In Praise of *VLSI Test Principles and Architectures: Design for Testability*

Testing techniques for VLSI circuits are today facing many exciting and complex challenges. In the era of large systems embedded in a single system-on-chip (SOC) and fabricated in continuously shrinking technologies, it is important to ensure correct behavior of the whole system. Electronic design and test engineers of today have to deal with these complex and heterogeneous systems (digital, mixed-signal, memory), but few have the possibility to study the whole field in a detailed and deep way. This book provides an extremely broad knowledge of the discipline, covering the fundamentals in detail, as well as the most recent and advanced concepts.

It is a textbook for teaching the basics of fault simulation, ATPG, memory testing, DFT and BIST. However, it is also a complete testability guide for an engineer who wants to learn the latest advances in DFT for soft error protection, logic built-in self-test (BIST) for at-speed testing, DRAM BIST, test compression, MEMS testing, FPGA testing, RF testing, etc.

Michel Renovell, Laboratoire d'Informatique, de Robotique et de Microélectronique de Montpellier (LIRMM), Montpellier, France

This book combines in a unique way insight into industry practices commonly found in commercial DFT tools but not discussed in textbooks, and a sound treatment of the technical fundamentals. The comprehensive review of future test technology trends, including self-repair, soft error protection, MEMS testing, and RF testing, leads students and researchers to advanced DFT research.

Hans-Joachim Wunderlich, University of Stuttgart, Germany

Recent advances in semiconductor manufacturing have made design for testability (DFT) an essential part of nanometer designs. The lack of an up-to-date DFT textbook that covers the most recent DFT techniques, such as at-speed scan testing, logic built-in self-test (BIST), test compression, memory built-in self-repair (BISR), and future test technology trends, has created problems for students, instructors, researchers, and practitioners who need to master modern DFT technologies. I am pleased to find a DFT textbook of this comprehensiveness that can serve both academic and professional needs.

Andre Ivanov, University of British Columbia, Canada

This is the most recent book covering all aspects of digital systems testing. It is a "must read" for anyone focused on learning modern test issues, test research, and test practices.

Kewal K. Saluja, University of Wisconsin-Madison

Design for testability (DFT) can no longer be considered as a graduate-level course. With growing design starts worldwide, DFT must be also part of the undergraduate curriculum. The book's focus on VLSI test principles and DFT architectures, while deemphasizing test algorithms, is an ideal choice for undergraduate education. In addition, system-on-chip (SOC) testing is one among the most important technologies for the development of ultra-large-scale integration (ULSI) devices in the 21st century. By covering the basic DFT theory and methodology on digital, memory, as well as analog and mixed-signal (AMS) testing, this book further stands out as one best reference book that equips practitioners with testable SOC design skills.

Yihe Sun, Tsinghua University, Beijing, China

VLSI Test Principles and Architectures

The Morgan Kaufmann Series in Systems on Silicon
Series Editor: Wayne Wolf, Princeton University

The rapid growth of silicon technology and the demands of applications are increasingly forcing electronics designers to take a systems-oriented approach to design. This has led to new challenges in design methodology, design automation, manufacture and test. The main challenges are to enhance designer productivity and to achieve correctness on the first pass. *The Morgan Kaufmann Series in Systems on Silicon* presents high-quality, peer-reviewed books authored by leading experts in the field who are uniquely qualified to address these issues.

The Designer's Guide to VHDL, Second Edition
Peter J. Ashenden

The System Designer's Guide to VHDL-AMS
Peter J. Ashenden, Gregory D. Peterson, and Darrell A. Teegarden

Readings in Hardware/Software Co-Design
Edited by Giovanni De Micheli, Rolf Ernst, and Wayne Wolf

Modeling Embedded Systems and SoCs
Axel Jantsch

ASIC and FPGA Verification: A Guide to Component Modeling
Richard Munden

Multiprocessor Systems-on-Chips
Edited by Ahmed Amine Jerraya and Wayne Wolf

Comprehensive Functional Verification
Bruce Wile, John Goss, and Wolfgang Roesner

Customizable Embedded Processors: Design Technologies and Applications
Edited by Paolo Ienne and Rainer Leupers

Networks on Chips: Technology and Tools
Giovanni De Micheli and Luca Benini

Designing SOCs with Configured Cores: Unleashing the Tensilica Diamond Cores
Steve Leibson

VLSI Test Principles and Architectures: Design for Testability
Edited by Laung-Terng Wang, Cheng-Wen Wu, and Xiaoqing Wen

Contact Information

Charles B. Glaser
Senior Acquisitions Editor
Elsevier
(Morgan Kaufmann; Academic Press; Newnes)
(781) 313-4732
c.glaser@elsevier.com
http://www.books.elsevier.com

Wayne Wolf
Professor
Electrical Engineering, Princeton University
(609) 258-1424
wolf@princeton.edu
http://www.ee.princeton.edu/~wolf/

VLSI TEST PRINCIPLES AND ARCHITECTURES
DESIGN FOR TESTABILITY

Edited by

Laung-Terng Wang
Cheng-Wen Wu
Xiaoqing Wen

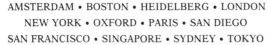

AMSTERDAM • BOSTON • HEIDELBERG • LONDON
NEW YORK • OXFORD • PARIS • SAN DIEGO
SAN FRANCISCO • SINGAPORE • SYDNEY • TOKYO

ELSEVIER

Morgan Kaufmann Publishers is an imprint of Elsevier

MORGAN KAUFMANN PUBLISHERS

Acquisitions Editor	Charles B. Glaser
Publishing Services Manager	George Morrison
Production Editor	Dawnmarie Simpson
Assistant Editor	Michele Cronin
Production Assistant	Melinda Ritchie
Cover Design	Paul Hodgson
Cover Illustration	©Dennis Harms/Getty Images
Composition	Integra Software Services
Technical Illustration	Integra Software Services
Copyeditor	Sarah Fortener
Proofreader	Phyllis Coyne et al. Proofreading Services
Indexer	Broccoli Information Management
Interior printer	The Maple-Vail Book Manufacturing Group
Cover printer	Phoenix Color Corporation

Morgan Kaufmann Publishers is an imprint of Elsevier.
500 Sansome Street, Suite 400, San Francisco, CA 94111

This book is printed on acid-free paper.

Library of Congress Cataloging-in-Publication Data
VLSI test principles and architectures: design for testability/edited by
    Laung-Terng Wang, Cheng-Wen Wu, Xiaoqing Wen.
        p. cm.
    Includes bibliographical references and index.
    ISBN-13: 978-0-12-370597-6 (hardcover: alk. paper)
    ISBN-10: 0-12-370597-5 (hardcover: alk. paper)
    1. Integrated circuits—Very large scale integration—Testing.   2. Integrated circuits—Very large
scale integration—Design.
I. Wang, Laung-Terng.   II. Wu, Cheng-Wen, EE Ph.D.   III. Wen, Xiaoqing.
TK7874.75.V587 2006
621.39'5—dc22

                                                    2006006869

ISBN 13: 978-0-12-370597-6
ISBN 10: 0-12-370597-5

For information on all Morgan Kaufmann publications,
visit our Web site at www.mkp.com or www.books.elsevier.com

CONTENTS

Preface   **xxi**

In the Classroom   **xxiv**

Acknowledgments   **xxv**

Contributors   **xxvii**

About the Editors   **xxix**

1 Introduction   **1**

Yinghua Min and Charles Stroud

1.1 Importance of Testing 1
1.2 Testing During the VLSI Lifecycle 2
    1.2.1 VLSI Development Process 3
        1.2.1.1 Design Verification 4
        1.2.1.2 Yield and Reject Rate 5
    1.2.2 Electronic System Manufacturing Process 6
    1.2.3 System-Level Operation 6
1.3 Challenges in VLSI Testing 8
    1.3.1 Test Generation 9
    1.3.2 Fault Models 11
        1.3.2.1 Stuck-At Faults 12
        1.3.2.2 Transistor Faults 15
        1.3.2.3 Open and Short Faults 16
        1.3.2.4 Delay Faults and Crosstalk 19
        1.3.2.5 Pattern Sensitivity and Coupling Faults 20
        1.3.2.6 Analog Fault Models 21
1.4 Levels of Abstraction in VLSI Testing 22
    1.4.1 Register-Transfer Level and Behavioral Level 22
    1.4.2 Gate Level 23
    1.4.3 Switch Level 24
    1.4.4 Physical Level 24

1.5		Historical Review of VLSI Test Technology	25
	1.5.1	Automatic Test Equipment	25
	1.5.2	Automatic Test Pattern Generation	27
	1.5.3	Fault Simulation	28
	1.5.4	Digital Circuit Testing	28
	1.5.5	Analog and Mixed-Signal Circuit Testing	29
	1.5.6	Design for Testability	29
	1.5.7	Board Testing	31
	1.5.8	Boundary Scan Testing	32
1.6		Concluding Remarks	33
1.7		Exercises	33
		Acknowledgments	34
		References	34

2 Design for Testability   37

Laung-Terng (L.-T.) Wang, Xiaoqing Wen, and Khader S. Abdel-Hafez

2.1		Introduction	37
2.2		Testability Analysis	40
	2.2.1	SCOAP Testability Analysis	41
		2.2.1.1 Combinational Controllability and Observability Calculation	41
		2.2.1.2 Sequential Controllability and Observability Calculation	43
	2.2.2	Probability-Based Testability Analysis	45
	2.2.3	Simulation-Based Testability Analysis	47
	2.2.4	RTL Testability Analysis	48
2.3		Design for Testability Basics	50
	2.3.1	*Ad Hoc* Approach	51
		2.3.1.1 Test Point Insertion	51
	2.3.2	Structured Approach	53
2.4		Scan Cell Designs	55
	2.4.1	Muxed-D Scan Cell	55
	2.4.2	Clocked-Scan Cell	56
	2.4.3	LSSD Scan Cell	57
2.5		Scan Architectures	59
	2.5.1	Full-Scan Design	59
		2.5.1.1 Muxed-D Full-Scan Design	59
		2.5.1.2 Clocked Full-Scan Design	62
		2.5.1.3 LSSD Full-Scan Design	62
	2.5.2	Partial-Scan Design	64
	2.5.3	Random-Access Scan Design	67
2.6		Scan Design Rules	70
	2.6.1	Tristate Buses	71
	2.6.2	Bidirectional I/O Ports	71

	2.6.3	Gated Clocks	71
	2.6.4	Derived Clocks	74
	2.6.5	Combinational Feedback Loops	74
	2.6.6	Asynchronous Set/Reset Signals	75
2.7	Scan Design Flow		76
	2.7.1	Scan Design Rule Checking and Repair	77
	2.7.2	Scan Synthesis	78
		2.7.2.1 Scan Configuration	79
		2.7.2.2 Scan Replacement	82
		2.7.2.3 Scan Reordering	82
		2.7.2.4 Scan Stitching	83
	2.7.3	Scan Extraction	83
	2.7.4	Scan Verification	84
		2.7.4.1 Verifying the Scan Shift Operation	85
		2.7.4.2 Verifying the Scan Capture Operation	86
	2.7.5	Scan Design Costs	86
2.8	Special-Purpose Scan Designs		87
	2.8.1	Enhanced Scan	87
	2.8.2	Snapshot Scan	88
	2.8.3	Error-Resilient Scan	90
2.9	RTL Design for Testability		92
	2.9.1	RTL Scan Design Rule Checking and Repair	93
	2.9.2	RTL Scan Synthesis	94
	2.9.3	RTL Scan Extraction and Scan Verification	95
2.10	Concluding Remarks		95
2.11	Exercises		96
Acknowledgments			99
References			99

3 Logic and Fault Simulation   105

Jiun-Lang Huang, James C.-M. Li, and Duncan M. (Hank) Walker

3.1	Introduction		106
	3.1.1	Logic Simulation for Design Verification	106
	3.1.2	Fault Simulation for Test and Diagnosis	107
3.2	Simulation Models		108
	3.2.1	Gate-Level Network	109
		3.2.1.1 Sequential Circuits	109
	3.2.2	Logic Symbols	110
		3.2.2.1 Unknown State u	111
		3.2.2.2 High-Impedance State Z	113
		3.2.2.3 Intermediate Logic States	114
	3.2.3	Logic Element Evaluation	114
		3.2.3.1 Truth Tables	115
		3.2.3.2 Input Scanning	115

		3.2.3.3	Input Counting	116
		3.2.3.4	Parallel Gate Evaluation	116
	3.2.4	Timing Models		118
		3.2.4.1	Transport Delay	118
		3.2.4.2	Inertial Delay	119
		3.2.4.3	Wire Delay	119
		3.2.4.4	Functional Element Delay Model	120
3.3	Logic Simulation			121
	3.3.1	Compiled-Code Simulation		121
		3.3.1.1	Logic Optimization	121
		3.3.1.2	Logic Levelization	123
		3.3.1.3	Code Generation	124
	3.3.2	Event-Driven Simulation		125
		3.3.2.1	Nominal-Delay Event-Driven Simulation	126
	3.3.3	Compiled-Code *Versus* Event-Driven Simulation		129
	3.3.4	Hazards		130
		3.3.4.1	Static Hazard Detection	131
		3.3.4.2	Dynamic Hazard Detection	132
3.4	Fault Simulation			132
	3.4.1	Serial Fault Simulation		133
	3.4.2	Parallel Fault Simulation		135
		3.4.2.1	Parallel Fault Simulation	135
		3.4.2.2	Parallel-Pattern Fault Simulation	137
	3.4.3	Deductive Fault Simulation		139
	3.4.4	Concurrent Fault Simulation		143
	3.4.5	Differential Fault Simulation		146
	3.4.6	Fault Detection		148
	3.4.7	Comparison of Fault Simulation Techniques		149
	3.4.8	Alternatives to Fault Simulation		151
		3.4.8.1	Toggle Coverage	151
		3.4.8.2	Fault Sampling	151
		3.4.8.3	Critical Path Tracing	152
		3.4.8.4	Statistical Fault Analysis	153
3.5	Concluding Remarks			154
3.6	Exercises			155
References				158

4 Test Generation   161

Michael S. Hsiao

4.1	Introduction		161
4.2	Random Test Generation		163
	4.2.1	Exhaustive Testing	166
4.3	Theoretical Background: Boolean Difference		166
	4.3.1	Untestable Faults	168

4.4   Designing a Stuck-At ATPG for Combinational Circuits 169
       4.4.1   A Naive ATPG Algorithm 169
                4.4.1.1   Backtracking 172
       4.4.2   A Basic ATPG Algorithm 173
       4.4.3   D Algorithm 177
       4.4.4   PODEM 182
       4.4.5   FAN 186
       4.4.6   Static Logic Implications 187
       4.4.7   Dynamic Logic Implications 191
4.5   Designing a Sequential ATPG 194
       4.5.1   Time Frame Expansion 194
       4.5.2   5-Valued Algebra Is Insufficient 196
       4.5.3   Gated Clocks and Multiple Clocks 197
4.6   Untestable Fault Identification 200
       4.6.1   Multiple-Line Conflict Analysis 203
4.7   Designing a Simulation-Based ATPG 207
       4.7.1   Overview 208
       4.7.2   Genetic-Algorithm-Based ATPG 208
                4.7.2.1   Issues Concerning the GA Population 212
                4.7.2.2   Issues Concerning GA Parameters 213
                4.7.2.3   Issues Concerning the Fitness Function 213
                4.7.2.4   CASE Studies 215
4.8   Advanced Simulation-Based ATPG 218
       4.8.1   Seeding the GA with Helpful Sequences 218
       4.8.2   Logic-Simulation-Based ATPG 222
       4.8.3   Spectrum-Based ATPG 225
4.9   Hybrid Deterministic and Simulation-Based ATPG 226
       4.9.1   ALT-TEST Hybrid 228
4.10   ATPG for Non-Stuck-At Faults 231
       4.10.1   Designing an ATPG That Captures Delay Defects 231
                4.10.1.1   Classification of Path-Delay Faults 233
                4.10.1.2   ATPG for Path-Delay Faults 236
       4.10.2   ATPG for Transition Faults 238
       4.10.3   Transition ATPG Using Stuck-At ATPG 240
       4.10.4   Transition ATPG Using Stuck-At Vectors 240
                4.10.4.1   Transition Test Chains via Weighted
                           Transition Graph 241
       4.10.5   Bridging Fault ATPG 244
4.11   Other Topics in Test Generation 246
       4.11.1   Test Set Compaction 246
       4.11.2   N-Detect ATPG 247
       4.11.3   ATPG for Acyclic Sequential Circuits 247
       4.11.4   IDDQ Testing 247
       4.11.5   Designing a High-Level ATPG 248
4.12   Concluding Remarks 248
4.13   Exercises 249
References 256

5 Logic Built-In Self-Test   263

Laung-Terng (L.-T.) Wang

5.1   Introduction 264
5.2   BIST Design Rules 266
       5.2.1   Unknown Source Blocking 267
                5.2.1.1   Analog Blocks 267
                5.2.1.2   Memories and Non-Scan Storage Elements 268
                5.2.1.3   Combinational Feedback Loops 268
                5.2.1.4   Asynchronous Set/Reset Signals 268
                5.2.1.5   Tristate Buses 269
                5.2.1.6   False Paths 270
                5.2.1.7   Critical Paths 270
                5.2.1.8   Multiple-Cycle Paths 270
                5.2.1.9   Floating Ports 270
                5.2.1.10   Bidirectional I/O Ports 271
       5.2.2   Re-Timing 271
5.3   Test Pattern Generation 271
       5.3.1   Exhaustive Testing 275
                5.3.1.1   Binary Counter 275
                5.3.1.2   Complete LFSR 275
       5.3.2   Pseudo-Random Testing 277
                5.3.2.1   Maximum-Length LFSR 278
                5.3.2.2   Weighted LFSR 278
                5.3.2.3   Cellular Automata 278
       5.3.3   Pseudo-Exhaustive Testing 281
                5.3.3.1   Verification Testing 282
                5.3.3.2   Segmentation Testing 287
       5.3.4   Delay Fault Testing 288
       5.3.5   Summary 289
5.4   Output Response Analysis 290
       5.4.1   Ones Count Testing 291
       5.4.2   Transition Count Testing 291
       5.4.3   Signature Analysis 292
                5.4.3.1   Serial Signature Analysis 292
                5.4.3.2   Parallel Signature Analysis 294
5.5   Logic BIST Architectures 296
       5.5.1   BIST Architectures for Circuits without Scan Chains 296
                5.5.1.1   A Centralized and Separate Board-Level
                           BIST Architecture 296
                5.5.1.2   Built-In Evaluation and Self-Test (BEST) 297
       5.5.2   BIST Architectures for Circuits with Scan Chains 297
                5.5.2.1   LSSD On-Chip Self-Test 297
                5.5.2.2   Self-Testing Using MISR and Parallel SRSG . . . 298
       5.5.3   BIST Architectures Using Register Reconfiguration 298
                5.5.3.1   Built-In Logic Block Observer 299

		5.5.3.2	Modified Built-In Logic Block Observer	300
		5.5.3.3	Concurrent Built-In Logic Block Observer	300
		5.5.3.4	Circular Self-Test Path (CSTP)	302
	5.5.4	BIST Architectures Using Concurrent Checking Circuits		303
		5.5.4.1	Concurrent Self-Verification	303
	5.5.5	Summary		304
5.6	Fault Coverage Enhancement			304
	5.6.1	Test Point Insertion		305
		5.6.1.1	Test Point Placement	306
		5.6.1.2	Control Point Activation	307
	5.6.2	Mixed-Mode BIST		308
		5.6.2.1	ROM Compression	308
		5.6.2.2	LFSR Reseeding	308
		5.6.2.3	Embedding Deterministic Patterns	309
	5.6.3	Hybrid BIST		309
5.7	BIST Timing Control			310
	5.7.1	Single-Capture		310
		5.7.1.1	One-Hot Single-Capture	310
		5.7.1.2	Staggered Single-Capture	311
	5.7.2	Skewed-Load		311
		5.7.2.1	One-Hot Skewed-Load	312
		5.7.2.2	Aligned Skewed-Load	312
		5.7.2.3	Staggered Skewed-Load	314
	5.7.3	Double-Capture		315
		5.7.3.1	One-Hot Double-Capture	315
		5.7.3.2	Aligned Double-Capture	316
		5.7.3.3	Staggered Double-Capture	317
	5.7.4	Fault Detection		317
5.8	A Design Practice			319
	5.8.1	BIST Rule Checking and Violation Repair		320
	5.8.2	Logic BIST System Design		320
		5.8.2.1	Logic BIST Architecture	320
		5.8.2.2	TPG and ORA	321
		5.8.2.3	Test Controller	322
		5.8.2.4	Clock Gating Block	323
		5.8.2.5	Re-Timing Logic	325
		5.8.2.6	Fault Coverage Enhancing Logic and Diagnostic Logic	325
	5.8.3	RTL BIST Synthesis		326
	5.8.4	Design Verification and Fault Coverage Enhancement		326
5.9	Concluding Remarks			327
5.10	Exercises			327
Acknowledgments				331
References				331

6 Test Compression   341

Xiaowei Li, Kuen-Jong Lee, and Nur A. Touba

6.1   Introduction 342
6.2   Test Stimulus Compression 344
       6.2.1   Code-Based Schemes 345
                6.2.1.1   Dictionary Code (Fixed-to-Fixed) 345
                6.2.1.2   Huffman Code (Fixed-to-Variable) 346
                6.2.1.3   Run-Length Code (Variable-to-Fixed) 349
                6.2.1.4   Golomb Code (Variable-to-Variable) 350
       6.2.2   Linear-Decompression-Based Schemes 351
                6.2.2.1   Combinational Linear Decompressors 355
                6.2.2.2   Fixed-Length Sequential
                           Linear Decompressors 355
                6.2.2.3   Variable-Length Sequential
                           Linear Decompressors 356
                6.2.2.4   Combined Linear and
                           Nonlinear Decompressors 357
       6.2.3   Broadcast-Scan-Based Schemes 359
                6.2.3.1   Broadcast Scan 359
                6.2.3.2   Illinois Scan 360
                6.2.3.3   Multiple-Input Broadcast Scan 362
                6.2.3.4   Reconfigurable Broadcast Scan 362
                6.2.3.5   Virtual Scan 363
6.3   Test Response Compaction 364
       6.3.1   Space Compaction 367
                6.3.1.1   Zero-Aliasing Linear Compaction 367
                6.3.1.2   X-Compact 369
                6.3.1.3   X-Blocking 371
                6.3.1.4   X-Masking 372
                6.3.1.5   X-Impact 373
       6.3.2   Time Compaction 374
       6.3.3   Mixed Time and Space Compaction 375
6.4   Industry Practices 376
       6.4.1   OPMISR+ 377
       6.4.2   Embedded Deterministic Test 379
       6.4.3   VirtualScan and UltraScan 382
       6.4.4   Adaptive Scan 385
       6.4.5   ETCompression 386
       6.4.6   Summary 388
6.5   Concluding Remarks 388
6.6   Exercises 389
Acknowledgments 390
References 391

7 Logic Diagnosis   397

Shi-Yu Huang

7.1   Introduction 397
7.2   Combinational Logic Diagnosis 401
    7.2.1   Cause–Effect Analysis 401
        7.2.1.1   Compaction and Compression of Fault Dictionary 403
    7.2.2   Effect–Cause Analysis 405
        7.2.2.1   Structural Pruning 407
        7.2.2.2   Backtrace Algorithm 408
        7.2.2.3   Inject-and-Evaluate Paradigm 409
    7.2.3   Chip-Level Strategy 418
        7.2.3.1   Direct Partitioning 418
        7.2.3.2   Two-Phase Strategy 420
        7.2.3.3   Overall Chip-Level Diagnostic Flow 424
    7.2.4   Diagnostic Test Pattern Generation 425
    7.2.5   Summary of Combinational Logic Diagnosis 426
7.3   Scan Chain Diagnosis 427
    7.3.1   Preliminaries for Scan Chain Diagnosis 427
    7.3.2   Hardware-Assisted Method 430
    7.3.3   Modified Inject-and-Evaluate Paradigm 432
    7.3.4   Signal-Profiling-Based Method 434
        7.3.4.1   Diagnostic Test Sequence Selection 434
        7.3.4.2   Run-and-Scan Test Application 434
        7.3.4.3   Why Functional Sequence? 435
        7.3.4.4   Profiling-Based Analysis 437
    7.3.5   Summary of Scan Chain Diagnosis 441
7.4   Logic BIST Diagnosis 442
    7.4.1   Overview of Logic BIST Diagnosis 442
    7.4.2   Interval-Based Methods 443
    7.4.3   Masking-Based Methods 446
7.5   Concluding Remarks 449
7.6   Exercises 450
Acknowledgments 453
References 454

8 Memory Testing and Built-In Self-Test   461

Cheng-Wen Wu

8.1   Introduction 462
8.2   RAM Functional Fault Models and Test Algorithms 463
    8.2.1   RAM Functional Fault Models 463
    8.2.2   RAM Dynamic Faults 465
    8.2.3   Functional Test Patterns and Algorithms 466
    8.2.4   March Tests 469

	8.2.5	Comparison of RAM Test Patterns	471
	8.2.6	Word-Oriented Memory	473
	8.2.7	Multi-Port Memory	473
8.3	RAM Fault Simulation and Test Algorithm Generation		475
	8.3.1	Fault Simulation	476
	8.3.2	RAMSES	477
	8.3.3	Test Algorithm Generation by Simulation	480
8.4	Memory Built-In Self-Test		488
	8.4.1	RAM Specification and BIST Design Strategy	489
	8.4.2	BIST Architectures and Functions	493
	8.4.3	BIST Implementation	495
	8.4.4	BRAINS: A RAM BIST Compiler	500
8.5	Concluding Remarks		508
8.6	Exercises		509
Acknowledgments			513
References			513

9   Memory Diagnosis and Built-In Self-Repair   517

Cheng-Wen Wu

9.1	Introduction		518
	9.1.1	Why Memory Diagnosis?	518
	9.1.2	Why Memory Repair?	518
9.2	Refined Fault Models and Diagnostic Test Algorithms		518
9.3	BIST with Diagnostic Support		521
	9.3.1	Controller	521
	9.3.2	Test Pattern Generator	523
	9.3.3	Fault Site Indicator (FSI)	524
9.4	RAM Defect Diagnosis and Failure Analysis		526
9.5	RAM Redundancy Analysis Algorithms		529
	9.5.1	Conventional Redundancy Analysis Algorithms	529
	9.5.2	The Essential Spare Pivoting Algorithm	531
	9.5.3	Repair Rate and Overhead	535
9.6	Built-In Self-Repair		537
	9.6.1	Redundancy Organization	537
	9.6.2	BISR Architecture and Procedure	538
	9.6.3	BIST Module	541
	9.6.4	BIRA Module	542
	9.6.5	An Industrial Case	545
	9.6.6	Repair Rate and Yield	548
9.7	Concluding Remarks		552
9.8	Exercises		552
Acknowledgments			553
References			553

10 Boundary Scan and Core-Based Testing   **557**

Kuen-Jong Lee

10.1 Introduction . . . : 558
    10.1.1 IEEE 1149 Standard Family 558
    10.1.2 Core-Based Design and Test Considerations 559
10.2 Digital Boundary Scan (IEEE Std. 1149.1) 561
    10.2.1 Basic Concept 561
    10.2.2 Overall 1149.1 Test Architecture and Operations 562
    10.2.3 Test Access Port and Bus Protocols 564
    10.2.4 Data Registers and Boundary-Scan Cells 565
    10.2.5 TAP Controller 567
    10.2.6 Instruction Register and Instruction Set 569
    10.2.7 Boundary-Scan Description Language 574
    10.2.8 On-Chip Test Support with Boundary Scan 574
    10.2.9 Board and System-Level Boundary-Scan Control
            Architectures 576
10.3 Boundary Scan for Advanced Networks (IEEE 1149.6) 579
    10.3.1 Rationale for 1149.6 579
    10.3.2 1149.6 Analog Test Receiver 581
    10.3.3 1149.6 Digital Driver Logic 581
    10.3.4 1149.6 Digital Receiver Logic 582
    10.3.5 1149.6 Test Access Port (TAP) 584
    10.3.6 Summary 585
10.4 Embedded Core Test Standard (IEEE Std. 1500) 585
    10.4.1 SOC (System-on-Chip) Test Problems 585
    10.4.2 Overall Architecture 587
    10.4.3 Wrapper Components and Functions 589
    10.4.4 Instruction Set 597
    10.4.5 Core Test Language (CTL) 601
    10.4.6 Core Test Supporting and System Test Configurations . . 603
    10.4.7 Hierarchical Test Control and Plug-and-Play 606
10.5 Comparisons between the 1500 and 1149.1 Standards 610
10.6 Concluding Remarks 611
10.7 Exercises 612
Acknowledgments 614
References 614

11 Analog and Mixed-Signal Testing   **619**

Chauchin Su

11.1 Introduction 619
    11.1.1 Analog Circuit Properties 620
            11.1.1.1 Continuous Signals 621

11.1.1.2   Large Range of Circuits 621
11.1.1.3   Nonlinear Characteristics 621
11.1.1.4   Feedback Ambiguity 622
11.1.1.5   Complicated Cause–Effect Relationship 622
11.1.1.6   Absence of Suitable Fault Model 622
11.1.1.7   Requirement for Accurate Instruments for
          Measuring Analog Signals 623
11.1.2   Analog Defect Mechanisms and Fault Models 623
11.1.2.1   Hard Faults 625
11.1.2.2   Soft Faults 625
11.2   Analog Circuit Testing 627
11.2.1   Analog Test Approaches 627
11.2.2   Analog Test Waveforms 629
11.2.3   DC Parametric Testing 631
11.2.3.1   Open-Loop Gain Measurement 632
11.2.3.2   Unit Gain Bandwidth Measurement 633
11.2.3.3   Common Mode Rejection Ratio Measurement . 634
11.2.3.4   Power Supply Rejection Ratio Measurement . . 635
11.2.4   AC Parametric Testing 635
11.2.4.1   Maximal Output Amplitude Measurement 636
11.2.4.2   Frequency Response Measurement 637
11.2.4.3   SNR and Distortion Measurement 639
11.2.4.4   Intermodulation Distortion Measurement 641
11.3   Mixed-Signal Testing 641
11.3.1   Introduction to Analog–Digital Conversion 642
11.3.2   ADC and DAC Circuit Structure 644
11.3.2.1   DAC Circuit Structure 646
11.3.2.2   ADC Circuit Structure 646
11.3.3   ADC/DAC Specification and Fault Models 647
11.3.4   IEEE 1057 Standard 652
11.3.5   Time-Domain ADC Testing 654
11.3.5.1   Code Bins 654
11.3.5.2   Code Transition Level Test (Static) 655
11.3.5.3   Code Transition Level Test (Dynamic) 655
11.3.5.4   Gain and Offset Test 656
11.3.5.5   Linearity Error and Maximal Static Error 657
11.3.5.6   Sine Wave Curve-Fit Test 658
11.3.6   Frequency-Domain ADC Testing 658
11.4   IEEE 1149.4 Standard for a Mixed-Signal Test Bus 658
11.4.1   IEEE 1149.4 Overview 659
11.4.1.1   Scope of the Standard 660
11.4.2   IEEE 1149.4 Circuit Structures 661
11.4.3   IEEE 1149.4 Instructions 665
11.4.3.1   Mandatory Instructions 665
11.4.3.2   Optional Instructions 665

11.4.4   IEEE 1149.4 Test Modes 666
      11.4.4.1   Open/Short Interconnect Testing 666
      11.4.4.2   Extended Interconnect Measurement 667
      11.4.4.3   Complex Network Measurement 671
      11.4.4.4   High-Performance Configuration 672
11.5   Concluding Remarks 673
11.6   Exercises 673
Acknowledgments 676
References 677

12   Test Technology Trends in the Nanometer Age   679

Kwang-Ting (Tim) Cheng, Wen-Ben Jone, and Laung-Terng (L.-T.) Wang

12.1   Test Technology Roadmap 680
12.2   Delay Testing 685
  12.2.1   Test Application Schemes for Testing Delay Defects 686
  12.2.2   Delay Fault Models 687
  12.2.3   Summary 690
12.3   Coping with Physical Failures, Soft Errors,
    and Reliability Issues 692
  12.3.1   Signal Integrity and Power Supply Noise 692
      12.3.1.1   Integrity Loss Fault Model 693
      12.3.1.2   Location 694
      12.3.1.3   Pattern Generation 694
      12.3.1.4   Sensing and Readout 695
  12.3.2   Parametric Defects, Process Variations, and Yield 696
      12.3.2.1   Defect-Based Test 697
  12.3.3   Soft Errors 698
  12.3.4   Fault Tolerance 701
  12.3.5   Defect and Error Tolerance 705
12.4   FPGA Testing 706
  12.4.1   Impact of Programmability 706
  12.4.2   Testing Approaches 708
  12.4.3   Built-In Self-Test of Logic Resources 708
  12.4.4   Built-In Self-Test of Routing Resources 709
  12.4.5   Recent Trends 710
12.5   MEMS Testing 711
  12.5.1   Basic Concepts for Capacitive MEMS Devices 711
  12.5.2   MEMS Built-In Self-Test 713
      12.5.2.1   Sensitivity BIST Scheme 713
      12.5.2.2   Symmetry BIST Scheme 713
      12.5.2.3   A Dual-Mode BIST Technique 714
  12.5.3   A BIST Example for MEMS Comb Accelerometers 716
  12.5.4   Conclusions 719

12.6   High-speed I/O Testing 719
    12.6.1   I/O Interface Technology and Trend 720
    12.6.2   I/O Testing and Challenges 724
    12.6.3   High-Performance I/O Test Solutions 725
    12.6.4   Future Challenges 726
12.7   RF Testing 728
    12.7.1   Core RF Building Blocks 729
    12.7.2   RF Test Specifications and Measurement Procedures . . . 730
        12.7.2.1   Gain 730
        12.7.2.2   Conversion Gain 731
        12.7.2.3   Third-Order Intercept 731
        12.7.2.4   Noise Figure 733
    12.7.3   Tests for System-Level Specifications 733
        12.7.3.1   Adjacent Channel Power Ratio 733
        12.7.3.2   Error Vector Magnitude, Magnitude Error, and
                    Phase Error 734
    12.7.4   Current and Future Trends 735
        12.7.4.1   Future Trends 736
12.8   Concluding Remarks 737
Acknowledgments 738
References 738

Index   **751**

PREFACE

Beginning with the introduction of commercial manufacturing of integrated circuits (ICs) in the early 1960s, modern electronics testing has a history of more than 40 years. The integrated circuit was developed in 1958, concurrently at Texas Instruments (TI) and Fairchild Semiconductor. Today, semiconductors lie at the heart of ongoing advances across the electronics industry. The industry enjoyed a banner year in 2005, with almost $230 billion in sales worldwide.

The introduction of new technologies, especially nanometer technologies with 90 nm or smaller geometry, has allowed the semiconductor industry to keep pace with increased performance-capacity demands from consumers. This has brightened the prospects for future industry growth; however, new technologies come with new challenges. Semiconductor test costs have been growing steadily. Test costs can now amount to 40% of overall product cost. In addition, product quality and yield could drop significantly if these chips are not designed for testability and thoroughly tested.

New problems encountered in semiconductor testing are being recognized quickly today. Because very-large-scale integration (VLSI) technologies drive test technologies, more effective test technologies are key to success in today's competitive marketplace. It is recognized that, in order to tackle the problems associated with testing semiconductor devices, it is necessary to attack them at earlier design stages. The field of design for testability (DFT) is a mature one today. Test cost can be significantly reduced by inserting DFT in earlier design stages; thus, it is important to expose students and practitioners to the most recent, yet fundamental, VLSI test principles and DFT architectures in an effort to help them design better quality products now and in the future that can be reliably manufactured in quantity.

In this context, it is important to make sure that undergraduates and practitioners, in addition to graduate students and researchers, are introduced to the variety of problems encountered in semiconductor testing and that they are made aware of the new methods being developed to solve these problems at earlier stages of design. A very important factor in doing so is to ensure that introductory textbooks for semiconductor testing are kept up to date with the latest process, design, and test technology advances.

This textbook is being made available with this goal in mind. It is a fundamental yet comprehensive guide to new DFT methods that will show readers how to design a testable and quality product, drive down test cost, improve product quality and yield, and speed up time-to-market and time-to-volume. Intended users of the book include undergraduates, engineers and engineering managers who have the need to know; it is not simply for graduate students and researchers. It focuses more on basic VLSI test concepts, principles, and DFT architectures and includes the latest advances that are in practice today, including *at-speed scan testing*, *test compression*, *at-speed built-in self-test* (BIST), *memory built-in self-repair* (BISR), and *test technology trends*. These advanced subjects are key to *system-on-chip* (SOC) designs in the nanometer age.

The semiconductor testing field is quite broad today, so the scope of this textbook is also broad, with topics ranging from *digital* to *memory* to *AMS* (analog and mixed-signal) *testing*. This book will allow the readers to understand fundamental VLSI test principles and DFT architectures and prepare them for tackling test problems caused by advances in semiconductor manufacturing technology and complex SOC designs in the nanometer era.

Each chapter of this book follows a specific template format. The subject matter of the chapter is first introduced, with a historical perspective provided, if needed. Then, related methods and algorithms are explained in sufficient detail while keeping the level of intended users in mind. Examples are taken from the current DFT tools, products, etc. Comprehensive reference sources are then provided. Each chapter (except Chapter 12) ends with a variety of exercises for students to solve to help them master the topic at hand.

Chapter 1 provides a comprehensive introduction to *semiconductor testing*. It begins with a discussion of the importance of testing as a requisite for achieving manufacturing quality of semiconductor devices and then identifies difficulties in *VLSI testing*. After the author explains how testing can be viewed as a design moving through different abstraction levels, a historical view of the *development of VLSI testing* is presented.

Chapter 2 is devoted to introducing the basic concepts of design for testability (DFT). *Testability analysis* to assess the testability of a logic circuit is discussed. *Ad hoc* and *structured approaches* to ease testing are then presented, which leads to *scan design*, a widely used DFT method in industry today. The remainder of the chapter is then devoted to *scan cell designs*, *scan architectures*, *scan design rules*, and *scan synthesis* and *verification*. Following a discussion of *scan cost issues*, *special-purpose scan designs* suitable for delay testing, system debug, and *soft error* protection, *RTL DFT techniques* are briefly introduced.

Chapter 3 and Chapter 4 are devoted to the familiar areas of logic/fault simulation and automatic test pattern generation (ATPG), respectively. Care is taken to describe methods and algorithms used in these two areas in an easy-to-grasp language while maintaining the overall perspective of VLSI testing.

Chapter 5 is completely devoted to logic *built-in self-test* (BIST). After a brief introduction, specific *BIST design rules* are presented. *On-chip test pattern generation* and *output response analysis* are then explained. The chapter puts great emphasis on documenting important *on-chip test pattern generation* techniques and *logic BIST architectures*, as these subjects are not yet well researched. *At-speed BIST techniques*, a key feature in this chapter, are then explained in detail. A *design practice* example provided at the end of the chapter invites readers to design a *logic BIST system*.

Chapter 6 then jumps into the most important test cost aspect of testability insertion into a *scan design*. How cost reduction can be achieved using test compression is discussed in greater detail. Representative, commercially available compression tools are introduced so readers (practitioners) can appreciate what is best suited to their needs.

Chapter 7 delves into the topic of logic diagnosis. Techniques for *combinational logic diagnosis* based on *cause–effect analysis*, *effect–cause analysis*, and *chip-level strategy* are first described. Then, innovative techniques for *scan chain diagnosis* and *logic BIST diagnosis* are explained in detail.

Chapter 8 and Chapter 9 cover the full spectrum of *memory test* and *diagnosis* methods. In both chapters, after a description of basic memory test and diagnosis concepts, *memory BIST* and *memory BISR* architectures are then explained in detail. *Memory fault simulation*, a unique topic, is also discussed in Chapter 8.

Chapter 10 covers *boundary scan* and *core-based testing* for board-level and system-level testing. The IEEE 1149 standard addresses boundary-scan-based testing; after a brief history, the *boundary-scan standards* (IEEE 1149.1 and 1149.6) are discussed. The newly endorsed *IEEE 1500 core-based testing standard* is then described.

Chapter 11 is devoted to analog and mixed-signal testing. Important *analog circuit properties* and their *defect mechanisms* and *fault models* are described first. Methods for analog circuit testing are then explained. Mixed-signal circuit testing is introduced by a discussion of *ADC/DAC testing*. The IEEE 1057 standard for *digitizing waveform recorders* is then explained. A related standard, IEEE 1149.4, and instructions for *mixed-signal test buses* are covered in detail. Special topics related to ADC/DAC testing, including *time-domain ADC testing* and *frequency-domain ADC testing*, are also touched on in this chapter.

Chapter 12 is devoted to test technology trends in the nanometer age. It presents an international test technology roadmap to put these new trends in perspective and predicts test technology needs in the coming 10 to 15 years, such as better methods for *delay testing*, as well as *coping with physical failures, soft errors*, and *reliability issues*. The emerging field of *FPGA* and *MEMS testing* is briefly touched upon before the chapter jumps into other modern topics such as *high-speed I/O testing* and *RF testing*.

In the Classroom

This book is designed to be used as a text for undergraduate and graduate students in computer engineering, computer science, or electrical engineering. It is also intended for use as a reference book for researchers and practitioners. The book is self-contained, with most topics covered extensively from fundamental concepts to current techniques used in research and industry. We assume that the students have had basic courses in logic design, computer science, and probability theory. Attempts are made to present algorithms, where possible, in an easily understood format.

In order to encourage self-learning, readers are advised to check the Elsevier companion Web site (www.books.elsevier.com/companions) to access up-to-date software and presentation slides, including errata, if any. Professors will have additional privileges to assess the solutions directory for all exercises given in each chapter by visiting www.textbooks.elsevier.com and registering a username and password.

Laung-Terng (L.-T.) Wang
Cheng-Wen Wu
Xiaoqing Wen

ACKNOWLEDGMENTS

The editors would like to acknowledge many of their colleagues who helped create this book. First and foremost are the 27 chapter/section contributors listed in the following section. Without their strong commitments to contributing the chapters and sections of their specialties to the book in a timely manner, it would not have been possible to publish this fundamental DFT textbook, which covers the most recent advances in VLSI testing and DFT architectures.

We also would like to give additional thanks to the reviewers of the book, particularly Prof. Fa Foster Dai (Auburn University), Prof. Andre Ivanov (University of British Columbia, Canada), Prof. Chong-Min Kyung (Korea Advanced Institute of Science and Technology, Korea), Prof. Adam Osseiran (Edith Cowan University, Australia), Prof. Sudhakar M. Reddy (University of Iowa), Prof. Michel Renovell (LIRMM, France), Prof. Kewal K. Saluja (University of Wisconsin–Madison), Prof. Masaru Sanada (Kochi University of Technology, Japan), Prof. Hans-Joachim Wunderlich (University of Stuttgart, Germany), Prof. Dong Xiang (Tsinghua University, China), Prof. Xiaoyang Zeng (Fudan University, China), Dwayne Burek (Magma Design Automation, Santa Clara, CA), Sachin Dhingra and Sudheer Vemula (Auburn University), Grady L. Giles (Advanced Micro Devices, Austin, TX), Dr. Yinhe Han and Dr. Huawei Li (Chinese Academy of Sciences, China), Dr. Augusli Kifli (Faraday Technology, Taiwan), Dr. Yunsik Lee (Korea Electronics Technology Institute, Korea), Dr. Samy Makar (Azul Systems, Mountain View, CA), Erik Jan Marinissen (Philips Research Laboratories, The Netherlands), Dr. Kenneth P. Parker (Agilent Technologies, Loveland, CO), Takeshi Onodera (Sony Corp. Semiconductor Solutions Network Co., Japan), Jing Wang and Lei Wu (Texas A&M University, College Station, TX), and Thomas Wilderotter (Synopsys, Bedminster, NJ), as well as all chapter/section contributors for cross-reviewing the manuscript. Special thanks also go to many colleagues at SynTest Technologies, Inc., including Dr. Ravi Apte, Jack Sheu, Dr. Zhigang Jiang, Zhigang Wang, Jongjoo Park, Jinwoo Cho, Jerry Lin, Paul Hsu, Karl Chang, Tom Chao, Feng Liu, Johnson Guo, Xiangfeng Li, Fangfang Li, Yiqun Ding, Lizhen Yu, Angelia Yu, Huiqin Hu, Jiayong Song, Jane Xu, Jim Ma, Sammer Liu, Renay Chang, and Teresa Chang and her lovely daughter, Alice Yu, all of whom helped review the manuscript, solve exercises, develop lecture slides, and draw/redraw figures and tables.

Finally, the editors would like to acknowledge the generosity of SynTest Technologies (Sunnyvale, CA) for allowing Elsevier to put an exclusive version of the company's most recent VLSI Testing and DFT software on the Elsevier companion Web site (www.books.elsevier.com/companions) for readers to use in conjunction with the book to become acquainted with DFT practices.

CONTRIBUTORS

Khader S. Abdel-Hafez, Director of Engineering (Chapter 2)
SynTest Technologies, Inc., Sunnyvale, California

Soumendu Bhattacharya, Post-Doctoral Fellow (Chapter 12)
School of Electrical and Computer Engineering, Georgia Institute of Technology, Atlanta, Georgia

Abhijit Chatterjee, Professor, IEEE Fellow; Prof. Chatterjee was elected to the honor in November 2006 (Chapter 12)
School of Electrical and Computer Engineering, Georgia Institute of Technology, Atlanta, Georgia

Xinghao Chen, Associate Professor (Chapter 2)
Department of Electrical Engineering, The Grove School of Engineering, City College and Graduate Center of The City University of New York, New York

Kwang-Ting (Tim) Cheng, Chair and Professor, IEEE Fellow (Chapter 12)
Department of Electrical and Computer Engineering, University of California, Santa Barbara, California

William Eklow, Distinguished Manufacturing Engineer (Chapter 10)
Cisco Systems, Inc., San Jose, California; Chair, IEEE 1149.6 Standard Committee

Michael S. Hsiao, Professor and Dean's Faculty Fellow; Prof. Hsio got promoted in July 2006. (Chapter 4)
Bradley Department of Electical and Computer Engineering, Virginia Tech, Blacksburg, Virginia

Jiun-Lang Huang, Assistant Professor (Chapter 3)
Graduate Institute of Electronics Engineering, National Taiwan University, Taipei, Taiwan

Shi-Yu Huang, Associate Professor (Chapter 7)
Department of Electrical Engineering, National Tsing Hua University, Hsinchu, Taiwan

Wen-Ben Jone, Associate Professor (Chapter 12)
Department of Electrical & Computer Engineering and Computer Science, University of Cincinnati, Cincinnati, Ohio

Rohit Kapur, Scientist, IEEE Fellow (Chapter 6)
Synopsys, Inc., Mountain View, California

Brion Keller, Senior Architect (Chapter 6)
Cadence Design Systems, Inc., Endicott, New York

Kuen-Jong Lee, Professor (Chapters 6 and 10)
Department of Electrical Engineering, National Cheng Kung University, Tainan, Taiwan

James C.-M. Li, Associate Professor; Prof. Li got promoted in 2006 (Chapter 3)
Graduate Institute of Electronics Engineering, National Taiwan University, Taipei, Taiwan

Mike Peng Li, Chief Technology Officer (Chapter 12)
Wavecrest Corp., San Jose, California

Xiaowei Li, Professor (Chapter 6)
Institute of Computing Technology, Chinese Academy of Sciences, Beijing, China

T.M. Mak, Senior Researcher (Chapter 12)
Intel Corp., Santa Clara, California

Yinghua Min, Professor Emeritus, IEEE Fellow (Chapter 1)
Institute of Computing Technology, Chinese Academy of Sciences, Beijing, China

Benoit Nadeau-Dostie, Chief Scientist (Chapter 6)
LogicVision, Inc., Ottawa, Ontario, Canada

Mehrdad Nourani, Associate Professor (Chapter 12)
Department of Electrical Engineering, University of Texas at Dallas, Richardson, Texas

Janusz Rajski, Chief Scientist and Director of DFT Engineering (Chapter 6)
Mentor Graphics Corp., Wilsonville, Oregon

Charles Stroud, Professor, IEEE Fellow (Chapters 1 and 12)
Department of Electrical and Computer Engineering, Auburn University, Auburn, Alabama

Chauchin Su, Professor (Chapter 11)
Department of Electrical and Control Engineering, National Chiao Tung University, Hsinchu, Taiwan

Nur A. Touba, Professor; Prof. Touba got promoted in july 2007 (Chapters 5, 6, and 7)
Department of Electrical and Computer Engineering, University of Texas, Austin, Texas

Erik H. Volkerink, Manager, Agilent Semiconductor Test Labs. (Chapter 6)
Agilent Technologies, Inc., Palo Alto, California

Duncan M. (Hank) Walker, Professor (Chapters 3 and 12)
Department of Computer Science, Texas A&M University, College Station, Texas

Shianling Wu, Vice President of Engineering (Chapter 2)
SynTest Technologies, Inc., Princeton Junction, New Jersey

About the Editors

Laung-Terng (L.-T.) Wang, Ph.D., founder and chief executive officer (CEO) of SynTest Technologies (Sunnyvale, CA), received his BSEE and MSEE degrees from National Taiwan University in 1975 and 1977, respectively, and his MSEE and EE Ph.D. degrees under the Honors Cooperative Program (HCP) from Stanford University in 1982 and 1987, respectively. He worked at Intel (Santa Clara, CA) and Daisy Systems (Mountain View, CA) from 1980 to 1986 and was with the Department of Electrical Engineering of Stanford University as Research Associate and Lecturer from 1987 to 1991. Encouraged by his advisor, Professor Edward J. McCluskey, a member of the National Academy of Engineering, he founded SynTest Technologies in 1990. Under his leadership, the company has grown to more than 50 employees and 250 customers worldwide. The design for testability (DFT) technologies Dr. Wang has developed have been successfully implemented in thousands of ASIC designs worldwide. He has filed more than 25 U.S. and European patent applications in the areas of scan synthesis, test generation, at-speed scan testing, test compression, logic built-in self-test (BIST), and design for debug and diagnosis, of which seven have been granted. Dr. Wang's work in at-speed scan testing, test compression, and logic BIST has proved crucial to ensuring the quality and testability of nanometer designs, and his inventions are gaining industry acceptance for use in designs manufactured at the 90-nanometer scale and below. He spearheaded efforts to raise endowed funds in memory of his NTU chair professor, Dr. Irving T. Ho, cofounder of the Hsinchu Science Park and vice chair of the National Science Council, Taiwan. Since 2003, he has helped establish a number of chair professorships, graduate fellowships, and undergraduate scholarships at Stanford University and National Taiwan University, as well as Xiamen University, Tsinghua University, and Shanghai Jiaotong University in China.

Cheng-Wen Wu, Ph.D., Dean and Professor of the College of Electrical Engineering and Computer Science (EECS), National Tsing Hua University, Taiwan, received his BSEE degree from National Taiwan University in 1981 and his MSEE and EE Ph.D. degrees from the University of California, Santa Barbara, in 1985 and 1987, respectively. He joined the faculty of the Department of Electrical Engineering, National Tsing Hua University, immediately after graduation. His research interests are in the areas of memory BIST and diagnosis, memory built-in self-repair (BISR), and security processor design with related system-on-chip test issues. He has published more than 200 journal and conference papers. Among the many honors and awards Dr. Wu has received is the Guo-Guang Sports Medal from the Ministry of Education, Taiwan, the nation's highest honor bestowed upon athletes; he was honored for being a pitcher and shortstop for the national Little League Baseball Team, which won the 1971 Little League World Series. Additional honors include the Distinguished Teaching Award from National Tsing Hua University in 1996, the Outstanding Academic Research Award from Taiwan's Ministry of Education in 2005, the Outstanding Contribution Award from the IEEE Computer Society in 2005, and Best Paper awards from the International Workshop on Design and Diagnostics of Electronic Circuits and Systems (DDECS) in 2002 and the Asia and South Pacific Design Automation Conference (ASP–DAC) in 2003. Dr. Wu has served on numerous program committees for IEEE-sponsored conferences, symposia, and workshops and currently chairs a test subcommittee of the IEEE Computer Society. He was elected an IEEE Fellow in 2003 and an IEEE Computer Society Golden Core Member in 2006.

Xiaoqing Wen, Ph.D., Associate Professor at the Graduate School of Computer Science and Systems Engineering, Kyushu Institute of Technology, Japan, received his B.E. degree in Computer Science and Engineering from Tsinghua University, China, in 1986; his M.E. degree in Information Engineering from Hiroshima University, Japan, in 1990; and his Ph.D. degree in Applied Physics from Osaka University, Japan, in 1993. He was an Assistant Professor at Akita University, Japan, from 1993 to 1997 and a Visiting Researcher at the University of Wisconsin–Madison from 1995 to 1996. From 1998 to 2003, he served as the chief technology officer (CTO) at SynTest Technologies (Sunnyvale, CA), where he conducted research and development. In 2004, Dr. Wen joined the Kyushu Institute of Technology. His research interests include design for testability (DFT), test compression, logic BIST, fault diagnosis, and low-power testing. He has published more than 50 journal and conference papers and has been a co-inventor with Dr. Laung-Terng Wang of more than 15 U.S. and European patent applications, of which seven have been granted. He is a member of the IEEE, the IEICE, and the REAJ.

CHAPTER 1

INTRODUCTION

Yinghua Min
Institute of Computing Technology, Chinese Academy of Sciences, Beijing, China

Charles Stroud
Electrical and Computer Engineering, Auburn University, Auburn, Alabama

ABOUT THIS CHAPTER

The introduction of *integrated circuits* (ICs), commonly referred to as *microchips* or simply *chips*, was accompanied by the need to test these devices. *Small-scale integration* (SSI) devices, with tens of transistors in the early 1960s, and *medium-scale integration* (MSI) devices, with hundreds of transistors in the late 1960s, were relatively simple to test. However, in the 1970s, *large-scale integration* (LSI) devices, with thousands and tens of thousands of transistors, created a number of challenges when testing these devices. In the early 1980s, *very-large-scale integration* (VLSI) devices with hundreds of thousands of transistors were introduced. Steady advances in VLSI technology have resulted in devices with hundreds of millions of transistors and many new testing challenges. This chapter provides an overview of various aspects of VLSI testing and introduces fundamental concepts necessary for studying and comprehending this book.

1.1 IMPORTANCE OF TESTING

Following the so-called **Moore's law** [Moore 1965], the scale of ICs has doubled every 18 months. A simple example of this trend is the progression from SSI to VLSI devices. In the 1980s, the term "VLSI" was used for chips having more than 100,000 transistors and has continued to be used over time to refer to chips with millions and now hundreds of millions of transistors. In 1986, the first megabit *random-access memory* (RAM) contained more than 1 million transistors. Microprocessors produced in 1994 contained more than 3 million transistors [Arthistory 2005]. VLSI devices with many millions of transistors are commonly used in today's computers and electronic appliances. This is a direct result of the steadily decreasing dimensions, referred to as **feature size**, of the transistors and interconnecting wires from tens of microns to tens of nanometers, with current submicron technologies based on a feature size of less than 100 nanometers (100 nm). The reduction in feature size has also resulted in increased operating frequencies and clock speeds; for example, in 1971, the first microprocessor ran at a clock frequency of 108 KHz, while current commercially available microprocessors commonly run at several gigahertz.

The reduction in feature size increases the probability that a manufacturing **defect** in the IC will result in a faulty chip. A very small defect can easily result in a faulty transistor or interconnecting wire when the feature size is less than 100 nm. Furthermore, it takes only one faulty transistor or wire to make the entire chip fail to function properly or at the required operating frequency. Yet, defects created during the manufacturing process are unavoidable, and, as a result, some number of ICs is expected to be faulty; therefore, testing is required to guarantee fault-free products, regardless of whether the product is a VLSI device or an electronic system composed of many VLSI devices. It is also necessary to test components at various stages during the manufacturing process. For example, in order to produce an electronic system, we must produce ICs, use these ICs to assemble **printed circuit boards** (PCBs), and then use the PCBs to assemble the system. There is general agreement with the **rule of ten**, which says that the cost of detecting a faulty IC increases by an order of magnitude as we move through each stage of manufacturing, from device level to board level to system level and finally to system operation in the field.

Electronic testing includes IC testing, PCB testing, and system testing at the various manufacturing stages and, in some cases, during system operation. Testing is used not only to find the fault-free devices, PCBs, and systems but also to improve production yield at the various stages of manufacturing by analyzing the cause of defects when faults are encountered. In some systems, periodic testing is performed to ensure fault-free system operation and to initiate repair procedures when faults are detected. Hence, VLSI testing is important to designers, product engineers, test engineers, managers, manufacturers, and end-users [Jha 2003].

1.2   TESTING DURING THE VLSI LIFECYCLE

Testing typically consists of applying a set of test stimuli to the inputs of the *circuit under test* (CUT) while analyzing the output responses, as illustrated in Figure 1.1 Circuits that produce the correct output responses for all input stimuli pass the test and are considered to be fault-free. Those circuits that fail to produce a correct response at any point during the test sequence are assumed to be faulty. Testing is

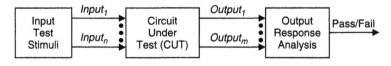

■ **FIGURE 1.1**

Basic testing approach.

performed at various stages in the lifecycle of a VLSI device, including during the VLSI development process, the electronic system manufacturing process, and, in some cases, system-level operation. In this section, we examine these various types of testing, beginning with the VLSI development process.

1.2.1 VLSI Development Process

The VLSI development process is illustrated in Figure 1.2, where it can be seen that some form of testing is involved at each stage of the process. Based on a customer or project need, a VLSI device requirement is determined and formulated as a design specification. Designers are then responsible for synthesizing a circuit that satisfies the design specification and for verifying the design. Design verification is a predictive analysis that ensures that the synthesized design will perform the required functions when manufactured. When a design error is found, modifications to the design are necessary and design verification must be repeated. As a result, design verification can be considered as a form of testing.

Once verified, the VLSI design then goes to fabrication. At the same time, test engineers develop a test procedure based on the design specification and fault models associated with the implementation technology. A **defect** is a flaw or physical imperfection that may lead to a fault. Due to unavoidable statistical flaws in the materials and masks used to fabricate ICs, it is impossible for 100% of any particular kind of IC to be defect-free. Thus, the first testing performed during the manufacturing process is to test the ICs fabricated on the wafer in order to determine which devices are defective. The chips that pass the wafer-level test are extracted and packaged. The packaged devices are retested to eliminate those devices that may have been damaged during the packaging process or put into defective packages. Additional testing is used to assure the final quality before going to market. This final testing includes measurement of such parameters as input/output timing

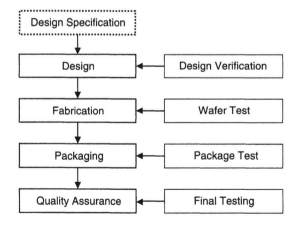

■ **FIGURE 1.2**

VLSI development process.

specifications, voltage, and current. In addition, burn-in or stress testing is often performed where chips are subjected to high temperatures and supply voltage. The purpose of burn-in testing is to accelerate the effect of defects that could lead to failures in the early stages of operation of the IC. ***Failure mode analysis*** (FMA) is typically used at all stages of IC manufacturing testing to identify improvements to processes that will result in an increase in the number of defect-free devices produced.

Design verification and yield are not only important aspects of the VLSI development process but are also important in VLSI testing. The following two subsections provide more detail on verification and yield, while their relationship to and impact on testing are discussed throughout this chapter.

1.2.1.1 Design Verification

A VLSI design can be described at different levels of abstraction, as illustrated in Figure 1.3. The design process is essentially a process of transforming a higher level description of a design to a lower level description. Starting from a design specification, a behavioral (architecture) level description is developed in ***very high speed integrated circuit hardware description language*** (VHDL) or Verilog or as a C program and simulated to determine if it is functionally equivalent to the specification. The design is then described at the ***register-transfer level*** (RTL), which contains more structural information in terms of the sequential and combinational logic functions to be performed in the data paths and control circuits. The RTL description must be verified with respect to the functionality of the behavioral description before proceeding with synthesis to the logical level.

A logical-level implementation is automatically synthesized from the RTL description to produce the gate-level design of the circuit. The logical-level implementation should be verified in as much detail as possible to guarantee the correct functionality of the final design. In the final step, the logical-level description must be transformed to a physical-level description in order to obtain the physical placement and interconnection of the transistors in the VLSI device prior to fabrication.

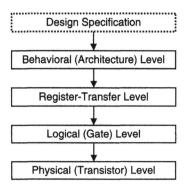

■ FIGURE 1.3

Design hierarchy.

This physical-level description is used to verify that the final design will meet timing and operating frequency specifications.

There are many tools available to assist in the design verification process including *computer-aided design* (CAD) synthesis and simulation tools, hardware emulation, and formal verification methods; however, design verification takes time, and insufficient verification fails to detect design errors. As a result, design verification is economically significant as it has a definite impact on time-to-market. It is interesting to note that many design verification techniques are borrowed from test technology because verifying a design is similar to testing a physical product. Furthermore, the test stimuli developed for design verification of the RTL, logical, and physical levels of abstraction are often used, in conjunction with the associated output responses obtained from simulation, to test the VLSI device during the manufacturing process.

1.2.1.2 Yield and Reject Rate

Some percentage of the manufactured ICs is expected to be faulty due to manufacturing defects. The **yield** of a manufacturing process is defined as the percentage of acceptable parts among all parts that are fabricated:

$$Yield = \frac{Number\ of\ acceptable\ parts}{Total\ number\ of\ parts\ fabricated}$$

There are two types of yield loss: catastrophic and parametric. Catastrophic yield loss is due to random defects, and parametric yield loss is due to process variations. Automation of and improvements in a VLSI fabrication process line drastically reduce the particle density that creates random defects over time; consequently, parametric variations due to process fluctuations become the dominant reason for yield loss.

When ICs are tested, the following two undesirable situations may occur:

1. A faulty device appears to be a good part passing the test.

2. A good device fails the test and appears as faulty.

These two outcomes are often due to a poorly designed test or the lack of *design for testability* (DFT). As a result of the first case, even if all products pass acceptance test, some faulty devices will still be found in the manufactured electronic system. When these faulty devices are returned to the IC manufacturer, they undergo FMA for possible improvements to the VLSI development and manufacturing processes. The ratio of field-rejected parts to all parts passing quality assurance testing is referred to as the **reject rate**, also called the **defect level**:

$$Reject\ rate = \frac{Number\ of\ faulty\ parts\ passing\ final\ test}{Total\ number\ of\ parts\ passing\ final\ test}$$

The reject rate provides an indication of the overall quality of the VLSI testing process [Bushnell 2000]. Generally speaking, a reject rate of 500 *parts per million*

(PPM) chips may be considered to be acceptable, while 100 PPM or lower represents high quality. The goal of **six sigma** manufacturing, also referred to as **zero defects**, is 3.4 PPM or less.

1.2.2   Electronic System Manufacturing Process

An electronic system generally consists of one or more units comprised of PCBs on which one or more VLSI devices are mounted. The steps required to manufacture an electronic system, illustrated in Figure 1.4, are also susceptible to defects. As a result, testing is required at these various stages to verify that the final product is fault-free. The PCB fabrication process is a photolithographic process similar in some ways to the VLSI fabrication process. The bare PCBs are tested in order to discard defective boards prior to assembly with expensive VLSI components. After assembly, including placement of components and wave soldering, the PCB is tested again; however, this time the PCB test includes testing of the various components, including VLSI devices, mounted on the PCB to verify that the components are properly mounted and have not been damaged during the PCB assembly process. Tested PCBs are assembled in units and systems that are tested before shipment for field operation, but unit- and system-level testing typically may not utilize the same tests as those used for the PCBs and VLSI devices.

1.2.3   System-Level Operation

When a manufactured electronic system is shipped to the field, it may undergo testing as part of the installation process to ensure that the system is fault-free before placing the system into operation. During system operation, a number of events can result in a system failure; these events include single-bit upsets, electromigration, and material aging. Suppose the state of system operation is represented as S, where $S = 0$ means the system operates normally and $S = 1$ represents a system failure. Then S is a function of time t, as shown in Figure 1.5.

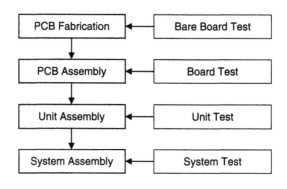

■ **FIGURE 1.4**

Manufacturing process.

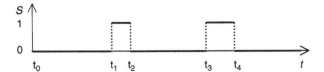

■ **FIGURE 1.5**

System operation and repair.

Suppose the system is in normal operation at $t = 0$, it fails at t_1, and the normal system operation is recovered at t_2 by some software modification, reset, or hardware replacement. Similar failure and repair events happen at t_3 and t_4. The duration of normal system operation (T_n), for intervals such as $t_1 - t_0$ and $t_3 - t_2$, is generally assumed to be a random number that is exponentially distributed. This is known as the **exponential failure law**. Hence, the probability that a system will operate normally until time t, referred to as **reliability**, is given by:

$$P(T_n > t) = e^{-\lambda t}$$

where λ is the **failure rate**. Because a system is composed of a number of components, the overall failure rate for the system is the sum of the individual failure rates (λ_i) for each of the k components:

$$\lambda = \sum_{i=0}^{k} \lambda_i$$

The ***mean time between failures*** (MTBF) is given by:

$$MTBF = \int_0^\infty e^{-\lambda t} dt = \frac{1}{\lambda}$$

Similarly, the ***repair time*** (R) is also assumed to obey an exponential distribution and is given by:

$$P(R > t) = e^{-\mu t}$$

where μ is the **repair rate**. Hence, the ***mean time to repair*** (MTTR) is given by:

$$MTTR = \frac{1}{\mu}$$

The fraction of time that a system is operating normally (failure-free) is the **system availability** and is given by:

$$System\ availability = \frac{MTBF}{MTBF + MTTR}$$

This formula is widely used in reliability engineering; for example, telephone systems are required to have system availability of 0.9999 (simply called **four nines**), while high-reliability systems may require seven nines or more.

Testing is required to ensure system availability. This testing may be in the form of **online testing** or **offline testing**, or a combination of both. Online testing is performed concurrently with normal system operation in order to detect failures as quickly as possible. Offline testing requires that the system, or a portion of the system, be taken out of service in order to perform the test. As a result, offline testing is performed periodically, usually during low-demand periods of system operation. In many cases, when online testing detects a failure, offline test techniques are then used for **diagnosis** (location and identification) of the failing replaceable component to improve the subsequent repair time. When the system has been repaired, the system, or portion thereof, is retested using offline techniques to verify that the repair was successful prior to placing the system back in service for normal operation.

The faulty components (PCBs, in most cases) replaced during the system repair procedure are sometimes sent to the manufacturing facility or a repair facility for further testing. This typically consists of board-level tests, similar to the board-level test used to test the manufactured PCBs. The goal in this case is to determine the location of the faulty VLSI devices on the PCB for replacement and repair. The PCB is then retested to verify successful repair prior to shipment back to the field for use as a replacement component for future system repairs. It should be noted that this PCB test, diagnosis, and repair scenario is viable only when it is cost effective, as might be the case with expensive PCBs. The important point to note is that testing goes on long after the VLSI development process and is performed throughout the life cycle of many VLSI devices.

1.3 CHALLENGES IN VLSI TESTING

The physical implementation of a VLSI device is very complicated. Figure 1.6 illustrates the microscopic world of the physical structure of an IC with six levels of interconnections and effective transistor channel length of $0.12\,\mu m$ [Geppert 1998]. Any small piece of dust or abnormality of geometrical shape can result in a defect. Defects are caused by process variations or random localized manufacturing imperfections. Process variations affecting transistor channel length, transistor threshold voltage, metal interconnect width and thickness, and intermetal layer dielectric thickness will impact logical and timing performance. Random localized imperfections can result in resistive bridging between metal lines, resistive opens in metal lines, improper via formation, etc.

Recent advances in physics, chemistry, and materials science have allowed production of nanometer-scale structures using sophisticated fabrication techniques. It is widely recognized that nanometer-scale devices will have much higher manufacturing defect rates compared to conventional complementary metal oxide semiconductor (CMOS) devices. They will have much lower current drive capabilities and will be much more sensitive to noise-induced errors such as **crosstalk**. They will

■ FIGURE 1.6

IBM CMOS integrated circuit with six levels of interconnections and effective transistor channel length of 0.12 μm [Geppert 1998].

be more susceptible to failures of transistors and wires due to soft (cosmic) errors, process variations, electromigration, and material aging. As the integration scale increases, more transistors can be fabricated on a single chip, thus reducing the cost per transistor; however, the difficulty of testing each transistor increases due to the increased complexity of the VLSI device and increased potential for defects, as well as the difficulty of detecting the faults produced by those defects. This trend is further accentuated by the competitive price pressures of the high-volume consumer market, as well as by the emergence of *system-on-chip* (SOC) implementations; mixed-signal circuits and systems, including *radiofrequency* (RF); and *microelectromechanical systems* (MEMSs).

1.3.1 Test Generation

A **fault** is a representation of a defect reflecting a physical condition that causes a circuit to fail to perform in a required manner. A **failure** is a deviation in the performance of a circuit or system from its specified behavior and represents an irreversible state of a component such that it must be repaired in order for it to provide its intended design function. A circuit **error** is a wrong output signal produced by a defective circuit. A circuit defect may lead to a fault, a fault can cause a circuit error, and a circuit error can result in a system failure.

To test a circuit with n inputs and m outputs, a set of input patterns is applied to the circuit under test (CUT), and its responses are compared to the known good responses of a fault-free circuit. Each input pattern is called a **test vector**. In order to completely test a circuit, many test patterns are required; however, it is difficult to know how many test vectors are needed to guarantee a satisfactory reject rate. If the CUT is an n-input combinational logic circuit, we can apply all 2^n possible input patterns for testing stuck-at faults; this approach is called **exhaustive testing**. If a circuit passes exhaustive testing, we might assume that the circuit does not contain functional faults, regardless of its internal structure. Unfortunately, exhaustive testing is not practical when n is large. Furthermore, applying all 2^n possible input patterns to an n-input sequential logic circuit will not guarantee that all possible states have been visited. However, this example of applying all possible input test patterns to an n-input combinational logic circuit also illustrates the basic idea of **functional testing**, where every entry in the truth table for the combinational logic circuit is tested to determine whether it produces the correct response. In practice, functional testing is considered by many designers and test engineers to be testing the CUT as thoroughly as possible in a system-like mode of operation. In either case, one problem is the lack of a quantitative measure of the defects that will be detected by the set of functional test vectors.

A more practical approach is to select specific test patterns based on circuit structural information and a set of **fault models**. This approach is called **structural testing**. Structural testing saves time and improves test efficiency, as the total number of test patterns is decreased because the test vectors target specific faults that would result from defects in the manufactured circuit. Structural testing cannot guarantee detection of all possible manufacturing defects, as the test vectors are generated based on specific fault models; however, the use of fault models does provide a quantitative measure of the fault-detection capabilities of a given set of test vectors for a targeted fault model. This measure is called **fault coverage** and is defined as:

$$Fault\ coverage = \frac{Number\ of\ detected\ faults}{Total\ number\ of\ faults}$$

It may be impossible to obtain a fault coverage of 100% because of the existence of undetectable faults. An undetectable fault means there is no test to distinguish the fault-free circuit from a faulty circuit containing that fault. As a result, the fault coverage can be modified and expressed as the **fault detection efficiency**, also referred to as the **effective fault coverage**, which is defined as:

$$Fault\ detection\ effeciency = \frac{Number\ of\ detected\ faults}{Total\ number\ of\ faults - number\ of\ undetectable\ faults}$$

In order to calculate fault detection efficiency, let alone reach 100% fault coverage, all of the undetectable faults in the circuit must be correctly identified, which is usually a difficult task. Fault coverage is linked to the yield and the defect level by the following expression [Williams 1981]:

$$Defect\ level = 1 - yield^{(1-fault\ coverage)}$$

From this equation, we can show that a PCB with 40 chips, each having 90% fault coverage and 90% yield, could result in a reject rate of 41.9%, or 419,000 PPM. As a result, improving fault coverage can be easier and less expensive than improving manufacturing yield because making yield enhancements can be costly; therefore, generating test stimuli with high fault coverage is very important.

Any input pattern, or sequence of input patterns, that produces a different output response in a faulty circuit from that of the fault-free circuit is a **test vector**, or sequence of test vectors, that will detect the faults. The goal of **test generation** is to find an efficient set of test vectors that detects all faults considered for that circuit. Because a given set of test vectors is usually capable of detecting many faults in a circuit, **fault simulation** is typically used to evaluate the fault coverage obtained by that set of test vectors. As a result, fault models are needed for fault simulation as well as for test generation.

1.3.2 Fault Models

Because of the diversity of VLSI defects, it is difficult to generate tests for real defects. Fault models are necessary for generating and evaluating a set of test vectors. Generally, a good fault model should satisfy two criteria: (1) It should accurately reflect the behavior of defects, and (2) it should be computationally efficient in terms of fault simulation and test pattern generation. Many fault models have been proposed [Abramovici 1994], but, unfortunately, no single fault model accurately reflects the behavior of all possible defects that can occur. As a result, a combination of different fault models is often used in the generation and evaluation of test vectors and testing approaches developed for VLSI devices.

For a given fault model there will be k different types of faults that can occur at each potential fault site ($k = 2$ for most fault models). A given circuit contains n possible fault sites, depending on the fault model. Assuming that there can be only one fault in the circuit, then the total number of possible single faults, referred to as the **single-fault model** or **single-fault assumption**, is given by:

$$Number\ of\ single\ faults = k \times n$$

In reality of course, multiple faults may occur in the circuit. The total number of possible combinations of multiple faults, referred to as the **multiple-fault model**, is given by:

$$Number\ of\ multiple\ faults = (k+1)^n - 1$$

In the multiple-fault model, each fault site can have one of k possible faults or be fault-free, hence the $(k+1)$ term. Note that the latter term in the expression (the "-1") represents the fault-free circuit, where all n fault sites are fault-free. While the multiple-fault model is more accurate than the single-fault assumption, the number of possible faults becomes impractically large other than for a small number of fault types and fault sites. Fortunately, it has been shown that high fault coverage obtained under the single-fault assumption will result in high fault coverage for the multiple-fault model [Bushnell 2000]; therefore, the single-fault assumption is typically used for test generation and evaluation.

Under the single-fault assumption, two or more faults may result in identical faulty behavior for all possible input patterns. These faults are called **equivalent faults** and can be represented by any single fault from the set of equivalent faults. As a result, the number of single faults to be considered for test generation for a given circuit is usually much less than $k \times n$. This reduction of the entire set of single faults by removing equivalent faults is referred to as **fault collapsing**. Fault collapsing helps to reduce both test generation and fault simulation times. In the following subsections, we review some well-known and commonly used fault models.

1.3.2.1   Stuck-At Faults

The stuck-at fault is a logical fault model that has been used successfully for decades. A stuck-at fault affects the state of logic signals on lines in a logic circuit, including *primary inputs* (PIs), *primary outputs* (POs), internal gate inputs and outputs, fanout stems (sources), and fanout branches. A stuck-at fault transforms the correct value on the faulty signal line to appear to be stuck at a constant logic value, either a logic 0 or a logic 1, referred to as *stuck-at-0* (SA0) or *stuck-at-1* (SA1), respectively.

Consider the example circuit shown in Figure 1.7, where the nine signal lines representing potential fault sites are labeled alphabetically. There are 18 (2×9) possible faulty circuits under the single-fault assumption. Table 1.1 gives the truth tables for the fault-free circuit and the faulty circuits for all possible single stuck-at faults. It should be noted that, rather than a direct short to a logic 0 or logic 1 value, the stuck-at fault is emulated by disconnection of the source for the signal and connection to a constant logic 0 or 1 value. This can be seen in Table 1.1, where SA0 on fanout branch line d behaves differently from SA0 on fanout branch line e, while the single SA0 fault on the fanout source line b behaves as if both fanout branches line d and line e are SA0.

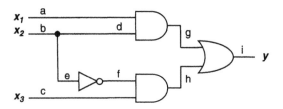

■ FIGURE 1.7

Example circuit.

TABLE 1.1 ■ Truth Tables for Fault-Free and Faulty Circuits of Figure 1.7

$x_1 x_2 x_3$	000	001	010	011	100	101	110	**111**
y	0	1	0	0	0	1	1	1
a SA0	0	1	0	0	0	1	0	0
a SA1	0	1	1	1	0	1	1	1
b SA0	0	1	0	1	0	1	0	1
b SA1	0	0	0	0	1	1	1	1
c SA0	0	0	0	0	0	0	1	1
c SA1	1	1	0	0	1	1	1	1
d SA0	0	1	0	0	0	1	0	0
d SA1	0	1	0	0	1	1	1	1
e SA0	0	1	0	1	0	1	1	1
e SA1	0	0	0	0	0	0	1	1
f SA0	0	0	0	0	0	0	1	1
f SA1	0	1	0	1	0	1	1	1
g SA0	0	1	0	0	0	1	0	0
g SA1	1	1	1	1	1	1	1	1
h SA0	0	0	0	0	0	0	1	1
h SA1	1	1	1	1	1	1	1	1
i SA0	0	0	0	0	0	0	0	0
i SA1	1	1	1	1	1	1	1	1

The truth table entries where the faulty circuit produces an output response different from that of the fault-free circuit are highlighted in gray. As a result, the input values for the highlighted truth table entries represent valid test vectors to detect the associated stuck-at faults. With the exception of line *d* SA1, line *e* SA0, and line *f* SA1, all other faults can be detected with two or more test vectors; therefore, test vectors 011 and 100 must be included in any set of test vectors that will obtain 100% fault coverage for this circuit. These two test vectors detect a total of ten faults, and the remaining eight faults can be detected with test vectors 001 and 110; therefore, this set of four test vectors obtains 100% single stuck-at fault coverage for this circuit.

Four sets of equivalent faults can be observed in Table 1.1. One fault from each set can be used to represent all of the equivalent faults in that set. Because there is a total of ten unique faulty responses to the complete set of input test patterns, then ten faults constitute the set of collapsed faults for the circuit. Stuck-at **fault**

collapsing typically reduces the total number of faults by 50 to 60% [Bushnell 2000]. Fault collapsing for stuck-at faults is based on the fact that a SA0 at the input to an AND (NAND) gate is equivalent to the SA0 (SA1) at the output of the gate. Similarly, a SA1 at the input to an OR (NOR) gate is equivalent to the SA1 (SA0) at the output of the gate. For an inverter, a SA0 (SA1) at the input is equivalent to the SA1 (SA0) at the output of the inverter. Furthermore, a stuck-at fault at the source (output of the driving gate) of a fanout-free net is equivalent to the same stuck-at fault at the destination (gate input being driven). Therefore, the number of collapsed stuck-at faults in any combinational circuit constructed from elementary logic gates (AND, OR, NAND, NOR, and inverter) is given by:

$$Number\ of\ collapsed\ faults = 2 \times (number\ of\ POs + number\ of\ fanout\ stems)$$
$$+\ total\ number\ of\ gate\ (including\ inverter)\ inputs$$
$$-\ total\ number\ of\ inverters$$

The example circuit in Figure 1.7 has one primary output and one fanout stem. The total number of gate inputs is 7, including the input to the one inverter; therefore, the number of collapsed faults $= 2 \times (1+1) + 7 - 1 = 10$. Note that single-input gates, including buffers, are treated the same as an inverter in the calculation of the number of collapsed faults because all faults at the input of the gate are equivalent to faults at the output.

A number of interesting properties are associated with detecting stuck-at faults in combinational logic circuits; for example, two such properties are described by the following theorems [Abramovici 1994]:

Theorem 1.1

A set of test vectors that detects all single stuck-at faults on all primary inputs of a fanout-free combinational logic circuit will detect all single stuck-at faults in that circuit.

Theorem 1.2

A set of test vectors that detect all single stuck-at faults on all primary inputs and all fanout branches of a combinational logic circuit will detect all single stuck-at faults in that circuit.

The stuck-at fault model can also be applied to sequential circuits; however, high fault coverage test generation for sequential circuits is much more difficult than for combinational circuits because, for most faults in a sequential logic circuit, it is necessary to generate sequences of test vectors. Therefore, DFT techniques are frequently used to ease sequential circuit test generation.

Although it is physically possible for a line to be SA0 or SA1, many other defects within a circuit can also be detected with test vectors developed to detect stuck-at faults. The idea of **N-detect** single stuck-at fault test vectors was proposed to detect more defects not covered by the stuck-at fault model [Ma 1995]. In an *N*-detect set of test vectors, each single stuck-at fault is detected by at least *N* different test vectors; however, test vectors generated using the stuck-at fault model do not necessarily guarantee the detection of all possible defects, so other fault models are needed.

1.3.2.2 Transistor Faults

At the switch level, a transistor can be **stuck-open** or **stuck-short**, also referred to as **stuck-off** or **stuck-on**, respectively. The stuck-at fault model cannot accurately reflect the behavior of stuck-open and stuck-short faults in CMOS logic circuits because of the multiple transistors used to construct CMOS logic gates. To illustrate this point, consider the two-input CMOS NOR gate shown in Figure 1.8. Suppose transistor N_2 is stuck-open. When the input vector $AB = 01$ is applied, output Z should be a logic 0, but the stuck-open fault causes Z to be isolated from ground (V_{SS}). Because transistors P_2 and N_1 are not conducting at this time, Z keeps its previous state, either a logic 0 or 1. In order to detect this fault, an ordered sequence of two test vectors $AB = 00 \rightarrow 01$ is required. For the fault-free circuit, the input 00 produces $Z = 1$ and 01 produces $Z = 0$ such that a falling transition at Z appears. But, for the faulty circuit, while the test vector 00 produces $Z = 1$, the subsequent test vector 01 will retain $Z = 1$ without a falling transition such that the faulty circuit behaves like a level-sensitive latch. Thus, a stuck-open fault in a CMOS combinational circuit requires a sequence of two vectors for detection rather than a single test vector for a stuck-at fault.

Stuck-short faults, on the other hand, will produce a conducting path between V_{DD} and V_{SS}. For example, if transistor N_2 is stuck-short, there will be a conducting path between V_{DD} and V_{SS} for the test vector 00. This creates a voltage divider at the output node Z where the logic level voltage will be a function of the resistances of the conducting transistors. This voltage may or may not be interpreted as an incorrect logic level by the gate inputs driven by the gate with the transistor fault; however, stuck-short transistor faults may be detected by monitoring the power supply current during steady state, referred to as I_{DDQ}. This technique of monitoring the steady-state power supply current to detect transistor stuck-short faults is referred to as **I_{DDQ} testing**.

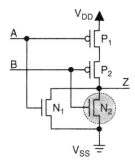

■ **FIGURE 1.8**

Two-input CMOS NOR gate.

The circuit in Figure 1.8 has a total of eight (2×4) possible single transistor faults; however, there are equivalent faults at the transistor level, as stuck-open faults in a group of series transistors (such as P_1 and P_2) are indistinguishable. The same holds true for stuck-short faults in a group of parallel transistors (such as N_1 and N_2); therefore, fault collapsing can be applied to transistor-level circuits [Stroud 2002]. The number of collapsed transistor faults in a circuit is given by:

$$Number\ of\ collapsed\ faults = 2 \times T - T_S + G_S - T_P + G_P$$

where T is the total number of transistors, T_S is the total number of transistors in series, G_S is the total number of groups of transistors in series, T_P is the total number of transistors in parallel, and G_P is the total number of groups of transistors in parallel. For the two-input NOR gate of Figure 1.8, there are four transistors $(T = 4)$, two transistors $(P_1$ and $P_2)$ in the only group of series transistors $(T_S = 2$ and $G_S = 1)$, and two transistors $(N_1$ and $N_2)$ in the only group of parallel transistors $(T_P = 2$ and $G_P = 1)$; hence, the number of collapsed faults is 6. The fault equivalence associated with the transistors can also be seen in Table 1.2, which gives the behavior of the fault-free circuit and each of the 8 possible faulty circuits under the single-fault assumption. Note that table entries labeled "last Z" indicate that the output node will retain its previous value and would require a two-test vector sequence for detection. Similarly, entries labeled "I_{DDQ}" indicate that the output node logic value will be a function of the voltage divider of the conducting transistors and can be detected by I_{DDQ} testing. Because both N_1 and N_2 stuck-short faults as well as P_1 and P_2 stuck-open faults can be tested by the same test set, the collapsed fault count is 6, as proven above.

1.3.2.3 Open and Short Faults

Defects in VLSI devices can include **opens** and **shorts** in the wires that interconnect the transistors forming the circuit. Opens in wires tend to behave like transistor

TABLE 1.2 ■ Truth Tables for Fault-Free and Faulty Circuits of Figure 1.8

AB	00	01	10	11
Z	1	0	0	0
N_1 stuck-open	1	0	Last Z	0
N_1 stuck-short	I_{DDQ}	0	0	0
N_2 stuck-open	1	Last Z	0	0
N_2 stuck-short	I_{DDQ}	0	0	0
P_1 stuck-open	Last Z	0	0	0
P_1 stuck-short	1	0	I_{DDQ}	0
P_2 stuck-open	Last Z	0	0	0
P_2 **stuck-short**	1	I_{DDQ}	0	0

stuck-open faults when the faulty wire segment is interconnecting transistors to form gates. On the other hand, opens tend to behave like stuck-at faults when the faulty wire segment is interconnecting gates. Therefore, a set of test vectors that provide high stuck-at fault coverage and high transistor fault coverage will also detect open faults; however, a resistive open does not behave the same as a transistor or stuck-at fault but instead affects the propagation delay of the signal path, as will be discussed in the next subsection.

A short between two elements is commonly referred to as a **bridging fault**. These elements can be transistor terminals or connections between transistors and gates. The case of an element being shorted to power (V_{DD}) or ground (V_{SS}) is equivalent to the stuck-at fault model; however, when two signal wires are shorted together, bridging fault models are required. In the first bridging fault model proposed, the logic value of the shorted nets was modeled as a logical AND or OR of the logic values on the shorted wires. This model is referred to as the **wired-AND/wired-OR** bridging fault model. The wired-AND bridging fault means the signal net formed by the two shorted lines will take on a logic 0 if either shorted line is sourcing a logic 0, while the wired-OR bridging fault means the signal net will take on a logic 1 if either of the two lines is sourcing a logic 1. Therefore, this type of bridging fault can be modeled with an additional AND or OR gate, as illustrated in Figure 1.9a, where A_S and B_S denote the sources for the two shorted signal nets and A_D and B_D

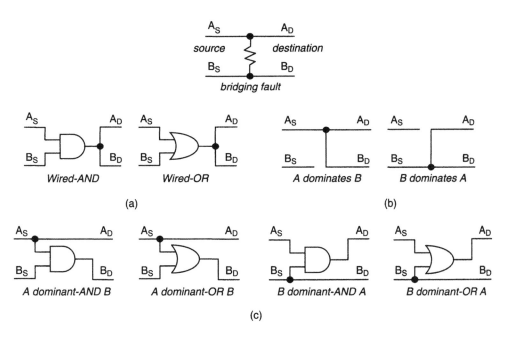

■ **FIGURE 1.9**

Bridging fault models.

TABLE 1.3 ■ Truth Tables for Bridging Fault Models of Figure 1.9

$A_S B_S$	0	0	0	1	1	0	1	1
$A_D B_D$	0	0	0	1	1	0	1	1
Wired-AND	0	0	0	0	0	0	1	1
Wired-OR	0	0	1	1	1	1	1	1
A dominates B	0	0	0	0	1	1	1	1
B dominates A	0	0	1	1	0	0	1	1
A dominant-AND B	0	0	0	0	1	0	1	1
B dominant-AND A	0	0	0	1	0	0	1	1
A dominant-OR B	0	0	0	1	1	1	1	1
B dominant-OR A	**0**	**0**	**1**	**1**	**1**	**0**	**1**	**1**

denote the destinations for the two nets. The truth tables for fault-free and faulty behavior are given in Table 1.3.

The wired-AND/wired-OR bridging fault model was originally developed for bipolar VLSI and does not accurately reflect the behavior of bridging faults typically found in CMOS devices; therefore, the **dominant bridging fault** model was proposed for CMOS VLSI where one driver is assumed to dominate the logic value on the two shorted nets. Two fault types are normally evaluated per fault site, where each driver is allowed to dominate the logic value on the shorted signal net (see Figure 1.9b). The dominant bridging fault model is more difficult to detect because the faulty behavior can only be observed on the dominated net, as opposed to both nets in the case of the wired-AND/wired-OR bridging fault model. However, it has been shown, and can be seen from the faulty behavior in Table 1.3, that a set of test vectors that detects all dominant bridging faults is also guaranteed to detect all wired-AND and wired-OR bridging faults.

The dominant bridging fault model does not accurately reflect the behavior of a resistive short in some cases. A recent bridging fault model has been proposed based on the behavior of resistive shorts observed in some CMOS VLSI devices [Stroud 2000]. In this fault model, referred to as the **dominant-AND/dominant-OR** bridging fault, one driver dominates the logic value of the shorted nets but only for a given logic value (see Figure 1.9c). While there are four fault types to evaluate for this fault model, as opposed to only two for the dominant and wired-AND/wired-OR models, a set of test vectors that detect all four dominant-AND/dominant-OR bridging faults will also detect all dominant and wired-AND/wired-OR bridging faults at that fault site.

Bridging faults commonly occur in practice and can be detected by I_{DDQ} testing. It has also been shown that many bridging faults are detected by a set of test vectors that obtains high stuck-at fault coverage, particularly with N-detect single stuck-at fault test vectors. In the presence of a bridging fault, a combinational logic circuit can have a feedback path and behave like a sequential logic circuit, making the testing problem more complicated. Another complication in test generation for bridging faults is the number of possible fault sites *versus* the number of realistic fault sites. While there are many signal nets in a VLSI circuit, it is impractical to evaluate detection of bridging faults between any possible pair of nets; for example, a circuit with N signal nets would have $N\text{-}choose\text{-}2 = N \times (N-1)/2$ possible fault sites, but a bridging fault between two nets on opposite sides of the device may not be possible. One solution to this problem is to extract likely bridging fault sites from the physical design after physical layout.

1.3.2.4 Delay Faults and Crosstalk

Fault-free operation of a logic circuit requires not only performing the logic function correctly but also propagating the correct logic signals along paths within a specified time limit. A **delay fault** causes excessive delay along a path such that the total propagation delay falls outside the specified limit. Delay faults have become more prevalent with decreasing feature sizes.

There are different delay fault models. In the **gate-delay fault** and the **transition fault** models, a delay fault occurs when the time interval taken for a transition from the gate input to its output exceeds its specified range. It should be noted that simultaneous transitions at inputs of a gate may change the gate delay significantly due to activation of multiple charge/discharge paths. The differences between the gate-delay and transition fault models will be discussed in more detail in Chapter 12. The other model is **path-delay fault**, which considers the cumulative propagation delay along a signal path through the CUT—in other words, the sum of all gate delays along the path; therefore, the path-delay fault model is more practical for testing than the gate-delay fault (or the transition fault) model. A critical problem encountered when dealing with path-delay faults is the large number of possible paths in practical circuits. This number, in the worst case, is exponential for the number of lines in the circuit, and in most practical cases the number of paths in a circuit makes it impossible to enumerate all path-delay faults for the purpose of test generation or fault simulation.

As with transistor stuck-open faults, delay faults require an ordered pair of test vectors to sensitize a path through the logic circuit and to create a transition along that path in order to measure the path delay. For example, consider the circuit in Figure 1.10, where the fault-free delay associated with each gate is denoted by the integer value label on that gate. The two test vectors, v_1 and v_2, shown in the figure

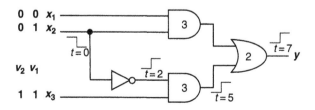

■ FIGURE 1.10

Path-delay fault test.

are used to test the path delay from input x_2, through the inverter and lower AND gate, to the output y. Assuming the transition between the two test vectors occurs at time $t = 0$, the resulting transition propagates through the circuit with the fault-free delays shown at each node in the circuit such that we expect to see the transition at the output y at time $t = 7$. A delay fault along this path would create a transition at some later time, $t > 7$. Of course, this measurement could require a high-speed, high-precision test machine.

With decreasing feature sizes and increasing signal speeds, the problem of modeling gate delays becomes more difficult. As technologies approach the deep submicron region, the portion of delay contributed by gates reduces while the delay due to interconnect becomes dominant. This is because the interconnect lengths do not scale in proportion to the shrinking area of transistors that make up the gates. In addition, if the operating frequencies also increase with scaling, then the on-chip inductances can play a role in determining the interconnect delay for long wide wires, such as those in clock trees and buses. However, wire delays can be taken into account in the path-delay fault model based on the physical layout, as interconnections are included in paths. As a result, it is no longer true that a path delay is equal to the sum of all delays of gates along the path.

The use of nanometer technologies increases cross-coupling capacitance and inductance between interconnects, leading to severe crosstalk effects that may result in improper functioning of a chip. Crosstalk effects can be separated to two categories: **crosstalk glitches** and **crosstalk delays**. A crosstalk glitch is a pulse that is provoked by coupling effects among interconnect lines. The magnitude of the glitch depends on the ratio of the coupling capacitance to the line-to-ground capacitance. When a transition signal is applied on a line that has a strong driver while stable signals are applied at other lines that have weaker drivers, the stable signals may experience coupling noise due to the transition of the stronger signal. Crosstalk delay is a signal delay that is provoked by the same coupling effects among interconnect lines, but it may be produced even if line drivers are balanced but have large loads. Because crosstalk causes a delay in addition to normal gate and interconnect delay, it is difficult to estimate the true circuit delay, which may lead to severe signal delay problems. Conventional delay fault analysis may be invalid if these effects are not taken into consideration based on the physical layout. Several design techniques, including physical design and analysis tools, are being developed to help design for margin and minimization of crosstalk problems; however, it may be impossible to anticipate in advance the full range of process variations and manufacturing defects that may significantly aggravate the cross-coupling effects. Hence, there is a critical need to develop testing techniques for manufacturing defects that produce crosstalk effects.

1.3.2.5 *Pattern Sensitivity and Coupling Faults*

Manufacturing defects can be of a wide variety and manifest themselves as faults that are not covered by the specific fault models for digital circuits discussed thus far. This is particularly true in the case of densely packed memories. In high-density RAMs, the contents of a cell or the ability of a memory cell to change can be influenced by the contents of its neighboring cells, referred to as a **pattern sensitivity fault**. A **coupling fault** results when a transition in one cell causes the content of another cell to change. Therefore, it is necessary when testing memories to add tests for pattern sensitivity and coupling faults in addition to stuck-at faults. Extensive work has been done on memory testing and many memory test algorithms have been proposed [van de Goor 1991] [Bushnell 2000]. One of the most efficient RAM test algorithms, in terms of test time and fault detection capability, currently in use is the March LR algorithm illustrated in Table 1.4. This algorithm has a test time on the order of $16N$, where N is the number of address locations, and is capable of detecting pattern sensitivity faults, intra-word coupling faults, and bridging faults in the RAM. For word-oriented memories, a *background data sequence* (BDS) must be added to detect faults within each word of the memory. The March LR with BDS shown in Table 1.4 is for a RAM with 2-bit words. In general, the number of BDSs $= \log_2(K) + 1$, where K is the number of bits per word.

1.3.2.6 Analog Fault Models

Analog circuits are constructed with passive and active components. Typical analog fault models include shorts, opens, and parameter variations in both active and passive components. Shorts and opens usually result in **catastrophic faults** that are relatively easy to detect. Parameter variations that cause components to be out of their tolerance ranges result in **parametric faults**. An active component can suffer from both *direct current* **(DC) faults** and *alternate current* **(AC) faults**. Op amps typically occupy a much larger silicon area in monolithic ICs than passive components and, hence, are more prone to manufacturing defects. As is the case with a catastrophic fault, a single parametric fault can result in a malfunctioning analog circuit; however, it is difficult to identify critical parameters and to supply a model of process fluctuations. Furthermore, because of the complex nature of analog circuits, a direct application of digital fault models, other than shorts and opens, is inadequate in capturing faulty behavior in analog circuits. It is also difficult to model all practical faults.

TABLE 1.4 ■ March LR RAM Test Algorithm

Test Algorithm	March Test Sequence
March LR w/o BDS	$\updownarrow$(w0); $\downarrow$(r0, w1); $\uparrow$(r1, w0, r0, r0, w1);
	$\uparrow$(r1, w0); $\uparrow$(r0, w1, r1, r1, w0); $\uparrow$(r0)
March LR with BDS	$\updownarrow$(w00); $\downarrow$(r00, w11); $\uparrow$(r11, w00, r00, r00, w11);
	$\uparrow$(r11, w00); $\uparrow$(r00, w11, r11, r11, w00);
	$\uparrow$(r00, w01, w10, r10); $\uparrow$(r10, w01, r01); $\uparrow$(r01)

Notation: w0 = write 0 (or all 0's); r1 = read 1 (or all 1's); $\uparrow$ = address up; $\downarrow$ = address down; $\updownarrow$ = address either way.

1.4   LEVELS OF ABSTRACTION IN VLSI TESTING

In the design hierarchy, a higher level description has fewer implementation details but more explicit functional information than a lower level description. As described in Section 1.2.1.1, the various levels of abstraction include behavioral (architecture), register-transfer, logical (gate), and physical (transistor) levels. The hierarchical design process lends itself to hierarchical test development, but the fault models described in the previous section are more appropriate for particular levels of abstraction. In this section, we discuss test generation and the use of fault models at these various levels of abstraction.

1.4.1   Register-Transfer Level and Behavioral Level

The demand for CAD tools for the design of digital circuits at high levels of abstraction has led to the development of synthesis and simulation technologies. The methodology in common practice today is to design, simulate, and synthesize *application-specific integrated circuits* (ASICs) of millions of gates at the RTL. So-called "black boxes" or *intellectual property* (IP) cores are often incorporated in VLSI design, especially in SOC design, for which there may be very little, if any, structural information. Traditional *automatic test pattern generation* (ATPG) tools cannot effectively handle designs employing blocks for which the implementation detail is either unknown or subject to change; however, several approaches to test pattern generation at the RTL have been proposed. Most of these approaches are able to generate test patterns of good quality, sometimes comparable to gate-level ATPG tools. It is the lack of general applicability that prevents these approaches from being widely accepted. Although some experimental results have shown that RTL fault coverage can be quite close to fault coverage achieved at the gate level when designs are completed and mapped to a technology library, it is unrealistic to expect that stuck-at fault coverage at the RTL will be as high as at the gate level [Min 2002].

To illustrate the importance of knowledge of the gate-level implementation on test generation, consider the two example circuits of Figure 1.11 which implement the following logic function, where x represents a "don't care" product term:

$$f = \bar{a}b\bar{c} + abc + xab\bar{c}$$

Because both circuits are valid implementations of the functional description, the gate-level implementation is not unique for a given RTL description. As a result, it may be difficult to generate tests at the RTL and achieve stuck-at fault coverage as high as at the gate level, as the stuck-at fault model is defined at the gate level. For example, if the "don't care" product term is assigned a logic 0, we obtain the logic equation along with resultant implementation and associated set of test vectors to detect all stuck-at faults shown in Figure 1.11a. If the "don't care" term is assigned a logic 1, on the other hand, we obtain the logic equation, gate-level implementation, and set of test vectors shown in Figure 1.11b. Note that the

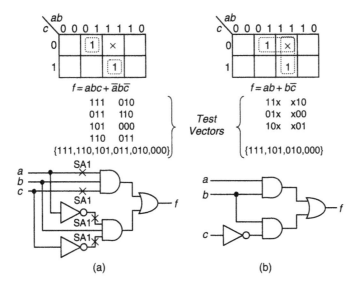

■ **FIGURE 1.11**

Example of different implementations and their test vectors.

set of test vectors for Figure 1.11b is a subset of those required for Figure 1.11a and, as a result, would not detect the four SA1 faults shown in the gate-level implementation of Figure 1.11a. This example can also be illustrated by considering Theorems 1.1 and 1.2. If ATPG assumes that a combinational logic circuit will be fanout free based on the functional description, it could produce test vectors to detect stuck-at faults for all primary inputs based on Theorem 1.1. Yet, if the synthesized circuit contains fanout stems, the set of test vectors produced by the APTG may not detect stuck-at faults on all fanout branches and, as a result of Theorem 1.2, may not detect all stuck-at faults in the circuit. Note that the four SA1 faults in Figure 1.11a not detected by the test vectors in Figure 1.11b are located on the additional fanout branches in Figure 1.11a. Therefore, if the ATPG is based on the functional description, test vectors can be generated based on assumptions that may not necessarily hold once the gate-level implementation is synthesized. Regardless, it is desirable to move ATPG operations toward higher levels of abstraction while targeting new types of faults in deep submicron devices. Because the main advantages of high-level approaches are compact test sets and reduced computation time, it is expected that this trend will continue.

1.4.2   Gate Level

For decades, traditional IC test generation has been at the gate level based on the gate-level netlist. The stuck-at fault model can easily be applied for which many ATPG and fault simulation tools are commercially available. Very often the stuck-at fault model is also employed to evaluate the effectiveness of the input stimuli used for simulation-based design verification. As a result, the design verification stimuli are often also used for fault detection during manufacturing testing. In addition to the stuck-at fault model, delay fault models and delay testing have been traditionally based on the gate-level description. While bridging faults can be modeled at the gate level, practical selection of potential bridging fault sites requires physical design information. The gate-level description has advantages of functionality and tractability because it lies between the RTL and physical levels; however, it is now widely believed that test development at the gate level is not sufficient for deep submicron designs.

1.4.3 Switch Level

For standard cell-based VLSI implementations, transistor fault models (stuck-open and stuck-short) can be applied and evaluated based on the gate-level netlist. When the switch-level model for each gate in the netlist is substituted, we obtain an accurate abstraction of the netlist used for physical layout. In addition, transmission gate and tristate buffer faults can also be tested at the switch level. For example, it may be necessary to place buffers in parallel for improved drive capabilities. In most gate-level models, these buffers will appear as a single buffer, but it is possible to model a fault on any of the multiple buffers at the switch level. Furthermore, a defect-based test methodology can be more effective with a switch-level model of the circuit as it contains more detailed structural information than a gate-level abstraction and will yield a more accurate defect coverage analysis. Of course, the switch-level description is more complicated than the gate-level description for both ATPG and fault simulation.

1.4.4 Physical Level

The physical level of abstraction is the most important for VLSI testing because it provides the actual layout and routing information for the fabricated device and, hence, the most accurate information for delay faults, crosstalk effects, and bridging faults. For deep submicron IC chips, in order to characterize electrical properties of interconnections, a distributed *resistance–inductance–capacitance* (RLC) model is based on the physical layout. This is then used to analyze and test for potential crosstalk problems. Furthermore, interconnect delays can be incorporated for more accurate delay fault analysis.

One solution to the problem of determining likely bridging fault sites is to extract the capacitance between the wires from the physical design after layout and routing [Maxwell 1994]. This provides an accurate determination of those wires that are adjacent and, therefore, likely to sustain bridging faults. In addition, the value of the capacitance between two adjacent wires is proportional to the distance between the wires and/or the length of adjacency. As a result, fault sites with the highest capacitance value can be targeted for test generation and evaluation as these sites have a higher probability of incurring bridging faults.

1.5 HISTORICAL REVIEW OF VLSI TEST TECHNOLOGY

VLSI testing includes two processes: test generation and test application. The goal of test generation is to produce test patterns for efficient testing. Test application is the process of applying those test patterns to the CUT and analyzing the output responses. Test application is performed by either *automatic test equipment* (ATE) or test facilities in the chip itself. This section gives a brief historical review of VLSI test technology development.

1.5.1 Automatic Test Equipment

Automatic test equipment (ATE) is computer-controlled equipment used in the production testing of ICs (both at the wafer level and in packaged devices) and PCBs. Test patterns are applied to the CUT and the output responses are compared to stored responses for the fault-free circuit. In the 1960s, when ICs were first introduced, it was foreseen that testing would become a bottleneck to high-volume production of ICs unless the tasks normally performed by technicians and laboratory instruments could be automated. An IC tester controlled by a minicomputer was developed in the mid-1960s, and the ATE industry was established. Since then, with advances in VLSI and computer technology, the ATE industry has developed electronic subassemblies (PCBs and backplanes), test systems, digital IC testers, analog testers, and SOC testers. A custom tester is often developed for testing a particular product, but a general-purpose ATE is often more flexible and enhances the productivity of high-volume manufacturing. Generally, ATE consists of the following parts:

1. *Computer*—A powerful computer is the heart of any ATE for central control and for making the test and measurement flexible for different products and different test purposes.

2. *Pin electronics and fixtures*—ATE architectures can be divided into two major subcomponents, the data generator and the pin electronics. The data generator supplies the input test vectors for the CUT, while the pin electronics are responsible for formatting these vectors to produce waveforms of the desired shape and timing. The pin electronics are also responsible for sampling the CUT output responses at the desired time. In order to actually touch the pads of an IC on a wafer or the pins of a packaged chip during testing, it is necessary to have a fixture with probes for each pin of the IC under test. Current VLSI devices may have over 1000 pins and require a tester with as many as 1024 pin channels. As a result, the pin electronics and fixtures constitute the most expensive part of the ATE.

3. *Test program*—In conjunction with the pin electronics, ATE contains waveform generators that are designed to change logic values at the setup and hold times associated with a given input pin. A test pattern containing logic 1's and 0's must be translated to these various timing formats. Also, ATE captures primary output responses, which are then translated to output vectors

for comparison with the fault-free responses. These translations and some environment settings are controlled by the central computer; therefore, a test program, usually written in a high-level language, becomes an important ingredient for controlling these translations and environment settings. Algorithmically generated test patterns may consist of subroutines, pattern and test routine calls, or sequenced events. The test program also specifies the timing format in terms of the tester **edge set**. An edge set is a data format with timing information for applying new data to a chip input pin and includes the input setup time, hold time, and the waveform type.

4. *Digital signal processor* (DSP)—Powerful 32-bit DSP techniques have been widely applied to analog testing for capturing analog characteristics at high frequencies. Digital signals are converted to analog signals and applied to the analog circuit inputs, while the analog output signals are converted to digital signals for response analysis by the DSP.

5. *Accurate DC and AC measurement circuitry*—ATE precision is a performance metric specifying the smallest measurement that the tester can resolve in a very low noise environment, especially for analog and mixed-signal testing. For example, a clock jitter (phase noise) of no more than 10 ps is required to properly test ICs that realize more than 100 Mb/s data rates. This requirement is even higher for today's high-performance ICs. The application of vectors to a circuit with the intent of verifying the timing compliance depends on the operational frequency of the ATE (*e.g.*, 200 MHz, 500 MHz, or 1 GHz). Ideally, the ATE operational frequency should be much higher than that of the ICs under test. Unfortunately, this is a difficult problem because the ATE itself is also constructed from ICs and limited by their maximum operating frequency.

Automatic test equipment can be very expensive. To satisfy the needs of advanced VLSI testing, the following features form the basis for keeping ATE costs under control:

1. *Modularization*—Modular systems give users the flexibility to purchase and use only those options that are suitable for the products under test.

2. *Configurability*—Test system configurability is essential for many test platforms. As testing needs change, users can reconfigure the test resources for particular products and continue to use the same basic framework.

3. *Parallel test capabilities*—Testing multiple devices in parallel improves the throughput and productivity of the ATE. Higher throughput means lower overall test cost.

4. *Third-party components*—The use of third-party hardware and software permits adopting the best available equipment and approaches, thus giving rise to competition that lowers test cost over time.

From a test economics point of view, there has been a systematic decrease in the capital cost of manufacturing a transistor over the past several decades as we continue to deliver more complex devices; however, testing capital costs per transistor have remained relatively constant. As a result, test costs are becoming an increasing portion of the overall industry capital requirement per transistor, to the extent that currently it costs almost as much to test as to manufacture a transistor.

From a test technology point of view on the other hand, ATE in the early 1980s had resolution capabilities well in excess of the component requirements. In 1985, for example, when testing a then fast 8-MHz 286 microprocessor, a 1-ns accuracy in the control of input signal transitions, referred to as **edge placement**, was available in ATE with very low yield loss due to tester tolerances. Later, for testing 700-MHz Pentium III microprocessors, only a 100-ps edge placement accuracy was available in ATE; thus, the hundredfold increase in CUT speed was accompanied by only a tenfold increase in the tester accuracy [Gelsinger 2000].

1.5.2 Automatic Test Pattern Generation

In the early 1960s, structural testing was introduced and the stuck-at fault model was employed. A complete ATPG algorithm, called the **D-algorithm**, was first published [Roth 1966]. The D-algorithm uses a logical value to represent both the "good" and the "faulty" circuit values simultaneously and can generate a test for any stuck-at fault, as long as a test for that fault exists. Although the computational complexity of the D-algorithm is high, its theoretical significance is widely recognized. The next landmark effort in ATPG was the **PODEM** algorithm [Goel 1981], which searches the circuit primary input space based on simulation to enhance computation efficiency. Since then, ATPG algorithms have become an important topic for research and development, many improvements have been proposed, and many commercial ATPG tools have appeared. For example, **FAN** [Fujiwara 1983] and **SOCRATES** [Schulz 1988] were remarkable contributions to accelerating the ATPG process. Underlying many current ATPG tools, a common approach is to start from a random set of test patterns. Fault simulation then determines how many of the potential faults are detected. With the fault simulation results used as guidance, additional vectors are generated for hard-to-detect faults to obtain the desired or reasonable fault coverage. The International Symposium on Circuits and Systems (ISCAS) announced combinational logic benchmark circuits in 1985 [Brglez 1985] and sequential logic benchmark circuits in 1989 [Brglez 1989] to assist in ATPG research and development in the international test community. A major problem in large combinational logic circuits with thousands of gates was the identification of undetectable faults. In the 1990s, very fast ATPG systems were developed using advanced high-performance computers which provided a speed-up of five orders of magnitude from the D-algorithm with 100% fault detection efficiency. As a result, ATPG for combinational logic is no longer a problem; however, ATPG for sequential logic is still difficult because, in order to propagate the effect of a fault to a primary output so it can be observed and detected, a state sequence must be traversed with the fault undertaken. For large sequential circuits, it is difficult to reach 100% fault coverage in reasonable computational time and cost unless DFT techniques are adopted [Breuer 1987].

1.5.3 Fault Simulation

A fault simulator emulates the target faults in a circuit in order to determine which faults are detected by a given set of test vectors. Because there are many faults to emulate for fault detection analysis, fault simulation time is much greater than that required for design verification. To accelerate the fault simulation process, improved approaches have been developed in the following order. **Parallel fault simulation** uses bit-parallelism of logical operations in a digital computer. Thus, for a 32-bit machine, 31 faults are simulated simultaneously. **Deductive fault simulation** deduces all signal values in each faulty circuit from the fault-free circuit values and the circuit structure in a single pass of true-value simulation augmented with the deductive procedure. **Concurrent fault simulation** is essentially an event-driven simulation to emulate faults in a circuit in the most efficient way. Hardware fault simulation accelerators based on parallel processing are also available to provide a substantial speed-up over purely software-based fault simulators.

For analog and mixed-signal circuits, fault simulation is traditionally performed at the transistor level using circuit simulators such as HSPICE. Unfortunately, analog fault simulation is a very time-consuming task and, even for rather simple circuits, a comprehensive fault simulation is normally not feasible. This problem is further complicated by the fact that acceptable component variations must be simulated along with the faults to be emulated, which requires many Monte Carlo simulations to determine whether the fault will be detected. Macro models of circuit components are used to decrease the long computation time. Fault simulation approaches using high-level simulators can simulate analog circuit characteristics based on differential equations but are usually avoided due to lack of adequate fault models.

1.5.4 Digital Circuit Testing

The development of digital circuit testing began with the introduction of the stuck-at fault model which was followed by the first bridging fault model, the transistor fault model, and finally by delay fault models. Digital testing now typically uses a combination of tests developed for different fault models because tests for any given fault model cannot assure the detection of all defects. For example, current testing practices by some manufacturers include stuck-at fault tests with 99% fault coverage in conjunction with path-delay fault tests with greater than 90% fault coverage.

Digital testing is also improved by monitoring the **quiescent power supply current** (I_{DDQ}). Normally, the leakage current of CMOS circuits under a quiescent state is very small and negligible. When a fault occurs, such as a transistor stuck-short or a bridging fault, and causes a conducting path from power to ground, it may draw an excessive supply current. I_{DDQ} testing became an accepted test method for the IC industry in the 1980s; however, normal fault-free I_{DDQ} has become quite large for current, complex VLSI devices due to the collective leakage currents of millions of transistors on a chip. This makes the detection of the additional I_{DDQ} current due to a single faulty transistor or bridging fault difficult; hence, I_{DDQ} testing is becoming ineffective.

A similar approach is **transient power supply current** (I_{DDT}) testing. When a CMOS circuit switches states, a momentary path is established between the supply lines V_{DD} and V_{SS} that results in a dynamic current I_{DDT}. The I_{DDT} waveform exhibits a spike every time the circuit switches with the magnitude and frequency components of the waveform dependent on the switching activity; therefore, it is possible to differentiate between fault-free and faulty circuits by observing either the magnitude or the frequency spectrum of I_{DDT} waveforms. Monitoring the I_{DDT} of a CMOS circuit may also provide additional diagnostic information about possible defects unmatched by I_{DDQ} and other test techniques [Min 1998]; however, I_{DDT} testing suffers many of the same problems as I_{DDQ} testing as the number of transistors in VLSI devices continues to grow.

1.5.5   Analog and Mixed-Signal Circuit Testing

Analog circuits are used in various applications, such as telecommunications, multimedia, and man–machine interfaces. Mixed-signal circuits include analog circuitry (*e.g.*, amplifiers, filters) and digital circuitry (*e.g.*, data paths, control logic), as well as *digital-to-analog converters* (DACs) and *analog-to-digital converters* (ADCs). Due to the different types of circuitry involved, several different schemes to test a mixed-signal chip are usually required. Test methods for analog circuitry and converters have not achieved maturity comparable to that for digital circuitry. Traditionally, the analog circuitry is tested by explicit functional testing to directly measure performance parameters, such as linearity, frequency response (phase and gain), or signal-to-noise ratio. The measured parameters are compared against the design specification tolerance ranges to determine if the device is faulty or operational within the specified limits. Long test application times and complicated test equipment are often required, making functional testing very expensive. Recently, defect-oriented test approaches based on fault models, similar to those used in digital testing (such as shorts and opens), have been investigated for reducing the cost for functional testing of the analog components and converters [Stroud 2002].

1.5.6   Design for Testability

Test engineers usually have to construct test vectors after the design is completed. This invariably requires a substantial amount of time and effort that could be avoided if testing is considered early in the design flow to make the design more testable. As a result, integration of design and test, referred to as *design for testability* (DFT), was proposed in the 1970s.

To structurally test circuits, we need to control and observe logic values of internal lines. Unfortunately, some nodes in sequential circuits can be very difficult to control and observe; for example, activity on the most significant bit of an n-bit counter can only be observed after 2^{n-1} clock cycles. Testability measures of controllability and observability were first defined in the 1970s [Goldstein 1979] to help find those parts of a digital circuit that will be most difficult to test and to assist in test pattern generation for fault detection. Many DFT techniques have been proposed since that time [McCluskey 1986]. DFT techniques generally fall into one of the following three categories: (1) **ad hoc DFT** techniques, (2) **level-sensitive scan design** (LSSD) or **scan design**, or (3) **built-in self-test** (BIST).

Ad hoc methods were the first DFT techniques introduced in the 1970s. The goal was to target only those portions of the circuit that would be difficult to test and to add circuitry to improve the controllability or observability. *Ad hoc* techniques typically use **test point** insertion to access internal nodes directly. An example of a test point is a multiplexer inserted to control or observe an internal node, as illustrated in Figure 1.12.

Level-sensitive scan design, also referred to as scan design, was the next, and most important, DFT technique proposed [Eichelberger 1977]. LSSD is latch based. In a flip-flop-based scan design, testability is improved by adding extra logic to each flip-flop in the circuit to form a shift register, or scan chain, as illustrated in Figure 1.13. During the scan mode, the scan chain is used to shift in (or scan in) a

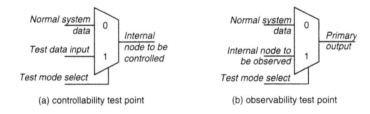

(a) controllability test point   (b) observability test point

■ **FIGURE 1.12**

Ad hoc DFT test points using multiplexers.

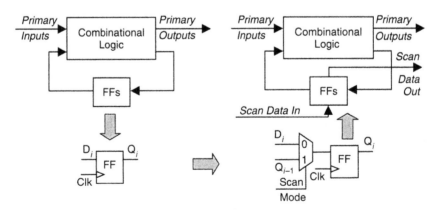

■ **FIGURE 1.13**

Transforming a sequential circuit for scan design.

test vector to be applied to the combinational logic. During one clock cycle in the system mode of operation, the test vector is applied to the combinational logic and the output responses are clocked into the flip-flops. The scan chain is then used in the scan mode to shift out (or scan out) the combinational logic output response to the test vector while shifting in the next test vector to be applied. As a result, LSSD reduces the problem of testing sequential logic to that of testing combinational logic and thereby facilitates the use of ATPG developed for combinational logic.

Built-in self-test was proposed around 1980 [Bardell 1982] [Stroud 2002] to integrate a ***test-pattern generator*** (TPG) and an ***output response analyzer*** (ORA) in the VLSI device to perform testing internal to the IC, as illustrated in Figure 1.14. Because the test circuitry resides with the CUT, BIST can be used at all levels of testing, from wafer through system-level testing.

1.5.7 Board Testing

Like the VLSI fabrication process, PCB manufacturing is a capital-intensive process with minimum human intervention. Once a high-volume batch has been started, the process is totally unmanned. Potential problems that could cause a line stoppage or poor yield are monitored throughout the process. In the 1970s and 1980s, PCBs were tested by probing the backs of the boards with probes (also called nails) in a bed-of-nails tester. The probes are positioned to contact various solder points on the PCB in order to force signal values at the component pins and monitor the output responses. Generally, a PCB tester is capable of performing both analog and digital functional tests and is usually designed to be modular and flexible enough to integrate different external instruments.

Two steps were traditionally taken before testing an assembled PCB. First, the **bare board** was tested for all interconnections using a PCB tester, primarily targeting shorts and opens. Next, the components to be assembled on the PCB were tested. After assembly, the PCB was tested by using a PCB tester. In the modern automated PCB production process, solder paste inspection, automated optical and x-ray inspections, and in-circuit (bed-of-nails) testing are used for quality control. With the advent of surface-mount devices on PCBs in the mid-1980s, problems arose for PCB in-circuit testing, as the pins of the package did not go through the board to guarantee contact sites on the bottom of the PCB. These problems were overcome with the introduction of **boundary scan**.

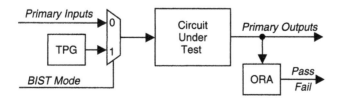

■ **FIGURE 1.14**

Basic BIST architecture.

1.5.8 Boundary Scan Testing

In the mid-1980s, the *Joint Test Action Group* (JTAG) proposed a boundary scan standard, approved in 1990 as IEEE Standard 1149.1 [IEEE 1149.1-2001]. Boundary scan, based on the basic idea of scan design, inserted logic to provide a scan path through all I/O buffers of ICs to assist in testing the assembled PCB. A typical boundary scan cell is illustrated in Figure 1.15 with regard to its application to a bidirectional I/O buffer. The scan chain provides the ability to shift in test vectors to be applied through the pad to the pins and interconnections on the PCB. The output responses are captured at the input buffers on other devices on the PCB and subsequently shifted out for fault detection. Thus, boundary scan provides access to the various signal nodes on a PCB without the need for physical probes.

The *test access port* (TAP) provides access to the boundary scan chain through a four-wire serial bus interface (summarized in Table 1.5) in conjunction with instructions transmitted over the interface. In addition to testing the interconnections on the PCB, the boundary scan interface also provides access to DFT features, such as LSSD or BIST, designed and implemented in the VLSI devices for board and system-level testing. The *boundary scan description language* (BSDL) provides a mechanism with which IC manufacturers can describe testability features in

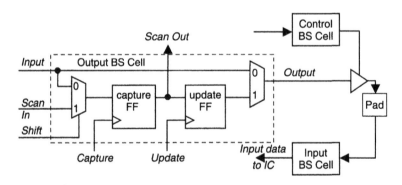

■ **FIGURE 1.15**

Basic boundary scan cell applied to a bidirectional buffer.

TABLE 1.5 ■ Boundary Scan Four-Wire Interface

BS pin	I/O	Function
TCK	Input	Test clock
TMS	Input	Test mode select
TDI	Input	Test data in
TDO	Output	Test data out

a chip [Parker 2001]. In 1999, another boundary scan standard, IEEE 1149.4, was adopted for mixed-signal systems; it defines boundary scan cells as well as a TAP for the analog portion of the device [IEEE 1149.4-1999] [Mourad 2000]. In 2003, an extended boundary scan standard for the I/O protocol of high-speed networks, namely 1149.6, was approved [IEEE 1149.6-2003].

System-on-chip implementations face test challenges in addition to those of normal VLSI devices. SOCs incorporate embedded cores that may be difficult to access during testing. The IEEE P1500 working group was approved in 1997 to develop a scalable wrapper architecture and access mechanism similar to boundary scan for enabling test access to embedded cores and the associated interconnect between embedded cores. This proposed P1500 test method, approved as an IEEE 1500 standard in 2005 [IEEE 1500-2005], is independent of the underlying functionality of the SOC or its individual embedded cores and creates the necessary testability requirements for detection and diagnosis of faults for debug and yield enhancement.

1.6 CONCLUDING REMARKS

This chapter provides an overview of VLSI testing as an area of both theoretical and great practical significance. The importance and challenges of VLSI testing at different abstraction levels were discussed along with a brief historical review of test technology development. New and continuing testing challenges, along with the critical mind of the test community, drive creative advances in test technology and motivate further developments for nanometer technology. Why do we need VLSI testing? How difficult is VLSI testing? What are the fundamental concepts and techniques for VLSI testing? Although many of these issues were briefly reviewed in this chapter, a more detailed discussion of these questions can be found in the following chapters of this book.

1.7 EXERCISES

1.1 **(Stuck-At Fault Models)** Consider the combinational logic circuit in Figure 1.16. How many possible single stuck-at faults does this circuit have? How many possible multiple stuck-at faults does this circuit have? How many collapsed single stuck-at faults does this circuit have?

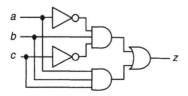

■ **FIGURE 1.16**

Circuit for Problem 1.1.

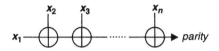

■ **FIGURE 1.17**

Circuit for Problem 1.4.

1.2 (**Bridging Fault Models**) Show an example where a combinational logic circuit will become a sequential circuit in the presence of a bridging fault.

1.3 (**Automatic Test-Pattern Generation**) Generate a minimum set of test vectors to completely test an n-input NAND gate under the single stuck-at fault model. How many test vectors are needed?

1.4 (**Automatic Test-Pattern Generation**) Generate a minimum set of test vectors to detect all single stuck-at faults for a cascade of $(n-1)$ exclusive-OR gates for an n-bit parity checker, as shown in Figure 1.17, where each exclusive-OR gate is implemented by elementary logic gates (AND, OR, NAND, NOR, NOT). How many test vectors are needed?

1.5 (**Mean Time between Failures**) The number of failures in 10^9 hours is a unit (abbreviated FITS) that is often used in reliability calculations. Calculate the MTBF for a system with 500 components where each component has a failure rate of 1000 FITS.

1.6 (**Mean Time to Repair**) On average, how long would it take to repair a system each year if the availability of the system is 99.999%?

1.7 (**Defect Level**) What percentage of all parts shipped will be defective if the yield is 50% and the fault coverage is 90% for the set of test vectors used to test the parts?

Acknowledgments

The authors wish to acknowledge the following people for their assistance during the preparation of this chapter: Dr. Huawei Li of the Institute of Computing Technology at the Chinese Academy of Sciences, Sachin Dhingra and Sudheer Vemula of the Department of Electrical and Computer Engineering at Auburn University, and Prof. Wen-Ben Jone of the Department of Electrical & Computer Engineering and Computer Science at the University of Cincinnati.

References

R1.0—Books

[Abramovici 1994] M. Abramovici, M. A. Breuer, and A. D. Friedman, *Digital Systems Testing and Testable Design*, IEEE Press, Piscataway, NJ, 1994 (revised printing).

[Breuer 1987] M. A. Breuer and A. D. Friedman, *Diagnosis and Reliable Design of Digital Systems*, Computer Science Press, 1987 (revised printing).

[Bushnell 2000] M. L. Bushnell and V. D. Agrawal, *Essentials of Electronic Testing for Digital, Memory and Mixed-Signal VLSI Circuits*, Springer Science, New York, 2000.

[IEEE 1149.4-1999] IEEE Std. 1149.4-1999, *IEEE Standard for a Mixed Signal Test Bus*, Institute of Electrical and Electronics Engineers, New York, 1999.

[IEEE 1149.1-2001] IEEE Std. 1149.1-2001, *IEEE Standard Test Access Port and Boundary Scan Architecture*, Institute of Electrical and Electronics Engineers, New York, 2001.

[IEEE 1149.6-2003] IEEE Std. 1149.6-2003, *IEEE Standard for Boundary-Scan Testing of Advance Digital Networks*, Institute of Electrical and Electronics Engineers, New York, 2003.

[IEEE 1500-2005] IEEE Std. 1500-2005, *IEEE Standard Testability Method for Embedded Core-Based Integrated Circuits*, Institute of Electrical and Electronics Engineers, New York, 2005.

[Jha 2003] N. K. Jha and S. K. Gupta, *Testing of Digital Systems*, Cambridge University Press, Cambridge, U.K., 2003.

[McCluskey 1986] E. J. McCluskey, *Logic Design Principles with Emphasis on Testable Semi-custom Circuits*, Prentice Hall, Englewood Cliffs, NJ, 1986.

[Min 1986] Y. Min, *Logical Circuit Testing* (in Chinese), China Railway Publishing House, Beijing, China, 1986.

[Mourad 2000] S. Mourad and Y. Zorian, *Principles of Testing Electronic Systems*, John Wiley & Sons, Somerset, NJ, 2000.

[Parker 2001] K. P. Parker, *The Boundary-Scan Handbook*, Kluwer Academic, Norwell, MA, 2001.

[Stroud 2002] C. E. Stroud, *A Designer's Guide to Built-In Self-Test*, Kluwer Academic, Norwell, MA, 2002.

[van de Goor 1991] A. J. van de Goor, *Testing Semiconductor Memories: Theory and Practice*, John Wiley & Sons, Chichester, U.K. 1991.

R1.1—Importance of Testing

[Arthistory 2005] Art History Club (http://www.arthistoryclub.com/art_history/Integrated_circuit #VLSI).

[Moore 1965] G. Moore, Cramming more components onto integrated circuits, *Electronics*, 38(8), 114–117, 1965.

R1.3—Challenges in VLSI Testing

[Geppert 1998] L. Geppert, Technology 1998 analysis and forecast: Solid state, *IEEE Spectr.*, 35(1), 23–28, 1998.

[Ma 1995] S. C. Ma, P. Franco, and E. J. McCluskey, An experimental chip to evaluate test techniques: Experimental results, in *Proc. Int. Test Conf.*, October 1995, pp. 663–672.

[Stroud 2000] C. Stroud, J. Emmert, and J. Bailey, A new bridging fault model for more accurate fault behavior, in *Pro. Automat. Test Conf. (AUTOTESTCON)*, September 2000, pp. 481–485.

[Williams 1981] T. Williams and N. Brown, Defect level as a function of fault coverage, *IEEE Trans. Comput.*, C-30(12), 987–988, 1981.

R1.4—Levels of Abstraction in VLSI Testing

[Maxwell 1994] P. Maxwell, R. Aitken, and L. Huismann, The effect on quality of non-uniform fault coverage and fault probability, in *Proc. Int. Test Conf.*, October 1994, pp. 739–746.

[Min 2002] Y. Min, Why RTL ATPG?, *J. Comput. Sci. Technol.*, 17(2), 113–117, 2002.

R1.5—Historical Review of VLSI Test Technology

[Bardell 1982] P. H. Bardell and W. H. McAnney, Self-testing of multiple logic modules, in *Proc. Int. Test Conf.*, October 1982, pp. 200–204.

[Brglez 1985] F. Brglez and H. Fujiwara, A neutral netlist of 10 combinational benchmark designs and a special translator in Fortran, in *Proc. Int. Symp. on Circuits and Systems*, June 1985, pp. 663–698.

[Brglez 1989] F. Brglez, D. Bryan, and K. Kozminski, Combinational profiles of sequential benchmark circuits, in *Proc. Int. Symp. on Circuits and Systems*, May 1989, pp. 1929–1934.

[Eichelberger 1977] E. B. Eichelberger and T. W. Williams, A logic design structure for LSI testability, in *Proc. Des. Automat. Conf.*, June 1977, pp. 462–468.

[Fujiwara 1983] H. Fujiwara and T. Shimono, On the acceleration of test generation algorithms, *IEEE Trans. Comput.*, C-32(12), 1137–1144, 1983.

[Gelsinger 2000] P. Gelsinger, Discontinuities driven by a billion connected machines, *IEEE Design Test Comput.*, 17(1), 7–15, 2000.

[Goel 1981] P. Goel, An implicit enumeration algorithm to generate tests for combinational logic circuits, *IEEE Trans. Comput.*, C-30(3), 215–222, 1981.

[Goldstein 1979] L. H. Goldstein, Controllability/observability analysis of digital circuits, *IEEE Trans. Circuits Syst.*, CAS-26(9), 685–693, 1979.

[Min 1998] Y. Min and Z. Li, IDDT testing *versus* IDDQ testing, *J. Electron. Testing: Theory Appl.*, 13(1), 15–55, 1998.

[Roth 1966] J. Roth, Diagnosis of automata failures: A calculus and a method, *IBM J. Res. Develop.*, 10(4), 278–291, 1966.

[Schulz 1988] M. H. Schulz, E. Trischler, and T. M. Serfert, SOCRATES: A highly efficient automatic test pattern generation system, *IEEE Trans. Computer-Aided Design*, CAD-7(1), 126–137, 1988.

CHAPTER 2

Design for Testability

Laung-Terng (L.-T.) Wang
SynTest Technologies, Inc., Sunnyvale, California

Xiaoqing Wen
Kyushu Institute of Technology, Fukuoka, Japan

Khader S. Abdel-Hafez
SynTest Technologies, Inc., Sunnyvale, California

ABOUT THIS CHAPTER

This chapter discusses *design for testability* (DFT) techniques for testing modern digital circuits. These DFT techniques are required in order to improve the quality and reduce the test cost of the digital circuit, while at the same time simplifying the test, debug and diagnose tasks. The purpose of this chapter is to provide readers with the knowledge to judge whether a design is implemented in a test-friendly manner and to recommend changes in order to improve the testability of the design for achieving the above-mentioned goals. More specifically, this chapter will allow readers to be able to identify and fix scan design rule violations and understand the basics for successfully converting a design into a scan design.

In this chapter, we first cover the basic DFT concepts and methods for performing testability analysis. Next, following a brief yet comprehensive summary of *ad hoc* DFT techniques, **scan design**, the most widely used structured DFT methodology, is discussed, including popular scan cell designs, scan architectures, scan design rules, scan design flow, and special-purpose scan designs. Finally, advanced DFT techniques for use at the *register-transfer level* (RTL) are presented in order to further reduce DFT design iterations and test development time.

2.1 INTRODUCTION

During the early stages of *integrated circuit* (IC) production history, design and test were regarded as separate functions, performed by separate and unrelated groups of engineers. During these early years, a design engineer's job was to implement the required functionality based on design specifications, without giving any thought to how the manufactured device was to be tested. Once the functionality was implemented, the design information was transferred to test engineers. A test engineer's job was to determine how to efficiently test each manufactured device within a reasonable amount of time, in order to screen out the parts that may contain manufacturing defects and ship all defect-free devices to customers. The final quality of the test was determined by keeping track of the number of defective parts shipped to the customers, based on customer returns. This product quality, measured in terms of defective *parts per million* (PPM) shipped, was a final test score for quantifying the effectiveness of the developed test.

While this approach worked well for small-scale integrated circuits that mainly consisted of combinational logic or simple finite-state machines, it was unable to keep up with the circuit complexity as designs moved from *small-scale integration* (SSI) to *very-large-scale integration* (VLSI). A common approach to test these VLSI devices during the 1980s relied heavily on fault simulation to measure the fault coverage of the supplied functional patterns. Functional patterns were developed to navigate through the long sequential depths of a design, with the goal of exercising all internal states and detecting all possible manufacturing defects. A **fault simulation** or **fault grading** tool was used to quantify the effectiveness of the functional patterns. If the supplied functional patterns did not reach the target fault coverage goal, additional functional patterns were further added. Unfortunately, this approach typically failed to improve the circuit's fault coverage beyond 80%, and the quality of the shipped products suffered.

Gradually, it became clear that designing devices without paying much attention to test resulted in increased test cost and decreased test quality. Some designs that were otherwise best in class with regard to functionality and performance failed commercially due to prohibitively high test cost or poor product quality. These problems have since led to the development and deployment of DFT engineering in the industry.

The first challenge facing DFT engineers was to find simpler ways of exercising all internal states of a design and reaching the target fault coverage goal. Various **testability measures** and *ad hoc* **testability enhancement** methods were proposed and used in the 1970s and 1980s to serve this purpose. These methods were mainly used to aid in the circuit's **testability** or to increase the circuit's **controllability** and **observability** [McCluskey 1986] [Abramovici 1994]. While attempts to use these methods have substantially improved the testability of a design and eased sequential *automatic test pattern generation* (ATPG), their end results at reaching the target fault coverage goal were far from being satisfactory; it was still quite difficult to reach more than 90% fault coverage for large designs. This was mostly due to the fact that, even with these testability aids, deriving functional patterns by hand or generating test patterns for a sequential circuit is a much more difficult problem than generating test patterns for a combinational circuit [Fujiwara 1982] [Bushnell 2000] [Jha 2002].

For combinational circuits, many innovative ATPG algorithms have been developed for automatically generating test patterns within a reasonable amount of time. Automatically generating test patterns for sequential circuits met with limited success, due to the existence of numerous internal states that are difficult to set and check from external pins. Difficulties in controlling and observing the internal states of sequential circuits led to the adoption of structured DFT approaches in which direct external access is provided for storage elements. These *reconfigured storage elements* with direct external access are commonly referred to as **scan cells**. Once the capability of controlling and observing the internal states of a design is added, the problem of testing the sequential circuit is transformed into a problem of testing the combinational logic, for which many solutions already existed.

Scan design is currently the most popular structured DFT approach. It is implemented by connecting selected storage elements of a design into multiple shift registers, called **scan chains**, to provide them with external access. Scan design accomplishes this task by replacing all selected storage elements with scan cells, each having one additional *scan input* (SI) port and one shared/additional *scan output* (SO) port. By connecting the SO port of one scan cell to the SI port of the next scan cell, one or more scan chains are created.

Since the 1970s, numerous scan cell designs and scan architectures have been proposed [Fujiwara 1985] [McCluskey 1986]. A design where all storage elements are selected for scan insertion is called a **full-scan design**. A design where almost all (*e.g.*, more than 98%) storage elements are selected is called an **almost full-scan design**. A design where some storage elements are selected and sequential ATPG is applied is called a **partial-scan design**. A partial-scan design where storage elements are selected in such a way as to break all *sequential feedback loops* [Cheng 1990] and to which combinational ATPG can be applied is further classified as a **pipelined, feed-forward**, or **balanced partial-scan design**. As silicon prices have continued to drop since the mid-1990s with the advent of deep submicron technology, the dominant scan architecture has shifted from partial-scan design to full-scan design.

In order for a scan design to achieve the desired PPM goal, specific circuit structures and design practices that can affect fault coverage must be identified and fixed. This requires compiling a set of **scan design rules** that must be adhered to. Hence, a new role of DFT engineer emerged, with responsibilities including identifying and fixing scan design rule violations in the design, inserting or synthesizing scan chains into the design, generating test patterns for the scan design, and, finally, converting the test patterns to *test programs* for test engineers to perform manufacturing testing on *automatic test equipment* (ATE). Since then, most of these DFT tasks have been automated.

In addition to being the dominant DFT architecture used for detecting manufacturing defects, scan design has become the basis of more advanced DFT techniques, such as logic *built-in self-test* (BIST) [Nadeau-Dostie 2000] [Stroud 2002] and test compression. Furthermore, as designs continue to move towards the nanometer scale, scan design is being utilized as a design feature, with uses varying from debug, diagnosis, and failure analysis to special applications, such as reliability enhancement against **soft errors** [Mitra 2005]. A few of these **special-purpose scan designs** are included in this chapter for completeness.

Recently, design for testability has started to migrate from the gate level to the *register-transfer level* (RTL). The motivation for this migration is to allow additional DFT features, such as logic BIST and test compression, to be integrated at the RTL, thereby reducing test development time and creating reusable and testable RTL cores. This further allows the integrated DFT design to go through synthesis-based optimization to reduce performance and area overhead.

2.2 TESTABILITY ANALYSIS

Testability is a relative measure of the effort or cost of testing a logic circuit. In general, it is based on the assumption that only primary inputs and primary outputs can be directly controlled and observed, respectively. *Testability* reflects the effort required to perform the main test operations of controlling internal signals from primary inputs and observing internal signals at primary outputs. **Testability analysis** refers to the process of assessing the testability of a logic circuit by calculating a set of numerical measures for each signal in the circuit.

One important application of testability analysis is to assist in the decision-making process during test generation. For example, if during test generation it is determined that the output of a certain AND gate must be set to 0, testability analysis can help decide which AND gate input is the easiest to set to 0. Another application is to identify areas of poor testability to guide testability enhancement, such as test point insertion, for improving the testability of the design. For this purpose, testability analysis is performed at various design stages so testability problems can be identified and fixed as early as possible.

Since the 1970s, many testability analysis techniques have been proposed [Rutman 1972] [Stephenson 1976] [Breuer 1978] [Grason 1979]. The **Sandia Controllability/Observability Analysis Program** (SCOAP) [Goldstein 1979] [Goldstein 1980] was the first topology-based program that popularized testability analysis applications. Enhancements based on SCOAP have also been developed and used to aid in test point selection [Wang 1984] [Wang 1985]. These methods perform testability analysis by calculating the **controllability** and **observability** of each signal line, where controllability reflects the difficulty of setting a signal line to a required logic value from primary inputs and observability reflects the difficulty of propagating the logic value of the signal line to primary outputs.

Traditionally, gate-level topological information of a circuit is used for testability analysis. Depending on the target application, deterministic or random testability measures are calculated. In general, **topology-based testability analysis**, such as SCOAP or probability-based testability analysis, is computationally efficient but can produce inaccurate results for circuits containing many reconvergent fanouts. **Simulation-based testability analysis**, on the other hand, can generate more accurate results by simulating the circuit behavior using deterministic, random, or pseudo-random test patterns but may require a long simulation time.

In this section, we first describe the method for performing SCOAP testability analysis. Next, probability-based testability analysis and simulation-based testability analysis are discussed. Finally, because the capability to perform testability analysis at the RTL is becoming increasingly important, we discuss how **RTL testability analysis** is performed.

2.2.1 SCOAP Testability Analysis

The SCOAP testability analysis program [Goldstein 1979] [Goldstein 1980] calculates six numerical values for each signal s in a logic circuit:

- $CC0(s)$—combinational 0-controllability of s
- $CC1(s)$—combinational 1-controllability of s
- $CO(s)$—combinational observability of s
- $SC0(s)$—sequential 0-controllability of s
- $SC1(s)$—sequential 1-controllability of s
- $SO(s)$—sequential observability of s

Roughly speaking, the three combinational testability measures (CC0, CC1, and CO) are related to the number of signals that must be manipulated in order to control or observe s from primary inputs or at primary outputs, whereas the three sequential testability measures (SC0, SC1, and SO) are related to the number of clock cycles required to control or observe s from primary inputs or at primary outputs [Bushnell 2000]. The values of controllability measures range between 1 and infinite, while the values of observability measures range between 0 and infinite. As a boundary condition, the CC0 and CC1 values of a primary input are set to 1, the SC0 and SC1 values of a primary input are set to 0, and the CO and SO values of a primary output are set to 0.

2.2.1.1 *Combinational Controllability and Observability Calculation*

The first step in SCOAP is to calculate the combinational controllability measures of all signals. This calculation is performed from primary inputs toward primary outputs in a breadth-first manner. More specifically, the circuit is levelized from primary inputs to primary outputs in order to assign a *level order* for each gate. The output controllability of each gate is then calculated in level order after the controllability measures of all of its inputs have been calculated. The rules for combinational controllability calculation are summarized in Table 2.1, where a 1 is added to each rule to indicate that a signal passes through one more level of logic gate. From this table, we can see that $CC0(s) \geq 1$ and $CC1(s) \geq 1$ for any signal s. A larger $CC0(s)$ or $CC1(s)$ value implies that it is more difficult to control s to 0 or 1 from primary inputs.

Once the combinational controllability measures of all signals are calculated, the combinational observability of each signal can be calculated. This calculation is also performed in a breadth-first manner while moving from primary outputs toward primary inputs. The rules for combinational observability calculation are summarized in Table 2.2, where a 1 is added to each rule to indicate that a signal passes through one more level of logic gate. From this table, we can see that $CO(s) \geq 0$ for any signal s. A larger $CO(s)$ value implies that it is more difficult to observe s at any primary output.

TABLE 2.1 ■ SCOAP Combinational Controllability Calculation Rules

	0-Controllability (Primary Input, Output, Branch)	1-Controllability (Primary Input, Output, Branch)
Primary Input	1	1
AND	*min* {input 0-controllabilities} + 1	$\sum$ (input 1-controllabilities) + 1
OR	$\sum$ (input 0-controllabilities) + 1	*min* {input 1-controllabilities} + 1
NOT	Input 1-controllability + 1	Input 0-controllability + 1
NAND	$\sum$ (input 1-controllabilities) + 1	*min* {input 0-controllabilities} + 1
NOR	*min* {input 1-controllabilities} + 1	$\sum$ (input 0-controllabilities) + 1
BUFFER	Input 0-controllability + 1	Input 1-controllability + 1
XOR	*min* {CC1(a) + CC1(b), CC0(a) + CC0(b)} + 1	*min* {CC1(a) + CC0(b), CC0(a) + CC1(b)} + 1
XNOR	*min* {CC1(a) + CC0(b), CC0(a) + CC1(b)} + 1	*min* {CC1(a) + CC1(b), CC0(a) + CC0(b)} + 1
Branch	Stem 0-controllability	Stem 1-controllability

Note: a and *b* are inputs of an XOR or XNOR gate.

TABLE 2.2 ■ SCOAP Combinational Observability Calculation Rules

	Observability (Primary Output, Input, Stem)
Primary Output	0
AND/NAND	$\sum$ (output observability, 1-controllabilities of other inputs) + 1
OR/NOR	$\sum$ (output observability, 0-controllabilities of other inputs) + 1
NOT/BUFFER	Output observability + 1
XOR/XNOR	*a*: $\sum$ (output observability, *min* {CC0(b), CC1(b)}) + 1
	b: $\sum$ (output observability, *min* {CC0(a), CC1(a)}) + 1
Stem	*min* {branch observabilities}

Note: a and *b* are inputs of an XOR or XNOR gate.

Figure 2.1 shows the combinational controllability and observability measures of a full-adder. The three-value tuple $v_1/v_2/v_3$ on each signal line represents the signal's 0-controllability (v_1), 1-controllability (v_2), and observability (v_3). The boundary condition is set by initializing the CC0 and CC1 values of the primary inputs A, B, and C_{in} to 1 and the CO values of the primary outputs Sum and C_{out} to 0. By applying the rules given in Tables 2.1 and 2.2 and starting with the given boundary condition, one can first calculate all combinational controllability measures forward and then calculate all combinational observability measures backward in level order.

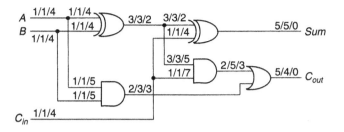

■ FIGURE 2.1

SCOAP full-adder example.

2.2.1.2 *Sequential Controllability and Observability Calculation*

Sequential controllability and observability measures are calculated in a similar manner as combinational measures, except that a 1 is not added as we move from one level of logic gate to another; rather, a 1 is added when a signal passes through a storage element. The difference is illustrated using the sequential circuit example shown in Figure 2.2, which consists of an AND gate and a positive-edge-triggered D flip-flop. The D flip-flop includes an active-high asynchronous reset pin r. SCOAP measures of a D flip-flop with a synchronous, as opposed to asynchronous, reset are shown in [Bushnell 2000].

First, we calculate the combinational and sequential controllability measures of all signals. In order to control signal d to 0, either input a or b must be set to 0. In order to control d to 1, both inputs a and b must be set to 1. Hence, the combinational and sequential controllability measures of signal d are:

$$CC0(d) = min\ \{CC0(a), CC0(b)\} + 1$$
$$SC0(d) = min\ \{SC0(a), SC0(b)\}$$
$$CC1(d) = CC1(a) + CC1(b) + 1$$
$$SC1(d) = SC1(a) + SC1(b)$$

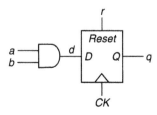

■ FIGURE 2.2

SCOAP sequential circuit example.

In order to control the data output q of the D flip-flop to 0, the data input d and the reset signal r can be set to 0 while applying a rising clock edge (a 0-to-1 transition) to the clock CK. Alternatively, this can be accomplished by setting r to 1 while holding CK at 0. Because a clock pulse is not applied to CK, a 1 is not added to the sequential controllability calculation in the second case; therefore, the combinational and sequential 0-controllability measures of q are:

$$CC0(q) = min \ \{CC0(d) + CC0(CK) + CC1(CK) + CC0(r), \ CC1(r) + CC0(CK)\}$$

$$SC0(q) = min \ \{SC0(d) + SC0(CK) + SC1(CK) + SC0(r) + 1, \ SC1(r) + SC0(CK)\}$$

Here, $CC0(q)$ measures how many signals in the circuit must be set to control q to 0, whereas $SC0(q)$ measures how many flip-flops in the circuit must be clocked to set q to 0. The only way to control the data output q of the D flip-flop to 1 is to set the data input d to 1 and the reset signal r to 0 while applying a rising clock edge to the clock CK. Hence,

$$CC1(q) = CC1(d) + CC0(CK) + CC1(CK) + CC0(r)$$

$$SC1(q) = SC1(d) + SC0(CK) + SC1(CK) + SC0(r) + 1$$

Next, we calculate the combinational and sequential observability measures of all signals. The data input d can be observed at q by holding the reset signal r at 0 and applying a rising clock edge to CK. Hence,

$$CO(d) = CO(q) + CC0(CK) + CC1(CK) + CC0(r)$$

$$SO(d) = SO(q) + SC0(CK) + SC1(CK) + SC0(r) + 1$$

The asynchronous reset signal r can be observed by first setting q to 1 and then holding CK at the inactive state 0. Again, a 1 is not added to the sequential controllability calculation because a clock pulse is not applied to CK:

$$CO(r) = CO(q) + CC1(q) + CC0(CK)$$

$$SO(r) = SO(q) + SC1(q) + SC0(CK)$$

There are two ways to indirectly observe the clock signal CK at q: (1) set q to 1, r to 0, and d to 0 and apply a rising clock edge at CK; or (2) set both q and r to 0, set d to 1, and apply a rising clock edge at CK. Hence,

$$CO(CK) = CO(q) + CC0(CK) + CC1(CK) + CC0(r)$$
$$+ min \ \{CC0(d) + CC1(q), \ CC1(d) + CC0(q)\}$$
$$SO(CK) = SO(q) + SC0(CK) + SC1(CK) + SC0(r)$$
$$+ min \ \{SC0(d) + SC1(q), \ SC1(d) + SC0(q)\} + 1$$

To observe an input of the AND gate at d requires setting the other input to 1; therefore, the combinational and sequential observability measures for both inputs a and b are:

$$CO(a) = CO(d) + CC1(b) + 1$$
$$SO(a) = SO(d) + SC1(b)$$
$$CO(b) = CO(d) + CC1(a) + 1$$
$$SO(b) = SO(d) + SC1(a)$$

It is important to note that controllability and observability measures calculated using SCOAP are heuristics and only approximate the actual testability of a logic circuit. When scan design is used, testability analysis can assume that all scan cells are directly controllable and observable. It was also shown in [Agrawal 1982] that SCOAP may overestimate testability measures for circuits containing many reconvergent fanouts; however, by being able to perform testability analysis in $O(n)$ computational complexity for n signals in a circuit, SCOAP provides a quick estimate of the circuit's testability that can be used to guide testability enhancement and test generation.

2.2.2 Probability-Based Testability Analysis

Topology-based testability analysis techniques, such as SCOAP, have been found to be extremely helpful in test generation, which is the main topic of Chapter 4. These testability measures are able to analyze the **deterministic testability** of the logic circuit in advance. On the other hand, in **logic built-in self-test** (BIST), which is the main topic of Chapter 5, random or pseudo-random test patterns are generated without specifically performing deterministic test pattern generation on any signal line. In this case, topology-based testability measures using signal probability to analyze the **random testability** of the circuit can be used [Parker 1975] [Savir 1984] [Seth 1985] [Jain 1985]. These measures are often referred to as **probability-based testability measures** or probability-based testability analysis techniques. For example, given a random input pattern, one can calculate three measures for each signal s in a combinational circuit as follows:

- $C0(s)$—probability-based 0-controllability of s
- $C1(s)$—probability-based 1-controllability of s
- $O(s)$—probability-based observability of s

Here, $C0(s)$ and $C1(s)$ are the probability of controlling signal s to 0 and 1 from primary inputs, respectively. $O(s)$ is the probability of observing signal s at primary outputs. These three probabilities range between 0 and 1. As a boundary condition, the C0 and C1 probabilities of a primary input are typically set to 0.5, and the O probability of a primary output is set to 1. For each signal s in the circuit, $C0(s) + C1(s) = 1$.

Many methods have been developed to calculate the probability-based testability measures. A simple method is given below, whose basic procedure is similar to the one used for calculating combinational testability measures in SCOAP except that different calculation rules are used. The rules for probability-based controllability and observability calculation are summarized in Tables 2.3 and 2.4, respectively. In Table 2.3, p_0 is the initial 0-controllability chosen for a primary input, where $0 < p_0 < 1$.

Compared to SCOAP testability measures, where non-negative integers are used, probability-based testability measures range between 0 and 1. The smaller a probability-based testability measure of a signal, the more difficult it is to control or observe the signal. Figure 2.3 illustrates the difference between SCOAP testability

TABLE 2.3 ■ Probability-Based Controllability Calculation Rules

	0-Controllability (Primary Input, Output, Branch)	1-Controllability (Primary Input, Output, Branch)
Primary Input	p_0	$p_1 = 1 - p_0$
AND	1 − (output 1-controllability)	$\prod$ (input 1-controllabilities)
OR	$\prod$ (input 0-controllabilities)	1 − (output 0-controllability)
NOT	Input 1-controllability	Input 0-controllability
NAND	$\prod$ (input 1-controllabilities)	1 − (output 0-controllability)
NOR	1 − (output 1-controllability)	$\prod$ (input 0-controllabilities)
BUFFER	Input 0-controllability	Input 1-controllability
XOR	1 − 1-controllability	$\sum (C1(a) \times CO(b), CO(a) \times C1(b))$
XNOR	1 − 1-controllability	$\sum (CO(a) \times CO(b), C1(a) \times C1(b))$
Branch	Stem 0-controllability	Stem 1-controllability

Note: a and b are inputs of an XOR or XNOR gate.

TABLE 2.4 ■ Probability-Based Observability Calculation Rules

	Observability (Primary Output, Input, Stem)
Primary Output	1
AND/NAND	$\prod$ (output observability, 1-controllabilities of other inputs)
OR/NOR	$\prod$ (output observability, 0-controllabilities of other inputs)
NOT/BUFFER	Output observability
XOR/XNOR	a: $\prod$ (output observability, max {0-controllability of b, 1-controllability of b})
	b: $\prod$ (output observability, max {0-controllability of a, 1-controllability of a})
Stem	max {branch observabilities}

Note: a and b are inputs of an XOR or XNOR gate.

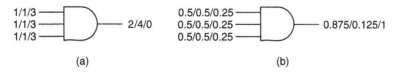

■ **FIGURE 2.3**

Comparison of SCOAP and probability-based testability measures: (a) SCOAP combinational measures, and (b) probability-based measures.

measures and probability-based testability measures of a three-input AND gate. The three-value tuple $v_1/v_2/v_3$ of each signal line represents the signal's 0-controllability (v_1), 1-controllability (v_2), and observability (v_3).

Signals with poor probability-based testability measures tend to be difficult to test with random or pseudo-random test patterns. The faults on these signal lines are often referred to as ***random-pattern resistant*** (RP-resistant) [Savir 1984]. That is, either the probability of these signals randomly receiving a 0 or 1 from primary inputs or the probability of observing these signals at primary outputs is low, assuming that all primary inputs have the equal probability of being set to 0 or 1 and are independent from each other.

The existence of such RP-resistant faults is the main reason why fault coverage using random or pseudo-random test patterns is low compared to using deterministic test patterns. In applications such as logic BIST, in order to solve this low fault coverage problem, test points are often inserted in the circuit to enhance the circuit's random testability. A few commonly used test point insertion techniques are discussed in Section 2.3. Interested readers can find more information in Chapter 5.

2.2.3 Simulation-Based Testability Analysis

In the calculation of SCOAP and probability-based testability measures as described above, only the topological information of a logic circuit is explicitly explored. These topology-based methods are static, in the sense that they do not use input test patterns for testability analysis. Their controllability and observability measures can be calculated in linear time, thus making them very attractive for applications that require fast testability analysis, such as test generation and logic BIST. However, the efficiency of these methods is achieved at the cost of reduced accuracy, especially for circuits that contain many reconvergent fanouts [Agrawal 1982].

As an alternative or supplement to static or topology-based testability analysis, dynamic or simulation-based methods that use input test patterns for testability analysis or testability enhancement can be performed through **statistical sampling**. Logic simulation and fault simulation techniques can be employed. Logic simulation and fault simulation are both covered in Chapter 3.

In statistical sampling, a sample set of input test patterns are selected that are either generated randomly or derived from a given pattern set, and logic simulation is conducted to collect the responses of all or part of signal lines of interest. The commonly collected responses are the number of occurrences of 0's, 1's, 0-to-1 transitions, and 1-to-0 transitions, which are then used to statistically profile the testability of a logic circuit. These data are then analyzed to find locations of poor testability. If a signal line exhibits only a few transitions or no transitions for the sample input patterns, it might be an indication that the signal likely has poor controllability.

In addition to logic simulation, fault simulation has also been used to enhance the testability of a logic circuit using random or pseudo-random test patterns. For example, a ***random resistant fault analysis*** (RRFA) method has been successfully applied to a high-performance microprocessor to improve the circuit's random testability in logic BIST [Rizzolo 2001]. This method is based on statistical data collected during fault simulation for a small number of random test patterns. Controllability and observability measures of each signal in the circuit are calculated using the probability models developed in the ***statistical fault analysis*** (STAFAN) algorithm [Jain 1985], which is described in Section 3.4.8 (STAFAN is the first method able to give reasonably accurate estimates of fault coverage in combinational circuits purely using input test patterns and without running fault simulation). With these data, RRFA identifies signals that are difficult to control or observe, as well as signals that are statistically correlated. Based on the analysis results, RRFA then recommends test points to be added to the circuit to improve the circuit's random testability.

Because it can take a long simulation time to run through all input test patterns, these simulation-based methods are in general used to guide testability enhancement in test generation or logic BIST when it is necessary to meet a very high fault coverage goal. This approach is crucial for life-critical and mission-critical applications, such as in the healthcare and defense/aerospace industries.

2.2.4 RTL Testability Analysis

The testability analysis methods discussed earlier are mostly used for logic circuits described at the gate level. Although they can be used to ease test generation and guide testability enhancement, testability enhancement at the gate level can be costly in terms of area overhead and possible performance degradation. In addition, it may require many DFT iterations and increase test development time. In order to address these problems, many **RTL testability analysis** methods have been proposed [Stephenson 1976] [Lee 1992] [Boubezari 1999].

The RTL testability analysis method described in [Lee 1992] can be used to improve data path testability. This method begins by building a **structure graph** to represent the data transfer within an RTL circuit, where each vertex represents a register, and each directed edge from vertex v_i to vertex v_j represents a functional block from register v_i to register v_j. The maximum level in a structure graph, referred to as the **sequential depth**, can be used to reflect the difficulty of testing the RTL circuit. This approach ignores all the details of the functional block.

The RTL testability analysis method discussed in [Boubezari 1999] can be used to improve the random-pattern testability of a scan-based logic BIST circuit, in which the outputs and inputs of all storage elements are treated as primary inputs and outputs, respectively. A ***directed acyclic graph*** (DAG) is constructed for each functional block in order to represent the flow of information and data dependencies. Each internal node of a DAG corresponds to a high-level operation (such as an arithmetic, relational, data transfer, and logical operation) of multiple bits, and each edge represents a signal, which can be composed of multiple bits. This modeling method keeps useful high-level information about a functional block while ignoring the details of the gate-level implementation. This information is then used to compute the 0-controllability, 1-controllability, and observability of each bit in a signal line.

As an example, consider the n-bit ripple-carry adder shown in Figure 2.4, which consists of n 1-bit full-adders. By considering the minterms leading to a 1 on the respective output, the probability-based 1-controllability measures of s_i and c_{i+1}, denoted by $C1(s_i)$ and $C1(c_{i+1})$, respectively, are calculated as follows [Boubezari 1999]:

$$C1(s_i) = \alpha + C1(c_i) - 2 \times (\alpha \times C1(c_i))$$

$$C1(c_{i+1}) = \alpha \times C1(c_i) + C1(a_i) \times C1(b_i)$$

where

$$\alpha = C1(a_i) + C1(b_i) - 2 \times C1(a_i) \times C1(b_i)$$

Here, α is the probability that $(a_i \oplus b_i) = 1$ and, consequently, $C1(s_i)$ is the probability that $(a_i \oplus b_i \oplus c_i) = 1$. By applying the above formulas from the leftmost full-adder toward the rightmost full-adder in the n-bit ripple-carry adder, the 1-controllability of each output is obtained. This calculation can be completed in linear time in terms of the number of inputs. The probability-based 0-controllability of each output l, denoted by $C0(l)$, in the n-bit ripple-carry adder is $1 - C1(l)$.

Next, we consider the probability-based observability of an input l on an output s_i, denoted by $O(l, s_i)$, in the n-bit ripple-carry adder. $O(l, s_i)$ is defined as the probability that a signal change on l will result in a signal change on s_i. According to the Boolean function of a 1-bit adder, the change on any input a_i, b_i, or c_i is always observable at s_i. Hence, we have:

$$O(a_i, s_i) = O(b_i, s_i) = O(c_i, s_i) = O(s_i)$$

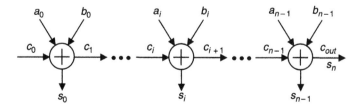

■ **FIGURE 2.4**

Ripple-carry adder composed of n full-adders.

where $i = 0, 1, \ldots, n-1$. On the other hand, the probability-based observability of an input l at stage i on an output s_k—$O(l, s_k)$, where $k > i$—depends on the propagation of the carry output from stage i to the output s_k. This calculation is left as a problem at the end of this chapter.

In general, RTL testability analysis can sometimes lead to more accurate results than gate-level testability analysis. The reason is that the number of reconvergent fanouts in an RTL model is usually much less than that in a gate-level model. RTL testability analysis is also more time efficient than gate-level testability analysis because an RTL model is much simpler than an equivalent gate-level model; however, the practical application of RTL testability analysis for testability enhancement in complex RTL designs remains a challenging research topic.

2.3 DESIGN FOR TESTABILITY BASICS

As discussed in the previous section, the testability of combinational logic decreases as the level of the combinational logic increases. A more serious issue is that good testability for sequential circuits is difficult to achieve. Because many internal states exist, setting a sequential circuit to a required internal state can require a very large number of input events. Furthermore, identifying the exact internal state of a sequential circuit from the primary outputs might require a very long **checking experiment**. Hence, a more structured approach for testing designs that contain a large amount of sequential logic is required as part of a methodical ***design for testability*** (DFT) approach [Williams 1983].

Initially, many ***ad hoc*** techniques were proposed for improving testability. These techniques relied on making local modifications to a circuit in a manner that was considered to result in testability improvement. While *ad hoc* DFT techniques do result in some tangible testability improvement, their effects are local and not systematic. Furthermore, these techniques are not methodical, in the sense that they have to be repeated differently on new designs, often with unpredictable results. Due to the *ad hoc* nature, it is also difficult to predict how long it would take to implement the required DFT features.

The **structured** approach for testability improvement was introduced to allow DFT engineers to follow a methodical process for improving the testability of a design. A structured DFT technique can be easily incorporated and budgeted for as part of the design flow and can yield the desired results. Furthermore, structured DFT techniques are much easier to automate. To date, ***electronic design automation*** (EDA) vendors have been able to provide sophisticated DFT tools to simplify and speed up DFT tasks. Scan design, which is the main topic in this chapter, has been found to be one of the most effective structured DFT methodologies for testability improvement. Not only can scan design achieve the targeted fault coverage goal, but it also makes DFT implementation in scan design manageable. In the following two subsections, we briefly introduce a few typical *ad hoc* DFT techniques, followed by a detailed description of the structured DFT approach, focusing specifically on scan design.

2.3.1 *Ad Hoc* Approach

The *ad hoc* approach involves using a set of design practice and modification guidelines for testability improvement. *Ad hoc* DFT techniques typically involve applying good design practices learned through experience or replacing a bad design practice with a good one. Table 2.5 lists some typical *ad hoc* techniques. In this subsection, we describe test point insertion, which is one of the most widely used *ad hoc* techniques. A few other techniques are further described in Section 2.6. Additional *ad hoc* techniques can be found in [Abramovici 1994].

TABLE 2.5 ■ Typical *Ad hoc* DFT Techniques

A1	Insert test points
A2	Avoid asynchronous set/reset for storage elements
A3	Avoid combinational feedback loops
A4	Avoid redundant logic
A5	Avoid asynchronous logic
A6	Partition a large circuit into small blocks

2.3.1.1 *Test Point Insertion*

Test point insertion (TPI) is a commonly used *ad hoc* DFT technique for improving the controllability and observability of internal nodes. Testability analysis is typically used to identify the internal nodes where ***test points*** should be inserted, in the form of control or observation points.

Figure 2.5 shows an example of observation point insertion for a logic circuit with three low-observability nodes. OP_2 shows the structure of an observation point that is composed of a multiplexer (MUX) and a D flip-flop. A low-observability node is connected to the 0 port of the MUX in an observation point, and all observation points are serially connected into an observation shift register using the 1 port of the MUX. An *SE* signal is used for MUX port selection. When *SE* is set to 0 and the clock *CK* is applied, the logic values of the low-observability nodes are captured into the D flip-flops. When *SE* is set to 1, the D flip-flops within OP_1, OP_2, and OP_3 operate as a shift register, allowing us to observe the captured logic values through *OP_output* during sequential clock cycles. As a result, the observability of the circuit nodes is greatly improved.

Figure 2.6 shows an example of control point insertion for a logic circuit with three low-controllability nodes. CP_2 shows the structure of a *control point* (CP) that is composed of a MUX and a D flip-flop. The original connection at a low-controllability node is cut, and a MUX is inserted between the source and destination ends. During normal operation, the ***test mode*** (*TM*) is set to 0 so that the value from the source end drives the destination end through the 0 port of the MUX.

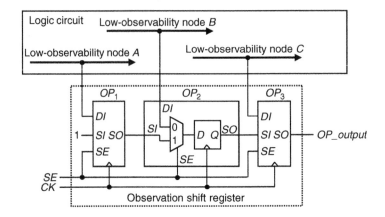

■ **FIGURE 2.5**

Observation point insertion.

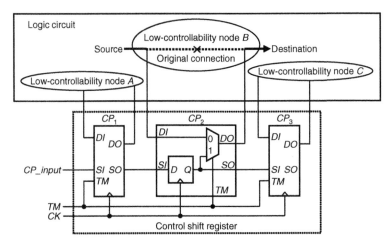

■ **FIGURE 2.6**

Control point insertion.

During test, *TM* is set to 1 so that the value from the D flip-flop drives the destination end through the 1 port of the MUX. The D flip-flops in CP_1, CP_2, and CP_3 are designed to form a shift register so the required values can be shifted into the flip-flops using *CP_input* and used to control the destination ends of low-controllability nodes. As a result, the controllability of the circuit nodes is dramatically improved. This, however, results in additional delay to the logic path. Hence, care must be taken not to insert control points on a critical path. Furthermore, it is preferable to add a *scan point*, which is a combination of a control point and an observation point, instead of a control point, as this allows us to observe the source end as well.

Some other test point designs are described in [Abramovici 1994] and [Nadeau-Dostie 2000]. In addition, test points can be shared among multiple internal nodes; for example, a network of XOR gates can be used to merge a few low-observability nodes together to share one observation point. This can potentially reduce the area overhead, although in some cases it might increase routing difficulty.

2.3.2 Structured Approach

The structured DFT approach attempts to improve the overall testability of a circuit with a *test-oriented design methodology* [Williams 1983] [McCluskey 1986]. This approach is methodical and systematic with much more predictable results.

Scan design, the most widely used structured DFT methodology, attempts to improve testability of a circuit by improving the controllability and observability of storage elements in a sequential design. Typically, this is accomplished by converting the sequential design into a scan design with three modes of operation: **normal mode**, **shift mode**, and **capture mode**. Circuit operations with associated clock cycles conducted in these three modes are referred to as normal operation, shift operation, and capture operation, respectively.

In normal mode, all test signals are turned off, and the scan design operates in the functional configuration. In both shift and capture modes, a **test mode** signal *TM* is often used to turn on all test-related fixes that are necessary to simplify the test, debug, and diagnosis tasks, improve fault coverage, and guarantee the safe operation of the circuit under test. These circuit modes and operations are distinguished using additional test signals or test clocks. The details are described in the following sections.

In order to illustrate how scan design works, consider the sequential circuit shown in Figure 2.7. This circuit contains combinational logic and three D flip-flops. Assume that a stuck-at fault f in the combinational logic requires the primary input X_3, flip-flop FF_2, and flip-flop FF_3 to be set to 0, 1, and 0, respectively, to capture the *fault effect* into FF_1. Because the values stored in FF_2 and FF_3 are not directly controllable from the primary inputs, a long sequence of operations may have to be applied in order to set FF_2 and FF_3 to the required values. Furthermore, in order to observe the fault effect on the captured value in flip-flop FF_1, a long *checking experiment* may be required to propagate the value of FF_1 to a primary output. From this example, it can be seen that the main difficulty in testing a sequential circuit stems from the fact that it is difficult to control and observe the internal state of the circuit.

Scan design, whose concept is illustrated in Figure 2.8, attempts to ease this difficulty by providing external access to selected storage elements in a design. This is accomplished by first converting selected storage elements in the design into **scan cells** and then stitching them together to form one or more shift registers, called **scan chains**. In the scan design illustrated in Figure 2.8, the n storage elements are now configured as a shift register in shift mode. Any test stimulus and test response can now be shifted into and out of the n scan cells in n clock cycles, respectively,

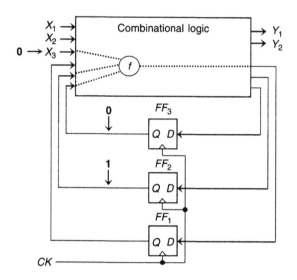

■ **FIGURE 2.7**

Difficulty in testing a sequential circuit.

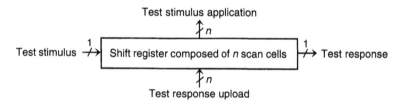

■ **FIGURE 2.8**

Scan design concept.

without having to resort to applying an exponential number of clock cycles to force all storage elements to a desired internal state. Hence, the task of detecting fault f in Figure 2.7 becomes a simple matter of: (1) switching to shift mode and shifting in the desired test stimulus, 1 and 0, to FF_2 and FF_3, respectively; (2) driving a 0 onto primary input X_3; (3) switching to capture mode and applying one clock pulse to capture the fault effect into FF_1; and, finally, (4) switching back to shift mode and shifting out the test response stored in FF_1, FF_2, and FF_3 for comparison with the expected response.

Because scan design provides access to internal storage elements, test generation complexity is reduced. In the following two sections, a number of popular scan cell designs and scan architectures for supporting scan design are described in more detail.

2.4 SCAN CELL DESIGNS

As mentioned in the previous section, in general, a scan cell has two different input sources that can be selected. The first input, *data input*, is driven by the combinational logic of a circuit, while the second input, *scan input*, is driven by the output of another scan cell in order to form one or more shift registers called **scan chains**. These scan chains are made externally accessible by connecting the scan input of the first scan cell in a scan chain to a primary input and the output of the last scan cell in a scan chain to a primary output.

Because there are two input sources in a scan cell, a selection mechanism must be provided to allow a scan cell to operate in two different modes: *normal/capture mode* and *shift mode*. In normal/capture mode, data input is selected to update the output. In shift mode, scan input is selected to update the output. This makes it possible to shift in an arbitrary test pattern to all scan cells from one or more primary inputs while shifting out the contents of all scan cells through one or more primary outputs. In this section, we describe three widely used scan cell designs: **muxed-D scan**, **clocked-scan**, and **level-sensitive scan design** (LSSD).

2.4.1 Muxed-D Scan Cell

The D storage element is one of the most widely used storage elements in logic design. Its basic function is to pass a logic value from its input to its output when a clock is applied. A D flip-flop is an edge-triggered D storage element, and a D latch is a level-sensitive D storage element. The most widely used scan cell replacement for the D storage element is the muxed-D scan cell. Figure 2.9a shows an **edge-triggered muxed-D scan cell** design. This scan cell is composed of a D flip-flop and a multiplexer. The multiplexer uses a *scan enable* (*SE*) input to select between the *data input* (*DI*) and the *scan input* (*SI*).

In normal/capture mode, *SE* is set to 0. The value present at the data input *DI* is captured into the internal D flip-flop when a rising clock edge is applied. In shift mode, *SE* is set to 1. The *SI* is now used to shift in new data to the D flip-flop while the content of the D flip-flop is being shifted out. Sample operation waveforms are shown in Figure 2.9b.

Figure 2.10 shows a **level-sensitive/edge-triggered muxed-D scan cell** design, which can be used to replace a D latch in a scan design. This scan cell is composed of a multiplexer, a D latch, and a D flip-flop. Again, the multiplexer uses a scan enable input *SE* to select between the data input *DI* and the scan input *SI*; however, in this case, shift operation is conducted in an edge-triggered manner, while normal operation and capture operation are conducted in a level-sensitive manner.

Major advantages of using muxed-D scan cells are their compatibility to modern designs using single-clock D flip-flops, and the comprehensive support provided by existing design automation tools. The disadvantage is that each muxed-D scan cell adds a multiplexer delay to the functional path.

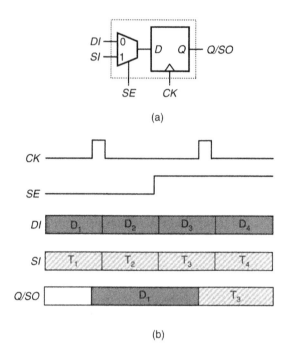

(a)

(b)

■ **FIGURE 2.9**

Edge-triggered muxed-D scan cell design and operation: (a) edge-triggered muxed-D scan cell, and (b) sample waveforms.

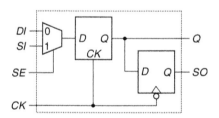

■ **FIGURE 2.10**

Level-sensitive/edge-triggered muxed-D scan cell design.

2.4.2 Clocked-Scan Cell

An edge-triggered ***clocked-scan cell*** can also be used to replace a D flip-flop in a scan design [McCluskey 1986]. Similar to a muxed-D scan cell, a clocked-scan cell also has a data input *DI* and a scan input *SI*; however, in the clocked-scan cell, input selection is conducted using two independent clocks, data clock *DCK* and shift clock *SCK*, as shown in Figure 2.11a.

In normal/capture mode, the data clock *DCK* is used to capture the value present at the data input *DI* into the clocked-scan cell. In shift mode, the shift clock *SCK* is used to shift in new data from the scan input *SI* into the clocked-scan cell, while

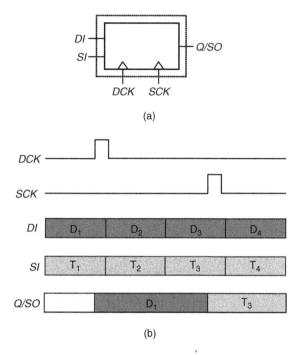

(a)

(b)

Clocked-scan cell design and operation: (a) clocked-scan cell, and (b) sample waveforms.

the current content of the clocked-scan cell is being shifted out. Sample operation waveforms are shown in Figure 2.11b.

As in the case of muxed-D scan cell design, a clocked-scan cell can also be made to support scan replacement of a D latch. The major advantage of using a clocked-scan cell is that it results in no performance degradation on the data input. The major disadvantage, however, is that it requires additional shift clock routing.

2.4.3   LSSD Scan Cell

While muxed-D scan cells and clocked-scan cells are generally used for edge-triggered, flip-flop-based designs, an LSSD scan cell is used for level-sensitive, latch-based designs [Eichelberger 1977] [Eichelberger 1978] [DasGupta 1982]. Figure 2.12a shows a polarity-hold **shift register latch** (SRL) design described in [Eichelberger 1977] that can be used as an LSSD scan cell. This scan cell contains two latches, a master two-port D latch L_1 and a slave D latch L_2. Clocks C, A, and B are used to select between the data input D and the scan input I to drive $+L_1$ and $+L_2$. In an LSSD design, either $+L_1$ or $+L_2$ can be used to drive the combinational logic of the design.

In order to guarantee race-free operation, clocks A, B, and C are applied in a nonoverlapping manner. In designs where $+L_1$ is used to drive the combinational

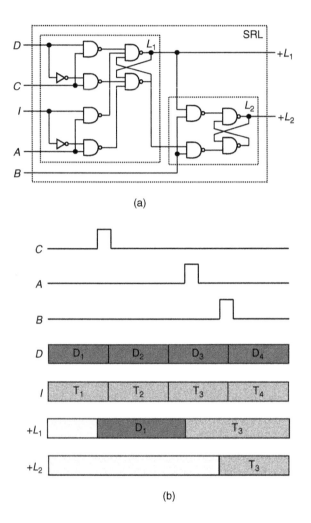

■ FIGURE 2.12

Polarity-hold SRL design and operation: (a) polarity-hold SRL, and (b) sample waveforms.

logic, the master latch L_1 uses the system clock C to latch system data from the data input D and to output this data onto $+L_1$. In designs where $+L_2$ is used to drive the combinational logic, clock B is used after clock C to latch the system data from latch L_1 and to output this data onto $+L_2$. In both cases, capture mode uses both clocks C and B to output system data onto $+L_2$. Finally, in shift mode, clocks A and B are used to latch scan data from the scan input I and to output this data onto $+L_1$ and then latch the scan data from latch L_1 and to output this data onto $+L_2$, which is then used to drive the scan input of the next scan cell. Sample operation waveforms are shown in Figure 2.12b.

The major advantage of using an LSSD scan cell is that it allows us to insert scan into a latch-based design. In addition, designs using LSSD are guaranteed to be race-free, which is not the case for muxed-D scan and clocked-scan designs. The major disadvantage, however, is that the technique requires routing for the additional clocks, which increases routing complexity.

2.5 SCAN ARCHITECTURES

In this section, we describe three popular scan architectures. These scan architectures include: (1) *full-scan design*, where all storage elements are converted into scan cells and combinational ATPG is used for test generation; (2) *partial-scan design*, where a subset of storage elements is converted into scan cells and sequential ATPG is typically used for test generation; and (3) *random-access scan design*, where a random addressing mechanism, instead of serial scan chains, is used to provide direct access to read or write any scan cell.

2.5.1 Full-Scan Design

In full-scan design, all storage elements are replaced with scan cells, which are then configured as one or more shift registers (also called *scan chains*) during the shift operation. As a result, all inputs to the combinational logic, including those driven by scan cells, can be controlled and all outputs from the combinational logic, including those driving scan cells, can be observed. The main advantage of full-scan design is that it converts the difficult problem of sequential ATPG into the simpler problem of combinational ATPG.

A variation of full-scan design, where a small percentage of storage elements (sometimes only a few) are not replaced with scan cells, is referred to as **almost full-scan design**. These storage elements are often left out of scan design for performance reasons, such as storage elements that are on critical paths, or for functional reasons, such as storage elements driven by a small clock domain that are deemed too insignificant to be worth the additional scan insertion effort. In this case, these storage elements may result in fault coverage loss.

2.5.1.1 *Muxed-D Full-Scan Design*

Figure 2.13 shows a sequential circuit example with three D flip-flops. The corresponding muxed-D full-scan circuit is shown in Figure 2.14a. The three D flip-flops, FF_1, FF_2, and FF_3, are replaced with three muxed-D scan cells, SFF_1, SFF_2, and SFF_3, respectively.

In Figure 2.14a, the data input *DI* of each scan cell is connected to the output of the combinational logic as in the original circuit. To form a scan chain, the scan inputs *SI* of SFF_2 and SFF_3 are connected to the outputs *Q* of the previous scan cells, SFF_1 and SFF_2, respectively. In addition, the scan input *SI* of the first scan cell SFF_1 is connected to the primary input *SI*, and the output *Q* of the last scan cell SFF_3 is connected to the primary output *SO*. Hence, in shift mode, *SE* is set to 1, and the scan cells operate as a single scan chain, which allows us to shift in any combination of logic values into the scan cells. In capture mode, *SE* is set to 0, and the

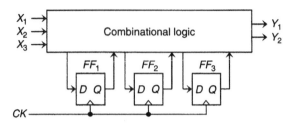

■ **FIGURE 2.13**

Sequential circuit example.

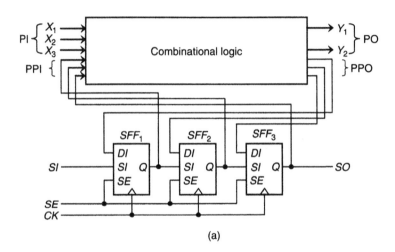

(a)

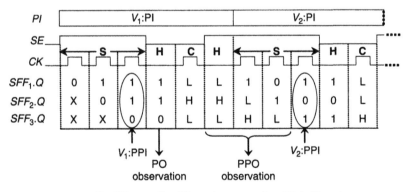

S: shift operation / **C**: capture operation / **H**: hold cycle

(b)

■ **FIGURE 2.14**

Muxed-D full-scan circuit and test operations: (a) muxed-D full-scan circuit, and (b) test operations.

scan cells are used to capture the test response from the combinational logic when a clock is applied.

In general, combinational logic in a full-scan circuit has two types of inputs: ***primary inputs*** (PIs) and ***pseudo primary inputs*** (PPIs). Primary inputs refer to the external inputs to the circuit, while pseudo primary inputs refer to the scan cell outputs. Both PIs and PPIs can be set to any required logic values. The only difference is that PIs are set directly in parallel from the external inputs, and PPIs are set serially through scan chain inputs. Similarly, the combinational logic in a full-scan circuit has two types of outputs: ***primary outputs*** (POs) and ***pseudo primary outputs*** (PPOs). Primary outputs refer to the external outputs of the circuit, while pseudo primary outputs refer to the scan cell inputs. Both POs and PPOs can be observed. The only difference is that POs are observed directly in parallel from the external outputs, while PPOs are observed serially through scan chain outputs.

Figure 2.14b shows a timing diagram to illustrate how the full-scan design is utilized to test the circuit shown in Figure 2.14a for stuck-at faults. During test, the test mode signal TM (not shown) is set to 1, in order to turn on all test-related fixes (see Table 2.6). Two test vectors, V_1 and V_2, are applied to the circuit. In order to apply V_1, SE is first set to 1 to operate the circuit in shift mode (marked by S in Figure 2.14b), and three clock pulses are applied to the clock CK. As a result, the PPI portion of V_1, denoted by V_1:PPI, is now applied to the combinational logic. A **hold cycle** is introduced between the shift and capture operations. During the hold cycle, SE is switched to 0 such that the muxed-D scan cells are operated in capture mode, and the PI portion of V_1, denoted by V_1:PI, is applied. The purpose of the hold cycle is to apply the PI portion of V_1 and to give enough time for the globally routed SE signal to settle from 1 to 0. At the end of the hold cycle, the complete test vector is now applied to the combinational logic, and the logic values at the primary outputs PO are compared with their expected values. Next, the capture operation is conducted (marked by C in Figure 2.14b) by applying one clock pulse to the clock CK in order to capture the test response of the combinational logic to V_1 into the scan cells. A second hold cycle is added in order to switch SE back to 1 and to observe the PPO value of the last scan cell at the SO output. Next, a new shift operation is conducted to shift out the test response captured in the scan cells serially through SO, while shifting in V_2:PPI, which is the PPI portion of the next test pattern V_2.

TABLE 2.6 ■ Circuit Operation Type and Scan Cell Mode

Circuit Operation Type	Scan Cell Mode	*TM*	*SE*
Normal	Normal	0	0
Shift operation	Shift	1	1
Capture operation	Capture	1	0

2.5.1.2 Clocked Full-Scan Design

Figure 2.15 shows a clocked full-scan circuit implementation of the circuit given in Figure 2.13. Clocked-scan cells are shown in Figure 2.11a. This clocked full-scan circuit is tested using shift and capture operations, similar to a muxed-D full-scan circuit. The main difference is how these two operations are distinguished. In a muxed-D full-scan circuit, a scan enable signal SE is used, as shown in Figure 2.14a. In the clocked full-scan circuit shown in Figure 2.15, these two operations are distinguished by properly applying the two independent clocks SCK and DCK during shift mode and capture mode, respectively.

2.5.1.3 LSSD Full-Scan Design

It is possible to implement LSSD full-scan designs, based on the polarity-hold SRL design shown in Figure 2.12a, using either a **single-latch design** or a **double-latch design**. In single-latch design [Eichelberger 1977], the output port $+L_1$ of the master latch L_1 is used to drive the combinational logic of the design. In this case, the slave latch L_2 is used only for scan testing. Because LSSD designs use latches instead of flip-flops, at least two system clocks C_1 and C_2 are required to prevent combinational feedback loops from occurring. In this case, combinational logic driven by the master latches of the first system clock C_1 are used to drive the master latches of the second system clock C_2, and vice versa. In order for this to work, the system clocks C_1 and C_2 should be applied in a nonoverlapping fashion. Figure 2.16a shows an LSSD single-latch design.

Figure 2.16b shows an example of LSSD **double-latch design** [DasGupta 1982]. In normal mode, the C_1 and C_2 clocks are used in a nonoverlapping manner, where the C_2 clock is the same as the B clock. The testing of an LSSD full-scan

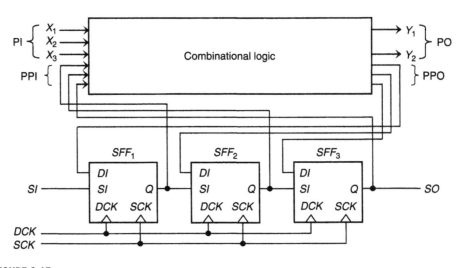

■ **FIGURE 2.15**

Clocked full-scan circuit.

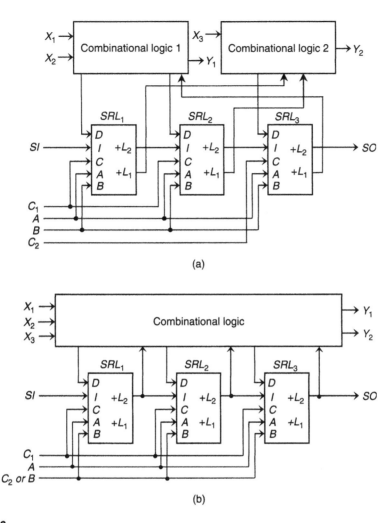

■ **FIGURE 2.16**

LSSD designs: (a) LSSD single-latch design, and (b) LSSD double-latch design.

circuit is conducted using shift and capture operations, similar to a muxed-D full-scan circuit. The main difference is how these two operations are distinguished. In a muxed-D full-scan circuit, a scan enable signal *SE* is used, as shown in Figure 2.14a. In an LSSD full-scan circuit, these two operations are distinguished by properly applying nonoverlapping clock pulses to clocks C_1, C_2, A, and B. During the shift operation, clocks A and B are applied in a nonoverlapping manner, and the scan cells $SRL_1 \sim SRL_3$ form a single scan chain from *SI* to *SO*. During the capture operation, clocks C_1 and C_2 are applied in a nonoverlapping manner to load the test response from the combinational logic into the scan cells.

As mentioned in Section 2.4.3, the operation of a polarity-hold SRL is race-free if clocks C and B as well as A and B are nonoverlapping. This characteristic is used to implement LSSD circuits that are guaranteed to have race-free operation in normal mode as well as in test mode. The required design rules [Eichelberger 1977] [Eichelberger 1978] are briefly summarized below:

- All storage elements must be polarity-hold latches.

- The latches are controlled by two or more nonoverlapping clocks such that any two latches where one feeds the other cannot have the same clock.

- A set of clock primary inputs must exist from which the clock ports of all SRLs are controlled either through a single clock tree or through logic that is gated by SRLs and/or non-clock primary inputs. In addition, the following three conditions should be satisfied: (1) all clock inputs to SRLs must be inactive when clock PIs are inactive, (2) the clock input to any SRL must be controlled from one or more clock primary inputs, and (3) no clock can be ANDed with another clock or its complement.

- Clock primary inputs must not feed the data inputs to SRLs either directly or through combinational logic.

- Each system latch must be part of an SRL, and each SRL must be part of a scan chain.

- A scan state exists under the following conditions: (1) each SRL or scan output SO is a function of only the preceding SRL or scan input SI in its scan chain during the scan operation, and (2) all clocks except the shift clocks are disabled at the SRL clock inputs.

2.5.2 Partial-Scan Design

Unlike full-scan design where all storage elements in a circuit are replaced with scan cells, **partial-scan design** only requires that a subset of storage elements be replaced with scan cells and connected into scan chains [Trischler 1980] [Abadir 1985] [Agrawal 1987] [Ma 1988] [Cheng 1989] [Saund 1997]. Partial-scan design was used in the industry long before full-scan design became the dominant scan architecture. It can also be implemented using muxed-D scan cells, clocked-scan cells, or LSSD scan cells. Depending on the structure of a partial-scan design, either combinational ATPG or sequential ATPG, both of which are described in Chapter 4, should be used.

An example of muxed-D partial-scan design is shown in Figure 2.17. In this example, a scan chain is constructed with two scan cells SFF_1 and SFF_3, while flip-flop FF_2 is left out. Because only one clock is used, typically sequential ATPG has to be used to control and observe the value of the non-scan flip-flop FF_2 through SFF_1 and SFF_3 in order to detect faults related to FF_2. This increases test generation complexity for partial-scan designs [Cheng 1995]. It is possible to reduce the test generation complexity by splitting the single clock into two separate clocks, one for controlling all scan cells, the other for controlling all non-scan storage elements;

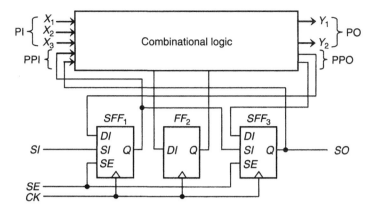

■ FIGURE 2.17

Partial-scan design.

however, this may result in the additional complexity of routing two separate clock trees during physical implementation.

In order to reduce the test generation complexity, many approaches have been proposed for determining the subset of storage elements for scan cell replacement. Scan cell selection can be conducted by using a functional partitioning approach, a pipelined or feed-forward partial-scan design approach, or a balanced partial-scan design approach.

In the **functional partitioning** approach, a circuit is viewed as being composed of a data path portion and a control portion. Typically, because storage elements on the data path portion cannot afford too much delay increase, especially when replaced with muxed-D scan cells, they are left out of the scan cell replacement process. On the other hand, storage elements in the control portion can be replaced with scan cells. This approach makes it possible to improve fault coverage while limiting the performance degradation due to scan design.

In the **pipelined** or **feed-forward partial-scan design** approach [Cheng 1990], a subset of storage elements to be replaced with scan cells is selected to make the sequential circuit feedback-free. This is accomplished by selecting the storage elements to break all sequential feedback loops so that test generation complexity is reduced and the silicon area overhead is kept low. In order to select these storage elements, a *structure graph* is first constructed for the sequential circuit, where each vertex represents a storage element and each directed edge from vertex v_i to vertex v_j represents a combinational logic path from v_i to v_j. For a feedback-free sequential circuit, the structure graph is a *directed acyclic graph*, where the maximum level in the structure graph is referred to as **sequential depth**. On the other hand, the structure graph of a sequential circuit containing feedback loops is a *directed cyclic graph* (DCG). Figure 2.18a shows a block diagram of a feedback-free sequential circuit; its corresponding structure graph is shown in Figure 2.18b with a sequential depth of 3.

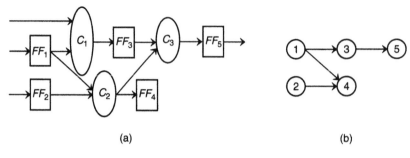

(a)   (b)

■ **FIGURE 2.18**

Sequential circuit and its structure graph: (a) sequential circuit, and (b) structure graph.

The sequential depth of a circuit is equal to the maximum number of clock cycles that must be applied in order to control and observe values to and from all non-scan storage elements. In a full-scan design, because all scan cells can be controlled and observed directly in shift mode, the sequential depth of a full-scan circuit is 0. Similarly, the sequential depth of a combinational logic block is also 0. In a partial-scan design, replacing a storage element with a scan cell is equivalent to removing its corresponding vertex from the structure graph.

In general, the difficulty of sequential ATPG is largely due to the existence of sequential feedback loops. By breaking all feedback loops, test generation for feedback-free sequential circuits becomes computationally efficient; hence, the scan cell selection problem can be expressed as finding the smallest set of vertices to break all feedback loops in a structure graph. The selected vertices are the storage elements that must be replaced with scan cells in order to produce a pipelined or feed-forward partial-scan design; however, a design can contain many self-loops or small loops. Breaking all feedback loops may result in large area overhead. The authors of [Cheng 1990] and [Agrawal 1995] have demonstrated that breaking only large loops, while keeping self-loops or small loops, can produce equally good results. As reported in [Cheng 1990], fault coverage as high as over 95% can be achieved by replacing roughly 25 to 50% of all storage elements with scan cells for a small design.

In the **balanced partial-scan design** approach, a target sequential depth (*e.g.*, 3 to 5) is used to further simplify the test generation process for the pipelined or feed-forward partial-scan design. In this approach, additional vertices are removed from the structure graph by replacing their corresponding storage elements with scan cells so the target sequential depth is met. By keeping the sequential depth under a small limit, one can apply combinational ATPG using multiple time frames to further increase the fault coverage of the design [Gupta 1990].

To summarize, the main advantage of partial-scan design is that it reduces silicon area overhead and performance degradation. The main disadvantage is that it can result in lower fault coverage and longer test generation time than a full-scan design. In practice, functional test vectors often have to be added in order to meet the target fault coverage goal. In addition, partial-scan design offers less support for debug, diagnosis, and failure analysis.

2.5.3 Random-Access Scan Design

Full-scan design and partial-scan design can be classified as **serial scan design**, as test pattern application and test response acquisition are both conducted serially through scan chains. The major advantage of serial scan design is its low routing overhead, as scan data is shifted through adjacent scan cells. Its major disadvantage, however, is that individual scan cells cannot be controlled or observed without affecting the values of other scan cells within the same scan chain. High switching activities at scan cells can cause excessive test power dissipation, resulting in circuit damage, low reliability, or even test-induced yield loss. *Random-access scan* (RAS) attempts to alleviate these problems by making each scan cell randomly and uniquely addressable, similar to storage cells in a *random-access memory* (RAM).

Traditional RAS design [Ando 1980] is illustrated in Figure 2.19. All scan cells are organized into a two-dimensional array, where they can be accessed individually for observing (reading) or updating (writing) in any order. This full-random access capability is achieved by decoding a full address with a row (X) decoder and a column (Y) decoder. A $\lceil \log_2 n \rceil$-bit address shift register, where n is the total number of scan cells, is used to specify which scan cell to access.

The RAS design significantly reduces test power dissipation and simplifies the process of performing delay tests because two independent test vectors can be applied consecutively. Its major disadvantage, however, is high overhead in scan cell design and routing required to set up the addressing mechanism. In addition,

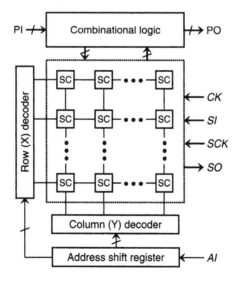

■ **FIGURE 2.19**

Traditional random-access scan architecture.

there is no guarantee that the test application time can be reduced if a large number of scan cells have to be updated for each test vector or the addresses of scan cells to be consecutively accessed have little overlap.

Recently, the ***progressive random-access scan*** (PRAS) design [Baik 2005] was proposed in an attempt to alleviate the problems associated with the traditional RAS design. The PRAS scan cell, as shown in Figure 2.20a, has a structure similar to that of a ***static random access memory*** (SRAM) cell or a grid-addressable latch [Susheel 2002], which has significantly smaller area and routing overhead than the traditional scan cell design [Ando 1980]. In normal mode, all horizontal row enable *RE* signals are set to 0, forcing each scan cell to act as a normal D flip-flop. In test mode, to capture the test response from D, the *RE* signal is set to 0 and a pulse is applied on clock Φ, which causes the value on D to be loaded into the scan cell. To read out the stored value of the scan cell, clock Φ is held at 1, the *RE* signal for the selected scan cell is set to 1, and the content of the scan cell is read out through the bidirectional scan data signals *SD* and $\overline{SD}$. To write or update a scan value into the scan cell, clock Φ is held at 1, the *RE* signal for the selected scan cell is set to 1, and the scan value and its complement are applied on *SD* and $\overline{SD}$, respectively.

The PRAS architecture is shown in Figure 2.20b, where rows are enabled in a fixed order, one at a time, by rotating a 1 in the row enable shift register. That is, it is only necessary to supply a column address to specify which scan cell in an enabled row to access. The length of the column address, which is $\lceil \log_2 m \rceil$ for a circuit with *m* columns, is considerably shorter than a full (row and column) address; therefore, the column address is provided in parallel in one clock cycle instead of providing a full address in multiple clock cycles. This reduces test application time. In order to minimize the need to shift out test responses, the scan cell outputs are compressed with a ***multiple-input signature register*** (MISR). More details on MISRs can be found in Section 5.4.3 of Chapter 5.

The test procedure of the PRAS design is shown in Figure 2.20c. For each test vector, the test stimulus application and test response compression are conducted in an interleaving manner when the test mode signal *TM* is enabled. That is, all scan cells in a row are first read into the MISR for compression simultaneously, and then each scan cell in the row is checked and updated if necessary. Repeating this operation for all rows compresses the test response to the previous test vector into the MISR and sets the next test vector to all scan cells. Next, *TM* is disabled and the normal clock is applied to conduct test response acquisition. It can be seen that the smaller the number of scan cells to be updated for each row, the shorter the test application time. This can be achieved by reducing the Hamming distance between the next test vector and the test response to the previous test vector. Possible solutions include test vector reordering and test vector modification [Baik 2004] [Baik 2005].

It was reported in [Baik 2005] that on average, PRAS design achieved a 37.1%, 64.9%, 85.9%, and 99.5% reduction in test data volume, test application time, peak switching activity, and average switching activity, respectively, when compared with full scan design for several benchmark circuits. The costs were a 25.6% increase in routing overhead and an 11.0% increase in area overhead. Similar results with a different RAS architecture were reported in [Mudlapur 2005]. These results indicate

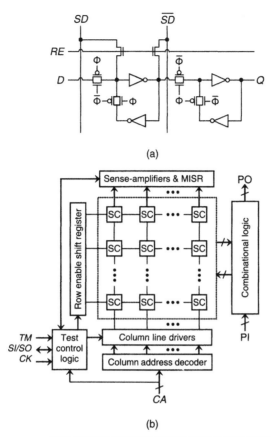

Progressive random-access scan design: (a) PRAS scan cell design, (b) PRAS architecture, and (c) PRAS test procedure.

that RAS design achieves significant reduction in test power dissipation, as well as a good reduction in test data volume and test application time. As test power and delay fault testing are becoming crucial issues in nanometer designs, the RAS approach represents a promising alternative to serial scan design and thus deserves further research.

2.6 SCAN DESIGN RULES

In order to implement scan into a design, the design must comply with a set of scan design rules [Cheung 1996]. In addition, a set of design styles must be avoided, as they may limit the fault coverage that can be achieved. A number of scan design rules that are required to successfully utilize scan and achieve the target fault coverage goal are listed in Table 2.7. In this table, a possible solution is recommended for each scan design rule violation. Scan design rules that are labeled "avoid" must be fixed throughout the shift and capture operations. Scan design rules that are labeled "avoid during shift" must be fixed only during the shift operation. Detailed descriptions are provided for some critical scan design rules.

TABLE 2.7 ■ Typical Scan Design Rules

Design Style	Scan Design Rule	Recommended Solution
Tristate buses	Avoid during shift	Fix bus contention during shift
Bidirectional I/O ports	Avoid during shift	Force to input or output mode during shift
Gated clocks (muxed-D full-scan)	Avoid during shift	Enable clocks during shift
Derived clocks (muxed-D full-scan)	Avoid	Bypass clocks
Combinational feedback loops	Avoid	Break the loops
Asynchronous set/reset signals	Avoid	Use external pins
Clocks driving data	Avoid	Block clocks to the data portion
Floating buses	Avoid	Add bus keepers
Floating inputs	Not recommended	Tie to V_{DD} or ground
Cross-coupled NAND/NOR gates	Not recommended	Use standard cells
Non-scan storage elements	Not recommended for full-scan design	Initialize to known states, bypass, or make transparent

2.6.1 Tristate Buses

Bus contention occurs when two bus drivers force opposite logic values onto a tristate bus, which can damage the chip. Bus contention is designed not to happen during the normal operation and is typically avoided during the capture operation, as advanced ATPG programs can generate test patterns that guarantee only one bus driver controls a bus. However, during the shift operation, no such guarantees can be made; therefore, certain modifications must be made to each tristate bus in order to ensure that only one driver controls the bus. For example, for the tristate bus shown in Figure 2.21a, which has three bus drivers (D_1, D_2, and D_3), circuit modification can be made as shown in Figure 2.21b, where EN_1 is forced to 1 to enable the D_1 bus driver, while EN_2 and EN_3 are set to 0 to disable both D_2 and D_3 bus drivers, when $SE = 1$.

In addition to bus contention, a bus without a pull-up, pull-down, or bus keeper may result in fault coverage loss. The reason is that the value of a floating bus is unpredictable, which makes it difficult to test for a stuck-at-1 fault at the enable signal of a bus driver. To solve this problem, a pull-up, pull-down, or bus keeper can be added. The bus keeper added in Figure 2.21b is an example of fixing this problem by forcing the bus to preserve the logic value driven onto it prior to when the bus becomes floating.

2.6.2 Bidirectional I/O Ports

Bidirectional I/O ports are used in many designs to increase the data transfer bandwidth. During the capture operation, a bidirectional I/O port is usually specified as being either input or output; however, conflicts may occur at a bidirectional I/O port during the shift operation. An example is shown in Figure 2.22a, where a bidirectional I/O port is used as an input and the direction control is provided by the scan cell. Because the output value of the scan cell can vary during the shift operation, the output tristate buffer may become active, resulting in a conflict if BO and the I/O port driven by the tester have opposite logic values. Figure 2.22b shows an example of how to fix this problem by forcing the tristate buffer to be inactive when $SE = 1$, and the tester is used to drive the I/O port during the shift operation. During the capture operation, the applied test vector determines whether a bidirectional I/O port is used as input or output and controls the tester appropriately.

2.6.3 Gated Clocks

Clock gating is a widely used design technique for reducing power by eliminating unnecessary storage element switching activity. An example is shown in Figure 2.23a. The clock enable signal (EN) is generated at the rising edge of CK and is loaded into the latch LAT at the failing edge of CK to become CEN. CEN is then used to enable or disable clocking for the flip-flop DFF. Although clock gating is a good approach for reducing power consumption, it prevents the clock ports of some flip-flops from being directly controlled by primary inputs. As a result, modifications are necessary to allow the scan shift operation to be conducted on these storage elements.

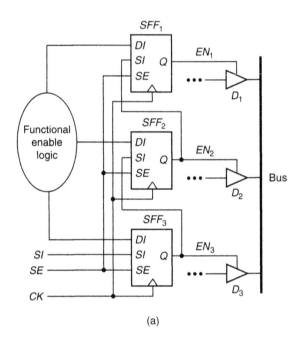

(a)

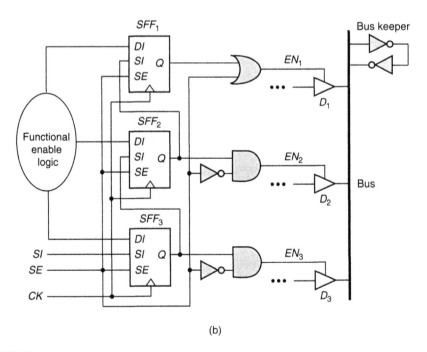

(b)

■ **FIGURE 2.21**

Fixing bus contention: (a) original circuit, and (b) modified circuit.

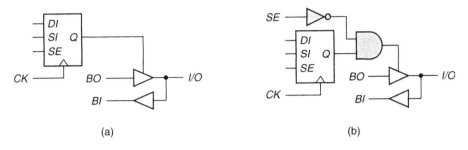

Fixing bidirectional I/O ports: (a) original circuit, and (b) modified circuit.

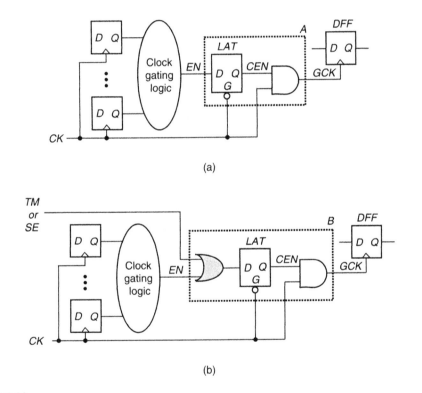

Fixing gated clocks: (a) original circuit, and (b) modified circuit.

The clock gating function should be disabled at least during the shift operation. Figure 2.23b shows how the clock gating can be disabled. In this example, an OR gate is used to force *CEN* to 1 using either the test mode signal *TM* or the scan enable signal *SE*. If *TM* is used, *CEN* will be held at 1 during the entire scan test operation (including the capture operation). This will make it impossible to detect faults in the clock gating logic, causing fault coverage loss. If *SE* is used, *CEN* will be held at 1 only during the shift operation but will be released during the capture operation; hence, higher fault coverage can be achieved but at the expense of increased test generation complexity.

2.6.4 Derived Clocks

A derived clock is a clock signal generated internally from a storage element or a clock generator, such as **phase-locked loop** (PLL), frequency divider, or pulse generator. Because derived clocks are not directly controllable from primary inputs, in order to test the logic driven by these derived clocks, these clock signals must be bypassed during the entire test operation. An example is illustrated in Figure 2.24a, where the derived clock *ICK* drives the flip-flops DFF_1 and DFF_2. In Figure 2.24b, a multiplexer selects *CK*, which is a clock directly controllable from a primary input, to drive DFF_1 and DFF_2 during the entire test operation when $TM = 1$.

2.6.5 Combinational Feedback Loops

Depending on whether the number of inversions on a combinational feedback loop is even or odd, it can introduce either sequential behavior or oscillation into a design. Because the value stored in the loop cannot be controlled or determined during test, this can lead to an increase in test generation complexity or fault coverage loss. Because combinational feedback loops are not a recommended design practice, the best way to fix this problem is to rewrite the RTL code generating the loop. In cases where this is not possible, a combinational feedback loop, as shown in Figure 2.25a, can be fixed by using a test mode signal *TM*. This signal permanently disables the loop throughout the entire shift and capture operations by inserting a scan point (*i.e.*, a combination of control and observation points) to break the loop, as shown in Figure 2.25b.

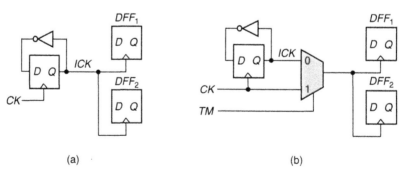

(a)   (b)

■ FIGURE 2.24

Fixing derived clocks: (a) original circuit, and (b) modified circuit.

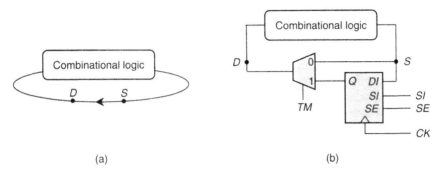

(a)   (b)

■ **FIGURE 2.25**

Fixing combinational feedback loops: (a) original circuit, and (b) modified circuit.

2.6.6   Asynchronous Set/Reset Signals

Asynchronous set/reset signals of scan cells that are not directly controlled from primary inputs can prevent scan chains from shifting data properly. In order to avoid this problem, it is required that these asynchronous set/reset signals be forced to an inactive state during the shift operation. These asynchronous set/reset signals are typically referred to as being sequentially controlled. An example of a sequentially controlled reset signal RL is shown in Figure 2.26a. A method for fixing this asynchronous reset problem using an OR gate with an input tied to the test mode signal TM is shown in Figure 2.26b. When $TM = 1$, the asynchronous reset signal RL of scan cell SFF_2 is permanently disabled during the entire test operation.

The disadvantage of using the test mode signal TM to disable asynchronous set/reset signals is that faults within the asynchronous set/reset logic cannot be tested. Using the scan enable signal SE instead of TM makes it possible to detect faults within the asynchronous set/reset logic, because during the capture operation

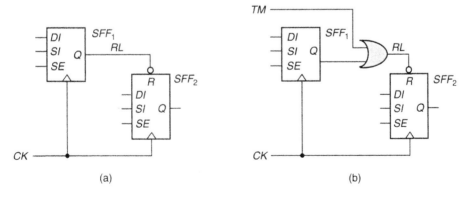

(a)   (b)

■ **FIGURE 2.26**

Fixing asynchronous set/reset signals: (a) original circuit, and (b) modified circuit.

($SE = 0$) these asynchronous set/reset signals are not forced to the inactive state. However, this might result in mismatches due to race conditions between the clock and asynchronous set/reset ports of the scan cells. A better solution is to use an independent reset enable signal RE to replace TM and to conduct test generation in two phases. In the first phase, RE is set to 1 during both shift and capture operations to test data faults through the DI port of the scan cells while all asynchronous set/reset signals are held inactive. In the second phase, RE is set to 1 during the shift operation and 0 during the capture operation without applying any clocks to test faults within the asynchronous set/reset logic.

2.7   SCAN DESIGN FLOW

Although conceptually scan design is not difficult to understand, the practice of inserting scan into a design in order to turn it into a scan design requires careful planning. This often requires many circuit modifications where care must be taken in order not to disrupt the normal functionality of the circuit. In addition, many physical implementation details must be taken into consideration in order to guarantee that scan testing can be performed successfully. Finally, a good understanding of scan design, with respect to which scan cell design and scan architecture to use, is required in order to better plan in advance which scan design rules must be complied with and which debug and diagnose features must be included to facilitate simulation, debug, and fault diagnosis [Crouch 1999].

The shift operation and the capture operation are the two key scan operations where care needs to be taken in order to guarantee that the scan design can operate properly. The shift operation, which is common to all scan designs, must be designed to perform successfully, regardless of the clock skew that exists within the same clock domain and between different clock domains. The capture operation is also common to all scan designs, albeit with more stringent scan design rules in some scan designs as compared to others. It must be designed such that the ATPG tool is able to correctly and deterministically predict the expected responses of the generated test patterns. This requires a basic understanding of the logic simulation and fault models used during ATPG, as well as the clocking scheme used during the capture operation.

A typical design flow for implementing scan in a sequential circuit is shown in Figure 2.27. In this figure, scan design rule checking and repair are first performed on a presynthesis RTL design or on a postsynthesis gate-level design, typically referred to as a **netlist**. The resulting design after scan repair is referred to as a **testable design**. Once all scan design rule violations are identified and repaired, scan synthesis is performed to convert the testable design into a scan design. The scan design now includes one or more scan chains for scan testing. A scan extraction step is used to further verify the integrity of the scan chains and to extract the final scan architecture of the scan chains for ATPG. Finally, scan verification is performed on both shift and capture operations in order to verify that the expected responses predicted by the zero-delay simulator used in test generation or fault simulation match with the full-timing behavior of the circuit

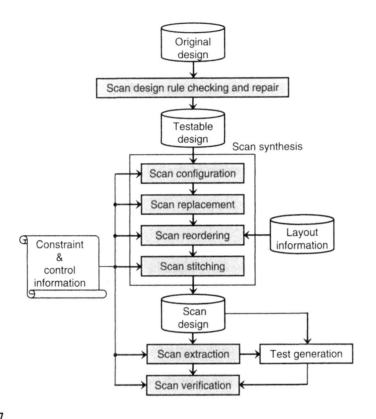

■ **FIGURE 2.27**

Typical scan design flow.

under test. The steps shown in the scan design flow are described in the following subsections in more detail.

2.7.1 Scan Design Rule Checking and Repair

The first step in implementing a scan design is to identify and repair all scan design rule violations in order to convert the original design into a testable design. Repairing these violations allows the testable design to meet target fault coverage requirements and guarantees that the scan design will operate correctly. These scan design rules were described in the previous section. In addition to these scan design rules, certain clock control structures may have to be added for at-speed delay testing. Typically, scan design rule checking is also performed on the scan design after scan synthesis to confirm that no new violations exist.

Upon successful completion of this step, the testable design must guarantee the correct shift and capture operations. During the shift operation, all clocks controlling scan cells of the design are directly controllable from external pins. The clock skew between adjacent scan cells must be properly managed in order not to cause any shift failure. During the capture operation, fixing all scan design rule violations should guarantee correctness for data paths that originate and terminate within the same clock domain. For data paths that originate and terminate in different clock domains, additional care must be taken in terms of the way the clocks are applied in order to guarantee the success of the capture operation. This is mainly due to the fact that the clock skew between different clock domains is typically large. A data path originating in one clock domain and terminating in another might result in a mismatch when both clocks are applied simultaneously, and the clock skew between the two clocks is larger than the data path delay from the originating clock domain to the terminating clock domain. In order to avoid the mismatch, the timing governing the relationship of such a data path shown in the following equation must be observed:

$$clock\ skew\ <\ data\ path\ delay + clock\text{-}to\text{-}Q\ delay\ \text{(originating clock)}$$

If this is not the case, a mismatch may occur during the capture operation. In order to prevent this from happening, clocks belonging to different clock domains can be applied sequentially (using the **staggered clocking** scheme), as opposed to simultaneously, such that any clock skew that exists between the clock domains can be tolerated during the test generation process. It is also possible to apply only one clock during each capture operation using the **one-hot clocking** scheme. On the other hand, a design typically contains a number of noninteracting clock domains. In this case, these clocks can be applied simultaneously, which can reduce the complexity and final pattern count of the pattern generation and fault simulation process. **Clock grouping** is a process used to identify all independent or noninteracting clocks that can be grouped and applied simultaneously.

An example of the clock grouping process is shown in Figure 2.28. This example shows the results of performing a circuit analysis operation on a testable design in order to identify all clock interactions, marked with an arrow, where a data transfer from one clock domain to a different clock domain occurs. As seen in Figure 2.28, the circuit in this example has seven clock domains ($CD_1 \sim CD_7$) and five crossing-clock-domain data paths ($CCD_1 \sim CCD_5$). From this example, it can be seen that CD_2 and CD_3 are independent from each other; hence, their related clocks can be applied simultaneously during test as CK_2. Similarly, clock domains CD_4 through CD_7 can also be applied simultaneously during test as CK_3. Therefore in this example, three grouped clocks instead of seven individual clocks can be used to test the circuit during the capture operation.

2.7.2 Scan Synthesis

When all the repairs have been made to the circuit, the scan synthesis flow is commenced. The scan synthesis flow converts a testable design into a scan design without affecting the functionality of the original design. Static analysis tools and equivalency checkers, which can compare the logic circuitry of two circuits under certain constraints, are typically used to verify that this is indeed the case. Depending on the type of scan cells used and the type of scan architecture implemented, minor modifications to the scan synthesis flow shown in Figure 2.27 may be necessary.

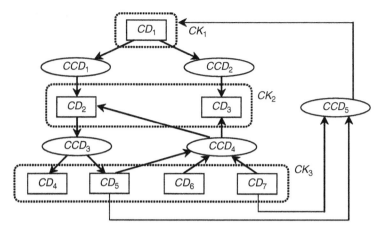

■ **FIGURE 2.28**

Clock grouping example.

During the 1990s, this scan synthesis operation was typically performed using a separate set of scan synthesis tools, which were applied after the logic synthesis tool had synthesized a gate-level netlist out of an RTL description of the design. More recently, these scan synthesis features are being integrated into the logic synthesis tools, and scan designs are synthesized automatically from the RTL. The process of performing scan synthesis during logic synthesis is often referred to as **one-pass synthesis** or **single-pass synthesis**. The scan synthesis flow shown in Figure 2.27 includes four separate steps: (1) scan configuration, (2) scan replacement, (3) scan reordering, and (4) scan stitching. Each of these steps is described below in more detail.

2.7.2.1   Scan Configuration

Scan configuration describes the initial step in scan chain planning, where the general structure of the scan design is determined. The main decisions that are made at this stage include: (1) the number of scan chains used; (2) the types of scan cells used to implement these scan chains; (3) storage elements to be excluded from the scan synthesis process; and (4) the way the scan cells are arranged within the scan chains.

The number of scan chains used is typically determined by analyzing the input and output pins of the circuit to determine how many pins can be allocated for the scan use. In order not to increase the number of pins of the circuit, which is typically limited by the size of the die, scan inputs and outputs are shared with existing pins during scan testing. In general, the larger the number of scan chains used, the shorter the time to perform test on the circuit. This is due to the fact that the maximum length of the scan chains dictates the overall test application time required to run each test pattern. One limitation that can preclude many scan chains from being used is the presence of high-speed I/O pads. The addition of any wire load to the high-speed I/O pad may adversely affect the timing of the design. An additional limitation is the number of tester channels available for scan testing.

The second issue regarding the types of scan cells to use typically depends on the process library. In general, for each type of storage element used, most process libraries have a corresponding scan cell type that closely resembles the functionality and timing of the storage element during normal operation.

The third issue relates to which storage elements to exclude from scan synthesis. This is typically determined by investigating parts of the design where replacing storage elements with functionally equivalent scan cells can adversely affect timing. Therefore, storage elements lying on the critical paths of a design where the timing margin is very tight are often excluded from the scan replacement step, in order to guarantee that the manufactured device will meet the restricted timing. In addition, certain parts of a design may be excluded from scan for many different reasons, including security reasons (*e.g.*, parts of a circuit that deal with encryption). In these cases, individual storage element types, individual storage element instances, or a complete section of the design can be specified as "don't scan."

The remaining issue is to determine how the storage elements are arranged within the scan chains. This typically depends on how the number of clock domains relates to the number of scan chains in the design. In general, a scan chain is formed out of scan cells belonging to a single clock domain. For clock domains that contain a large number of scan cells, several scan chains are constructed, and a scan-chain balancing operation is performed on the clock domain to reduce the maximum scan-chain length. Oftentimes, a clock domain will include both negative-edge and positive-edge scan cells. If the number of negative-edge scan cells in a clock domain is large enough to construct a separate scan chain, then these scan cells can be allocated as such. In cases where a scan chain has to include both negative-edge and positive-edge scan cells, all negative-edge scan cells are arranged in the scan chains such that they precede all positive-edge scan cells in order to guarantee that the shift operation can be performed correctly.

Figure 2.29a shows an example of a circuit structure comprising a negative-edge scan cell followed by a positive-edge scan cell. The associated timing diagram, shown in Figure 2.29b, illustrates the correct shift timing of the circuit structure. During each shift clock cycle, Y will first take on the state X at the rising CK edge before X is loaded with the SI value at the falling CK edge. If we accidentally place the positive-edge scan cell before the negative-edge scan cell, both scan cells will always incorrectly contain the same value at the end of each shift clock cycle.

In cases where scan chains must include scan cells from several different clock domains, a lock-up latch is inserted between adjacent cross-clock-domain scan cells to guarantee that any clock skew between the clocks can be tolerated. Clock skew between different clock domains is expected, as clock skew is controlled within a clock domain to remain below a certain threshold, but not controlled across different clock domains. As a result, a race caused by hold time violation could occur between these two scan cells if a lock-up latch is not inserted.

Figure 2.30a shows an example of a circuit structure having a scan cell SCp belonging to clock domain CK_1 driving a scan cell SCq belonging to clock

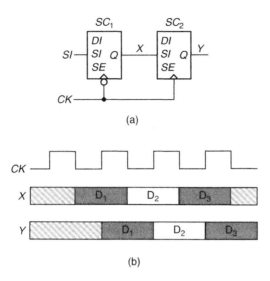

Mixing negative-edge and positive-edge scan cells in a scan chain: (a) circuit structure, and (b) timing diagram.

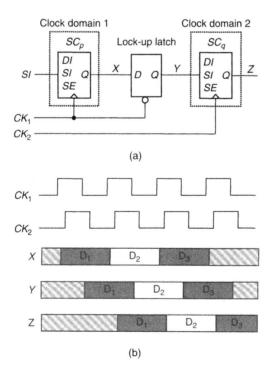

Adding a lock-up latch between cross-clock-domain scan cells: (a) circuit structure, and (b) timing diagram.

domain CK_2 through a lock-up latch. The associated timing diagram is shown in Figure 2.30b, where CK_2 arrives after CK_1, to demonstrate the effect of clock skew on cross-clock-domain scan cells. During each shift clock cycle, X will first take on the SI value at the rising CK_1 edge, then Z will take on the Y value at the rising CK_2 edge. Finally, the new X value is transferred to Y at the falling CK_1 edge to store the SCp contents. If CK_2 arrives earlier than CK_1, Z will first take on the Y value at the rising CK_2 edge. Then, X will take on the SI value at the rising CK_1 edge. Finally, the new X value is transferred to Y at the falling CK_1 edge to store the SCp contents. In both cases, the lock-up latch design in Figure 2.30a allows correct shift operation regardless of whether CK_2 arrives earlier or later than CK_1. It is important to note that this scheme works only when the clock skew between CK_1 and CK_2 is less than the width (duty cycle) of the clock pulse. If this is not the case, then slowing down the shift clock frequency or enlarging the duty cycle of the shift clock can guarantee that this approach will work for any amount of clock skew. Other lock-up latch and lock-up flip-flop designs can also be used.

Once the clock structure of the scan chains is determined, it is still necessary to determine which scan cells should be stitched together into one scan chain and the order in which these scan cells should be placed. In some scan synthesis flows, a preliminary layout placement is used to allocate scan cells to different scan chains belonging to the same clock domain. Then, the best order in which to stitch these scan cells within the scan chains is determined in order to minimize the scan routing required to connect the output of each scan cell to the scan input of the next scan cell. In cases where a preliminary placement is not available, scan cells can be assigned to different scan chains based on an initial floor plan of the testable design, by grouping scan cells in proximate regions of the design together. Once the final placement is determined, the scan chains can then be reordered and stitched, and the scan design is modified based on the new scan chain order.

2.7.2.2 Scan Replacement

After scan configuration is complete, **scan replacement** replaces all original storage elements in the testable design with their functionally equivalent scan cells. The testable design after scan replacement is often referred to as a **scan-ready design**. Functionally equivalent scan cells are the scan cells that most closely match power, speed, and area requirements of the original storage elements. The scan inputs of these scan cells are often tied to the scan outputs of the same scan cell to prevent floating inputs from being present in the circuit. These connections are later removed during the scan stitching step. In cases where one-pass or single-pass synthesis is used, scan replacement is transparent to tool users. Recently, some RTL scan synthesis tools have implemented scan replacement at the RTL, even before going to the logic/scan synthesis tool, in order to reflect the scan design changes in the original RTL design.

2.7.2.3 Scan Reordering

Scan reordering refers to the process of reordering scan cells in scan chains, based on the physical scan cell locations, in order to minimize the amount of interconnect wires used to implement the scan chains. During design implementation, if the physical location of each scan cell instance is not available, a "random" scan order based purely on the module-level and bus-level connectivity of the testable design can be used. However, if a preliminary placement is available, scan cells can be assigned to different scan chains based on the initial floor plan of the design. Only after the final placement process of the physical implementation is performed on this testable design is the physical location of each scan cell instance taken into consideration. During the routing process of the physical implementation, scan reordering can be performed using *intra-scan-chain reordering, inter-scan-chain reordering*, or a combination of both. **Intra-scan-chain reordering**, in which scan cells are reordered only within their respective scan chains, does not reorder any scan cells across clock or clock-polarity boundaries. **Inter-scan-chain reordering**, in which scan cells are reordered among different scan chains, must make sure that the clock structure of the scan chains is preserved. In both intra-scan-chain reordering and inter-scan-chain reordering, care must be also taken to limit the minimum distance between scan cells to avoid timing violations that can destroy the integrity of the shift operation.

Advanced techniques have also been proposed to further reduce routing congestion while avoiding timing violations during the shift operation [Duggirala 2002] [Duggirala 2004]. For deep submicron circuits, the capacitance of the scan chain interconnect must also be taken into account to guarantee correct shift operation [Barbagallo 1996].

2.7.2.4 Scan Stitching

Finally, the **scan stitching** step is performed to stitch all scan cells together to form scan chains. Scan stitching refers to the process of connecting the output of each scan cell to the scan input of the next scan cell, based on the scan order specified above. An additional step is also performed by connecting the scan input of the first scan cell of each scan chain to the appropriate scan chain input port and the scan output of the last scan cell of each scan chain to the appropriate scan chain output port to make the scan chains externally accessible. In cases where a shared I/O port is used to connect to the scan chain input or the scan chain output, additional signals must be connected to the shared I/O port to guarantee that it always behaves as either input or output, respectively, throughout the shift operation. As mentioned earlier, it is important to avoid using high-speed I/O ports as scan chain inputs or outputs, as the additional loading could result in a degradation of the maximum speed at which the device can be operated. In addition to stitching the existing scan cells, lock-up latches or lock-up flip-flops are often inserted during the scan stitching step for adjacent scan cells where clock skew may occur. These lock-up latches or lock-up flip-flops are then stitched between adjacent scan cells.

2.7.3 Scan Extraction

When the scan stitching step is complete, the scan synthesis process is complete. The original design has now been converted into a scan design; however, an additional step is often performed to verify the integrity of the scan chains, especially if any design changes are made to the scan design. **Scan extraction** is the process used for extracting all scan cell instances from all scan chains specified in the scan design. This procedure is performed by tracing the design for each scan chain to verify that all the connections are intact when the design is placed in shift mode. Scan extraction can also be used to prepare for the test generation process to identify the scan architecture of the design in cases where this information is not otherwise available.

2.7.4 Scan Verification

When the physical implementation of the scan design is completed, including placement and routing of all the cells of the design, a timing file in *standard delay format* (SDF) is generated. This timing file resembles the timing behavior of the manufactured device. This is then used to verify that scan testing can be successfully performed on the manufactured scan design.

Other than the trivial problems of scan chains being incorrectly stitched, verification errors during the shift operation are typically due to hold time violations between adjacent scan cells, where the data path delay from the output of a driving scan cell to the scan input of the following scan cell is smaller than the clock skew that exists between the clocks driving the two scan cells. In cases where the two scan cells are driven by the same clock, this may indicate a failure of the *clock tree synthesis* (CTS) process in guaranteeing that the clock skew between scan cells belonging to the same clock domain be kept at a minimum. In cases where the two scan cells are driven by different clocks, this may indicate a failure of inserting a required lock-up latch between the scan cells of the two different clock domains.

Apart from clock skew problems, other scan shift problems can occur. Often, they stem from (1) an incorrect scan initialization sequence that fails to put the design into test mode; (2) incomplete scan design rule checking and repair, where the asynchronous set/reset signals of some scan cells are not disabled during shift operation or the gated/generated clocks for some scan cells are not properly enabled or disabled; or (3) incorrect scan synthesis, where positive-edge scan cells are placed before negative-edge scan cells.

Scan capture problems typically occur due to mismatches between the zero-delay model used in the test generation and fault simulation tool, and the full-timing behavior of the real device. In these cases, care must be taken during the scan design and test application process to: (1) provide enough clock delay between the supplied clocks such that the clock capture order becomes deterministic, and (2) prevent simultaneous clock and data switching events from occurring. Failing to take clock events into proper consideration can easily result in a breakdown of the zero-delay (cycle-based) simulator used in the test generation and fault simulation process. More detailed information regarding scan verification of the shift and capture operations is described below.

2.7.4.1 *Verifying the Scan Shift Operation*

Verifying the scan shift operation involves performing **flush tests** using a full-timing logic simulator during the shift operation. A flush test is a shift test where a selected flush pattern is shifted all the way through the scan chains in order to verify that the same flush pattern arrives at the end of the scan chains at the correct clock cycle. For example, a scan chain containing 1000 scan cells requires 1000 shift cycles to be applied to the scan chain for the selected flush pattern to begin arriving at the scan output. If the data arrive early by a number of shift cycles, this may indicate that a similar number of hold time problems exist in the circuit.

To detect clock skew problems between adjacent scan cells, the selected flush pattern is typically a pattern that is capable of providing both 0-to-1 and 1-to-0 transitions to each scan cell. In order to ensure that a 0-to-0 or 1-to-1 transition of a scan cell does not corrupt the data, the selected flush pattern is further extended to provide these transitions. A typical flush pattern that is used for testing the shift operation is "01100," which includes all four possible transitions. Different flush patterns can also be used for debugging different problems, such as the all-zero and all-one flush patterns used for debugging stuck-at faults in the scan chain.

Because observing the arrival of the data on the scan chain output cannot pinpoint the exact location of any shift error in a faulty scan chain, flush testbenches are typically created to observe the values at all internal scan cells to identify the locations at which the shift errors exist. By using this technique, the faulty scan chain can be easily and quickly diagnosed and fixed during the scan shift verification process; for example:

- Scan hold time problems that exist between scan cells belonging to different clock domains indicate that a lock-up latch may be missing. Lock-up latches should be inserted between these adjacent scan cells.

- Scan hold time and setup time problems that exist between scan cells belonging to the same clock domain indicate that the CTS process was not performed correctly. In this case, either CTS has to be redone or additional buffers need to be inserted between the failing scan cells to slow down the path.

- Scan hold time problems due to positive-edge scan cells followed by negative-edge scan cells indicate that the scan chain order was not performed correctly. Lock-up flip-flops rather than lock-up latches can be inserted between these adjacent scan cells or the scan chains may have to be reordered by placing all negative-edge scan cells before all positive-edge scan cells.

An additional approach to scan shift verification that has become more popular in recent years involves performing *static timing analysis* (STA) on the shift path in shift mode. In this case, the STA tool can immediately identify the locations of all adjacent scan cells that fail to meet timing. The same solutions mentioned earlier are then used to fix problems identified by the STA tool.

2.7.4.2   Verifying the Scan Capture Operation

Verifying the scan capture operation involves simulating the scan design using a full-timing logic simulator during the capture operation. This is used to identify the location of any failing scan cells where the captured response does not match the expected response predicted by the zero-delay logic simulator used in test generation or fault simulation. To reduce simulation time, a **broadside-load** testbench is often used, where a test pattern is loaded directly into all scan cells in the scan chains and only the capture cycle is simulated. Because the **broadside-load test** does not involve any shift cycle in the test pattern, broadside-load testbenches often include at least one shift cycle in the capture verification testbench to ensure that each test pattern can at least shift once. This requires loading the test pattern into the outputs of the previous scan cells, rather than directly into the outputs of the current scan cells. In addition, verifying the scan capture operation often includes a *serial simulation*, in which a limited number of test patterns, typically three to five or as many as can be simulated within a reasonable time, are simulated. In this serial simulation, a test pattern is simulated exactly how it would be applied on the tester by shifting in each pattern serially through the scan chains inputs. Next, a capture cycle is applied. The captured response is then shifted out serially to verify that the complete scan chain operation can be performed successfully.

As mentioned before, mismatches in the capture cycle indicate that the zero-delay simulation model used by the test generator and fault simulator failed to capture all the details of the actual timing occurring in the device. Debugging these types of failures is tedious and may involve observing all signals of the mismatching scan cells as well as signal lines (also called nets) driving these scan cells. One brute-force method commonly used by designers for removing these mismatches is to mask off the locations by changing the expected response of the mismatching location into an unknown (X) value. A new approach that has become more popular is to use the static timing analysis tool for both scan shift and scan capture verification.

2.7.5   Scan Design Costs

The price of converting a design into a scan design involves numerous costs, including area overhead cost, I/O pin cost, performance degradation cost, and design effort cost. However, these costs are far outweighed by the benefits of scan, in terms of the increased testability, lower test development cost, higher product quality with a smaller number of defective parts shipped, and reduced fault diagnosis and failure analysis time. As a result, implementing scan on a design has become almost mandatory. The costs of implementing scan are summarized below:

- *Area overhead cost*—This cost comes primarily in two forms. The first is the scan cell overhead cost due to the replacement of a storage element with a scan cell. The second is the routing cost, which is caused by additional routing of the scan chains, the scan enable signal, and additional shift clocks. Layout-based scan reordering techniques typically do a good job of reducing the overhead due to scan chain routing.

- *I/O pin cost*—Scan design typically requires a dedicated test mode pin to indicate when scan testing is performed. Some designers have been able to get around this need by developing an initialization sequence that is capable of putting the design into test mode. Additional I/O cost is due to the possible performance degradation of pins where scan inputs and scan outputs are shared.

- *Performance degradation cost*—The additional scan input of a scan cell may require placing an additional delay on the functional path. The effects of this delay can be alleviated by embedding the scan replacement step in logic/scan synthesis such that the logic optimization process can be aggressively performed to reduce the effect of the added delay.

- *Design effort cost*—Implementing scan requires additional steps to be added to the typical design flow to perform scan design rule checking and repair, scan synthesis, scan extraction, and scan verification. Additional effort may also be required by the layout engineers in order to perform global routing of the scan enable signal or additional shift clocks, which must be designed to reach all scan cells in the design while having the ability to switch value within a reasonable time. As mentioned before, this cost is far outweighed by the savings in test development efforts that would otherwise have to be performed.

2.8   SPECIAL-PURPOSE SCAN DESIGNS

As discussed above, scan design allows us to use a small external interface to control and observe the states of scan cells in a design which dramatically simplifies the task of test generation. In addition, scan design can be used to reduce debug and diagnosis time and facilitate failure analysis by giving access to the internal states of the circuit. A few other scan methodologies have been proposed for special-purpose testing. In this section, we describe three special-purpose scan designs—namely, enhanced scan, snapshot scan, and error-resilient scan—used for delay testing, system debug, and soft error protection, respectively.

2.8.1   Enhanced Scan

Testing for a delay fault requires applying a pair of test vectors in an at-speed fashion. This is used to generate a logic value transition at a signal line or at the source of a path, and the circuit response to this transition is captured at the circuit's operating frequency. Applying an arbitrary pair of vectors as opposed to a functionally dependent pair of vectors, generated through the combinational logic of the circuit under test, allows us to maximize the delay fault detection capability. This can be achieved using **enhanced scan** [Malaiya 1983] [Glover 1988] [Dervisoglu 1991].

Enhanced scan increases the capacity of a typical scan cell by allowing it to store two bits of data that can be applied consecutively to the combinational logic driven

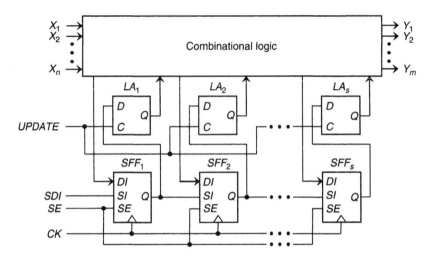

■ FIGURE 2.31

Enhanced-scan architecture.

by the scan cells. For a muxed-D scan cell or a clocked-scan cell, this is achieved through the addition of a D latch.

Figure 2.31 shows a general enhanced-scan architecture using muxed-D scan cells. In this figure, in order to apply a pair of test vectors $<V_1, V_2>$ to the design, the first test vector V_1 is first shifted into the scan cells ($SFF_1 \sim SFF_s$) and then stored into the additional latches ($LA_1 \sim LA_s$) when the *UPDATE* signal is set to 1. Next, the second test vector V_2 is shifted into the scan cells while the *UPDATE* signal is set to 0, in order to preserve the V_1 values in the latches ($LA_1 \sim LA_s$). Once the second vector V_2 is shifted in, the *UPDATE* signal is applied to change V_1 to V_2 while capturing the output response at-speed into the scan cells by applying *CK* after exactly one clock cycle.

The main advantage of enhanced scan is that it allows us to achieve high delay fault coverage, by applying any arbitrary pair of test vectors, that otherwise would have been impossible. The disadvantages, however, are that each enhanced-scan cell requires an additional scan-hold D latch and that maintaining the timing relationship between *UPDATE* and *CK* for at-speed testing may be difficult. An additional disadvantage is that many **false paths**, instead of functional data paths, may be activated during test, causing an **over-test** problem. In order to reduce over-test, the conventional **launch-on-shift** (also called *skewed-load* in [Savir 1993]) and **launch-on-capture** (also called *broad-side* in [Savir 1994] or *double-capture* in Chapter 5) delay test techniques using normal scan chains can be used. These conventional delay test techniques are described in more detail in Chapters 4 and 5.

2.8.2 Snapshot Scan

Snapshot scan is used to capture a snapshot of the internal states of the storage elements in a design at any time without having to disrupt the functional operation

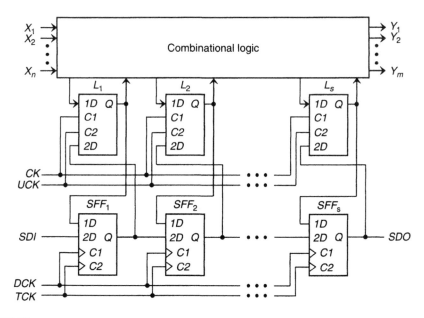

■ **FIGURE 2.32**

Scan-set architecture.

of the circuit. This is done by adding a scan cell to each storage element of interest in the circuit. These scan cells are connected as one or more scan chains that can be used to shift in and shift out any required test data or internal state snapshot of the design. A snapshot scan design technique, called **scan set**, was proposed in [Stewart 1978]. An example of scan-set architecture implemented by adding clocked-scan cells to the system latches (two-port D latches) for snapshot scan is shown in Figure 2.32.

In this figure, four different operations are possible: (1) Test data can be shifted into and out of the scan cells ($SFF_1 \sim SFF_s$) from the SDI and SDO pins, respectively, using TCK. (2) The test data can be transferred to the system latches ($L_1 \sim L_s$) in parallel through their $2D$ inputs using UCK. (3) The system latch contents can be loaded into the scan flip-flops through their $1D$ inputs using DCK. (4) The circuit can be operated in normal mode using CK to capture the values from the combinational logic into the system latches ($L_1 \sim L_s$).

During normal (system) operation, the contents of the system latches can be captured into the scan flip-flops any time DCK is applied. The captured response stored in the scan cells ($SFF_1 \sim SFF_s$) can then be shifted out for analysis. This provides a powerful means of getting a snapshot of the system status that is very helpful in system debug. It is also possible to shift in test data to the system latches to ease fault diagnosis and failure analysis when UCK is applied to the system latches. In addition, by adding observation scan cells that are connected to specific circuit nodes, the scan-set technique makes it possible to capture the logic value at any circuit node of interest and to shift it out for observation. As a result, the observability at nonstorage circuit nodes can be dramatically improved. Hence, the scan-set technique can significantly improve the circuit's diagnostic resolution and silicon debug capability. These advantages have made the approach attractive to high-performance and high-complexity designs [Kuppuswamy 2004], despite the increased area overhead. The technique has also been extended to the LSSD architecture [DasGupta 1981].

2.8.3   Error-Resilient Scan

Soft errors are transient **single-event upsets** (SEUs) caused by various types of radiation. Cosmic radiation has long been regarded as the major source of soft errors, especially in memories [May 1979], and chips used in space applications typically use parity or **error-correcting code** (ECC) for soft error protection. As circuit features begin to shrink into the nanometer ranges, error-causing activation energies are reduced. As a result, terrestrial radiation, such as alpha particles from the packaging materials of a chip, is also beginning to cause soft errors with increasing frequency. This has created reliability concerns, especially for microprocessors, network processors, high-end routers, and network storage components. **Error-resilient scan**, proposed in [Mitra 2005], can also be used to allow scan design to protect a device from soft errors during normal system operation.

Error-resilient scan is based on the observation that soft errors either: (1) occur in memories and storage elements and manifest themselves by flipping their stored states, or (2) result in a transient fault in a combinational gate, as caused by an ion striking a transistor within the combinational gate, and can be captured by a memory or storage element [Nicolaidis 1999]. Data from [Mitra 2005] show that combinational gates and storage elements contribute to a total of 60% of the *soft error rate* (SER) of a design manufactured using current state-of-the-art technology *versus* 40% for memories. Hence, it is no longer enough to consider soft error protection only for memories without considering any soft error protection for storage elements, as well.

Figure 2.33 shows an error-resilient scan cell design [Mitra 2005] that reduces the impact of soft errors affecting storage elements by more than 20 times. This scan cell consists of a system flip-flop and a scan portion, each comprised of a one-port D latch and a two-port D latch, a C-element, and a bus keeper. This scan cell supports two operation modes: system mode and test mode.

In test mode, *TEST* is set to 1, and the C-element acts as an inverter. During the shift operation, a test vector is shifted into latches *LA* and *LB* by alternately applying clocks *SCA* and *SCB* while keeping *CAPTURE* and *CLK* at 0. Then, the *UPDATE* clock is applied to move the content of *LB* to PH_1. As a result, a test vector is written into the system flip-flop. During the capture operation, *CAPTURE* is first set to 1, and then the functional clock *CLK* is applied which captures the circuit response to the test vector into the system flip-flop and the scan portion simultaneously. The circuit response is then shifted out by alternately applying clocks *SCA* and *SCB* again.

In system mode, *TEST* is set to 0, and the C-element acts as a hold-state comparator. The function of the C-element is shown in Table 2.8. When inputs O_1

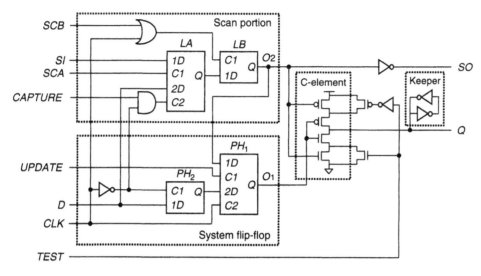

Error-resilient scan cell.

TABLE 2.8 ■ C-Element Truth Table

O_1	O_2	Q
0	0	1
1	1	0
0	1	Previous value retained
1	0	Previous value retained

and O_2 are unequal, the output of the C-element keeps its previous value. During this mode, a 0 is applied to the *SCA*, *SCB*, and *UPDATE* signals, and a 1 is applied to the *CAPTURE* signal. This converts the scan portion into a master-slave flip-flop that operates as a shadow of the system flip-flop. That is, whenever the functional clock *CLK* is applied, the same logic value is captured into both the system flip-flop and the scan portion. When *CLK* is 0, the outputs of latches PH_1 and *LB* hold their previous logic values. If a soft error occurs either at PH_1 or at *LB*, O_1 and O_2 will have different logic values. When *CLK* is 1, the outputs of latches PH_2 and *LA* hold their previous logic values, and the logic values drive O_1 and O_2, respectively. If a soft error occurs either at PH_2 or at *LA*, O_1 and O_2 will have different logic values. In both cases, unless such a soft error occurs after the correct logic value passes through the C-element and reaches the keeper, the soft error will not propagate to the output Q and the keeper will retain the correct logic value at Q.

Error-resilient scan is one of the first online test techniques developed for soft error protection. While the error-resilient scan cell requires more test signals, clocks, and area overhead than conventional scan cells, the technique paves the way to develop more advanced error-resilient and error-tolerant scan and logic BIST architectures to cope with the physical failures of the nanometer age.

2.9 RTL DESIGN FOR TESTABILITY

During the 1990s, the testability of a circuit was primarily assessed and improved at the gate level. The reason was because the circuits were not too large that the logic/scan synthesis process took an unreasonable amount of time. As device size grows toward tens to hundreds of millions of transistors, tight timing, potential yield loss, and low power issues begin to pose serious challenges. When combined with increased core reusability and time-to-market pressure, it is becoming imperative that most, if not all, testability issues be fixed at the RTL. This allows the logic/scan synthesis tool and the physical synthesis tool, which takes physical layout information into consideration, to optimize area, power, and timing after DFT repairs are made. Fixing DFT problems at the RTL also allows designers to create testable RTL cores that can be reused without having to repeat the DFT checking and repair process for a number of times.

Figure 2.34 shows a design flow for performing testability repair at the gate level. It is clear that performing testability repair at the gate level introduces a loop in the design flow that requires repeating the time-consuming logic synthesis process every time testability repair is made. This makes it attractive to attempt to perform testability checking and repair at the RTL instead so testability violations can be detected and fixed at the RTL, as shown in Figure 2.35, without having to repeat the logic synthesis process.

An additional benefit of performing testability repair at the RTL is that it allows scan to be more easily integrated with other advanced DFT features implemented at the RTL, such as memory BIST, logic BIST, test compression, boundary scan, and *analog and mixed-signal* (AMS) BIST. This allows us to perform all testability integration at the RTL, as opposed to the current practices of integrating the

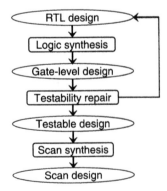

■ FIGURE 2.34

Gate-level testability repair design flow.

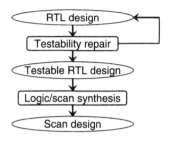

■ **FIGURE 2.35**

RTL testability repair design flow.

advanced DFT features at the RTL, and later integrating them with scan at the gate level. In the following, we describe the RTL DFT problems by focusing mainly on scan design.

Some modern synthesis tools now incorporate testability repair and scan synthesis as part of the logic synthesis process, such that a testable design free of scan rule violations is generated automatically. In this case, if the DFT fixes made are acceptable and do not have to be incorporated into the RTL, the flow can proceed directly to test generation and scan verification.

2.9.1 RTL Scan Design Rule Checking and Repair

In order to perform scan design rule checking and repair at the RTL, a **fast synthesis** step of the RTL is usually performed first. In fast synthesis, combinational RTL code is mapped onto combinational primitives and high-level models, such as adders and multipliers. This allows us to identify all possible scan design rule violations and infer all storage elements in the RTL design.

Static solutions for identifying testability problems at the RTL without having to perform any test vector simulation or dynamic solutions that simulate the structure of the design through the RTL have been developed. These solutions allow us to identify almost all testability problems at the RTL. While a few testability problems remain that can be identified only at the gate level, this approach does reduce the number of iterations involving logic synthesis, as shown in Figure 2.35. In addition, it has become common to add scan design rules as part of RTL "lint" tools that check for good coding and reusability styles, as well as user-defined coding style rules [Keating 1999]. To further optimize testability results, *clock grouping* can also be performed at the RTL as part of scan design rule checking [Wang 2005a].

Automatic methods for repairing RTL testability problems have also been developed [Wang 2005a]. An example of this is shown in Figure 2.36. The RTL code shown in Figure 2.36a, which is written in the Verilog *hardware description language* (HDL) [IEEE 1463-2001], represents a generated clock. In this example, a flip-flop *clk_*15 can be inferred, whose value is driven to 1 when a counter value q is equal to "1111." The output of this flip-flop is then used to trigger the second "always" statement, where an additional flip-flop can be inferred. Figure 2.36b

```
always @(posedge clk)
  if (q ==4'b1111)
    clk_15 <= 1;
  else
    begin
      clk_15 <= 0;
      q <= q + 1;
    end
always @(posedge clk_15)
  d < = start;
```

(a)

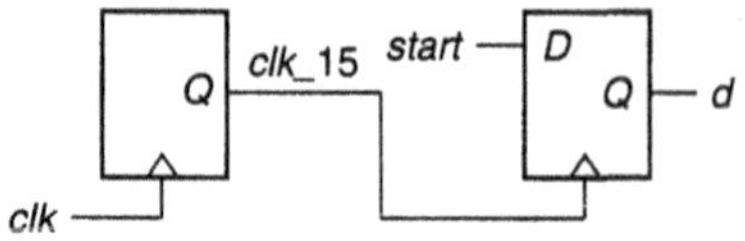

(b)

```
always @(posedge clk)
  if(q == 4'b1111)
    clk_15 <= 1;
  else
    begin
      clk_15 <= 0;
      q <= q + 1;
    end
assign clk_test = (TM)? clk : clk_15;
always @(posedge clk_test)
  d <= start;
```

(c)

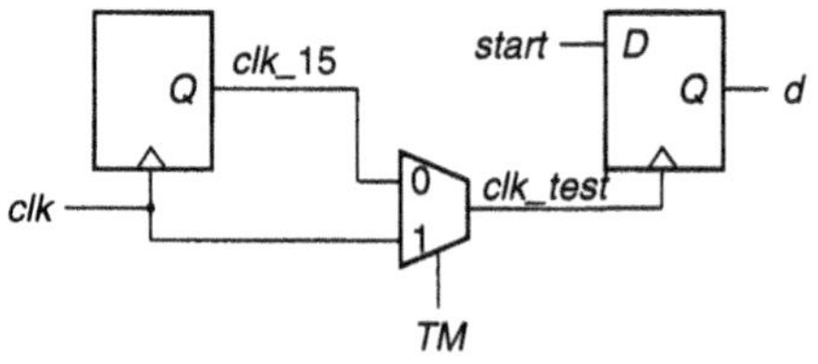

(d)

■ **FIGURE 2.36**

Automatic repair of a generated clock violation at the RTL: (a) generated clock (RTL code), (b) generated clock (schematic), (c) generated clock repair (RTL code), and (d) generated clock repair (schematic).

shows a schematic of the flip-flop generating the *clk*_15 signal, as well as the flip-flop driven by the generated clock, which is likely to be the structure synthesized out of the RTL using a logic synthesis tool. This scan design rule violation can be fixed using the test mode signal *TM* by modifying the RTL code as shown in Figure 2.36c. The schematic for the modified RTL code is shown in Figure 2.36d.

2.9.2 RTL Scan Synthesis

When storage elements have been identified during RTL scan design rule checking, either **RTL scan synthesis** or **pseudo RTL scan synthesis** can be performed. In RTL scan synthesis, the scan synthesis step as described in Section 2.7.2 is performed. The only difference is that the scan equivalent of each storage element does not refer to a library cell but to an RTL structure that is equivalent to the original storage element in normal mode. In this case, the scan chains are inserted into the RTL design. In pseudo RTL scan synthesis, the scan synthesis step is not performed; only pseudo primary inputs and pseudo primary outputs are specified and stitched to primary inputs and primary outputs, respectively. This approach is becoming more appealing to designers nowadays, because it can cope with many advanced DFT structures, such as logic BIST and test compression, where scan chains are driven internally by additional test structures synthesized at the RTL. Once all advanced DFT structures are inserted at the RTL, a one-pass or single-pass synthesis step is performed using the RTL design flow, as shown in Figure 2.35.

Several additional steps are actually performed in order to identify the storage elements in the RTL design. First, all clocks are identified, either explicitly by tracing from specified clock signal names, or implicitly by analyzing the sensitivity list of all "always" blocks. When the clocks have been identified, all registers, each consisting of one or more storage elements in the RTL design, are inferred by analyzing all "assign" statements to determine which assignments can be mapped onto a register while keeping track of the clock domain to which each register belongs. In addition, the clock polarity of each register is determined.

When all registers have been identified and each converted into its scan equivalent at the RTL, the next step is to stitch these individual scan cells into one or more scan chains. One approach is to allocate scan cells to different scan chains based on the driving clocks and to stitch all scan cells within a scan chain in a random fashion [Aktouf 2000]. Although this approach is simple and straightforward, it can introduce wiring congestion as well as high interconnect area overhead. In order to solve these issues, it is better to take full advantage of the rich functional information available at the RTL [Roy 2000] [Huang 2001]. Because storage elements are identified as registers as opposed to a large number of unrelated individual storage elements, it is beneficial to connect the scan cells (which are scan equivalence of these storage elements) belonging to the same register sequentially in a scan chain. This has been found to dramatically reduce wiring congestion and interconnect area overhead.

2.9.3 RTL Scan Extraction and Scan Verification

In order to verify the scan-inserted RTL design (also called *RTL scan design*), both scan extraction and scan verification must be performed. Scan extraction relies on performing fast synthesis on the RTL scan design. This generates a software model where scan extraction can be performed by tracing the scan connections of each scan chain in a similar manner as scan extraction from a *gate-level scan design*. Scan verification relies on a flush testbench that is used to simulate flush tests on the RTL scan design. Because the inputs and outputs of the RTL scan design should match the inputs and outputs of its gate-level scan design, the same flush testbench can be used to verify the scan operation for both RTL and gate-level designs. It is also possible to apply broadside-load tests for verifying the scan capture operation at the RTL. In this case, either random test patterns or deterministic test patterns generated at the RTL can be used [Ghosh 2001] [Ravi 2001] [Zhang 2003].

2.10 CONCLUDING REMARKS

Design for testability (DFT) has become vital for ensuring product quality. Over the past decades, we have seen DFT engineering evolve in order to bridge the gap between design engineering and test engineering. An early task of DFT engineering was to quantify testability. This led to the development of testability analysis, used to identify design areas of poor controllability and observability. These techniques have since proven effective in test generation, logic built-in self-test (BIST), and fault coverage estimation.

When it was recognized that generating test patterns for a sequential circuit was a much more difficult problem than generating test patterns for a combinational circuit, *ad hoc* DFT techniques were proposed but were met with limited success. Scan design, which has proven to be the most powerful DFT technique ever invented, allowed the transformation of sequential circuit testing into combinational circuit testing and has since become an industry standard.

In this chapter, we have presented a comprehensive discussion of scan design. This included scan cell designs, scan architectures, scan design rules, and a typical scan design flow. The RTL DFT techniques that include RTL testability analysis and RTL design for testability were briefly touched upon; these techniques are used to guide testability enhancement and enable DFT integration at the RTL. Finally, we examined promising random-access scan architecture along with a number of special-purpose scan designs, hoping to shed some light on future DFT research.

As we continue to move towards even smaller geometries, new design and test challenges have started to evolve. Novel and advanced DFT architectures will be required to further reduce test power, test data volume, and test application time. We anticipate that advanced at-speed scan and logic BIST architectures [Wang 2005b], low-power scan and logic BIST architectures [Girard 2002] [Wen 2005], and novel error-resilient and error-tolerant architectures [Breuer 2004] will be of growing importance in the coming decades to help us cope with the physical failures of the nanometer design era.

2.11 EXERCISES

2.1 **(Testability Analysis)** Calculate the SCOAP controllability and observability measures for a three-input XOR gate and for its NAND–NOR implementation.

2.2 **(Testability Analysis)** Use the rules given in Tables 2.3 and 2.4 to calculate the probability-based testability measures for a three-input XNOR gate and for its NAND–NOR implementation. Assume that the probability-based controllability values at all primary inputs and the probability-based observability value at the primary output are 0.5 and 1, respectively.

2.3 **(Testability Analysis)** Solve Problem 2.2 again for the full-adder circuit shown in Figure 2.1.

2.4 **(Testability Analysis)** Calculate the combinational observability of input a_i at output s_k, denoted by $O(a_i, s_k)$, where $k > i$, for the n-bit ripple-carry adder shown in Figure 2.4.

2.5 **(*Ad Hoc* Technique)** Use an example to show why a combinational feedback loop in a combinational circuit can cause low testability.

2.6 **(Test Point Insertion)** Show an implementation where a single observation point is used to observe the three low-observability nodes A, B, and C in Figure 2.5 using XOR gates.

2.7 **(Clocked-Scan Cell)** Show a possible gate-level implementation of the clocked-scan cell shown in Figure 2.11a.

2.8 **(LSSD Scan Cell)** Show a possible CMOS implementation of the LSSD scan cell shown in Figure 2.12a.

2.9 **(Full-Scan Design)** Calculate the number of clock cycles required for testing a full-scan design with n test vectors. Assume that the full-scan design has m scan chains, each having the same length L, and that scan testing is conducted in the way shown in Figure 2.14b.

2.10 **(Full-Scan Design)** Explain the main differences between an LSSD single-latch design and an LSSD double-latch design.

2.11 **(Random-Access Scan)** Assume that a sequential circuit with n storage elements has been reconfigured as a full-scan design as shown in Figure 2.14a and a random-access scan design as shown in Figure 2.19. In addition, assume that the full-scan circuit has m balanced scan chains and that a test vector v_i is currently in the scan cells of both scan designs. Now consider the application of the next test vector v_{i+1}. Assume that v_{i+1} is different in d bits. Calculate the number of clock cycles required for applying v_{i+1} to the full-scan design and the random-access scan design, respectively.

2.12 **(Combinational Feedback Loop)** Show an algorithm that checks whether a sequential circuit contains combinational feedback loops.

2.13 **(Lock-Up Latch)** Suppose that a scan chain is configured as $SI \rightarrow SFF_1 \rightarrow SFF_2 \rightarrow SFF_3 \rightarrow SFF_4 \rightarrow SFF_5 \rightarrow SO$, where SFF_1 through SFF_5 are muxed-D scan cells, and SI and SO are the scan input pin and scan output pin, respectively. Suppose that this scan chain fails scan shift verification in which the flush test sequence $<t_1t_2t_3t_4t_5> = <01010>$ is applied but the response sequence is $<r_1r_2r_3r_4r_5> = <01100>$. Identify the scan flip-flops that may have caused this failure, and show how to fix this problem by using a lock-up latch.

2.14 **(Lock-Up Latch)** A scan chain may contain both positive-edge-triggered and negative-edge-triggered muxed-D scan cells. If, by accident, all positive-edge-triggered scan flip-flops are placed before all negative-edge-triggered muxed-D scan cells, show how to stitch them into one single scan chain. (*Hint*: Positive-edge-triggered muxed-D scan cells and negative-edge-triggered muxed-D scan cells should be placed in two separate sections.)

2.15 **(Lock-Up Latch)** Refer to Figure 2.30. The scheme works only when the clock skew between CK_1 and CK_2 is less than the width (duty cycle) of the clock pulse. If CK_2 is delayed more than the duty cycle of CK_1 (*i.e.*, CK_1 and CK_2 become nonoverlapping), show whether or not it is possible to stitch the two cross-clock-domain scan cells into one single scan chain using a lock-up latch. If not, can it be done using a lock-up flip-flop instead?

2.16 **(Scan Stitching)** Use examples to show why a scan chain may not be able to perform the shift operation properly if two neighboring scan cells in the scan chain are too close to or too far from each other. Also describe how to solve these problems.

2.17 **(Test Signal)** Describe the difference between the test mode signal *TM* and the scan enable signal *SE* used in scan testing.

2.18 **(Clock Grouping)** Show an algorithm to find the smallest number of clock groups in clocking grouping.

2.19 **(RTL Testability Enhancement)** Read the following Verilog HDL code and draw its schematic. Then determine if there is any scan design rule violation. If there is any violation, modify the RTL code to fix the problem, then draw the schematic of the modified RTL code.

```
reg [3:0] tri_en;
always @(posedge clk)
begin
    case (bus_sel)
    0: tri_en[0] = 1'b1;
    1: tri_en[1] = 1'b1;
    2: tri_en[2] = 1'b1;
    3: tri_en[3] = 1'b1;
    endcase
end
assign dbus = (tri_en[0])? d1 : 8'bz;
assign dbus = (tri_en[1])? d2 : 8'bz;
assign dbus = (tri_en[2])? d3 : 8'bz;
assign dbus = (tri_en[3])? d4 : 8'bz;
```

2.20 **(A Design Practice)** Use the scan design rule checking programs and user's manuals contained on the companion Web site to show if you can detect any asynchronous set/reset signal violations and bus contention. Try to redesign a Verilog circuit to include such violations. Then, fix the violations by hand, and see whether the problems disappear.

2.21 **(A Design Practice)** Use the scan synthesis programs and user's manuals contained on the companion Web site to convert the two ISCAS-1989 benchmark circuits s27 and s38417 [Brglez 1989] into scan designs. Perform scan extraction and then run Verilog flush tests and broadside-load tests on the scan designs to verify whether the generated testbenches pass Verilog simulation.

Acknowledgments

The authors wish to thank Prof. Xinghao Chen of The City College and Graduate Center of The City University of New York for contributing the Testability Analysis section, and Shianling Wu of SynTest Technologies for contributing a portion of the RTL Design for Testability section, as well as Dwayne Burek of Magma Design Automation, Dr. Augusli Kifli of Faraday Technology, Thomas Wilderotter of Synopsys, and Xiangfeng Li and Fangfang Li of SynTest Technologies for reviewing the text and providing valuable comments.

References

R2.0—Books

[Abramovici 1994] M. Abramovici, M. A. Breuer, and A. D. Friedman, *Digital Systems Testing and Testable Design*, IEEE Press, Piscataway, NJ, 1994.

[Bushnell 2000] M. L. Bushnell and V. D. Agrawal, *Essentials of Electronic Testing for Digital, Memory and Mixed-Signal VLSI Circuits*, Springer Science, New York, 2000.

[Crouch 1999] A. Crouch, *Design for Test for Digital IC's and Embedded Core Systems*, Prentice Hall, Upper Saddle River, NJ, 1999.

[Fujiwara 1985] H. Fujiwara, *Logic Testing and Design for Testability*, The MIT Press, Cambridge, MA, 1985.

[IEEE 1463-2001] *IEEE Standard Description Language Based on the Verilog Hardware Description Language*, IEEE Std. 1463-2001, Institute of Electrical and Electronics Engineers, New York, 2001.

[Jha 2002] N. K. Jha and S. K. Gupta, *Testing of Digital Systems*, Cambridge University Press, London, 2002.

[Keating 1999] M. Keating and P. Bricaud, *Reuse Methodology Manual for System-on-a-Chip Designs*, Kluwer Academic, Boston, MA, 1999.

[McCluskey 1986] E. J. McCluskey, *Logic Design Principles: With Emphasis on Testable Semi-custom Circuits*, Prentice Hall, Englewood Cliffs, NJ, 1986.

[Nadeau-Dostie 2000] B. Nadeau-Dostie, *Design for At-Speed Test, Diagnosis and Measurement*, Kluwer Academic, Boston, MA, 2000.

[Stroud 2002] C. E. Stroud, *A Designer's Guide to Built-In Self-Test*, Kluwer Academic, Boston, MA, 2002.

R2.1—Introduction

[Cheng 1990] K.-T. Cheng and V. D. Agrawal, A partial scan method for sequential circuits with feedback, *IEEE Trans. Comput.*, 39(4), 544–548, 1990.

[Fujiwara 1982] H. Fujiwara and S. Toida, The complexity of fault detection problems for combinational circuits, *IEEE Trans. Comput.*, C-31(6), 555–560, 1982.

[Mitra 2005] S. Mitra, N. Seifert, M. Zhang and K. Kim, Robust system design with built-in soft-error resilience, *IEEE Compt.*, 38(2), 43–52, 2005.

R2.2—Testability Analysis

[Agrawal 1982] V. D. Agrawal and M. R. Mercer, Testability measures: What do they tell us?, in *Proc. Int. Test Conf.*, November 1982, pp. 391–396.

[Boubezari 1999] S. Boubezari, E. Cerny, B. Kaminska, and B. Nadeau-Dostie, Testability analysis and test-point insertion in RTL VHDL specifications for scan-based BIST, *IEEE Trans. Comput.-Aided Des.*, 18(9), 1327–1340, 1999.

[Breuer 1978] M. A. Breuer, New concepts in automated testing of digital circuits, in *Proc. EEC Symp. on CAD of Digital Electronic Circuits and Systems*, 1978, pp. 69–92.

[Grason 1979] J. Grason, TMEAS: A testability measurement program, in *Proc. Des. Automat. Conf.*, June 1979, pp. 156–161.

[Goldstein 1979] L. H. Goldstein, Controllability/observability analysis of digital circuits, *IEEE Trans. Circuits Syst.*, CAS-26(9), 685–693, 1979.

[Goldstein 1980] L. H. Goldstein and E. L. Thigpen, SCOAP: Sadia Controllability/Observability Analysis Program, in *Proc. Des. Automat. Conf.*, June 1980, pp. 190–196.

[Jain 1985] S. K. Jain and V. D. Agrawal, Statistical fault analysis, *IEEE Des. Test Comput.*, 2(2), 38–44, 1985.

[Lee 1992] T.-C. Lee, W. H. Wolf, N. K. Jha, and J. M. Acken, Behavioral synthesis for easy testability in data path allocation, in *Proc. Int. Test Conf.*, October 1992, pp. 29–32.

[Rizzolo 2001] R. F. Rizzolo, B. F. Robbins, and D. G. Scott, A hierarchical approach to improving random pattern testability on IBM eServer z900 chips, in *Digest of Papers: North Atlantic Test Workshop*, May 2001, pp. 84–89.

[Rutman 1972] R. A. Rutman, Fault detection test generation for sequential logic heuristic tree search, *IEEE Computer Group Repository*, Paper No. R-72-187, 1972.

[Parker 1975] K. P. Parker and E. J. McCluskey, Probability treatment of general combinational networks, *IEEE Trans. Comput.*, 24(6), 668–670, 1975.

[Savir 1984] J. Savir, G. S. Ditlow, and P. H. Bardell, Random pattern testability, *IEEE Trans. Comput.*, C-33(1), 79–90, 1984.

[Seth 1985] S. C. Seth, L. Pan, and V. D. Agrawal, PREDICT: Probabilistic estimation of digital circuit testability, in *Proc. Int. Symp. on Fault-Tolerant Computing*, June 1985, pp. 220–225.

[Stephenson 1976] J. E. Stephenson and J. Garson, A testability measure for register transfer level digital circuits, in *Proc. Int. Symp. on Fault-Tolerant Computing*, June 1976, pp. 101–107.

[Wang 1984] L.-T. Wang and E. Law, Daisy Testability Analyzer (DTA), in *Proc. Int. Conf. on Computer-Aided Design*, November 1984, pp. 143–145.

[Wang 1985] L.-T. Wang and E. Law, An enhanced Daisy Testability Analyzer (DTA), in *Proc. Automat. Test. Conf.*, October 1985, pp. 223–229.

R2.3—Design for Testability Basics

[Williams 1983] T. W. Williams and K. P. Parker, Design for testability: A survey, *Proc. IEEE*, 71(1), 98–112, 1983.

R2.4—Scan Cell Designs

[DasGupta 1982] S. DasGupta, P. Goel, R. G. Walther, and T. W. Williams, A variation of LSSD and its implications on design and test pattern generation in VLSI, in *Proc. Int. Test Conf.*, November 1982, pp. 63–66.

[Eichelberger 1977] E. B. Eichelberger and T. W. Williams, A logic design structure for LSI testability, in *Proc. Des. Automat. Conf.*, June 1977, pp. 462–468.

[Eichelberger 1978] E. B. Eichelberger and T. W. Williams, A logic design structure for LSI testability, *J. Des. Automat. Fault-Tolerant Comput.*, 2(2), 165–178, 1978.

R2.5—Scan Architectures

[Abadir 1985] M. S. Abadir and M. A. Breuer, A knowledge-based system for designing testable VLSI chips, *IEEE Design Test Comput.*, 2(4), 56–68, 1985.

[Agrawal 1987] V. D. Agrawal, K.-T. Cheng, D. D. Johnson, and T. Lin, A complete solution to the partial scan problem, in *Proc. Int. Test Conf.*, September 1987, pp. 44–51.

[Agrawal 1995] V. D. Agrawal, Special issue on partial scan methods, Vol. 7, No. 1/2, *Journal of Electronic Testing: Theory and Applications (JETTA)*, Aug.–Oct., 7(1/2), 1995.

[Ando 1980] H. Ando, Testing VLSI with random access scan, in *Proc. COMPCON*, February 1980, pp. 50–52.

[Arslan 2004] B. Arslan and A. OIn raioglu, Test cost reduction through a reconfigurable scan architecture, in *Proc. Int. Test Conf.*, November 2004, pp. 945–952.

[Baik 2004] D. Baik, S. Kajihara, and K. K. Saluja, Random access scan: A solution to test power, test data volume and test time, in *Proc. Int. Conf. on VLSI Design*, January 2004, pp. 883–888.

[Baik 2005] D. Baik and K. K. Saluja, Progressive random access scan: A simultaneous solution to test power, test data volume and test time, in *Proc. Int. Test Conf.*, November 2005, Paper 15.2 (10 pp.).

[Cheng 1989] K.-T. Cheng and V. D. Agrawal, An economical scan design for sequential logic test generation, in *Proc. Int. Symp. on Fault-Tolerant Computing*, June 1989, pp. 28–35.

[Cheng 1990] K.-T. Cheng and V. D. Agrawal, A partial scan method for sequential circuits with feedback, *IEEE Trans. Comput.*, 39(4), 544–548, 1990.

[Cheng 1995] K.-T. Cheng, Single-clock partial scan, *IEEE Des. Test Comput.*, 12(2), 24–31, 1995.

[Gupta 1990] R. Gupta, R. Gupta, and M. A. Breuer, The ballast methodology for structured partial scan design, *IEEE Trans. Comput.*, 39(4), 538–544, 1990.

[Ma 1988] H.-K. Ma, S. Devadas, A. R. Newton, and A. Sangiovanni-Vincentelli, An incomplete scan design approach to test generation for sequential machines, in *Proc. Int. Test Conf.*, September 1988, pp. 730–734.

[Mudlapur 2005] A. S. Mudlapur, V. D. Agrawal, and A. D. Singh, A random access scan architecture to reduce hardware overhead, in *Proc. Int. Test Conf.*, November 2005, Paper 15.1 (9 pp.).

[Saund 1997] G. S. Saund, M. S. Hsiao, and J. H. Patel, Partial scan beyond cycle cutting, in *Proc. Int. Symp. on Fault-Tolerant Computing*, 1997, pp. 320–328.

[Susheel 2002] T. G. Susheel, J. Chandra, T. Ferry, and K. Pierce, ATPG based on a novel grid-addressable latch element, in *Proc. Des. Automat. Conf.*, June 2002, pp. 282–286.

[Trischler 1980] E. Trischler, Incomplete scan path with an automatic test generation methodology, in *Proc. Int. Test Conf.*, November 1980, pp. 153–162.

[Williams 1983] T. W. Williams and K. P. Parker, Design for testability: A survey, *Proc. IEEE*, 71(1), 98–112, 1983.

R2.6—Scan Design Rules

[Cheung 1996] B. Cheung and L.-T. Wang, The seven deadly sins of scan-based designs, *Integrated Syst. Des.*, August 1996 (www.eetimes.com/editorial/1997/test9708.html).

R2.7—Scan Design Flow

[Barbagallo 1996] S. Barbagallo, M. Bodoni, D. Medina, F. Corno, P. Prinetto, and M. Sonza Reorda, Scan insertion criteria for low design impact, in *Proc. VLSI Test Symp.*, April 1996, pp. 26–31.

[Duggirala 2002] S. Duggirala, R. Kapur, and T. W. Williams, System and Method for High-Level Test Planning for Layout, U.S. Patent No. 6,434,733, August 13, 2002.

[Duggirala 2004] S. Duggirala, R. Kapur, and T. W. Williams, System and Method for High-Level Test Planning for Layout, U.S. Patent No. 6,766,501, July 20, 2004.

R2.8—Special-Purpose Scan Designs

[Cheng 1991] K. Cheng, S. Devadas, and K. Keutzer, A partial enhanced-scan approach to robust delay fault test generation for sequential circuits, in *Proc. Int. Test Conf.*, October 1991, pp. 403–410.

[DasGupta 1981] S. DasGupta, R. G. Walther, T. W. Williams, and E. B. Eichelberger, An enhancement to LSSD and some applications of LSSD in reliability, availability and serviceability, in *Proc. Int. Symp. on Fault-Tolerant Computing*, June 1981, pp. 32–34.

[Dervisoglu 1991] B. I. Dervisoglu and G. E. Strong, Design for testability, using scan path techniques for path-delay test and measurement, in *Proc. Int. Test Conf.*, October 1991, pp. 365–374.

[Glover 1988] C. T. Glover and M. R. Mercer, A method of delay fault test generation, in *Proc. Des. Automat. Conf.*, June 1988, pp. 90–95.

[Kuppuswamy 2004] R. Kuppuswamy, P. DesRosier, D. Feltham, R. Sheikh, and P. Thadikaran, Full hold-scan systems in microprocessors: Cost/benefit analysis, *Intel Technol. J.*, 8(1), February 18, 2004.

[Malaiya 1983] Y. K. Malaiya and R. Narayanaswamy, Testing for timing faults in synchronous sequential integrated circuits, in *Proc. Int. Test Conf.*, October 1983, pp. 560–571.

[May 1979] T. C. May and M. H. Woods, Alpha-particle-induced soft errors in dynamic memories, *IEEE Trans. Electron Devices*, ED-26(1), 2–9, 1979.

[Mitra 2005] S. Mitra, N. Seifert, M. Zhang, and K. Kim, Robust system design with built-in soft-error resilience, *IEEE Comput.*, 38(2), 43–52, 2005.

[Nicolaidis 1999] M. Nicolaidis, Time redundancy based soft-error tolerance to rescue nanometer technologies, in *Proc. VLSI Test Symp.*, April 1999, pp. 86–94.

[Savir 1993] J. Savir and S. Patil, Scan-based transition test, *IEEE Trans. Comput.-Aided Des.*, 12(8), 1232–1241, 1993.

[Savir 1994] J. Savir and S. Patil, Broad-side delay test, *IEEE Trans. Comput.-Aided Design*, 13(8), 1057–1064, 1994.

[Stewart 1978] J. H. Stewart, Application of scan/set for error detection and diagnostics, in *Proc. Semiconductor Test Conf.*, October 1978, pp. 152–158.

R2.9—RTL Design for Testability

[Aktouf 2000] C. Aktouf, H. Fleury, and C. Robach, Inserting scan at the behavioral level, *IEEE Des. Test Comput.*, 17(3), 34–42, 2000.

[Ghosh 2001] I. Ghosh and M. Fujita, Automatic test pattern generation for functional register-transfer level circuits using assignment decision diagrams, *IEEE Trans. Comput.-Aided Des.*, 20(3), 402–415, 2001.

[Huang 2001] Y. Huang, C. C. Tsai, N. Mukherjee, O. Samoan, W.-T. Cheng, and S. M. Reddy, On RTL scan design, in *Proc. Int. Test Conf.*, November 2001, pp. 728–737.

[Ravi 2001] S. Ravi and N. Jha, Fast test generation for circuits with RTL and gate-level views, in *Proc. Int. Test Conf.*, November 2001, pp. 1068–1077.

[Roy 2000] S. Roy, G. Guner, and K.-T. Cheng, Efficient test mode selection and insertion for RTL-BIST, in *Proc. Int. Test Conf.*, October 2000, pp. 263–272.

[Wang 2005a] L.-T. Wang, A. Kifli, F.-S. Hsu, S.-C. Kao, X. Wen, S.-H. Lin, and H.-P. Wang, Computer-Aided Design System to Automate Scan Synthesis at Register-Transfer Level Test, U.S. Patent No. 6,957,403, October 18, 2005.

[Zhang 2003] L. Zhang, I. Ghosh, and M. S. Hsiao, Efficient sequential ATPG for functional RTL circuits, in *Proc. Int. Test Conf.*, 2003, pp. 290–298.

R2.10—Concluding Remarks

[Breuer 2004] M. Breuer, S. Gupta, and T. M. Mak, Defect and error tolerance in the presence of massive numbers of defects, *IEEE Des. Test Comput.*, May/June, 216–227, 2004.

[Girard 2002] P. Girard, Survey of low-power testing of VLSI circuits, *IEEE Des. Test Comput.*, 19(3), 82–92, 2002.

[Wang 2005b] L.-T. Wang, X. Wen, P.-C. Hsu, S. Wu, and J. Guo, At-speed logic BIST architecture for multi-clock designs, in *Proc. Int. Conf. on Computer Design*, October 2005, pp. 475–478.

[Wen 2005] X. Wen, H. Yamashita, S. Kajihara, L.-T. Wang, K. Saluja, and K. Kinoshita, On low-capture-power test generation for scan testing, in *Proc. VLSI Test Symp.*, May 2005, pp. 265–270.

CHAPTER 3

LOGIC AND FAULT SIMULATION

Jiun-Lang Huang
National Taiwan University, Taipei, Taiwan

James C.-M. Li
National Taiwan University, Taipei, Taiwan

Duncan M. (Hank) Walker
Texas A&M University, College Station, Texas

ABOUT THIS CHAPTER

Simulation is a powerful set of techniques that are used heavily in digital circuit verification, test development, design debug, and diagnosis. During the design stage, logic simulation is performed to help verify whether the design meets its specifications and contains any design errors. It also helps locate these design errors that escape to fabrication during design debug. In test development, faulty circuit behavior is simulated with a set of test patterns to assess the pattern quality and guide further pattern development. Simulation of faulty circuits is referred to as fault simulation and is also used during fault diagnosis, where test results are used to locate manufacturing defects within the hardware.

This chapter begins with a discussion of logic simulation. After an introduction to the logic circuit models, the popular compiled-code and event-driven logic simulation techniques are described. This is followed by a description of hazards, the undesirable transient pulses (glitches) that can occur in circuits, what causes them, and how they can be detected during logic simulation. The second half of the chapter discusses fault simulation. Although fault simulation is rooted in logic simulation, many techniques have been developed to quickly simulate all possible faulty behaviors. A discussion of the serial, parallel, deductive, concurrent, and differential fault simulation techniques is followed by qualitative comparisons between their advantages and drawbacks. The chapter concludes with alternative techniques to fault simulation. These techniques trade accuracy for reduced execution time which is crucial for managing the complexity of large designs. By working through this chapter, the reader will learn about the major logic and fault simulation techniques. This background will be valuable in selecting the simulation methodology that best meets the design needs.

3.1   INTRODUCTION

Simulation is the process of predicting the behavior of a circuit design before it is physically built. For digital circuits, simulation serves dual purposes. First, during the design stage, **logic simulation** helps the designer verify that the design conforms to the functional specifications. Second, during test development, **fault simulation** is used to simulate faulty circuits. (For this reason, logic simulation is generally referred to as fault-free simulation.) Given a set of test patterns, fault simulation determines its efficiency in detecting the modeled faults of interest. Furthermore, fault simulation is also an important component of *automatic test pattern generator* (ATPG) programs.

3.1.1   Logic Simulation for Design Verification

The main application of logic simulation is **design verification**, the process of verifying the correctness of a digital design prior to its physical realization in the form of silicon, a *printed circuit board* (PCB), or even a system. To manage growing design complexity, logic simulation or design verification is generally performed at each design stage, ranging from the behavioral down to the switch level. During each design stage, the design is described in a suitable description language that captures the required functional specification for fulfilling the design goal of that stage.

In general, design verification begins at the **behavioral level** or *electronic system level* (ESL). At this level, the behavioral model of the target design is described in ESL languages such as C/C++, SystemC [SystemC 2006], and SystemVerilog [SystemVerilog 2006]. Once the behavioral model has been verified to an acceptable confidence level, the verification process moves to the *register-transfer level* (RTL) design stage. The circuit at this stage is described in *hardware description languages* (HDLs) (*e.g.*, Verilog [IEEE 1463-2001] [Thomas 2002] and VHDL [IEEE 1076-2002]), in terms of blocks such as registers, counters, data processing units, and controllers, as well as the data/control flow between these blocks. Because ESL/RTL verification usually does not involve detailed timing analysis, design verification of the ESL or RTL is also referred to as **functional verification** [Wile 2005].

Logic/scan synthesis comes into play after the RTL design stage. The **gate-level** netlist of the RTL design that includes scan cells is synthesized from logic elements provided in a cell library. For high-performance designs, the **switch-level** model may be employed for the timing-critical portions. A switch-level network is described as the interconnection of MOS switches. Finally, at the transistor level, the circuit is described as interconnections of devices such as transistors, resistors, and capacitors. The **transistor-level** description provides the most accurate model for the design under development, but transistor-level simulation is much slower than gate-level simulation. Thus, transistor-level simulation is usually only used for characterizing cell libraries, including SRAMs and DRAMs. For digital system designs, in general, logic simulation at the gate level suffices.

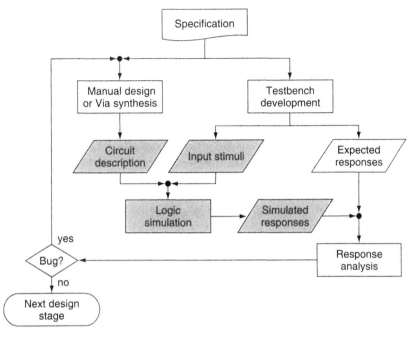

Logic simulation for design verification.

The flow of using logic simulation for digital circuit design verification is shown in Figure 3.1. The functional specification documents the required functionality and performance for the target design. During each design stage, a corresponding circuit description that contains ESL code for the behavioral design, HDL code for the RTL design, a netlist for the gate-level design, or SPICE models for the switch- and transistor-level design is generated in conformance with the given specification. To ensure conformance, verification testbenches consisting of a set of input stimuli and expected output responses are created. The logic simulator then takes the circuit description and the input stimuli as inputs and produces the simulated responses. Any discrepancy between the simulated and expected responses (detected by the response analysis process) indicates the existence of a design bug. The circuit is then redesigned or modified until no more design errors exist. The design process then advances to the next design stage.

3.1.2 Fault Simulation for Test and Diagnosis

The major difference between logic simulation and fault simulation lies in the nature of the nonidealities they deal with. Logic simulation is intended for identifying design errors using the given specifications or a known good design as the reference. Design errors may be introduced by human designers or EDA tools and should be caught prior to physical implementation. Fault simulation, on the other hand, is concerned with the behavior of fabricated circuits as a consequence of inevitable fabrication process imperfections. Manufacturing defects (*e.g.*, wire shorts and opens), if present, may cause the circuits to behave differently from the expected behavior. Fault simulation generally assumes that the design is functionally correct.

The capability of fault simulation to predict the faulty circuit behavior is of great importance for test and diagnosis. First, fault simulation rates the effectiveness of a set of test patterns in detecting manufacturing defects. The quality of a test set is expressed in terms of **fault coverage**, the percentage of modeled faults that causes the design to exhibit observable erroneous responses if the test set is applied. In practice, the designer employs the **fault simulator** to evaluate the fault coverage of a set of input stimuli (test vectors or test patterns) with respect to the modeled faults of interest. Because fault simulation concerns the fault coverage of a test set rather than the detection of design bugs, it is also termed **fault grading**. Low fault coverage test patterns will jeopardize the manufacturing test quality and eventually lead to unacceptable field returns from customers. Second, fault simulation helps identify undetected faults which is especially important when the achieved fault coverage is unacceptable. In this case, either the designer or the ATPG has to generate additional test vectors to improve the fault coverage (*i.e.*, to detect the undetected faults). Third, fault simulation allows one to compress the test set without sacrificing fault coverage. As part of the **test compaction** process, fault simulation identifies redundant test patterns, which are discarded with no negative impact on the fault coverage. With the above capabilities and applications, fault simulation is one of the crucial components of ATPG. In fact, implementation of an ATPG program usually starts with the fault simulator. Finally, fault simulation assists **fault diagnosis**, which determines the type and location of faults that best explain the faulty circuit behavior of the device under diagnosis. The fault simulation results are compared against the observed circuit responses to identify the most likely faults. The fault type and location information can then be used as a starting point for locating the defects that cause the circuit malfunction.

Although fault simulation can also be used to fault-grade analog and mixed-signal circuits, this chapter will only focus on the most popular fault simulation techniques for digital circuits. Readers interested in analog and mixed-signal testing should refer to Chapter 11.

3.2 SIMULATION MODELS

In this section, we discuss the gate-level circuit simulation models for combinational and sequential networks, which have widespread acceptance in the integrated circuit testing community. Gate-level circuit descriptions contain sufficient circuit structure information necessary to capture the effects of many realistic manufacturing defects. On the other hand, the abstraction level of gate-level models is high enough to permit development of efficient simulation techniques.

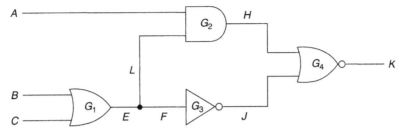

■ FIGURE 3.2

The gate-level model of the combinational circuit *N*.

3.2.1   Gate-Level Network

A gate-level network is described as the interconnections of logic gates, which are circuit elements that realize Boolean operations or expressions. The available gates to realize a Boolean expression range from the standard gates (AND, OR, NOT, NAND, and NOR) to complex gates such as XOR and XNOR. For example, the combinational circuit N[1] in Figure 3.2 is composed of an OR gate (G_1), an AND gate (G_2), an inverter (G_3), and a NOR gate (G_4). The Boolean expression associated with the network can be obtained after a few Boolean algebraic manipulations[2]:

$$K = (A \cdot E + E')'$$
$$= (A + E')'$$
$$= A' \cdot (B + C)$$

3.2.1.1   Sequential Circuits

Most logic designs are **sequential circuits**, which differ from combinational circuits in that their outputs depend on both the current and past input values; that is, they have memories. Sequential circuits are divided into two categories: synchronous and asynchronous. Here, we limit our discussion to synchronous circuits due to their widespread acceptance.

Figure 3.3 illustrates the Huffman model of a synchronous sequential circuit. The sequential circuit is comprised of two parts: the combinational logic and the flip-flops *synchronized* by a common clock signal. The inputs to the combinational logic consist of the ***primary inputs*** (PIs) $x_1, x_2, \ldots, x_n$ and the flip-flop outputs $y_1, y_2, \ldots, y_l$, also called the ***pseudo primary inputs*** (PPIs) to the combinational logic. The outputs are comprised of the ***primary outputs*** (POs) $z_1, z_2, \ldots, z_m$ and the flip-flop inputs $Y_1, Y_2, \ldots, Y_l$, also called the ***pseudo primary outputs*** (PPOs) to the combinational logic. Assuming that the flip-flops are edge triggered, upon

---

[1] Circuit N will be the example network throughout this chapter, unless specifically mentioned.
[2] The three basic Boolean operations (*i.e.*, AND, OR, and NOT) are represented by the multiplication ($\cdot$), addition ($+$), and prime ($'$) operators, respectively.

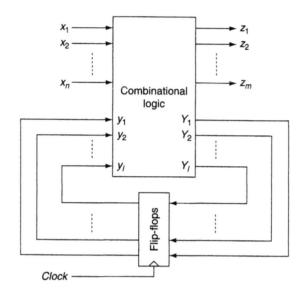

■ **FIGURE 3.3**

The Huffman model of a sequential circuit.

the active clock transition the states of all the flip-flops are updated according to the PPO values at that time and the flip-flop characteristic functions (*e.g.*, $y_i = Y_i$ for a D flip-flop).

In the gate-level description, a flip-flop may be modeled as a functional block or as the interconnections of logic gates. Figure 3.4 shows the NAND implementation of the positive-edge-triggered D flip-flop and its functional symbol. Besides data (*D*) and clock (*Clock*) inputs, the D flip-flop also has active low asynchronous preset (*PresetB*) and clear (*ClearB*) inputs. Its outputs are the uncomplemented (*Q*) and complemented (*QB*) data.

3.2.2 Logic Symbols

The basic mathematics for most digital systems is the two-valued Boolean algebra (referred to as Boolean algebra hereafter for convenience). In Boolean algebra, a variable can assume only one of the two values, *true* or *false*, which are represented by the two symbols "1" and "0," respectively. Note that "1" and "0" here do not represent numerical quantities. Physical representations of the two symbols depend on the logic family of choice. Consider the most popular CMOS logic as an example; the two symbols "1" and "0" represent two distinct voltage levels, V_{dd} and ground,[3] respectively. Whether a signal's value is 1 or 0 depends on which voltage source it is connected to.[4]

---

[3] Assume that positive logic is used.
[4] In the following discussion, it is assumed that the CMOS logic family is chosen.

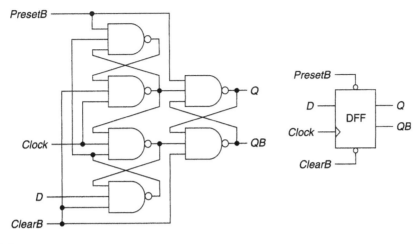

Positive-edge-triggered D flip-flop.

In addition to 1 and 0, logic simulators often include two more symbols: u (unknown) and Z (high-impedance); the former represents the uncertain circuit behavior, and the latter helps resolve the behavior of tristate logic. For cases when $0, 1, u$, and Z are insufficient to meet the required simulation accuracy, intermediate logic states that incorporate both value and strength may be utilized.

3.2.2.1 *Unknown State* u

Almost all practical digital circuits contain memory elements (*e.g.*, flip-flops and memories) to store the circuit state; however, when these circuits are powered up, the initial states of their memory elements are usually unknown. To handle such situations, the logic symbol u is introduced to indicate an *unknown* logic value. By associating u with a signal, we mean that the signal is 1 or 0, but we are not sure which one is the actual value.

Basic Boolean operations for **ternary logic** (0, 1, and u) are straightforward. First, the three symbols are viewed as three sets of symbols: 0 as {0}, 1 as {1}, and u as {0, 1}. Then, the outcome of a ternary logic operation is the union of the results obtained by applying the same operation to the elements of the sets; for example, the result of $0 \cdot u$ is derived as follows:

$$0 \cdot u = \{0\} \cdot \{0, 1\}$$
$$= \{0 \cdot 0, 0 \cdot 1\}$$
$$= \{0, 0\}$$
$$= \{0\}$$
$$= 0$$

TABLE 3.1 ■ Basic Boolean Operations for Ternary Logic

AND	0	1	u	OR	0	1	u	NOT	0	1	u
0	0	0	0	0	0	1	u		1	0	u
1	0	1	u	1	1	1	1				
u	0	u	u	u	u	1	u				

The input/output relationships of the three basic Boolean operations using ternary logic are summarized in Table 3.1. From Table 3.1, one can observe that for an AND operation, the output is determined if one of the inputs is 0. Thus, we say that 0 is the **controlling value** of the AND operation. Similarly, 1 is the controlling value of an OR operation.

Simulation based on ternary logic is pessimistic; it may report that a signal is unknown when in fact its value can be uniquely determined as 0 or 1 [Breuer 1972]. To illustrate the information loss caused by ternary logic, the example circuit N is redrawn in Figure 3.5. Let the input vector be $ABC = 1u0$. Ternary logic simulation (Figure 3.5a) will report that the output K is unknown; however, recall that $ABC = 1u0$ represents two possibilities: $ABC = 100$ and 110. Figure 3.5b shows the simulation results for both cases using binary logic; K equals 0 regardless of the value of B, be it 0 or 1. Apparently, ternary logic simulation causes information loss in this example.

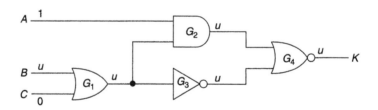

(a) Ternary logic simulation: $K = u$

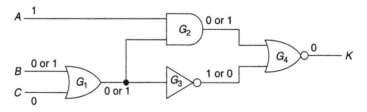

(b) Enumerate all possible cases ($B = 0$ and 1): $K = 0$

■ **FIGURE 3.5**

Information loss caused by ternary logic.

To resolve the problem of information loss, one would have to assign to each flip-flop a unique unknown symbol u_i and associate with u_i the following rules:

$$NOT(u_i) = u_i'$$
$$NOT(u_i') = u_i$$
$$u_i \cdot u_i' = 0$$
$$u_i + u_i' = 1$$

Let us revisit the example in Figure 3.5. Based on the above rules, the output of G_3 will be u' instead of u, and finally one has $K = 0$, the correct answer. The problem with this approach is that signals that are affected by multiple unknown symbols have to be expressed as Boolean expressions of u_i's. As the number of unknown symbols grows, the required symbolic simulation becomes cumbersome.

3.2.2.2 High-Impedance State Z

Until now, the logic signal states that we have discussed are 1 and 0, indicating that the signal is connected to either V_{dd} or ground. (The unknown symbol indicates uncertainty; however, the signal of interest is still 1 or 0.) In addition to 1 or 0, tristate gates have a third, high-impedance state, denoted by logic symbol Z. Tristate gates permit several gates to time-share a common wire, called a *bus*. A signal is in the Z state if it is connected to neither V_{dd} nor ground.

Figure 3.6 depicts a typical bus application. In this example, three bus drivers (G_1, G_2, and G_3) drive the bus wire y. Each driver G_i is controlled by an *enable* signal e_i, and its output o_i is determined as follows:

$$o_i = \begin{cases} x_i & \text{if } e_i = 1 \\ Z & \text{if } e_i = 0 \end{cases}$$

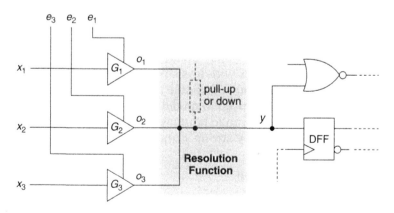

■ FIGURE 3.6

Tristate circuits.

That $o_i = Z$ indicates that G_i has no effect on the bus wire y, leaving the control to other drivers.

For the bus to function correctly, there should not be more than one active tristate control at a time. If multiple drivers are enabled and they intend to drive the bus to the same value, the bus wire is assigned the active drivers' output value; however, if at least two drivers drive the bus wire to opposite binary values, a **bus conflict** occurs. Such situations may cause the circuit to be permanently damaged. Finally, if no driver is activated, the bus is in a **floating state** because it is not connected to V_{dd} or ground. A pull-up or down network that connects the bus wire to V_{dd} or ground via a resistor may be added to provide a default 1 or 0 logic value (Figure 3.6); otherwise, the bus wire will retain its previous value as a result of trapped charge in the parasitic wire capacitance.

In addition to design errors, abnormal bus states could occur during testing when the circuit is not in its normal operating environment and may receive *illegal* input sequences; for example, e_1, e_2, and e_3 may come from the outputs of flip-flops fed by mutual exclusion logic. However, during test, the flip-flops may have random values scanned into them, producing a bus conflict.

To facilitate logic simulation of tristate buses, one may insert a **resolution function** into the circuit description for each bus wire (Figure 3.6). When the simulator encounters a bus signal, the resolution function will check the outputs (and other necessary information) of all the drivers to determine the bus signal. Depending on the simulation requirement, the accuracy of the resolution functions varies. In the simplest form, it may report the occurrence of a bus conflict. To achieve higher simulation accuracy, more sophisticated resolution functions utilize multiple-valued logic systems to represent intermediate logic states.

3.2.2.3 Intermediate Logic States

To model the intermediate logic states that may occur in tristate buses, switch-level networks, and defective circuits, logic simulators employ multiple-valued logic systems that include symbols carrying information of not only signal values but also strengths.

Consider the 21-valued logic system in [Miczo 2003]. Six symbols are used to represent six distinct logic levels: strong, weak, and floating 1's and 0's. The strong 1 and 0 are the same as the 1 and 0 that we have been using. The weak signals, on the other hand, drive circuit nodes with less strength and are overridden by strong signals. Floating signals denote trapped charge and are the weakest. Besides the six logic levels, 15 symbols are introduced to model uncertain circuit behavior. Each of the symbols corresponds to a subrange bounded by a pair of the 6 logic levels. For example, the subrange bounded by strong 1 and 0 denotes the most uncertainty.

3.2.3 Logic Element Evaluation

Logic element evaluation (or gate evaluation) is the process of computing the output of a logic element based on its current input and state values. The choice of evaluation technique depends on the considered logic symbols and the types and models of the logic elements.

3.2.3.1 Truth Tables

Using the truth table is the most straightforward way to evaluate logic elements. Assuming only binary values, an n-input combinational logic element requires a 2^n-entry truth table to store the output value with respect to all possible input combinations. (For a sequential element, n corresponds to the number of its input and state variables.) In practice, the truth table is stored in an array of size 2^n. To access the array, the values of the n input variables are packed in a word that serves as the index to access the array. For example, consider the array T_{NAND3} to store the truth table of a three-input NAND gate. Then, the output value with respect to input pattern 010 is obtained by:

$$T_{NAND3}[010_2] = T_{NAND3}[2]$$

where the subscript 2 indicates the binary number system.

For a multivalued logic system with k symbols, the required array size for an n-input element is calculated as follows. Let m be the number of bits needed to code the k logic symbols; that is, m is the smallest integer such that $2^m \geq k$. The n input values will be packed into an $m \cdot n$-bit word; therefore, the array size is 2^{mn}, although only k^n entries are needed. For example, a nine-valued logic system requires four bits to code the nine symbols (i.e., $m = 4$). For a five-input element, an array of size $2^{4 \times 5} = 2^{20}$ is needed to store the $9^5 = 19{,}683$ truth table entries. Truth-table-based logic element evaluation techniques are fast; however, their usage is limited because the required memory grows exponentially with respect to the number of gate inputs.

3.2.3.2 Input Scanning

Recall that the outputs of AND and OR gates (and similarly NAND and NOR gates) can be determined if any of their inputs has a controlling value. The idea of input scanning is to scan through the inputs and determine the corresponding output based on the presence of the controlling and unknown values in the gate input list.

In addition to the controlling value, denoted by c, we need the **inversion value**, denoted by i, to characterize the AND, OR, NAND, and NOR gates. The c and i parameters of these gates are summarized in Table 3.2. The input scanning algorithm determines the gate output value according to the following rules:

1. If any of the inputs is the controlling value, the gate output is $c \oplus i$.

2. Otherwise, if any of the inputs is u, the gate output is u.

3. Otherwise, the gate output is $c' \oplus i$.

TABLE 3.2 ■ The c and i Values of Basic Gates

	c	i
AND	0	0
OR	1	0
NAND	0	1
NOR	1	1

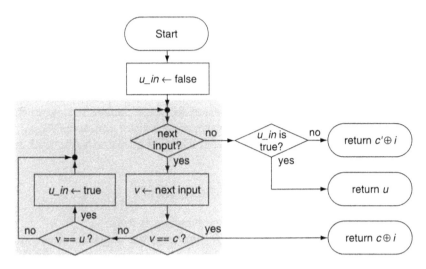

■ FIGURE 3.7

The input scanning algorithm.

The input scanning algorithm flow is depicted in Figure 3.7. The scanning process (the shaded region) detects the existence of controlling and unknown inputs. If an unknown input is encountered, the u_in variable is set to true. On the other hand, once a controlling input is detected, the algorithm will exit the loop and return $c \oplus i$. If there is no controlling input, the output value depends on whether there is any unknown input.

3.2.3.3 Input Counting

Examining the input scanning algorithm, one can observe that knowing the number of controlling and unknown inputs is sufficient to evaluate the output of AND, OR, NAND, and NOR gates. Based on this observation, the input counting algorithm maintains, for each gate, the number of controlling and unknown inputs, denoted by c_count and u_count, respectively. During logic simulation, the two counts are updated if the value of any gate input changes. Consider the NAND gate as an example. If one of its inputs switches from 0 to u, then c_count will be decremented and u_count incremented. Finally, the same rules as those for the input scanning algorithm are applied to determine the output value.

3.2.3.4 Parallel Gate Evaluation

One way to speed up logic simulation is to implement simulation concurrency on the host computer. Because modern computers process data in the unit of a word, usually 32- or 64-bits wide, one can store in a single word multiple copies of a signal (with respect to different input vectors) and process them at the same time. This is referred to as **parallel simulation** or **bitwise parallel simulation**.

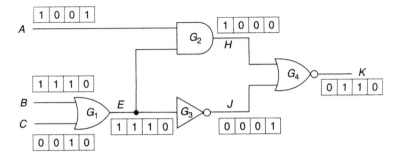

■ **FIGURE 3.8**

Parallel gate evaluation.

Figure 3.8 depicts how parallel simulation is realized to simulate circuit N with binary logic on a computer with a 4-bit word. Because one bit is sufficient to code binary logic symbols, four vectors can be stored in a word and processed in parallel. In this example, the four input vectors to be simulated are $ABC = \{(110),(010),(011),(100)\}$, and next to each signal is the 4-bit data word that stores the values corresponding to the four input vectors. Bitwise logic operations are performed to evaluate the gate outputs.

Parallel simulation is more complicated for multi-valued logic. Consider the ternary logic for which two bits are needed to code the three symbols. One possible coding scheme is:

$$v_0 = (00)$$
$$v_1 = (11)$$
$$v_u = (01)$$

Assume that the word width of the host computer is w. For each signal, two words, denoted by X_1 and X_2 for signal X, are allocated to store w signal values, with X_1 storing the first bit of each symbol and X_2 storing the second bit. Under this symbol coding and packing scheme, the AND and OR operations can be realized by directly applying the same bitwise operation. For example, evaluation of an AND gate with inputs A and B and output C is performed as follows:

$$C_1 = \text{AND}(A_1, B_1)$$
$$C_2 = \text{AND}(A_2, B_2)$$

If $A = 00$ and $B = 11$, then $C = 00$. If $A = 01$ and $B = 11$, then $C = 01$. The complement operation (say, $C = A'$), on the other hand, is realized by:

$$C_1 = \text{NOT}(A_2)$$
$$C_2 = \text{NOT}(A_1)$$

Interchanging A_1 and A_2 ensures that the inversion of an unknown is still unknown.

3.2.4 Timing Models

Delay is a fact of life for all electrical components, including logic gates and inter-connection wires. In this section, we discuss the commonly used gate and wire delay models.

3.2.4.1 Transport Delay

The **transport delay** refers to the time duration it takes for the effect of gate input changes to appear at gate outputs. Several transport delay models characterize this phenomenon from different aspects. The **nominal delay** model specifies the same delay value for the output rising and falling transitions and thus is also referred to as the **transition-independent delay** model. Consider the AND gate G in Figure 3.9 as an example. Here B is fixed at 1; thus, the output of G is only affected by A. Assuming that G has a nominal delay of $d_N = 2$ ns and A is pulsed to 1 for 1 ns, the corresponding simulation result is shown in Figure 3.9a. Using the nominal delay model, the output waveform at F is simply a version of A delayed by 2 ns.

For cases where the rising and falling times are different (*e.g.*, the pull-up and pull-down transistors of the gate have different driving strengths), one may opt for the **rise/fall delay** model. In Figure 3.9b, the setup is the same as that in Figure 3.9a, except that the rise/fall delay model is employed instead; the rise and fall delays are $d_r = 2$ ns and $d_f = 1.5$ ns, respectively. Due to the difference between the two delays, the duration of the output pulse shrinks from 1 to 0.5 ns.

If the gate transport delay cannot be uniquely determined (*e.g.*, due to process variations), one may employ the **min–max delay** model. In the min–max delay

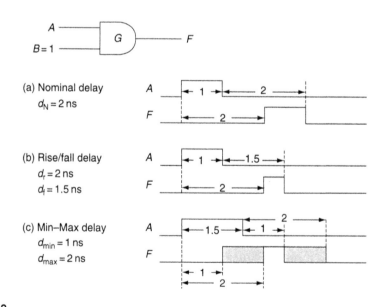

(a) Nominal delay
$d_N = 2$ ns

(b) Rise/fall delay
$d_r = 2$ ns
$d_f = 1.5$ ns

(c) Min–Max delay
$d_{min} = 1$ ns
$d_{max} = 2$ ns

■ **FIGURE 3.9**

Transport delay models.

model, the minimum and maximum gate delays (d_{min} and d_{max}) are specified to represent the ambiguous time interval in which the output change may occur. In Figure 3.9c, the minimum and maximum delays are 1 and 2 ns, respectively, and a 1.5-ns pulse is applied at A. In response to the delay uncertainty, two ambiguous intervals (the shaded regions), corresponding to the rising and falling transitions, are observed at output F. Within the two ambiguous intervals, the exact output value is unknown.

Note that one may combine the min–max and rise/fall delay models to represent more complicated delay behaviors.

3.2.4.2 Inertial Delay

The **inertial delay** is defined as the minimum input pulse duration necessary for the output to switch states. Pulses shorter than the inertial delay cannot pass through the circuit element. The inertial delay models the limited bandwidth of logic gates. Figure 3.10 illustrates this filtering effect. Assume that the AND gate has an inertial delay of 1.5 ns and a nominal delay of 3 ns. Let us fix B at 1 and apply a pulse on A. In Figure 3.10a, the 1-ns pulse is filtered and the output remains at a constant 0. In Figure 3.10b, the pulse is long enough (2 ns) and an output pulse is observed 3 ns later.

3.2.4.3 Wire Delay

In the past, when gate delays dominated circuit delay, the interconnection wires were regarded as ideal conductors with no signal propagation delay. In reality, wires are three-dimensional structures that are inherently resistive and capacitive. Furthermore, they may interact with neighboring conductors to form mutual capacitance. Figure 3.11a illustrates the distributed RLC model of a metal wire. In the

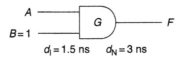

(a) Pulse duration less than d_I

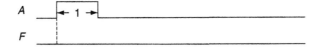

(b) Pulse duration longer than d_I

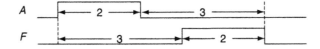

■ **FIGURE 3.10**

Inertial delay.

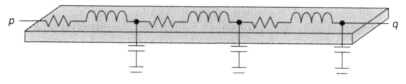

(a) Distributed wire delay model

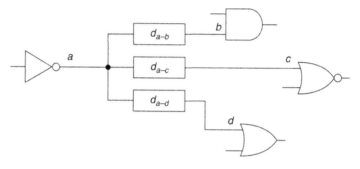

(b) Fanout delay modeling

■ **FIGURE 3.11**

Wire delay model.

presence of the passive components, it takes finite time, called the **propagation delay**, for a signal to travel from point p to point q.

In general, wire delays are specified for each connected gate output and gate input pair because the physical distances and thus the propagation delays between the driver and receiver gates vary. In Figure 3.11b, the inverter output a branches out to drive three gates. To model the wire delays associated with the three signal paths, one may insert delay elements d_{a-b}, d_{a-c}, and d_{a-d} into the fanout branches. For convenience, wire delays may also be viewed as the receiver gate input delays and become part of the receiver gate delay model.

Thanks to the advance of integrated-circuit fabrication technology, continuous device scaling has significantly reduced gate delays; however, wire delays do not benefit as much from device scaling. As a result, wire delays have replaced gate delays as the dominant performance-limiting factor. The challenge of wire delay modeling is that accurate delay values are not available until the physical design stage when the functional blocks are placed and signal nets are routed. Very often, the designers have to go back to earlier design stages to fix the timing violations, a time-consuming process.

3.2.4.4 *Functional Element Delay Model*

Functional elements, such as flip-flops, have more complicated behaviors than simple logic gates and require more sophisticated timing models. In Table 3.3, the I/O delay model of the positive-edge-triggered D flip-flop (Figure 3.4) is depicted.

TABLE 3.3 ■ The D Flip-Flop I/O Delay Model

| Input Condition | | | | Present State | Outputs | | Delays (ns) | | |
D	Clock	PresetB	ClearB	q	Q	QB	to Q	to QB	Comments
X	X	↓	1	0	↑	↓	1.6	1.8	Asynchronous preset
X	X	1	↓	1	↓	↑	1.8	1.6	Asynchronous clear
1	↑	1	1	0	↑	↓	2	3	$Q: 0 \rightarrow 1$
0	↑	1	1	1	↓	↑	3	2	$Q: 1 \rightarrow 0$

Note: X indicates "don't care."

Take the asynchronous preset operations (second row) as an example. Regardless of the *Clock* and *D* values, if the current flip-flop state (q) is 0 and *ClearB* remains 1, changing *PresetB* from 1 to 0 (denoted by the down arrow) will cause output transitions at Q and QB after 1.6 and 1.8 ns, respectively. Besides the input-to-output transport delay, the flip-flop timing model usually contains timing constraints, such as setup/hold times and inertial delays for each input.

3.3 LOGIC SIMULATION

In this section, we will discuss two commonly used gate-level logic simulation methodologies: compiled-code and event-driven. The reader should note that, although not included in this chapter, hardware emulation and acceleration approaches are often employed to speed up the logic simulation process, especially for large designs.

3.3.1 Compiled-Code Simulation

The idea of **compiled-code simulation** is to translate the logic network into a series of machine instructions that model the functions of the individual gates and interconnections between them. The compiled-code simulation flow is illustrated in Figure 3.12a. In each clock cycle, the compiled code program together with the input pattern is executed in the host machine. The simulation results are displayed or stored for later analysis. The code generation flow is depicted in Figure 3.12b. Note that logic optimization and levelization are performed prior to the actual code generation process.

3.3.1.1 Logic Optimization

The purpose of **logic optimization** is to enhance the simulation efficiency. A typical optimization process consists of the following transformation [Wang 1987]:

1. Remove gate inputs that are tied to noncontrolling values (Figure 3.13a).

2. Convert a one-input gate into an inverter or buffer (Figure 3.13b).

3. Remove a gate with one or more inputs tied to its controlling value, and replace the gate's output with 1 or 0 (Figure 3.13c).

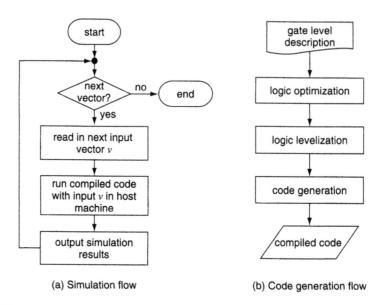

(a) Simulation flow   (b) Code generation flow

■ FIGURE 3.12

Compiled code simulation.

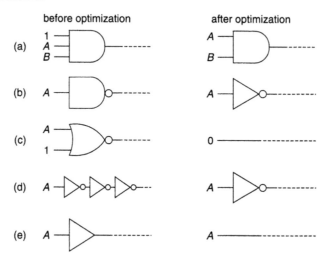

■ FIGURE 3.13

Logic optimization.

4. Replace three consecutive inverters with a single one (Figure 3.13d); this case is common in clock trees.

5. Replace a buffer with a single wire (Figure 3.13e).

6. Remove logic gates that drive unobservable or floating outputs.

Because each gate corresponds to one or more statements in the compiled code, logic optimization reduces the program size and execution time.

3.3.1.2 Logic Levelization

To avoid unnecessary computations, logic gates must be evaluated in an order such that a gate will not be evaluated until all its driving gates have been evaluated. For circuit N, the evaluation order:

$$G_1 \rightarrow G_2 \rightarrow G_3 \rightarrow G_4$$

satisfies this requirement. For most networks, there exists more than one evaluation order that meets the requirement; for example, for N:

$$G_1 \rightarrow G_3 \rightarrow G_2 \rightarrow G_4$$

The **logic levelization** algorithm shown in Figure 3.14 can be utilized to produce the desired gate evaluation order. At the beginning of the algorithm, all the PIs are assigned level 0, and all the PI fanout gates are appended to the first-in/first-out queue Q that stores the gates to be processed. While Q is non-empty, the first gate g in Q is popped out. If all the driving gates of g are levelized and the maximum level is l, g is assigned level $l+1$ and all of the fanout gates of g are appended to Q; otherwise, g is put back in Q to be processed later. The levelization process repeats until Q is empty. Note that for gates assigned the same level, their order

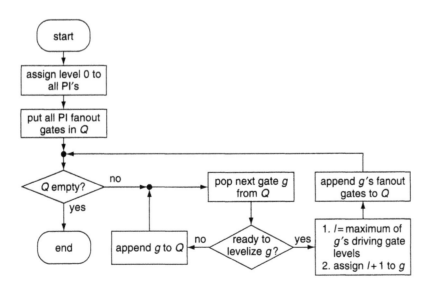

■ **FIGURE 3.14**

The logic levelization algorithm.

TABLE 3.4 ■ The Levilization Process of Circuit N

Step	A	B	C	G_1	G_2	G_3	G_4	Q
0	0	0	0					$<G_2, G_1>$
1	0	0	0					$<G_1, G_2>$
2	0	0	0	1				$<G_2, G_3>$
3	0	0	0	1	2			$<G_3, G_4>$
4	0	0	0	1	2	2		$<G_4>$
5	0	0	0	1	2	2	3	$< >$

of evaluation does not matter. This levelization process is also referred to as **rank ordering**.

The levelization process for circuit N is shown step by step in Table 3.4. At the beginning, PIs are assigned level 0, and their fanout gates G_1 and G_2 are appended to Q. In step 1, G_2 is not ready and put back to Q because G_1 is not levelized yet. In step 2, G_1 is assigned level 1 because it is driven by level 0 PIs only. At the end of the process, the following orders are produced:

$$G_1 \rightarrow G_2 \rightarrow G_3 \rightarrow G_4$$
$$G_1 \rightarrow G_3 \rightarrow G_2 \rightarrow G_4$$

3.3.1.3 Code Generation

Depending on performance, portability, and maintainability needs, different code generation techniques may be used [Wang 1987]. Three approaches for code generation are described below:

- **Approach 1—High-level programming language source code.** The network to be simulated is described in a high-level programming language, such as C. The advantage is that it is easier to debug and can be ported to any target machine that has a C compiler. The compilation time could be a severe limitation for fault simulators that require recompilation for each faulty circuit.

- **Approach 2—Native machine code.** This approach generates the target machine code directly without the need of compilation, which makes it a more viable solution to fault simulation. High simulation efficiency can be achieved if code optimization techniques are utilized to maximize the usage of the target machine's data registers.

- **Approach 3—Interpreted code.** In this approach, the target machine is a software emulator. During simulation, the instructions are interpreted and executed one at a time. This approach offers the best portability and maintainability at the cost of reduced performance.

Shown below is the pseudo code for circuit N. In the actual implementation, each statement is replaced with the corresponding language constructs or machine instructions, depending on the adopted code generation approach:

```
while(true) do
    read(A, B, C);
    E←OR(B, C);
    H←AND(A, E);
    J←NOT(E);
    K←NOR(H, J);
end
```

Compiled-code simulation is most effective when binary logic simulation suffices. In such cases, machine instructions are readily available for Boolean operations (*e.g.*, AND, OR, and NOT). Its main limitations include its incapability of timing modeling and low simulation efficiency. The compiled-code simulation methodology cannot handle gate and wire delay models. As a result, it fails to detect timing problems such as glitches and race conditions. The low efficiency of compiled-code simulation is because the *entire* network is evaluated for each input vector, despite the fact that in general only 1 to 10% of input signals change values between consecutive vectors.

3.3.2 Event-Driven Simulation

In contrast to compiled-code simulation, **event-driven simulation** exhibits high simulation efficiency by performing gate evaluations only when necessary. We will use Figure 3.15 to illustrate the event-driven simulation concept. In this example, two consecutive input patterns $ABC = 001$ and 111 are applied to circuit N and the corresponding signal values are shown. Note that the application of the second vector does not change the input of G_3, so G_3 is not evaluated for the second vector. In event-driven simulation, the switching of a signal's value is called an **event**, and an event-driven simulator monitors the occurrences of events to determine which gates to evaluate.

Figure 3.16 depicts the zero-delay event-driven simulation flow. (A zero-delay simulation is one in which gates and interconnect are assumed to have zero delay.) At the beginning of the simulation flow, the initial signal values, which may be given

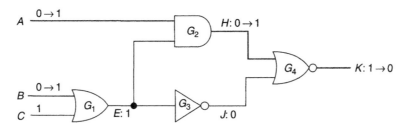

■ **FIGURE 3.15**

Signal transitions between consecutive inputs.

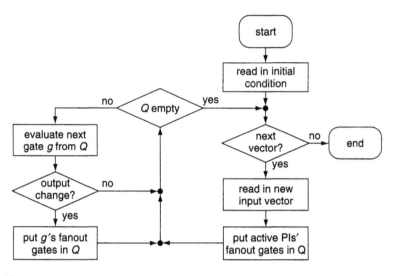

■ **FIGURE 3.16**

Zero-delay event-driven simulation.

or simply unknown, are read in and assigned. Then, a new input vector is loaded and the primary inputs at which events occur (called active PIs) are identified. To propagate the events toward primary outputs, gates driven by active primary inputs are put in the event queue Q, which stores the gates to be evaluated. As long as Q is not empty, a gate g is popped from Q and evaluated. If the output of g changes (*i.e.*, a new event occurs), the fanout gates of g are placed in Q. When Q becomes empty, the simulation for the current input vector is finished, and the simulator proceeds to process the next input vector.

Doing only the necessary work, event-driven simulation is more efficient than compiled-code simulation. Besides simulation efficiency, the biggest advantage of event-driven simulation is its capability to simulate any delay model.

3.3.2.1 Nominal-Delay Event-Driven Simulation

The scheduler is an important component of an event-drive simulator. It keeps track of event occurrences and schedules the necessary gate evaluations. For zero-delay simulation, the event queue is a good enough scheduler because timing is not considered. For nominal-delay simulation, however, a more sophisticated scheduler is required to determine not only which gates to evaluate but also when to evaluate them. Because events must be evaluated in chronological order, the scheduler is implemented as a priority queue.

Figure 3.17 depicts one possible priority queue implementation for a nominal delay event-driven simulator. In the priority queue, the vertical list is an ordered list that stores the time stamps when events occur. Attached to each time stamp t_i is a horizontal list of events that occur at time t_i. During simulation, a new event that will occur at time t_i is appended to the event list of time stamp t_i. For

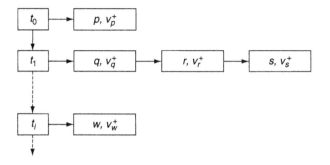

Priority queue event scheduler.

example, in Figure 3.17, the value of signal w will switch to v_w^+ at t_i. If t_i is not in the time stamp list yet, the scheduler will first place it in the list according to the chronological order.

For the priority queue scheduler in Figure 3.17, the time needed to locate a time stamp to insert an event grows with the circuit size. To improve the event scheduler efficiency, one may use, instead of a linked list, an array of evenly spaced time stamps. Although some entries in the array may have empty event lists, the overall search time is reduced because the target time stamp can be indexed by its value. Further enhancement is possible with the concept of **timing wheel** [Ulrich 1969]. Let the time resolution be one time unit and the array size M. A time stamp that is d time units ahead of current simulation time (with array index i) is stored in the array and indexed by $(i+d) \bmod^5 M$ if d is less than M; otherwise, it is stored in an overflow remote event list similar to that is shown in Figure 3.17. Remote event lists are brought into the timing wheel once their time stamps are within $M-1$ time units from current simulation time.

A two-pass strategy for nominal delay event-driven simulation is depicted in Figure 3.18. When there are still future time stamps to process, the event list L_E of next time stamp t is retrieved. L_E is processed in a two-pass manner. In pass one (the left shaded box), the simulator determines the set of gates to be evaluated. The notation (g, v_g^+) indicates that the output of gate g is to become v_g^+. For each event (g, v_g^+), if v_g^+ is the same as g's current value v_g, this event is false and is discarded. On the other hand, if $v_g^+ \neq v_g$ (i.e., (g, v_g^+) is a valid event), then v_g is updated to v_g^+, and the fanout gates of g are appended to the activity list L_A. In the second pass (the right shaded box), gates are evaluated and new events are scheduled. While the activity list L_A is non-empty, a gate g is retrieved and evaluated. Let the evaluation result be v_g^+. The scheduler will schedule the new event (g, v_g^+) at time stamp $t +$ delay(g), where delay(g) denotes the nominal delay of gate g. The two-pass strategy avoids repeated evaluation of gates with events on multiple inputs.

---

[5] "mod" denotes modulo operation. The array is referred to as the timing wheel due to the modulo-M-induced circular structure.

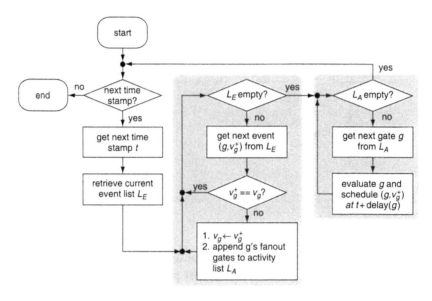

■ **FIGURE 3.18**

Two-pass event-driven simulation strategy.

In the following, we will use circuit N to demonstrate the two-pass event-driven strategy. In this example, the nominal delays for G_1, G_2, G_3, and G_4 are 8, 8, 4, and 6 ns, respectively, and there are four input events (see Figure 3.19): $(A, 1, 0)$, $(C, 0, 2)$, $(B, 0, 4)$, and $(A, 0, 8)$, where the notation (w, v'_w, t) represents the event that signal w switches to v'_w at time t. The simulation progress is shown in Table 3.5.

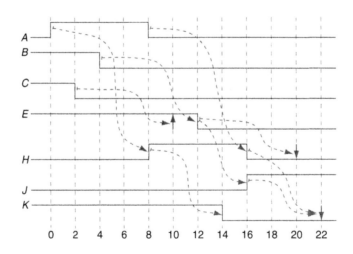

■ **FIGURE 3.19**

Flow of events and voided events.

TABLE 3.5 ■ Two-Pass Event-Driven Simulation

Time	L_E	L_A	Scheduled Events
0	{(A, 1)}	{G_2}	{(H, 1, 8)}
2	{(C, 0)}	{G_1}	{(E, 1, 10)}
4	{(B, 0)}	{G_1}	{(E, 0, 12)}
8	{(A, 0), (H, 1)}	{G_2, G_4}	{(H, 0, 16), (K, 0, 14)}
10	{(E, 1)}		
12	{(E, 0)}	{G_2, G_3}	{(H, 0, 20), (J, 1, 16)}
14	{(K, 0)}		
16	{(H, 0), (J, 1)}	{G_4}	{(K, 0, 22)}
20	{(H, 0)}		
22	{(K, 0)}		

At time 0, there is only one primary input event $(A, 1)$. Because A drives G_2, G_2 is added to activity list L_A. Evaluation of G_2 returns $H = 1$; therefore, the event $(H, 1)$ is scheduled at time 8 (*i.e.*, 8 ns, the delay of G_2 after the current time.) At time stamps 2 and 4, the two input events at C and B are processed in the same way. There are two events at time 8: the input event $(A, 0)$ and the scheduled event $(H, 1)$ from time stamp 0. As both events are valid, the two affected gates, G_2 and G_4, are put in L_A for evaluation. The corresponding events $(H, 0)$ and $(K, 0)$ are scheduled at time 16 and 14, respectively. Note that the event $(E, 1)$ at time 10 is false because it does not cause a signal transition; therefore, no gate evaluation is performed.

In Figure 3.19, the detailed signal waveforms are drawn to illustrate the flow of events and the unnecessarily scheduled false events: $(E, 1, 10)$, $(H, 0, 20)$, and $(K, 0, 22)$. One way to avoid false events is to compare the gate evaluation result with the *last scheduled value* of that gate. A new event is scheduled only if the two values differ.

3.3.3 Compiled-Code *versus* Event-Driven Simulation

Compiled-code and event-driven simulation each have their advantages and disadvantages. Compiled-code simulation is good for **cycle-based simulation**, where only the circuit behavior at the end of each clock cycle is of interest and zero-delay simulation can be used. Compiled-code simulation is also good for hiding the details of a simulation model, such as a processor core. Compiled-code simulation is also good when the circuit activity is high or when bitwise parallel simulation is used. The overhead of compilation restricts compiled-code simulation to applications where a large number of input vectors will be simulated. Event-driven simulation is the best approach for implementing general delay models, and detecting hazards. It is also the best approach for circuits with low activity, such as low-power circuits that employ clock gating. Event-driven simulation is also the best approach during circuit debug, when frequent edit-simulate-debug cycles occur and simulation startup time is important.

3.3.4 Hazards

Because of the difference in delays along reconvergent signal paths, input transitions may cause unwanted transient pulses or glitches, called **hazards**, to appear at internal signals or primary outputs. We will use circuit N to illustrate the cause of hazards. In this example, the inverter has a nominal delay of 3 ns, and the other gates have nominal delays of 2 ns. At first, the input vector to circuit N is $ABC = 110$ and the output value is $K = 0$. After circuit N stabilizes, the second input vector $ABC = 100$ is applied. Without considering the gate delays, the simulator will report that K remains unchanged; however, as shown in Figure 3.20, a delay-aware simulator will reveal the existence of a spurious one pulse at K, called a static 0-hazard.

Hazards are divided into two categories: static and dynamic. A **static hazard** refers to the transient pulse on a signal line whose static value does not change. Depending on what the signal's static value is, a static hazard may be a **static 1-hazard** or a **static 0-hazard**. A **dynamic hazard**, on the other hand, refers to the transient pulse during a 0-to-1 or 1-to-0 transition. Figure 3.21 illustrates the possible outputs of a network with hazards. In the figures, only one hazard pulse is shown, but in general there can be multiple pulses. The presence of hazards may cause a sequential network to malfunction. Following the above example, if the

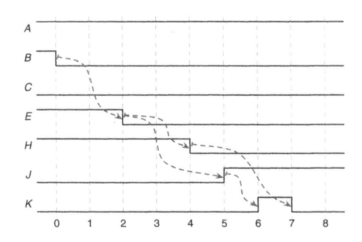

■ FIGURE 3.20

Static 0-hazard.

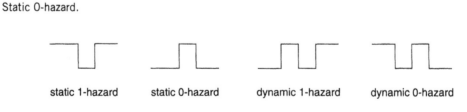

static 1-hazard   static 0-hazard   dynamic 1-hazard   dynamic 0-hazard

■ FIGURE 3.21

Types of hazards.

output signal K is connected to the active high clear input of a flip-flop, the flip-flop may be erroneously cleared by the 1 spike.

Hazard detection is straightforward if the network timing information is available and supported by the simulator; however, the accuracy of this approach suffers from gate delay deviations caused by process variations. In the following, we discuss multivalued logic-based hazard detection techniques that perform worst-case hazard analysis regardless of the timing model.

3.3.4.1 Static Hazard Detection

Recall that hazards are caused by the difference of delays associated with reconvergent paths (*e.g.*, $E \rightarrow H \rightarrow K$ and $E \rightarrow J \rightarrow K$ in circuit N). (The event flow corresponding to the two paths are shown in Figure 3.20.) One must therefore analyze the transient behavior of the network for hazard detection; however, without the correct delay information, it is impossible to predict the exact moment at which a signal transition occurs. One solution to this difficulty is to model the network's transient behavior by associating an uncertainty interval to each input signal transition [Yoeli 1964] [Eichelberger 1965]; that is, a $0 \rightarrow 1$ transition becomes $0 \rightarrow u \rightarrow 1$. (Similarly, a $1 \rightarrow 0$ transition becomes $1 \rightarrow u \rightarrow 0$.) Because $0u1$ may be 001 or 011 (a slower and a faster transition, respectively), the added u signifies the fact that we do not know exactly when the transition occurs.

Let $V^1 = v_1^1 v_2^1 \dots v_n^1$ and $V^2 = v_1^2 v_2^2 \dots v_n^2$ be two consecutive input vectors. The extra input vector $V^+ = v_1^+ v_2^+ \dots v_n^+$ that models the transition uncertainty is obtained in the following way:

$$v_i^+ = \begin{cases} v_i^1 & \text{if } v_i^1 = v_i^2 \\ u & \text{if } v_i^1 \neq v_i^2 \end{cases}$$

When V^+ is available, the modified input sequence $V^1 V^+ V^2$ is simulated. If the $0u0$ or $1u1$ pattern is observed at any primary output, the static hazard is detected. Note that the above method performs a worst-case analysis independent of the delay model.

Now, let us apply this procedure to circuit N with input sequences $V^1 = 110$ and $V^2 = 100$. Following the above procedure, one has $V^+ = 1u0$. Simulating the $V^1 V^+ V^2$ sequence (using ternary logic) reports that $K = 0u0$; thus, a static 0-hazard is detected in this example, which agrees with the simulation results in Figure 3.20.

Based on the same idea, a simulator may utilize the **six-valued logic** to detect static hazards [Hayes 1986]. The symbols and interpretations of the six-valued logic are listed in Table 3.6. The results of Boolean operations on the six symbols can be obtained by applying the same operation bitwise. For example, the outcome of $AND(F, 1^*)$ is derived as follows:

$$AND(F, 1^*) = AND(\{1u0\}, \{1u1\})$$

$$= \{1u0\}$$

$$= F$$

TABLE 3.6 ■ Multivalued Logic for Hazard Detection

Symbol	Interpretation	Six-Valued Logic	Eight-Valued Logic
0	Static 0	{000}	{0000}
1	Static 1	{111}	{1111}
R	Rise transition	{001,011} = 0u1	{0001,0011,0111}
F	Fall transition	{100,110} = 1u0	{1110,1100,1000}
0*	Static 0-hazard	{000,010} = 0u0	{0000,0100,0010,0110}
1*	Static 1-hazard	{111,101} = 1u1	{1111,1011,1101,1001}
R*	Dynamic 1-hazard		{0001,0011,0101,0111}
F*	Dynamic 0-hazard		{1000,1010,1100,1110}

3.3.4.2 Dynamic Hazard Detection

A dynamic hazard causes an unwanted pulse to appear during a 0-to-1 or 1-to-0 transition. To detect dynamic hazards, four-bit sequences are necessary. The **eight-valued logic** [Hayes 1986] that covers all the 4-bit sequences necessary for dynamic hazard detection is shown in Table 3.6. Compared to six-valued logic, two symbols R* and F* are added to denote the dynamic 1- and 0-hazard, respectively. The result of a Boolean operation on the eight-valued logic symbols is the union of the results obtained by applying the same operation to all possible sequence pairs of the two operands. For example, the process of deriving OR(0*, F) is shown below:

$$OR\,(0^*, F) = OR\left(\left\{\begin{array}{l}0000\\0100\\0010\\0110\end{array}\right\}, \left\{\begin{array}{l}1110\\1100\\1000\end{array}\right\}\right) = \left\{\begin{array}{l}1110\\1100\\1000\\1010\end{array}\right\} = F^*$$

3.4 FAULT SIMULATION

Fault simulation is a more challenging task than logic simulation due to the added dimension of complexity; that is, the behavior of the circuit containing all the modeled faults must be simulated. When simulating one fault at a time, the amount of computation is approximately proportional to the circuit size, the number of test patterns, and the number of modeled faults. Because the number of modeled faults is roughly proportional to the circuit size, the overall time complexity of fault simulation is O(pn^2), for p test patterns and n logic gates, which becomes infeasible for large circuits.

To improve fault simulation performance, various fault simulation techniques have been developed. In the following sections, we restrict our discussion to the single stuck-at fault model and illustrate the key fault simulation techniques. Before introducing these techniques, we would like to clarify terminology. Although the terms "test vectors" and "test patterns" are interchangeable in most cases, for the subject of logic simulation the term "test vectors" is preferred, because test vectors are mostly written by human designers for design verification. For fault simulation, on the other hand, the term "test patterns" is used, as the fault simulators frequently work with ATPG to grade test patterns.

3.4.1 Serial Fault Simulation

Serial fault simulation is the simplest fault simulation technique. It consists of fault-free and faulty circuit simulations. Initially, fault-free logic simulation is performed on the original circuit to obtain the fault-free output responses. The fault-free responses are stored and later employed to determine whether a test pattern can detect a fault or not. After fault-free simulation, a serial fault simulator simulates faults one at a time. For each fault, **fault injection** is first performed, which modifies the original circuit to mimic the circuit behavior in the presence of the fault. Then, the faulty circuit is simulated to derive the faulty responses of the current fault with respect to the given test patterns. This process repeats until all faults in the fault list have been simulated.

The serial fault simulation process is demonstrated using the example circuit N. In this example, the fault list is comprised of two faults, A stuck-at one (denoted by f) and J stuck-at zero (denoted by g), which are depicted in Figure 3.22. Note that, although both faults are drawn in the figure, only one fault is present at a time under the single stuck-at fault model. The test set consists of three test patterns (denoted by P_1, P_2, and P_3 and shown in the "Input" columns of Table 3.7).

The serial fault simulator starts from fault-free simulation. The fault-free responses are $K_{good} = \{1, 1, 0\}$ for input patterns P_1, P_2, and P_3, respectively. After the fault-free responses are available, fault f is processed; fault injection is achieved by forcing A to a constant one and the obtained faulty circuit is simulated. The circuit responses for fault f are $K_f = \{\mathbf{0}, \mathbf{0}, 0\}$ with respect to the three input patterns. Compared with the fault-free responses (the "Output" column in Table 3.7), it is observed that patterns P_1 and P_2 detect fault f but pattern P_3 does not. After fault f has been simulated, circuit N is restored by removing fault f. The next fault, g, is then injected by forcing J to zero. Simulation of the resulting faulty circuit is then

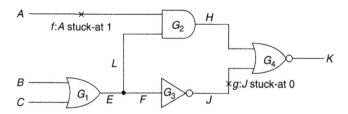

■ **FIGURE 3.22**

An example circuit with two faults.

TABLE 3.7 ■ Serial Fault Simulation Results for Figure 3.22

Pattern No.	Input			Internal					Output		
	A	B	C	E	F	L	J	H	K_{good}	K_f	K_g
P_1	0	1	0	1	1	1	0	0	1	**0**	1
P_2	0	0	1	1	1	1	0	0	1	**0**	1
P_3	1	0	0	0	0	0	1	0	0	0	**1**

performed to obtain the faulty outputs $K_g = \{1, 1, \underline{1}\}$ (also listed in Table 3.7). Fault g is detected by pattern P_3 but not P_1 and P_2.

In this example, nine simulation runs are performed: three fault-free and six faulty circuit simulations. These nine simulation runs can be divided into three **simulation passes**. In each simulation pass, either the fault-free or the faulty circuit is simulated for the whole test pattern set; thus, the first simulation pass consists of fault-free simulations for P_1, P_2, and P_3, and the second and third passes correspond to the faulty circuit simulations of faults f and g, respectively, for P_1, P_2, and P_3.

By careful inspection of the simulation results in Table 3.7, one can observe that, if we are only concerned with the set of faults that is detected by the test set $\{P_1, P_2, P_3\}$, simulations of the faulty circuit with fault f for patterns P_2 and P_3 are redundant because f is already detected by P_1. (It is assumed that the test patterns are simulated in the order P_1, P_2, and then P_3.) Halting simulation of detected faults is called **fault dropping**. For the purpose of fault grading, fault dropping dramatically improves fault simulation performance, as most faults are detected after relatively few test patterns have been applied. Fault dropping, however, should be avoided in fault diagnosis applications in which the entire fault simulation results are usually required to facilitate the identification of the fault type and location.

The simplified serial fault simulation flow is depicted in Figure 3.23. Prior to fault simulation, *fault collapsing* is executed to reduce the size of the fault list, denoted by F. Fault-free simulation is then performed for all test patterns to obtain the correct responses O_{good}. The algorithm then proceeds to fault simulation. For each fault f in F, if there exists a test pattern whose output response O_f differs from that of the corresponding good circuit O_{good}, f is removed from F, indicating that it is detected. When all patterns have been simulated, the remaining faults in F are the undetected faults.

The major advantage of serial fault simulation is its ease of implementation; a regular logic simulator plus fault injection and output comparison procedures will suffice. In addition, serial fault simulation can handle a wide range of fault models, as long as the fault effects can be properly injected into the circuit. The major disadvantage of serial fault simulation is its low performance. As is discussed in the following sections, practical fault simulation techniques exploit parallelism or similarities among the faulty circuits to speed up the fault simulation process.

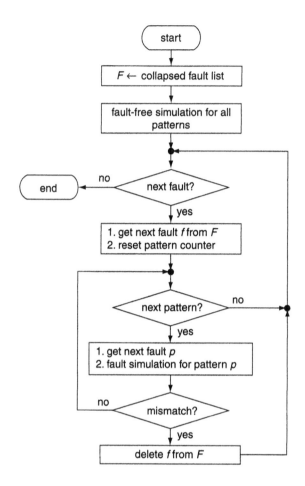

■ FIGURE 3.23

The serial fault simulation algorithm flow.

3.4.2 Parallel Fault Simulation

Similar to parallel logic simulation, fault simulation can take advantage of the bitwise parallelism inherent in the host computer to reduce fault simulation time. For example, in a 32-bit wide CPU, logic operations (AND, OR, or XOR) can be performed on all 32 bits at once. There are two ways to realize bitwise parallelism in fault simulation: parallelism in faults and parallelism in patterns. These two approaches are referred to as **parallel fault simulation** and **parallel pattern fault simulation**.

3.4.2.1 Parallel Fault Simulation

Parallel fault simulation was proposed as early as the 1960s [Seshu 1965]. Assuming that binary logic is utilized, one bit is sufficient to store the logic value of a signal.

Thus, in a host computer using w-bit wide data words, each signal is associated with a data word of which $w-1$ bits are allocated for $w-1$ faulty circuits and the remaining bit is reserved for the fault-free circuit. This way, $w-1$ faulty and one fault-free circuit can be processed in parallel using bitwise logic operations which correspond to a speedup factor of approximately $w-1$ compared to serial fault simulation. A fault is detected if its bit value differs from that of the fault-free circuit at any of the outputs.

We will reuse the example from serial fault simulation to illustrate the parallel fault simulation process. Assuming that the width of a computer word is three bits, the first bit stores the *fault-free* (FF) circuit response, and the second and third bits store the faulty responses in the presence of faults f and g, respectively. The simulation results are shown in Table 3.8. Because fault f, A stuck-at one, uses the second bit, it is injected by forcing the second bit of the data word of signal A to 1 during fault simulation (shown in the "A_f" column with the forced value underlined; the "A" column corresponds to the fault-free case). Similarly, the "J_g" column depicts how fault g is injected by forcing the third bit to 0.

As we have mentioned, parallel fault simulation is performed using bitwise logic operations. For example, the logic value of signal H is obtained by a bitwise AND operation on the data words of signals A and L (A, L, and H are circled in Table 3.8). The faulty response of the first pattern is {1, **0**, 1}. This means that fault f is detected (the second bit) but fault g (the third bit) is not. Similarly, the outputs of P_2 and P_3 are {1, **0**, 1} and {0, 0, **1**}, respectively. In this example, three simulations (in one simulation pass) are performed. Compared to serial fault simulation, which requires nine simulations, parallel fault simulation saves two-thirds of the simulation time.

To perform parallel fault simulation using regular parallel logic simulators, one may inject the faults by adding extra logic gates. Figure 3.24 shows how this is done for faults f and g in N. To inject f, a stuck-at one fault, an OR gate (G_f) is

TABLE 3.8 ■ Parallel Fault Simulation for Figure 3.22

		Input				Internal						Output
		A	A_f	B	C	E	F	L	J	J_g	H	K
P_1	FF	0	0	1	0	1	1	1	0	0	0	1
	f	**0**	**1**	1	0	1	1	1	0	0	**1**	**0**
	g	0	0	1	0	1	1	1	0	0	0	1
P_2	FF	0	0	0	1	1	1	1	0	0	0	1
	f	0	**1**	0	1	1	1	1	0	0	**1**	**0**
	g	0	0	0	1	1	1	1	0	0	0	1
P_3	FF	1	1	0	0	0	0	0	1	1	0	0
	f	1	1	0	0	0	0	0	1	1	0	0
	g	1	1	0	0	0	0	0	1	**0**	0	**1**

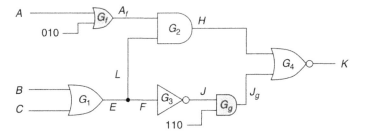

Fault injection for parallel fault simulation.

inserted. To force the second bit of A_f to one without affecting the other two bits, the side input of G_f is set to be 010. Note that the injection of fault f does not affect the fault-free circuit and the faulty circuit with fault g. Similarly, injecting fault g, a stuck-at zero fault, is achieved by adding the AND gate G_g and setting its side input to be 110.

Note that the parallel fault simulation technique is applicable to the unit or zero delay models only. More complicated delay models cannot be modeled because several faults are evaluated at the same time. Furthermore, a simulation pass cannot terminate unless all the faults in this pass are detected. For example, we cannot drop fault f alone after simulating pattern P_1 because fault g is not detected yet. Parallel fault simulation is best used for simulating the beginning of the test pattern sequence, when a large number of faults are detected by each pattern.

3.4.2.2 Parallel-Pattern Fault Simulation

Bitwise parallelism can be used to simulate test patterns in parallel. For a host computer with a w-bit data width, the signal values for a sequence of w test patterns are packed into a data word. For the fault-free or faulty circuit, w test patterns can be simulated in parallel by utilizing bitwise logic operations. This approach was first reported in [Waicukauski 1985], in which it is called **parallel-pattern single-fault propagation** (PPSFP), as one fault at a time is simulated. This approach is especially useful for combinational circuits or full-scan sequential circuits.

In PPSFP, logic simulations on the fault-free circuit are first performed on the first w test patterns, and the circuit outputs are recorded. Then, the faults are simulated one at a time on these w test patterns. For each fault, the simulation results are compared with the correct responses to determine if the fault is detected. Simulation continues until the fault is detected or all the test patterns are simulated. The faulty circuit is restored to its original state and the next fault is processed. The same procedure repeats until all faults in the fault list are simulated.

The PPSFP results of the fault simulation example are shown in Table 3.9. The "Fault-free" row lists the fault-free simulation results. Note that the three patterns are packed into one single word and thus are evaluated simultaneously using bitwise logic operations. The "f" row represents the simulation results with fault f injected.

TABLE 3.9 ■ PPSFP for Figure 3.22

		Input			Internal					Output
		A	**B**	**C**	**E**	**F**	**L**	**J**	**H**	**K**
Fault-free	P_1	0	1	0	1	1	1	0	0	1
	P_2	0	0	1	1	1	1	0	0	1
	P_3	1	0	0	0	0	0	1	0	0
f	P_1	1	1	0	1	1	1	0	1	**0**
	P_2	1	0	1	1	1	1	0	1	**0**
	P_3	1	0	0	0	0	0	1	0	0
g	P_1	0	1	0	1	1	1	0	0	1
	P_2	0	0	1	1	1	1	0	0	1
	P_3	1	0	0	0	0	0	**0**	0	**1**

In PPSFP, faults are injected by activating rising or falling events, depending on the stuck-at value, at the faulty signal. Thus, fault f, A stuck-at one, is injected by activating two rising events on input A. The faulty responses are {**0**, **0**, 0} which indicates that fault f is detected by the first and second patterns but not the third one. After fault f is simulated, fault f is removed by activating two falling events on input A at patterns P_1 and P_2. Then, fault g is injected by activating one falling event on signal J at pattern P_3. Three simulation runs are carried out.

Figure 3.25 illustrates the simplified PPSFP flow. Again, fault collapsing is first executed to obtain the collapsed fault list F. Then, the first w patterns are simulated on the fault free circuit in parallel and the good outputs (O_{good}) are stored. Then, each fault f in fault list F is simulated one by one using the same w test patterns. A fault is dropped and not simulated against the remaining test patterns if its output response O_f is different from O_{good}. To fault simulate the next fault, the fault effect of the current fault is removed and the next fault is injected. This process continues until all faults are either detected or simulated against all test patterns. If the number of test patterns is not an even multiple of the machine word width, only part of the machine word is used when simulating this last batch of patterns.

Parallel-pattern single-fault propagation is best suited for simulation of test patterns that come later in the test sequence, where the fault drop rate per pattern is lower. Parallel fault simulation does not work well in this situation because it cannot terminate a simulation pass until all $w - 1$ faults being processed are detected. PPSFP is not suitable for sequential circuits because the circuit state for test pattern i in the w-bit word is dependent on the previous $i - 1$ patterns in the word, and this state is not available when the patterns are processed in parallel.

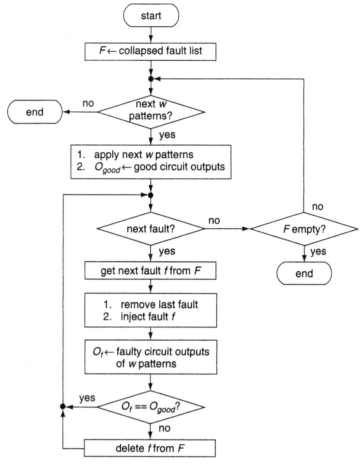

■ FIGURE 3.25

The PPSFP flowchart.

3.4.3 Deductive Fault Simulation

Deductive fault simulation [Armstrong 1972], unlike the fault simulation techniques described above, takes a very different approach; it is based on logic reasoning rather than simulation. For a given test pattern, deductive simulation identifies, all at once, the faults that can be detected. Deductive fault simulation can be very fast because only fault-free simulations have to be performed.

In deductive fault simulation, a fault list (L_x) is associated with a signal x. L_x is the set of faults that causes x to differ from its fault-free value. Figure 3.26 shows the fault list of each signal with respect to test pattern P_1. Fault $A/1$ appears in L_A because its presence causes the value of primary input A to deviate from its correct value of zero. Fault $A/0$ is not in the fault list because the value of A remains correct when the fault $A/0$ is present. The fault lists for inputs B and C are derived in the

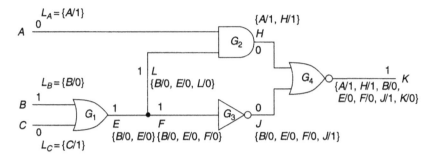

■ **FIGURE 3.26**

Deductive fault simulation (P_1).

same way. Based on logic reasoning, the process of deriving the fault list of a gate output from those of the gate inputs is called **fault list propagation**; for example, the fault list of gate output E is the union of the fault list of B and the $E/0$ fault. Clearly, the $E/0$ fault should be included in L_E as the correct value of E is one. On the other hand, because the fault-free value of C is a noncontrolling value of G_1, the fault effect of each fault in L_B will propagate to E (which causes E to be 1); therefore, all faults in L_B are propagated to L_E. L_C is not propagated to the gate output because the other input B holds the controlling value (one) of gate G_1.

Similarly, the fault list L_E is propagated to signals L and F. The fanout branches do nothing but add faults $L/0$ and $F/0$ to L_L and L_F, respectively. The fault list of gate output H contains $A/1$ and $H/1$; the fault list of A is propagated through G_2 because L is one, and the fault list of L is discarded because A is zero. Finally, the fault list of primary output K is the union of the fault lists of the two gate inputs; that is, $L_K = L_H \cup L_J = \{A/1, H/1, B/0, E/0, F/0, J/1, K/0\}$ because both gate inputs of G_4 are zeros; all the fault effects at the gate inputs are propagated to the gate output. By definition, we can conclude that pattern P_1 detects the seven faults in L_K. From this simple example, we can see the advantage of deductive fault simulation—all faults detected by a test pattern are obtained in one fault list propagation pass. Note that, for ease of explanation, no fault collapsing is performed in this example. In practice, however, the faults are collapsed before deductive fault simulation and only the collapsed faults are considered during fault list propagation.

In Figure 3.27, the deductive fault simulation results for test pattern P_2 are shown. The notable difference is that those faults previously detected by pattern P_1 are dropped and not taken into account. The fault list of K indicates that one more fault, $C/0$, is detected by P_2. The fault simulation results for pattern P_3 are depicted in Figure 3.28. Three more faults $\{F/1, J/0, K/1\}$ are detected.

Figure 3.29 illustrates the deductive fault simulation flow. For each test pattern, fault-free simulation is first performed to obtain the correct values of each signal. Fault list propagation is then conducted. A fault is detected and removed from the fault list if it appears in any primary output's fault list. The same process repeats until all test patterns are simulated or all faults are detected.

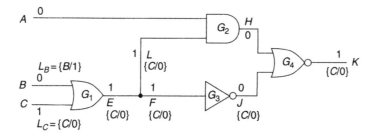

■ FIGURE 3.27

Deductive fault simulation (P_2).

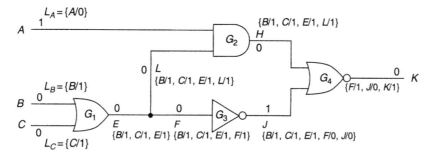

■ FIGURE 3.28

Deductive fault simulation (P_3).

Although in our simple example, the fault list propagation rules are demonstrated only for two-input gates, they can be generalized to multiple input gates. Let I and z be the set of gate inputs and the gate output, respectively. Equation 3.1 shows the fault list propagation rule when all gate inputs hold noncontrolling values:

$$L_z = \left(\bigcup_{j \in I} L_j \right) \cup \{z/(c \oplus i)\} \tag{3.1}$$

In Equation 3.1, c and i are the controlling and inversion values of the gate. (See Table 3.2 for the c and i values of basic gates.) Because no controlling value appears in the gate inputs, the fault lists at the inputs are propagated to the fault list of the gate output L_z, represented by the term $\bigcup_{j \in I} L_j$. At the same time, the correct value of z is $c \oplus i$; therefore, the fault z stuck-at $c \oplus i$, denoted by $z/(c \oplus i)$, is added to L_z. (Recall that $(c \oplus i')' = c \oplus i$.) According to the rule, the fault list of the NOR gate G_4 in Figure 3.26 is simply $L_K = L_H \cup L_J \cup \{K/0\}$.

For cases where at least one gate input holds the controlling value, the fault list propagation rule is depicted in Equation 3.2, where S and $I - S$ stand for the sets of

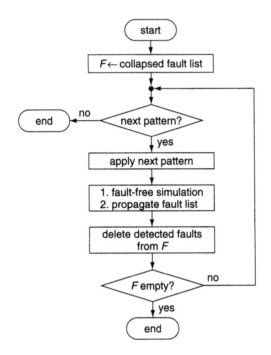

Deductive fault simulation flowchart.

gate inputs that hold the controlling and noncontrolling values, respectively, and the minus sign represents the set difference operation:

$$L_z = \left[\left(\bigcap_{j \in S} L_j\right) - \left(\bigcup_{j \in I-S} L_j\right)\right] \cup \{z/c \oplus i'\} \qquad (3.2)$$

The term $\left(\bigcap_{j \in S} L_j\right) - \left(\bigcup_{j \in I-S} L_j\right)$ represents the set of faults in the gate input fault lists that will propagate to the gate output. First, a fault cannot be observed unless it appears in every fault list of gate inputs in S, represented by the term $\bigcap_{j \in S} L_j$; otherwise, some gate inputs will retain the controlling value and block the fault effect propagation. Second, the fault lists of the noncontrolling gate inputs (i.e., $I - S$) cannot propagate to the gate output, represented by the $\bigcup_{j \in I-S} L_j$ term and the set difference operation, because these faults prevent the gate output from being changed. Applying Equation 3.2 to the NOR gate G_4 in Figure 3.28, one has $L_K = (L_J - L_H) \cup \{K/1\}$; the faults in L_H are taken out of L_J because flipping H does not change the value of output K.

Although deductive fault simulation is efficient in that it processes all faults at the same time, it has several limitations. The first problem is that unknown values are not easily handled. For each unknown value, both cases must be considered (i.e., when the unknown is a controlling or noncontrolling value). The logic reasoning becomes even more complicated if more than one unknown appears. See

[Abramovici 1994] for more detailed discussions of this problem. The second problem is that deductive fault simulation is only suitable for the zero-delay timing model, because no timing information is considered during the deductive fault propagation process. Finally, deductive fault simulation has a potential memory management problem. Because the size of fault lists cannot be predicted in advance, there can be a large variation in memory requirements during algorithm execution.

3.4.4 Concurrent Fault Simulation

Because a fault only affects the logic in the fanout cone from the fault site, the good circuit and faulty circuits typically only differ in a small region. **Concurrent fault simulation** exploits this fact and simulates only the differential parts of the whole circuit [Ulrich 1974]. Concurrent fault simulation is essentially an event-driven simulation with the fault-free circuit and faulty circuits simulated altogether.

In concurrent fault simulation, every gate has a **concurrent fault list**, which consists of a set of **bad gates.** A bad gate of gate x represents an imaginary copy of gate x in the presence of a fault. Every bad gate contains a fault index and the associated gate I/O values in the presence of the corresponding fault. Initially, the concurrent fault list of gate x contains **local faults** of gate x. The local faults of gate x are faults on the inputs or outputs of gate x. As the simulation proceeds, the concurrent fault list contains not only local faults but also faults propagated from previous stages. Local faults of gate x remain in the concurrent fault list of gate x until they are detected.

Figure 3.30 illustrates the concurrent simulation of the example circuit for test pattern P_1. For clear illustration, we demonstrate three faults in this example: A stuck-at one, C stuck-at zero, and J stuck-at zero faults. The concurrent

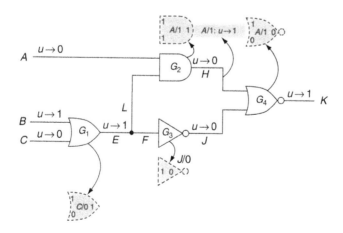

■ FIGURE 3.30

Concurrent fault simulation (P_1).

fault lists with bad gates in gray are drawn beside the good gates. The fault indices are labeled in the middle of bad gates and their associated bad gate I/O values are labeled beside their I/O pins. The fault list of G_1, G_2, and G_3 initially contains their local faults: $C/0$, $A/1$, and $J/0$. When we apply the first pattern, three events occur in the primary inputs: $u \rightarrow 0$ on A, $u \rightarrow 1$ on B, and $u \rightarrow 0$ on C. They are **good events** because they happen in the good circuit. The output of good gate G_1 changes from unknown to one. In the presence of fault $C/0$, the output of faulty G_1 is the same as that of good G_1. A bad gate is **invisible** if its faulty output is the same as the good output. The bad gates C/0 and J/0 are both invisible so they are not propagated to the subsequent stages.

The output of G_2 changes from unknown to zero. In the presence of fault $A/1$, the faulty output changes from unknown to 1. Because the faulty output differs from the good output, bad gate $A/1$ becomes visible. A bad gate is **visible** if its faulty output is different from the good output. The visible bad gate $A/1$ creates a bad event $u \rightarrow 1$ on net H (in gray). A **bad event** does not occur in the good circuit; it only occurs in the faulty circuit of the corresponding fault. A new copy of bad gate $A/1$ is added to the concurrent fault list of G_4 because it has one input different from the good gate. It is said that bad gate $A/1$ **diverges from** its good gate. Finally, fault $A/1$ is detected because the faulty output K is different from the good output. At this time, we could drop detected fault $A/1$ but we keep it for illustration purposes.

Figure 3.31 illustrates the concurrent fault simulation for test pattern P_2. Two good events occur in this figure: $0 \rightarrow 1$ on C and $1 \rightarrow 0$ on B. The bad gate $C/0$, which

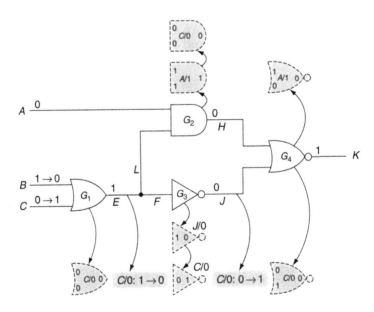

■ **FIGURE 3.31**

Concurrent fault simulation (P_2).

was invisible in pattern P_1, now becomes **newly visible.** The newly visible bad gate creates a bad event—net E falls to zero, which in turn creates two divergences in G_2 and G_3. The former is invisible but the latter creates a bad event—net J rises to one. Finally, the concurrent fault list of G_4 contains two bad gates; both faults $A/1$ and $C/0$ are detected. Again, we keep $A/1$ and $C/0$ faults for demonstrating the simulation of pattern P_3.

   For the last test pattern P_3 (Figure 3.32), two good events occur at primary inputs A and C. The bad gate $C/0$ now becomes invisible. The bad gate $C/0$ is deleted from the concurrent fault list of G_3. A bad gate **converges to** its good gate if it is not a local fault and its I/O values are identical to those of the good gate. Similarly, the other bad gates of $C/0$ also converge to G_2 and G_4. Note that bad gate $C/0$ does not converge to G_1 because it is a local fault for G_1. The bad gate $A/1$ can be examined in the same way. For gate G_3, although the faulty output of bad gate $J/0$ does not change, the good event $0 \rightarrow 1$ on J makes bad gate $J/0$ newly visible. The newly visible event (in gray) is propagated to G_4 and a new bad gate $J/0$ diverges from G_4. Eventually, the fault $J/0$ is detected by pattern P_3.

   Figure 3.33 shows a simplified concurrent fault simulation flowchart. The fault simulator applies one pattern at a time. The concurrent fault simulation is an event-driven simulation with both good events and bad events simulated at the same time. The events on the gate inputs are first analyzed. A good event affects both good and bad gates but a bad event only affects bad gates of the corresponding fault. After the analysis, events are then executed. The diverged bad gates and converged bad gates are added to or deleted from the fault list, respectively. Determining whether

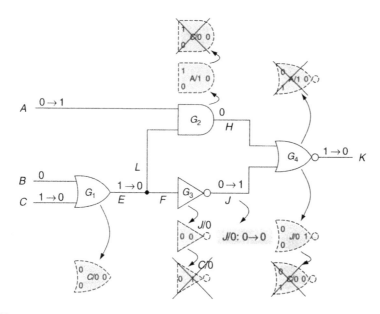

■ **FIGURE 3.32**

Concurrent fault simulation (P_3).

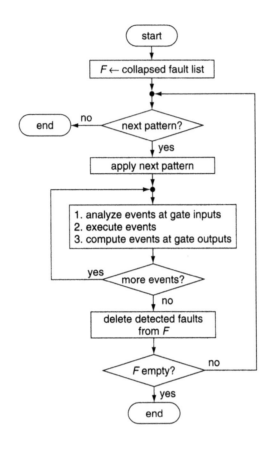

■ **FIGURE 3.33**

Concurrent fault simulation flowchart.

a bad gate diverges or converges depends on three factors: the visibility, the bad event, and the concurrent fault list (see [Abramovici 1994] for more details). After the event execution, new events are computed at the gate output. If an event reaches the primary outputs, detected faults can be removed from concurrent fault lists of all gates. This process repeats until there are no more test patterns or no undetected faults.

3.4.5 Differential Fault Simulation

Concurrent fault simulation constructs the state of the faulty circuit from that of the same faulty circuit of the previous test pattern. Concurrent fault simulation has a potential memory problem because the size of the concurrent fault list changes at run time. In contrast, the single fault propagation technique constructs the state of the faulty circuit from that of the good circuit. For sequential circuits, the single fault propagation technique would require a large overhead to store

	P_1	P_2	$\ldots$	P_i	P_{i+1}	$\ldots$	P_n
Good	G_1	G_2	$\ldots$	G_i	G_{i+1}	$\ldots$	G_n
f_1	$F_{1,1}$	$F_{1,2}$	$\ldots$	$F_{1,i}$	$F_{1,i+1}$	$\ldots$	$F_{1,n}$
f_2	$F_{2,1}$	$F_{2,2}$	$\ldots$	$F_{2,i}$	$F_{2,i+1}$	$\ldots$	$F_{2,n}$
$\ldots$	$\ldots$	$\ldots$	$\ldots$	$\ldots$	$\ldots$	$\ldots$	$\ldots$
f_k	$F_{k,1}$	$F_{k,2}$	$\ldots$	$F_{k,i}$	$F_{k,i+1}$	$\ldots$	$F_{k,n}$
f_{k+1}	$F_{k+1,1}$	$F_{k+1,2}$	$\ldots$	$F_{k+1,i}$	$F_{k+1,i+1}$	$\ldots$	$F_{k+1,n}$
$\ldots$	$\ldots$	$\ldots$	$\ldots$	$\ldots$	$\ldots$	$\ldots$	$\ldots$
f_m	$F_{m,1}$	$F_{m,2}$	$\ldots$	$F_{m,i}$	$F_{m,i+1}$	$\ldots$	$F_{m,n}$

■ **FIGURE 3.34**

Differential fault simulation.

the states of the good circuit. Neither of the above two techniques is good for sequential fault simulation. Differential fault simulation combines the merits of concurrent fault simulation and single fault propagation techniques [Cheng 1989]. The idea is to simulate in turn every faulty circuit by tracking only the difference between a faulty circuit and the last simulated one. An event-driven simulator can easily implement differential fault simulation with the differences injected as events.

Figure 3.34 illustrates how differential fault simulation works. First, the first pattern P_1 is simulated on the good circuit G_1 and the good primary outputs are stored. Then a faulty circuit ($F_{1,1}$) is simulated with fault f_1 injected as an event. The first subscript indicates the fault and the second subscript indicates the pattern. The difference of states between G_1 and $F_{1,1}$ is stored. Note that only the states of memory elements, such as flip-flops, are stored, so the memory required is small compared to concurrent fault simulation. If the primary outputs of $F_{1,1}$ and G_1 are not the same, then fault f_1 is detected. Following F_1 the second faulty circuit ($F_{2,1}$) is simulated with f_1 removed and f_2 injected. Similarly, the difference of states between F_1 and F_2 is stored. The above process continues until pattern P_1 has been simulated for all faults (f_1 to f_m).

Following the first pattern, the state of the good circuit (G_2) is restored and the second pattern P_2 is applied. After the fault-free simulation, the primary outputs of G_2 are stored. The state of faulty circuit $F_{1,2}$ is restored by injecting the difference of G_1 and $F_{1,1}$. The fault f_1 is again injected as an event. The differential fault simulation for P_2 is the same as that of pattern P_1. Differential fault simulation goes in the direction of the arrows in Figure 3.34: $G_i, F_{1,i}, F_{2,i}, \ldots, F_{m,i}, G_{i+1}, F_{1,i+1}, \ldots$.

Figure 3.35 shows a simplified flowchart for differential fault simulation. For every test pattern, a fault-free simulation is performed first, then the faulty circuits are simulated one after another. The states of every circuit are restored from the last simulation. If the faulty circuit outputs are different from the good outputs, the fault is detected and dropped. The state difference of every circuit is stored. With fault dropping, the state difference of the dropped fault must be accumulated into

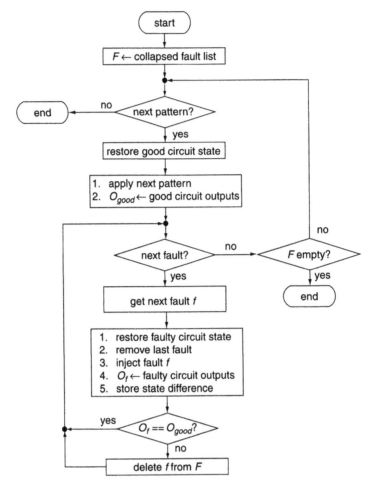

■ **FIGURE 3.35**

Differential fault simulation flowchart.

the state differences of its next undetected fault. This process repeats until there are no test patterns or no undetected faults.

The problem with differential fault simulation is that the order of events caused by fault sites is not the same as the order of the timing of their occurrence. If the circuit behavior depends on the gate delay of the circuit, the timing information of every event must be included. This solution, however, can potentially require high memory consumption.

3.4.6 Fault Detection

In the previous sections, we defined fault detection as an output value being different from the good value. In the simple example we used to illustrate fault simulation techniques, making the fault detection decision is easy because the faults are **hard detected**; that is, the outputs of the fault-free and faulty circuits are either 1 or 0 and are different. In practical cases, the fault detection decision is more difficult. For example, consider the stuck-at-zero fault that occurs at the enable input of a tristate buffer. With its enable input forced to 0, the tristate buffer's output is in a floating state. It is unclear whether the fault is detected, because the logic value of a floating signal may be the same as the correct value by accident; however, if the fault is simulated against many test patterns, it is very likely that it will eventually be detected. For this reason, some fault simulators regard this kind of fault as **potentially detected**. Faults that cause the circuit to oscillate (called **oscillation faults**) also complicate the fault detection decision because it is impossible to predict the faulty circuit outputs. Finally, some faults may cause the faulty circuit behavior to deviate significantly from the correct behavior—for example, stuck-at faults on clock signals. Called **hyperactive faults**, this type of fault makes the fault simulation process extremely time and memory consuming, due to the large number of differences between the good and faulty circuit. Hyperactive faults are in general easily detected, so they are regarded as detected without actual fault simulation, to avoid memory explosion in the fault simulator.

3.4.7 Comparison of Fault Simulation Techniques

The reader may have realized that the major concerns of fault simulation techniques are the simulation speed and the required memory. In practice, factors such as multivalued logic simulation capability, delay model simulation capability, functional model simulation compatibility, and sequential fault simulation capability should be considered as well. Also, the choice of the most suitable fault simulation technique depends on the system memory space, the simulation time constraint, the presence of unknown or high-impedance states, the delay model, the circuit characteristics (sequential or combinational), and the presence of functional level descriptions in the circuit. In the following, we make a qualitative comparison of the previously discussed fault simulation techniques.

In terms of simulation speed, it is apparent that serial fault simulation is the slowest among all the techniques. Deductive fault simulation can be faster than parallel fault simulation as their complexities are $O(n^2)$ and $O(n^3)$, respectively [Goel 1980], where n is the number of logic gates in a circuit. There is no direct comparison between the deductive and concurrent fault simulation techniques. It is suspected, however, that the latter is faster than the former because concurrent fault simulation only deals with the "active" parts of the circuit that are affected by faults. Deductive fault simulation, in contrast, performs deduction on the entire circuit whenever the input patterns change. Differential fault simulation is shown to be up to twelve times faster than concurrent fault simulation and PPSFP [Cheng 1989].

Memory usage is in general not a problem for serial fault simulation because it deals with one fault at a time. Similarly, parallel fault simulation and PPSFP do not require much more memory than the fault-free simulation. The memory requirement of deductive fault simulation, in contrast, can be a problem because the fault lists are dynamically created at run time and their sizes are difficult to predict prior to simulation. Concurrent fault simulation has even more severe memory problems than deductive fault simulation because the concurrent fault list is larger than the deductive fault list. Furthermore, the I/O values of every bad gate in concurrent fault simulation must be recorded. Differential fault simulation relieves the memory management problem of concurrent fault simulation because only the difference in flip-flips is stored.

When the unknown (X) or high-impedance (Z) values are present in the circuit, multivalued fault simulation becomes necessary. Serial fault simulation has no problem in handling multivalued fault simulation because it can be realized with a regular logic simulator. In contrast, to exploit bitwise word parallelism, it is more difficult for parallel fault simulation or PPSFP to handle X or Z. Deductive fault simulation, as mentioned earlier, becomes awkward in the presence of X and Z. In concurrent fault simulation, dealing with multivalued simulations is straightforward because every bad gate is evaluated in the same way as in the fault-free simulation. Finally, differential fault simulation can simulate X or Z without a problem as it is based on event-driven simulation.

From the aspect of delay and functional modeling capability, serial fault simulation does not encounter any difficulty. Parallel fault simulation and PPSFP cannot take delay or functional models into account as they pack the information of multiple faults or test patterns into the same word and rely on bitwise logic operations. Based on logic deduction, a deductive fault simulator can deal with neither delay nor functional models. Being event driven, both concurrent and differential fault simulation techniques are capable of handling functional models; however, only the former is able to process circuit delays.

When sequential circuits are of concern, serial as well as parallel fault simulation techniques do not have a problem. The PPSFP technique, however, is not suited for sequential circuit simulation because a large memory space is required to store the states of the fault-free circuit. Deductive fault simulation might get very complicated because sequential circuits usually contain many unknowns. Concurrent and differential fault simulations are able to perform sequential fault simulation without difficulty.

Based on the above discussions, PPSFP and concurrent fault simulation techniques are currently the most popular fault simulation techniques for combinational (full-scan) circuits. On the other hand, differential and concurrent fault simulation techniques have been widely adopted for sequential circuits. Algorithm switching has also been employed to improve performance. Parallel fault simulation can be used when the fault drop rate per test pattern is high, and then PPSFP is employed when more patterns are required to drop each fault.

Even for fault simulation techniques that are efficient in time and memory, the problems of memory explosion and long simulation time still exist as the complexity of integrated circuits continues to grow. To overcome the memory problem, the **multiple-pass fault simulation** approach is often adopted. The idea of multiple-pass fault simulation is to partition the faults into small groups, each of which is simulated independently. If the faults are well partitioned, multiple-pass fault simulation prevents the memory explosion problem. To further reduce the fault simulation time, **distributed fault simulation** approaches may be employed.

Distributed fault simulation divides the entire fault simulation into smaller tasks, each of which is performed independently on a separate processor.

3.4.8 Alternatives to Fault Simulation

Because fault simulation is very time consuming and difficult for large circuits, alternatives to avoid "true" fault simulation have been developed. These alternatives require only one fault-free simulation or very few fault simulations, so the run time is significantly reduced. The alternatives give approximate fault coverage numbers. It should be noted that these alternatives are probably acceptable if the purpose of fault simulation is to estimate the quality of test patterns (*i.e.*, fault grading). These alternatives are probably not acceptable when it comes to diagnosis. This is because diagnosis requires exact information about which patterns detect which faults. (Please see Chapter 7 for more detailed information about diagnostic fault simulation.)

3.4.8.1 Toggle Coverage

Toggle coverage is a popular technique to evaluate the quality of test patterns because it requires only one single fault-free simulation. There are two definitions for toggling. The relaxed definition says that a net is toggled if its value has been set to 0 and 1 (the order does not matter) during the fault-free simulation. The stringent definition requires that the net have both a 0-to-1 transition and a 1-to-0 transition (the order does not matter) during the fault-free simulation. Both definitions can be used to calculate the toggle coverage. The toggle coverage is the number of toggled nets over the number of total nets in the circuit. Please note that toggling a net does not guarantee its fault propagation so we do not know the relationship between the toggle coverage and the fault coverage.

3.4.8.2 Fault Sampling

The fault sampling technique was proposed to simulate only a sampled group of faults [Butler 1974]. The real fault coverage is approximated by the simulation result of the sampled group of faults. Fault sampling is like polling before an election. The error of the polling depends on two factors: (1) the sample size, and (2) whether the sample is biased or not.

Let M be the total number of faults in the circuit and K be the number of faults detected by the test set. The true fault coverage is therefore $FC = K/M$. Suppose that m is the number of sampled faults, and k is the number of sampled faults detected in the simulation. The estimated fault coverage is $fc = k/m$.

Based on probability theory, the random variable k follows the hypergeometric distribution. When M is much greater than m, random variable k can be approximated by a normal random variable, of which the mean is $\mu_k = mK/M = mFC$; therefore, the mean of the simulated fault coverage is $\mu_{fc} = \mu_k/m = FC$. The standard deviation σ of fc is approximately $\sqrt{FC(1-FC)/m}$. From the normal distribution assumption, we know that the confidence level of the $\pm 3\sigma$ interval is 99.7%. This means that the probability that the mean of simulated fault coverage μ_{fc} falls in the $\pm 3\sigma$ interval of the true fault coverage FC is 99.7%.

3.4.8.3 Critical Path Tracing

Critical path tracing is another alternative to fault simulation [Abramovici 1984]. Given a test pattern t, net x has a **critical value** v if and only if the x stuck-at v' fault is detected by t. A net that has a critical value is a **critical net**. The **critical path** is a path that consists of nets with critical values. Tracing the critical path from PO to PI gives a list of critical nets and hence a list of detected faults.

Critical path tracing is demonstrated in Figure 3.36. All the critical values are circled. The primary output K is certainly critical, as any change in K is observed. Both gate inputs H and J of gate G_4 are critical because flipping either one of them would change the primary output K. It can be seen that E, F, A, and B are all critical. Note that L is not critical, because changing L would not change the primary output. After the critical path tracing, seven critical nets are identified and their associated faults $\{A/1, H/1, B/0, E/0, F/0, J/1, K/0\}$ are detected.

Special attention is needed when fanout branches reconverge. Figure 3.37 shows the example circuit for pattern P_3. As is the case in pattern P_1, nets K, J, and F

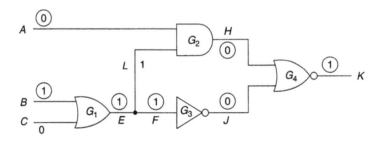

■ **FIGURE 3.36**

Critical path tracing (P_1).

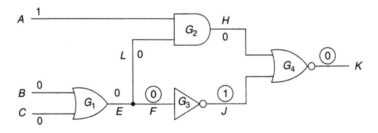

■ **FIGURE 3.37**

Critical path tracing (P_3).

are critical nets; however, L and E are not critical because changing their values does not affect the circuit output. The critical path tracing is stopped due to the reconvergence of fanout branches L and F. Eventually, faults $\{F/1, J/0, K/1\}$ are detected. One solution to this fanout reconvergence is to partition the circuit into fanout-free subcircuits. The detailed implementation of the critical path tracing can be found in [Abramovici 1984]. A modified critical path tracing technique that is linear time, exact, and complete can be found in [Wu 2005].

3.4.8.4   Statistical Fault Analysis

Instead of performing actual fault simulation, the **statistical fault analysis** (STAFAN) approach proposes to use probability theory to estimate the expected value of fault coverage [Jain 1985]. The **detectability** of fault f (d_f) is the probability that fault f is detected by a random pattern. STAFAN calculates the detectability of a fault by two numbers: controllability and observability. The **1-controllability** of net x, $C1(x)$, is the probability of setting net x to 1 by a random pattern. The **0-controllability** of net x, $C0(x)$, is the probability of setting net x to 0 by a random pattern. STAFAN runs one fault-free simulation and keeps track of the number of 1's and 0's of every net. After the simulation, $C1(x)$ is the number of 1's divided by the number of patterns, and $C0(x)$ is the number of 0's divided by the number of patterns.

The **observability** of net x, $O(x)$, represents the probability that the given patterns propagate the fault effect on net x to the primary outputs. During the fault-free simulation, STAFAN counts the number of times that every gate input is sensitized to its gate output. The sensitization probability, $S(x)$, is then obtained by dividing the sensitization count of gate input x by the number of test patterns. The observability of primary outputs is 1, because fault effects on primary outputs will certainly be observed. The observability of a gate input x is $S(x)$ times the observability of its gate output. The observability of every net can be calculated from primary outputs to primary inputs.

The observability calculation becomes complicated in the presence of fanout branches. The lower bound of the observability of a fanout stem is the maximum value of the observability of its fanout branches. The upper bound of a fanout stem is the "union" of the observability of its fanout branches. This upper bound assumes that observing the fault effect via each fanout branch is independent. For example, the observability of a fanout stem with two branches is $O(x) = O(x_1) + O(x_2) - O(x_1)O(x_2)$, where x_1 and x_2 are the fanout branches of x. Finally, the observability of a fanout stem is a linear combination of its upper bound and its lower bound. In the presence of fanout reconvergence, the independent observation of fanout branches is not a valid assumption.

Eventually, the detectability (d_f) of the net x stuck-at zero fault is $C1(x)$ times $O(x)$. The detectability of the net x stuck-at one fault is $C0(x)$ times $O(x)$. Given a set of n independent patterns, the probability of detecting fault f is $d_f^n = 1 - (1 - d_f)^n$. The expected fault coverage is the summation of d_f^n of all faults in the circuit over the number of total faults. Statistical data show that more than 91% of faults that have detectability higher than 0.9 are actually detected, while less than 25% of faults that have a detectability lower than 0.1 are actually detected for single stuck-at fault test sets.

3.5 CONCLUDING REMARKS

We have presented two fundamental subjects, logic simulation and fault simulation, that are important for readers to design quality digital circuits. Logic simulation checks whether the design will behave as predicted before its physical implementation is built, while fault simulation tells us in advance how effective the given test pattern set is in detecting faults.

For logic simulation, event-driven simulation that can take timing (delay) models and sequential circuit behavior into consideration is the technique most widely used in commercially available logic simulators. Examples of logic simulators include Verilog-XL, NC-Verilog (both from Cadence [Cadence 2006]), ModelSim (from Mentor Graphics [Mentor 2006]), and VCS (from Synopsys [Synopsys 2006]). These logic simulators can accept gate-level models as well as RTL and behavioral descriptions of the circuits written in hardware description languages, such as Verilog and VHDL, both IEEE standards. HDLs are beyond the scope of this book but are important for digital designers to learn. More detailed descriptions of both languages can be found in books or Web sites, such as [Palnitkar 1996], http://www.verilog.com, and http://www.verilog.net.

For fault simulation, both event-driven simulation and compiled-code simulation techniques can be found in commercially available electronic design automation applications. The fault simulators can be standalone tools or can be used as an integrated feature in the ATPG programs. As a standalone tool, concurrent fault simulation using the event-driven simulation technique is used in Verifault-XL (from Cadence) and TurboFault and TurboScan (both from SynTest [SynTest 2006]). As an integrated feature in ATPG, bitwise parallel simulation using the compiled-code simulation technique is widely used in Encounter Test (from Cadence), FastScan (from Mentor Graphics), and TetraMAX (from Synopsys).

As we move to the nanometer age, we have begun to see nanometer designs that contain hundreds of millions of transistors. We anticipate that the semiconductor industry will completely adopt the scan methodology for quality considerations. As a result, it is becoming imperative that advanced techniques for both logic simulation and fault simulation be developed to address the high-performance and high-capacity issues, in particular, for addressing new fault models, such as transition faults [Waicukauski 1986], path-delay faults [Schulz 1989], and bridging faults [Li 2003]. At the same time, more innovations are needed in developing advanced concurrent fault simulation techniques, as designs today that are based on the scan methodology are still not 100% scan testable. Fault simulation using functional patterns remains important in order to meet excellent quality and parts-per-million defect level goals.

3.6   EXERCISES

3.1   (Parallel Gate Evaluation) Consider a logic simulator with four logic symbols
(0, 1, u, and Z) that are coded as follows:

$$v_0 = (00)$$
$$v_1 = (11)$$
$$v_u = (01)$$
$$v_Z = (10)$$

Assume that the host computer has a word width of w. To simulate w input
vectors in parallel, two words (X_1 and X_2) are allocated for each signal X to
store the first and second bits of the logic symbol codes, respectively.

(a)   Derive the gate evaluation procedures for AND, OR, and NOT operations.

(b)   Derive the evaluation procedures for complex gates such as a 2-to-1 multiplexer, XOR, and tristate buffer.

Note that the simulator is based on ternary logic; therefore, Z-to-u conversions
may be necessary to convert Z inputs to u's prior to gate evaluations.

3.2   (Timing Models) For circuit M shown in Figure 3.38, complete the following
timing diagram (Figure 3.39) with respect to each timing model given below:

(a)   *Nominal delay*—Two-input gate, 1 ns; three-input gate, 1.2 ns; inverter,
0.6 ns.
Inertial delay—All gates, 0.3 ns.

(b)   *Rise delay*—Two-input gate, 0.8 ns; three-input gate, 1 ns; inverter, 0.6 ns.
Fall delay—Two-input gate, 1 ns; three-input gate, 1.2 ns; inverter, 0.8 ns.

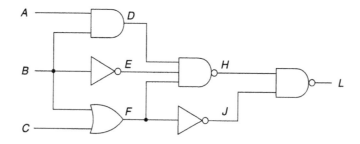

■ **FIGURE 3.38**

Example circuit M.

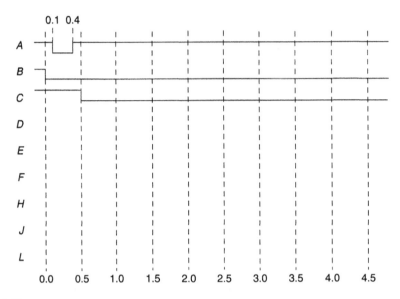

■ **FIGURE 3.39**

The timing diagram.

(c) *Minimum delay*—Two-input gate, 0.8 ns; three-input gate, 1 ns; inverter, 0.6 ns.
Maximum delay—Two-input gate, 1 ns; three-input gate, 1.2 ns; inverter, 0.8 ns.

3.3 **(Compiled-Code Simulation)** Apply logic levelization on circuit *M* given in Figure 3.38. Assign a level number to each gate starting from level 1 at the primary inputs. Assume that a target machine can only support basic logic operations using two-input AND/OR and inversion. What is the pseudo code for circuit *M* if it is to be simulated in the target machine?

3.4 **(Event-Driven Simulation)** Redo Problem 3.2a using the nominal-delay event-driven simulation technique. Show all events and activity lists of each time stamp.

3.5 **(Hazard Detection)** Use eight-valued logic to detect static and dynamic hazards in circuit *M* in response to an input change of *ABC* from {101} to {010}.

3.6 **(Hazard Detection)** For the circuit and test patterns given in Figure 3.40 below, determine whether there is a static or dynamic hazard, assuming there are no faults present in the design.

3.7 **(Parallel-Pattern Single-Fault Propagation)** For the circuit and two given stuck-at faults shown in Figure 3.40, use the parallel-pattern single-fault propagation fault simulation technique to identify which faults can be detected by the given test patterns.

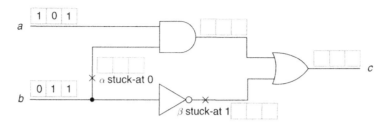

■ **FIGURE 3.40**

Example circuit.

■ **FIGURE 3.41**

Circuit for Problem 3.9.

3.8 **(Parallel Fault Simulation)** Repeat Problem 3.7 using parallel fault simulation.

3.9 **(Deductive Fault Simulation)** Write the fault list propagation rule for the three-input NOR gate given in Figure 3.41.

3.10 **(Deductive Fault Simulation)** Repeat Problem 3.7 using deductive fault simulation.

3.11 **(Concurrent Fault Simulation)** Repeat Problem 3.7 using concurrent fault simulation.

3.12 **(Critical Path Tracing)** For the circuit in Problem 3.7, circle all the critical values for the three test patterns. What faults are detected?

3.13 **(A Design Practice)** Repeat Problem 3.7 using the logic simulation program provided on the Web site. What are the correct outputs of the circuit?

3.14 **(A Design Practice)** Repeat Problem 3.7 using the fault simulation program provided on the Web site. What is the fault coverage of this test set?

3.15 **(A Design Practice)** For the circuit given in Problem 3.2, use any commercially available logic simulator, such as Verilog-XL, VCS, or ModelSim, to simulate the circuit behavior. Show the correct outputs of the circuit on a waveform display. Do they agree with your answers?

3.16 **(A Design Practice)** For the circuit given in Problem 3.7, use the fault simulation program (TurboFault) provided on the Web site to simulate the faulty output in the presence of fault α. Is the fault detected?

References

R3.0—Books

[Abramovici 1994] M. Abramovici, M. A. Breuer, and A. D. Friedman, *Digital Systems Testing and Testable Design*, IEEE Press, Piscataway, NJ, 1994 (revised printing).

[Miczo 2003] A. Miczo, *Digital Logic Testing and Simulation*, 2nd ed., John Wiley & Sons, Hoboken, NJ, 2003.

[Palnitkar 1996] S. Palnitkar, *Verilog HDL: A Guide to Digital Design and Synthesis*, Sunsoft, Mountain View, CA, 1996.

[Thomas 2002] D. E. Thomas and P. R. Moorby, *The Verilog Hardware Description Language*, Springer Science, New York, 2002.

[IEEE 1076-2002] *IEEE Standard VHDL Language Reference Manual*, IEEE Std. 1076-2002, Institute of Electrical and Electronics Engineers, New York, 2002.

[IEEE 1463-2001] *IEEE Standard Description Language Based on the Verilog Hardware Description Language*, IEEE Std. 1463-2001), Institute of Electrical and Electronics Engineers, New York, 2001.

[Wile 2005] B. Wile, J. C. Goss, and W. Roesner, *Comprehensive Functional Verification*, Morgan Kaufmann, San Francisco, CA, 2005.

R3.1—Introduction

[SystemC 2006] SystemC (http://www.systemc.org).

[SystemVerilog 2006] SystemVerilog (http://systemverilog.org).

R3.2—Simulation Models

[Breuer 1972] M. A. Breuer, A note on three valued logic simulation, *IEEE Trans. Comput.*, C-21(4), 399–402, 1972.

R3.3—Logic Simulation

[Eichelberger 1965] E. B. Eichelberger, Hazard detection in combinational and sequential circuits, *IBM J. Res. Develop.*, 9(9), 90–99, 1965.

[Hayes 1986] J. P. Hayes, Uncertainty, energy, and multiple-valued logics, *IEEE Trans. Comput.*, C-35(2), 107–114, 1986.

[Ulrich 1969] E. G. Ulrich, Exclusive simulation of activity in digital networks, *Commun. ACM*, 12(2), 102–110, 1969.

[Wang 1987] L.-T. Wang, N. E. Hoover, E. H. Porter, and J. J. Zasio, SSIM: A software levelized compiled-code simulator, in *Proc. Des. Automat. Conf.*, June 1987, pp. 2–8.

[Yoeli 1964] M. Yoeli and S. Rinon, Applications of ternary algebra to the study of static hazards, *J. ACM*, 11(1), 84–97, 1964.

R3.4—Fault Simulation

[Abramovici 1984] M. Abramovici, P. R. Menon, and D. T. Miller, Critical path tracing: An alternative to fault simulation, *IEEE Des. Test Comput.*, 1(1), 83–93, 1984.

[Armstrong 1972] D. B. Armstrong, A deductive method for simulating faults in logic circuits, *IEEE Trans. Comput.*, C-21(5), 464–471, 1972.

[Butler 1974] T. T. Butler, T. G. Hallin, J. J. Kulzer, and K. W. Johnson, LAMP: Application to switching system development, *Bell System Tech. J.*, 53, 1535–1555, 1974.

[Cheng 1989] W. T. Cheng and M. L. Yu, Differential fault simulation: A fast method using minimal memory, in *Proc. Des. Automat. Conf.*, June 1989, pp. 424–428.

[Goel 1980] P. Goel, Test generation cost analysis and projections, in *Proc. Des. Automat. Conf.*, June 1980, pp. 77–84.

[Jain 1985] S. K. Jain and V. D. Agrawal, Statistical fault analysis, *IEEE Des. Test Comput.*, 2(1), 38–44, 1985.

[Seshu 1965] S. Sesuh and D. N. Freeman, On improved diagnosis program, *IEEE Trans. Electron. Comput.*, EC-14(1), 76–79, 1965.

[Ulrich 1974] E. G. Ulrich and T. Baker, Concurrent simulation of nearly identical digital networks, *IEEE Trans. Comput.*, 7(4), 39–44, 1974.

[Waicukauski 1985] J. A. Waicukauski, E. B. Eichelberger, D. O. Forlenza, E. Lindbloom, and T. McCarthy, Fault simulation for structured VLSI, *Proc. VLSI Syst. Des.*, 6(12), 20–32, 1985.

[Waicukauski 1986] J. A. Waicukauski, E. Lindbloom, B. K. Rosen, and V. S. Iyengar, Transition fault simulation by parallel pattern single fault propagation, in *Proc. IEEE Int. Test Conf.*, September 1986, pp. 542–549.

[Wu 2005] L. Wu and D. M. H. Walker, A fast algorithm for critical path tracing in VLSI digital circuits, in *Proc. Int. Symp. on Defect and Fault Tolerance in VLSI Systems*, October 2005, pp. 178–186.

R3.5—Concluding Remarks

[Cadence 2006] Cadence Design Systems (http://www.cadence.com).

[Li 2003] Z. Li, X. Lu, W. Qiu, W. Shi, and D. M. H. Walker, A circuit level fault model for resistive bridges, *ACM Trans. Des. Automat. Electron. Sys. (TODAES)*, 8(4), 546–559, 2003.

[Mentor 2006] Mentor Graphics (http://www.mentor.com).

[Schulz 1989] M. Schulz, F. Fink, and K. Fuchs, Parallel pattern fault simulation of path delay faults, in *Proc. Des. Automat. Conf.*, June 1989, pp. 357–363.

[Synopsys 2006] Synopsys (http://www.synopsys.com).

[SynTest 2006] SynTest Technologies (http://www.syntest.com).

[Waicukauski 1986] J. A. Waicukauski, E. Lindbloom, B. K. Rosen, and V. S. Iyengar, Transition fault simulation by parallel pattern single fault propagation, in *Proc. IEEE Int. Test Conf.*, September 1986, pp. 542–549.

CHAPTER 4

TEST GENERATION

Michael S. Hsiao
Virginia Tech, Blacksburg, Virginia

ABOUT THIS CHAPTER

Test generation is the task of producing an effective set of vectors that will achieve high fault coverage for a specified fault model. While much progress has been made over the years in *automatic test pattern generation* (ATPG), this problem remains an extremely difficult one. Without powerful ATPGs, chips will increasingly depend on *design for testability* (DFT) techniques to alleviate the high cost of generating vectors. This chapter deals with the fundamental issues behind the design of an ATPG, as well as the underlying learning mechanisms that can improve the overall performance of ATPG.

This chapter is organized as follows. First, an overview of the problem of test generation is given, followed by random test generation. Next, deterministic algorithms for test generation for stuck-at faults are explained, including techniques that enhance the deterministic engines such as static and dynamic learning. Simulation-based test generation is covered next, where genetic algorithms are used to derive intelligent vectors. Test generation for other fault models such as delay faults is explained, including ATPG for path-delay faults and transition faults. A brief discussion on bridging faults is also included. Finally, advanced test generation topics are briefly discussed.

4.1 INTRODUCTION

Due to the imperfect manufacturing process, defects may be introduced during fabrication, resulting in chips that could potentially malfunction. The objective of test generation is the task of producing a set of test vectors that will uncover any defect in a chip. Figure 4.1 illustrates a high-level concept of test generation. In this figure, the circuit at the top is defect free, and for any defective chip which is functionally different from the defect-free one there must exist some input that can differentiate the two. Generating effective test patterns efficiently for a digital circuit is thus the goal of any *automatic test pattern generation* (ATPG) system.

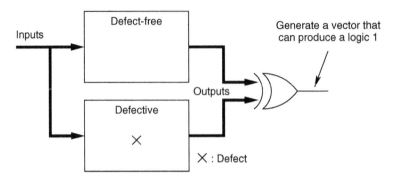

■ **FIGURE 4.1**

Conceptual view of test generation.

As this problem is extremely difficult, ***design for testability*** (DFT) methods have been frequently used to relieve the burden on the ATPG. In this sense, a powerful ATPG can be regarded as the holy grail in testing, with which all DFT methods could potentially be eliminated. In other words, if the ATPG engine is capable of delivering high-quality test patterns that achieve high fault coverages and small test sets, DFT would no longer be necessary. This chapter thus deals with the algorithms and inner workings of an automatic test pattern generator. Both the underlying theory and the implementation details are covered.

As it is difficult and unrealistic to generate vectors targeting all possible defects that could potentially occur during the manufacturing process, automatic test generators operate on an abstract representation of defects referred to as *faults*. The single stuck-at fault model is one of the most popular fault models and is discussed first in this chapter, followed by discussion of test generation for other fault models. In addition, only a single fault is assumed to be present in the circuit to simplify the test generation problem.

Consider the single stuck-at fault model: Any fault simply denotes that a circuit node is tied to logic 1 or logic 0. Figure 4.2 shows a circuit with a single stuck-at fault in which signal d is tied to logic 0 ($d/0$). A logic 1 must be applied from the primary inputs of the circuit to node d if there is to be a difference between the fault-free (or good) circuit and the circuit with the stuck-at fault present. Next, in order to observe the effect of the fault, a logic 1 must be applied to signal c so if

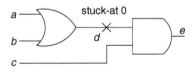

■ **FIGURE 4.2**

Example of a single stuck-at fault.

the fault $d/0$ is present it can be detected at the output e. Test generation attempts to generate test vectors for every possible fault in the circuit. In this example, in addition to the $d/0$ fault, faults such as $a/1, b/1, c/0$, etc. are also targeted by the test generator. As some of the fault in the circuit can be logically equivalent, no test can be obtained to distinguish between them. Thus, equivalent fault collapsing is often used to identify equivalent faults *a priori* in order to reduce the number of faults that must be targeted [Abramovici 1994] [Bushnell 2000] [Jha 2003]. Subsequently, the ATPG is only concerned with generating test vectors for each fault in the collapsed fault list.

4.2 RANDOM TEST GENERATION

Random test generation (RTG) is one of the simplest methods for generating vectors. Vectors are randomly generated and fault-simulated (or fault-graded) on the *circuit under test* (CUT). Because no specific fault is targeted, the complexity of RTG is low. However, the disadvantages of RTG are that the test set size may grow to be very large and the fault coverage may not be sufficiently high, due to difficult-to-test faults.

In RTG, logic values are randomly generated at the primary inputs, with equal probability of assigning a logic 1 or logic 0 to each primary input. Thus, the random vectors are uniformly distributed in the test set. Note that the random test set is not truly random because a pseudo-random number generator is generally used. In other words, the random test set can be repeated with the same pseudo-random number generator. Nevertheless, the vectors generated hold the necessary statistical properties of a random vector set.

The **level of confidence** one can have on a random test set T can be measured as the probability that T can detect all the stuck-at faults in the circuit. For N random vectors, the **test quality** t_N indicates the probability that all detectable stuck-at faults are detected by these N random vectors. Thus, the test quality of a random test set highly depends on the circuit under test.

Consider a circuit with an eight-input AND gate (or equivalently a cone of seven two-input AND gates), illustrated in Figure 4.3. While achieving a logic 0 at the output of the AND gate is easy, getting a logic 1 is difficult. A logic 1 would require all the inputs to be at logic 1. If the RTG assigns each primary input with an equal probability of logic 0 or logic 1, the chance of getting eight logic 1's simultaneously would only be $0.5^8 = 0.0039$. In other words, the AND gate output stuck-at-0 fault would be difficult to test by the RTG. Such faults are called **random-pattern resistant faults**.

As discussed earlier, the quality of a random test set depends on the underlying circuit. More random-pattern resistant faults will more likely reduce the quality of the random test set.

To tackle the problem of targeting random-pattern resistant faults, biasing is required so the input vectors are no longer viewed as uniformly distributed. Consider the same eight-input AND gate example again. If each input of the AND gate has a much higher probability of receiving a logic 1, the probability of getting a logic 1 at the

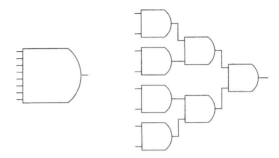

■ **FIGURE 4.3**

Two equivalent circuits.

output of the AND gate significantly increases. For example, if each input has a 75% probability of receiving a logic 1, then getting a logic 1 at the output of the AND gate now becomes $0.75^8 = 0.1001$, rather than the previous 0.0039.

Determining the optimal bias values for each primary input is not an easy task. Thus, rather than trying to obtain the optimal set of values, the objective is frequently to increase the probabilities for those difficult-to-control and difficult-to-observe nodes in the circuit. For instance, suppose a circuit has an eight-input AND gate; any fault that requires the AND gate output equal to logic 1 for detection will be considered difficult to test. It would then be beneficial to attempt to increase the probability of obtaining a logic 1 at the output of this AND gate.

Another issue regarding random test generation is the number of random vectors needed. Given a circuit with n primary inputs, there are clearly 2^n possible input vectors. One can express the probability of detecting fault f by any random vector to be:

$$d_f = \frac{T_f}{2^n}$$

where T_f is the set of vectors that can detect fault f. Consequently, the probability that a random vector will not detect f (*i.e.*, f escapes a random vector) is:

$$e_f = 1 - d_f$$

Therefore, given N random vectors, the probability that none of the N vectors detects fault f is:

$$e_f^N = (1 - d_f)^N$$

In other words, the probability that at least one out of N vectors will detect fault f is:

$$1 - (1 - d_f)^N$$

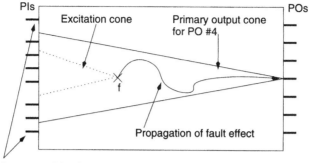

Inputs outside of PO cone are not needed for detection of fault f

■ **FIGURE 4.4**

Detection of a fault.

If the detection probability, d_f, for the hardest fault is known, N can be readily computed by solving the following inequality:

$$1 - (1 - d_f)^N \geq p$$

where p is the probability that N vectors should detect fault f.

If the detection probability is not known, it can be computed directly from the circuit. The detection probability of a fault is directly related to: (1) the controllability of the line that the fault is on, and (2) the observability of the fault-effect to a primary output. The controllability and observability computations have been introduced previously in the chapter on design for testability.

It is worth noting that the minimum detection probability of a detectable fault f can be determined by the output cone in which f resides. In fact, if f is detectable, it must be excited and propagated to at least one primary output, as illustrated in Figure 4.4. It is clear that all the primary inputs necessary to excite f and propagate the fault-effect must reside in the cone of the output to which f is detected. Thus, the detection probability for f is at least $(0.5)^m$, where m is the number of primary inputs in the cone of the corresponding primary output. Taking this concept a step further, the detection probability of the most difficult fault can be obtained with the following lemma [David 1976] [Shedletsky 1977].

Lemma 1

In a combinational circuit with multiple outputs, let n_{max} be the number of primary inputs that can lead to a primary output. Then, the detection probability for the most difficult detectable fault, d_{min}, is:

$$d_{min} \geq (0.5)^{n_{max}}$$

Proof

The proof follows from the preceding discussion.

4.2.1   Exhaustive Testing

If the combinational circuit has few primary inputs, **exhaustive testing** may be a viable option, where every possible input vector is enumerated. This may be superior to random test generation as RTG can produce duplicated vectors and may miss certain ones.

In circuits where the number of primary inputs is large, exhaustive testing becomes prohibitive. However, based on the results of Lemma I, it may be possible to partition the circuit and only exhaust the input vectors within each cone for each primary output. This is called **pseudo-exhaustive testing**. In doing so, the number of input vectors can be drastically reduced. When enumerating the input vectors for a given primary output cone, the values for the primary inputs that are outside the cone are simply assigned random values. Therefore, if a circuit has three primary outputs, each of which has a corresponding primary output cone. Note that these three primary output cones may overlap. Let n_1, n_2, and n_3 be the number of primary inputs corresponding to these three cones. Then the number of pseudo-exhaustive vectors is simply at most $2^{n_1} + 2^{n_2} + 2^{n_3}$.

4.3   THEORETICAL BACKGROUND: BOOLEAN DIFFERENCE

Consider the circuit shown in Figure 4.5. Let the target fault be the stuck-at-0 fault on primary input y. Recall the high-level concept of test generation illustrated in Figure 4.1, where the objective is to distinguish the fault-free circuit from the faulty circuit. In the example circuit shown in Figure 4.5, the faulty circuit is the circuit with y stuck at 0.

Note that the circuit output can be expressed as a Boolean formula:

$$f = xy + \bar{y}z$$

Let f' be the faulty circuit with the fault $y/0$ present. In other words,

$$f' = f(y = 0).$$

In order to distinguish the faulty circuit f' from the fault-free counterpart f, any input vector that can make $f \oplus f' = 1$ would suffice. Furthermore, as the aim is test

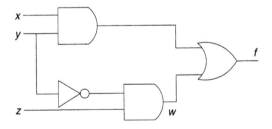

■ FIGURE 4.5

Example circuit to illustrate the concept of Boolean difference.

generation, the target fault must be excited. In this example, the logic value on primary input y must be logic 1 to excite the fault $y/0$. Putting these two conditions together, the following equation is obtained:

$$y \cdot f(y = 1) \oplus f(y = 0) = 1. \tag{4.1}$$

Note that $f(y = 1) \oplus f(y = 0)$ indicates the exclusive-or operation on the two functions $f(y = 1)$ and $f(y = 0)$; it evaluates to logic 1 if and only if the two functions evaluate to opposing values. In terms of ATPG, this is synonymous to propagating the fault effect at node y to the primary output f. Therefore, any input vector on primary inputs x, y, and z that can satisfy Equation (4.1) is a valid test vector for fault $y/0$:

$$y \cdot f(y = 1) \oplus f(y = 0) = y \cdot (x \oplus z)$$

$$= y \cdot (x\bar{z} + \bar{x}z)$$

$$= xy\bar{z} + \bar{x}yz$$

In this running example, the two vectors $xyz = \{110, 011\}$ are candidate test vectors for fault $y/0$.

Formally, $f(y = 1) \oplus f(y = 0)$ is called the **Boolean difference** of f with respect to y and is often written as:

$$\frac{df}{dy} = f(y = 1) \oplus f(y = 0).$$

In general, if f is a function of $x_1, x_2, \ldots, x_n$, then:

$$\frac{df}{dx_i} = f(x_1, x_2, \ldots, x_i, \ldots, x_n) \oplus f(x_1, x_2, \ldots, \bar{x}_i, \ldots, x_n)$$

In terms of test generation, for any target fault on some fault α/v, the set of all vectors that can *propagate* the fault-effect to the primary output f is then those vectors that can satisfy:

$$\frac{df}{d\alpha} = 1$$

(Note that this is independent of the polarity of the fault, whether it is stuck-at-0 or stuck-at-1.) Next, the constraint that the fault must be excited, α set to value $\bar{v}$, must be added. Subsequently, the set of test vectors that can detect the fault becomes all those input values that can satisfy the following equation:

$$(\alpha = \bar{v}) \cdot \frac{df}{d\alpha} = 1 \tag{4.2}$$

Consider the same circuit shown in Figure 4.5 again. Suppose the target fault is $w/0$. The same analysis can be performed for this new fault. The set of test vectors that can detect $w/0$ is simply:

$$w \cdot \frac{df}{dw} = 1$$

$$\Rightarrow \quad w \cdot (f(w = 1) \oplus f(w = 0)) = 1$$

$$\Rightarrow \quad w \cdot (1 \oplus xy) = 1$$
$$\Rightarrow \quad w \cdot (\overline{xy}) = 1$$
$$\Rightarrow \quad w \cdot (\overline{x} + \overline{y}) = 1$$
$$\Rightarrow \quad w\overline{x} + w\overline{y} = 1$$

Now, w can be expanded from the circuit shown in the figure to be $w = \overline{y} \cdot z$. Plugging this into the equation above gives us:

$$w \cdot \overline{x} + w \cdot \overline{y} = 1$$
$$\Rightarrow \quad \overline{y} \cdot z \cdot \overline{x} + \overline{y} \cdot z \cdot \overline{y} = 1$$
$$\Rightarrow \quad \overline{x} \cdot \overline{y} \cdot z + \overline{y} \cdot z = 1$$
$$\Rightarrow \quad \overline{y} \cdot z = 1$$

Therefore, the set of vectors that can detect $w/0$ is {001, 101}.

4.3.1 Untestable Faults

If the target fault is untestable, it would be impossible to satisfy Equation 4.2. Consider the circuit shown in Figure 4.6. Suppose the target fault is $z/0$. Then the set of vectors that can detect $z/0$ are those that can satisfy:

$$z \cdot \frac{df}{dz} = 1$$
$$\Rightarrow \quad z \cdot (f(z = 1) \oplus f(z = 0)) = 1$$
$$\Rightarrow \quad z \cdot (xy \oplus xy) = 1$$
$$\Rightarrow \quad z \cdot 0 = 1$$
$$\Rightarrow \quad \text{UNSATISFIABLE}$$

In other words, there exists no input vectors that can satisfy $z \cdot \dfrac{df}{dz} = 1$, indicating that the fault $z/0$ is untestable.

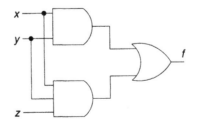

■ FIGURE 4.6

Example circuit for an untestable fault.

4.4 DESIGNING A STUCK-AT ATPG FOR COMBINATIONAL CIRCUITS

In deterministic ATPG algorithms, there are two main tasks. The first is to excite the target fault, and the second is to propagate the fault-effect to a primary output. Because the logic values in both the fault-free and faulty circuits are needed, composite logic values are used. For each signal in the circuit, the values v/v_f are needed, where v denotes the value for the signal in the fault-free circuit, and v_f represents the value in the corresponding faulty circuit. Whenever $v = v_f$, v is sufficient to denote the signal value. To facilitate the manipulation of such composite values, a 5-valued algebra was proposed [Roth 1966], in which the five values are 0, 1, X, D, and $\overline{D}$; 0, 1, and X are the conventional values found in logic design for true, false, and "don't care." D represents the composite logic value 1/0 and $\overline{D}$ represents 0/1. Boolean operators such as AND, OR, NOT, XOR, etc., can work on the 5-valued algebra as well. The simplest way to perform Boolean operations is to represent each composite value into the v/v_f form and operate on the fault-free value first, followed by the faulty value. For example, 1 AND D is 1/1 AND 1/0. AND-ing the fault-free values yields 1 AND $1 = 1$, and AND-ing the faulty values yields 1 AND $0 = 0$. So the result of the AND operation is $1/0 = D$. As another example,

$$D \text{ OR } \overline{D} = 1/0 \text{ OR } 0/1$$
$$= 1/1$$
$$= 1$$

Tables 4.1, 4.2, and 4.3 show the AND, OR, and NOT operations for the 5-valued algebra, respectively. Operations on other Boolean conjunctives can be constructed in a similar manner.

4.4.1 A Naive ATPG Algorithm

A very simple and naive ATPG algorithm is shown in Algorithm 1, where combinational circuits with fanout structures can be handled.

TABLE 4.1 ■ AND Operation

AND	0	1	D	$\overline{D}$	X
0	0	0	0	0	0
1	0	1	D	$\overline{D}$	X
D	0	D	D	0	X
$\overline{D}$	0	$\overline{D}$	0	$\overline{D}$	X
X	0	X	X	X	X

TABLE 4.2 ■ OR Operation

OR	0	1	D	$\bar{D}$	X
0	0	1	D	$\bar{D}$	X
1	1	1	1	1	1
D	D	1	D	1	X
$\bar{D}$	$\bar{D}$	1	1	$\bar{D}$	X
X	X	1	X	X	X

TABLE 4.3 ■ NOT Operation

NOT	
0	1
1	0
D	$\bar{D}$
$\bar{D}$	D
X	X

Algorithm 1 Naive ATPG (C, f)

1: **while** a fault-effect of f has not propagated to a PO and all possible vector combinations have not been tried **do**
2:     pick a vector, v, that has not been tried;
3:     fault simulate v on the circuit C with fault f;
4: **end while**

Note that in an ATPG, the worst-case computational complexity is exponential, as all possible input patterns may have to be tried before a vector is found or that the fault is determined to be undetectable. One may go about line #2 of the algorithm in an intelligent fashion, so a vector is not simply selected indiscriminately. Whether or not intelligence is incorporated, some mechanism is needed to account for those attempted input vectors so no vector would be repeated. If it is possible to deduce some knowledge during the search for the input vector, the ATPG may be able to mark a set of solutions as tried and thus reduce the remaining search space. For instance, after attempting a number of input vectors, this naive ATPG realizes that any input vector with the first primary input set to logic 0 cannot possibly detect the target fault and it can safely mark all vectors with the first primary input equal to 0 as a tried input vector. Subsequently, only those vectors with the first primary input set to 1 will be selected.

In certain cases, it may not be possible for the ATPG to deduce that all vectors with some primary input set to a given logic value definitely do not qualify to be solution vectors. However, it may be able to make an intelligent guess that input vectors with primary input #i set to some specific logic value are more likely to lead to a solution. In such a case, the ATPG would make a **decision** on primary input #i. Because the decision may actually be wrong, the ATPG may eventually have to alter its decision, trying the vectors that have the opposite Boolean value on primary input #i.

The process of making decisions and reversing decisions will result in a **decision tree**. Each node in the decision tree represents a decision variable. If only two choices are possible for each decision variable, then the decision tree is a binary tree. However, there may be cases where multiple choices are possible in a general search tree.

Figure 4.7 shows an example decision tree. While this figure only allows decisions to be made at the primary inputs, in general this may not be the case. This is used simply to allow the reader to have a clearer picture of the concept behind decision trees. At each decision, the search space is halved. For example, if the circuit has n primary inputs, then there is a total of 2^n possible vectors in the solution space. After a decision is made, the solution spaces under the two branches of a decision node are disjoint. For instance, the space under the decision $a = 1$ does not contain any vectors with $a = 0$. Note that the decision tree for a solution vector may not require the ATPG to *exhaustively* enumerate every possible vector; rather, it *implicitly* enumerates the vectors. If a solution vector exists, there must be a path along the decision tree that leads to the solution. On the other hand, if the fault is undetectable, every path in the decision tree would lead to no solution.

It is important to note that a fault may be detected without having made all decisions. For example, the circuit nodes that do not play a role in exciting or propagating the fault would not have to be included in the decision process. Likewise,

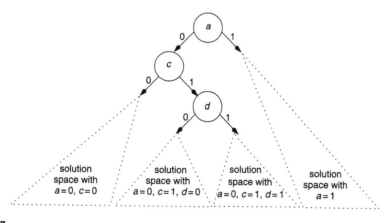

■ **FIGURE 4.7**

An example decision tree.

it may not require all decision variables before the ATPG can determine that it is on the wrong path. For example, if a certain path already sets a value on the fault site such that the fault is not excited, then no value combination on the remaining decision variables can help to excite and propagate the fault. Using Figure 4.7 as an example again, suppose the path $a = 0$, $c = 1$, $d = 1$ cannot excite the target fault α. Then, the rest of the decision variables, $b, e, f, \ldots$, cannot undo the effect rendered by $a = 0$, $c = 1$, $d = 1$.

4.4.1.1 Backtracking

Whenever a conflict is encountered (*i.e.*, a path segment leading to no solution), the search must not continue searching along that path, but must go back to some earlier point and re-decide on a previous decision. If only two choices are possible for a decision variable, then some previous decision needs to be reversed, if the other branch has not been explored before. This reversal of decision is called a **backtrack**. In order to keep track of where the search spaces have been explored and avoid repeating the search in the same spaces, the easiest mechanism is to reverse the most recent decision made. When reversing any decision, the signal values implied by the assignment of the previous decision variable must be undone.

Consider the decision tree illustrated in Figure 4.8 as an example. Suppose the current decisions made so far are $a = 0$, $c = 1$, $d = 0$, and this causes a conflict in detecting the target fault. Then, the search must reverse the most recently made decision, which is $d = 0$. When reversing $d = 0$ to $d = 1$, all values resulted from $d = 0$ must be first undone. Then, the search continues with the path $a = 0$, $c = 1$, $d = 1$. If the reversal of a decision also caused a conflict (in this case, reversing $d = 0$ also caused a conflict), then it means $a = 0$, $c = 1$ actually cannot lead to any solution vector that can detect the target fault. The backtracking mechanism would then take the search to the previous decision and attempt to reverse that decision. In the running example, it would undo the decision on d, assigning d to "don't care," followed by reversing of the decision $c = 1$ and searching the portion of the search space under $a = 0$, $c = 0$. Finally, if there is no previous decision that can be reversed, the ATPG concludes that the target fault is undetectable.

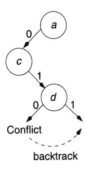

■ **FIGURE 4.8**

Backtrack on a decision.

Technically, whenever a decision is reversed, say $d = 0$ is reversed to $d = 1$ as shown in Figure 4.8, $d = 1$ is no longer a decision; rather, it becomes an implied value by a subset of the previous decisions made. The exact subset of decisions that implied $d = 1$ can be computed by a **conflict analysis** [Marques-Silva 1999b]. However, the details of conflict analysis are beyond the scope of this chapter and are thus omitted. The reader can refer to [Marques-Silva 1999b] for details of this mechanism. In addition, intelligent conflict analysis can also allow for **nonchronological backtracking**.

4.4.2 A Basic ATPG Algorithm

Given a target fault g/v in a fanout-free combinational circuit C, a simple procedure to generate a vector for the fault is shown in Algorithm 2, where JustifyFanoutFree() and PropagateFanoutFree() are both recursive functions.

Algorithm 2 Basic Fanout Free ATPG (C, g/v)

1: initialize circuit by setting all values to X;
2: JustifyFanoutFree($C, g, \bar{v}$); /* excite the fault by justifying line g to $\bar{v}$ */
3: PropagateFanoutFree(C, g); /* propagate fault-effect from g to a PO */

The JustifyFanoutFree(g, v) function recursively justifies the predecessor signals of g until all signals that need to be justified are indeed justified from the primary inputs. The simple outline of the JustifyFanoutFree routine is listed in Algorithm 3. In line #10 of the algorithm, controllability measures can be used to select the best input to justify. Selecting a good gate input may help to reach a primary input sooner.

Consider the circuit C shown in Figure 4.9. Suppose the objective is to justify $g = 1$. According to the above algorithm, the following sequence of recursive calls to JustifyFanoutFree() would have been made:

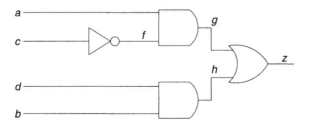

■ **FIGURE 4.9**

Example fanout-free circuit.

Algorithm 3 JustifyFanoutFree(C, g, v)

1: $g = v$;
2: **if** gate type of $g ==$ primary input **then**
3:    return;
4: **else if** gate type of $g ==$ AND gate **then**
5:    **if** $v == 1$ **then**
6:       **for all** inputs h of g **do**
7:          JustifyFanoutFree($C, h, 1$);
8:       **end for**
9:    **else** $\{v == 0\}$
10:       $h =$ pick one input of g whose value $== X$;
11:       JustifyFanoutFree($C, h, 0$);
12:    **end if**
13: **else if** gate type of $g ==$ OR gate **then**
14:    . . .
15: **end if**

---

call #1: JustifyFanoutFree($C, g, 1$)
call #2: JustifyFanoutFree($C, a, 1$)
call #3: JustifyFanoutFree($C, f, 1$)
call #5: JustifyFanoutFree($C, c, 0$)

After these calls to JustifyFanoutFree(), $abcd = 1X0X$ is an input vector that can justify $g = 1$.

Consider another circuit C shown in Figure 4.10. Note that the circuit is not fanout-free, but the above algorithm will still work for the objective of trying to justify the signal $g = 1$. According to the algorithm, the following sequence of calls to the JustifyFanoutFree function would have been made:

call #1: JustifyFanoutFree($C, g, 1$)
call #2: JustifyFanoutFree($C, a, 1$)
call #3: JustifyFanoutFree($C, f, 1$)
call #4: JustifyFanoutFree($C, d, 0$)
call #5: JustifyFanoutFree($C, c, 0$)

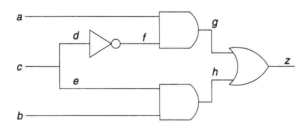

■ **FIGURE 4.10**

Example circuit with a fanout structure.

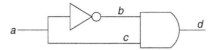

■ **FIGURE 4.11**

Circuit with a constant circuit node.

After these five calls to JustifyFanoutFree(), $abc = 1X0$ is an input vector that can justify $g = 1$.

Note that in a fanout-free circuit, the JustifyFanoutFree() routine will *always* be able to set g to the desired value v and no conflict will ever be encountered. However, this is not always true for circuits with fanout structures. This is because in circuits with fanout branches, two or more signals that can be traced back to the same fanout stem are **correlated**, and setting arbitrary values on these correlated signals may not always be possible. For example, in the simple circuit shown in Figure 4.11, justifying $d = 1$ is impossible, as it requires both $b = 1$ and $c = 1$, thereby causing a conflict on a.

Consider again the circuit shown in Figure 4.10. Suppose the objective is to set $z = 0$. Based on the JustifyFanoutFree() algorithm, it would first justify both $g = 0$ and $h = 0$. Now, for justifying $g = 0$, suppose it picks the signal f for justifying the objective $g = 0$; it would eventually assign $c = 1$ through the recursive JustifyFanoutFree() function. Next, for justifying $h = 0$, it no longer can choose $e = 0$ as a viable option, because choosing $e = 0$ will eventually cause a **conflict** on signal c. In other words, a different **decision** has to be made for justifying $h = 0$. In this case, $b = 0$ should be chosen. While this example is very simple, it illustrates the possibility of making poor decisions, causing potential **backtracks** in the search. In the rest of this chapter, more discussion on avoiding conflicts will be covered.

In the above running example, suppose the target fault is $g/0$, and JustifyFanoutFree(C, g, 1) would have successfully excited the fault. With the fault $g/0$ excited, the next step is to propagate the fault-effect to a primary output. Similar to the JustifyFanoutFree() function, PropagateFanoutFree() is a recursive function as well, where the fault-effect is propagated one gate at a time until it reaches a primary output. Algorithm 4 illustrates the pseudo-code for one possible implementation of the propagate function.

Again, although the PropagateFanoutFree() routine is meant for fanout-free circuits, it is sufficient for the running example. Using the PropagateFanoutFree() function on the fault-effect D at signal g, listed in Algorithm 3, the following calls to the JustifyFanoutFree and PropagateFanoutFree functions would have been made:

   call #1: PropagateFanoutFree(C, g)
   call #2: JustifyFanoutFree(C, h, 0)
   call #3: JustifyFanoutFree(C, b, 0)
   call #4: PropagateFanoutFree(C, z)

Algorithm 4 PropagateFanoutFree(C, g)

```
 1: if g has exactly one fanout then
 2:     h = fanout gate of g;
 3:     if none of the inputs of h has the value of X then
 4:        backtrack;
 5:     end if
 6: else {g has more than one fanout}
 7:     h = pick one fanout gate of g that is unjustified;
 8: end if
 9: if gate type of h == AND gate then
10:     for all inputs, j, of h, such that j ≠ g do
11:        if the value on j == X then
12:           JustifyFanoutFree(C, j, 1);
13:        end if
14:     end for
15: else if gate type of h == OR gate then
16:     for all inputs, j, of h, such that j ≠ g do
17:        if the value on j == X then
18:           JustifyFanoutFree(C, j, 0);
19:        end if
20:     end for
21: else if gate type of h == ... gate then
22:     ...
23: end if
24: PropagateFanoutFree(C, h);
```

Because the fault-effect has successfully propagated to the primary output z, the fault $g/0$ is detected, with the vector $abc = 100$.

The reader may notice that once $g/0$ has been excited, it is also propagated to z as well, because $c = 0$ also has made $h = 0$. In other words, the JustifyFanoutFree(C, h, 0) step is unnecessary. However, this is only possible if logic simulation or implication capability is embedded in the BasicFanoutFreeATPG() algorithm. For this discussion, it is not assumed that logic simulation is included.

Using the same circuit shown in Figure 4.10, consider the fault $g/1$. The Basic-FanoutFreeATPG() algorithm will again be used to generate a test vector for this fault. In this case, the ATPG first attempts to justify $g = 0$, followed by propagating the fault-effect to z. During the justification of $g = 0$, the ATPG can pick either a or f as the next signal to justify. At this point, the ATPG must make a **decision**. Testability measures discussed in an earlier chapter can be used as a guide to make more intelligent decisions. In this example, choosing a is considered to be better than f, because choosing a requires no additional decisions to be made. Note that testability measures only serve as a guide to decision selection; they do not guarantee that the guidance will always lead to better decision selection.

It is important to note that in circuits with fanout structures, because the simple JustifyFanoutFree() and PropagateFanoutFree() functions described above are meant for fanout-free circuits, will not always be applicable as illustrated in some of the examples above due to potential conflicts. In order to generate test vectors for general combinational circuits, there must be mechanisms that will allow the ATPG to avoid conflicts, as well as get out of a conflict when a conflict is encountered. To do so, the corresponding decision tree must be constructed during the search for a solution vector, and backtracks must be enforced for any conflict encountered. The following sections describe a few ATPG algorithms.

4.4.3 D Algorithm

The D algorithm was proposed to tackle the generation of vectors in general combinational circuits [Roth 1966] [Roth 1967]. As indicated by the name of the algorithm, the D algorithm tries to propagate a D or $\overline{D}$ of the target fault to a primary output. Note that because each detectable fault can be excited, a fault-effect can always be created. In the following discussion, propagation of the fault-effect will take precedence over the justification of the signals. This allows for enhanced efficiency of the algorithm as well as for simpler discussion.

Before proceeding to discussing the details of the D algorithm, two important terms should be defined: the **D-frontier** and the **J-frontier**. The D-frontier consists of all the gates in the circuit whose output value is x and a fault-effect (D or $\overline{D}$) is at one or more of its inputs. In order for this to occur, one or more inputs of the gate must have a "don't care" value. For example, at the start of the D algorithm, for a target fault f there is exactly one D (or $\overline{D}$) placed in the circuit corresponding to the stuck-at fault. All other signals currently have a "don't care" value. Thus, the D-frontier consists of the successor gate(s) from the line with the fault f. Two scenarios of a D-frontier are illustrated in Figure 4.12. Clearly, at any time if the D-frontier is empty, the fault no longer can be detected. For example, consider Figure 4.12a. If the bottom input of gate a is assigned a value of 0, the output of gate a will become 0, and the D-frontier now becomes empty. At this time, the search must backtrack and try a different search path.

The J-frontier consists of all the gates in the circuit whose output values are known (can be any of the five values in the 5-valued logic) but is not justified by its inputs. Figure 4.13 illustrates an example of a J-frontier. Thus, in order to detect the target fault, all gates in the J-frontier must be justified; otherwise, some gates in the J-frontier must have caused a conflict, where these gates cannot be justified to the desired values.

Having discussed the two fundamental concepts of the D-frontier and the J-frontier, the explanation for the D algorithm can begin. The D algorithm begins by trying to propagate the initial D (or $\overline{D}$) at the fault site to a primary output. For example, in Figure 4.14, the propagation routine will set all the side inputs of the path necessary (gates $a \rightarrow b \rightarrow c$) to propagate the fault-effect to the respective noncontrolling values. These side input gates, namely x, y, and z, thus form the J-frontier as they are not currently justified. And as the D is propagated to the primary output, the D-frontier eventually becomes the output gate.

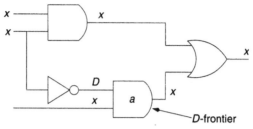

(a) *D*-frontier contains one gate

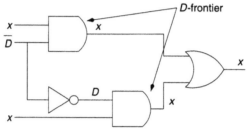

(b) *D*-frontier contains two gates

■ **FIGURE 4.12**

Illustrations of *D*-frontier.

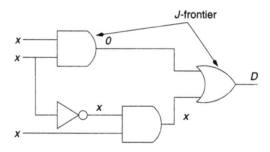

■ **FIGURE 4.13**

Illustration of *J*-frontier.

Whenever there are paths to choose from in advancing the *D*-frontier, observability values can be used to select the corresponding gates. However, this does not guarantee that the more observable path will definitely lead to a solution.

When a D or a $\bar{D}$ has reached a primary output, all the gates in the *J*-frontier must now be justified. This is done by advancing the *J*-frontier backward by placing predecessor gates in the *J*-frontier such that they justify the previous unjustified gates. Similar to propagation of the fault-effect, whenever a conflict occurs, a backtrack must be invoked. In addition, at each step, the *D*-frontier must be checked so the D (or $\bar{D}$) that has reached a primary output is still there. Otherwise, the search returns

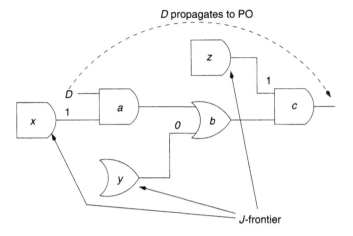

■ **FIGURE 4.14**

Propagation of D- and J-frontier.

to the propagation phase and attempts to propagate the fault-effect to a primary output again. The overall procedure for the D algorithm is shown in Algorithms 5 and 6.

Note that the above procedure has not incorporated any intelligence in the decision-making process. In other words, sometimes it may be possible to determine that some value assignments are not justifiable, given the current circuit state. For instance, consider the circuit fragment shown in Figure 4.15. Justifying gate $a = 1$ and gate $b = 0$ is not possible because $a = 1$ requires both of its inputs set to logic 1, while $b = 0$ requires both of its inputs set to logic 0. Noting such conflicting scenarios early can help to avoid future backtracks. Such knowledge can be incorporated into line #1 of the D-Alg-Recursion() shown in Algorithm 6. In particular, static and dynamic implications can be used to identify such potential conflicts, and they are used extensively to enhance the performance of the D algorithm (as

---

Algorithm 5 D-Algorithm(C, f)

1: initialize all gates to don't-cares;
2: set a fault-effect (D or $\bar{D}$) on line with fault f and insert it to the D-frontier;
3: J-frontier $= \phi$;
4: result $=$ D-Alg-Recursion(C);
5: **if** result $==$ success **then**
6:     print out values at the primary inputs;
7: **else**
8:     print fault f is untestable;
9: **end if**

Algorithm 6 D-Alg-Recursion(*C*)

 1: **if** there is a conflict in any assignment or D-frontier is Ø **then**
 2:    return failure;
 3: **end if**
 4: /* first propagate the fault-effect to a PO */
 5: **if** no fault-effect has reached a PO **then**
 6:    **while** not all gates in D-frontier has been tried **do**
 7:       g = a gate in D-frontier that has not been tried;
 8:       set all unassigned inputs of g to non-controlling value and add them to the J-frontier;
 9:       result = D-Alg-Recursion(*C*);
10:       **if** result == success **then**
11:          return (success);
12:       **end if**
13:    **end while**
14:    return (failure);
15: **end if** {fault-effect has reached at least one PO}
16: **if** J-frontier is Ø **then**
17:    return (success);
18: **end if**
19: g = a gate in J-frontier;
20: **while** g has not been justified **do**
21:    j = an unassigned input of g;
22:    set j = 1 and insert j = 1 to J-frontier;
23:    result = D-Alg-Recursion(*C*);
24:    **if** result == success **then**
25:       return (success);
26:    **else** try the other assignment
27:       set j = 0;
28:    **end if**
29: **end while**
30: return(failure);

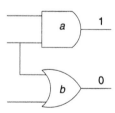

■ **FIGURE 4.15**

Conflict in the justification process.

---

well as other ATPG algorithms). The implications of these procedures are discussed later in this chapter.

Consider the multiplexer circuit shown in Figure 4.10. If the target fault is *f* stuck-at-0, then, after initializing all gate values to *x*, the *D* algorithm places a *D* on line f. The algorithm then tries to propagate the fault-effect to z. First it will place $a = 1$ in the J-frontier, followed by $h = 0$ in the J-frontier. At this time, the fault-effect has reached the primary output. Now, the ATPG tries to justify all unjustified values in the J-frontier. Because a is a primary input, it is already justified. The other signals in the J-frontier are $f = D$ and $h = 0$. For $f = D, d = 0$, thereby making $c = 0$. For $h = 0$, either $e = 0$ or $b = 0$ is sufficient. Whichever one it picks, the search process will terminate, as a solution has been found.

Consider the same multiplexer circuit (Figure 4.10) again. Suppose the target fault now is f stuck-at-1. Following the similar discussion as the previous target fault $f/0$, the algorithm initializes the circuit and places a $\overline{D}$ on f. Next, to propagate the fault-effect to a primary output, it likewise inserts $a = 1$ and $h = 0$ into the J-frontier. Now, the ATPG needs to justify all the gates in the J-frontier, which includes $a = 1$, $f = \overline{D}$, and $h = 0$. Because a is a primary output, it is already justified. For $f = \overline{D}$, $d = 1$. For $h = 0$, suppose it selects $e = 0$. At this time, the J-frontier consists of two gate values: $d = 1$ and $e = 0$. No value assignment on c can satisfy both $d = 1$ and $e = 0$; therefore, a conflict has occurred, and backtrack on the previous decision is needed. The only decision that has been made is $e = 0$ for $h = 0$, as there were two choices possible for justifying $h = 0$. At this time, the value on e is reversed, and $b = 0$ is added to the J-frontier. The process continues and all gate values in the J-frontier can be successfully justified, ending the process with the vector $abc = 101$.

Note that, in the above example, if some learning procedure (such as implications) is present, the decision for $h = 0$ would not result in $e = 0$, because the ATPG would have detected that $e = 0$ would conflict with $d = 1$. This knowledge could potentially improve the performance of the ATPG, which will be discussed later in this chapter.

Consider another example circuit shown in Figure 4.16. Suppose the target fault is $g/1$. After circuit initialization, the D algorithm places a $\overline{D}$ on g. Now, the J-frontier consists of $g = \overline{D}$ and the D-frontier consists of h. In order to advance the D-frontier, f is set to logic 1; $f = 1$ is added to the J-frontier, and the D-frontier is now i. Next, to propagate the fault-effect to the output, $c = 1$ is added to the J-frontier. At this time, the fault-effect has been propagated to the output, and the task is to justify the signal values in the J-frontier: $\{g = \overline{D}, f = 1, c = 1\}$. To justify $g = \overline{D}$, two choices are possible: $a = 0$ or $b = 0$. If $a = 0$ is selected, it is necessary to justify $f = 1, b = 1$. Finally, $c = 1$ remains in the J-frontier which is still unjustified. At this time, a contradiction has occurred ($a = 0$ and $c = 1$), and the search reverses its last decision, changing $a = 0$ to $a = 1$. The search discovers that this reversal also causes a conflict. Thus, a backtrack occurs where line b is chosen instead

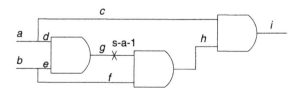

■ **FIGURE 4.16**

Example circuit.

of a for the previous decision, so a is reset to "don't care." By assigning $b = 0$, a conflict is observed. Reversing b also cannot justify all the J-frontier. At this time, backtracking on b leads to no prior decisions. Thus, target fault $g/1$ is declared to be untestable.

4.4.4 PODEM

In the D algorithm, the decision space encompasses the entire circuit. In other words, every internal gate could be a decision point. However, noting that the end result of any ATPG algorithm is to derive a solution vector at the primary inputs and that the number of primary inputs generally is much fewer than the total number of gates, it may be possible to arrive at a very different ATPG algorithm that makes decisions only at primary inputs rather than at internal nodes of the circuit.

PODEM [Goel 1981] is based on this notion and makes decisions only at the primary inputs. Similar to the D algorithm, a D-frontier is kept. However, because decisions are made at the primary inputs, the J-frontier is unnecessary. At each step of the ATPG search process, it checks if the target fault is excited. If the fault is excited, it then checks if there exists an X-path from at least one fault-effect in the D-frontier to a primary output, where an X-path is a path of "don't care" values from the fault-effect to a primary output. If no X-path exists, it means that all the fault-effects in the D-frontier are blocked, as illustrated in Figure 4.17, where both possible propagation paths of the D have been blocked. Otherwise, PODEM will pick the best X-path to propagate the fault-effect. Note that if the target fault has not been excited, the first steps of PODEM will be to excite the fault.

The basic flow of PODEM is illustrated in Algorithms 7 and 8. It is also based on a branch-and-bound search, but the decisions are limited to the primary inputs. All internal signals obtain their logic values via logic simulation (or implications) from the decision points. As a result, no conflict will ever occur at the internal signals of the circuit. The only possible conflicts in PODEM are either (1) the target fault is not excited, or (2) the D-frontier becomes empty. In either of these cases, the search must backtrack.

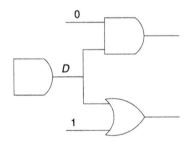

■ **FIGURE 4.17**

No X path.

Algorithm 7 PODEM(C, f)

1: initialize all gates to don't-cares;
2: D-frontier $= \emptyset$;
3: result = PODEM-Recursion(C);
4: **if** result == success **then**
5:    print out values at the primary inputs;
6: **else**
7:    print fault f is untestable;
8: **end if**

Algorithm 8 PODEM-Recursion(C)

1: **if** fault-effect is observed at a PO **then**
2:    return (success);
3: **end if**
4: (g, v) = getObjective(C);
5: (pi, u) = backtrace(g, v);
6: logicSimulate_and_imply(pi, u);
7: result = PODEM-Recursion(C);
8: **if** result == success **then**
9:    return(success);
10: **end if**
11: /* backtrack */
12: logicSimulate_and_imply($pi, \bar{u}$);
13: result = PODEM-Recursion(C);
14: **if** result == success **then**
15:    return(success);
16: **end if**
17: /* bad decision made at an earlier step, reset pi */
18: logicSimulate_and_imply(pi, x);
19: return(failure);

According to the algorithm in **PODEM**, the search starts by picking an objective, and it backtraces from the objective to a primary input via the best path. Controllability measures can be used here to determine which path is regarded as the best. Gradually more primary inputs will be assigned logic values. At any time the target fault becomes unexcited or the D-frontier becomes empty, a bad decision must have been made, and reversal of some previously decisions is needed. The backtracking mechanism proceeds by reversing the most recent decision. If reversing the most recent decision also causes a conflict, the recursive algorithm will continue to backtrack to earlier decisions, until no more reversals are possible, at which time the fault is determined to be undetectable.

Three important functions in PODEM-Recursion() are getObjective(), backtrace(), and logicSimulate_and_imply(). The getObjective() function returns the next objective the ATPG should try to justify. Before the target fault has been excited, the objective is simply to set the line on which the target fault resides to the value opposite to the stuck value. Once the fault is excited, the getObjective() function selects the best fault-effect from the D-frontier to propagate. The pseudo-code for getObjective() is shown in Algorithm 9.

Algorithm 9 getObjective(C)

1: **if** fault is not excited **then**
2:    return $(g, \bar{v})$;
3: **end if**
4: $d =$ a gate in D-frontier;
5: $g =$ an input of d whose value is x;
6: $v =$ non-controlling value of d;
7: return (g, v);

The backtrace() function returns a primary input assignment from which there is a path of unjustified gates to the current objective. Thus, backtrace() will never traverse through a path consisting of one or more justified gates. From the objective's point of view, the getObjective() function returns an objective, say $g = v$, which means the current value of g is "don't care." If g were set to $\bar{v}$, $g = v$ would have never been selected as an objective, as it conflicts with gate g's current value. Now, if $g = x$ currently, and the objective is to set $g = v$, there must exist a path of unjustified gates from at least one primary input to g. This backtrace() function can simply be implemented as a loop from the objective to some primary inputs through a path of "don't cares." Algorithm 10 shows the pseudo-code for the backtrace() routine.

Finally, the logicSimulate_and_imply() function can simply be a regular logic simulation routine. The added imply is used to derive additional implications, if any, that can enhance the getObjective() routine later on.

Consider the multiplexer circuit shown in Figure 4.10 again. Consider the target fault f stuck-at-0. First, PODEM initializes all gate values to x. Then, the first

---

Algorithm 10 backtrace(C)

1: $i = g$;
2: num_inversion $= 0$;
3: **while** $i \neq$ primary input **do**
4:    $i =$ an input of i whose value is x;
5:    **if** i is an inverted gate type **then**
6:       num_inversion++;
7:    **end if**
8: **end while**
9: **if** num_inversion $==$ odd **then**
10:    $v = \bar{v}$;
11: **end if**
12: return(i, v);

TABLE 4.4 ■ PODEM Objectives and Decisions for f Stuck-At-0

getObjective()	backtrace()	logicSim()	D-frontier
$f = 1$	$c = 0$	$d = 0, f = D,$ $e = 0, h = 0$	g
$a = 1$	$a = 1$	$g = D, z = D$	$f/0$ detected

objective would be to set $f = 1$. The backtrace routine selects $c = 0$ as the decision. After logic simulation, the fault is excited, together with $e = h = 0$. The D-frontier at this time is g. The next objective is to advance the D-frontier, thus getObjective() returns $a = 1$. Because a is already a primary input, backtrace() will simply return $a = 1$. After simulating $a = 1$, the fault-effect is successfully propagated to the primary output z, and PODEM is finished with this target fault with the computed vector $abc = 1X0$. Table 4.4 shows the series of objectives and backtraces for this example.

Consider the circuit shown in Figure 4.11. Suppose the target fault is b stuck-at-0. After circuit initialization, the first objective is $b = 1$ to excite the fault. The backtrace() returns $a = 0$. After logic simulation, although the target fault is excited, there is no D-frontier, because $c = d = 0$. At this time, PODEM reverses its last decision $a = 0$ to $a = 1$. After logic simulating $a = 1$, the target fault is not excited and the D-frontier is still empty. PODEM backtracks but there is no prior decision point. Thus, it concludes that fault $b/0$ is undetectable. Table 4.5 shows the steps made for this example, and Figure 4.18 shows the corresponding decision tree.

Consider again the circuit shown in Figure 4.16 with the target fault $g/1$. After circuit initialization, the first objective is to excite the fault; in other words, the objective is $g = 0$. The backtrace() function backtraces from the objective backward to a primary input via a path of "don't cares." Suppose the backtrace reaches $a = 0$. After logic simulation, $g = 0$, $c = d = 0$, and $i = 0$. The D-frontier is h. However, note that there is no path of "don't cares" from any fault-effect in the D-frontier to a primary output! If the PODEM algorithm is modified to check that any objective has at least a path of "don't cares" to one or more primary outputs, some needless

TABLE 4.5 ■ PODEM Objectives and Decisions for b Stuck-At-0

getObjective()	backtrace()	logicSim()	D-frontier
$b = 1$	$a = 0$	$b = 1, c = 0, d = 0$	$\emptyset$
$a = 1$ (reversal)	—	$b = 0, c = 1, d = 0$	$\emptyset$

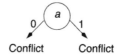

■ **FIGURE 4.18**

Decision tree for fault $b/0$.

TABLE 4.6 ■ PODEM Objectives and Decisions for g Stuck-At-1

getObjective()	backtrace()	logicSim()	D-frontier
$g = 0$	$a = 0$	$g = D, c = 0$ $d = 0, i = 0$	h (but no X-path to PO)
$a = 1$ (reversal)	—	$c = 1, d = 1$	∅

searches can be avoided. For instance, in this example, if the next objective was $f = 1$, even after the decision of $b = 1$ is made, the target fault still would not have been detected, as there was no path to propagate the fault-effect to a primary output even before the decision $b = 1$ was made. In other words, the search could immediately backtrack on the first decision $a = 0$. In this case, $a = 1$, and the objective is still $g = 0$. Backtrace() will now return $b = 0$. After logic simulation, $g = 0, c = 1, f = 0, h = 0, i = 0$. Again, there is no propagation path possible. As there is no earlier decision to backtrack to, the ATPG concludes that fault $g/1$ is untestable. Table 4.6 shows the steps for this example.

4.4.5 FAN

While PODEM reduces the number of decision points from the number of gates in the circuit to the number of primary inputs, it still can make an excessive number of decisions. Furthermore, because PODEM targets one objective at a time, the decision process may sometimes be too localized and miss the global picture. The FAN (Fanout-Oriented TG) algorithm [Fujiwara 1983] extends the PODEM-based algorithm to remedy these shortcomings.

To reduce the number of decision points, FAN first identifies the **headlines** in the circuit, which are the output signals of fanout-free regions. Due to the fanout-free nature of each cone, all signals outside the cone that do not conflict with the headline assignment would never require a conflicting value assignment on the primary inputs of the corresponding fanin cone. In other words, any value assignment on the headline can always be justified by its fanin cone. This allows the backtrace() function to backtrace to either headlines or primary inputs. Because each headline has a corresponding fanin cone with several primary inputs, this allows the number of decision points to be reduced.

Consider the circuit shown in Figure 4.19. If the current objective is to set $z = 1$, the corresponding decision tree based on the PODEM algorithm will involve many decisions at the primary inputs, such as $a = 1, c = 1, d = 1, e = 1, f = 1$. On the other hand, the decision based on the FAN algorithm is significantly smaller, involving only two decisions: $x = 1$ and $y = 1$. If $z = 1$ was not the first objective, there would have been other decisions made earlier. In other words, if there was a poor decision made in an earlier step, PODEM would need to reverse and backtrack many more decisions compared to FAN.

The next improvement that FAN makes over PODEM is the simultaneous satisfaction of multiple objectives, as opposed to only one target objective at each step. Consider the circuit fragment shown in Figure 4.20. Without taking into account multiple objectives, the backtrace() routine may choose the easier path in trying

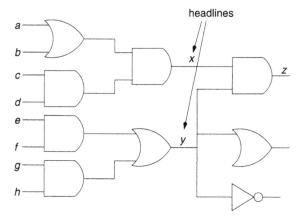

■ **FIGURE 4.19**

Circuit with identified headlines.

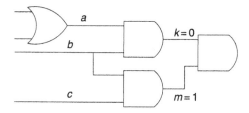

■ **FIGURE 4.20**

Multiple backtrace to avoid potential conflicts.

to justify $k = 0$. The easier path may be through the fanout stem b. However, this would cause a conflict later on with the other objective $m = 1$. In FAN, multiple objectives are taken into account, and the backtrace routine scores the nodes visited from each objective in the current set of objectives. The nodes along the path with the best scores are chosen. In this example, $a = 0$ will be chosen rather than $b = 0$, even if $a = 0$ is less controllable.

4.4.6 Static Logic Implications

Logic implications capture the effect of assigning logic values on other gate values in a circuit. They can be extremely helpful for the ATPG to make better decisions, reduce the number of backtracks, etc. Over the past few decades, logic implications have been applied and shown their effectiveness in several areas relevant to testing. They include test-pattern-generation [Schulz 1988] [El-Maleh 1998] [Tafertshofer 2000], logic and fault simulation [Kajihara 2004], fault diagnosis [Amyeen 1999], logic verification [Paul 2000] [Marques-Silva 1999a] [Arora 2004], logic optimization [Ichihara 1997] [Kunz 1997], and untestable fault identification [Iyer 1996a] [Iyer 1996b] [Peng 2000] [Hsiao 2002] [Syal 2004].

A powerful implication engine can have a significant impact on the performance of ATPG algorithms. Thus, much effort has been invested over the years in the efficient computation of implications. The quality of implications was improved with the computation of indirect implications in **SOCRATES** [Schulz 1988]. Static learning was extended to dynamic learning in [Schulz 1989] and [Kunz 1993], where some nodes in the circuit already had value assignments during the learning process. A 16-valued logic was introduced by Cox *et al.* [Rajski 1990], and reduction lists were used to dynamically determine the gate values. Chakradhar *et al.* proposed a transitive closure procedure based on the implication graph. Recursive learning was later proposed by Kunz *et al.* [Kunz 1994] in which a complete set of pair-wise implications could be computed. In order to keep the computational costs low, a small recursion depth can be enforced in the recursive learning procedure. Finally, implications to capture time frame information in sequential circuits in a graphical representation were proposed in [Zhao 2001] to compactly store the implications in sequential circuits.

All of the aforementioned techniques require the proper understanding of logic implications. As indicated earlier, logic implications identify the effect of asserting logic values on gates in a circuit. Static logic implications, in particular, can be computed as a one-time process before ATPG begins. At the end of the process, relationships among a subset of signals in the circuit would have been learned. Static logic implications have been categorized into direct, indirect, and extended backward implications. Direct implications for a gate g simply denote logic relationships immediately on a circuit gate. On the other hand, indirect and extended backward implications require circuit simulation and the application of transition and contrapositive properties. Because they are more involved, they help to identify the logical effect of asserting a value on g with nodes in the circuit that may not be directly connected to g. The following terminology is used for the discussion on logic implications:

1. $[N, v, t]$: Assign logic value v to gate N in time frame t. In combinational circuits, t is equal to 0 and can thus be dropped from the expression; that is, if $t = 0$, $[N, v, t]$ is rewritten as $[N, v]$.

2. $[N, v, t_1] \rightarrow [M, w, t_2]$: Assigning logic value v to gate N in time frame t_1 would imply a logic value w to gate M in time frame t_2.

3. $Impl[N, v, t]$: The set of all implications resulting from the value assignment of logic value v to gate N in time frame t. For $t = 0$, $Impl[N, v, t]$ is simply represented as $Impl[N, v]$.

Consider an AND gate and its implication graph, shown in Figure 4.21. Because the simple AND gate has three corresponding signals, a, b, and c, the associated implication graph has six nodes. An edge in the implication graph indicates the implication relationship. For example, $c = 1$ has two implications: $b = 1$ and $a = 1$.

The following example will explain further the concepts of direct, indirect, and extended backward implications. Note that the static logic implications are

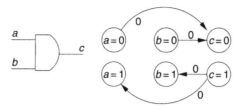

■ FIGURE 4.21

Example of an implication graph.

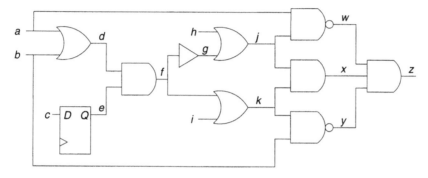

■ FIGURE 4.22

Sequential circuit fragment.

applicable to both combinational and sequential circuits. Given the sequential circuit fragment shown in Figure 4.22, consider gate $f = 1$:

1. *Direct implications*: A logic value of 1 on gate f would directly imply $g = k = 1$ because they are directly connected to gate f. In addition, $f = 1 \rightarrow d = 1$ and $e = 1$. Thus, the set $\{(f, 1, 0), (g, 1, 0), (k, 1, 0), (d, 1, 0), (e, 1, 0)\}$ is the set of direct implications for $f = 1$. Similarly, direct implications associated with $g = 1$ can be computed to be $\{(g, 1, 0), (j, 1, 0), (f, 1, 0)\}$. These implications are stored in the form of a graph, where each node represents a gate (with a logic value). A directed edge between two nodes represents an implication, and a weight along an edge represents the relative time frame associated with the implication. Figure 4.23 shows the graphical representation of a portion of direct implications for $f = 1$ in this example. The complete set of implications resulting from setting $f = 1$ can be obtained by traversing the graph rooted at node $f = 1$. Computing the set of all nodes reachable from this root node $(f = 1)$ (transitive closure on $f = 1$) would return the set $Impl[f = 1]$. Thus, the complete set of direct implications using the implication graph shown in the figure for $f = 1$ is: $\{(f, 1, 0), (d, 1, 0), (e, 1, 0), (g, 1, 0), (k, 1, 0), (j, 1, 0), (c, 1, -1)\}$

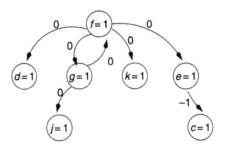

■ FIGURE 4.23

Portion of implication graph for $f = 1$.

2. *Indirect implications*: Note that neither $j = 1$ nor $k = 1$ implies a logic value on gate x individually. However, if they are taken collectively, they imply $x = 1$. Thus, indirectly, $f = 1$ would imply $x = 1$. This is an indirect implication of $f = 1$, and it can be computed by performing a logic simulation on the current set of implications of the root node on the circuit. In this example, by inserting the implications of $f = 1$ into the circuit, followed by a run of logic simulation, $x = 1$ would be obtained as a result. This new implication is then added as an additional outgoing dashed edge from $f = 1$ in the implication graph as shown in Figure 4.24. Another nontrivial implication that can be inferred from each indirect implication is based on the contrapositive law. According to the contrapositive law, if $[N, v] \rightarrow [M, w, t_1]$, then $[M, \overline{w}] \rightarrow [N, \overline{v}, -t_1]$. Because $[f, 1] \rightarrow [x, 1, 0]$, by the contrapositive law, $[x, 0] \rightarrow [f, 0, 0]$.

3. *Extended backward* (EB) *implications*: Extended backward implications aim to increase the number of implications for any single node by exploring the unjustified implied nodes in the implication list. Using the same circuit shown in Figure 4.22 again, in the implication list of $f = 1$, $d = 1$ is an unjustified gate because none of d's inputs has been implied to a value of logic 1. Thus, d is a candidate for the application of extended backward implications. To obtain extended backward implications on d, a transitive closure is first performed

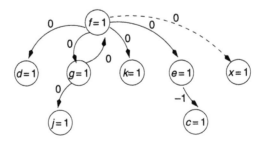

■ FIGURE 4.24

Adding indirect implications for $f = 1$.

for each of its unspecified inputs. In this case, $impl[a = 1]$ and $impl[b = 1]$ are first computed. The implications of $f = 1$ are logic simulated *together* with each of d's unspecified input's implication sets in turn, creating a set of newly found logic assignments for each input of the chosen unjustified gate. For this example, when the implications of $(a = 1)$ and $(f = 1)$ are simulated, the new assignments (set_a) found include $(w, 0, 0)$ and $(z, 0, 0)$. Similarly, for the combined implication set of $(b = 1)$ and $(f = 1)$, the new assignments (set_b) found include $(y, 0, 0)$ and $(z, 0, 0)$. All logic assignments that are not already in $Impl[f = 1]$ which are common to set_a and set_b are the extended backward implications. These new implications are added as new edges to the original node $f = 1$. In this running example, because $(z, 0, 0)$ is common in set_a and set_b, it is a new implication. The corresponding new implication graph is illustrated in Figure 4.25, where the new implication is shown as a dotted edge.

4.4.7 Dynamic Logic Implications

While static implications are computed one time for the entire circuit, dynamic implications are performed during the ATPG process. At a given step in the ATPG process, some signals in the circuit would have taken on values, including D or $\bar{D}$. This set of values may imply other signals which are currently unassigned to necessary value assignments. In general, dynamic implications work locally around assigned signals to see if any implication can be derived. For instance, consider the simple AND gate $c = a \cdot b$. According to static logic implications, $c = 0$ does not imply any value on either a or b. However, if $a = 1$ has been assigned by the current decision process, then $c = 0$ would imply $b = 0$. This can be deduced readily. The implicant, $b = 0$, may be propagated further to imply other signals.

The concept of direct, indirect, and extended backward implications can be applied in dynamic implications as well. Consider the circuit shown in Figure 4.26. Suppose $c = 1$ has been achieved by the decision process. Then, in order to achieve $z = 0$, either d must be 0 or e must be 0. For $d = 0$, both a and b must be 0. On the other hand, for $e = 0$, since $c = 1$, the only way for $e = 0$ is that b be assigned to 0.

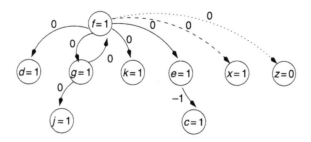

■ **FIGURE 4.25**

Adding extended backward implications for $f = 1$.

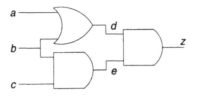

■ **FIGURE 4.26**

Dynamic implications.

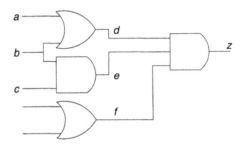

■ **FIGURE 4.27**

Another dynamic implications example.

The intersection of $\{a = 0, b = 0\}$ and $\{b = 0\}$ is $\{b = 0\}$. In other words, the dynamic implication for $z = 0$ given $c = 1$ is $b = 0$.

Dynamic implications can also be applied to signals with a fault-effect. For instance, consider the circuit shown in Figure 4.27. Suppose there is a D on signal b, and this fault-effect is the only one for the current target fault. Then, in order to propagate the fault-effect to the primary output z, $f = 1$ is a necessary condition. This dynamic implication can be obtained via the following analysis. For $b = D$ to propagate to z, either $a = 0$ or $c = 1$ is needed, resulting in a fault effect at signal d or e. Regardless of which path the fault effect propagates, signal $f = 1$ is a necessary condition for the fault effect to propagate to z. Such an observation was made in [Akers 1976] [Fujiwara 1983].

The work in [Hamzaoglu 1999] extended this concept of dynamic implications a step further. Suppose the D-frontier for the current target fault consists of gates $g_1, g_2, \ldots, g_n$. By a similar analysis as the previous example shown in Figure 4.27, each gate $g_i \in D$-frontier would have a set of necessary assignments, A_i. Clearly, the necessary assignment for any single fault-effect may not be necessary for detecting the target fault. However, in order to propagate the fault effect to a primary output, at least one fault effect in the D-frontier must be sensitized to the output. Subsequently, the intersection of all the necessary assignments for each of the gates in the D-frontier would be the set of required assignments for detection of the target fault.

In other words, $\cap_{\forall\ g_i \in D\text{-}frontier}\ A_i$ is the set of necessary assignments for detecting the target fault.

Finally, another form of dynamic learning consists of finding a partial circuit decomposition in the form of a frontier called the *evaluation frontier* (or E-frontier for short) [Giraldi 1990]. The idea behind this is that at any point in the decision process there exists a frontier of evaluated gates, and that the same frontier may be achieved by a different set of decision variables. For instance, three value assignments are possible to achieve the output of an AND gate set to logic 0. Each frontier can be associated with an edge in the decision tree. Suppose a set of E-frontiers has been learned for fault f_i and the corresponding decision tree for f_i is available. Now, for a different fault f_j, if a similar E-frontier is obtained, where the E-frontier has at least one fault effect as illustrated in Figure 4.28, the subtree for f_j's decision tree could be directly copied from the subtree in f_i's decision tree, to which the E-frontier was mapped. Note that the set of current primary input assignments is sufficient to justify the E-frontier, and all nodes to the right of the E-frontier are all "don't cares." In this figure, the only primary inputs that could have been used to propagate the fault effect are a, b, and m. If there was an assignment on these three primary inputs that was able to propagate the D for fault f_i to a primary output, then the same assignment would be able to propagate the D for f_j as it had the same E-frontier. In other words, the decision variables in the subtree corresponding to this point in the decision process consisted of only these three variables outside the

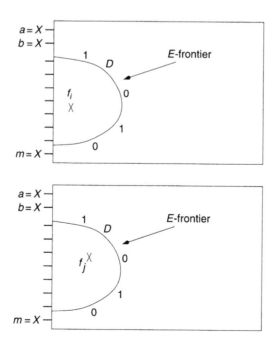

■ **FIGURE 4.28**

Example of evaluation frontier.

E-frontier. Stated differently, the propagation of the fault effect could directly be borrowed from a previous fault. The same concept can be extended to untestable faults as well.

4.5 DESIGNING A SEQUENTIAL ATPG

4.5.1 Time Frame Expansion

Test generation for sequential circuits bears much similarity with that for combinational circuits. However, one vector may be insufficient to detect the target fault, because the excitation and propagation conditions may necessitate some of the flip-flop values to be specified at certain values.

The general model for a sequential circuit is shown in Figure 4.29, where flip-flops constitute the memory/state elements of the design. All the flip-flops receive the same clock signal, so no multiple clocks are assumed in the circuit model. Figure 4.30 illustrates an example of a sequential circuit which is unrolled into several time frames, also called an **iterative logic array** of the circuit. For each time frame, the flip-flop inputs from the previous time frame are often referred to as **pseudo primary inputs** with respect to that time frame, and the output signals to feed the flip-flops to the next time frame are referred to as **pseudo primary outputs**. Note that in any unrolled circuit, a target fault is present in every time frame.

When the test generation begins, the first time frame is referred to as time frame 0. An ATPG search similar to a combinational circuit is carried out. At the end of

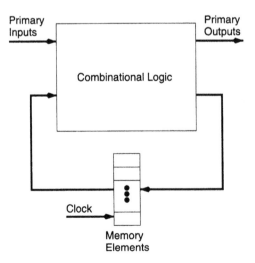

■ **FIGURE 4.29**

Model of a sequential circuit.

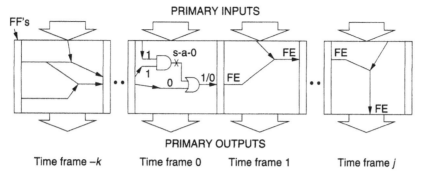

PRIMARY INPUTS

PRIMARY OUTPUTS

Time frame −k   Time frame 0   Time frame 1   Time frame j

■ **FIGURE 4.30**

An ILA model.

the search, a combinational vector is derived, where the input vector consists of primary inputs and pseudo primary inputs. The fault-effect for the target fault may be sensitized to either a primary output of the time frame or a pseudo primary output. If at least one pseudo primary input has been specified, then the search must attempt to justify the needed flip-flop values in time frame −1. Similarly, if fault-effects only propagate to pseudo primary outputs, the ATPG must try to propagate the fault-effects across time frame +1. Note that this results in a **test sequence** of vectors. As opposed to combinational circuits, where a single vector is sufficient to detect a detectable fault, in sequential circuits a test sequence is often needed.

One question naturally arises: Should the ATPG first attempt the fault excitation via several time frames −1, −2, etc., or should the ATPG attempt to propagate the fault-effect through time frames 1, 2, etc.? It can be observed that in propagating the fault-effect in time frame 1, the search may place additional values on the flip-flops between the boundary of time frames 0 and 1. These added constraints propagate backward and may add additional values needed at the pseudo primary inputs at time frame 0. In other words, if the ATPG first justifies the pseudo primary inputs at time frame 0, it would have missed the additional constraints placed by the propagation. Therefore, the ATPG first tries to propagate the fault-effect to a primary output via several time frames, with all the intermediate flip-flop values propagated back to time frame 0. Then, the ATPG proceeds to justify all the pseudo primary input values at time frame 0.

While easy to understand, the process can be very complex. For example, if the fault-effect has propagated forward for three time frames: time frames 1, 2, and 3. Now in time frame 4, suppose the ATPG successfully propagates the fault-effect to a primary output (*i.e.*, it has derived a vector at time frame 4), it must go back to time frame 3 to make sure the values assigned to the flip-flops at the boundary between time frames 3 and 4 are indeed possible. It must perform this check for time frames 2, 1, and 0. If at any time frame a conflict occurs, the vector derived at time frame 4 is actually invalid, as it is not justifiable from the previous vectors.

At this time, a backtrack occurs in time frame 4 and the ATPG must try to find a different solution vector #4. This process is repeated.

One way to reduce the complexity discussed above is to try to propagate the fault-effect in an unrolled circuit instead of propagating the fault-effect time frame by time frame. In doing so, a k-frame combinational circuit is obtained, say $k = 256$, and the ATPG views the entire 256-frame circuit as one large combinational circuit. However, the ATPG must keep in mind that the target fault is present in all 256 time frames. This eliminates the need to check for state boundary justifiability and allows the ATPG to propagate the fault-effect across multiple time frames at a time.

When the fault-effect has been propagated to at least one primary output, the pseudo primary inputs at time frame 0 must be justified. Again, the justification can be performed in a similar process of viewing an unrolled 256-frame circuit. As before, the ATPG must ensure that the fault is present in every time frame of the unrolled circuit.

4.5.2 5-Valued Algebra Is Insufficient

Because the fault is present in every time frame, it makes value justification tricky. For example, when justifying the pseudo primary input vector $01X1$, is it sufficient to obtain the fault-free values of $01X1$, or do the corresponding faulty values on these inputs need to be at the same logic values as fault-free values? If the faulty values can be different from the fault-free values, the 5-valued logic would be insufficient for this task [Muth 1976]. In other words, to justify a fault-free value of 1, the corresponding faulty value could be X and the justified state may still be sufficient to propagate the fault-effect to the primary output. Consider the circuit shown in Figure 4.31a. In the one-time-frame illustration of the sequential circuit, the target fault is $b/0$. Because the fault is present in every time frame of the unrolled ILA, the fault-free and faulty values arriving at the flip-flops in the previous time frame may be different. Taking this into consideration, it may be possible to obtain a value of 1/0 or 1/1 at signal a. However, when looking at this target fault, either 1/0 or 1/1 would be able to successfully propagate the fault effect at b to the output of the AND gate. Therefore, $a = 1/X$ is a sufficient condition and should be returned by the getObjective() function of the ATPG. If $a = 1/1$ were the objective returned by the getObjective() function, it may not be possible to derive this value from the flip-flops, thus over-constraining the search space. By a similar discussion, the $b/1$ fault shown in Figure 4.31b only requires $a = X/1$ in order to propagate the fault effect.

HITEC [Niermann 1991] is a popular sequential test generator that performs the search similar to the discussed methodologies with a 9-valued algebra. In addition, it uses the concept of **dominators** to help reduce the search complexity. A dominator for a target fault is a gate in the circuit through which the fault-effect must traverse [Kirkland 1987]. Therefore, for a given target fault, all inputs of any dominator gate that are not in the fanout cone of the fault must be assigned to noncontrolling values in order to detect the fault.

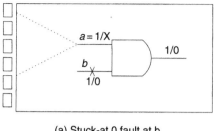

(a) Stuck-at 0 fault at b

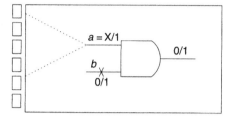

(b) Stuck-at 1 fault at *b*

■ **FIGURE 4.31**

The need for 9-valued algebra in sequential circuits.

The concept of controllability and observability metrics can be extended to sequential circuits such that the backtrace routine would prefer to backtrace toward primary inputs and those easy-to-justify flip-flops. Using sequential testability metrics allows the ATPG to narrow the search space by favoring the easy-to-reach states and avoiding getting into difficult-to-justify states.

The computational complexity of a sequential ATPG is intuitively higher than that of the combinational ATPG. Therefore, aggressive learning can help to reduce the computational cost. For instance, if a known subset of unreachable states is available, this information can be used to allow the ATPG to backtrack much sooner when an intermediate state is unreachable. This can avoid successive justification of an unreachable state. Likewise, if a justification sequence has been successfully computed for state S before, and a different target fault requires the same state S, the previous justification sequence can be used to guide the search. Note that, because the target faults are different, the justification sequence may not simply be copied from the solution for one fault to another.

4.5.3   Gated Clocks and Multiple Clocks

All the algorithms for sequential ATPG thus far assumed the sequential circuit has a single global clock. This assumption is simple as all memory elements (flip-flops) switch synchronously at every clock; however, in modern digital systems, this assumption is often not true. For instance, gated clocks (illustrated in Figure 4.32a) and multiple clocks (Figure 4.32b) are becoming mainstream. Gated clocks are

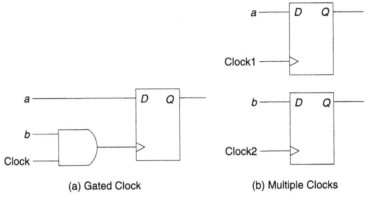

(a) Gated Clock   (b) Multiple Clocks

■ **FIGURE 4.32**

Non-traditional clocking schemes.

mostly used for power savings, such that not all memory elements will switch at every clock. On the other hand, multiple clocks benefit performance, power, and design as blocks can be partitioned to different clock domains.

If circuit modification is not possible, ATPG should be designed to perform some circuit modeling as a preprocessing step to ease the ATPG process. Actually, this is the approach taken by most current EDA vendors today. In other words, instead of designing new ATPG algorithms that can handle designs with gated clocks and multiple clocks, it may be easier to slightly modify the circuit such that the original circuit is transformed to one that uses only a single, global clock such that the transformed circuit is functionally equivalent to the original design. For instance, consider the gated clock case. The memory element that depends on a gated clock can easily be modified to one that depends on a single global clock by adding a small multiplexer, as shown in Figure 4.33. In the top half of the figure, the gated clock with signal *b* is easily transformed to the one shown on the right. The lower portion of the figure shows an example where the clock signal is an arbitrary internal signal; this also can be transformed in a similar manner. Note that the transformed designs shown on the right are functionally equivalent to the original designs.

Likewise, for a circuit with multiple clocks, a transformation is possible with similar design changes. Figure 4.34 illustrates the modification. The modified design is one where the clock is modified. This can further be converted by adding a multiplexer as done in the gated-clock scenario so the resulting design has a single global clock. In particular, the "new a" and "new b" signals can be converted to those having MUX-based inputs, as shown in Figure 4.33. The Clock1 and Clock2 signals may be used as the select signals for the multiplexers.

After a circuit with gated and/or multiple clocks has been modified, conventional stuck-at ATPG algorithms (combinational or sequential) will be readily applicable.

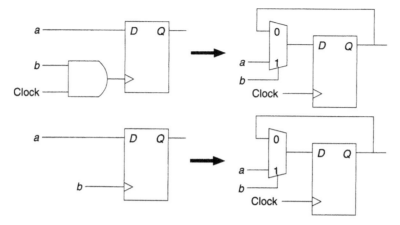

■ **FIGURE 4.33**

Transformation of gated clock.

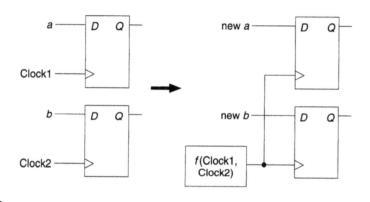

■ **FIGURE 4.34**

Transformation of multiple clocks.

Note, however, fault models other than the stuck-at model may not necessarily benefit from this transformation.

Finally, alternatives to the above MUX-based modifications are possible for handling designs with multiple clocks. They include the one-hot or the staggered clocking schemes. The details of the clocking are described in Section 5.7. One-hot clocking gives better fault coverage, but it suffers from potential large test sets. On the other hand, staggered clocking results in slightly lower fault coverages, but it can be applied using a combinational ATPG with circuit expansion. Sequential ATPG may be used as well, but it may incur longer execution times.

In addition to one-hot and staggered clocking, simultaneous clocking allows for all clocks to be run at the same time, but it marks unknowns (X's) between clock domains and uses one-time-frame combinational ATPG. EDA vendors tend to start with staggered or simultaneous clocking schemes, then a one-hot clocking scheme is used to detect any remaining faults [Wang 2003].

4.6 UNTESTABLE FAULT IDENTIFICATION

Untestable faults are faults for which there exists no test pattern that can both excite the fault and propagate its fault-effect to a primary output. Thus, a fault may be untestable for any of the following three reasons:

- The conditions necessary to excite the fault are not possible.

- The conditions necessary to propagate the fault-effect to a primary output are not possible.

- The conditions for fault excitation and fault propagation cannot be simultaneously satisfied.

In combinational circuits, untestable faults are due to redundancies in the circuit, while in sequential circuits untestable faults may also result from the presence of unreachable states or impossible state transitions.

From an ATPG's point of view, the presence of untestable faults in a design can degrade the performance of the ATPG tool. When considering untestable faults, an ATPG engine must exhaust the entire search space before declaring such faults as untestable. Thus, the performance of ATPG engines (as well as fault-simulators) can be enhanced if knowledge of untestable faults is available *a priori*. In other words, untestable faults can first be filtered from the fault list and the tools work only on the remaining faults, which could be much fewer than the original number of faults. There are additional benefits of untestable fault identification: Untestable faults in the form of redundancies increase the chip area; they may also increase the power consumption and the propagation delays through a circuit [Friedman 1967]. The presence of an untestable fault can potentially prevent the detection of other faults in the circuit [Friedman 1967]. Finally, untestable faults may result in unnecessary yield loss in scan-based testing. This is because even though the circuit remains fully operational in the presence of untestable faults, scan-based testing may detect such faults and reject the chip. As a result, significant effort has been invested in the efficient identification of untestable faults.

The techniques that have been proposed in the past for untestable fault identification can be classified into fault-oriented methods based on deterministic ATPG [Cheng 1993] [Agrawal 1995] [Reddy 1999], fault-independent methods [Iyer 1996a] [Iyer 1996b] [Hsiao 2002] [Syal 2003] [Syal 2004], and hybrid methods [Peng 2000]. The fault-independent methods generally are based on conflict analysis. While the deterministic ATPG-based methods outperform fault-independent methods for smaller circuits, the computational complexity of deterministic ATPGs makes them impractical for large circuits. On the other hand, conflict-based analysis targets the identification of untestable faults that require a conflicting scenario in the circuit. These methods do not target specific faults, thus they are fault-independent approaches. FIRE [Iyer 1996a] is a technique to identify untestable faults based on conflict analysis. While the theory can be applicable to any conflicting scenario, only single-line conflicts were implemented in FIRE. The basic idea behind FIRE is very simple. Because it is impossible for a single line to take on both logic values 0 and 1 simultaneously, logic values 0 and 1 set on any signal would clearly be a conflicting scenario. Subsequently, any fault that requires a signal set to both logic values 0 and 1 for its detection would be untestable. In order to reduce the computational cost, FIRE restricts its search to only fanout stems instead of every gate in the circuit.

In the single-line conflict analysis, for each gate in the circuit, the following two sets are computed:

- S_0—Set of faults not detectable when signal $g = 0$.

- S_1—Set of faults not detectable when signal $g = 1$.

Essentially, all the faults in each set S_i require $g = \bar{i}$ for their detection. Thus, any fault that is in the intersection of sets S_0 and S_1 would be untestable because it requires conflicting values on g as necessary conditions for its detection. The following example illustrates the single line conflict analysis.

Consider the circuit shown in Figure 4.35. During static learning, the implications for every gate can be computed as discussed earlier in this chapter. For example, $Impl[b, 1, 0] = \{(b, 1, 0), (b_1, 1, 0), (b_2, 1, 0), (d, 1, 0), (x, 0, 0), (z, 0, 0)\}$.

- **Faults unexcitable due to $b = 1$:**

  With $b = 1$, it would not be possible to set line $d = 0$, as $\{b = 1\} \rightarrow \{d = 1\}$. Thus, fault $d/1$ would be unexcitable when $b = 1$. In other words, this fault requires $b = 0$ as a necessary condition for its detection. Essentially, *if $[k, v, t] \in Impl[N, w]$, then fault k/v would be unexcitable in time frame t with $N = w$ in the reference time frame 0*. Similarly, faults $b/1, b_1/1, b_2/1, d/1, x/0, z/0$ would be unexcitable with $b = 1$.

- **Faults unobservable due to $b = 1$:**

  Because $\{b = 1\} \rightarrow \{x = 0\}$, line y is blocked. Hence, faults $y/0$ and $y/1$ would require $b = 0$ as a necessary condition for their detection. Similarly, any faults

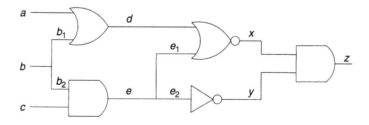

■ **FIGURE 4.35**

Example of single line conflict analysis.

appearing on lines a, e_1, e_2, etc., would also be blocked due to implications of $b = 1$. The unobservable information could be propagated backwards until a fanout stem is reached at which the faults on the fanout stem may no longer be unobservable. The condition for checking if the fanout stem is unobservable is to see if the stem can reach any of the blocking conditions for each of the fanout branches. For instance, using the circuit illustrated in Figure 4.35, if $a = 1$ and $c = 0$, both fanout branchs $b1$ and $b2$ would be unobservable. However, because the fanout stem b cannot reach any of the conditions for blocking any of the branches (*i.e.*, the blocking condition for $b1$ is $a = 1$ and the blocking condition for $b2$ is $c = 0$), stem b would still be unobservable. The complete set of faults that cannot be propagated due to $b = 1$ is:

$$\{a/0, a/1, e_1/0, e_1/1, y/0, y/1, e_2/0, e_2/1, e/0, e/1, c/0, c/1, b_2/0, b_2/1\}.$$

Thus, S_1 is the union of the two sets computed above:

$$S_1 = \{b/1, b_1/1, b_2/1, d/1, x/0, z/0, a/0, a/1, e_1/0, e_1/1, y/0, y/1, e_2/0, e_2/1,$$
$$e/0, e/1, c/0, c/1, b_2/0, b_2/1\}.$$

Now, consider the implications of $b = 0$:

$$Impl[b, 0, 0] = \{(b, 0, 0), (b_1, 0, 0), (b_2, 0, 0), (e, 0, 0), (e_1, 0, 0), (e_2, 0, 0), (y, 1, 0)\}.$$

Similar to the analysis performed for $b = 1$, faults that are unexcitable and unobservable due to $b = 0$ can be computed, resulting in:

$$S_0 = \{b/0, b_1/0, b_2/0, e/0, e_1/0, e_2/0, y/1, c/0, c/1\}$$

Now that both S_1 and S_0 are computed, any fault that is in the intersection of the two sets would be untestable. In this example, $S_0 \cap S_1 = \{b_2/0, e/0, e_1/0, e_2/0, y/1, c/0, c/1\}$. These faults are untestable because they require a conflicting assignment on line b ($b = 1$ and $b = 0$ simultaneously) as a necessary condition for their detection.

In a follow-up work to FIRE, FIRES [Iyer 1996b] targeted untestable faults in sequential circuits based on single-line conflicts. In addition, FILL and FUNI [Long 2000] adapted the concept of single-line conflicts to multiple nodes on the state variables (flip-flops) because any illegal state in sequential circuits is considered an impossible value assignment. As a result, any fault that requires an illegal state necessary for its detection would be untestable. A **binary decision diagram** (BDD)-based approach is used to identify illegal states, and FUNI [Long 2000] utilized this illegal state space information to identify untestable faults. MUST [Peng 2000] was built over the framework of FIRES as a hybrid approach (fault-oriented and fault-independent) to identify untestable faults; however, the memory requirement for MUST can be quadratic in the number of signals in the circuit. Next, Hsiao presented a fault-independent technique to identify untestable faults using multiple-node impossible value combinations [Hsiao 2002]. Finally, the concept of multiple-node conflicts is extended in [Syal 2004] to identify more untestable faults. The underlying concept of multiple line conflict is discussed next.

4.6.1   Multiple-Line Conflict Analysis

The application of logic implications to quickly identifying untestable faults is evident from the previous example. However, it is restricted to single-line conflicts. The application of logic implications to the identification of untestable faults can be taken to the next level, where impossible value combinations on multiple signals in the circuit are used as conflicting scenarios. These impossible value combinations are then used to identify untestable faults.

Finding trivial conflicting value assignments from the implication graph is easy, but it will not help to find more untestable faults because the single-line conflict approach has already taken these conflicts into account. For instance, if the implication set $impl[x,0]$ includes $[y,1]$, then the pair $\{[x, 0], [y, 0]\}$ naturally forms a conflicting value assignment. However, in the original FIRE algorithm, if Set_0 and Set_1 have been computed to be the faults that require $x = 0$ and $x = 1$, respectively, then Set_1 already contains all the faults that require $y = 0$ to be testable. This can be explained as follows: Because the set of faults that require $y = 0$ are obtained as those undetectable due to the value assignments in $impl[y,1]$, and because $y = 0 \rightarrow x = 1$, by the contrapositive law $x = 0 \rightarrow y = 1$ can be obtained. Thus, $impl[y, 1] \subseteq impl[x, 0]$. This leads to the following observation: The set of faults requiring $x = 1$, set_1 (*i.e.*, undetectable computed from $impl[x,0]$) must contain every fault that requires $y = 0$ as well.

Consequently, methods that can quickly identify non-trivial impossible combinations are needed in order to find more untestable faults. Finding arbitrary value conflicts in the circuit can be computationally expensive, thus any algorithm must limit the search for conflicting value assignments to computationally feasible approaches. In [Hsiao 2002], the impossible value assignments are limited to those associated with a single Boolean gate, making the algorithm of $O(n)$ complexity, where n is the number of gates in the circuit.

Consider the AND gate and its implication graph again, shown in Figure 4.36. When considering a single-line-conflict algorithm, there are three such cases for the AND gate: $\{a = 0, a = 1\}$, $\{b = 0, b = 1\}$, and $\{c = 0, c = 1\}$. (Recall that identification of undetectable faults when $a = 0$ requires $impl[a = 1]$, as described earlier.) By traversing the implication graph, the impossible value combination imposed by the conflicting line assignment $\{a = 0, a = 1\}$ includes the set $\{a = 0, a = 1, c = 0\}$. Similarly, one can obtain the sets of impossible value combination for conflicting

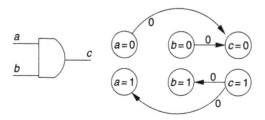

■ **FIGURE 4.36**

AND gate example.

line assignments $\{b = 0, b = 1\}$ and $\{c = 0, c = 1\}$ as $\{b = 0, b = 1, c = 0\}$ and $\{c = 0, c = 1, a = 1, b = 1\}$, respectively.

Note that there exist other sets of impossible value combinations not covered by any of these three single-line conflicts. Not all remaining conflicting combinations are nontrivial. For example, consider the conflicting scenario $\{a = 0, c = 1\}$. This is a trivial value conflict because $[a = 0] \rightarrow [c = 0]$ and $[c = 1] \rightarrow [a = 1]$. Therefore, $\{a = 0, c = 1\}$ is already covered by the single-line conflicts $\{a = 0, a = 1\}$ and $\{c = 0, c = 1\}$.

There exists a conflicting assignment that is not covered by any single-line conflicts: $\{a = 1, b = 1, c = 0\}$. In order to compute the corresponding impossible value assignment set, it is necessary to compute the following implications: $impl[a = 0]$, $impl[b = 0]$, and $impl[c = 1]$. By traversing the implication edges in the graph, the impossible value assignment set $\{a = 0, b = 0, c = 0, c = 1, a = 1, b = 1\}$ is obtained. This set has not been covered in any of the previous impossible value assignment sets, and hence the value set $\{a = 1, b = 1, c = 0\}$ may be used for obtaining additional untestable faults that require this conflict.

Impossible value combinations for other gate primitives and/or gates with different number of inputs can be derived in a similar manner.

The technique would then identify the value combination of $\{a = 1, b = 1, c = 0\}$ as impossible to achieve, and then untestable faults would be identified by creating the following sets:

- S_0—Set of faults not detectable when signal $a = 0$.

- S_1—Set of faults not detectable when signal $b = 0$.

- S_2—Set of faults not detectable when signal $c = 1$.

The faults in S_0, S_1, and S_2 require $a = 1$, $b = 1$, and $c = 0$, respectively, as necessary conditions for their detection. Then, the intersection of S_0, S_1, and S_2 would represent the set of untestable faults due to this conflicting value assignment.

Because the aim is to identify as many nontrivial conflicting value assignments as possible, which leads to untestable faults, the new approach of maximizing local impossibilities is performed on top of the single-line conflict FIRE algorithm, which is described below in Algorithm 11.

In this algorithm, the implication graph is first constructed, with indirect implications computed and added to the graph. Then, a single-line conflict FIRE algorithm is performed (line #3). Next, for each set of conflicting values not covered by the single-line conflict for each *gate*, the set of faults untestable due to such conflicts is computed. Because the algorithm on maximizing local value impossibilities is performed once for each gate, the complexity is kept linear in the size of the circuit. For large circuits, the number of additional untestable faults identified can be significant.

Maximizing local impossibilities can be extended further so the conflicting value assignments are no longer local to a Boolean gate. Consider the circuit shown in Figure 4.37. In [Hsiao 2002], the technique would identify the value combination of $\{g = 1, h = 1, z = 0\}$ as impossible to achieve, and then untestable faults would be identified correspondingly.

Algorithm 11 MaxLocalConflicts()

1: construct implication graph (learn any additional implications via extended backward impl, etc.);
2: **for** each line *l* in circuit **do**
3:   identify all untestable faults using the single-line-conflict FIRE algorithm;
4: **end for**
5: /* maximizing impossibilities algorithm */
6: **for** for each gate *g* in circuit **do**
7:   SIV = set of impossible value combinations not yet covered for gate *g*;
8:   $i = 0$;
9:   **for** each value assignment $(a = v)$ in SIV **do**
10:     set_i = faults requiring $a = \bar{v}$ to be detectable;
11:     $i = i + 1$;
12:   **end for**
13:   untestable_faults = untestable_faults $\cup (\cap_{v_i} set_i)$;
14: **end for**

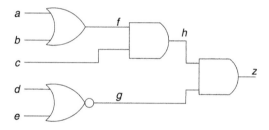

■ FIGURE 4.37

Circuit to illustrate multi-node impossible combination.

Now, it is interesting to note from Figure 4.37 that the value combination $\{(d = 0, e = 0), (f = 1, c = 1), z = 0\}$ also forms a conflicting value assignment. In addition, because $Impl[f, 0, 0] \supset Impl[h, 0, 0]$ and $Impl[c, 0, 0] \supset impl[h, 0, 0]$, the set of faults untestable due to $f = 0$ and $c = 0$ could potentially be greater than that due to $h = 0$. Similarly, the set of faults that can be identified as untestable due to $d = 1$ and $e = 1$ could be greater than that untestable due to $g = 0$. Consequently, the set of untestable faults identified using this new conflicting combination could be greater than that identified with the original conflicting value set $\{g = 1, h = 1, z = 0\}$.

This comes at a small price: The number of sets for which intersection must be performed for the conflict $\{(f = 1, c = 1), (d = 0, e = 0), z = 0\}$ would be greater than that for $\{g = 1, h = 1, z = 0\}$. However, because set intersection can be performed on the fly (with the faults computed for each implication set), the intersection operation can be aborted as soon as the intersection becomes empty. A larger conflicting value set might hurt the performance if each set intersection remains non-empty until the last intersection is performed and the intersection becomes empty only after the last intersection operation. However, this does not happen often, and the computational overhead due to a bigger set of impossible value combinations remains acceptable.

Before proceeding to the algorithm, the following terms are first defined:

Definition 1

Nonterminating necessary condition set (NTC): NTC for an assignment $x = v$ is defined as the set of value assignments $\{a_i = w_i | w_i \in \{0, 1\}\}$ that are necessary to achieve $x = v$. However, there may exist other assignments that are necessary to achieve some or all conditions in NTC.

For example, in Figure 4.37, $h = 1$ and $g = 1$ are necessary for $z = 1$. However, there exist assignments ($f = 1$, $c = 1$, $d = 1$, and $e = 1$) that are necessary to achieve $h = 1$ and $g = 1$. Thus, $h = 1$, $g = 1$ forms the NTC for $z = 1$.

Definition 2

Terminating necessary condition set (TNC): TNC for an assignment $x = v$ is the set of value assignments $\{a_i = w_i | w_i \in \{0, 1\}\}$ necessary to achieve $x = v$ such that there exist no additional assignments that are necessary to achieve any conditions in this set.

For example, in Figure 4.37, $f = 1$, $c = 1$, $d = 0$, and $e = 0$ form the TNC for $z = 1$.

According to Definitions 1 and 2, the set of conflicting conditions obtained in [Hsiao 2002] would correspond to the NTC set for a gate $x = v$. These conflicting conditions would form the TNC only if $|NTC| = |TNC|$ for any $x = v$.

In the new approach [Syal 2004], the TNC for any assignment $x = v$ is first identified (rather than the NTC). Then the set $\{TNC, x = \bar{v}\}$ forms a conflicting assignment. As the size of the TNC is greater than that of NTC, the new approach may take more execution time than that taken by the previous approach in [Hsiao 2002], but the following definition and corresponding lemma guarantee that the new approach always identifies at least as many (and potentially more) untestable faults as identified in [Hsiao 2002].

Definition 3

Related elements: Gates a and b are related elements if there exists at least one topological path from a to b.

Lemma 2

If two related elements a and b exist such that the assignment $a = v$ is a part of TNC for a gate $g = u$ and $b = w$ is a part of NTC for the same gate $g = u$, then $impl[a, \bar{v}] \supseteq impl[b, \bar{w}]$.

Algorithm 12 Multi-Line-Conflicts()

1: construct implication graph;
2: /* identification of impossible combinations */
3: **for** each gate assignment $g = val$ **do**
4:     identify the TNC for $g = val$;
5:     Impossible Combination (IC) set = TNC, $g = \overline{val}$;
6:     $i = 0$, $S_{untest} = \emptyset$;
7:     **for** each assignment $a = w$ in IC **do**
8:         S_i = fault untestable with $a = \overline{w}$;
9:         **if** $i == 0$ **then**
10:             $S_{untest} = S_{untest} \cup S_i$;
11:         **else**
12:             $S_{untest} = S_{untest} \cap S_i$;
13:         **end if**
14:         **if** $S_{untest} = \emptyset$ **then**
15:             break;
16:         **else**
17:             $i++$;
18:         **end if**
19:     **end for**
20: **end for**

Proof

Because $b = w$ is not a terminating necessary condition for $g = u$, there must exist some necessary conditions to achieve $b = w$. Now, because $a = v$ is a terminating condition for $g = u$ and because a and b are related, then $a = v$ must be a part of the conditions necessary to set $b = w$. This means that in order to set $b = w$, gate a must be set to v, or in other words, $[b, w, 0] \rightarrow [a, v, 0]$. By contrapositive law, $[a, \overline{v}, 0] \rightarrow [b, \overline{w}, 0]$. Thus, $impl[a, \overline{v}, 0] \supseteq impl[b, \overline{w}, 0]$.

Thus, according to Lemma 2, the implications of the complement of all elements in a TNC are a superset of the complemented related elements in a NTC for any given assignment. Therefore, the set of untestable faults obtained using TNCs is always a superset of those using NTCs as used in [Hsiao 2002]. The complete algorithm to identify untestable faults using a multiple-line conflict analysis is shown in Algorithm 12.

4.7 DESIGNING A SIMULATION-BASED ATPG

In this section, we will discuss how simulation, as opposed to deterministic algorithms, can be used for generating test vectors. This section begins with an overview of how simulation can be used to guide the test generation process and then discusses how tests can be generated in specific frameworks, such as genetic algorithms, state partitioning, spectrum, etc.

4.7.1   Overview

As we have already seen earlier in this chapter, the random test generator is a simple type of simulation-based ATPG. The vectors are randomly generated and simulated on the circuit under test, and any vector that is capable of detecting new faults is added to the test set. While this concept is relatively simple, its applicability is limited as random ATPG cannot generate vectors that target hard faults.

Simulation-based test generators were first proposed in 1962 by Seshu and Freeman [Seshu 1962]. Subsequently, several other simulation-based test generators have been developed, including [Breuer 1971], [Schnurmann 1975], [Lisanke 1987], [Wunderlich 1990], [Snethen 1977], and [Agrawal 1989]. Each of these test generators will be described in the following discussion.

Random vectors are simulated and selected using a fault simulator in [Breuer 1971]. Weighted random test generators were introduced in [Schnurmann 1975], [Lisanke 1987], and [Wunderlich 1990], in which each bit is generated with a biased coin (as opposed to an unbiased one in the simple random test pattern generator), and high fault coverages were reported for combinational circuits. Specific faults are targeted in the test generators proposed in [Snethen 1977] and [Agrawal 1989], and the ATPGs only considered vectors of Hamming distance equal to one between consecutive vectors. In other words, any two successive vectors can differ in only a single bit. Finally, cost functions computed during fault simulation were used to evaluate the generated vectors in [Agrawal 1989].

While these aforementioned simulation-based ATPGs were able to reduce the test generation time, the test sets generated were typically much longer than those generated by deterministic test generators. In addition, in sequential circuits, many difficult-to-test faults were frequently aborted. Finally, even when simulation-based test generators can be effective in detecting hard faults, simulation-based algorithms, per their nature, cannot detect untestable faults. In this regard, deterministic algorithms will be needed to uncover any faults that are untestable.

4.7.2   Genetic-Algorithm-Based ATPG

A simple *genetic algorithm* (GA) can be used for the generation of individual test vectors for combinational as well as sequential circuits. In a typical GA, a **population of individuals** (or chromosomes) is defined, where each individual is a candidate solution for the problem at hand. As the individual represents a test vector for combinational circuit test generation, each character in the individual is mapped to a primary input. If a binary coding is used, the individual simply represents a test vector. Each individual is associated with a **fitness**, which measures the quality of this individual for solving the problem. In the test generation context, this fitness measures how good the candidate individual is for detecting the faults. The fitness evaluation can simply be computed by logic or fault simulation. Based on the evaluated fitness, the evolutionary processes of **selection**, **crossover**, and **mutation** are used to generate a new population from the existing population. The process is repeated until the fitness of the best individual cannot be improved or is satisfactory.

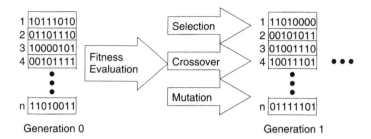

■ FIGURE 4.38

GA framework.

One simple application of GAs for test generation is to select the best test vectors for each GA run. A simple view of a GA framework is illustrated in Figure 4.38. The test generator starts with a random population of n individuals, and a (fault) simulator is used to calculate the fitness of each individual. The best test vector evolved in any generation is selected and added to the test set. Then, the fault set is updated by removing the detected faults by the added vector(s). The GA process repeats itself until no more faults can be detected.

Because a new random population is used initially, the GA process may not guarantee that a successful vector can be found. Likewise, in sequential circuits, a number of vectors may be necessary to drive the circuit to a state before the fault can be excited. Therefore, a *progress limit* should be used to limit the amount of execution allowed before the the entire process stops. When the population does not start with a right combination of individuals, the GA process may not result in an effective test vector. When this happens, the GA is reinitialized with a new random population, and a new GA attempt proceeds. This overall procedure is shown in Algorithm 13.

Note that in this procedure, the GA operators of selection, crossover, and mutation are applied in each iteration. Rather than exposing the reader to the numerous schemes for each GA operator, the following discussion will focus on the classic methods. First, for the selection operator, two popular schemes are often used: binary tournament selection and roulette wheel selection. In binary tournament selection, to select one individual from the population, two individuals are first randomly chosen from the population, and the one with the greater fitness value is selected as a parent individual. This is repeated to select a second parent. Note that, because a comparison is made in the process, selection is biased toward the more fit individuals. In the roulette wheel selection scheme, the n individuals in the population are mapped onto n slots on the wheel, where the size of each slot corresponds to the fitness of the individual, as illustrated in Figure 4.39. Thus, when the roulette wheel is spun, the position where the marker lands will determine the individual selected. Note that both roulette wheel and binary tournament selections may be conducted with or without replacement. When no replacement is used, the individuals selected are not put back into the population for the subsequent selection. In other words, an individual will not be selected more than once as a

---

Algorithm 13 Simple_GA_ATPG

```
 1: test set T = Ø;
 2: while there is improvement do
 3:     initialize a random GA currentPopulation;
 4:     compute fitness of currentPopulation;
 5:     for i = 1 to maxGenerations do
 6:         add the best individual to test set T;
 7:         nextPopulation = Ø;
 8:         for j = 1 to populationSize/2 do
 9:             select parent₁ and parent₂ from currentPopulation;
10:             crossover(parent₁, parent₂, child₁, child₂);
11:             mutate(child₁);
12:             mutate(child₂);
13:             place child₁ and child₂ to nextPopulation;
14:         end for
15:         compute fitness of nextPopulation;
16:         currentPopulation = nextPopulation;
17:     end for
18: end while
```

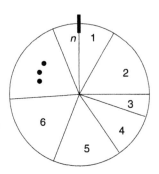

■ **FIGURE 4.39**

Roulette wheel selection.

candidate parent individual. Finally, when comparing the effectiveness of roulette wheel with binary tournament selections, the notion of **selection pressure** is necessary. Selection pressure is the driving force that determines the convergence rate of the GA population, in which the population converges to n identical (or very similar) individuals. Note that fast convergence may not necessarily lead to a better solution. Roulette wheel selection with replacement results in a higher selection pressure than binary tournament selection when there are some highly fit individuals in the population. On the other hand, when individuals' fitnesses have a small variance, binary selection will apply a higher selection pressure.

The next GA operator to be discussed is the crossover operator. Again, the discussion will focus on classic crossover techniques. In essence, once two parent individuals are selected, crossover is applied to the two parent individuals to produce two children individuals, where each child inherits parts of the chromosomes from each parent. The idea behind crossover is that the building blocks from two different solutions are combined in a random fashion to produce two new solutions. Intuitively, a more fit individual contains more valuable building blocks when compared with a less fit individual. Because the selection biases toward more fit individuals, the building blocks from the more fit parents are passed down to the next generation. When the valuable building blocks from different fit parents are combined, more fit individuals may result. In the following, one-point, two-point, and uniform crossover are explained.

Suppose the length of an individual is λ. In one-point crossover, the two parents are crossed at a randomly chosen point, r, between 1 and $\lambda - 1$. Consequently, the first child inherits the first r bits from parent #1 and the final $\lambda - r$ bits from parent #2, while the second child inherits the first r bits from parent #2 and the final $\lambda - r$ bits from parent #1. Table 4.7 illustrates an example of the one-point crossover scheme. The vertical line in the table indicates the crossover point.

Similar to the one-point crossover, two-point crossover works in a similar fashion except that two points are chosen instead of one. The portion of the parent individuals between the two points are swapped to produce the new individuals. Table 4.8 illustrates an example for the two-point crossover scheme. Again, the vertical lines indicate the crossover point.

Finally, in uniform crossover, a crossover mask is first generated randomly, and the bits between the two parent individuals are swapped whenever the corresponding bit position in the crossover mask is one. Table 4.9 illustrates an example for the uniform crossover scheme.

The reader is encouraged to try applying crossover on individuals over a few generations to see how new individuals may be produced, similar to the examples illustrated here.

TABLE 4.7 ■ One-Point Crossover

Parent #1	110011001100	110011001100
Parent #2	101010101010	101010101010
Child #1	110011001100	101010101010
Child #2	101010101010	110011001100

TABLE 4.8 ■ Two-Point Crossover

Parent #1	11001100	11001100	11001100
Parent #2	10101010	10101010	10101010
Child #1	11001100	10101010	11001100
Child #2	10101010	11001100	10101010

TABLE 4.9 ■ Uniform Crossover

Mask	01001110010001001110101
Parent #1	11001100110011001100
Parent #2	10101010101010101010
Child #1	10001010100010010101000
Child #2	11101100111011011001110

TABLE 4.10 ■ Mutating Bit Position #3

Before mutation	11001100110011001100
After mutation	11101100110011001100

The third GA operator to be discussed is the mutation operator. It allows the child individual to vary slightly from the two parents it had inherited. The mutation operator simply selects a random bit position in an individual and flips its logic value with a mutation probability. An example of mutating bit #3 is shown in Table 4.10. Let μ be the mutation probability. If μ is too small, children individuals that are produced after crossover may rarely see any mutation. In other words, it is less likely that new genes (building blocks) will be produced. On the other hand, if μ is too large, too much random perturbation may occur, and the resemblance may soon be lost after a few generations.

4.7.2.1 Issues Concerning the GA Population

The population size should be a function of the individual length. In sequential circuits, the individual length is equal to the number of primary inputs in the circuit multiplied by the test sequence length. The population size may be increased from time to time to increase the diversity of the individuals, thereby helping to expand the search space.

One pertinent issue in the GA population is the encoding of the individuals: whether a binary or nonbinary coding should be used. In a binary coding, the individual is simply the test vector itself (or a sequence of vectors in the case of sequential circuits). Thus, the GA operates directly on that string. For instance, bitwise crossover and bitwise mutation can be used. (Bitwise mutation is simply the flipping of a single bit in the vector.) On the other hand, in a nonbinary coding, several bits are combined and represented by a separate character in the alphabet, and the GA operates on the individual as a string of characters. Special operators are needed for the nonbinary alphabet. For example, crossover can now occur only at multi-bit boundaries, and mutation involves replacing a given character in an individual with a randomly generated character. Finally, in a nonbinary coding, a larger population size and mutation rate may be required to ensure adequate diversity.

Obtaining a compact test set is another concern; thus, an accurate fitness measure is needed. As fault simulation is used to compute the fitness, computation of the fitnesses in each GA generation can be costly. To reduce this cost, approximation can be used, in which a fault sample from the complete fault list may be used. In doing so, fault simulation only has to consider the faults in the sample rather than the entire fault list.

Another method to reduce the execution time is to use overlapping populations in the GA. In overlapping populations, some individuals from the parent generation are copied over to the offspring generation. Therefore, only a fraction of the population is replaced in each generation.

The success of using GAs to obtain the desired solution depends also on how the GA parameters are chosen. First and foremost, the population size of the GA should be such that adequate diversity is represented. In the context of test generation, certain values on specific primary inputs may be necessary to excite a fault. If no individual in the initial population has this specific combination, then none of the strings in the population would have been able to excite the fault. As the number of bit combinations increases exponentially with the vector length, the population size should be large enough to appropriately reflect the embedded diversity. However, the population should not be too large to the extent that the cost of evaluating the fitnesses of individual becomes infeasible. These two factors must be carefully considered when determining the GA population size.

4.7.2.2 Issues Concerning GA Parameters

The first GA parameter to be considered is the number of generations necessary to achieve a desirable solution. Similar to population size, the number of generations necessary to obtain an individual with a specific bit pattern requires the GA designer's attention. For instance, if the target fault demands a pattern of "1011" among four bits in the vector, and if this pattern is absent in the initial population, it may take several generations before an individual arrives at this pattern. The number of generations is also related to the population size. Larger populations will likely require more generations to allow for more diverse pairs of individuals to be as parent individuals. Thus, it may suffice to have a small population and a small number of generations to target the easy to detect faults and then increase both the population size and the number of generations when targeting the more difficult faults.

The next two parameters are the crossover and mutation probabilities. A crossover probability of 1.0 means that two parent individuals are always crossed so that two children individuals are produced from the parents. Mutation is used to introduce added diversity. A population after several generations will be more likely to have individuals that are more fit than those in the initial population. As the more fit individuals may have similarities, mutation can randomly flip certain bits among the individuals to decrease their similarity. However, mutation can also destroy those good patterns already achieved in some individuals. Thus, an appropriate mutation probability is needed to achieve an appropriate balance.

4.7.2.3 Issues Concerning the Fitness Function

How the fitness values are computed for the individuals in the population is a very important concern, as the search critically relies on the fitness values. An ill-defined fitness metric can mislead the GA to arrive at a suboptimal solution, or even no solution at all. For instance, a population whose individuals' fitnesses are similar will not allow the selection process to identify more highly fit individuals to act as parents. Furthermore, without a metric, the individuals may become indistinguishable even when they really are distinguishable. For example, if the fitness function is simply a binary function, where an individual's fitness is equal to one if the target fault is excited and zero otherwise, this will result in many individuals with fitness equal to zero if they do not excite the target fault. It is clear, however, that some individuals may be closer to exciting the target fault than others. However, the aforementioned binary fitness function would prevent the GA from distinguishing those more fit individuals from the less fit ones.

At the start of the ATPG process, there may exist many easy-to-detect faults; therefore, it may be advantageous to first detect them before targeting the harder faults. In this regard, dividing the ATPG process into different phases would be desirable. As an example, CONTEST [Agrawal 1989] targets test generation in three phases, each having its own distinct fitness measure.

A two-stage ATPG process is described here. In the first stage, the aim is to detect as many faults as possible. The fitness function could simply be the number of faults detected. This fitness metric allows the GA to bias toward those vectors that could potentially detect more faults. One can refine this fitness function to become:

$$Fitness = \alpha \times detected + \beta \times excited$$

In this case, individuals that detect the same number of faults may be distinguished.

Initially, when there are still many easy faults undetected, many individuals will have high fitness values. As vectors are added to the test set and detected faults removed from the fault list, the average fitness of individuals will be expected to come down. When this occurs, it will become increasingly difficult for the GA to distinguish good individuals from the less fit individuals, as discussed earlier. Consequently, the ATPG enters the second stage, where the goal is targeting individual faults instead.

In the second stage, each GA process targets a specific fault. Thus, the fitness function should also be adjusted similarly for this purpose. The fitness ought to measure how close the individual is to exciting the fault, as well as how close it is to propagating the fault-effect to a primary output. For measuring how close the individual is to exciting the fault, one can check the number of necessary value assignments. For instance, suppose the target fault is $h/0$ at the output of AND gate h, as illustrated in Figure 4.40. Then, an individual that sets both inputs of h to logic 0 (Case 2 in the figure) is further away from exciting the fault than another individual that sets one input to logic 1 (Case 1 in the figure). For measuring how close the individual propagates a fault effect to a primary output, the fitness can measure the number of D or $\overline{D}$ present in the circuit generated by the individual, together with the observability value associated with the lines to which the $D(\overline{D})$ has propagated.

For sequential circuits, it may be appropriate to have a stage zero where the goal is to initialize all the flip-flops. The fitness of an individual is then simply the

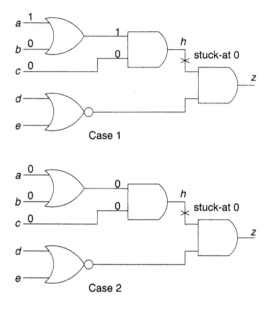

■ **FIGURE 4.40**

Fitness measure on how close a fault is excited.

number of additional flip-flops set to logic zero or logic one, as done in [Rudnick 1994a] [Rudnick 1994b]. Note that only logic simulation is needed in this stage. For subsequent stages, the fitness function may take into account the fault-effects that propagate to flip-flops as well, as it may take several time frames in order to propagate the fault-effect to a primary output.

As calculating the fitnesses of individuals dominates the computational cost of the GA, care must be taken when designing the fitness functions. Data structures that allow for fast access to the fault-free and faulty values in the circuit, for instance, would be desired. When fitness value calculation becomes prohibitive, one may reduce the cost by estimating the fitness instead of computing the exact fitness. In stage one, for example, fault samples may be used instead of simulating all faults. Also, counting the number of events in logic simulation may be used to estimate the number of faults excited; this may eliminate the high cost of fault simulation. When using such fitness estimates, one must be aware of the potential loss in the quality of the derived solution and that the final fault coverage may also be reduced.

4.7.2.4 CASE Studies

In the GA-based ATPG by Srinivas and Patnaik [Srinivas 1993], combinational circuits were targeted. Each individual represented a test vector. The fitness function accounted for excitation of the fault and propagation of the fault effect. Depending on the fitness of an individual, different crossover and mutation rates were used. While test sets were large, high coverages were obtained.

Genetic algorithms based on [Holland 1975] were used in CRIS [Saab 1992] in which two individuals were evolved in each generation, which replaced the least fit individuals with some probability. The fitness measure was based on the fault-free activities in the internal nodes in the circuit. This allowed the fitness evaluation to be simple, as only logic simulation is required, thus significantly reducing the computation costs. The circuit is divided into various partitions based on each primary output, and the fitness function favors those individuals that produce similar levels of activity in each partition. It has been presumed that vectors that induce high levels of activity are expected to result in higher fault coverage. As the fitness metric is an estimate of fault coverage, the resulting test sets are longer and may have lower fault coverage compared to deterministic test generators.

GATTO [Prinetto 1994] targeted sequential circuits and was based on GAs in the fault propagation phase during the test generation process. First, a reset state was assumed and random vectors were generated from the reset state until at least one fault had been excited. Then, for a limited group of excited faults (up to at most 64 faults), the GAs were used to propagate them toward the primary outputs or flip-flops. If any fault-effect reached a primary output, the corresponding test sequence was added to the test set. If the GAs were unsuccessful in propagating the fault-effects to a primary output, the GA stopped and started over from the reset state to obtain a different set of excited faults. GATTO was able to achieve higher fault coverages compared to CRIS for some circuits.

A GA-based combinational test generator was developed by Pomeranz and Reddy [Pomeranz 1997] in which problem-specific information was used. In this case, circuit information played a significant role. For instance, primary inputs lying in the same cone of logic were grouped together, and crossover was limited to primary input group boundaries. This enforced that fault excitation and propagation information from effective individuals are preserved during and after crossover operation. The grouping of the primary inputs was done carefully so that no group was either too large or too small. Note that, because a primary input can belong to multiple groups, care must be taken when copying bit values from a parent individual to a child individual. Uniform crossover was used, and only individuals that were shown to improve the fault coverage were added to the GA population. Further, the population size increased dynamically and the GA process terminated when all faults were detected or a given number of iterations had been reached.

A three-phased sequential test generator based on GAs was developed in GATEST [Rudnick 1994a] [Rudnick 1994b] that is based on the PROOFS sequential circuit fault simulator [Niermann 1992]. Table 4.11 shows the population sizes and mutation probabilities used in GATEST as a function of the vector length. Tournament selection without replacement and uniform crossover are used. In the initial phase of GATEST, the aim is to initialize all the flip-flops. Thus, the fitness metric measures the number of new flip-flops initialized to a known value from a previously unknown state. In this phase, only logic simulation is needed. When all flip-flops have been initialized, GATEST exits phase one and enters the second phase. In phase two, the goal is to detect as many faults as possible in any GA attempt. So the fitness is simply the number of faults detected by the candidate individual and the number of faults excited and propagated to flip-flops, with more emphasis placed

TABLE 4.11 ■ GA Parameter Values

Vector Length (L)	Population Size	Mutation Probability
<4	8	1/8
4–16	16	1/16
17–49	16	1/L
50–63	24	1/L
64–99	24	1/64
>99	32	1/64

on fault detection. Phase two continues until no more faults can be detected, at which point GATEST enters phase three. Similar to phase two, phase three aims to detect as many faults as possible, except that the fitness function now accounts for the fault-free and faulty circuit activities in addition to fault detection and propagation. Individuals that induce more activity would have higher fitness values. GATEST allows for phase three to exit and return to phase two when vectors are found to detect additional faults. Finally, in phase four, sequences of vectors are used as individuals, and the fitness function is similar to phase two, except that the test sequence length is also factored in. The fitness of a candidate test for each phase is calculated as follows:

- Phase 1—Fitness is a function of total new flip-flops initialized.

- Phase 2—Fitness is a function of the number of faults detected and the number of faults propagated to flip-flops.

- Phase 3—Fitness is a function of the number of faults detected, the number of faults propagated to flip-flops, and the number of fault-free and faulty circuit events.

- Phase 4—Fitness is a function of the number of faults detected and the number of faults propagated to flip-flops for a test sequence.

Because one fault is targeted at a time and the majority of time spent by the GA is in the fitness evaluation, parallelism among the individuals can be exploited. Parallel-fault simulation [Abramovici 1994] [Bushnell 2000] [Jha 2003] is used to speed up the process.

High fault coverages and compact test sets have been obtained by GATEST for combinational circuits. For some circuits, however, deterministic ATPGs could achieve higher coverage in much less time. For sequential circuits, the number of faults detected is either greater than or equal to that of deterministic test generators for most circuits, and the test set sizes are much shorter. In most cases, GATEST takes only a fraction of the execution time compared to deterministic test generators. Thus, GATEST can be used as a preprocessor in test generation to screen out many faults before applying a more expensive deterministic test generator.

4.8 ADVANCED SIMULATION-BASED ATPG

4.8.1 Seeding the GA with Helpful Sequences

Genetic algorithms have been shown to be effective for test generation in the above discussion. However, for some difficult faults, the previous GA-based methods may still underperform the deterministic ATPGs. For such faults, it may be helpful to embed certain individuals in the initial population to guide the GA. This is called **seeding**.

For example, suppose a fault has been excited and propagated to one or more flip-flops in a sequential circuit, and now the GA attempts to drive the fault-effect from those flip-flops to a primary output. If there are previously known sequences that were successful in propagating fault-effects from a similar set of flip-flops, then seeding these sequences into the initial population may tremendously help the GA.

The DIGATE [Hsiao 1996a] [Hsiao 1998] and the STRATEGATE test generators [Hsiao 1997] [Hsiao 2000] aggressively apply seeding of useful sequences for the GA. When there are no such sequences available, both DIGATE and STRATEGATE try to genetically engineer such sequences. For example, initially, there are no known sequences that could propagate a fault-effect from any flip-flop to a primary output. So the test generator generates some of these sequences in a preprocessing step. Essentially, propagating a fault-effect from a flip-flop to a primary output is the same as trying to distinguish between two sets of states in the circuit. Two states, S_1 and S_2, are said to be distinguishable if there exists a finite sequence T such that the output sequence observed by applying T starting at state S_1 differs from the output sequence observed by applying T starting at state S_2. If such a sequence T exists, T is a candidate distinguishing sequence for states S_1 and S_2. Figure 4.41 illustrates an example of a distinguishing sequence for a state pair. The sequence of four vectors, '1001, 0101, 1011, 0111', distinguishes the state pair (11010, 11000).

In the context of test generation, consider a sequential circuit with five flip-flops, ff_1 through ff_5. Suppose a fault has been excited and propagated to ff_4, and suppose the fault-free state at this time is $ff_1 \ldots ff_5 = 11010$. Then, the faulty state must be 11000, in which the faulty value at ff_4 differs from the fault-free value. Thus if a sequence exists that can distinguish the state pair (11010, 11000), by definition of a distinguishing sequence, it would be able to produce different output sequences starting from these two states. In other words, the fault-effect at ff_4 would likely be propagated to at least one primary output by the application of this sequence. Note that this sequence may not detect the fault because the faulty circuit is slightly different from the fault-free circuit. Therefore, the test generator is trying to distinguish the state 11010 in the fault-free circuit from the state 11000 in the faulty circuit. Nevertheless, for most cases, the distinguishing sequence is effective in propagating the fault-effect to a primary output.

Generating distinguishing sequences for sequential circuits can be a very difficult task. As the main target is test generation, the underlying ATPG ought not spend too much time on generating distinguishing sequences, but the focus should be on generating those sequences that are sufficient to detect the set of hard faults. In other words, to facilitate a fast generation of distinguishing sequences, one cannot

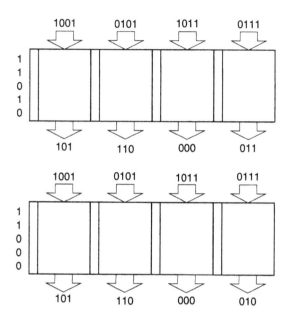

1001 0101 1011 0111

1 1 0 1 0

101 110 000 011

1001 0101 1011 0111

1 1 0 0 0

101 110 000 010

■ **FIGURE 4.41**

A distinguishing sequence that distinguishes states 11010 and 11000.

afford to generate a distinguishing sequence for each possible state pair. Rather, the search may simply be on finding those distinguishing sequences that are applicable for distinguishing many pairs of states. Using the above five-flip-flop circuit example again, if a distinguishing sequence exists that can distinguish all pairs of states that differ in ff_4, this sequence would be a powerful distinguishing sequence for many pairs of states. Although this type of distinguishing sequence can be computed prior to the start of test generation, such sequences may not exist for every flip-flop in the circuit. Thus, less powerful distinguishing sequences are also captured during test generation dynamically. However, less powerful sequences may only be applicable when the circuit is in a specific state.

In both **DIGATE** and **STRATEGATE**, distinguishing sequences are generated both statically and dynamically during test generation with the help of the GA, and these sequences are used as seeds for the GA whenever they are applicable to propagate fault-effects from flip-flops to primary outputs. If a fault is excited and propagated to multiple flip-flops, all relevant distinguishing sequences corresponding to these flip-flops are seeded. Whenever newly distinguishing sequences are learned, they are recorded and saved for future use. To avoid having a huge database of distinguishing sequences, the list of distinguishing sequences is pruned dynamically such that less useful sequences are removed from the database.

Results of DIGATE show very high fault coverages compared with previous GA-based ATPGs. For those faults that have been excited and propagated to at least one flip-flop, many of them would be detected via the help of the genetically engineered distinguishing sequences. Note that generation of distinguishing sequences on the fault-free machine is possible using binary decision diagrams instead of GAs, as has been done in [Park 1995] for the purpose of test generation. However, no pruning of sequences was performed, and no procedure for modifying the sequences was available to handle faulty circuits.

Despite the high coverages achieved by DIGATE, for some faults that were not activated to any flip-flop, seeding of distinguishing sequences would not be useful. These faults are those difficult-to-activate faults. They require specific states and justification sequences to arrive at those states in order for the faults to be excited and propagated to one or more flip-flops. For a number of circuits, previous GA-based methods, including DIGATE, achieved low fault coverages due to the lack of specific state justification successes with regard to exciting the difficult-to-activate faults. The difference in fault coverages for some of these circuits was up to 30%. Even when a GA was specifically targeted at state justification such as in [Rudnick 1995], the simple fitness function used was insufficient to successfully justify these states.

Storing the complete state information for large sequential circuits is impractical, as there could potentially be 2^n states for circuits with n flip-flops. Likewise, keeping a database of sequences capable of reaching each reachable state is infeasible. To tackle this problem, the STRATEGATE test generator [Hsiao 1997] [Hsiao 2000] was built on top of DIGATE for this very purpose. STRATEGATE uses the *linear* list of states obtained by the test vectors generated during ATPG to guide state justification. Thus, the storage requirement is only on the order of the number of test vectors rather than exponential based on the number of flip-flops.

To facilitate the state justification, the set of visited states is stored in a table, together with the corresponding list of vectors that took the circuit to the state, as shown in Figure 4.42. During state justification, the aim is to engineer a sequence that will justify the target state from the current state. At any given time during ATPG, the current state reached by the current set of vectors in the test set is the starting state. Suppose the current state has been reached at the end of vectors i, k, and m. When justifying states that have been visited before, the target state is the state reached at the end of vectors j and l in Figure 4.42. Either sequence

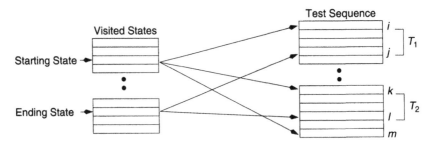

■ **FIGURE 4.42**

Data structure for dynamic state traversal.

$[i+1, \ldots, j]$ or sequence $[k+1, \ldots, l]$ would suffice to drive the circuit to the target ending state.

However, if the target state has not been visited before, STRATEGATE tries to genetically engineer a sequence that can reach it. Note that a sequence that correctly justifies one portion of the desired state may simultaneously set a different value on the other portions, resulting in conflicts. Nevertheless, the justification sequences for each partial state may be viewed as a set of partial solutions for finding the justification sequence for the complete target state. In other words, the important information about justifying specific portions of a state is intrinsically embedded in each partial solution, and this information may be extremely helpful to the GA in deriving the complete solution.

Based on the above discussion, during the state justification phase for a new state, STRATEGATE first gathers the set of *ending states* that closely match (*i.e.*, are similar to) the target state from the visited state table. Then, the sequences corresponding to these states are seeded in the GA in an attempt to engineer a valid justification sequence for the target state. Consider the example illustrated in Figure 4.43 in which the state **1X0X10** has to be justified. Sequence T_1 successfully justifies all but the third flip-flop value; on the other hand, sequence T_2 justifies all but the final flip-flop value. As explained previously, these two sequences (T_1 and T_2) may provide important information for reaching the complete solution, T_3, which justifies the complete state. T_1 and T_2 are thus used as seeds for the GA in an attempt to genetically engineer the sequence T_3 in the faulty circuit. Because the GA performs simulation in the presence of the fault to derive a sequence, any sequence derived will still be valid. Note that the GA may still abort on the state justification step, in which it fails to justify the target state. When this happens, the GA enters the single-time-frame mode, which is discussed next.

Essentially, the single-time-frame phase divides the state justification into two steps. First, it attempts to derive a single-time-frame vector (consisting of the primary input and flip-flop values) that can excite the fault and propagate its fault-effect to at least one flip-flop. Then, it tries to justify the state in the flip-flop portion of the single-time-frame vector from the current state. Because an unjustifiable state is undesirable, the fitness function also uses the dynamic controllability values of the flip-flops to guide the search toward more easily justifiable states. Note that the state portion of the vector is relaxed (some values are relaxed to "don't cares")

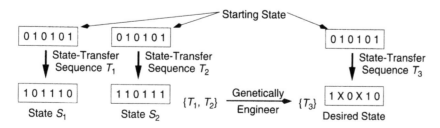

■ **FIGURE 4.43**

Genetic justification of desired state.

to ease the burden of state justification. The relaxed state is ensured by the engine such that the target fault is still excited and propagated.

Even though STRATEGATE may not justify every required state, the embedded dynamic state traversal for state justification allows it to close the 30% gap in fault coverage difference among those circuits where previous GA-based approaches failed. For other circuits, STRATEGATE has been able to achieve extremely high fault coverages compared to other simulation-based and deterministic test generators. The STRATEGATE test sets are often more compact than those obtained by deterministic test generators, even when higher fault coverages are achieved. The test sets are more compact than those obtained by CRIS or DIGATE for most circuits. Finally, simulation-based test generation can be applied to design validation rather than manufacturing test, such as the work reported in [Hsiao 2001], [Sheng 2002], and [Wu 2004].

4.8.2 Logic-Simulation-Based ATPG

The GA-based ATPGs discussed thus far use repeated fault simulation runs to gather information related to targeted faults to guide the search for test sequences. As fault simulation can be significantly more computationally intensive compared to logic simulation, approaches that use logic simulation rather than fault simulation have been proposed.

Logic-simulation-based test generators usually target some inherent "property" in the fault-free circuit and try to derive test vectors that exercise these properties. It has been brought up earlier in the chapter that the CRIS test generator attempts to maximize the circuit activity (events in logic simulation), as it has been observed that circuit activities are correlated to fault excitation. In another approach, LOCSTEP [Pomeranz 1995] made the observation that test sequences for sequential circuits achieved higher coverage when more states are visited. This is because difficult-to-test faults often require difficult-to-reach states in order to be excited, propagated, etc. Thus, LOCSTEP tries to maximize the number of new states visited. Because no fault simulation is invoked to remove the detected faults, and also because the number of reachable states can be huge in large sequential circuits, the number of vectors can potentially grow to be very large. In addition, the fault coverage obtained can be inferior to that achieved by fault-simulation-based test generators. Finally, in other ATPGs that target some properties, such as in [Guo 1999] and [Giani 2001], compaction is used to identify useful vectors that may be repeated to detect additional faults. However, repeated applications of fault simulation are necessary in test set compaction. More discussion on the use of compaction for test generation is provided later in the chapter.

As logic-simulation-based ATPGs do not call fault simulation on a regular basis, we may end up with a large number of vectors, where many vectors may not contribute to detection of new faults. The reason why the indiscriminate addition of vectors may not contribute to the detection of new faults can be explained by the following: Because some flip-flops belong to the data path and others to the controller of the circuit, maximizing the number of new states on the data path generally will not play as significant a role as maximizing those on the controller.

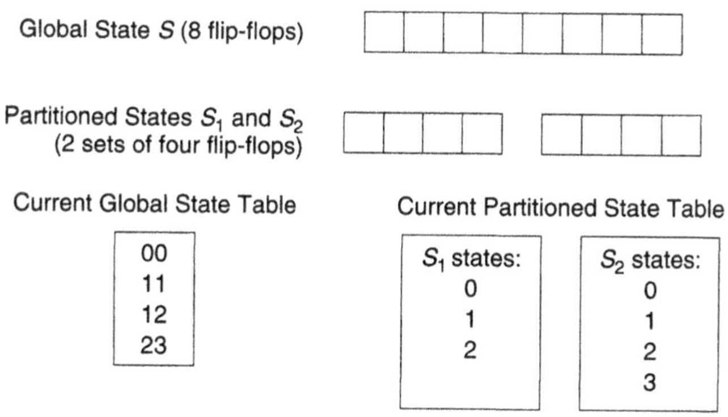

Global State S (8 flip-flops)

Partitioned States S_1 and S_2
(2 sets of four flip-flops)

Current Global State Table

| 00 |
| 11 |
| 12 |
| 23 |

Current Partitioned State Table

S_1 states:	S_2 states:
0	0
1	1
2	2
	3

■ **FIGURE 4.44**

State partition examples.

Different states on the datapath generally map to different operand values for the functional units in the design, while different states in the controller dictate different modes of operation for the circuit. This implies that the underlying ATPG should not treat the entire state as one entity. In other words, treating the entire state as one entity may mislead the test generator, particularly by the "noise" from those *unimportant states*. Thus, *partitioning* of state will help to weed out the noise. State partitioning can remove the noise and provide better guidance in the search space, as shown in Figure 4.44.

In Figure 4.44, consider a circuit with eight flip-flops. Let the global state, S, be partitioned into two partial states, S_1 and S_2, where the value of each partial state can be expressed in a hexadecimal number; for example, a partial state "1010" appearing on partition S_1 is represented by the hexadecimal value 'A'. The same notation is used for S_2. For the global state S, a pair (S_1, S_2) is used to represent its value. For example, the global state "0101 1010" is represented as $(5, A)$.

Assume that the current test set has traversed the following global states in the circuit: $(0, 0)$, $(1, 1)$, $(1, 2)$, $(2, 3)$. Correspondingly, the distinct partial states visited on S_1 and S_2 are $\{0, 1, 2\}$ and $\{0, 1, 2, 3\}$, respectively. Based on this partial state information, the following two scenarios can occur.

First, suppose there are two new candidate sequences that drive the circuit from the current state to two new, but different, global states: $(2, 1)$ and $(3, 1)$. While both states are new, it may be possible that one is better than the other. If no distinction is made about these two global states, the test generator would simply pick one randomly. Now, considering state partition as discussed before, the two states can be differentiated by noting that $(3, 1)$ may be a more useful state because 3 brings a new state in partition S_1, while both 2 and 1 have been reached in the two separate partitions already.

For the second scenario, suppose the two different global states reached by the two candidate sequences are $(3,0)$ and $(2,4)$, and, similar to the first scenario, both states are new global states; in addition, 3 is a new partial state on S_1 and 4 is a new partial state on S_2. In other words, both states bring something new. However, if different weights are imposed on different state partitions, it may be possible to differentiate them. A partition has a greater weight if it is deemed to have greater influence on the circuit. Suppose the weight assigned to S_1 is greater than S_2, then (3, 1) will be favored.

Based on the above discussion, a new test generator was proposed in [Sheng 2002b] that uses logic simulation as the core engine in the test generation process, in addition to state partitioning. Ideally, a clear distinction between control path flip-flops and the datapath flip-flops is desired. However, this may be difficult if the higher levels of the circuit description are unavailable. Without complete knowledge of controller and datapath, the partitioning is done in a different manner. One possibility is to partition all the circuit's flip-flops based on the controllability values of the flip-flops. Flip-flops with similar ranges of controllability values are grouped together. The reason behind this grouping is based on the observation that in a given circuit, some state variables will be less controllable than others. Thus, less activity will occur in those less-controllable flip-flops. In order to traverse more useful states, it would be desirable to stimulate more activity on those less active flip-flops. By grouping them together, any new partial state value reached in that group will be regarded as valuable. Other partitioning methods exist, such as using the circuit's structure to group those flip-flops that belong in the same output cone, etc.

In addition to state partitioning, the search must avoid repeated visits of certain types of states, such as reset states. A technique called **reset signal masking** was proposed in [Sheng 2002b] for this very task. It is based on the following observation. Digital circuit designers often put reset or partial reset input signals in circuits for **design for testability** (DFT) purposes. When the circuit is extensively reset or partially reset, the chance of visiting new states is significantly reduced. Therefore, identifying the signals that can reset some of the flip-flops is necessary. Then, during test generation, those primary input values that can reset some (or all) flip-flops are avoided. This is the idea behind reset signal masking. Consider a state space in which the circuit is currently traversing, illustrated in Figure 4.45. In this figure, circles denote states, and edges denote transitions between the states. Generally speaking, the circuit visits a set of easy states initially (which may contain some reset or synchronizing states) such as those states in the dotted region of the figure. Then, this set of reached states grows gradually as more states are visited. As the goal is to expand the state space beyond the current frontier, the search must avoid repeating the visit of any previously visited states, including reset states. Using Figure 4.45 again, states A, B, and C are some of the states currently at the frontier of the reached state space. Consider the frontier state B. In order to avoid going back to a previously visited state, say A, the search must place constraints on the primary inputs so that returning to state A will not occur.

The overall test generation procedures that incorporate reset signal masking and state partitioning are given in Algorithm 14.

In this algorithm, static partitioning is used to obtain initial state partition. The *stop condition* is either 100,000 vectors have been generated or the execution time has reached a preset value. This is an efficient yet simple sequential circuit test generator based on logic simulation and circuit state partitioning. Very high

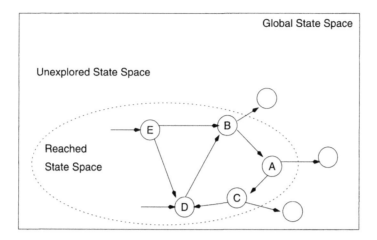

■ **FIGURE 4.45**

State space exploration.

Algorithm 14 LogicSimATPG

1: Identify reset signal masking for each primary input;
2: Partition the flip-flops (e.g., based on their controllability values.);
3: **while** stop condition not satisfied **do**
4:   Generate test vectors that expand the search space the most using reset signal masking information and partition information;
5:   Re-partition the flip-flops if desired;
6: **end while**

---

fault coverage has been obtained for large sequential circuits with significantly less computational effort. For some circuits, the highest fault coverage was obtained compared with existing deterministic and simulation-based approaches.

4.8.3 Spectrum-Based ATPG

Similar to logic-simulation-based ATPG, spectrum-based ATPG tries to seek embedded properties in the fault-free circuit that can help with the test generation process. For spectrum-based ATPG, the underlying sequential circuit is viewed as a black-box system that is identifiable and predictable from its input/output signals, rather than the traditional view as a netlist of primitives. In studying a signal, the foremost concern is the predictability of the signal. If the signal is predictable, then a portion of it can be used to represent and reconstruct its entirety. Testing of sequential systems, then, becomes a problem of constructing a set of waveforms which when applied at the primary inputs of the circuit can achieve high fault coverages by exciting and propagating many faults in the circuit.

In order to capture the spectral characteristics of a signal, a clean representation for that signal is desired (wider spectra lead to more unpredictable or random signals). Thus, any embedded noise should be filtered from the signal. In the context of test generation, static test set compaction can be viewed as a filter as it reduces the size of the test set by removing any unnecessary vectors while retaining the useful ones that achieve the same fault coverage as the original uncompacted test set. In other words, static test set compaction *filters* unwanted noise from the derived test vectors, leaving a cleaner signal (narrower spectrum) for analysis. Vectors that are generated from the narrow spectrum may have better fault detection characteristics.

Frequency decomposition is the most commonly used technique in signal processing. A signal can be projected to a set of independent waveforms that have different frequencies. In the work by Giani *et al.* [Giani 2001], Hadamard transform is used to perform frequency decomposition. The reader is referred to the cited work for details of Hadamard transform, as it is beyond the scope of this chapter.

The overall framework of the spectrum-based test generation procedure is relatively straightforward. Initially (iteration 0), the test set simply consists of random vectors. A call to static compaction will filter any unnecessary vectors such that no fault coverage is lost. Then, using the Hadamard transform on the obtained compacted test set, each primary input is analyzed and the dominant frequency components for each primary input are identified. Next, test vectors are generated based on this identified spectrum. Any spectrum can be represented as a linear combination of the basis vectors. Then, test vectors can be generated by spanning the likely vector space using only the basis vectors. This process is repeated until either a desired fault coverage is obtained or a maximum number of iterations is reached. This approach has consistently achieved very high fault coverages and small vector sets in short execution times for most sequential benchmark circuits.

4.9 HYBRID DETERMINISTIC AND SIMULATION-BASED ATPG

As both deterministic and simulation-based test generators have their own merits, in terms of coverage, execution time, test set size, etc., it may be beneficial to combine the two different types of ATPGs together. In general, deterministic ATPGs are better suited for control-dominant circuits, while simulation-based ATPGs perform better on data-dominant designs. In addition, circuits with many untestable faults should not be handled by simulation-based test generators, unless the untestable faults are first identified and removed from the fault list.

A simple combination of the two approaches would be to start with a fast run of a simulation-based test generator, followed by a deterministic test generator to improve the fault coverage and to identify untestable faults. For instance, a quick run of a random TPG would remove many of the easy-to-detect faults, leaving the deterministic ATPG only those more difficult and untestable faults. The CRIS-hybrid test generator [Saab 1994] is also based on this notion. It switches from simulation-based to deterministic test generation when a fixed number of test vectors have been generated by the simulation-based ATPG without improving the fault coverage. During the deterministic ATPG phase, in addition to generation of vectors for some undetected faults, some untestable faults are also identified. Simulation-based test generation may resume after a test sequence is obtained from the deterministic procedure.

There are of course other methods of combining simulation-based and deterministic algorithms for test generation. The GA–HITEC hybrid test generator [Rudnick 1995] uses deterministic algorithms for fault excitation and propagation and a GA for state justification. Deterministic procedures for state justification are used if the GA is unsuccessful. Instead of targeting one group of faults at a time, GA-HITEC targets one fault at a time, as is generally done in deterministic ATPGs.

This particular method of combination in GA-HITEC is based on the observation that deterministic algorithms for combinational circuit test generation have proven to be more effective than genetic algorithms [Rudnick 1994]. Furthermore, in sequential circuits, state justification using deterministic approaches is known to be very difficult and is vulnerable to many backtracks, leading to excessive execution times. Therefore, it makes sense to include the deterministic algorithm for fault excitation and propagation, while the GA is used for state justification. Note that this approach cannot identify some untestable faults.

In GA-HITEC, a fault is taken as a target. Then, the fault is excited by the deterministic engine, followed by propagation to a primary output, perhaps through several time frames, also by the deterministic engine. Through this process, several primary inputs and flip-flop variables at time frame 0 would have been chosen as decision points, as illustrated in Figure 4.46. The decisions made on the flip-flops

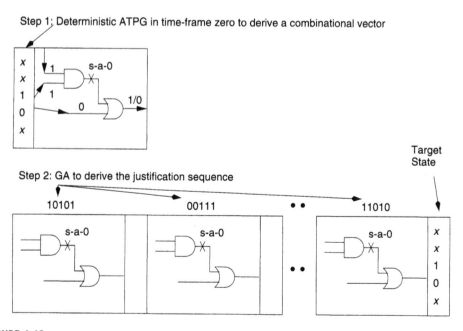

■ **FIGURE 4.46**

Test generation using GA for state justification.

at time frame 0 now become the target state to be justified. The GA is invoked at this time to generate a justification sequence for the target state. If a sequence is found that justifies the target state, then this sequence is concatenated with the vectors derived for fault excitation and propagation, and the complete test sequence is added to the test set. Faults that are detected by this sequence are removed from the fault-list. On the other hand, if a justification sequence cannot be found by the GA, then backtracks are made in the fault propagation phase in the deterministic test generator, and attempts are made to justify any new state.

In the state justification phase, the GA evolves over four full generations for each target state. Each individual in the population represents a candidate sequence of vectors. A small population size of 64 is used, and the number of generations is limited to four to reduce the execution time. The population size is doubled and the number of generations increased to eight later for the more difficult faults. The sequence length is also doubled.

During the GA search for a justification sequence, both fault-free and faulty states are checked for each individual in the population. Note that this check is done for every vector in an individual, which contains several vectors. Thus, if a match is found, the length of the actual justification sequence may be less than the length of the individual. For the purposes of the GA, the fitness function simply measures how closely the final state matches the target state:

$$fitness = \frac{9}{10}(\#\,flip\text{-}flops\ matched\ in\ fault\text{-}free\ circuit)$$

$$+ \frac{1}{10}(\#\,flip\text{-}flop\ matches\ in\ faulty\ circuit).$$

A flip-flop is said to be matched if the value achieved is the same as the target value. If the target value is "don't care," it is considered matched as well. Therefore, if both the fault-free and faulty states match, the fitness will equal the total number of flip-flops in the circuit. Note that unequal weights are given to the fault-free and faulty states. Changing the weights will alter the search by placing emphases differently. Again, during fitness evaluation, parallelism among the individuals can be exploited, where 32 individuals may be simulated together to reduce the computational cost.

Results of GA-HITEC have demonstrated that higher fault coverages can be obtained as compared to pure deterministic HITEC for many circuits. Similar numbers of untestable faults were identified as well.

4.9.1 ALT-TEST Hybrid

ALT-TEST [Hsiao 1996b] is another hybrid approach that combines a GA-based test generator and a deterministic engine. HITEC [Niermann 1991] is again used as the deterministic test generator in ALT-TEST. The number of calls to the deterministic test generator is very few, which differs significantly from the CRIS hybrid, where hundreds of calls to the deterministic engines were made. ALT-TEST alternates repeatedly between GA-based and deterministic test generation.

A fast run of a GA-based test generator is followed by a run of a deterministic test generator that targets faults that were left undetected by the previous GA-based test generator. Any successful sequences derived by the deterministic test generator are used as seeds for the successive GA-based ATPG run. The test sequences derived by the deterministic engine typically will traverse previously unvisited states. Thus, the deterministic test generator may be viewed as an external engine whose purpose is mainly to guide the GA to new state spaces of the circuit that have not been visited. By visiting new state spaces, the test generator can maximize the search space. Furthermore, the use of a deterministic test generator also helps to identify any untestable faults, thus saving the computational effort in the GA runs on those faults that could never be detected.

The test generation process in ALT-TEST is divided into three stages; each of the three stages is composed of alternating phases of GA-based and HITEC test generation. The first stage attempts to detect as many faults as possible from the fault list. The second stage tries to maximize the number of visited states and propagate fault effects to flip-flops. Finally, the third stage tries to both detect the remaining faults and visit new states. In each of the three stages, the GA is first run until little or no more improvement is obtained, then the deterministic approach is used to target undetected faults. A stage is finished when no more improvements are made for the remaining undetected faults. The pseudo code for test generation within a stage is given in Algorithm 15.

In each GA run, the initial population consists of: (1) the best sequence from the previous GA run or the deterministic engine, (2) the sequences having fitness

---

Algorithm 15 ALT-TEST

```
 1: while there is improvement in this stage do
 2:     /* GA-based test generation */
 3:     while there is improvement in the GA phase do
 4:         for all undetected faults, in groups of 31 faults do
 5:             select next 31 undetected faults as target faults;
 6:             best-individual = GA-evolve();
 7:             add best-individual to test set;
 8:             seed the next GA population;
 9:             compute improvement;
10:         end for
11:     end while
12:     /* deterministic test generation */
13:     select the hard-to-detect faults;
14:     best-sequence = deterministic-ATPG(hard faults);
15:     if a best-sequence is found then
16:         add best-sequence to test set;
17:     end if
18:     seed best-sequence into the next GA;
19:     compute improvement;
20: end while
```

---

values greater than or equal to one-half of the best fitness from the previous GA run, and (3) random individuals to fill the entire population if needed. Instead of targeting individual faults, the GA tries to detect as many faults as possible by any individual. Because the target is to generate a sequence that can detect as many faults as possible, parallel-fault simulation (on 31 faults) is used during fitness evaluation; 31 faults are used instead of 32 due to the nature of the embedded fault simulator. A set of 31 undetected faults in the fault list are selected as target faults. All individuals in the population would then target the same group of 31 faults.

For successive GA runs, faults are chosen cleverly so that efforts can be reduced. For instance, if the best sequence added to the test set detected a total of 20 faults, it may have also excited and propagated some faults to one or more flip-flops at the end of that sequence. These activated faults should be placed in the next targeted fault group, as faults that have propagated to at least one flip-flop are deemed to have a greater chance of being detected. If the more than 31 faults have propagated to the flip-flops, preference is given to those that have propagated to more flip-flops. On the other hand, if fewer than 31 faults have propagated to the flip-flops, the remaining faults in the group are filled from the undetected fault list.

After the GA phase, the deterministic test generator is activated that targets the difficult-to-detect faults identified by the previous GA run. A difficult-to-detect fault is one that has never been detected by any of the test sequences added to the test set. The sequence generated by the deterministic test generator is also seeded to the next GA run with hopes that it can help to expand the search space.

As the GA and deterministic phases alternate, the number of faults detected as a function of time will experience periodic jumps, as illustrated in Figure 4.47 for

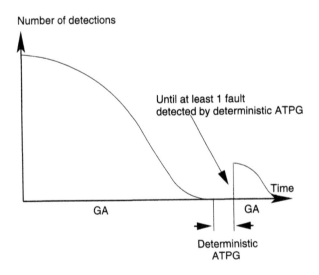

■ FIGURE 4.47

Number of detections in alternating phases.

a case in which the deterministic ATPG successfully finds a sequence for at least one fault.

The fitness functions for the three separate stages of ALT-TEST depend on a variety of parameters. Because each stage targets a different goal, the set of parameters that control the search will differ as well. The parameters that can affect the fitness of an individual include the following:

P_1—Number of faults detected
P_2—Number of flip-flops to which fault effects have arrived
P_3—Number of new states visited
P_4—Number of flip-flops set to their difficult-to-control values

It can be seen that parameters P_3 and P_4 contribute to the expansion of the searched state space. While P_3 explicitly aims for visitation of more states, P_4 guides the search by favoring sequences that are able to set the difficult-to-control flip-flops to values that have not yet been encountered. Consequently, a new state is likely to be visited. All four parameters are given different weights across the three stages of ALT-TEST:

Stage 1—Fitness $= 0.8P_1 + 0.1P_2 + 0.1(P_3 + P_4).$
Stage 2—Fitness $= 0.1P_1 + 0.45P_2 + 0.45(P_3 + P_4).$
Stage 3—Fitness $= 0.4P_1 + 0.2P_2 + 0.4(P_3 + P_4).$

In the first stage, because most faults have not yet been detected, the aim is thus to detect as many faults as possible, which makes the parameter P_1 the most dominant. At the end of the first stage, little improvement in fault detection is observed, indicating that the fitness function is no longer effective. In other words, it is unlikely to detect faults without other ingredients added. Therefore, in the second stage, maximizing visitation of new states and fault-effect propagation to flip-flops becomes the aim. By doing this, the search tries to expand the state space together with those states that can still excite and propagate the fault-effects. Finally, in the third stage of ALT-TEST, the focus is shifted once again. Now the target is detecting the remainder of the faults that have been difficult to detect by the GA and the deterministic engine in the previous two stages. Therefore, the fitness function weights fault detections and new state identifications more heavily. ALT-TEST achieves high coverages compared with GA-HITEC for many circuits, with the ability to identify untestable faults.

4.10 ATPG FOR NON-STUCK-AT FAULTS

4.10.1 Designing an ATPG That Captures Delay Defects

Today's integrated circuits are seeing an escalating clock rate, shrinking dimensions, increasing chip density, etc. Consequently, there arises a class of defects that would affect the functionality of the design if the chip is run at a high speed. In other words, the design is functionally correct when it is operated at a slow clock. This type of defect is referred to as a *delay defect*. While the conventional stuck-at testing can catch some delay defects, the stuck-at fault model is insufficient to model delay defects satisfactorily. This has prompted engineers and researchers to propose a variety of methods and fault models for detecting speed failures. Among the fault models are the *transition fault* [Levendel 1986] [Waicukauski 1987] [Cheng 1993], the *path-delay fault* [Smith 1985], and the *segment delay fault* [Heragu 1996]. This section is devoted to path-delay fault test generation.

The path-delay fault model considers the cumulative effect of the delays along a specific combinational path in the circuit. If the cumulative delay in a faulty circuit exceeds the clock period for the path, then the test pattern that can exercise this path will fail the chip. The segment delay fault model targets path segments instead of complete paths.

Because a transition has to be launched in order to propagate across a given path, two vectors are needed. The first vector initializes the circuit nodes, and the second vector launches a transition at the start of a path and ensures that the transition is propagated along the given path. Given a path P, a signal is an **on-input** of P if it is on P. Conversely, a signal is an **off-input** of P if it is an input to a gate in P but is not an on-input of P. A path-delay fault can be a rising fault, where a rising transition is at the start of the path, or a falling fault, where a falling transition is at the start of the path. The rising and falling path-delay faults are denoted with the up-arrow ↑ and the down-arrow ↓ before path P, respectively. For example, $\uparrow g_1 g_4 g_7$ is a rising path that traverses through gates g_1, g_4, and g_7.

Delay tests can be applied three different ways: **launch-on-capture** (also called broadside or double-capture) [Savir 1994], **launch-on-shift** (also called skewed-load) [Savir 1993], and **enhanced-scan** [Dervisoglu 1991]. In launch-on-capture-based testing, the first n-bit vector is scanned in to the circuit with n scan flip-flops at a slow speed, followed by another clock which creates the transition. Finally, an at-speed functional clock is applied that captures the response. Thus, only one vector has to be stored per test, and the second vector is directly derived from the initial vector by pulsing the clock. In launch-on-shift-based testing, the first $n-1$ bits of an n-bit vector are shifted in at a slow speed. The final n^{th} shift is performed, and it is also used to launch the transition. This is followed by an at-speed quick capture. Similar to launch-on-capture, only one vector has to be stored per test, as the second vector is simply the shifted version of the first vector. Finally, in enhanced-scan testing, both vectors in the vector pair (V_1, V_2) have to be stored in the tester memory. The first vector is loaded into the scan chain, followed by its immediate application to initialize the circuit under test. Next, the second vector is scanned in, followed by an immediate application and capture of the response. Note that the node values in the circuit is preserved during the shifting-in of the second vector V_2. In order to achieve this, a hold-scan design [Dervisoglu 1991] is required.

Because both launch-on-capture and launch-on-shift place constraints on what the second vector can be, they will achieve lower fault coverage when compared with enhanced-scan. However, enhanced-scan comes at a price of hold-scan cells, which consume more chip area. This may not be viewed as a huge negative in microprocessors and some custom-designed circuits because hold-scan cells are used to prevent the combinational logic from seeing the values being shifted. This is done because the intermediate state of the scan cells may cause contention in some of the signals in the logic, as well as reducing the power consumption in the combinational logic during the shifting of the data in scan cells. In addition, hold-scan cells also help increase the diagnostic capability on failing chips in which the data captured in the scan chain can be retrieved.

In terms of test data volume, enhanced-scan tests may actually require less storage to achieve the same delay fault coverage. In other words, for launch-on-capture or launch-on-shift to achieve the same level of fault coverage, many more patterns may have to be applied.

Unlike stuck-at faults, where a fault is either detected or not detected by a given test vector, a path-delay fault may be detected by different test patterns (consisting of two vectors) with differing levels of quality. In other words, some test patterns can detect a path-delay fault only with certain restrictions in place. Higher quality test patterns place more restrictions on sensitization of the path. On the other hand, similar to stuck-at faults, some paths may be untestable if the sensitization requirement for a given path is not satisfiable.

For designs with two interactive clock domains, modifications can be made to allow for test. For example, the following at-speed delay test approaches can be used for both launch-on-capture and launch-on-shift architectures: one-hot double-capture, aligned double-capture, and staggered double-capture [Bhawmik 1997] [Wang 2002]. Details of these architectures are further explained in Chapter 5.

4.10.1.1 Classification of Path-Delay Faults

Given the above discussion, the path-delay faults can be classified into several categories. A path P is said to be **statically sensitizable** if all the off-inputs of P can be justified to their corresponding noncontrolling values for some test vector. If all of the off-inputs of a path P cannot be justified to the respective noncontrolling values, P is said to be a **statically unsensitizable path**. A **false path** is a path such that no transition can propagate from the start to the end of the path. A path is false because the values necessary on the off-inputs of the path are not realizable by the circuit [Devadas 1993]. Note that a false path is always statically unsensitizable, but not vice versa. Figure 4.48 illustrates an example of a statically unsensitizable path $\downarrow abce$, as signal d cannot be at the noncontrolling value in the second vector. But this path is not false, because a transition can propagate from a to the end of the path, e, via a multi-path from a to e.

A path P is **single-path sensitizable** if the values of the off-inputs in P can be settled to their noncontrolling values in both vectors. This is the most stringent requirement, thus there are few paths that would satisfy this condition.

In order to detect a delay defect along a path, it may be possible to relax the constraint according to the single-path sensitizability. In other words, it may be possible to detect the delay fault without having all off-inputs set to noncontrolling values. Consider the circuit illustrated in Figure 4.49. The falling path $\downarrow bdfg$ is the target path. Note that the value for signals a and e can be relaxed in the first vector such that the transition on b can still be propagated to g. This is because the

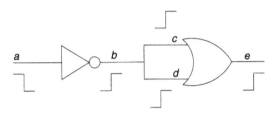

Path: ↓ *abce*

■ **FIGURE 4.48**

A statically unsensitizable but not false path.

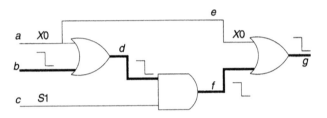

■ **FIGURE 4.49**

A robustly testable path.

propagation of the falling transition from *b* to *d* is independent of the value of *a* in the first vector (and similarly for the transition from *f* to *g*). On the other hand, a steady 1 (S1) is needed for both the first and second vectors on signal *c*. Relaxing the value in the first vector could block the transition from *d* to *f*.

The target path in the above example is said to be **robustly testable**. More specifically, the path is testable irrespective of other delay faults in the circuit [Smith 1985] [Lin 1987]. In the same running example shown in Figure 4.49, if the "don't care" value in the first vector for signal *a* is a logic 1, the transition on *b* would still be propagated to *d*, as the transition on *d* depends on the later of the two transitions. In short, a delay on *a* will not prevent the target path from being detected.

Given the above discussion, the value criteria for each off-input of *P* for a robustly testable path are as follows:

- When the corresponding on-input of *P* has a controlling to noncontrolling transition, the value in the first vector for the off-input can be "don't care," with the value for the off-input as a noncontrolling value in the second vector.

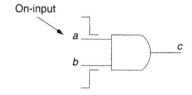

■ **FIGURE 4.50**

Sensitization criteria.

■ When the corresponding on-input of P has a noncontrolling to controlling transition, the values for the off-input must be a steady noncontrolling value for both vectors.

Because a robust test for a path P can detect P irrespective of any other delay faults in the circuit, they are the most desirable tests. For most circuits, however, the number of robustly testable paths is small. Thus, for those **robustly untestable paths**, less restrictive tests must be sought.

Consider the AND gate shown in Figure 4.50. Suppose signal a is the on-input and b is the off-input along some path. In the robust sensitization criterion, because the on-input is going from a noncontrolling to a controlling value, the off-input must be at steady noncontrolling value. As discussed before, such a restriction will ensure that the path going through a, if testable, will be tested irrespective of any other delay faults in the circuit. However, if such a test is not possible, one may wish to relax the condition such that the target path is the only faulty path in the circuit. In other words, if the off-input b is not late, then it may be possible to relax the steady noncontrolling value somewhat. In this example, if the transition on the target path through a is late, and if the transition on the off-input is on time, then the output c will still have a faulty value. Therefore, it may be sufficient to require the values of $X1$ for b instead of a steady 1. This sensitization condition is called the **nonrobust sensitization** condition. Figure 4.51 illustrates an example of a path, $\uparrow bcdf$, that is robustly untestable but is nonrobustly testable. Note that in a robustly testable path, a transition is present at every gate along the path. On the other hand, in a nonrobustly testable path, some transitions may be lost along the path. In the example shown in Figure 4.51, the transition is lost at f.

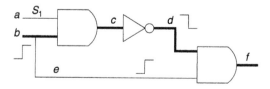

■ **FIGURE 4.51**

A nonrobust path.

Given the above discussion, the value criteria for each off-input of P for a nonrobustly testable path are as follows:

- Irrespective of the transition on the on-input, the value in the first vector for the off-input can be "don't care," with the value for the off-input as a noncontrolling value in the second vector.

There are other classes of path-delay faults, such as validatable nonrobustly testable path-delay faults, functional sensitizable path-delay faults, multi-path-delay faults, *etc*. They are not included in this discussion.

4.10.1.2   ATPG for Path-Delay Faults

Unlike stuck-at test generation, where only one vector is necessary and the value on any signal can be 0, 1, D, $\overline{D}$, or X, in path-delay fault test generation, two vectors are required, and the vector pair only has to ensure that a transition is launched at the start of the path and that the off-inputs satisfy the conditions specified by the robust or non-robust tests.

For a given target path P, a test pattern (of two vectors) can be generated for P. One can go about generating the two vectors separately, or a different logic system may be used such that the two vectors can be derived simultaneously with a single copy of the circuit. When the two vectors are generated separately, each signal can be 0, 1, or X. The vector-pair generated must ensure that a transition is launched at the start of P and that all off-inputs adhere to the robust or nonrobust conditions specified by the test. The value justification of the off-inputs is similar to the multi-objective value justification in stuck-at ATPG.

If two vectors are to be generated together, a value system different from the three-value system can be used to represent values over two vectors [Lin 1987]. In this case, a signal can be any of the following:

- $S0$—Initial and final values are both logic 0.
- $S1$—Initial and final values are both logic 1.
- $U0$—Initial logic can be either 0 or 1, but final value is logic 0.
- $U1$—Initial logic can be either 0 or 1, but final value is logic 1.
- XX—Both initial and final values are "don't cares."

Boolean operators also work on this new system of values. For example, Tables 4.12, 4.13, and 4.14 show the AND, NOT, and OR operations over these five values, respectively. Such tables can be generated for other Boolean operations as well. Using the new 5-valued system, conventional ATPG algorithms can be applied to generate path-delay tests.

Because many paths overlap and there may be an exponential number of paths, it may be possible and helpful to reuse some of the knowledge gained from targeting other paths. For instance, if it is known that $a = 1$ and $b = 0$ is not possible, then any path that requires this combination would be untestable. Likewise, if two paths share a segment, the two test patterns for the two paths may share bits in

TABLE 4.12 ■ AND Operation

AND	S0	U0	S1	U1	XX
S0	S0	S0	S0	S0	S0
U0	S0	U0	U0	U0	U0
S1	S0	U0	S1	U1	XX
U1	S0	U0	U1	U1	XX
XX	S0	U0	XX	XX	XX

TABLE 4.13 ■ NOT Operation

NOT	
S0	S1
U0	U1
S1	S0
U1	U0
XX	XX

TABLE 4.14 ■ OR Operation

OR	S0	U0	S1	U1	XX
S0	S0	U0	S1	U1	XX
U0	U0	U0	S1	U1	XX
S1	S1	S1	S1	S1	S1
U1	U1	U1	S1	U1	U1
XX	XX	XX	S1	U1	XX

common. In RESIST [Fuchs 1994], this concept is taken into account such that a recursion-based ATPG algorithm searches starting from a primary input. The search progresses by targeting paths that differ only in the last segment; in other words, they share the same initial subpath.

Essentially, RESIST starts at each primary input, and the circuit is traversed in a depth-first fashion. Once a complete path P from a primary input to primary output has been tested, a backtrack is invoked and a different path P_2 that differs from P in only one segment is tried. At the end, all the paths in the output cone of the starting primary input would have been considered. In doing so, the decision tree can be shared among different paths and knowledge is reused. Likewise, if a subpath is found to be untestable, all paths that share the same initial subpath would be untestable. Then, RESIST repeats for another primary input until all primary inputs have been processed.

RESIST is efficient because it incorporates knowledge into the ATPG process, and paths are handled such that much knowledge can be carried over from one path to the next.

4.10.2 ATPG for Transition Faults

If robust tests were possible for all the paths in a circuit, we would not need any additional test vectors for capturing the delay defects. However, because very few paths are robustly testable, there will be some delays that cannot be captured by either robust or nonrobust path-delay fault tests. Consider the situation where some small delay defects are distributed inside a circuit. If the circuit nodes lie on a robustly untestable path or a less critical path, then the path-delay fault test vectors may miss those faults. The segment delay fault model might also miss the faults because there might not be a path along which the effect may be propagated.

A transition fault at node g assumes a delay defect is present at node g such that the propagation of the transition at g will not reach the flip-flop or primary output within the clock period. While the path-delay fault model considers the cumulative effect of the delays along a specific path, the transition fault model does not specify the path through which the fault is to be excited or propagated. Today, the transition fault model is the most practical as the number of transition faults is linear to the number of circuit nodes and commercial tools are available for computing such tests. On the other hand, the number of path-delay faults is exponential to the number of circuit lines, which makes critical path analysis and identification procedures necessary. Finally, transition tests have been generated to improve the detection of speed failures in microprocessors [Tendulkar 2002] as well as *application-specific integrated circuits* (ASICs) [Hsu 2001]. These reasons make transition faults popular in industry.

Similar to the stuck-at fault model, two transition faults are possible at each node of the circuit: *slow-to-rise* and *slow-to-fall*. A test pattern for a transition fault consists of a pair of vectors {V1, V2} where V1 (called the *initial vector*) is required to set the target node to an initial value, and V2 (called the *test vector*) is required to launch the corresponding transition at the target node and also propagate the fault effect to a primary output [Waicukauski 1987] [Savir 1993].

Lemma 3

A transition fault can be launched robustly or nonrobustly, or in neither way, through the segment PI-fault site.

Proof

Consider a slow-to-rise transition fault at the output of the OR gate in Figure 4.52. This transition can only be launched by having rising transitions at both inputs of this gate. Hence, neither of two paths passing through the OR gate can be robustly or nonrobustly tested.

Lemma 4

A detectable transition fault can be detected by a robust segment or nonrobust segment, or in neither way, starting from the fault site to a primary output.

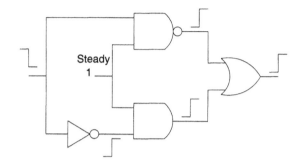

■ **FIGURE 4.52**

Slow-to-rise transition at the input of an OR gate.

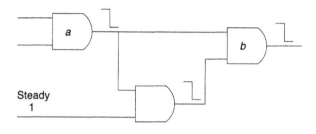

■ **FIGURE 4.53**

Slow-to-fall transition propagation example.

Proof

Consider the circuit shown in Figure 4.53. A slow-to-fall transition fault at *a* is propagated to the primary output and hence detected, but neither path from *a* to *b* is robustly or nonrobustly testable due to off-path inputs at gate *b*.

The two above lemmas conclude that both the launching and propagation of a transition fault can be done through multiple paths and none of the paths may be tested robustly or nonrobustly [Gupta 2004]. Hence, there are some faults that can be missed by the path-delay fault model and can only be captured by the transition fault model. But, for small delay defects, an enhanced transition fault model is needed to properly address the aforementioned issues.

Transition tests can also be applied in three different ways as for the other delay fault models discussed earlier: **launch-on-capture**, **launch-on-shift**, and **enhanced-scan**. As with path-delay tests, because both launch-on-capture and launch-on-shift place constraints on what the second vector can be, they will achieve lower transition fault coverage when compared with enhanced-scan.

4.10.3 Transition ATPG Using Stuck-At ATPG

A transition fault can be modeled as two stuck-at faults. Thus, one can view testing transition faults as testing two stuck-at faults. For example, a transition fault a slow-to-rise can be modeled as exciting the fault $a/1$ in the first time frame and detecting the fault $a/0$ in the second time frame. In other words, exciting $a/1$ requires setting $a = 0$, and testing for $a/0$ requires setting $a = 1$ and propagating its effect to an observable point.

With enhanced-scan, because the two vectors are not correlated, these two vectors can be generated independently. For launch-on-capture or launch-on-shift, the two time frames must be handled together. In the launch-on-capture-based test scheme, one may view the excitation of the fault in the first time frame as a constraint for the ATPG for detecting the fault in the second time frame. In other words, for testing the transition fault a slow-to-rise, the stuck-at fault $a/0$ is the target fault in the right (second) time frame of the two-time-frame unrolled circuit. A stuck-at ATPG is invoked to detect $a/0$ with the constraint that the signal a in the first time frame must be set to logic 0. A slow-to-fall transition can be modeled in a similar manner.

On the other hand, in launch-on-shift, the two time frames are related in a different way. The flip-flops of the second time frame are fed by a shifted version of the first time frame. Other than that, the ATPG setup is similar to the launch-on-capture-based test.

4.10.4 Transition ATPG Using Stuck-At Vectors

For enhanced-scan testing, because both vectors must be stored in the test equipment memory, there is considerable redundancy in the information stored. Consider the test sequence shown in Table 4.15. In this test sequence, V_2 and V_3 are used several times. Ideally, it would be desirable to store only one copy of V_2 and V_3. However, storing only one copy of a vector would require the ATE to have the ability to index and reuse the vector in a random order; this functionality is currently unavailable. Limited reuse of the information stored, however, may be possible. Whenever two copies of the same vector are stored in consecutive locations in the tester memory, it may be possible to store only one copy and scan in the vector as many times as needed during consecutive scan cycles. Thus, with the running example, the sequence $\{V_1, V_2*, V_3*, V_4, V_3, V_5, V_1, V_3\}$ can be restored and the vectors marked with the asterisk ($*$) are scanned in twice. Information about vectors that must be scanned in multiple times is stored in the control memory of the tester. In this example, only 8 out of 10 vectors have to be stored.

In this running example, the tester memory requirement was reduced at a price. Because vectors that are scanned in repeatedly do not form a regular pattern, the control memory necessary to store the asterisks can be costly. To avoid such a cost

TABLE 4.15 ■ Example Test Sequence

V_1	V_2	V_2	V_3	V_3	V_4	V_3	V_5	V_1	V_3

we can do the following: Apply each vector twice except for the first and the last vectors stored in the tester memory. Consider the sequence $\{V_1, V_2, V_3, V_4, V_1, V_3, V_5\}$. Because all but the first and the last vectors are applied twice, the set of transition test patterns we obtain would be (V_1, V_2), (V_2, V_3), (V_3, V_4), (V_4, V_1), (V_1, V_3), and (V_3, V_5). This set of patterns includes all the test patterns of Table 4.15, plus the additional (V_4, V_1). This example shows that 7 vectors (instead of 10 vectors) can be sufficient to apply all the needed transition tests. Such sequences where all but the first and the last vectors are applied twice are called **transition test chains**.

4.10.4.1 *Transition Test Chains via Weighted Transition Graph*

Because transition faults and stuck-at faults are closely related, it may be possible to construct transition test sets directly from stuck-at test sets using the concept of transition test chains [Liu 2005].

A *weighted transition graph algorithm* is used to construct transition test chains from a given stuck-at test set. Rather than computing a set of vector pairs and chaining them together as alluded to in the above examples, the weighted transition graph algorithm maps the chains onto a graph traversal problem. The algorithm first builds a weighted directed graph called the *weighted transition-pattern graph*. In this graph, each node represents a vector in the stuck-at test set; a directed edge from node V_i to node V_j denotes the transition test pattern (V_i, V_j); and its weight indicates the number of transition faults that can be detected by (V_i, V_j). This may potentially result in a complete graph, where a directed edge exists between every pair of vertices. In order to reduce the graph size, a subset of the transition faults that were not detected by the application of the original stuck-at test set may be considered instead of considering all transition faults. The graph construction procedure is described in Algorithm 16.

For example, consider a circuit with five gates, 10 stuck-at faults, and a stuck-at test set consisting of 4 vectors V_1, V_2, V_3, and V_4. Let the excitation and detection dictionary obtained be as shown in Table 4.16. Assuming the test set order is $\{V_1, V_2, V_3, V_4\}$, then only 3 of the 10 transition faults can be detected, namely c slow-to-fall, e slow-to-fall, and c slow-to-rise. However, using Table 4.16, the non-consecutive vectors can be combined to detect additional transition faults: (V_1, V_3) can detect a slow-to-fall; (V_3, V_1) detects a slow-to-rise; (V_1, V_4) detects d slow-to-fall; (V_4, V_2) detects d slow-to-rise; (V_4, V_1) detects a slow-to-rise, b slow-to-fall; and (V_2, V_4) detects b slow-to-rise, e slow-to-rise, and d slow-to-fall. The corresponding weighted transition graph is shown in Figure 4.54.

The weighted transition graph has a nice property that allows for formulation of the following theorem.

Theorem 1

Faults detected by pattern (V_i, V_j) and pattern (V_j, V_k) are mutually exclusive.

Proof

This is proved by contradiction. Without loss of generality, suppose a fault f *slow-to-fall* is detected by (V_i, V_j). Thus, V_i must excite f *s-a-0*(sets line f to 1) and V_j must

---

Algorithm 16 WeightedTransitionGraphConstruction(T)

Require: stuck-at test set $T = \{T_1 \ldots T_N\}$
  1: Perform transition fault simulation using pairs of vectors $\in T\{(T_1, T_2),$
     $(T_2, T_3). \ldots (T_{N-1}, T_N)\}$;
  2: $U_T =$ the set of undetected transition faults;
  3: $U_s = \emptyset$;
  4: **if** transition fault X slow-to-rise (or slow-to-fall) $\in U_T$ **then**
  5:    $U_s = U_s \cup \{X/0, X/1\}$;
  6: **end if**
  7: Perform stuck-at fault simulation without fault dropping using the stuck-at test set T
     on only the stuck-at faults in U_s.
  8: **for** each stuck-at fault $f \in U_s$ **do**
  9:    record the vectors in T that can excite f and the vectors that can detect f;
 10: **end for**
 11: **for** each vector $v \in T$ **do**
 12:    record the faults excited and detected by v;
 13: **end for**
 14: **for all** vector pairs T_i and T_j **do**
 15:    Insert a directed edge from V_i to V_j if test pattern (T_i, T_j) detects at least one
        transition fault in U_T;
 16:    weight of inserted directed edge = number of transition faults detected
        by (T_i, T_j);
 17: **end for**

TABLE 4.16 ■ Fault Dictionary Without Fault Dropping

Vectors	Excited Faults	Detected Faults
V_1	$a/0, b/1, c/1, d/0, e/0$	$a/0, b/1$
V_2	$b/1, c/0, d/0, e/1$	$c/0, d/0, e/1$
V_3	$a/1, c/1,$	$a/1, c/1$
V_4	$a/1, b/0, d/1, e/0$	$b/0, d/1, e/0$

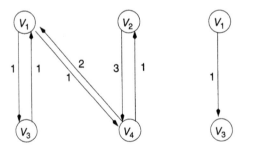

(a) Original graph   (b) Updated graph

■ FIGURE 4.54

Weighted transition-pattern graph example.

detect f s-a-1. If (V_j, V_k) also detects the transition fault f *slow-to-fall*, the vector V_j must set line f to 1, which is a contradiction.

A Eulerian trail in a graph is a path such that each edge in the graph is traversed exactly once. Using this concept, a Eulerian trail in the transition-pattern graph traverses all the (non-zero weight) edges in the graph exactly once. It is tempting to conclude such a Euler trail in the weighted transition is the best transition test chain. However, the Eulerian trail assumes that the edge weights in the graph are static. This may not be true in the case of selecting test patterns, in which some transition faults may be detectable by different patterns. For example, if edge (V_i, V_j) is traversed (*i.e.*, test (V_i, V_j) is selected), then a number of transition faults would have been detected by this test pattern. This also means that the weights on other edges should be modified because some other test patterns may have detected similar faults as well. Some of the edge weights may even become zero. Based on Theorem 1, edges whose weights will not change are only those originating from V_j. Thus, after a directed edge (V_i, V_j) is selected, all other directed edges that do not start at V_j must have their edge weights updated. This, however, would involve extensive fault simulation, which could be expensive. To reduce the cost of fault simulation, the edge weights could be updated periodically instead of after traversing each edge. A simple algorithm is outlined in Algorithm 17 where T is the transition test chain computed by the algorithm from the given stuck-at test set.

The idea behind this algorithm is as follows: First identify a test sequence of length 3 that can cover the most faults by traversing the weighted transition pattern graph. For example, in the original weighted transition pattern graph shown in Figure 4.54a (V_2, V_4, V_1) is the maximal-weight test chain of length 3. After traversing this chain, five transition faults (a slow-to-rise, b slow-to-rise, b slow-to-fall, d slow-to-fall, and e slow-to rise) are detected. The updated graph is shown in Figure 4.54b. It should be noted that in addition to removing the edges (V_2, V_4) and (V_4, V_1), two other edges $((V_1, V_4)$ and $(V_3, V_1))$ are also removed from the graph. This is because

---

Algorithm 17 TransitionTestGeneration

Require: $T = \{V_1, \ldots, V_n\}$;
  1: G = WeightedTransitionGraphConstruction(T);
  2: **while** transition FC < 100% AND number of iterations < N **do**
  3:    identify edge $(V_i, V_j) \in G$ with the largest weight;
  4:    append vectors V_i and V_j to test set T;
  5:    **for all** edges that start at V_j **do**
  6:     search for an edge (V_j, V_k) with the largest weight;
  7:    **end for**
  8:    append vector V_k to test set T;
  9: **end while**
10: return (T);

---

the fault d slow-to-fall, which can be detected by (V_1, V_4), has already been detected by selecting the test chain (V_2, V_4, V_1). Therefore, the edge (V_1, V_4) can be removed from the weighted pattern graph because it no longer contributes to the detection of other transition faults. Similar argument can be made for the edge (V_3, V_1). Finally, all the seven originally undetected faults in Table 4.16 are detected with the test chain $\{V_2, V_4, V_1, V_3, V_4, V_2\}$.

4.10.5 Bridging Fault ATPG

Recall that bridging faults are those faults that involve a short between two signals in the circuit. Given a circuit with n signals, there are potentially $n \times (n-1)$ possible bridging faults. However, practically, only those signals that are locally close on the die are more likely to be bridged. Therefore, the total number of bridging faults can be reduced to be linear in the number of signals in the circuit.

Consider two signals x and y in the circuit that are bridged. This bridging fault will not be excited unless different values are placed on x and y. Note that the actual voltage at x and y may be different due to the resistance value of the bridge. Subsequently, the logic that takes x as its input may interpret the logic value differently from the logic that takes y as its input. In order to reduce the complexity, five common bridging fault models are often used:

- AND bridge—The faulty value of the bridge for x' and y' is taken to be the logical AND of x and y in the original fault-free circuit.

- OR bridge—The faulty value of the bridge for x' and y' is taken to be the logical OR of x and y in the original fault-free circuit.

- x DOM y bridge—x dominates y; in other words, the faulty value of the bridge for both x' and y' is taken to be the logic value of x in the fault-free circuit.

- x DOM1 y bridge—x dominates y if $x = 1$; in other words, the faulty value of x' is unaffected, but the faulty value for y' is taken to be the logical OR of x and y in the fault-free circuit.

- x DOM0 y bridge—x dominates y if $x = 0$; in other words, the faulty value of x' is unaffected, but the faulty value for y' is taken to be the logical AND of x and y in the fault-free circuit.

Figure 4.55 illustrates the faulty circuit models corresponding to each of these five bridge types.

If there exists a path between x and y, then the bridging fault is said to be a **feedback bridging fault**. Otherwise, it is a **non-feedback bridging fault**. Figure 4.56 illustrates a feedback bridging fault. In this figure, if $abc = 110$, then in the fault-free circuit $z = 0$. If the bridge is an AND-bridge, then a cycle would result. In other words, a becomes 0 and in turn makes $z = 1$. And because $a = 1$ initially, it will again try to drive $z = 0$, resulting in an infinite loop around the bridge. For the following discussion, only nonfeedback bridging faults will be considered.

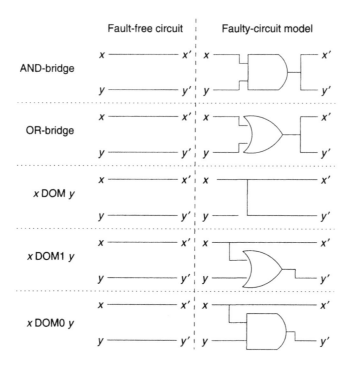

■ **FIGURE 4.55**

Bridging fault models.

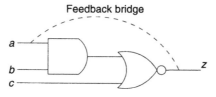

■ **FIGURE 4.56**

A feedback bridging fault.

Testing for bridging faults is similar to a constrained stuck-at ATPG. In other words, when testing for the AND-bridge(x, y), either (1) $x/0$ has to be detected with $y = 0$ or (2) $y/0$ has to be detected with $x = 0$ [Williams 1973]. A conventional stuck-at ATPG can be modified to handle the added constraint. Likewise, the ATPG can be modified for other bridging fault types.

4.11   OTHER TOPICS IN TEST GENERATION

4.11.1   Test Set Compaction

The vectors generated by any ATPG may include too many vectors. In other words, it may be possible to reduce the length of the test set without compromising on the fault coverage. Test compaction can be performed either statically or dynamically. Static compaction attempts to combine and remove certain vectors after the test set has been generated by an ATPG. Dynamic compaction, on the other hand, is integrated within the ATPG, in which the ATPG tries to generate vectors such that each vector detects as many faults as possible [Pomeranz 1991] [Rudnick 1999]. Obviously, static compaction can be performed even after dynamic compaction has been used.

Static compaction for combinational test vectors involves the selection of the minimal number of vectors that can detect all faults. Essentially, it is based on a covering algorithm, in which a matrix is constructed where the rows denote the vectors, and the columns denote the faults. A 1 is placed in element (i, j) of the matrix if vector i detects fault j. This matrix can be constructed by fault simulating the test set without fault dropping. Then, the compaction is set up as a covering problem with the following goal: Select a set of rows such that all columns (faults) are covered. For example, Table 4.17 shows such a matrix. In this example, vector v_2 is unnecessary because the faults that it detects can be detected by other vectors in the test set. Furthermore, vector v_4 is an **essential vector**, as it detects one or more faults that cannot be detected by any other vector.

Another form of static compaction is also possible in which compatible vectors are identified in a test set made up of incompletely specified vectors. For instance, vectors 11X1 and X101 are compatible. These two vectors can thus be combined, and one vector is sufficient.

Dynamic compaction, on the other hand, tries to intelligently fill in the "don't care" bits in the vectors such that more undetected faults can be detected. For example, when targeting fault f_i, the vector $1x10x$ may be sufficient. By filling the two "don't care" bits in a clever manner, more faults could be detected.

Compaction for sequential circuits is more involved, since removing a vector may not be permitted. Instead, any removal of a vector or sequence of vectors must be validated with a fault simulation to ensure that fault coverage is retained.

TABLE 4.17 ■ Combinational Test Compaction Matrix

	f_1	f_2	f_3	f_4	f_5	f_6
v_1	X		X		X	
v_2					X	X
v_3	X			X		X
v_4		X	X	X	X	

4.11.2 *N*-Detect ATPG

In order to enhance the quality of a test set, one may wish to derive different test sets targeting different fault models as an attempt to capture potential defects that could arise. However, this requires multiple ATPG engines, each targeting a different fault model. While this may be theoretically possible, it may not be possible in practice.

Instead, to increase the coverage of all possible defects, one may generate a test set that achieves multiple detections of every fault under a given fault model. A fault is detected multiple times if it is detected with different vectors. By exciting the fault and propagating the fault effect different ways, it is hoped that any defect locally close to the target fault will have an increased change of being detected [Franco 1995] [Chang 1998] [Dworak 2000]. For instance, detection of stuck-at fault $a/0$ with $b = 1$ will not have detected the AND-bridge fault between a and b. However, a different test that detects $a/0$ with $b = 0$ would have excited the bridge by setting $a = 1$ and $b = 0$.

In an N-detect setup, each fault must be targeted multiple times by an ATPG. In other words, all vectors generated that could detect a target fault are marked, and a fault is removed from further consideration when it has been detected N times. It has been shown that the size of an n-detect test set grows approximately linearly with respect to N [Reddy 1997].

4.11.3 ATPG for Acyclic Sequential Circuits

An acyclic sequential circuit is a circuit whose S-graph has no cycles. In such circuits, the sequential circuit may be transformed into a combinational circuit by unrolling the sequential circuit k time frames, where k is the sequential depth of the design [Kim 2001]. With the unrolled circuit, the circuit is inherently combinational and sequential ATPG is no longer needed. However, a fault in the original design may become a multiple stuck-at fault in the unrolled circuit, and the combinational ATPG must handle the multiple fault in order to detect the corresponding fault in the original sequential circuit. Several classes of acyclic circuits are studied in [Kim 2001], and different approaches to handling the test generation problem are reported.

4.11.4 IDDQ Testing

Unlike the test generation methods discussed thus far, which are focused on driving specific voltage values to circuit nodes and observing the voltage levels at the observable points such as the primary outputs, IDDQ testing targets the current drawn in the fabricated chips. Given a good chip, an expected current can be measured for a small set of input vectors. On defective chips, the currents drawn may differ drastically. For example, consider a circuit with a p-transistor of an inverter that is always on. Then, whenever the n-transistor of the inverter is switched on, a power-to-ground short is created, and the measured current could surge. Measuring the current is much slower than measuring the voltage, thus much fewer vectors can be considered. Further, the noise in current measurement must also be dealt with to ensure the quality of the test application.

4.11.5 Designing a High-Level ATPG

Because of the exponentially complex nature of ATPG, its performance can be severely limited to the size of the circuit. As a result, conventional gate-level ATPG may produce unsatisfactory results for large circuit sizes. On the other hand, higher level ATPGs have the advantages of fewer circuit primitives and easier access to circuit functional information that may enhance the ATPG effort.

The circuit is first given in a high-level description such as VHDL, Verilog, or SystemC. Then, the design is read in and an intermediate representation is constructed. Similar to gate-level ATPGs, the representation allows the high-level ATPG to traverse through the circuit and make decisions on the search. However, because the signals may not be Boolean, value justification and fault-effect propagation must work on the integer level. Backtracking mechanisms also have to be modified.

An alternative to testing the design at the high level structurally is testing the design with its *finite state machine* (FSM) as the circuit description. FSM-based testing relies on the traversal of states and transitions in the FSM description. Given a state diagram or flow-table of the FSM, any fault in the design will map to an error in the FSM, where the error could be a wrong transition, a wrong output, etc. Based on the FSM, sequences of vectors can be generated to traverse the state diagram. With an initializing sequence, the FSM can be driven to a know state. Transfer sequences are used to traverse the FSM. In addition, a distinguishing sequence is used to ensure that the circuit has indeed arrived at the desired state. Each state transition ($S_i \rightarrow S_j$) in the FSM is targeted one at a time in the following steps:

Step 1—Go to state S_i from the current state.

Step 2—Apply the input vector that takes the circuit from S_i to S_j.

Step 3—Apply the distinguishing sequence to check if the circuit is indeed in state S_j. Note that after the application of the distinguishing sequence the circuit may no longer be in state S_j.

While this approach is simple, it may not be scalable when the FSM is enormous. Furthermore, the test set generated by traversing the FSM may be very large.

Some success has been reported on some high-level ATPGs, where new value logic has been proposed. Nevertheless, high-level ATPGs remain an area of research in the days to come.

4.12 CONCLUDING REMARKS

This chapter describes in detail the underlying theory and implementation of an ATPG engine. It starts out with random TPG, followed by deterministic ATPG for combinational circuits, where branch-and-bound search is used. Several algorithms are laid out with specific examples given. Next, sequential ATPG is discussed where a combinational ATPG is extended to a 9-valued logic. Untestable fault identification is covered in detail where static logic implications are aggressively applied to help quickly identify untestable faults. Simulation-based ATPG is explained with particular emphasis on genetic-algorithm-based approaches. ATPG for non-stuck-at faults is also covered, with emphasis on those fault models that address delay defects, such as the path-delay fault and the transition fault. Finally, additional topics are briefly addressed that relate to the topic of test generation.

4.13 EXERCISES

4.1 (Random Test Generation) Given a circuit with three primary outputs, x, y, and z, the fanin cone of x is $\{a, b, c\}$, the fanin cone of y is $\{c, d, e, f\}$, and the fanin cone of z is $\{e, f, g\}$. Devise a pseudo-exhaustive test set for this circuit. Is this test set the minimal pseudo-exhaustive test set?

4.2 (Random Test Generation) Using the circuit shown in Figure 4.10, compute the detection probabilities for each of the following faults:

a. $e/0$

b. $e/1$

c. $c/0$

4.3 (Boolean Difference) Using the circuit shown in Figure 4.10, compute the set of all vectors that can detect each of the following faults using Boolean difference:

a. $e/0$

b. $e/1$

c. $c/0$.

4.4 (Boolean Difference) Using the circuit shown in Figure 4.16, compute the set of all vectors that can detect each of the following faults using Boolean difference:

a. $a/1$

b. $d/1$

c. $g/1$

4.5 (Boolean Difference) Using the circuit shown in Figure 4.35, compute the set of all vectors that can detect each of the following faults using Boolean difference:

a. $a/1$

b. $b1/1$

c. $e/0$

d. $e2/1$

4.6 **(Boolean Difference)** Assume a single-output combinational circuit, where the output is denoted as f. If two faults, α and β, are indistinguishable, it means that there does not exist a vector that can detect only one and not the other. Show that $f_\alpha \oplus f_\beta = 0$ if they are indistinguishable.

4.7 **(D Algorithm)** Construct the table for the XNOR operation for the 5-valued logic similar to Tables 4.1, 4.2, and 4.3.

4.8 **(D Algorithm)** Using the circuit shown in Figure 4.35, use the D algorithm to compute a vector for the fault $b/1$. Repeat for the fault $e/0$.

4.9 **(D Algorithm)** Consider a three-input AND gate g. Suppose $g \in D$-frontier. What are all the possible value combinations the three inputs of g can take such that g is a valid D-frontier?

4.10 **(PODEM)** Repeat Problem 4.8 using PODEM instead of the D algorithm.

4.11 **(PODEM)** Using the circuit shown in Figure 4.22, compute the vector that can detect the fault $f/0$. Note that even though the circuit is sequential it can be viewed as a combinational circuit because the D flip-flop does not have an explicit feedback.

4.12 **(Static Implications)** Using the circuit shown in Figure 4.22 and given the fact that the implications of $f = 1$ are shown in Figure 4.25, how could you use this information as multiple objectives to speed up the test generation for the fault $f/0$?

4.13 **(Static Implications)** Construct the static implication graph for the circuit shown in Figure 4.57 with only indirect implications. Based on the implication graph:

    a. What are all the implications for $g = 0$?

    b. What are all the implications for $f = 0$?

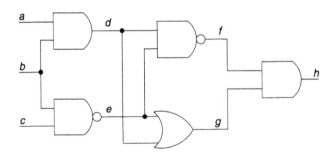

■ **FIGURE 4.57**

Example circuit.

4.14 **(Static Implications)** Construct the static implication graph for the circuit shown in Figure 4.58 by considering:

   a. Only direct implications

   b. Direct and indirect implications, including those obtained by the contra-positive law

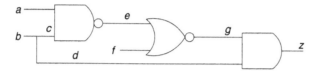

■ **FIGURE 4.58**

Example circuit.

4.15 **(Dynamic Implications)** Consider the circuit shown in Figure 4.58. Suppose justifying $e = 1$ via $a = 0$ is not possible due to some prespecified constraints. Perform all dynamic implications for all signals based on the knowledge of this constraint.

4.16 **(Dynamic Implications)** Prove that two faults, f and g, in a combinational circuit with the same E-frontier that has at least one D or $\bar{D}$, can be propagated to a primary output the same way.

4.17 **(Untestable Fault Identification)** Consider the circuit shown in Figure 4.58.

   a. Compute the static logic implications of $b = 0$.

   b. Compute the static logic implications of $b = 1$.

   c. Compute the set of faults that are untestable when $b = 0$.

   d. Compute the set of faults that are untestable when $b = 1$.

   e. Compute the set of untestable faults based on the stem analysis of b.

4.18 **(PODEM)** Consider the circuit shown in Figure 4.58, and use PODEM to generate a vector for each of the following faults:

   a. $c/0$

   b. $c/1$

   c. $d/0$

   d. $d/1$

4.19 (Untestable Fault Identification) Consider the circuit shown in Figure 4.59.

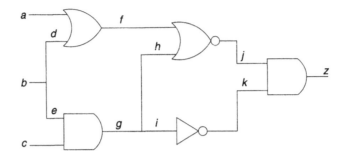

■ FIGURE 4.59

Example circuit.

    a. Compute the static logic implications of $b = 0$.

    b. Compute the static logic implications of $b = 1$.

    c. Compute the set of faults that are untestable when $b = 0$.

    d. Compute the set of faults that are untestable when $b = 1$.

    e. Compute the set of untestable faults based on the stem analysis of b.

4.20 (PODEM) Consider the circuit shown in Figure 4.59, and use PODEM to generate a vector for each of the following faults:

    a. $k/1$

    b. $k/0$

    c. $g/1$

    d. $g/0$

4.21 (Untestable Fault Identification) Prove that any fault that is combinationally untestable is also sequentially untestable.

4.22 (FAN) Consider the circuit shown in Figure 4.19. Suppose the constraint that $y = 1 \rightarrow x = 0$ is given. How could one use this knowledge to reduce the search space when trying to generate vectors in the circuit? For example, suppose the target fault is $y/0$.

4.23 (Sequential ATPG) Consider the circuit shown in Figure 4.60. The target fault is $a/0$.

    a. Generate a test sequence for the target fault using only 5-valued logic.

    b. Generate a test sequence for the target fault using 9-valued logic.

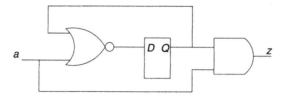

■ **FIGURE 4.60**

Example sequential circuit.

4.24 **(Sequential ATPG)** Given a sequential circuit, is it possible that two stuck-at faults, $a/0$ and $a/1$, are both detected by the same vector v_i in a test sequence $v_0, v_1, \ldots, v_k$?

4.25 **(Sequential ATPG)** Consider the sequential circuit shown in Figure 4.61. If the initial state is $de = 00$, what is the set of reachable states? Draw the corresponding state diagram for the finite state machine.

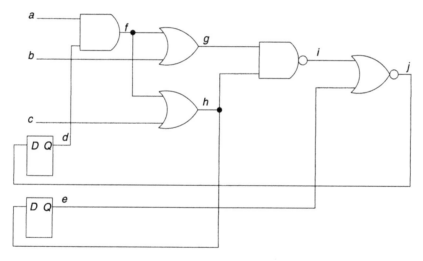

■ **FIGURE 4.61**

Example sequential circuit.

4.26 **(Sequential ATPG)** Consider an *iterative logic array* (ILA) expansion of a sequential circuit, where the initial pseudo primary inputs are fully controllable. Show that the states reachable in successive time frames of the ILA shrink monotonically.

4.27 **(Simulation-Based ATPG)** Design a simple genetic-algorithm based ATPG for combinational circuits. Design the fitness to be the number of faults detected. Adjust the GA parameters to observe the effectiveness of the test generator.

4.28 **(Simulation-Based ATPG)** Design a simple genetic-algorithm-based ATPG for sequential circuits, where an individual is a concatenation of several vectors. Design the fitness to be the number of faults detected. Adjust the GA parameters to observe the effectiveness of the test generator.

4.29 **(Advanced Simulation-Based ATPG)** Illustrate an example where a sequence that is able to propagate a fault-effect from a flip-flop FF_i to a primary output for fault f_1 cannot propagate a fault effect at the same flip-flop FF_i for a different fault f_j.

4.30 **(Hybrid ATPG)** Consider a fault f that is aborted by both deterministic and simulation-based test generators.

   a. What characteristics can be said for f considering that it is aborted by a deterministic ATPG?

   b. What characteristics can be said for f considering that it is aborted by a simulation-based ATPG?

   c. Suppose a hybrid ATPG detects f; what synergy is explored to detect f?

4.31 **(Path-Delay ATPG)** Consider the circuit fragment shown in Figure 4.62.

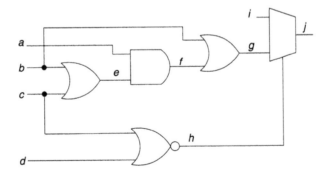

■ **FIGURE 4.62**

Example circuit.

   a. Generate all paths in this circuit. How many paths are there in this circuit?

   b. Which paths are functionally unsensitizable?

   c. For those sensitizable paths, which ones are robustly testable, and which ones are nonrobustly testable?

4.32 **(Path-Delay ATPG)** Given a combinational circuit with the knowledge of the implication $a = 1 \rightarrow b = 1$. How can this knowledge be used to deduce certain paths are unsensitizable?

4.33 **(Path-Delay ATPG)** Construct the table for the XNOR operation for the 5-valued system similar to Tables 4.12, 4.13, and 4.14.

4.34 **(Path-Delay ATPG)** Consider a full-scan circuit. Discuss how incidental detection of a sequentially untestable path-delay fault in the full-scan mode can lead to yield loss.

4.35 **(Transition Test Chains)** Consider the dictionary of excited and detected stuck-at faults of a test set shown in Table 4.18. Construct the smallest set of vectors that can detect as many transition faults as possible using only these seven stuck-at vectors.

TABLE 4.18 ■ Fault Dictionary Without Fault Dropping

Vectors	Excited Faults	Detected Faults
V_1	$a/0, b/0, c/0, d/0$	$e/1, f/1$
V_2	$c/0, f/0, g/0, h/0$	$e/1, f/1$
V_3	$d/0, e/0, h/0, i/0$	$a/1, b/1, c/1, f/1, g/1$
V_4	$a/0, b/0, g/0, i/0$	$d/1, e/1, f/1$
V_5	$c/0, d/0, g/0$	$a/1, d/1, h/1, i/1$
V_6	$d/0, e/0, i/0$	$a/1, b/1, c/1, f/1, g/1$
V_7	$b/0, g/0$	$e/1, i/1$

4.36 **(Bridging Faults)** Consider a bridging fault between the outputs of an AND gate $x = ab$ and an OR gate $y = c + d$. What values to $abcd$ would induce the largest current in the bridge?

4.37 **(A Design Practice)** Use the pseudo-random pattern generator and the ATPG program provided online to generate test sets for a number of combinational benchmark circuits. Compare and contrast the execution time and fault coverage obtained by the random TPG and the ATPG. What benefits does each have?

4.38 **(A Design Practice)** Repeat Problem 4.37 for sequential benchmark circuits.

4.39 **(A Design Practice)** Use the pseudo-random pattern generator provided online to generate test sets for a number of combinational benchmark circuits. Then, use the ATPG program also provided online to generate test vectors only for those undetected faults by the random vectors. Compare and contrast the execution time and fault coverage obtained by such a combined random and deterministic approach.

4.40 **(A Design Practice)** Repeat Problem 4.39 for sequential benchmark circuits.

References

R4.1—Introduction

[Abramovici 1994] M. Abramovici, M. A. Breuer, and A. D. Friedman, *Digital Systems Testing and Testable Design*, IEEE Press, Piscataway, NJ, 1994.

[Bushnell 2000] M. L. Bushnell and V. D. Agrawal, *Essentials of Electronic Testing for Digital, Memory, and Mixed-Signal VLSI Circuits*, Springer, New York, 2000.

[Holland 1975] J. H. Holland, *Adaptation in Natural and Artificial Systems*, University of Michigan Press, Ann Arbor, MI, 1975.

[Jha 2003] N. Jha and S. Gupta, *Testing of Digital Systems*, Cambridge University Press, Cambridge, U.K., 2003.

R4.2—Random Test Generation

[David 1976] R. David and G. Blanchet, About random fault detection of combinational networks, *IEEE Trans. Comput.*, C-25(6), 659–664, 1976.

[Shedletsky 1977] J. J. Shedletsky, Random testing: practicality vs. verified effectiveness, in *Proc. Int. Conf on Fault-Tolerant Computing*, June 1977, pp. 175–179.

R4.4—Designing a Stuck-At ATPG for Combinational Circuits

[Akers 1976] S. B. Akers, A logic system for fault test generation, *IEEE Trans. Comput.*, C-25(6), 620–630, 1976.

[Amyeen 1999] M. E. Amyeen, W. K. Fuchs, I. Pomeranz, and V. Boppana, Implications and evaluation techniques for proving fault equivalence, in *Proc. IEEE VLSI Test Symp.*, April 1999, pp. 201–207.

[Arora 2004] R. Arora and M. S. Hsiao, Enhancing SAT-based bounded model checking using sequential logic implications, in *Proc. IEEE Int. Conf on VLSI Design*, January 2004, pp. 784–787.

[El-Maleh 1998] A. El-Maleh, M. Kassab, and J. Rajski, A fast sequential learning technique for real circuits with application to enhancing ATPG performance, in *Proc. Design Automation Conf.*, June 1998, pp. 625–631.

[Fujiwara 1983] H. Fujiwara and T. Shimono, On the acceleration of test generation algorithms, *IEEE Trans. Comput.*, C-32(12), 1137–1144, 1983.

[Giraldi 1990] J. Giraldi and M. L. Bushnell, EST: the new frontier in automatic test pattern generation, in *Proc. Design Automation Conf.*, June 1990, pp. 667–672.

[Goel 1981] P. Goel, An implicit enumeration algorithm to generate tests for combinational logic circuits, *IEEE Trans. Comput.*, C-30(3), 215–222, 1981.

[Hamzaoglu 1999] I. Hamzaoglu and J. H. Patel, New techniques for deterministic test pattern generation, *J. Electron. Testing: Theory Appl.*, 15(1–2), 63–73, 1999.

[Hsiao 2002] M. S. Hsiao, Maximizing impossibilities for untestable fault identification, in *Proc. IEEE Design, Automation, and Test in Europe Conf.*, March 2002, pp. 949–953.

[Ichihara 1997] H. Ichihara and K. Kinoshita, On acceleration of logic circuits optimization using implication relations, in *Proc. Asian Test Symp.*, November 1997, pp. 222–227.

[Iyer 1996a] M. A. Iyer and M. Abramovici, FIRE: a fault independent combinational redundancy algorithm, *IEEE Trans. VLSI Syst.*, 4(2), 295–301, 1996.

[Iyer 1996b] M. A. Iyer, D. E. Long, and M. Abramovici, Identifying sequential redundancies without search, in *Proc. of Design Automation Conf.*, June 1996, pp. 457–462.

[Kajihara 2004] S. Kajihara, K. K. Saluja, and S. M. Reddy, Enhanced 3-valued logic/fault simulation for full scan circuits using implicit logic values, in *Proc. IEEE European Test Symp.*, May 2004, pp. 108–113.

[Kunz 1993] W. Kunz and D. K. Pradhan, Accelerated dynamic learning for test pattern generation in combinational circuits, *IEEE Trans. Comput.-Aided Des.*, 12(5), 684–694, 1993.

[Kunz 1994] W. Kunz and D. K. Pradhan, Recursive learning: a new implication technique for efficient solutions to CAD problems—test, verification, and optimization, *IEEE Trans. Comput.-Aided Des.*, 13(9), 1149–1158, 1994.

[Kunz 1997] W. Kunz, D. Stoffel, and P. R. Menon, Logic optimization and equivalence checking by implication analysis, *IEEE Trans. Comput.-Aided Des.*, 16(3), 266–281, 1997.

[Marques-Silva 1999a] J. Marques-Silva and T. Glass, Combinational equivalence checking using satisfiability and recursive learning, in *Proc. Design, Automation, and Test in Europe Conf.*, 145–149, 1999.

[Marques-Silva 1999b] J. P. Marques-Silva and K. A. Sakallah, GRASP: a search algorithm for propositional satisfiability, *IEEE Trans. Comput.*, 48(5), 506–521, 1999.

[Paul 2000] D. Paul, M. Chatterjee, and D. K. Pradhan, VERILAT: verification using logic augmentation and transformations, *IEEE Trans. Comput.-Aided Des.*, 19(9), 1041–1051, 2000.

[Peng 2000] Q. Peng, M. Abramovici, and J. Savir, MUST: multiple-stem analysis for identifying sequentially untestable faults, in *Proc. IEEE Int. Test Conf.*, October 2000, pp. 839–846.

[Rajski 1990] J. Rajski and H. Kox, A method to calculate necessary assignments in ATPG, in *Proc. IEEE Int. Test Conf.*, October 1990, pp. 25–34.

[Roth 1966] J. P. Roth, Diagnosis of automata failures: a calculus and a method, *IBM J. R&D*, 10(4), 278–291, 1966.

[Roth 1967] J. P. Roth, W. G. Bouricius, and P. R. Schneider, Programmed algorithms to compute tests to detect and distinguish between failures in logic circuits, *IEEE Trans. Electron. Comput.*, EC-16(10), 567–579, 1967.

[Schulz 1988] M. H. Schulz, E. Trischler, and T. M. Sarfert, SOCRATES: a highly efficient automatic test pattern generation system, *IEEE Trans. Comput.-Aided Des.*, 7(1), 126–137, 1988.

[Schulz 1989] M. H. Schulz and E. Auth, Improved deterministic test pattern generation with applications to redundancy identification, *IEEE Trans. Comput.-Aided Des.*, 8(7), 811–816, 1989.

[Syal 2004] M. Syal and M. S. Hsiao, Untestable fault identification using recurrence relations and impossible value assignments, in *Proc. IEEE Int. Conf. on VLSI Design*, January 2004, pp. 481–486.

[Tafertshofer 2000] P. Tafertshofer, A. Ganz, and K. J. Antreich, IGRAINE: an implication graph-based engine for fast implication, justification, and propagation, *IEEE Trans. Comput.-Aided Des.*, 19(8), 907–927, 2000.

[Zhao 2001] J. Zhao, J. A. Newquist, and J. H. Patel, A graph traversal based framework for sequential logic implication with an application to C-cycle redundancy identification, in *Proc. IEEE Int. Conf. on VLSI Design*, January 2001, pp. 163–169.

R4.5—Designing a Sequential ATPG

[Kirkland 1987] T. Kirkland and M. R. Mercer, A topological search algorithm for ATPG, in *Proc. IEEE Design Automation Conf.*, June 1987, pp. 502–508.

[Muth 1976] P. Muth, A nine-valued circuit model for test generation, *IEEE Trans. Comput.*, C-25(6), 630–636, 1976.

[Niermann 1991] T. M. Niermann and J. H. Patel, HITEC: a test generation package for sequential circuits, in *Proc. European Design Automation Conf.*, February 1991, pp. 214–218.

[Wang 2003] L.-T. Wang, K.S. Abdel-Hafez, X. Wen, B. Sheu, and S.-M. Wang, Smart ATPG (Automatic Test Pattern Generation) for Scan-Based Integrated Circuits, U.S. Patent Application No. 20050262409, May 23, 2003 (allowed May 2, 2006).

R4.6—Untestable Fault Identification

[Agrawal 1995] V. D. Agrawal and S. T. Chakradhar, Combinational ATPG theorems for identifying untestable faults in sequential circuits, *IEEE Trans. Comput.-Aided Des.*, 14(9), 1155–1160, 1995.

[Cheng 1993] K. T. Cheng, Redundancy removal for sequential circuits without reset states, *IEEE Trans. CAD*, 12(1), 13–24, 1993.

[Friedman 1967] A. D. Friedman, Fault detection in redundant circuits, *IEEE Trans. Electron. Comput.*, EC-16, 99–100, 1967.

[Hsiao 2002] M. S. Hsiao, Maximizing impossibilities for untestable faul1 identification, in *Proc. Design, Automation, and Test in Europe Conf.*, March 2002, pp. 949–953.

[Iyer 1996a] M. A. Iyer and M. Abramovici, FIRE: a fault independent combinational redundancy algorithm, *IEEE Trans. VLSI Syst.*, 4(2), 295–301, 1996.

[Iyer 1996b] M. A. Iyer, D. E. Long, and M. Abramovici, Identifying sequential redundancies without search, in *Proc. IEEE Design Automation Conf.*, June 1996, pp. 457–462.

[Long 2000] D. E. Long, M. A. Iyer, and M. Abramovici, FILL and FUNI: algorithms to identify illegal states and sequentially untestable faults, *ACM Trans. Des. Automat. Electron. Syst.*, 5(3), 631–657, 2000.

[Peng 2000] Q. Peng, M. Abramovici, and J. Savir, MUST: multiple-stem analysis for identifying sequentially untestable faults, in *Proc. IEEE Int. Test Conf.*, October 2000, pp. 839–846.

[Reddy 1999] S. M. Reddy, I. Pomeranz, X. Lim, and N. Z. Basturkmen, New procedures for identifying undetectable and redundant faults in synchronous sequential circuits, in *Proc. IEEE VLSI Test Symp.*, April 1999, pp. 275–281.

[Syal 2003] M. Syal and M. S. Hsiao, A novel, low-cost algorithm for sequentially untestable fault identification, in *Proc. IEEE Design, Automation, and Test in Europe Conf.*, March 2003, pp. 316–321.

[Syal 2004] M. Syal and M. S. Hsiao, Untestable fault identification using recurrence relations and impossible value assignments, in *Proc. IEEE Int. Conf. on VLSI Design*, January 2004, pp. 481–486.

R4.7—Designing a Simulation-Based ATPG

[Agrawal 1989] V. D. Agrawal, K. T. Cheng, and P. Agrawal, A directed search method for test generation using a concurrent simulator, *IEEE Trans. Comput.-Aided Des.*, 8(2), 131–138, 1989.

[Breuer 1971] M. A. Breuer, A random and an algorithmic technique for fault detection test generation for sequential circuits, *IEEE Trans. Comput.*, 20(11), 1364–1370, 1971.

[Holland 1975] J. H. Holland, *Adaptation in Natural and Artificial Systems*, University of Michigan Press, Ann Arbor, MI, 1975.

[Lisanke 1987] R. Lisanke, F. Brglez, A. J. Degeus, and D. Gregory, Testability-driven random test-pattern generation, *IEEE Trans. Comput.-Aided Des.*, 6(6), 1082–1087, 1987.

[Niermann 1992] T. M. Niermann, W.-T. Cheng, and J. H. Patel, PROOFS: A *Fast*, Memory-Efficient Sequential Circuit Fault Simulator, *IEEE Trans. Comput.-Aided Des.*, 11(2), 198–207, February 1992.

[Pomeranz 1997] I. Pomeranz and S. M. Reddy, On improving genetic optimization based test generation, in *Proc. European Design and Test Conf.*, February 1997, pp. 506–511.

[Prinetto 1994] P. Prinetto, M. Rebaudengo, and M. Sonza Reorda, An automatic test pattern generator for large sequential circuits based on genetic algorithms, in *Proc. IEEE Int. Test Conf.*, October 1994, pp. 240–249.

[Rudnick 1994a] E. M. Rudnick, J. G. Holm, D. G. Saab, and J. H. Patel, Application of simple genetic algorithms to sequential circuit test generation, in *Proc. European Design and Test Conf.*, February 1994, pp. 40–45.

[Rudnick 1994b] E. M. Rudnick, J. H. Patel, G. S. Greenstein, and T. M. Niermann, Sequential circuit test generation in a genetic algorithm framework, in *Proc. Design Automation Conf.*, June 1994, pp. 698–704.

[Saab 1992] D. G. Saab, Y. G. Saab, and J. A. Abraham, CRIS: a test cultivation program for sequential VLSI circuits, in *Proc. IEEE Int. Conf. on Comput.-Aided Des.*, November 1992, pp. 216–219.

[Schnurmann 1975] H. D. Schnurmann, E. Lindbloom, and R. G. Carpenter, The weighted random test-pattern generator, *IEEE Trans. Comput.*, 24(7), 695–700, 1975.

[Seshu 1962] S. Seshu and D. N. Freeman, The diagnosis of synchronous sequential switching systems, *IEEE Trans. Electron. Comput.*, 11, 459–465, 1962.

[Snethen 1977] T. J. Snethen, Simulator-oriented fault test generator, in *Proc. IEEE Design Automation Conf.*, June 1977, pp. 88–93.

[Srinivas 1993] M. Srinivas and L. M. Patnaik, A simulation-based test generation scheme using genetic algorithms, in *Proc. IEEE Int. Conf. on VLSI Design*, January 1993, pp. 132–135.

[Wunderlich 1990] H.-J. Wunderlich, Multiple distributions for biased random test patterns, *IEEE Trans. Comput.-Aided Des.*, 9(6), 584–593, 1990.

R4.8—Advanced Simulation-Based ATPG

[Giani 2001] A. Giani, S. Sheng, M. S. Hsiao, and V. Agrawal, Efficient spectral techniques for sequential ATPG, in *Proc. IEEE Design, Automation, and Test in Europe Conf.*, March 2001, pp. 204–208.

[Guo 1999] R. Guo, S. M. Reddy, and I. Pomeranz, Proptest: a property based test pattern generator for sequential circuits using test compaction, in *Proc. Design Automation Conf.*, June 1999, pp. 653–659.

[Hsiao 1996a] M. S. Hsiao, E. M. Rudnick, and J. H. Patel, Automatic test generation using genetically engineered distinguishing sequences, in *Proc. IEEE VLSI Test Symp.*, April 1996, pp. 216–223.

[Hsiao 1997] M. S. Hsiao, E. M. Rudnick, and J. H. Patel, Sequential circuit test generation using dynamic state traversal, in *Proc. European Design and Test Conf.*, February 1997, pp. 22–28.

[Hsiao 1998] M. S. Hsiao, E. M. Rudnick, and J. H. Patel, Application of genetically engineered finite-state-machine sequences to sequential circuit ATPG, *IEEE Trans. Comput.-Aided Des.*, 17(3), 239–254, 1998.

[Hsiao 2000] M. S. Hsiao, E. M. Rudnick, and J. H. Patel, Dynamic state traversal for sequential circuit test generation, *ACM Trans. Des. Automat. Electron. Sys.*, 5(3), 548–565, 2000.

[Hsiao 2001] M. S. Hsiao and J. Jain, Practical use of sequential ATPG for model checking: going the extra mile does pay off, in *Proc. Int. Workshop on High Level Design Validation and Test*, November 2001, pp. 39–44.

[Park 1995] J. Park, C. Oh, and M. R. Mercer, Improved sequential ATPG based on functional observation information and new justification methods, in *Proc. European Design and Test Conf.*, February 1995, pp. 262–266.

[Pomeranz 1995] 1. Pomeranz and S. M. Reddy, LOCSTEP: a logic simulation based test generation procedure, in *Proc. Fault Tolerant Computing Symp.*, June 1995, pp. 110–119.

[Rudnick 1995] E. M. Rudnick and J. H. Patel, Combining deterministic and genetic approaches for sequential circuit test generation, in *Proc. IEEE Design Automation Conf.*, June 1995, pp. 183–188.

[Sheng 2002] S. Sheng, K. Takayama, and M. S. Hsiao, Effective safety property checking based on simulation-based ATPG, in *Proc. IEEE Design Automation Conf.*, June 2002, pp. 813–818.

[Sheng 2002b] S. Sheng and M. S. Hsiao, Efficient sequential test generation based on logic simulation, *IEEE Des. Test Comput.*, 19(5), 56–64, 2002.

[Wu 2004] Q. WU and M. S. Hsiao, Efficient ATPG for design validation based on partitioned state exploration histories, in *Proc. IEEE VLSI Test Symp.*, April 2004, pp. 389–394.

R4.9—Hybrid Deterministic and Simulation-Based ATPG

[Hsiao 1996b] M. S. Hsiao, E. M. Rudnick, and J. H. Patel, Alternating strategies for sequential circuit ATPG, in *Proc. European Design and Test Conf.*, February 1996, pp. 368–374.

[Niermann 1991] T. M. Niermann and J. H. Patel, HITEC: a test generation package for sequential circuits, in *Proc. European Design Automation Conf.*, February 1991, pp. 214–218.

[Rudnick 1994] E. M. Rudnick, J. G. Holm, D. G. Saab, and J. H. Patel, Application of simple genetic algorithms to sequential circuit test generation, in *Proc. European Design and Test Conf.*, February 1994, pp. 40–45.

[Rudnick 1995] E. M. Rudnick and J. H. Patel, Combining deterministic and genetic approaches for sequential circuit test generation, in *Proc. Design Automation Conf.*, June 1995, pp. 183–188.

[Saab 1994] D. G. Saab, Y. G. Saab, and J. A. Abraham, Iterative [simulation-based genetics + deterministic techniques] = complete ATPG, in *Proc. IEEE Int. Conf. on Comput.-Aided Des.*, November 1994, pp. 40–43.

R4.10—ATPG for Non-Stuck-At Faults

[Bhawmik 1997] S. Bhawmik, Method and Apparatus for Built-In Self-Test with Multiple Clock Circuits, U.S. Patent No. 5,680,543, 1997.

[Cheng 1993] K. T. Cheng, S. Devadas, and K. Keutzer, Delay-fault test generation and synthesis for testability under a standard scan design methodology, *IEEE Trans. Comput.-Aided Des.*, 12(8), 1217–1231, 1993.

[Dervisoglu 1991] B. Dervisoglu and G. Stong, Design for testability: using scanpath techniques for path-delay test and measurement, in *Proc. IEEE Int. Test Conf.*, October 1991, pp. 365–374.

[Devadas 1993] S. Devadas, K. Keutzer, and S. Malik, Computation of floating mode delay in combinational circuits: theory and algorithms, *IEEE Trans. Comput.-Aided Des.*, 12(12), 1913–1923, 1993.

[Fuchs 1994] K. Fuchs, M. Pabst, and T. Rossel, RESIST: a recursive test pattern generation algorithm for path delay faults considering various test classes, *IEEE Trans. Comput.-Aided Des.*, 13(12), 1550–1562, 1994.

[Gupta 2004] P. Gupta and M. S. Hsiao, ALAPTF: a new transition fault mode l and the ATPG algorithm, in *Proc. IEEE Int. Test Conf.*, October 2004, pp. 1053–1060.

[Heragu 1996] K. Heragu, J. H. Patel, and V. D. Agrawal, Segment delay faults: a new fault model, in *Proc. IEEE VLSI Test Symp.*, April 1996, pp. 32–39.

[Hsu 2001] F. F. Hsu, K. M. Butler, and J. H. Patel, A case study of the Illinois scan architecture, in *Proc. IEEE Int. Test Conf.*, October 2001, pp. 538–547.

[Levendel 1986] Y. Levendel and P. Menon, Transition faults in combinational circuits: input transition test generation and fault simulation, in *Proc. Fault Tolerant Computing Symp.*, July 1986, pp. 278–283.

[Lin 1987] C. J. Lin and S. M. Reddy, On delay fault testing in logic circuits, *IEEE Trans. Comput.-Aided Des.*, 6(5), 694–703, 1987.

[Liu 2005] X. Liu, M. S. Hsiao, S. Chakravarty, and P. J. Thadikaran, Efficient techniques for transition testing, *ACM Trans. Des. Automat. Electron. Syst.*, 10(2), 258–278, 2005.

[Savir 1993] J. Savir and S. Patil, Scan-based transition test, *IEEE Trans. Comput.-Aided Des.*, 12(8), 1232–1241, 1993.

[Savir 1994] J. Savir and S. Patil, On broad-side delay test, in *Proc. IEEE VLSI Test Symp.*, April 1994, pp. 284–290.

[Smith 1985] G. L. Smith, Model for delay faults based upon paths, in *Proc. IEEE Int. Test Conf.*, October 1985, pp. 342–349.

[Tendulkar 2002] N. Tendulkar, R. Raina, R. Woltenburg, X. Lin, B. Swanson, and G. Aldrich, Novel techniques for achieving high at-speed transition fault coverage for Motorola's microprocessors based on PowerPC instruction set architecture, in *Proc. IEEE VLSI Test Symp.*, April 2002, pp. 3–8.

[Waicukauski 1987] J. A. Waicukauski, E. Lindbloom, B. K. Rosen, and V. S. Iyengar, Transition fault simulation, *IEEE Des. Test Comput.*, 4, 32–38, 1987.

[Wang 2002] L.-T. Wang, P.-C. Hsu, S.-C. Kao, M.-C. Lin, H.-P. Wang, H.-J. Chao, and X. Wen, Multiple-Capture DFT System for Detecting or Locating Crossing Clock-Domain Faults

During Self-Test or Scan-Test, U.S. Patent Application No. 20,020,120,896, August 29, 2002.

[Williams 1973] M. J. Y. Williams and J. B. Angel, Enhancing testability of large-scale integrated circuits via test points and additional logic, *IEEE Trans. Comput.*, C-22(1), 46–60, 1973.

R4.11—Other Topics in Test Generation

[Chang 1998] J. T.-Y. Chang, C.-W. Tseng, C.-M. J. Li, M. Purtell, and E. J. McCluskey, Analysis of pattern-dependent and timing-dependent failures in an experimental test chip, in *Proc. IEEE Int. Test Conf.*, October 1998, pp. 184–193.

[Dworak 2000] J. Dworak, M. R. Grimaila, S. Lee, L.-C. Wang, and M. R. Mercer, Enhanced DO-RE-ME based defect level prediction using defect site aggregation-MPG-D, in *Proc. IEEE Int. Test Conf.*, October 2000, pp. 930–939.

[Franco 1995] S. C. Ma, P. Franco, and E. J. McCluskey, An experimental chip to evaluate test technique experiment results, in *Proc. IEEE Int. Test Conf.*, October 1995, pp. 663–672.

[Kim 2001] Y. C. Kim, V. D. Agrawal, and K. K. Saluja, Combinational test generation for various classes of acyclic sequential circuits, *Proc. IEEE Int. Test Conf.*, October 2001, pp. 1078–1087.

[Pomeranz 1991] I. Pomeranz, L. N. Reddy, and S. M. Reddy, COMPACTEST: a method to generate compact test sets for combinatorial circuits, *Proc. IEEE Int. Test Conf.*, October 1991, pp. 194–203.

[Reddy 1997] S. M. Reddy, I. Pomeranz, and S. Kajihara, Compact test sets for high defect coverage, *IEEE Trans. Comput.-Aided Des.*, 16(8), 923–930, 1997.

[Rudnick 1999] E. M. Rudnick and J. H. Patel, Efficient techniques for dynamic test sequence compaction, *IEEE Trans. Comput.-Aided Des.*, 48(3), 323–330, 1999.

CHAPTER 5

LOGIC BUILT-IN SELF-TEST

Laung-Terng (L.-T.) Wang
SynTest Technologies, Inc., Sunnyvale, California

ABOUT THIS CHAPTER

Logic *built-in self-test* (BIST) is a *design for testability* (DFT) technique in which a portion of a circuit on a chip, board, or system is used to test the digital logic circuit itself. Logic BIST is crucial for many applications, in particular for life-critical and mission-critical applications. These applications commonly found in the aerospace/defense, automotive, banking, computer, healthcare, networking, and telecommunications industries require on-chip, on-board, or in-system self-test to improve the reliability of the entire system, as well as the ability to perform remote diagnosis.

This chapter first introduces the basic concepts and design rules of logic BIST. Next, we focus on a number of test pattern generation and output response analysis techniques suitable for BIST implementations. Test pattern generation techniques discussed include exhaustive testing, pseudo-random testing, and pseudo-exhaustive testing. Output response analysis techniques discussed include ones count testing, transition count testing, and signature analysis. Specific logic BIST architectures along with methods to further improve the circuit's fault coverage are then described, including the industry's widely used **STUMPS** architecture.

Finally, various BIST timing control diagrams are shown to illustrate how to test faults in a scan-based design containing multiple clock domains. This is particularly important for slow-speed testing of structural faults, such as stuck-at faults and bridging faults, as well as at-speed testing of timing-related delay faults, such as path-delay faults and transition faults.

A primary objective of this chapter is to enable the reader to design a logic BIST system comprised of a test pattern generator, output response analyzer, and logic BIST controller; therefore, we include a design practice example at the end of the chapter and show all necessary steps to arrive at the logic BIST system design, verify its correctness, and improve its fault coverage.

5.1   INTRODUCTION

With recent advances in semiconductor manufacturing technology, the production and usage of ***very-large-scale integration*** (VLSI) circuits has run into a variety of testing challenges during wafer probe, wafer sort, pre-ship screening, incoming test of chips and boards, test of assembled boards, system test, periodic maintenance, repair test, etc. Traditional test techniques that use ***automatic test pattern generation*** (ATPG) software to target single faults for digital circuit testing have become quite expensive and can no longer provide sufficiently high fault coverage for deep submicron or nanometer designs from the chip level to the board and system levels.

One approach to alleviate these testing problems is to incorporate ***built-in self-test*** (BIST) features into a digital circuit at the design stage [McCluskey 1986] [Abramovici 1994] [Bushnell 2000] [Mourad 2000] [Stroud 2002] [Jha 2003]. With logic BIST, circuits that generate test patterns and analyze the output responses of the functional circuitry are embedded in the chip or elsewhere on the same board where the chip resides.

There are two general categories of BIST techniques for testing random logic: (1) online BIST and (2) offline BIST. A general form of logic BIST techniques is shown in Figure 5.1 [Abramovici 1994].

Online BIST is performed when the functional circuitry is in normal operational mode. It can be done either *concurrently* or *nonconcurrently*. In **concurrent online BIST**, testing is conducted simultaneously during normal functional operation. The functional circuitry is usually implemented with coding techniques or with duplication and comparison [Abramovici 1994]. When an *intermittent* or *transient* error is detected, the system will correct the error on the spot, **rollback** to its previously stored system states, and repeat the operation, or generate an interrupt signal for repeated failures. These techniques are discussed in more detail in Chapter 12. In **nonconcurrent online BIST**, testing is performed when the functional circuitry is in idle mode. This is often accomplished by executing diagnosis software routines (macrocode) or diagnosis firmware routines (microcode) [Abramovici 1994]. The test process can be interrupted at any time so that normal operation can resume.

Offline BIST is performed when the functional circuitry is not in normal mode. This technique does not detect any *real-time errors* but is widely used in the industry

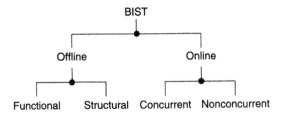

■ **FIGURE 5.1**

Logic BIST techniques.

for testing the functional circuitry at the system, board, or chip level to ensure product quality.

Functional offline BIST performs a test based on the functional specification of the functional circuitry and often employs a functional or high-level fault model. Normally such a test is implemented as diagnostic software or firmware.

Structural offline BIST performs a test based on the structure of the functional circuitry. There are two general classes of structural offline BIST techniques: (1) **external BIST**, in which test pattern generation and output response analysis is done by circuitry that is separate from the functional circuitry being tested, and (2) **internal BIST**, in which the **functional storage elements** are converted into test pattern generators and output response analyzers. Some external BIST schemes test sequential logic directly by applying test patterns at the inputs and analyzing the responses at its outputs. Such techniques are often used for board-level and system-level self-test. The BIST schemes discussed here all assume that the functional storage elements of the circuit are converted into a scan chain or multiple scan chains for combinational circuit testing. Such schemes are much more common than those that involve sequential circuit testing and are the primary focus of this chapter.

Figure 5.2 shows a typical logic BIST system using the *structural offline BIST* technique. The ***test pattern generator*** (TPG) automatically generates test patterns for application to the inputs of the ***circuit under test*** (CUT). The ***output response analyzer*** (ORA) automatically compacts the output responses of the CUT into a *signature*. Specific BIST timing control signals, including scan enable signals and clocks, are generated by the **logic BIST controller** for coordinating the BIST operation among the TPG, CUT, and ORA. The logic BIST controller provides a pass/fail indication once the BIST operation is complete. It includes comparison logic to compare the *final signature* with an embedded *golden signature*, and often comprises **diagnostic logic** for fault diagnosis. As compaction is commonly used for output response analysis, it is required that all storage elements in the TPG,

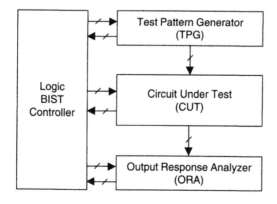

■ **FIGURE 5.2**

A typical logic BIST system.

CUT, and ORA be initialized to known states prior to self-test, and no unknown (X) values be allowed to propagate from the CUT to the ORA. In other words, the CUT must comply with additional *BIST-specific design rules*.

There are a number of advantages to using the *structural offline BIST* technique rather than conventional scan:

- BIST can be made to effectively test and report the existence of errors on the board or system and provide diagnostic information as required; it is always available to run the test and does not require the presence of an external tester.

- Because BIST implements most of the tester functions on-chip, the origin of errors can be easily traced back to the chip; some defects are detected without being modeled by software. *N*-detect, a method for detecting a fault *N*times, is done automatically. At-speed testing, which is inherent in BIST, can be used to detect many delay faults.

- Test costs are reduced due to reduced test time, tester memory requirements, or tester investment costs, as most of the tester functions reside on-chip itself.

However, there are also disadvantages associated with this approach. More stringent **BIST-specific design rules** are required to deal with unknown (X) sources originating from analog blocks, memories, non-scan storage elements, asynchronous set/reset signals, tristate buses, false paths, and multiple-cycle paths, to name a few. Also, because pseudo-random patterns are mostly used for BIST pattern generation, additional test points (including control points and observation points) may have to be added to improve the circuit's fault coverage.

While *BIST-specific design rules* are required and the BIST fault coverage may be lower than that using scan, BIST does eliminate the expensive process of software test pattern generation and the huge test data volume necessary to store the output responses for comparison. More importantly, a circuit embedded with BIST circuitry can be easily tested after being integrated into a system. Periodic in-system self-test, even using test patterns with less than perfect fault coverage, can diagnose problems down to the level where the BIST circuitry is embedded. This allows system repair to become trivial and economical.

5.2 BIST DESIGN RULES

Logic BIST requires much more stringent design restrictions when compared to conventional scan. While many *scan design rules* discussed in Chapter 2 are optional for scan designs, they are mandatory for BIST designs. The major logic BIST design restriction relates to the propagation of unknown (X) values. Because any unknown (X) value that propagates directly or indirectly to the output response analyzer will corrupt the *signature* and cause the BIST design to malfunction, no unknown (X) values can be tolerated. This is different from scan designs where unknown (X) values present in a scan design only result in fault coverage degradation. Therefore, when designing a logic BIST system, it is essential that the circuit under test meet all scan design rules and BIST-specific design rules, called *BIST design rules*. The process of taking a scan-based design and making it meet all additional *BIST-specific design rules* turns the design into a BIST-ready core.

5.2.1   Unknown Source Blocking

There are many unknown (X) sources in a CUT or BIST-ready core. Any unknown (X) source in the BIST-ready core, which is capable of propagating its unknown (X) value to the ORA directly or indirectly, must be blocked and fixed using a DFT repair approach often called **X-bounding** or **X-blocking**. Figure 5.3 shows a few of the more typically used X-bounding methods for blocking an unknown (X) source: The **0-control point** forces an X source to 0; the **1-control point** controls the X source to 1; the **bypass logic** allows the output of the X source to receive both 0 and 1 from a *primary input* (PI) or an internal node; the **control-only scan point** drives both 0 and 1 through a storage element, such as D flip-flop; and, finally, the **scan point** can capture the X-source value and drive both 0 and 1 through a scan cell, such as scan D flip-flop or *level-sensitive scan design* (LSSD) *shift register latch* (SRL) [Eichelberger 1977].

Depending on the nature of each unknown (X) source, several X-bounding methods can be appropriate for use. The most common problems inherent in these approaches include: (1) that they might increase the area of the design, and (2) that they might impact timing.

5.2.1.1   Analog Blocks

Examples of analog blocks are *analog-to-digital converters* (ADCs). Any analog block output that can exhibit unknown (X) behavior during test has to be forced to a

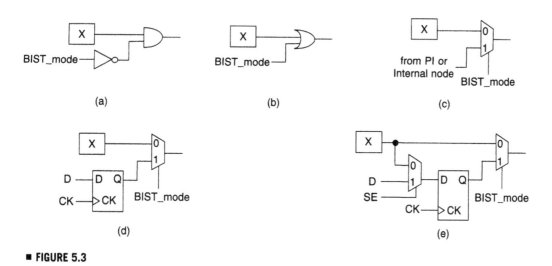

■ **FIGURE 5.3**

Typical X-bounding methods for blocking an unknown (X) source: (a) 0-control point; (b) 1-control point; (c) bypass logic; (d) control-only scan point; and (e) scan point.

known value. This can be accomplished by adding a 0-control point, 1-control point, bypass logic, or control-only scan point. We recommend the latter two approaches because they yield higher fault coverage than the former two approaches.

5.2.1.2 Memories and Non-Scan Storage Elements

Examples of memories are *dynamic random-access memories* (DRAMs), *static random-access memories* (SRAMs), or flash memories. Examples of non-scan storage elements are D flip-flops or D latches. Bypass logic is typically used to block each unknown (X) value originating from a memory or non-scan storage element. Another approach is to use an initialization sequence to set a memory or non-scan storage element to a known state. This is typically done to avoid adding delay to critical (functional) paths. Care must be taken to ensure that the stored state is not corrupted throughout the BIST operation.

5.2.1.3 Combinational Feedback Loops

All combinational feedback loops must be avoided. If they are unavoidable, then each loop must be broken using a 0-control point, a 1-control point, or a scan point. We recommend adding scan points because they yield higher fault coverage than the other approaches.

5.2.1.4 Asynchronous Set/Reset Signals

As indicated in Chapter 2, asynchronous set or reset can destroy the data during shift operation if a pattern causes the set/reset signal to become active. The asynchronous set or reset can be disabled using an external set/reset disable (RE) pin (see Figure 2.26). This set/reset disable pin must be set to 1 during shift operation. This may become cumbersome for BIST applications where there is a need to use the pin for other purposes. Thus, we recommend using the existing scan enable (SE) signal to protect each shift operation and adding a set/reset clock point (SRCK) on each set/reset signal to test the set/reset circuitry, as shown in Figure 5.4 [Abdel-Hafez 2004].

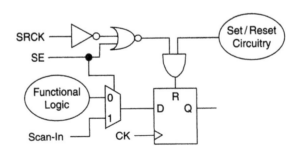

■ **FIGURE 5.4**

Set/reset clock point for testing a set/reset-type scan cell.

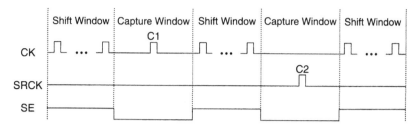

■ FIGURE 5.5

Example timing control diagram for testing data and set/reset faults.

In addition, we recommend testing all data and set/reset faults using two separate BIST sessions, as shown in Figure 5.5. The timing diagram in this figure is used for testing a circuit having one system clock (CK) and one added set/reset clock. To test data faults in the functional logic, a clock pulse C1 is triggered from CK while SRCK is held inactive in one capture window. Similarly, to test set/reset faults in the set/reset circuitry, C2 is enabled while CK is held inactive in another capture window. Using this approach, we can avoid races and hazards and prevent data in scan cells from being destroyed by the set/reset signals.

5.2.1.5   *Tristate Buses*

Bus contention occurs when two drivers force different values on the same bus which can damage the chip; hence, it is important to prevent bus conflicts during normal operation as well as shift operation [Cheung 1996]. For BIST applications, since pseudo-random patterns are commonly used, it is also crucial to protect the capture operation [Al-Yamani 2002]. To avoid potential bus contention, it is best to resynthesize each bus with multiplexers. If this is impractical, make sure only one tristate driver is enabled at any given time. The **one-hot decoder** shown in Figure 5.6 is an example of a circuit that can ensure that only one driver is selected during each shift or capture operation.

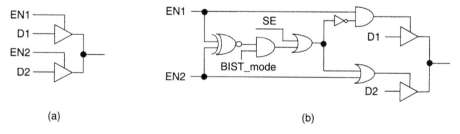

(a)   (b)

■ FIGURE 5.6

A one-hot decoder for testing a tristate bus with two drivers: (a) tristate bus, and (b) one-hot decoder.

5.2.1.6　False Paths

False paths are not normal functional paths. They do no harm to the chip during normal operation; however, for delay fault testing, a pseudo-random pattern might adversely attempt to test a selected false path. Because false paths are not exercised during normal circuit operation, they typically do not meet timing specifications, which can result in a mismatch during logic BIST delay fault testing. To avoid this potential problem, we recommend adding a 0-control point or 1-control point to each false path.

5.2.1.7　Critical Paths

Critical paths are timing-sensitive functional paths. Because the timing of these paths is critical, no additional gates are allowed to be added to the path, to prevent increasing the delay of the critical path. In order to remove an unknown (X) value from a critical path, we recommend adding an extra input pin to a selected combinational gate, such as an inverter, NAND gate, or NOR gate, on the critical path to minimize the added delay. The combinational gate is then converted to an embedded 0-control point or embedded 1-control point as shown in Figure 5.7, where an inverter is selected for adding the extra input.

5.2.1.8　Multiple-Cycle Paths

Multiple-cycle paths are normal functional paths but data are expected to arrive after two or more cycles. Similar to false paths, they can cause mismatches if exercised during delay fault testing, as they are intended to be tested in one cycle. To avoid this potential problem, we recommend adding a 0-control point or 1-control point to each multiple-cycle path or holding certain scan cell output states to avoid those multiple-cycle paths.

5.2.1.9　Floating Ports

Neither **primary inputs** (PIs) nor **primary outputs** (POs) can be floating. These ports must have a proper connection to power (V_{DD}) or ground (V_{SS}). Also, floating inputs to any internal modules must be avoided. This has a potential chance to propagate unknown (X) values to the ORA.

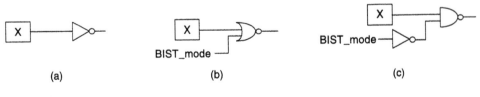

(a)　　　　　　　　　(b)　　　　　　　　　(c)

■ FIGURE 5.7

Embedded control points for testing a critical path having an inverter: (a) inverter; (b) embedded 0-control point; and (c) embedded 1-control point.

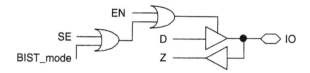

■ **FIGURE 5.8**

Forcing a bidirectional port to output mode.

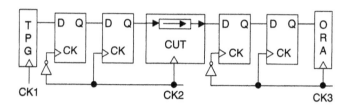

■ **FIGURE 5.9**

Re-timing logic among the TPG, CUT, and ORA.

5.2.1.10 Bidirectional I/O Ports

Bidirectional I/O ports are commonly used in a design. For BIST operations, make sure to fix the direction of each bidirectional I/O port to either input or output mode. Figure 5.8 shows an example of forcing a bidirectional I/O port to output mode.

5.2.2 Re-Timing

Because the TPG and the ORA are typically placed far from the CUT, races and hazards caused by clock skews may occur between the TPG and the (scan chain) inputs of the CUT as well as between the (scan chain) outputs of the CUT and the ORA. To avoid these potential problems and ease physical implementation, we recommend adding **re-timing logic** between the TPG and the CUT and between the CUT and the ORA. The re-timing logic should consist of at least one negative-edge **pipelining register** (D flip-flop) and one positive-edge pipelining register (D flip-flop). Figure 5.9 shows an example of re-timing logic among the TPG, CUT, and ORA, using two pipelining registers on each end. Note that the three clocks (CK1, CK2, and CK3) could belong to one clock tree.

5.3 TEST PATTERN GENERATION

For logic BIST applications, in-circuit TPGs constructed from *linear feedback shift registers* (LFSRs) are most commonly used to generate test patterns or test sequences for exhaustive testing, pseudo-random testing, and pseudo-exhaustive testing. **Exhaustive testing** always guarantees 100% single-stuck and multiple-stuck fault coverage. This technique requires all possible 2^n test patterns to be applied to an *n*-input combinational *circuit under test* (CUT), which can take too long for combinational circuits where *n* is huge; therefore, **pseudo-random testing** [Bardell 1987] is often used for generating a subset of the 2^n test patterns and uses fault simulation to calculate the exact fault coverage. In some cases, this might become quite time consuming, if not infeasible. In order to eliminate the need for fault simulation while at the same time maintaining 100% single-stuck fault coverage, we can use **pseudo-exhaustive testing** [McCluskey 1986] to generate 2^w or $2^k - 1$ test patterns, where $w < k < n$, when each output of the *n*-input combinational CUT at most depends on *w* inputs. For testing delay faults, hazards must also be taken into consideration.

Standard LFSR

Figure 5.10 shows an *n*-stage **standard LFSR**. It consists of *n* D flip-flops and a selected number of *exclusive-OR* (XOR) gates. Because XOR gates are placed on the external feedback path, the standard LFSR is also referred to as an **external-XOR LFSR** [Golomb 1982].

Modular LFSR

Similarly, an *n*-stage **modular LFSR** with each XOR gate placed between two adjacent D flip-flops, as shown in Figure 5.11, is referred to as an **internal-XOR LFSR** [Golomb 1982]. The modular LFSR runs faster than its corresponding standard LFSR, because each stage introduces at most one XOR-gate delay.

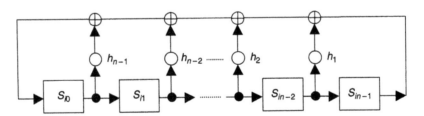

■ **FIGURE 5.10**

An *n*-stage (external-XOR) standard LFSR.

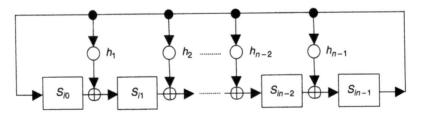

■ **FIGURE 5.11**

An *n*-stage (internal-XOR) standard LFSR.

LFSR Properties

The internal structure of the n-stage LFSR in each figure can be described by specifying a **characteristic polynomial** of degree n, $f(x)$, in which the symbol h_i is either 1 or 0, depending on the existence or absence of the feedback path, where:

$$f(x) = 1 + h_1 x + h_2 x^2 + \cdots + h_{n-1} x^{n-1} + x^n$$

Let S_i represent the contents of the n-stage LFSR after ith shifts of the initial contents, S_0, of the LFSR, and let $S_i(x)$ be the polynomial representation of S_i. Then, $S_i(x)$ is a polynomial of degree $n-1$, where:

$$S_i(x) = S_{i0} + S_{i1} x + S_{i2} x^2 + \cdots + S_{in-2} x^{n-2} + S_{in-1} x^{n-1}$$

If T is the smallest positive integer such that $f(x)$ divides $1 + x^T$, then the integer T is called the **period** of the LFSR. If $T = 2^n - 1$, then the n-stage LFSR generating the **maximum-length sequence** is called a **maximum-length LFSR**. For example, consider the four-stage standard and modular LFSRs shown in Figure 5.12a and Figure 5.12b, below. The characteristic polynomials, $f(x)$, used to construct both LFSRs are $1 + x^2 + x^4$ and $1 + x + x^4$, respectively.

The test sequences generated by each LFSR, when its initial contents, S_0, are set to {0001} or $S_0(x) = x^3$, are listed in Figures 5.12c and 5.12d, respectively. Because

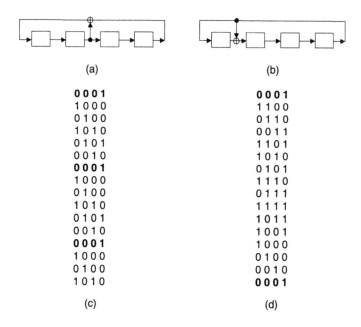

(a) (b)

(c)	(d)
0 0 0 1	**0 0 0 1**
1 0 0 0	1 1 0 0
0 1 0 0	0 1 1 0
1 0 1 0	0 0 1 1
0 1 0 1	1 1 0 1
0 0 1 0	1 0 1 0
0 0 0 1	0 1 0 1
1 0 0 0	1 1 1 0
0 1 0 0	0 1 1 1
1 0 1 0	1 1 1 1
0 1 0 1	1 0 1 1
0 0 1 0	1 0 0 1
0 0 0 1	1 0 0 0
1 0 0 0	0 1 0 0
0 1 0 0	0 0 1 0
1 0 1 0	**0 0 0 1**

■ **FIGURE 5.12**

Example four-stage test pattern generators (TPGs): (a) four-stage standard LFSR; (b) four-stage modular LFSR; (c) test sequence generated by (a); and (d) test sequence generated by (b).

the first test sequence repeats after 6 patterns and the second test sequence repeats after 15 patterns, the LFSRs have periods of 6 and 15, respectively. This further implies that $1+x^6$ can be divided by $1+x^2+x^4$, and $1+x^{15}$ can be divided by $1+x+x^4$.

Define a **primitive polynomial** of degree n over **Galois field** $GF(2)$, $p(x)$, as a polynomial that divides $1+x^T$, but not $1+x^i$, for any integer $i < T$, where $T = 2^n - 1$ [Colomb 1982]. A primitive polynomial is **irreducible**. Because $T = 15 = 2^4 - 1$, the characteristic polynomial, $f(x) = 1+x+x^4$, used to construct Figure 5.12b is a primitive polynomial; thus, the modular LFSR is a maximum-length LFSR. Let:

$$r(x) = f(x)^{-1} = x^n f(x^{-1})$$

Then $r(x)$ is defined as a **reciprocal polynomial** of $f(x)$ [Peterson 1972]. A reciprocal polynomial of a primitive polynomial is also a primitive polynomial. Thus, the reciprocal polynomial of $f(x) = 1+x+x^4$ is also a primitive polynomial, with $p(x) = r(x) = 1+x^3+x^4$.

Hybrid LFSR

Let a polynomial over $GF(2)$, $a(x) = 1+b(x)+c(x)$, be said to be **fully decomposable** if both $b(x)$ and $c(x)$ have no common terms and there exists an integer j such that $c(x) = x^j b(x)$, where $j \geq 1$. If $f(x)$ is fully decomposable such that:

$$f(x) = 1+b(x)+x^j b(x)$$

then a **(hybrid) top–bottom LFSR** [Wang 1988a] can be constructed using the connection polynomial:

$$s(x) = 1+ {}^{\wedge}x^j +x^j b(x)$$

where ${}^{\wedge}x^j$ indicates that the XOR gate with one input taken from the jth stage output of the LFSR is connected to the feedback path, not between stages. Similarly, if $f(x)$ is fully decomposable such that:

$$f(x) = b(x)+x^j b(x)+x^n$$

then a **(hybrid) bottom–top LFSR** [Wang 1988a] can be constructed using the connection polynomial:

$$s(x) = b(x)+ {}^{\wedge}x^{n-j} +x^n$$

It was shown in [Wang 1988a] that if top–bottom LFSRs exist for a characteristic polynomial, $f(x)$, then bottom–top LFSRs will exist for its reciprocal polynomial, $r(x)$. Assume that a standard or modular LFSR uses m XOR gates, where m is an odd number. If its characteristic polynomial, $f(x)$, is fully decomposable, then a hybrid LFSR can be realized with only $(m+1)/2$ XOR gates. Figure 5.13 shows two example five-stage hybrid LFSRs each using two, rather than three, XOR gates.

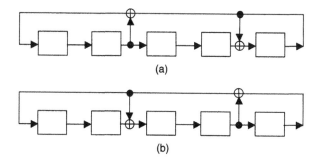

■ **FIGURE 5.13**

Hybrid LFSRs: (a) five-stage top–bottom LFSR using $s(x) = 1 + {}^\wedge x^2 + x^4 + x^5$ for $f(x) = 1 + x^2 + x^3 + x^4 + x^5$, and (b) five-stage bottom–top LFSR using $s(x) = 1 + x^2 + {}^\wedge x^4 + x^5$ for $f(x) = 1 + x + x^2 + x^3 + x^5$.

Table 5.1 lists a set of primitive polynomials of degree n up to 100 [Bardell 1987]. A different set was given in [Wang 1988a]. Each polynomial can be used to construct minimum-length LFSRs in standard, modular, or hybrid form. For primitive polynomials of degree up to 300, consult [Bardell 1987].

5.3.1 Exhaustive Testing

Exhaustive testing requires applying 2^n exhaustive patterns to an n-input combinational *circuit under test* (CUT). Any **binary counter** can be used as an *exhaustive pattern generator* (EPG) for this purpose; however, because the order of generation of the inputs is not important, it may be more efficient to use an autonomous, maximum-length LFSR that can cycle through all states. To do this, it is necessary to modify the LFSR so that the all-zero state is included [McCluskey 1981] [McCluskey 1986]. A general procedure for constructing modified (maximum-length) LFSRs that include the all-zero state is given in [Wang 1986b]. These modified LFSRs are called *complete LFSRs* (CFSRs).

5.3.1.1 Binary Counter

Figure 5.14 shows an example of a 4-bit binary counter design [Wakerly 2000] for testing a four-input combinational CUT. Binary counters are simple to design but require more hardware than LFSRs.

5.3.1.2 Complete LFSR

Figures 5.15a and 5.15b show two complete LFSRs for testing the four-input CUT. Each figure is reconfigured from a four-stage maximum-length LFSR such that the resulting standard or modular CFSR has period 16. In each CFSR, an XOR gate is inserted into the last stage of the LFSR, and a NOR gate (with $n - 1 = 3$ inputs from the first $n - 1$ stages of the LFSR) is used as a **zero-detector**. With this reconfiguration, both CFSRs insert the all-zero state right after state {0001} is reached.

TABLE 5.1 ■ Primitive Polynomials of Degree *n* up to 100

n	Exponents	*n*	Exponents	*n*	Exponents	*n*	Exponents
1	0	26	8 7 1 0	51	16 15 1 0	76	36 35 1 0
2	1 0	27	8 7 1 0	52	3 0	77	31 30 1 0
3	1 0	28	3 0	53	16 15 1 0	78	20 19 1 0
4	1 0	29	2 0	54	37 36 1 0	79	9 0
5	2 0	30	16 15 1 0	55	24 0	80	38 37 1 0
6	1 0	31	3 0	56	22 21 1 0	81	4 0
7	1 0	32	28 27 1 0	57	7 0	82	38 35 3 0
8	6 5 1 0	33	13 0	58	19 0	83	46 45 1 0
9	4 0	34	15 14 1 0	59	22 21 1 0	84	13 0
10	3 0	35	2 0	60	1 0	85	28 27 1 0
11	2 0	36	11 0	61	16 15 1 0	86	13 12 1 0
12	7 4 3 0	37	12 10 2 0	62	57 56 1 0	87	13 0
13	4 3 1 0	38	6 5 1 0	63	1 0	88	72 71 1 0
14	12 11 1 0	39	4 0	64	4 3 1 0	89	38 0
15	1 0	40	21 19 2 0	65	18 0	90	19 18 1 0
16	5 3 2 0	41	3 0	66	10 9 1 0	91	84 83 1 0
17	3 0	42	23 22 1 0	67	10 9 1 0	92	13 12 1 0
18	7 0	43	6 5 1 0	68	9 0	93	2 0
19	6 5 1 0	44	27 26 1 0	69	29 27 2 0	94	21 0
20	3 0	45	4 3 1 0	70	16 15 1 0	95	11 0
21	2 0	46	21 20 1 0	71	6 0	96	49 47 2 0
22	1 0	47	5 0	72	53 47 6 0	97	6 0
23	5 0	48	28 27 1 0	73	25 0	98	11 0
24	4 3 1 0	49	9 0	74	16 15 1 0	99	47 45 2 0
25	3 0	50	27 26 1 0	75	11 10 1 0	100	37 0

Note: "24 4 3 1 0" means $p(x) = x^{24} + x^4 + x^3 + x^1 + x^0 = x^{24} + x^4 + x^3 + x + 1$.
Source: P. H. Bardell *et al.*, *Built-In Test for VLSI: Pseudorandom Techniques*, John Wiley & Sons, Somerset, NJ, 1987.

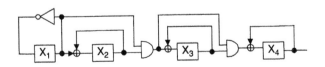

■ **FIGURE 5.14**

Example binary counter as EPG.

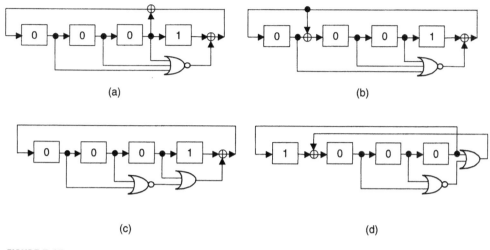

■ **FIGURE 5.15**

Example complete LFSRs (CFSRs) as EPGs: (a) four-stage standard CFSR; (b) four-stage modular CFSR; (c) minimized version of (a); and (d) minimized version of (b).

It is possible to further minimize both CFSR designs. For any standard CFSR, this can be by Boolean minimization, as shown in Figure 5.15c. For any modular CFSR, the minimization is done by replacing the XOR gate at the last stage by an OR gate and reconnecting the tap to the OR-gate output, as shown in Figure 5.15d. With this arrangement, the modular CFSR inserts the all-zero state right after state {1000} is reached and then switches to state {0100} on the next clock.

If further minimization is necessary, then using the hybrid LFSR scheme presented above can save about half of the XOR gates required for the feedback connection.

Exhaustive testing guarantees that all detectable, combinational faults (those that do not change a combinational circuit into a sequential circuit) will be detected. This approach is especially useful for circuits where the number of inputs, n, is a small number (*e.g.*, 20 or less). When n is larger than 20, the test time may be prohibitively long and is thus not recommended. The following techniques are aimed at reducing the number of test patterns. They are recommended when exhaustive testing is impractical.

5.3.2 Pseudo-Random Testing

One approach that can reduce test length but sacrifices the circuit fault coverage uses a ***pseudo-random pattern generator*** (PRPG) for generating a pseudo-random sequence of test patterns [Bardell 1987] [Rajski 1998] [Bushnell 2000] [Jha 2003]. **Pseudo-random testing** has the advantage of being applicable to both sequential and combinational circuits; however, there are difficulties in determining the required test length and fault coverage. Schemes to estimate the random test length required to achieve a certain level of fault detection or obtain a certain defect level can be found in [Savir 1984b], [Williams 1985], [Chin 1987], [Wagner 1987], and [Seth 1990]. Its effectiveness has been reported in [Lisanke 1987] and [Wunderlich 1988].

5.3.2.1 Maximum-Length LFSR

Maximum-length LFSRs are commonly used for pseudo-random pattern generation. Each LFSR produces a sequence with 0.5 probability of generating 1's (or with probability distribution 0.5) at every output. The **LFSR pattern generation technique** that uses these LFSRs in standard, modular, or hybrid form to generate patterns for the entire design has the advantage of being very easy to implement. The major problem with this approach is that some circuits may be *random-pattern resistant* (RP-resistant) [Savir 1984a]; that is, either the probability of certain nodes randomly receiving a 0 or 1 or the probability of observing certain nodes at the circuit outputs is low, assuming equi-probable inputs. For example, consider a five-input OR gate. The probability of applying an all-zero pattern to all inputs is 1/32. This makes it difficult to test the RP-resistant OR-gate output stuck-at-1.

5.3.2.2 Weighted LFSR

It is possible to increase fault coverage (and detect most RP-resistant faults) in RP-resistant designs. A **weighted pattern generation technique** employing an LFSR and a combinational circuit was first described in [Schnurmann 1975]. The combinational circuit inserted between the output of the LFSR and the CUT is to increase the frequency of occurrence of one logic value while decreasing the other logic value. This approach may increase the probability of detecting those faults that are difficult to detect using the typical LFSR pattern generation technique.

Implementation methods for realizing this scheme are further discussed in [Chin 1984]. The weighted pattern generation technique described in that paper modifies the maximum-length LFSR to produce an equally weighted distribution of 0's and 1's at the input of the CUT. It skews the LFSR probability distribution of 0.5 to either 0.25 or 0.75 to increase the chance of detecting those faults that are difficult to detect using just a 0.5 distribution. Better fault coverage was also found in [Wunderlich 1987] where probability distributions in a multiple of 0.125 (rather than 0.25) are used. For some circuits, several programmable probabilities or weight sets are required in order to further increase each circuit's fault coverage [Waicukauski 1989] [Bershteyn 1993] [Kapur 1994] [Lai 2005]. Additional discussions on weighted pattern generation can be found in the books [Rajski 1998] and [Bushnell 2000]. Figure 5.16 shows a four-stage weighted (maximum-length) LFSR with probability distribution 0.75 [Chin 1984].

5.3.2.3 Cellular Automata

Cellular automata were first introduced in [Wolfram 1983]. They yielded better randomness property than LFSRs [Hortensius 1989]. The *cellular-automaton-based* (or CA-based) *pseudo-random pattern generator* (PRPG) "is attractive for BIST applications" [Khara 1987] [Gloster 1988] [Wang 1989] [van Sas 1990] because it: (1) provides patterns that look more random at the circuit inputs, (2) has

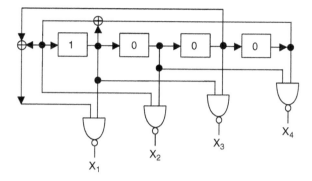

■ **FIGURE 5.16**

Example weighted LFSR as PRPG.

higher opportunity to reach very high fault coverage in a circuit that is RP-resistant, and (3) has implementation advantages as it only requires adjacent neighbor communication (no global feedback, unlike the modular LFSR case).

A ***cellular automaton*** (CA) is a collection of **cells** with forward and backward connections. A general structure is shown in Figure 5.17a. Each cell can only connect to its local neighbors (adjacent left and right cells). The connections are expressed as **rules**; each rule determines the next state of a cell based on the state of the cell and its neighbors. Assume cell i can only talk with its neighbors, $i-1$ and $i+1$. Define:

$$Rule\ 90:\ x_i(t+1) = x_{i-1}(t) + x_{i+1}(t)$$

and

$$Rule\ 150:\ x_i(t+1) = x_{i-1}(t) + x_i(t) + x_{i+1}(t)$$

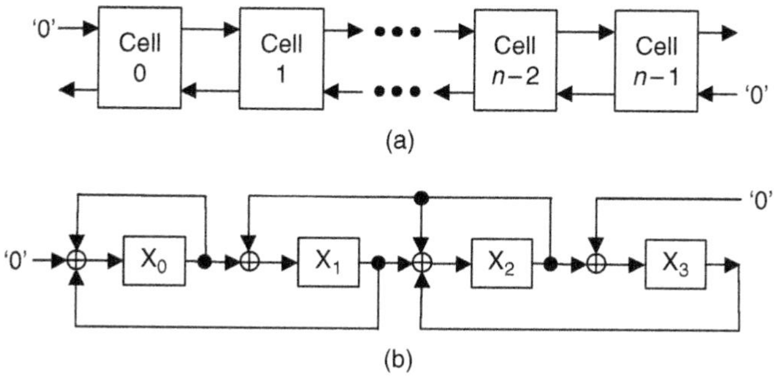

```
0 0 0 1
0 0 1 0
0 1 1 1
1 1 1 1
0 0 1 1
0 1 0 1
1 0 0 0
1 1 0 0
0 1 1 0
1 1 0 1
0 1 0 0
1 0 1 0
1 0 1 1
1 0 0 1
1 1 1 0
```

(a)

(b)

(c)

■ **FIGURE 5.17**

Example cellular automation (CA) as PRPG: (a) general structure of an *n*-stage CA; (b) four-stage CA; and (c) test sequence generated by (b).

Then the two rules, *rule 90* and *rule 150*, can be established based on the following state transition table:

$x_{i-1}(t)x_i(t)x_{i+1}(t)$	111	110	101	100	011	010	001	000
Rule 90: $x_i(t+1)$	0	1	0	1	1	0	1	0

$$2^6 + 2^4 + 2^3 + 2^1 = 90$$

Rule 150: $x_i(t+1)$	1	0	0	1	0	1	1	0

$$2^7 + 2^4 + 2^2 + 2^1 = 150$$

The terms *rule 90* and *rule 150* were derived from their decimal equivalents of the binary code for the next state of cell i [Hortensius 1989]. Figure 5.17b shows an example of a four-stage CA generated by alternating rules 150 (on even cells) and 90 (on odd cells). Similar to the four-stage modular LFSR given in Figure 5.12b, the four-stage CA generates a *maximum-length sequence* of 15 distinct states, as listed in Figure 5.17c.

It has been shown in [Hortensius 1989] that, by combining cellular automata rules 90 and 150, an n-stage CA can generate a maximum-length sequence of $2^n - 1$. The construction rules for $4 \leq n \leq 53$ can be found in [Hortensius 1989] and are listed in Table 5.2.

[Serra 1990] and [Slater 1990] demonstrated an isomorphism between a one-dimensional linear cellular automaton and a maximum-length LFSR having the same number of stages; however, state sequencing may still differ between the CA and the LFSR. CAs have much less shift-induced bit value correlation (only on those left-/right-edge-cells built with *rule 90*) than LFSRs. The LFSR, however, can be made more random by using a *linear phase shifter* [Das 1990].

The CA-based PRPG can be programmed as a **universal CA** for generating different orders of test sequences. A **universal CA-cell** for generating patterns based on *rule 90* or *rule 150* is given in Figure 5.18 [Wang 1989]. When the RULE150_SELECT signal is set to 1, the universal CA-cell will behave as a *rule 150* cell; otherwise, it will act as a *rule 90* cell. This universal CA structure is useful for BIST applications

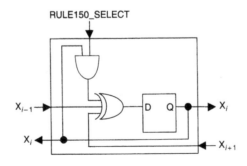

■ **FIGURE 5.18**

A universal CA-cell structure.

TABLE 5.2 ■ Construction Rules for Cellular Automata of Length n up to 53

n	Rule[a]	n	Rule[a]
4	05	29	2,512,712,103
5	31	30	7,211,545,075
6	25	31	04,625,575,630
7	152	32	10,602,335,725
8	325	33	03,047,162,605
9	625	34	036,055,030,672
10	0,525	35	127,573,165,123
11	3,252	36	514,443,726,043
12	2,525	37	0,226,365,530,263
13	14,524	38	0,345,366,317,023
14	17,576	39	6,427,667,463,554
15	44,241	40	00,731,257,441,345
16	152,525	41	15,376,413,143,607
17	175,763	42	11,766,345,114,746
18	252,525	43	035,342,704,132,622
19	0,646,611	44	074,756,556,045,302
20	3,635,577	45	151,315,510,461,515
21	3,630,173	46	0,112,312,150,547,326
22	05,252,525	47	0,713,747,124,427,015
23	32,716,432	48	0,606,762,247,217,017
24	77,226,526	49	02,675,443,137,056,631
25	136,524,744	50	23,233,006,150,544,226
26	132,642,730	51	04,135,241,323,505,027
27	037,014,415	52	031,067,567,742,172,706
28	0,525,252,525	53	207,121,011,145,676,625

[a] Rule is given in octal format. For $n = 7$, Rule $= 152 = 001,101,010 = 1,101,010$, where "0" denotes a *rule 90* cell and "1" denotes a *rule 150* cell, or vice versa.

where it is required to obtain very high fault coverage for RP-resistant designs or detect additional classes of faults.

5.3.3 Pseudo-Exhaustive Testing

Another approach to reduce the test time to a practical value while retaining many of the advantages of exhaustive testing is the **pseudo-exhaustive test technique**. It applies fewer than 2^n test patterns to an n-input combinational CUT. The technique depends on whether any output is driven by all of its inputs. If none of the outputs depends on all inputs, a **verification test approach** proposed in [McCluskey 1984] can be used to test these circuits. In circuits where there is one output that

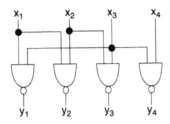

■ **FIGURE 5.19**

An $(n, w) = (4, 2)$ CUT.

depends on all inputs or the test time using verification testing is still too long, a **segmentation test approach** must be used [McCluskey 1981]. Pseudo-exhaustive testing guarantees single-stuck fault coverage without any detailed circuit analysis.

5.3.3.1 Verification Testing

Verification testing [McCluskey 1984] divides the circuit under test into m cones, where m is the number of outputs. It is based on backtracing from each circuit output to determine the actual number of inputs that drive the output. Each cone will receive exhaustive test patterns, and all cones are tested concurrently.

Assume the combinational CUT has n inputs and m outputs. Let w be the maximum number of input variables upon which any output of the CUT depends. Then, the n-input m-output combinational CUT is defined as an (n, w) CUT, where $w < n$. Figure 5.19 shows an $(n, w) = (4, 2)$ CUT that will be used as an example for designing the ***pseudo-exhaustive pattern generators*** (PEPGs).

Syndrome Driver Counter

The first method for **pseudo-exhaustive pattern generation** was proposed in [Savir 1980]. **Syndrome driver counters** (SDCs) are used to generate test patterns [Barzilai 1981]. The SDC can be a binary counter, a maximum-length LFSR, or a complete LFSR. This method checks whether some circuit inputs can share the same test signal. If $n - p$ inputs, $p < n$, can share the **test signals** with the other p inputs, then the circuit can be tested exhaustively with these p inputs. In this case, the test length becomes 2^p if $p = w$ or $2^p - 1$ if $p > w$. Figure 5.20a shows a three-stage SDC used to test the circuit given in Figure 5.19. Because both inputs x_1 and x_4 do not drive the same output, one test signal can be used to drive both inputs. In this case, p is 3, and the test length becomes $2^3 - 1 = 7$. Designs based on the SDC method for in-circuit test pattern generation are simple. The problem with this method is that when p is close to n, it may still take too long to test the circuit.

Constant-Weight Counter

To resolve the test length problem, a pattern generation technique using ***constant-weight counters*** (CWCs) was proposed in [McCluskey 1982] and [Tang 1983]. Constant-weight counters are constructed using **constant-weight code** or

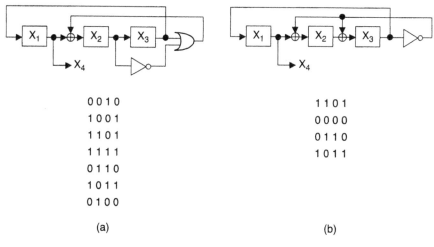

0 0 1 0
1 0 0 1
1 1 0 1
1 1 1 1
0 1 1 0
1 0 1 1
0 1 0 0

(a)

1 1 0 1
0 0 0 0
0 1 1 0
1 0 1 1

(b)

■ **FIGURE 5.20**

Example SDC and CWC as PEPGs: (a) three-stage syndrome driver counter, and (b) three-stage constant-weight counter.

M-out-of-*N* *code*. An *M*-out-of-*N* code contains a set of *N*-bit codewords, each having exactly *M* 1's. Figure 5.20b shows a three-stage constant-weight counter for generating a combination of 2-out-of-3 code and 0-out-of-3 code. The constant-weight test set is shown to be a minimum-length test set for many circuits [McCluskey 1982]; however, for circuits requiring higher *M*-out-of-*N* codes (*e.g.*, a 10-out-of-20 code), CWCs can become very costly to implement.

Combined LFSR/SR

An alternative to the high implementation cost of CWCs that sacrifices the minimum test length requirement was proposed in [Barzilai 1983] and [Tang 1984]. A **combined LFSR/SR** approach using a combination of an LFSR and a *shift register* (SR) is used for pattern generation. Figure 5.21a shows a four-stage combined LFSR/SR and its generated test sequence. We can see any two outputs of the LFSR/SR contain four input combinations, {00, 01, 10, 11}, and hence each output cone of the (4, 2) CUT is tested exhaustively.

The method is most effective when w is much less than n (*e.g.*, $w < n/2$); however, it usually requires at least two seeds (starting patterns). A similar method using maximum-length LFSRs for pseudo-exhaustive pattern generation was given in [Lempel 1985], [Chen 1986], [Golan 1988], and [Wang 1999]. The input register to the combinational CUT is reconfigured as a shift register during self-test. [Wang 1999] proposed inserting an AND gate and a toggle flip-flop between the maximum-length LFSR and the SR to reduce shift power. Test patterns are shifted in from the LFSR. Designs based on this method are simple but require more test patterns than when using other schemes.

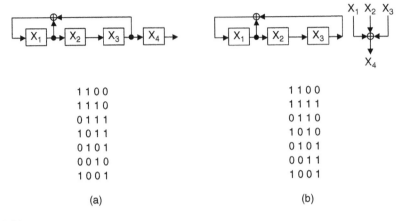

```
1100            1100
1110            1111
0111            0110
1011            1010
0101            0101
0010            0011
1001            1001

 (a)             (b)
```

■ **FIGURE 5.21**

Example combined LFSR/SR and combined LFSR/PS as PEPGs: (a) four-stage combined LFSR/SR, and (b) three-stage combined LFSR/PS.

Combined LFSR/PS

This multiple seed problem can be solved using linear sums [Akers 1985] or linear codes [Vasanthavada 1985]. A **combined LFSR/PS** approach using a combination of an LFSR and a *linear phase shifter* (PS) is used for pattern generation, where the *linear phase shifter* comprises a network of XOR gates. Figure 5.21b shows a three-stage combined LFSR/PS and its associated test sequence. Once again, because any two outputs contain all four combinations, {00, 01, 10, 11}, this (4, 2) CUT can be tested exhaustively. Test lengths derived with this method are very close to the LFSR/SR approach, but this method uses at most two seeds; however, the implementation cost for most circuits, as in the case of using CWCs, is still high.

Condensed LFSR

The multiple seed and implementation cost problems can be solved by using the **condensed LFSR** approach proposed in [Wang 1986a]. Condensed LFSRs are constructed based on **linear codes** [Peterson 1972]. An (n, k) linear code over $GF(2)$ generates a code space C containing 2^k distinct codewords (n-tuples) with the following property: if $c_1 \in C$ and $c_2 \in C$, then $c_1 + c_2 \in C$. Define an (n, k) condensed LFSR as an n-stage modular LFSR with period $2^k - 1$. A condensed LFSR for testing an (n, w) CUT is constructed by first computing the smallest integer k such that:

$$w \leq \lceil k/(n-k+1) \rceil + \lfloor k/(n-k+1) \rfloor$$

where $\lceil x \rceil$ denotes the smallest integer equal to or greater than the real number x, and $\lfloor y \rfloor$ denotes the largest integer equal to or smaller than the real number y.

Then, by using:

$$f(x) = g(x)p(x) = (1 + x + x^2 + \cdots + x^{n-k})p(x)$$

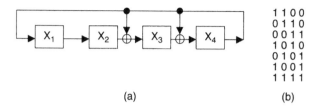

(a)   (b)

■ **FIGURE 5.22**

Example condensed LFSR as PEPG: (a) a (4, 3) condensed LFSR, and (b) test sequence generated by (a).

an (n, k) condensed LFSR can be realized, where $g(x)$ is a **generator polynomial** of degree $n-k$ generating the (n, k) linear code, and $p(x)$ is a primitive polynomial of degree k.

Consider the $(n, k) = (4, 3)$ condensed LFSR shown in Figure 5.22a used to test the $(n, w) = (4, 2)$ CUT. Because $n = 4$ and $w = 2$, we obtain $k = 3$ and $(n-k) = 1$. Selecting $p(x) = 1 + x + x^3$, we have $f(x) = (1+x)(1+x+x^3) = 1 + x^2 + x^3 + x^4$. Figure 5.22b lists the generated period-7 test sequence. It is important to note that the seed polynomial $S_0(x)$ of the LFSR must be divisible by $g(x)$. In the example, we set $S_0(x) = g(x) = 1 + x$, or S_0 to {1100}.

For any given (n, w) CUT, this method uses at most two seeds and has shown to be effective when $w \geq n/2$. Designs based on this method are simple; however, this technique uses more patterns than the LFSR/SR approach when $w < n/2$.

Cyclic LFSR

One approach for reducing the test length when $w < n/2$ is to use **cyclic LFSRs** for test pattern generation [Chen 1987] [Wang 1988b] [Wang 1988c]. Define an (n, k) cyclic LFSR as an n-stage LFSR with period $2^k - 1$. Cyclic LFSRs are based on **cyclic codes** [Peterson 1972]. An (n, k) cyclic code over $GF(2)$ contains 2^k distinct codewords (n-tuples) with the following property: If an n-tuple is a codeword, then the n-tuple obtained by rotating the codeword one bit to the right is also a codeword. Cyclic codes are a subset of linear codes. Each cyclic code has a *minimum distance* or *weight d* [Peterson 1972].

A cyclic code does not exist for every integer n; it exists for every $n' = 2^b - 1, b > 1$. To exhaustively test any (n, w) CUT using cyclic codes, one must start with the smallest integer $n' \geq n$. This method: (1) finds a generator polynomial $g(x)$ of largest degree k' (or smallest degree k) for generating an $(n', k') = (n', n'-k)$ cyclic code that divides $1 + x^{n'}$ and has a design distance $d \geq w + 1$, from any coding theory book such as [Peterson 1972]; (2) uses $h(x) = (1+x^{n'})/g(x)$ to generate an (n', k) cyclic code, which is the **dual code** of the $(n', n'-k)$ cyclic code generated by $g(x)$ [Hsiao 1977], and to construct an (n', k) *cyclic LFSR* using:

$$f(x) = h(x)p(x) = (1+x^{n'})p(x)/g(x)$$

where $h(x)$ is the **parity-check polynomial** of $g(x)$ that satisfies $g(x)h(x) = 1 + x^{n'}$; and, finally, (3) shortens this (n', k) cyclic LFSR to an (n, k) cyclic LFSR by deleting

TABLE 5.3 ■ Generator Polynomials for Given n', k', and d

n'	k'	d	$g(x)$
7	4	3	$1+x+x^3$
7	3	4	$(1+x)(1+x+x^3)$
7	1	6	$(1+x^7)/(1+x)$
15	11	3	$1+x+x^4$
15	10	4	$(1+x)(1+x+x^4)$
15	7	5	$(1+x+x^4)(1+x+x^2+x^3+x^4)$
15	6	6	$(1+x)(1+x+x^4)(1+x+x^2+x^3+x^4)$
15	5	7	$(1+x+x^4)(1+x+x^2+x^3+x^4)(1+x+x^2)$
15	4	8	$(1+x)(1+x+x^4)(1+x+x^2+x^3+x^4)(1+x+x^2)$
15	2	10	$(1+x)(1+x+x^4)(1+x+x^2+x^3+x^4)(1+x^3+x^4)$
15	1	14	$(1+x^{15})/(1+x)$

the rightmost, middle, or leftmost $n'-n$ stages from the (n', k) cyclic LFSR. It was demonstrated in [Wang 1988b] that deleting the middle $n'-n$ stages from the (n', k) cyclic LFSR yields the lowest overhead.

Table 5.3 shows a partial list of (n', k') cyclic codes generated by $g(x)$. It was taken from Appendix D in [Peterson 1972]. Assume that an $(n, w) = (8, 3)$ CUT is to be tested. Because a cyclic code does not exist for $n = 8$, we must choose an (n', k') cyclic code with the smallest integer n' and largest integer k' that has a design distance $d \geq w+1 = 4$ where $n' > n$. From Table 5.3, we obtain $n' = 15$ and $k' = 10$. This allows us to build an $(n', k') = (n', n'-k) = (15, 10)$ cyclic code with $g(x) = (1+x)(1+x+x^4) = 1+x^2+x^4+x^5$, or an $(n', k) = (15, 5)$ dual code using $h(x) = (1+x^{15})/g(x)$. Selecting $p(x) = 1+x^2+x^3+x^4+x^5$ from [Peterson 1972], we can then construct an $(n', k) = (15, 5)$ cyclic LFSR with $f(x) = h(x)p(x) = 1+x^3+x^5+x^8+x^9+x^{11}+x^{12}+x^{13}+x^{15}$. This cyclic LFSR uses seven XOR gates.

Figure 5.23 shows an $(n, k) = (8, 5)$ cyclic LFSR obtained by picking the first six stages and the last two stages of the (15, 5) cyclic LFSR that uses the least number of XOR gates. The figure was derived according to [Wang 1988b], who described a procedure for choosing the stage positions to be deleted. The initial state {10100100} shown in the figure is set to its corresponding state, $S_0(x)$, of the

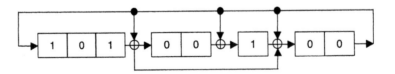

■ **FIGURE 5.23**

Example (8, 5) cyclic LFSR as PEPG.

(15, 5) cyclic LFSR, where $S_0(x) = h(x) = (1+x^{15})/g(x) = 1+x^2+x^5+x^6+x^8+x^9+x^{10}$. This (8, 5) cyclic LFSR has period 31 and only uses three XOR gates.

It was shown in [Wang 1987] that when $n = 2^b, b > 2$, an (n, k) cyclic LFSR can produce a longer test length than using the combined LFSR/PS approach. In this case, an $(n, k-s)$ **shortened cyclic LFSR** can be employed, where s is the number of information bits to be deleted from the (n, k) cyclic code, $1 \leq s < k < n$. The (8, 4) shortened cyclic LFSR shown in Figure 5.24 uses eight XOR gates, but its test length has been reduced from 31 (in the cyclic LFSR case) to 15.

Compatible LFSR

Recall from [Savir 1980] that a p-stage syndrome driver counter can test each output cone of an (n, w) CUT exhaustively, $w \leq p < n$, if $n-p$ inputs can share *test signals* with the other p inputs. This means that the SDC can detect all single-stuck-at and multiple-stuck-at faults within each output cone. If we only consider single-stuck-at faults, [Chen 1998] shows that additional inputs can be further combined using a **mapping logic** without losing any single-stuck fault coverage. This method requires finding the **compatibility classes** for all inputs and may require a detailed fault simulation. The l-to-n mapping logic, $l < p$, can be implemented using simple buffers or inverters. Additional decoders [Chakrabarty 1997] or a more general combinational circuit [Hamzaoglu 2000] can also be used. An l-stage **compatible LFSR**, which is a combination of an l-stage LFSR and an l-to-n mapping logic, can now further reduce the test length for some (n, w) CUTs.

Consider the $(n, w) = (5, 4)$ CUT shown in Figure 5.25a [Jha 2003]. Because x_1 and x_5 do not drive the same output, we obtain $p = 4$. A four-stage SDC generating $2^p - 1 = 15$ patterns is required to detect all single-stuck-at and multiple-stuck-at faults within each output cone; however, when only single-stuck-at faults are considered, the two-stage compatible LFSR shown in Figure 5.25b generating $2^2 = 4$ patterns can be used to detect all faults [Chen 1998].

5.3.3.2 Segmentation Testing

There are circuits where either the test length using the previous techniques is still too long or an output depends on all circuit inputs. For these circuits, a pseudo-exhaustive test is still possible, but it is necessary to resort to a partitioning or segmentation technique. Such a procedure is described in [McCluskey 1981]. This technique relies on exhaustive testing by dividing the circuit into segments or partitions in order to avoid excessively long test sequences. This is referred to

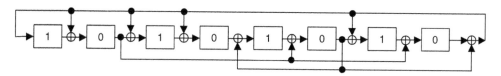

■ **FIGURE 5.24**

Example (8, 4) shortened cyclic LFSR as PEPG.

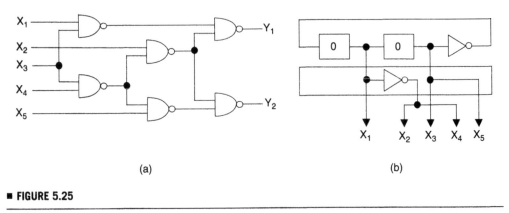

Example compatible LFSR as PEPG: (a) $(n, w) = (5, 4)$ CUT, and (b) two-stage compatible LFSR.

as *segmentation testing*. It differs from previous partitioning techniques in that the partitions may divide the signal paths through the circuit rather than only separating the signal paths from each other.

There are two techniques that can be used to achieve this partitioning: hardware partitioning and sensitized partitioning [Bozorgui-Nesbat 1980] [McCluskey 1981]. **Hardware partitioning** is based on inserting multiplexers and connecting the embedded inputs and outputs of the subcircuit to those primary inputs and outputs that are not used by the subcircuit under test. **Sensitized partitioning** refers to the technique in which circuit partitioning and subcircuit isolation can be achieved by applying the appropriate input patterns to some of the input lines.

Partitioning the circuit into several subcircuits and exhaustively testing each such subcircuit greatly simplifies the testing of the overall circuit; however, partitioning VLSI circuits is an *NP-complete* problem [Patashnik 1983].

Hardware partitioning using multiplexers can reduce the operating speed of a circuit and is costly to implement. Sensitized partitioning does not alter the functional circuitry and is therefore the preferred technique. This technique has been used in [Udell 1986] to develop pseudo-exhaustive test patterns. A *reconfigurable counter* that automatically generates these test patterns was given in [Udell 1987].

5.3.4 Delay Fault Testing

The BIST pattern generation techniques described above mainly target structural faults, such as stuck-at faults and bridging faults, which can be detected with one-pattern vectors. For delay faults requiring two-pattern vectors for testing, these methods do not provide adequate fault coverage. In this section, we discuss a few approaches that can be used for delay fault testing.

Unlike structural fault testing that requires an exhaustive one-pattern set of 2^n test patterns, an exhaustive two-pattern set of $2^n(2^n - 1)$ patterns is required to test delay faults in an n-input CUT exhaustively. This means that, for delay fault testing, one must use a **test pattern generator** (TPG) with $2n$ or more stages. A maximum-length LFSR having $2n$ stages is called a **double-length LFSR** [Jha 2003].

[Furuya 1991] has shown that when all even or odd stage outputs (called **even taps** or **odd taps**) of a $2n$-stage double-length LFSR are connected to the n-input CUT, the LFSR can generate $2^{2n} - 1$ vectors to test the CUT exhaustively. While all delay faults are tested exhaustively, there is a potential problem that the test set could cause **test invalidation** due to hazards present in the design [Bushnell 2000]. Test invalidation or hazards can occur when more than one circuit inputs change values. There are also risks that the power consumption during at-speed BIST can exceed the power rating of the chip or package. Increased average power can cause heating of the chip and increased peak power can produce noise-related failures [Bushnell 2000] [Girard 2002].

To solve these problems, it is important to generate *single-input change* (SIC) or one-transition patterns. [Breuer 1987] shows that when two-transition patterns are applied at the circuit inputs, no additional paths are tested for delay faults which could not be tested by one-transition test pairs. Due to complementary converging effects, increasing the number of simultaneous input transitions applied may lead to a reduction in the number of complete transition paths tested. [Breuer 1987] further proved that a delay fault TPG will require one-transition patterns not more than $2n2^n + 1$ but not less than $n2^n + 1$.

A **Gray code counter**, comprised of a binary counter or maximum-length LFSR, as well as a **Johnson counter** or **ring counter** are commonly used for this purpose [Breuer 1987] [Bushnell 1995] [Virazel 2002]. An example delay fault TPG for testing an n-input CUT is shown in Figure 5.26 [Bushnell 2000]. The standard maximum-length LFSR can cycle through $2^n - 1$ states. The n-bit Johnson counter can generate $2n$ one-transition patterns. By properly selecting the control signal TESTTYPE, the delay fault TPG can generate $2n(2^n - 1)$ one-transition patterns for delay fault testing or $2^n - 1$ patterns for stuck fault testing.

5.3.5 Summary

While many advantages for using exhaustive or pseudo-exhaustive testing exist, pseudo-random testing is still the most practical and commonly used technique for BIST pattern generation. Because this scheme generally leads to lower fault coverage, it is often required to augment pseudo-random test patterns, particularly for

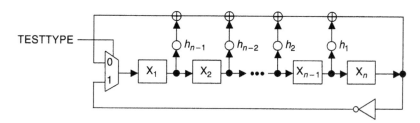

■ **FIGURE 5.26**

Example delay fault TPG as PRPG or PEPG.

life-critical and mission-critical applications. Methods that have been proposed for fault coverage enhancement include inserting test points or embedding deterministic patterns to the circuit under test. These approaches are discussed extensively in Section 5.6.

5.4 OUTPUT RESPONSE ANALYSIS

In the previous chapters where we discussed logic and fault simulation and test generation, our assumption was that output responses coming out of the circuit under test are compared directly on a tester. For BIST operations, it is impossible to store all output responses on-chip, on-board, or in-system to perform bit-by-bit comparison. An *output response analysis* technique must be employed such that output responses can be compacted into a **signature** and compared with a *golden signature* for the fault-free circuit either embedded on-chip or stored off-chip.

Compaction differs from *compression* in that compression is loss-less, while compaction is lossy. **Compaction** is a method for dramatically reducing the number of bits in the original circuit response during testing in which some information is lost. **Compression** is a method for reducing the number of bits in the original circuit response in which no information is lost, such that the original output sequence can be fully regenerated from the compressed sequence [Bushnell 2000]. Because all output response analysis schemes involve information loss, they are referred to as *output response compaction*; however, there is no general consensus in academia yet as to when the terms "compaction" or "compression" are to be used. For example, in the random-access scan architecture described in Chapter 2, the authors prefer to use the term "compression" for output response analysis; however, for output response analysis throughout the remainder of the book we will refer to the lossy compression as "compaction."

In this section, we present three different output response compaction techniques: (1) **ones count testing**, (2) **transition count testing**, and (3) **signature analysis**. We also describe the architectures of the *output response analyzers* (ORAs) that are used. The signature analysis technique is described in more detail, as it is the most popular compaction technique in use today.

When using compaction, it is important to ensure that the faulty and fault-free signatures are different. If they are the same, the faults can go undetected. This situation is referred to as **error masking**, and the erroneous output response is said to be an **alias** of the correct output response [Abramovici 1994]. It is also important to ensure that none of the output responses contain an unknown (X) value. If an unknown value is generated and propagated directly or indirectly to the *output response analyzer* (ORA), then the ORA can no longer function reliably. Therefore, it is necessary to fix all unknown (X) propagation problems to ensure that the logic BIST system will operate correctly by using the X-bounding techniques discussed in Section 5.2.

5.4.1 Ones Count Testing

Assume that the CUT has only one output and the output contains a stream of L bits. Let the fault-free output response, R_0, be $\{r_0 r_1 r_2 \ldots r_{L-1}\}$. The **ones count test technique** will only require a counter to count the number of 1's in the bit stream. For example, if $R_0 = \{0101100\}$, then the signature or ones count of R_0, $OC(R_0)$, is 3. If fault f_1 present in the CUT causes an erroneous response $R_1 = \{1100110\}$, then it will be detected because $OC(R_1) = 4$; however, fault f_2 causing $R_2 = \{0101010\}$ will not be detected because $OC(R_2) = OC(R_0) = 3$. Let the fault-free signature or ones count be m. There will be $C(L, m)$ possible ways having m 1's in an L-bit stream. Assuming all faulty sequences are equally likely to occur as the response of the CUT, the **aliasing probability** or **masking probability** of using ones count testing having m 1's [Savir 1985] can be expressed as:

$$P_{OC}(m) = (C(L, m) - 1)/(2^L - 1)$$

In the previous example, where $m = OC(R_0) = 3$ and $L = 7$, $P_{OC}(m) = 34/127 = 0.27$. Figure 5.27 shows the ones count test circuit for testing the CUT with T patterns. The number of stages in the counter design must be equal to or greater than $\lceil \log_2(L+1) \rceil$.

5.4.2 Transition Count Testing

The theory behind transition count testing is similar to that for ones count testing, except that the signature is defined as the number of 0-to-1 and 1-to-0 transitions. The **transition count test technique** [Hayes 1976] simply requires using a D flip-flop and an XOR gate connected to a ones counter (see Figure 5.28), to count the

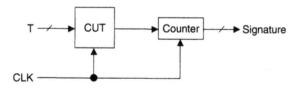

■ **FIGURE 5.27**

Ones counter as ORA.

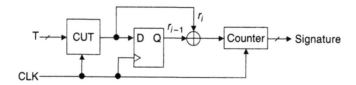

■ **FIGURE 5.28**

Transition counter as ORA.

number of transitions in the output data stream. Consider the example given above. Because $R_0 = \{0101100\}$, the signature or transition count of R_0, $TC(R_0)$, will be 4. Assume that the initial state of the D flip-flop, r_{-1}, is 0. Fault f_1 causing an erroneous response $R_1 = \{1100110\}$ will not be detected because $TC(R_1) = TC(R_0) = 4$, but fault f_2 causing $R_2 = \{0101010\}$ will be detected because $TC(R_2) = 6$.

Let the fault-free signature or transition count be m. Because a given L-bit sequence R_0 that starts with $r_0 = 0$ has $L-1$ possible transitions, the number of sequences with m transitions can be given by $C(L-1, m)$. Because R_0 can also start with $r_0 = 1$, there will be a total of $2C(L-1, m)$ possible ways to have m 0-to-1 and 1-to-0 transitions in an L-bit stream. Assuming all faulty sequences are equally likely to occur as the response of the CUT, the *aliasing probability* or *masking probability* of using transition count testing having m transitions [Savir 1985] is:

$$P_{TC}(m) = (2C(L-1, m) - 1)/(2^L - 1)$$

In the previous example, where $m = TC(R_0) = 4$ and $L = 7$, $P_{TC}(m) = 29/127 = 0.23$. Figure 5.28 shows the transition count test circuit. The number of stages in the counter design must be equal to or greater than $\lceil \log_2(L+1) \rceil$.

5.4.3 Signature Analysis

Signature analysis is the most popular response compaction technique used today. The compaction scheme, based on **cyclic redundancy checking** (CRC) [Peterson 1972], was first developed in [Benowitz 1975]. Hewlett-Packard commercialized the first logic analyzer (the HP 5004A Signature Analyzer) based on the scheme and referred to it as **signature analysis** [Frohwerk 1977]. In this section, we discuss two signature analysis schemes: (1) **serial signature analysis** for compacting responses from a CUT having a single output, and (2) **parallel signature analysis** for compacting responses from a CUT having multiple outputs.

5.4.3.1 Serial Signature Analysis

Consider the n-stage **single-input signature register** (SISR) shown in Figure 5.29. This SISR uses an additional XOR gate at the input for compacting an L-bit

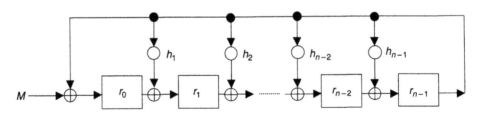

■ FIGURE 5.29

An n-stage single-input signature register (SISR).

output sequence, M, into the *modular LFSR*. Let $M = \{m_0 m_1 m_2 \ldots m_{L-1}\}$, and define:

$$M(x) = m_0 + m_1 x + m_2 x^2 + \cdots + m_{L-1} x^{L-1}$$

After shifting the L-bit output sequence, M, into the modular LFSR, the contents (remainder) of the SISR, R, is given as $\{r_0 r_1 r_2 \ldots r_{n-1}\}$, or:

$$r(x) = r_0 + r_1 x + r_2 x^2 + \cdots + r_{n-1} x^{n-1}$$

The SISR is basically a *CRC code generator* [Peterson 1972] or a *cyclic code checker* [Benowitz 1975]. Let the *characteristic polynomial* of the modular LFSR be $f(x)$. [Peterson 1972] has shown that the SISR performs polynomial division of $M(x)$ by $f(x)$, or:

$$M(x) = q(x)f(x) + r(x)$$

The final state or **signature** in the SISR is the *polynomial remainder*, $r(x)$, of the division. Consider the four-stage SISR given in Figure 5.30 using $f(x) = 1 + x + x^4$. Assuming $M = \{10011011\}$, we can express $M(x) = 1 + x^3 + x^4 + x^6 + x^7$. Using polynomial division, we obtain $q(x) = x^2 + x^3$ and $r(x) = 1 + x^2 + x^3$ or $R = \{1011\}$. The remainder $\{1011\}$ is equal to the *signature* derived from Figure 5.30a when the SISR is first initialized to a *starting pattern* (*seed*) of $\{0000\}$.

Now, assume fault f_1 produces an erroneous output stream $M' = \{11001011\}$ or $M'(x) = 1 + x + x^4 + x^6 + x^7$, as given in Figure 5.30b. Using polynomial division, we obtain $q'(x) = x^2 + x^3$ and $r'(x) = 1 + x + x^2$ or $R' = \{1110\}$. Because the faulty signature R', $\{1110\}$, is different from the fault-free signature R, $\{1011\}$, fault f_1 is detected. For fault f_2 with $M'' = \{11001101\}$ or $M''(x) = 1 + x + x^4 + x^5 + x^7$ as given in

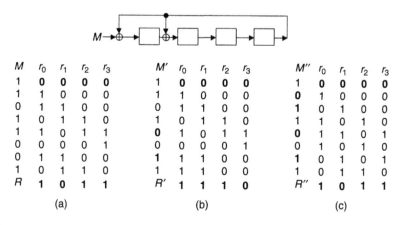

M	r_0	r_1	r_2	r_3		M'	r_0	r_1	r_2	r_3		M''	r_0	r_1	r_2	r_3
1	0	0	0	0		1	0	0	0	0		1	0	0	0	0
1	1	0	0	0		1	1	0	0	0		0	1	0	0	0
0	1	1	0	0		0	1	1	0	0		1	0	1	0	0
1	0	1	1	0		1	0	1	1	0		1	1	0	1	0
1	1	0	1	1		0	1	0	1	1		0	1	1	0	1
0	0	0	0	1		0	0	0	0	1		0	1	0	1	0
0	1	1	0	0		1	1	1	0	0		1	0	1	0	1
1	0	1	1	0		1	1	1	0	0		1	0	1	1	0
R	1	0	1	1		R'	1	1	1	0		R''	1	0	1	1
	(a)						(b)						(c)			

■ **FIGURE 5.30**

A four-stage SISR: (a) fault-free signature; (b) signature for fault f_1; and (c) signature for fault f_2.

Figure 5.30c, we have $q''(x) = x + x^3$ and $r''(x) = 1 + x^2 + x^3$ or $R'' = \{1011\}$. Because $R'' = R$, fault f_2 is not detected.

The *fault detection* or *aliasing* problem of an SISR can be better understood by looking at the *error sequence E* or *error polynomial E(x)* of the fault-free sequence M and a faulty sequence M'. Define $E = M + M'$, or:

$$E(x) = M(x) + M'(x)$$

If $E(x)$ is not divisible by $f(x)$, then all faults generating the faulty sequence M' will be detected; otherwise, these faults are not detected. Consider fault f_1 again. We obtain $E = \{01010000\} = M + M' = \{10011011\} + \{11001011\}$ or $E(x) = x + x^3$. Because $E(x)$ is not divisible by $f(x) = 1 + x + x^4$, fault f_1 is detected. Consider fault f_2 again. We have $E = \{01010110\} = M + M'' = \{10011011\} + \{11001101\}$ or $E(x) = x + x^3 + x^5 + x^6$. Because $f(x)$ divides $E(x)$—that is, $E(x) = (x + x^2)f(x)$—fault f_2 is not detected.

Assume that the SISR consists of n stages. For a given L-bit sequence, $L > n$, there are $2^{(L-n)}$ possible ways of producing an n-bit signature of which one is the correct signature. Because there are a total of $2^L - 1$ erroneous sequences in an L-bit stream, the *aliasing probability* using an n-stage SISR for serial signature analysis (SSA) is:

$$P_{SSA}(n) = (2^{(L-n)} - 1)/(2^L - 1)$$

If $L \gg n$, then $P_{SSA}(n) \approx 2^{-n}$. When $n = 20$, $P_{SSA}(n) < 2^{-20} = 0.0001\%$.

5.4.3.2 Parallel Signature Analysis

A common problem when using ones count testing, transition count testing, and serial signature analysis is the excessive hardware cost required to test an m-output CUT. It is possible to reduce the hardware cost by using an m-to-1 multiplexer, but this increases the test time m times. Consider the n-stage **multiple-input signature register** (MISR) shown in Figure 5.31. The MISR uses n extra XOR gates for compacting n L-bit output sequences, M_0 to M_{n-1}, into the modular LFSR simultaneously.

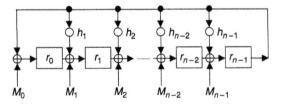

■ FIGURE 5.31

An n-stage multiple-input signature register (MISR).

[Hassan 1984] has shown that the n-input MISR can be remodeled as a single-input SISR with *effective input sequence* $M(x)$ and *effective error polynomial* $E(x)$ expressed as:

$$M(x) = M_0(x) + xM_1(x) + \cdots + x^{n-2}M_{n-2}(x) + x^{n-1}M_{n-1}(x)$$

and

$$E(x) = E_0(x) + xE_1(x) + \cdots + x^{n-2}E_{n-2}(x) + x^{n-1}E_{n-1}(x)$$

Consider the four-stage MISR shown in Figure 5.32 using $f(x) = 1 + x + x^4$. Let $M_0 = \{10010\}$, $M_1 = \{01010\}$, $M_2 = \{11000\}$, and $M_3 = \{10011\}$. From this information, the signature R of the MISR can be calculated as $\{1011\}$. Using $M(x) = M_0(x) + xM_1(x) + x^2M_2(x) + x^3M_3(x)$, we obtain $M(x) = 1 + x^3 + x^4 + x^6 + x^7$ or $M = \{10011011\}$, as shown in Figure 5.33. This is the same data stream we used in the SISR example in Figure 5.30a. Therefore, $R = \{1011\}$.

Assume there are mL-bit sequences to be compacted in an n-stage MISR, where $L > n \geq m \geq 2$. The *aliasing probability* for **parallel signature analysis** (PSA) now becomes:

$$P_{PSA}(n) = (2^{(mL-n)} - 1)/(2^{mL} - 1)$$

If $L \gg n$, then $P_{PSA}(n) \approx 2^{-n}$. When $n = 20$, $P_{PSA}(n) < 2^{-20} = 0.0001\%$. The result suggests that $P_{PSA}(n)$ mainly depends on n, when $L \gg n$. Hence, increasing the number of MISR stages or using the same MISR but with a different $f(x)$ can substantially reduce the *aliasing probability* [Hassan 1984] [Williams 1987].

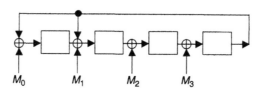

$M_0 \qquad M_1 \qquad M_2 \qquad M_3$

■ **FIGURE 5.32**

A four-stage MISR.

M_0	10010
M_1	01010
M_2	11000
M_3	10011
M	10011011

■ **FIGURE 5.33**

An equivalent M sequence.

5.5 LOGIC BIST ARCHITECTURES

Several architectures for incorporating **offline BIST** techniques into a design have been proposed. These architectures generally fall into four categories: (1) those that assume no special structure to the circuit under test, (2) those that make use of scan chains in the circuit under test, (3) those that configure the scan chains for test pattern generation and output response analysis, and (4) those that use the **concurrent checking** (implicit test) circuitry of the design.

In this section, we only discuss a few representative BIST architectures on each category. A more comprehensive survey can be found in [McCluskey 1985] and [Abramovici 1994]. To preserve integrity and continuity, we adopt the same naming convention used in [Abramovici 1994].

5.5.1 BIST Architectures for Circuits without Scan Chains

The first BIST architecture uses a pseudo-random pattern generator as well as a single-input signature register or multiple-input signature register for testing a combinational or sequential circuit that does not assume any special structure. This architecture is often used at the board or system level. Hewlett-Packard was among the first companies to adopt this BIST architecture for board-level fault diagnosis [Frohwerk 1977].

5.5.1.1 A Centralized and Separate Board-Level BIST Architecture

Figure 5.34a shows a BIST architecture described in [Benowitz 1975]. This is referred to as a ***centralized and separate board-level BIST architecture*** (CSBL) [Abramovici 1994]. Two LFSRs and two multiplexers are added to the circuit. The first ***multiplexer*** (MUX) selects either the primary inputs (PIs) or the PRPG outputs to drive the circuit under test (CUT). The CUT is typically a ***sequential circuit*** (S) but can be a ***combinational circuit*** (C) as well. The second multiplexer routes the primary outputs (POs) of the circuit to the SISR. Additional circuitry (not shown in the figure) is used to compare the final signature of the SISR with an embedded golden signature (*known good signature*). It also provides a pass/fail indication once the test is complete.

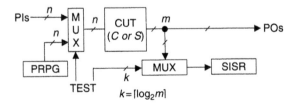

■ **FIGURE 5.34**

The centralized and separate board-level BIST (CSBL) architecture.

5.5.1.2 *Built-In Evaluation and Self-Test (BEST)*

A similar BIST architecture is described in [Perkins 1980]. This architecture makes use of a PRPG and a MISR that are external to the chip but could be located on the same board. The logic being tested on the chip is typically a sequential circuit but can be a combinational circuit as well. Pseudo-random patterns are applied in parallel from the PRPG to the chip's primary inputs, and a MISR is used to compact the chip's output responses. Both PRPG and MISR can also be embedded inside the chip. This architecture, referred to as a ***built-in evaluation and self-test*** (BEST), is shown in Figure 5.35.

5.5.2 BIST Architectures for Circuits with Scan Chains

For designs that incorporate scan chains, it is possible to make use of this scan architecture for the BIST circuitry. The resulting BIST architecture is generally referred to as a **test-per-scan** BIST system [Bushnell 2000].

5.5.2.1 *LSSD On-Chip Self-Test*

A BIST architecture that makes use of scan chains for the BIST circuitry was proposed in [Eichelberger 1983] and shown in Figure 5.36. It was called an ***LSSD on-chip self-test*** (LOCST) architecture [Abramovici 1994]. In addition to the internal scan chain comprised of LSSD *shift register latches* (SRLs), an external scan chain comprised of all primary inputs and primary outputs of the circuit under test (CUT) is required. The external scan chain input is connected to the scan output

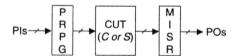

■ FIGURE 5.35

The built-in evaluation and self-test (BEST) architecture.

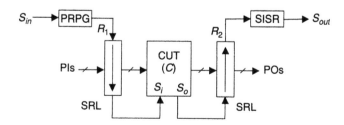

■ FIGURE 5.36

The LSSD on-chip self-test (LOCST) architecture.

of the internal scan chain. Pseudo-random patterns are generated by a PRPG and are shifted into the combined scan chain. The system clocks are triggered and the contents of the scan chain latches are shifted out to a SISR. The final signature is then compared in the SISR with a precomputed fault-free signature in order to generate a pass/fail error signal. The scan output is also connected to a pin so, in case of a failure, intermediate signatures can be examined externally for diagnosis purposes.

5.5.2.2 Self-Testing Using MISR and Parallel SRSG

A similar design was presented in [Bardell 1982]. This design, shown in Figure 5.37, contains an PRPG (parallel *shift register sequence generator* [SRSG]) and a MISR. The scan chains are loaded in parallel from the PRPG. The system clocks are then triggered and the test responses are shifted to the MISR for compaction. New test patterns are shifted in at the same time while test responses are being shifted out. This BIST architecture is referred to as *self-testing using MISR and parallel SRSG* (STUMPS) [Bardell 1982]. Due to the ease of integration with traditional scan architecture, the STUMPS architecture is the only BIST architecture widely used in industry to date. In order to further reduce the lengths of the PRPG and MISR and improve the randomness of the PRPG, a STUMPS-based architecture that includes an optional linear phase shifter and an optional linear phase compactor is often used in industrial applications [Nadeau-Dostie 2000] [Cheon 2005]. The linear phase shifter and linear phase compactor typically comprise a network of XOR gates. Figure 5.38 shows the STUMPS-based architecture.

5.5.3 BIST Architectures Using Register Reconfiguration

A concern with BIST designs is the amount of test time required. One technique for reducing the test time is to make use of the storage elements already in the design for both test generation and response analysis. The storage elements are redesigned so they can function as pattern generators or signature analyzers for test purposes. This BIST architecture is generally referred to as a **test-per-clock** BIST system [Bushnell 2000].

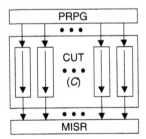

■ **FIGURE 5.37**

The self-testing using MISR and parallel (STUMPS) architecture.

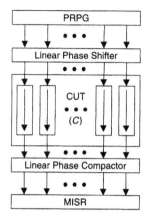

■ **FIGURE 5.38**

A STUMPS-based architecture.

5.5.3.1 Built-In Logic Block Observer

The architecture described in [Könemann 1979] and [Könemann 1980] applies to circuits that can be partitioned into independent modules (logic blocks). Each module is assumed to have its own input and output registers (storage elements), or such registers are added to the circuit where necessary. The registers are redesigned so that for test purposes they act as PRPGs for test generation or MISRs for signature analysis. The redesigned register is referred to as a ***built-in logic block observer*** (BILBO).

The BILBO is operated in four modes: normal mode, scan mode, test generation or signature analysis mode, and reset mode. A typical three-stage BILBO that is reconfigurable into an TPG or a MISR during self-test is shown in Figure 5.39. It

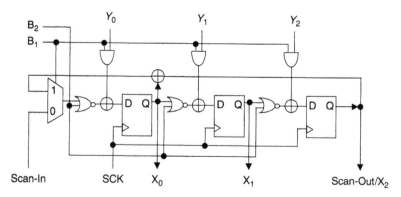

■ **FIGURE 5.39**

A three-stage built-in logic block observer (BILBO).

is controlled by two control inputs B_1 and B_2. When both control inputs B_1 and B_2 are equal to 1, the circuit functions in normal mode with the inputs Y_i gated directly into the D flip-flops. When both control inputs are equal to 0, the BILBO is configured as a shift register. Test data can be shifted in via the serial scan-in port or shifted out via the serial scan-out port. Setting $B_1 = 1$ and $B_2 = 0$ converts the BILBO into a MISR. It can then be used in this configuration as a TPG by holding every Y_i input to 1. The BILBO is reset after a system clock is triggered when $B_1 = 0$ and $B_2 = 1$.

This technique is most suitable for testing circuits, such as RAMs, ROMs, or bus-oriented circuits, where input and output registers of the partitioned modules can be reconfigured independently. For testing finite-state machines or pipelined-oriented circuits, as shown in Figure 5.40, the signature data from the previous module must be used as test patterns for the next module, because the test generation and signature analysis modes cannot be separated. In this case, a detailed fault simulation is required to achieve 100% single-stuck fault coverage.

5.5.3.2 Modified Built-In Logic Block Observer

One technique that overcomes the above BILBO problem is described in [McCluskey 1981]. It uses an additional control input to separate test generation from signature analysis. Such a *modified BILBO* (MBILBO) design is shown in Figure 5.41. The modification is obtained from the original BILBO by adding one more OR gate to each Y_i input. The control input B_3 is always set to 0 except when the register has to be configured into a TPG. In that case, B_3 is set to 1. For testing the pipelined-oriented circuit shown in Figure 5.40b, the MBILBO cells can now be used, and CC1 and CC2 will be tested alternatively. However, this approach still cannot test the finite-state machine of Figure 5.40a exhaustively, because the receiving BILBO cell must be always in signature analysis mode.

5.5.3.3 Concurrent Built-In Logic Block Observer

The above BILBO and MBILBO problems can be resolved by using the *concurrent BILBO* (CBILBO) approach [Wang 1986c]. It uses two storage elements to perform test generation and signature analysis simultaneously. A **CBILBO** design is shown

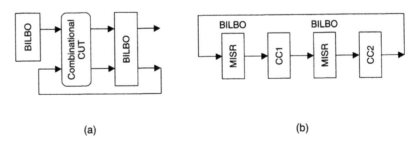

(a)　　　　　　　　　　　　　　　　　　(b)

■ **FIGURE 5.40**

BILBO architectures: (a) for testing a finite-state machine, and (b) for testing a pipelined-oriented circuit.

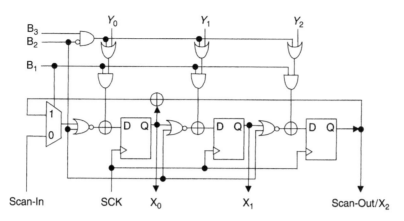

A three-stage modified built-in logic block observer (MBILBO).

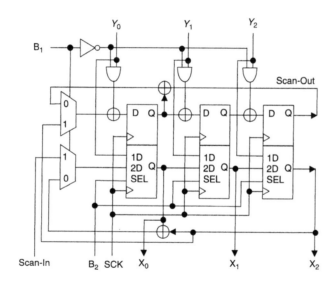

A three-stage concurrent BILBO (CBILBO).

in Figure 5.42, where only three modes of operation are considered: normal, scan, and test generation and signature analysis. When $B_1 = 0$ and $B_2 = 1$, the upper D flip-flops act as a MISR for signature analysis, whereas the lower two-port D flip-flops form a TPG for test generation. Because signature analysis is separated from test generation, an exhaustive or pseudo-exhaustive pattern generator (EPG/PEPG) can now be used for test generation; therefore, no fault simulation is required and it is possible to achieve 100% single-stuck fault coverage using the CBILBO architectures shown in Figure 5.43. However, the hardware cost associated with

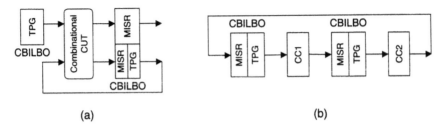

(a)   (b)

■ **FIGURE 5.43**

CBILBO architectures: (a) for testing a finite-state machine, and (b) for testing a pipelined-oriented circuit.

using the CBILBO approach is generally higher than for the BILBO or MBILBO approach. A gate-level design of the CBILBO one-cell structure using D latches is given in [Wang 1986c]. A CMOS version of the CBILBO structure can be found in [Liu 1987].

5.5.3.4 Circular Self-Test Path (CSTP)

The hardware cost can be substantially reduced using the **circular self-test path** (CSTP) architecture [Krasniewski 1989] shown in Figure 5.44a. In the CSTP configuration, all primary inputs and primary outputs are reconfigured as *external scan cells*. They are connected to the internal scan cells to form a circular path. If the entire circular path has n scan cells, then it corresponds to a MISR with characteristic polynomial $f(x) = 1 + x^n$.

During self-test, all primary inputs are connected as a shift register, whereas all internal scan cells and primary outputs are reconfigured as a MISR. The MISR consists of a number of *self-test cells* connected in series, where in self-test mode,

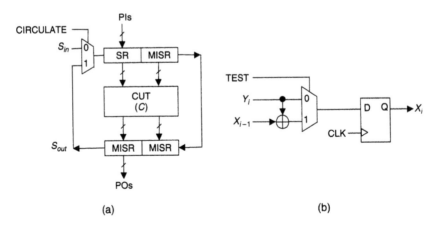

(a)   (b)

■ **FIGURE 5.44**

The circular self-test path (CSTP) architecture: (a) CSTP architecture, and (b) self-test cell.

each self-test cell takes as input from an XOR gate output of input Y_i and its previous scan cell output X_{i-1}, as shown in Figure 5.44b. One requirement for the CSTP design is that all registers must be initialized to known states prior to self-test. After initialization of all registers, the circuit runs for a number of clock cycles and then the final signature is read out for analysis. Because the characteristic polynomial, $f(x) = 1 + x^n$, is nonlinear, the CSTP design can lead to low fault coverage [Stroud 1988] [Pilarski 1992] [Carletta 1994] [Touba 2002].

The CSTP architecture is similar to the **simultaneous self-test** architecture [Bardell 1982] and the **circular BIST** architecture [Stroud 1988]. The primary differences in the three architectures are the functional modes of operation supported by the registers. Both *simultaneous self-test* and *circular BIST* architectures included scan chain capabilities. As a result, the fault coverage obtained during BIST operation could be augmented with additional scan vectors in the event that low fault coverage was obtained for a given application.

5.5.4   BIST Architectures Using Concurrent Checking Circuits

For systems that include concurrent checking circuits, it is possible to use the circuitry to verify the output response during explicit (offline) testing; hence, the need to implement a separate response analysis circuit, such as a MISR, is avoided.

5.5.4.1   Concurrent Self-Verification

A BIST architecture shown in Figure 5.45, *concurrent self-verification* (CSV), was described in [Sedmak 1979] and [Sedmak 1980]. A PRPG is applied to the functional circuitry (CUT) and the duplicate circuitry. The duplicate circuitry is realized in complementary form to reduce design and common-mode faults. Because the

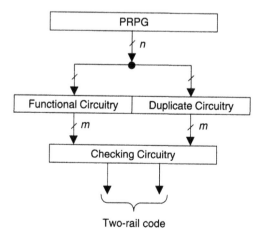

■ **FIGURE 5.45**

The concurrent self-verification (CSV) architecture.

TABLE 5.4 ■ Representative Logic BIST Architectures

Architecture	Level	TPG	ORA	Circuit	BIST
CSBL	B or C	PRPG	SISR	C or S	Test-per-clock
BEST	B or C	PRPG	MISR	C or S	Test-per-clock
LOCST	C	PRPG	SISR	C	Test-per-scan
STUMPS	B or C	PRPG	MISR	C	Test-per-scan
BILBO	C	PRPG	MISR	C	Test-per-clock
MBILBO	C	PRPG	MISR	C	Test-per-clock
CBILBO	C	EPG/PEPG	MISR	C	Test-per-clock
CSTP	C	PRPG	MISR	C or S	Test-per-clock
CSV	C	PRPG	Checker	C or S	Test-per-clock

checking circuitry recommended involves comparing the outputs of the two implementations, this technique avoids the aliasing problem and consequent loss of effective fault coverage. The checking circuitry is a **totally self-checking two-rail checker** [Abramovici 1994].

5.5.5 Summary

Many logic BIST architectures have been proposed in the 1980s. In this section, we have presented a number of representative BIST architectures for testing combinational or sequential circuits at the board or chip level. Table 5.4 shows some of the main attributes of the BIST architectures presented. A BIST technique that can be used for testing sequential circuits (S) can also be used for testing combinational circuits (C). Similarly, a BIST technique suitable for board-level testing (B) is also applicable for chip-level testing (C).

The CBILBO architecture is the only architecture that can be used for exhaustive or pseudo-exhaustive testing. The CSV architecture is the only architecture that does not require an additional SISR or MISR for output response analysis. Due to its ease of integration with traditional scan architecture, the STUMPS architecture is the only architecture widely used in industry to date; however, because pseudo-random patterns are used, fault coverage is still a concern. This has prevented the technique from being accepted across all industries.

5.6 FAULT COVERAGE ENHANCEMENT

In *pseudo-random testing*, the fault coverage is limited by the presence of *random-pattern resistant* (RP-resistant) faults. If the fault coverage is not sufficient, then three approaches can be used to enhance the fault coverage: (1) **test point insertion**,

(2) **mixed-mode BIST**, and (3) **hybrid BIST**. The first two approaches are applicable for in-field coverage enhancement, and the third approach is applicable for manufacturing coverage enhancement.

Test point insertion adds control points and observation points for providing additional controllability and observability to improve the detection probability of RP-resistant faults so they can be detected during pseudo-random testing. Mixed-mode BIST involves supplementing the pseudo-random patterns with some deterministic patterns that detect RP-resistant faults and are generated using on-chip hardware. When BIST is performed during manufacturing test where a tester is present, hybrid BIST involves combining BIST and external testing by supplementing the pseudo-random patterns with deterministic data from the tester to improve the fault coverage. This third option is not applicable when BIST is used in the field, as the tester is not present. Each of these approaches is described in more detail in the following subsections.

5.6.1 Test Point Insertion

Test points can be used to increase the circuit's fault coverage to a desired level. Figure 5.46 shows two typical types of test points that can be inserted. A **control point** can be connected to a primary input, an existing scan cell output, or a dedicated scan cell output. An **observation point** can be connected to a primary output through an additional multiplexer, an existing scan cell input, or a dedicated scan cell input.

Figure 5.47b shows an example where one control point and one observation point are inserted to increase the detection probability of a six-input AND-gate given in Figure 5.47a. By splitting the six-input AND gate into two fewer-input AND gates and placing a control point and an observation point between the two fewer-input AND gates, we can increase the probability of detecting faults in the original six-input AND gate (*e.g.*, output Y stuck-at-0 and any input X_i stuck-at-1), thereby making the circuit more RP-testable. After the test points are inserted, the most difficult fault to detect is the bottom input of the four-input AND gate stuck-at-1. In that case, one of inputs X_1, X_2, and X_3 must be 0, the control point must be

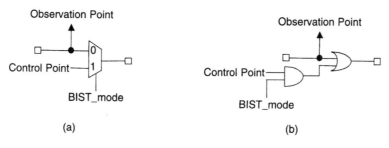

(a)　　　　　　　　　　　　　(b)

■ **FIGURE 5.46**

Typical test points inserted for improving a circuit's fault coverage: (a) test point with a multiplexer, and (b) test point with AND–OR gates.

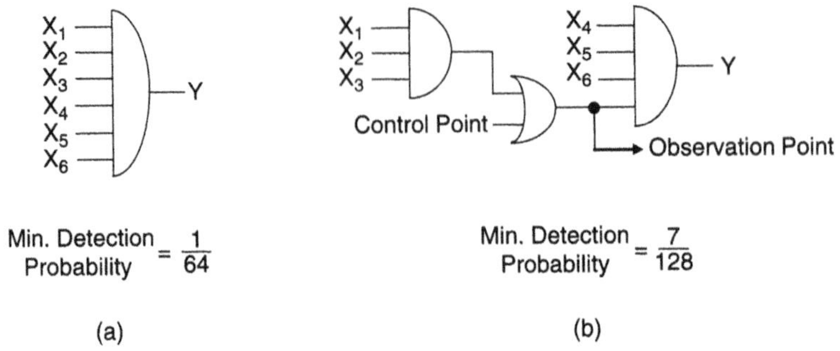

■ FIGURE 5.47

Example of inserting test points to improve detection probability: (a) an output RP-resistant stuck-at-0 fault, and (b) example inserted test points.

0, and all inputs X_4, X_5, and X_6 must be 1, resulting in a detection probability of $7/128 \, (= 7/8 \times 1/2 \times 1/2 \times 1/2 \times 1/2)$.

5.6.1.1   Test Point Placement

Because test points add area and performance overhead, an important issue for test point insertion is where to place the test points in the circuit to maximize the coverage and minimize the number of test points required. Note that it is not sufficient to only use observation points, as some faults require control points in order to be detected. Optimal placement of test points in circuits with reconvergent fanout has been shown to be NP-complete [Krishnamurthy 1987]. Several approximation techniques for placement of test points have been developed. They can be categorized depending on whether they use fault simulation or testability measures to guide them.

Fault simulation guided techniques require that the TPG is known ahead of time and can be simulated to determine the exact set of patterns that will be applied during self-test. Given this set of patterns, fault simulation is used to identify which faults will not be detected during self-test. Test points are then inserted to enable those faults to be detected. The technique in [Iyengar 1989] uses fault simulation to identify gates that block fault propagation and then inserts test points to allow propagation. The technique in [Touba 1996] uses path tracing to identify a set of test point solutions for each undetected fault, and then a covering algorithm is used to select the smallest set of test points that will allow detection of all faults. A limitation of fault simulation guided techniques is that the TPG must be known ahead of time which is not always the case, especially for cores that may be used in different *system-on-chips* (SOCs) with different BIST controllers. Also, if there are any late engineering changes that alter the set of patterns that are applied during self-test, then the fault coverage may be reduced.

Testability measure guided techniques avoid these problems because they do not require any knowledge of the TPG. They focus on improving the detection probability of RP-resistant faults which is approximated with testability measures. The gradient technique in [Seiss 1991] forms a cost function based on the ***controlla-bility/observability program*** (COP) testability measures [Brglez 1984] and then computes, in linear time, the gradient of the function with respect to each possible test point. The gradients are used to approximate the global testability impact for inserting a particular test point. Based on these approximations, the test point that has maximum benefit is inserted and the COP testability measures are recomputed. The process continues iteratively adding additional test points until the testability is satisfactory. Methods for speeding up this process are described in [Tsai 1998], where a hybrid cost function is used to estimate the actual cost function, and in [Nakao 1997], where several techniques including simultaneous selection of test points and candidate reduction are used. The technique in [Tamarapalli 1996] uses *probabilistic fault simulation*, which provides greater accuracy than COP testability measures, to guide the selection of test points to maximize the number of faults that exceed a specified detection probability threshold. In [Boubezari 1999], testability measures are computed and test points are inserted at the RTL. This has the advantage of allowing RTL synthesis procedures to take the test points into consideration when optimizing the design. In [Touba 1999], a logic synthesis procedure is described which uses *testability-driven factoring* combined with test point insertion to automatically synthesize random-pattern testable circuits. In [Xiang 2005], observation points are inserted in the scan chains and multiple capture cycles are used during shift operation.

One important concern with inserting test points is the impact on performance. If a test point adds delay on a critical timing path, then the timing requirements may not be satisfied. *Timing-driven test point insertion* techniques have been developed to address this problem. The technique in [Tsai 1998] computes the timing slack of each node and eliminates any node whose slack is not sufficiently long as a candidate for test point insertion. As test points are inserted, the slack information is updated. Because test points are not permitted in some locations due to timing constraints, the number of test points that is inserted to achieve sufficient fault coverage may be increased.

5.6.1.2 *Control Point Activation*

Once the test points have been inserted, the logic that drives the control points must be designed. When a control point is activated, it forces the logic value at a particular node in the circuit to a fixed value. During normal operation, all control points must be deactivated. During testing, there are different strategies as to when and how the control points are activated. One approach is **random activation**, where the control points are driven by the pseudo-random generator. The drawback of this approach is that when a large number of control points are inserted, they can interfere with each other and may not improve the fault coverage as much as desired. An alternative to *random activation* is to use **deterministic activation**. The technique in [Tamarapalli 1996] divides the BIST into phases and deterministically activates some subset of the control points in each phase. The technique in [Touba 1996] uses *pattern decoding logic* to activate the control points only for certain patterns where they are needed to detect RP-resistant faults.

5.6.2 Mixed-Mode BIST

A major drawback of test point insertion is that it requires modifying the circuit under test. In some cases this is not possible or not desirable (*e.g.*, for hard cores, macros, hand-crafted designs, or legacy designs). An alternative way to improve fault coverage without modifying the CUT is to use **mixed-mode BIST**. Pseudo-random patterns are generated to detect the RP-testable faults, and then some additional deterministic patterns are generated to detect the RP-resistant faults. There are a number of ways for generating deterministic patterns on-chip. Three approaches are described below.

5.6.2.1 ROM Compression

The simplest approach for generating deterministic patterns on-chip is to store them in a *read-only memory* (ROM). The problem with this approach is that the size of the required ROM is often prohibitive. Several **ROM compression** techniques have been proposed for reducing the size of the ROM in [Agarwal 1981], [Aboulhamid 1983], [Dandapani 1984], and [Edirisooriya 1992].

5.6.2.2 LFSR Reseeding

Instead of storing the test patterns themselves in a ROM, techniques have been developed for storing LFSR seeds that can be used to generate the test patterns [Könemann 1991]. The LFSR that is used for generating the pseudo-random patterns is also used for generating the deterministic patterns by reseeding it with computed seeds. The seeds can be computed with linear algebra as described in [Könemann 1991]. Because the seeds are smaller than the test patterns themselves, they require less ROM storage. One problem is that for an LFSR with a fixed characteristic (feedback) polynomial, it may not always be possible to find a seed that will efficiently generate the required deterministic test patterns. A solution to that problem was proposed in [Hellebrand 1995a] in which a ***multiple-polynomial LFSR*** (MP-LFSR), as illustrated in Figure 5.48, is used. An MP-LFSR is an LFSR

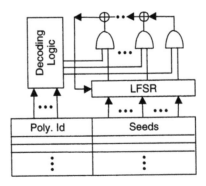

■ **FIGURE 5.48**

Reseeding with multiple-polynomial LFSR.

with a reconfigurable feedback network. A *polynomial identifier* is stored with each seed to select the characteristic polynomial that will be used for that seed. Techniques for further reductions in storage can be achieved by using variable-length seeds [Rajski 1998], a special ATPG algorithm [Hellebrand 1995b], folding counters [Liang 2001], and seed encoding [Al-Yamani 2005].

5.6.2.3 Embedding Deterministic Patterns

A third approach for mixed-mode BIST is to embed the deterministic patterns in the pseudo-random sequence. Many of the pseudo-random patterns generated during pseudo-random testing do not detect any new faults, so some of those "useless" patterns can be transformed into deterministic patterns that detect RP-resistant faults [Touba 1995]. This can be done by adding *mapping logic* between the scan chains and the CUT [Touba 1995] or in a less intrusive way by adding the *mapping logic* at the inputs to the scan chains to either perform bit-fixing [Touba 2001] or bit-flipping [Kiefer 1998]. Figure 5.49 shows a bit-flipping BIST scheme taken from [Kiefer 1998]. A bit-flipping function detects these "useless" patterns and maps them to deterministic patterns through the use of an XOR gate that is inserted between the LFSR and each scan chain.

5.6.3 Hybrid BIST

For manufacturing fault coverage enhancement where a tester is present, deterministic data from the tester can be used to improve the fault coverage. The simplest approach is to perform top-up ATPG for the faults not detected by BIST to obtain a set of deterministic test patterns that "top-up" the fault coverage to the desired level and then store those patterns directly on the tester. In a system-on-chip, test scheduling can be done to overlap the BIST run time with the transfer time for loading the deterministic patterns from the tester [Sugihara 1998] [Jervan 2003]. More elaborate hybrid BIST schemes have been developed which attempt to store the deterministic patterns on the tester in a compressed form and then make use of the existing BIST hardware to decompress them. Such techniques are described in [Das 2000], [Dorsch 2001], [Ichino 2001], [Krishna 2003], [Wohl 2003], [Jas 2004], and [Lei 2005]. More discussions on test compression can be found in Chapter 6.

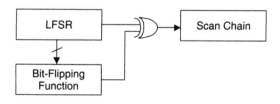

■ **FIGURE 5.49**

Bit-flipping BIST.

5.7   BIST TIMING CONTROL

While logic BIST can be used to reduce test costs by moving most of the tester functionality onto the circuit under test, its real value is in providing **at-speed testing** for high-speed and high-performance circuits. These circuits often contain multiple clock domains, each running at a frequency that is either synchronous or asynchronous to the other clock domains.

The most critical yet difficult part of using logic BIST is how to test intra-clock-domain faults and inter-clock-domain faults thoroughly and efficiently with a proper capture-clocking scheme. An **intra-clock-domain fault** originates at one clock domain and terminates at the same clock domain. An **inter-clock-domain fault** originates at one clock domain but terminates at another clock domain.

There are three basic capture-clocking schemes that can be used for testing multiple clock domains: (1) single-capture, (2) skewed-load, and (3) double-capture. We will illustrate with BIST timing control diagrams how to test synchronous and asynchronous clock domains using these schemes.

Two clock domains are said to be **synchronous** if the active edges of both clocks controlling the two clock domains can be aligned precisely or triggered simultaneously. Two clock domains are said to be **asynchronous** if they are not synchronous. Throughout this section, we will assume that a STUMPS-based architecture is used and that each clock domain contains one test clock and one scan enable signal. The faults we will consider include **structural faults**, such as stuck-at faults and bridging faults, as well as timing-related **delay faults**, such as path-delay faults and transition faults.

5.7.1   Single-Capture

Single-capture is a **slow-speed test** technique in which only one capture pulse is applied to each clock domain. It is the simplest for testing all intra-clock-domain and inter-clock-domain structural faults. There are two approaches that can be used: (1) one-hot single-capture, and (2) staggered single-capture.

5.7.1.1   One-Hot Single-Capture

Using the **one-hot single-capture** approach, a capture pulse is applied to only one clock domain during each capture window, while all other test clocks are held inactive. A sample timing diagram is shown in Figure 5.50. In the figure, because only one capture pulse (C1 or C2) is applied during each capture window, this scheme can only test intra-clock-domain and inter-clock-domain structural faults. The main advantage of this approach is that the designer does not have to worry about clock skews between the two clock domains during self-test, as each clock domain is tested independently. The only requirement is that delays $d1$ and $d2$ be properly adjusted; hence, this approach can be used for **slow-speed testing** of both synchronous and asynchronous clock domains. Another benefit of using this approach is that a single, slow-speed *global scan enable* (GSE) signal can be used

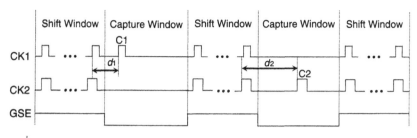

■ FIGURE 5.50

One-hot single-capture.

for driving both clock domains, which makes it easy to integrate with scan. A major drawback is longer test time, as all clock domains have to be tested one at a time.

5.7.1.2 Staggered Single-Capture

The long test time problem using one-hot single-capture can be solved using the staggered single-capture approach [Wang 2006]. A sample timing diagram is shown in Figure 5.51. In this approach, capture pulses C1 and C2 are applied in a sequential or staggered order during the capture window to test all intra-clock-domain and inter-clock-domain structural faults in the two clock domains. For clock domains that are synchronous, adjusting $d2$ will allow us to detect inter-clock-domain delay faults between the two clock domains at-speed. In addition, because $d1$ and $d3$ can be as long as desired, a single, slow-speed GSE signal can be used. This significantly simplifies the logic BIST physical implementation for designs with multiple clock domains. There may be some structural fault coverage loss between clock domains if the ordered sequence of capture clocks is fixed for all capture cycles.

5.7.2 Skewed-Load

Skewed-load is an **at-speed delay test** technique in which a last shift pulse followed immediately by a capture pulse, running at the test clock's operating

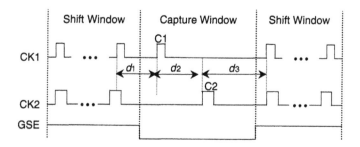

■ FIGURE 5.51

Staggered single-capture.

frequency, are used to launch the transition and capture the output response [Savir 1993]. It is also referred to as **launch-on-shift**. This technique addresses the intra-clock-domain delay fault detection problem which cannot be tested using single-capture schemes. Skewed-load uses the value difference between the last shift pulse and the next-to-last-shift pulse to launch the transition and uses the capture pulse to capture the output response. In order for the last shift pulse to launch the transition, the scan enable signal associated with the clock domain must be able to switch operations from shift to capture in one clock cycle. There are three approaches that can be used: (1) one-hot skewed-load, (2) aligned skewed-load, and (3) staggered skewed-load.

5.7.2.1 One-Hot Skewed-Load

Similar to one-hot single-capture, the **one-hot skewed-load** approach tests all clock domains one by one [Bhawmik 1997]. A sample timing diagram is shown in Figure 5.52. The main differences are: (1) It applies shift-followed-by-capture pulses (S1-followed-by-C1 or S2-followed-by-C2) to detect intra-clock-domain delay faults, and (2) each scan enable signal (SE1 or SE2) must switch operations from shift to capture within one clock cycle ($d1$ or $d2$). Thus, this approach can only be used for **at-speed testing** of intra-clock-domain delay faults in both synchronous and asynchronous clock domains. The disadvantages are: (1) It cannot be used to detect inter-clock-domain delay faults, (2) it has a long test time, and (3) it is incompatible with scan, as a single, slow-speed GSE signal can no longer be used.

5.7.2.2 Aligned Skewed-Load

The disadvantages of one-hot skewed-load can be resolved by using the aligned skewed-load scheme. One **aligned skewed-load** approach that aligns all capture edges together is illustrated in Figure 5.53 [Nadeau-Dostie 1994] [Nadeau-Dostie 2000]. The approach is referred to as **capture aligned skewed-load**. The major advantage of using this approach is that all intra-clock-domain and inter-clock-domain faults can be tested. The arrows shown in Figure 5.53 indicate the delay

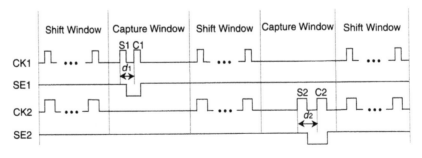

■ **FIGURE 5.52**

One-hot skewed-load.

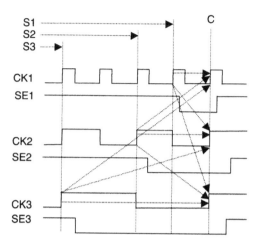

■ **FIGURE 5.53**

Capture aligned skewed-load.

faults that can be tested. For example, the three arrows from S1 (CK1) to C are used to test all intra-clock-domain delay faults in the clock domain controlled by CK1 and all inter-clock-domain delay faults from CK1 to CK2 and CK3. The remaining six arrows shown from S2 (CK2) to C and from S3 (CK3) to C are used to test all the remaining delay faults.

Because the active edges (rising edges) of the three capture pulses (see dash line C) must be aligned precisely, the circuit must contain one reference clock, and the frequency of all remaining test clocks must be derived from the reference clock. In the example given here, CK1 is the reference clock operating at the highest frequency, and CK2 and CK3 are derived from CK1 and designed to operate at 1/2 and 1/4 the frequency, respectively; therefore, this approach is only applicable for **at-speed testing** of intra-clock-domain and inter-clock-domain delay faults in synchronous clock domains.

A similar **aligned skewed-load** approach that aligns all last shift edges, rather than capture edges, is shown in Figure 5.54 [Hetherington 1999] [Rajski 2003]. This approach is referred to as **launch aligned skewed-load**. Similar to capture aligned skewed-load, it is also only applicable for **at-speed testing** of intra-clock-domain and inter-clock-domain delay faults in synchronous clock domains.

Consider the three clock domains, driven by CK1, CK2, and CK3, again. The eight arrows among the dash line S and the three capture pulses (C1, C2, and C3) indicate the intra-clock-domain and inter-clock-domain delay faults that can be tested. Unlike Figure 5.53, however, in order to test the inter-clock-domain delay faults from CK1 to CK3, a special shift pulse S1 (when SE1 is set to 1) is required. As this method requires a much more complex timing-control diagram, a **clock suppression** circuit is used to enable or disable selected shift or capture pulses [Rajski 2003]. The dotted clock pulses shown in the figure indicate the suppressed shift pulses.

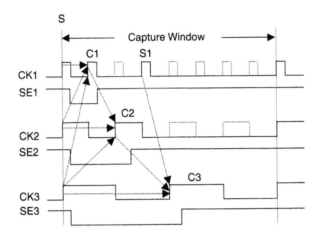

Launch aligned skewed-load.

5.7.2.3 *Staggered Skewed-Load*

While the above aligned skewed-load approaches can test all intra-clock-domain and inter-clock-domain faults in synchronous clock domains, their physical implementation is extremely difficult. There are two main reasons. First, in order to effectively align all active edges in either capture or last shift, the circuit must contain a reference clock. This reference clock must operate at the fastest clock frequency, and all other clock frequencies must be derived from the reference clock; such designs rarely exist. Second, for any two edges that cannot be aligned precisely due to clock skews, we must either resort to a one-hot skewed-load approach or add capture-disabling circuitry on the functional data paths of the two clock domains to prevent the cross-domain logic from interacting with each other during capture. This increases the circuit overhead, degrades the functional circuit performance, and reduces the ability to test inter-clock-domain faults.

The **staggered skewed-load** approach shown in Figure 5.55 relaxes these conditions [Wang 2005b]. For test clocks that cannot be precisely aligned, a delay $d3$ is inserted, to eliminate the clock skew interaction between the two clock domains. The two last shift pulses (S1 and S2) are used to create transitions at the outputs of some scan cells, and the output responses to these transitions are captured by the following two capture pulses (C1 and C2), respectively. Both delays $d1$ and $d2$ are set to their respective clock domains' operating frequencies; hence, this scheme can be used to test all intra-clock-domain faults and inter-clock-domain structural faults in **asynchronous clock domains**. A problem still exists, as each clock domain requires an at-speed scan enable signal, which complicates physical implementation.

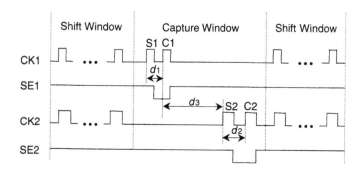

■ **FIGURE 5.55**

Staggered skewed-load.

5.7.3 Double-Capture

The physical implementation difficulty using skewed-load can be resolved by using the double-capture scheme. **Double-capture** is another **at-speed test** technique in which two consecutive capture pulses are applied to launch the transition and capture the output response. It is also referred to as **broad-side** [Savir 1994] or **launch-on-capture**. The double-capture scheme can achieve **true at-speed test** quality for intra-clock-domain and inter-clock-domain faults in any synchronous or asynchronous design and that is easy for physical implementation. Here, true at-speed testing is meant to: (1) allow detection of intra-clock-domain faults within each clock domain at its own operating frequency and detection of inter-clock-domain structural faults or delay faults, depending on whether the circuit under test is synchronous, asynchronous, or a mix of both; and (2) ease physical implementation for seamless integration with the conventional scan/ATPG technique.

5.7.3.1 One-Hot Double-Capture

Similar to one-hot skewed-load, the **one-hot double-capture** approach tests all clock domains one by one. A sample timing diagram is shown in Figure 5.56. The

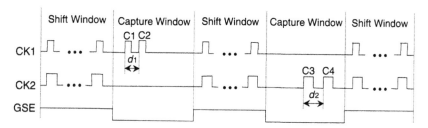

■ **FIGURE 5.56**

One-hot double-capture.

main differences are: (1) Two consecutive capture pulses are applied (C1-followed-by-C2 or C3-followed-by-C4) at their respective clock domains' frequencies (of period $d1$ or $d2$) to test intra-clock-domain delay faults, and (2) a single, slow-speed GSE signal is used to drive both clock domains. Hence, this scheme can be used for **true at-speed testing** of intra-clock-domain delay faults in both synchronous and asynchronous clock domains. Two drawbacks remain: (1) It cannot be used to detect inter-clock-domain delay faults, and (2) it has a long test time.

5.7.3.2 Aligned Double-Capture

The drawbacks of the one-hot double-capture scheme can be resolved by using an **aligned double-capture** approach. Similar to the aligned skewed-load approach, the aligned double-capture scheme allows all intra-clock-domain faults and inter-clock-domain faults to be tested [Wang 2006]. The main differences are: (1) Two consecutive capture pulses are applied, rather than shift-followed-by-capture pulses, and (2) a single, slow-speed GSE signal is used. Figures 5.57 and 5.58 show two

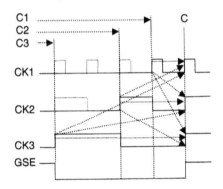

■ **FIGURE 5.57**

Capture aligned double-capture.

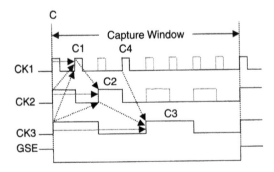

■ **FIGURE 5.58**

Launch aligned double-capture.

sample timing diagrams. This scheme can be used for **true at-speed testing** of **synchronous clock domains**. One major drawback is that precise alignment of the capture pulses is still required. This complicates physical implementation for designs with asynchronous clock domains.

5.7.3.3 Staggered Double-Capture

The capture alignment problem in the aligned double-capture approach can finally be relaxed by using the **staggered double-capture** scheme [Wang 2005a] [Wang 2006]. A sample timing diagram is shown in Figure 5.59. During the capture window, two capture pulses are generated for each clock domain. The first two capture pulses (C1 and C3) are used to create transitions at the outputs of some scan cells, and the output responses to the transitions are captured by the second two capture pulses (C2 and C4), respectively. Both delays $d2$ and $d4$ are set to their respective domains' operating frequencies. Because $d1$, $d3$, and $d5$ can be adjusted to any length, we can simply use a single, slow-speed GSE signal for driving all clock domains; hence, **true at-speed testing** is guaranteed using this approach for asynchronous clock domains. Because a single GSE signal is used, this scheme significantly eases physical implementation and allows us to integrate logic BIST with scan/ATPG easily in order to improve the circuit's manufacturing fault coverage.

5.7.4 Fault Detection

Scan ATPG and logic BIST are currently the two most widely used *structural offline test* techniques for improving a circuit's fault coverage and product quality. Unfortunately, 100% single-stuck fault coverage using scan ATPG does not guarantee perfect product quality (*i.e.*, no **test escapes**) [McCluskey 2000] [Li 2002]. Recent investigations reported in [McCluskey 2004] further revealed that only 5% (15) of the 324 defective ELF35 chips contained defects that acted as single-stuck-at faults, while 35% (41) of the 116 defective Murphy chips acted as single-stuck-at faults. The remaining defects were (1) timing dependent, (2) sequence dependent,

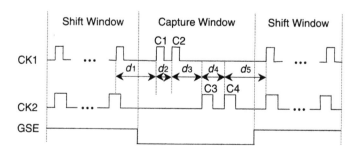

■ **FIGURE 5.59**

Staggered double-capture.

or (3) attributed to timing-independent, non-single-stuck-at faults, such as multiple-stuck-at faults or non-feedback bridging faults.

Possible causes of timing-dependent defects are *resistive opens*, connections that have significant higher resistance than intended or transistors with lower drive than designed for. Possible causes of sequence-dependent defects are: (1) a defect that acts like a stuck-open fault [Li 2002] or (2) one that causes a feedback bridging fault. The paper found that all test sets using 15-detect (an **N-detect** method for detecting a single stuck-at fault multiple times) [Ma 1995] or ***transition faults propagated to all reachable outputs*** (TARO) [Tseng 2001] resulted in zero to three test escapes on both devices. This suggests that in order to screen out more defects delay fault testing is no longer optional. Test patterns targeting single-stuck-at faults multiple times or all possible transition paths (not just critical paths) are also required. Logic BIST automatically addresses these issues as it is able to detect defects that cannot be modeled for scan ATPG.

Intra-clock-domain delay fault testing is relatively easy using any of the above skewed-load and double-capture timing control schemes. Inter-clock-domain delay fault testing, however, is more complex. [Wang 2005a] conducted an experiment that showed that applying a single capture pulse to each clock domain, rather than two pulses, as shown in Figure 5.60, yields the highest fault coverage. In this figure, *d* has to be set correctly to detect inter-clock-domain faults between the two clock domains.

Table 5.5 shows the type of faults that can be detected and the design styles that must be adopted when using the above timing control schemes. Each scheme has its advantages and disadvantages; for example,

1. One-hot single-capture yields the highest fault coverage for both intra-clock-domain and inter-clock-domain structural faults.

2. Staggered single-capture yields the highest fault coverage for inter-clock-domain delay faults.

3. One-hot skewed-load may yield the highest fault coverage for intra-clock-domain delay faults but may over-test the circuit by creating more invalid states in the functional circuitry than the one-hot double-capture approach.

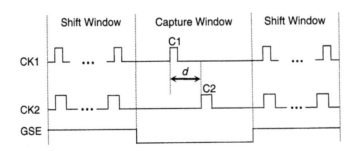

■ **FIGURE 5.60**

Inter-clock-domain fault detection.

TABLE 5.5 ■ Fault Detection Capability

Capture-Clocking Scheme	Intra-Structural	Intra-Delay	Inter-Structural	Inter-Delay	Sync. Design	Async. Design
One-hot single-capture	✓	–	✓	–	✓	✓
Staggered single-capture	✓	–	✓	✓	✓	✓
One-hot skewed-load	✓	✓	✓	–	✓	✓
Aligned skewed-load	✓	✓	✓	✓	✓	–
Staggered skewed-load	✓	✓	✓	✓	✓	✓
One-hot double-capture	✓	✓	✓	–	✓	✓
Aligned double-capture	✓	✓	✓	✓	✓	–
Staggered double-capture	✓	✓	✓	✓	✓	✓

4. Aligned skewed-load and aligned double-capture are best suited for testing synchronous clock domains.

5. Staggered skewed-load and staggered double-capture are best suited for testing asynchronous clock domains.

To summarize, a hybrid double-capture scheme using staggered double-capture and aligned double-capture seems to be the preferred scheme for **true at-speed testing** of designs having a number of synchronous and asynchronous clock domains. This hybrid approach makes physical implementation easier than other approaches and allows for seamless integration with any conventional scan/ATPG technique to further improve the circuit's fault coverage.

5.8   A DESIGN PRACTICE

In this section, we show an example of designing a logic BIST system for testing a scan-based design (core) comprised of two clock domains using s38417 and s38584. The two clock domains are taken from the ISCAS-1989 benchmark circuits [Brglez 1989] and their statistics are shown in Table 5.6. The design we consider is described at the *register-transfer level* (RTL). We show you all the necessary steps to arrive at the logic BIST system design, verify its correctness, and improve its fault coverage.

TABLE 5.6 ■ Design Statistics

Clock Domain	No. of PIs	No. of POs	No. of Flip-Flops	No. of Gates
CD1 (s38417)	28	106	1636	22179
CD2 (s38584)	12	278	1452	19253

5.8.1 BIST Rule Checking and Violation Repair

The first step is to perform logic BIST design rule checking on the RTL design. All DFT rule violations of the *scan design rules* and *BIST-specific design rules* provided in Section 2.6 of Chapter 2 and Section 5.2 must be repaired. Once all DFT rule violations are repaired, the design should meet all scan and logic BIST design rules. In addition, we should be aware of the following design parameters:

- The number of test clocks present in the design, each used for controlling one clock domain.

- The number of set/reset clocks present in the design to be used for breaking all asynchronous set/reset loops.

In the example given above, the design contains two test clocks and does not require any additional set/reset clock. The new RTL design (core) after BIST rule repair is performed is referred to as an *RTL BIST-ready core*.

5.8.2 Logic BIST System Design

The second step is to design the logic BIST system at the RTL. The decisions that need to be made at this stage include:

- The type of logic BIST architecture to adopt.

- The number of PRPG–MISR (or PEPG–MISR) pairs to use.

- The length of each PRPG–MISR (or PEPG–MISR) pair.

- The faults to be tested and BIST timing control diagrams to be used for testing these faults.

- The types of optional logic to be added in order to ease physical implementation and facilitate debug and diagnosis, as well as improve the circuit's fault coverage.

5.8.2.1 Logic BIST Architecture

In accordance with the logic BIST architectures summarized in Table 5.4, we choose to implement a STUMPS-based architecture, as it is easy to integrate with scan/ATPG and is the architecture widely used in the industry. We recommend using one PRPG–MISR pair for each clock domain, whenever possible, as the resulting BIST architecture is easier to debug. In addition, the use of one PRPG–MISR pair for each clock domain can eliminate the need for additional design efforts for managing clock skews between interactive clock domains, even when they operate at the same frequency. If it is required to use a single PRPG–MISR pair to test multiple clock domains, these clock domains should be placed within physical proximity in order to simplify physical implementation. An example logic BIST system based on the STUMPS architecture for testing the design given in Table 5.6 is shown in Figure 5.61.

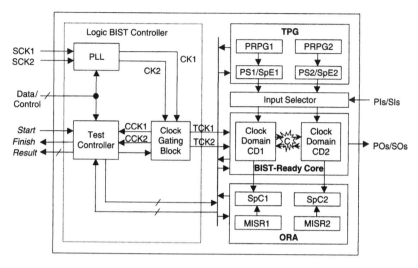

■ **FIGURE 5.61**

A logic BIST system for testing a design with two cores.

The BIST architecture used for testing the BIST-ready core consists of a TPG for generating test stimuli, an input selector for providing pseudo-random or ATPG patterns to the core-under-test, an ORA for compacting the test responses, and a logic BIST controller for coordinating the overall BIST operation. The logic BIST controller consists of a test controller and a clock gating block. The test controller initiates the BIST operation upon receiving a *Start* signal, issues a *Finish* signal once the BIST operation is complete, and reports the pass/fail status of the test through the *Result* bus. The clock gating block accepts internal PLL clocks (CK1 and CK2) derived from external functional clocks (SCK1 and SCK2) and generates the required test clocks (TCK1 and TCK2) and controller clocks (CCK1 and CCK2) for controlling the BIST-ready core and test controller, respectively. During normal functional operation, both CK1 and CK2 can run faster or slower than SCK1 and SCK2, respectively.

5.8.2.2 *TPG and ORA*

Next, we need to determine the length of each PRPG–MISR pair. Using a separate PRPG–MISR pair for each clock domain allows us to reduce the length of each PRPG and MISR. In the example shown in Figure 5.61, the linear phase shifters, PS1 and PS2, and space expanders, SpE1 and SpE2, can be used to further reduce the length of the PRPGs, whereas the space compactors, SpC1 and SpC2, can be used to further reduce the length of the MISRs. Each space expander or space compactor typically consists of an XOR-gate tree.

Now, suppose we decide to: (1) synthesize the two clock domains, CD1 and CD2, each with 20 balanced scan chains; (2) run 100,000 pseudo-random patterns to obtain very high BIST fault coverage by adding additional test points; and

(3) perform top-up ATPG after BIST to further increase the circuit's fault coverage. Because CD1 has 28 PIs, a logical conclusion would be to expect the length of the PRPG1 to be 48 for using a 48-stage PRPG to drive 28 PIs and 20 scan chains. Because we plan to perform top-up ATPG, which requires sharing 20 out of the 28 PIs with **scan inputs** (SIs), and another 20 POs with **scan outputs** (SOs), another possible length for the PRPG1 would be 28. What we need to determine is whether a 28-stage PRPG, constructed from a maximum-length LFSR or CA, is adequate for generating the required 100,000 pseudo-random patterns.

For CD1 with 20 balanced scan chains, 82 shift clock pulses are required (1636 flip-flops/20 scan chains) to scan-in a single pseudo-random pattern. This means that a total of 8.2 million shift clock pulses are required to scan-in all 100,000 patterns. This number is much smaller than the 256 million ($2^{28} - 1$) patterns generated using a 28-stage maximum-length LFSR or CA for the PRPG1. From Table 5.1, we choose a 28-stage maximum-length LFSR with characteristic polynomial $f(x) = 1 + x^3 + x^{28}$.

A similar analysis applies for CD2. The main difference is that CD2 has 12 PIs. Suppose we pick 10 out of the 12 PIs to share with 10 SIs for top-up ATPG. We will need to use a 10-to-20 space expander (SpE2) for driving the 20 scan chains and a 20-to-10 space compactor (SpC2) for driving the 10 SOs. Because testing this clock domain requires a total of 7.3 million ($1452/20 \times 100,000$) shift clock pulses, we need to use at least a 23-stage maximum-length LFSR or CA as PRPG2 to drive the 12 PIs. From Table 5.1, we choose a 25-stage maximum-length LFSR with characteristic polynomial $f(x) = 1 + x^3 + x^{25}$.

As indicated in Section 5.4.3, each MISR can cause an *aliasing* problem, but the problem is of less concern when the MISR length is greater than 20. Because CD1 and CD2 both have 106 and 278 POs, we choose a 106-to-27 space compactor (SpC1) and a 278-to-35 space compactor (SpC2), respectively. Thus, we will use a 47-stage MISR and a 45-stage MISR to compact the test responses from both CD1 and CD2, respectively, where 47 = 27 (shared POs) + 20 (SOs) and 45 = 35 (shared POs) + 10 (SOs). From Table 5.1, we choose to implement the 47-stage MISR with $f(x) = 1 + x^5 + x^{47}$ and the 45-stage MISR with $f(x) = 1 + x + x^3 + x^4 + x^{45}$. Table 5.7 shows the decisions made for each PRPG–MISR pair so far.

5.8.2.3 Test Controller

The test controller plays a central role in coordinating the overall BIST operation. In general, a *finite-state machine* written at the RTL is used to implement the test

TABLE 5.7 ■ PRPG–MISR Choices

Clock Domain	No. of Scan Chains	No. of Shared SIs or SOs	Maximum Scan Chain Length	PRPG Length	MISR Length
CD1 (s38417)	20	20	82	28	47
CD2 (s38584)	20	10	73	25	45

controller for interfacing with all external signals, such as *Start*, *Finish*, and *Result*, and generating the required timing control signals for controlling each PRPG–MISR pair and the BIST-ready core. Comparison logic is included in the test controller to compare the *final signature* with an embedded *golden signature*.

Often, these interface signals are controlled through an IEEE 1149.1 boundary-scan-standard-based **test access port** (TAP) **controller**. In this case, all signals can be assessed through the TAP: **test data in** (TDI), **test data out** (TDO), **test clock** (TCK), and **test mode select** (TMS). Optionally, an IEEE 1500 standard-based **wrapper** may be also embedded in each selected clock domain. Both IEEE standards are discussed extensively in Chapter 10.

In order to test structural faults in the BIST-ready core, we choose the *staggered single-capture* approach rather than the *one-hot single-capture* approach. The slow-speed timing control diagram is shown in Figure 5.62, where test clocks TCK1 and TCK2 are staggered and generated by the clock gating block shown in Figure 5.61.

In order to test delay faults in the BIST-ready core, we choose the *staggered double-capture* approach if CD1 and CD2 are asynchronous or the *aligned double-capture* approach if they are synchronous. This is due to the fact that either approach allows us to operate a GSE signal at slow speed for driving all clock domains simultaneously, in both BIST and scan ATPG modes. The at-speed timing control diagrams using the *staggered double-capture* and *launch aligned double-capture* schemes are shown in Figures 5.63 and 5.64, respectively.

5.8.2.4 *Clock Gating Block*

In order to generate an ordered sequence of *single-capture* or *double-capture* clocks, *clock suppression* [Rajski 2003] [Wang 2004], *daisy-chain clock triggering*, or *token-ring clock enabling* [Wang 2005a] can be used. The clock suppression scheme typically requires using a reference clock operating at the highest frequency. Daisy-chain clock triggering means that a completion of one event automatically triggers the next event, as the arrows show in Figure 5.65. The only difference between daisy-chain clock triggering and token-ring clock enabling is that the former uses a clock edge to trigger the next event, while the latter uses a signal level to enable the next event.

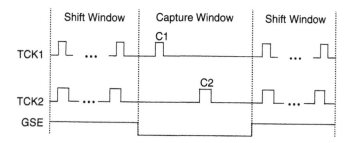

■ **FIGURE 5.62**

Slow-speed timing control using staggered single-capture.

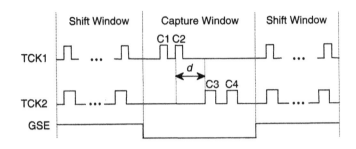

■ **FIGURE 5.63**

At-speed timing control using staggered double-capture.

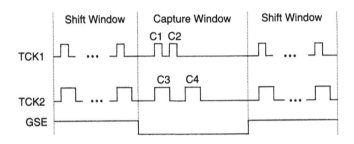

■ **FIGURE 5.64**

At-speed timing control using launch aligned double-capture.

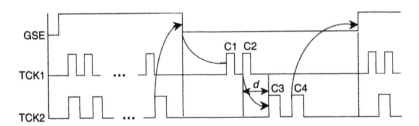

■ **FIGURE 5.65**

Daisy-chain clock triggering.

Figure 5.66 shows a daisy-chain clock-triggering circuit for generating the *staggered double-capture* waveform given in Figure 5.65. When the BIST mode is activated, the SE1/SE2 generators and two-pulse controllers will generate the required scan enable and double-capture clock pulses, per the arrows shown in Figure 5.65. Each SE1/SE2 can be treated as a GSE signal for CD1/CD2.

Figure 5.67 shows a clock suppression circuit for generating the *launch aligned double-capture* waveform given in Figure 5.64. This circuit uses a reference clock (CK1) to program the capture window. The contents of the 8-bit shift register are preset to {0011,1111} during each shift window. Due to its programmability, the

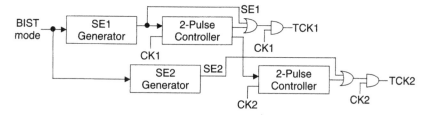

■ **FIGURE 5.66**

A daisy-chain clock triggering circuit for generating the waveform given in Figure 5.65.

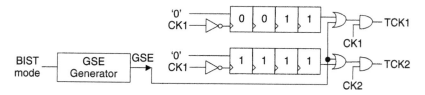

■ **FIGURE 5.67**

A clock suppression circuit for generating the waveform given in Figure 5.64.

approach can also be used to generate timing waveforms for testing asynchronous designs. One major requirement is that we guarantee that the delay measured by the number of reference clock pulses will be longer than delay d between C2 and C3, as shown in Figure 5.63.

5.8.2.5 Re-Timing Logic

The main difference between ATE-based scan testing and logic BIST is that the latter requires that more complex BIST circuitry be implemented on the functional circuitry. Successfully completing the physical implementation of the functional circuitry of a high-speed and high-performance design is a challenge in itself. If the BIST circuitry adds a large number of timing critical signals and requires strict clock-skew management, the physical implementation of logic BIST can become extremely difficult; therefore, we recommend adding two pipelining registers (see Figure 5.9) between each PRPG and the BIST-ready core and two additional pipelining registers between the BIST-ready core and each MISR. In this case, the maximum scan chain length for each clock domain, CD1 or CD2, is effectively increased by 2, not 4.

5.8.2.6 Fault Coverage Enhancing Logic and Diagnostic Logic

The drawback to using pseudo-random patterns is that the circuit may not meet the target fault coverage goal. In order to improve the circuit's fault coverage, we recommend adding extra test points and additional logic for top-up ATPG support at the RTL. A general rule of thumb is to add one extra test point every 1000

TABLE 5.8 ■ Example Test Modes Supported by the Logic BIST System

Test Mode	CD1 Effective Chain Count	CD2 Effective Chain Count
Normal	0	0
BIST	20	20
ATPG	20	10
ATPG compression	20	20
Serial debug and diagnosis	1	1

gates. For top-up ATPG support, the inserted logic includes an input selector for selecting test patterns either from the PRPGs or PIs/SIs, as shown in Figure 5.61, as well as circuitry for reconfiguring the scan chains to perform top-up ATPG in: (1) ATPG mode, or (2) ATPG compression mode, which is discussed in more detail in Chapter 6.

We also recommend including **diagnostic logic** in the RTL BIST code to facilitate debug and diagnosis. One simple approach is to connect all PRPG–MISR pairs (and all scan chains) as a serial scan chain and make them externally accessible. (Please refer to Chapter 7 for more advanced BIST diagnosis techniques.) Table 5.8 summarizes all possible test modes of the logic BIST system along with the effective scan chain counts for each test mode.

5.8.3   RTL BIST Synthesis

Once all decisions regarding the logic BIST architecture are made, it is time to create the RTL logic BIST code. At this stage, it is possible to either design the logic BIST system by hand or generate the RTL code automatically using a (commercially available) RTL logic BIST tool. In either case, the number of scan chains for each clock domain should be specified along with the names of their associated scan inputs and scan outputs without inserting the actual scan chains into the circuit. The scan synthesis task can be handled as part of the general synthesis task, implemented using any commercially available synthesis tool for converting the RTL BIST-ready core and the logic BIST system into a gate-level netlist.

5.8.4   Design Verification and Fault Coverage Enhancement

Finally, the synthesized netlist must be verified with functional or timing verification to ensure that the logic BIST system functions as intended. If any pattern mismatch occurs, the problem must be identified and resolved. Next, fault simulation must be performed on the pseudo-random patterns generated by the TPG in order to determine the circuit's fault coverage. If the circuit does not reach the target fault coverage goal, additional test points should be inserted or top-up ATPG should be used. The extra test points that were added in advance at the RTL design should allow you to achieve the target fault coverage goal; otherwise, the test point insertion and fault simulation process may have to be repeated until the final fault

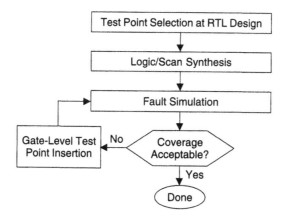

■ **FIGURE 5.68**

Fault simulation and test point insertion flow.

coverage goal is reached. Once this process is complete, the *golden signature* can be either recorded to be compared externally or hard-coded into the comparison logic. The fault simulation and test point insertion flow are illustrated in Figure 5.68.

5.9 CONCLUDING REMARKS

Bardell and McAnney [Bardell 1982] are among the pioneers who have proposed a widely adopted logic BIST architecture, called STUMPS, for scan-based designs. The acceptance of this STUMPS architecture is mostly due to the ease of integration of the BIST circuitry into a scan design; however, the efforts required to implement the BIST circuitry and the loss of the fault coverage for using pseudo-random patterns have prevented the STUMPS-based logic BIST architecture from being widely used across all industries. As the semiconductor manufacturing technology moves into the nanometer design era, it remains to be seen how the CBILBO-based architecture proposed by Wang and McCluskey [Wang 1986c], which can always guarantee 100% single stuck-at fault coverage and has the ability of running 10 times more BIST patterns than the STUMPS-based architecture, will perform. Challenges lie ahead with regard to whether or not pseudo-exhaustive testing will become a preferred BIST pattern generation technique.

5.10 EXERCISES

5.1 (BIST Design Rules) A scan design can contain many asynchronous set/reset signals that may require adding two or more set/reset clock points to break all ripple set/reset loops. A ripple set/reset loop is a combinational feedback loop.

Assume that the design now contains two system clocks (CK1 and CK2) and two set/reset clocks (SRCK1 and SRCK2). Derive two BIST timing control diagrams, including a scan enable (SE) signal, to test all data faults and set/reset faults controlled by these four clocks. Explain which timing control diagram can detect more faults.

5.2 (BIST Design Rules) Design a one-hot decoder for testing a tristate bus with four independent tristate drivers in BIST mode.

5.3 (BIST Design Rules) Design an X-bounding circuit for improving the fault coverage of a bidirectional I/O port by forcing it to input mode during BIST operation.

5.4 (Complete LFSR) Insert a zero-state into each hybrid LFSR given in Figures 5.13a and 5.13b. Minimize each modified hybrid LFSR, called complete LFSR, so it contains the least number of gates. What is the period of each complete LFSR?

5.5 (Weighted LFSR) Design a four-stage weighted LFSR with each output having a different weight of 0.75, 0.5, 0.25, or 0.125.

5.6 (Cellular Automata) Prove why the cellular automaton of length 5 given in Table 5.2 can generate a maximum-length sequence of $2^5 - 1$. Derive the construction rules for cellular automata of lengths 54 and 55.

5.7 (Condensed LFSR) Design an (n, k) condensed LFSR to test each output cone of an $(n, w) = (8, 3)$ CUT exhaustively. Compare its pros and cons with the (8, 5) cyclic LFSR given in Figure 5.23.

5.8 (Cyclic LFSR) Derive how the (8, 5) cyclic LFSR given in Figure 5.23 is shortened from the (15, 5) cyclic LFSR with $g(x) = (1 + x)(1 + x + x^4) = 1 + x^2 + x^4 + x^5$, $h(x) = (1 + x^{15})/g(x)$, $p(x) = 1 + x^2 + x^3 + x^4 + x^5$, and $f(x) = h(x)p(x) = 1 + x^3 + x^5 + x^8 + x^9 + x^{11} + x^{12} + x^{13} + x^{15}$.

5.9 (Shortened Cyclic LFSR) Assume that the number of information bits to be deleted (s) is 1. Derive how the (8, 4) shortened cyclic LFSR given in Figure 5.24 is shortened from the $(n - s, k - s) = (15 - 1, 5 - 1) = (14, 4)$ shortened cyclic LFSR with $g(x) = (1 + x)(1 + x + x^4) = 1 + x^2 + x^4 + x^5$, $p(x) = 1 + x + x^4$, $h(x) = (1 + x^{15})/g(x)$, and $f(x) = h(x)p(x)$.

5.10 (Compatible LFSR) Mark all collapsed single stuck-at faults in Figure 5.25a with up and down arrows. Mark faults detected by each test pattern $(X_1, X_3) = \{00, 01, 10, 11\}$ given in Figure 5.25b.

5.11 (Ones Count Testing *versus* Transition Count Testing) Assume a fault-free output response $R_0 = \{01101111\}$ and a faulty response $R_1 = \{00110001\}$. Compute the ones-count and transition-count signatures; indicate which compaction scheme can detect the faulty response, and show the aliasing probability using either compaction scheme.

5.12 **(Serial Signature Analysis)** Compute the signature of the SISR using $f(x) = 1 + x + x^4$ given in Figure 5.30 for a faulty sequence $M' = \{11111111\}$. Explain why M' is detected or not detected.

5.13 **(Parallel Signature Analysis)** Let $M'_0 = \{00010\}$, $M'_1 = \{00010\}$, $M'_2 = \{11100\}$, and $M'_3 = \{10000\}$. Compute the signature of the MISR using $f(x) = 1 + x + x^4$ given in Figure 5.32 and explain why M' is detected or not detected.

5.14 **(BILBO *versus* MBILBO *versus* CBILBO)** Discuss further the advantages and disadvantages of using the BILBO, modified BILBO (MBILBO), and concurrent BILBO (CBILBO) approaches for testing a pipeline-oriented circuit, from the points of view of hardware cost, test time, and fault coverage.

5.15 **(STUMPS *versus* BILBO)** Compare the performance of a STUMPS design and a BILBO design. Assume that both designs operate at 200 MHz and the circuit under test has 100 scan chains each having 1000 scan cells. Compute the test time for each design when 100,000 test patterns are to be applied. In general, the shift (scan) speed is much slower than a circuit's operating speed. Assume that the shift speed is 20 MHz, and compute the test time for the STUMPS design again. Explain further why the STUMPS-based architecture is gaining more popularity than the BILBO-based architecture.

5.16 **(Test Point Insertion)** For the circuit shown in Figure 5.47, calculate the detection probabilities, before and after test point insertion, for a stuck-at-0 fault present at input X_3 and then for a stuck-at-1 fault at input X_6.

5.17 **(Test Point Insertion)** For the circuit shown in Figure 5.69, insert two test points so the minimum detection probability for any fault in the circuit is greater than or equal to 1/16 and draw the resulting circuit. Assume control points are randomly activated.

5.18 **(Aligned Skewed-Load *versus* Aligned Double-Capture)** Assume there are four synchronous clock domains each controlled by a capture clock, CK1, CK2, CK3, or CK4, and each is operated at a frequency $F1 = 2 \times F2 = 4 \times F3 = 8 \times F4$. Derive BIST timing control diagrams using aligned skewed-load and aligned double-capture to test all intra-clock-domain and inter-clock-domain delay faults. Specify by arrows the delay faults that can be detected in the diagram.

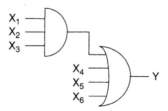

■ **FIGURE 5.69**

An example circuit for Problem 5.17.

5.19 **(Staggered Skewed-Load *versus* Staggered Double-Capture)** Assume there are four asynchronous clock domains each controlled by a capture clock, CK1, CK2, CK3, or CK4, and each is operated at a frequency F1 > F2 > F3 > F4. Derive BIST timing control diagrams using staggered skewed-load and staggered double-capture to test all intra-clock-domain and inter-clock-domain delay faults. Specify by arrows the delay faults that can be detected in the diagram.

5.20 **(Hybrid Double-Capture)** Assume there are four mixed synchronous and asynchronous clock domains each controlled by a capture clock, CK1, CK2, CK3, and CK4, operating at F1 = 100 MHz, F2 = 50 MHz, F3 = 60 MHz, and F4 = 30 MHz, respectively. Derive a BIST timing control diagram using a hybrid double-capture scheme comprised of staggered double-capture and aligned double-capture to test all intra-clock-domain and inter-clock-domain delay faults. Specify by arrows the delay faults that can be detected in the diagram.

5.21 **(A Design Practice)** Use the logic BIST programs and user's manuals contained on the companion Web site to design the logic BIST system using staggered double-capture for the circuit given in Section 5.8. Report the circuit's BIST fault coverage every 10,000 increments up to 100,000 pseudo-random patterns.

5.22 **(A Design Practice)** Repeat Problem 5.21, but instead implement the two pseudo-random pattern generators, PRPG1 and PRPG2, with a 28-stage CA and a 25-stage CA, respectively, using the construction rules given in Table 5.2. Explain why the CA-based logic BIST system can or cannot reach higher BIST fault coverage than the LFSR-based logic BIST system given in Problem 5.21.

5.23 **(A Design Practice)** Use the ATPG programs and user's manuals contained on the Web site to report the circuit's ATPG fault coverage when the logic BIST system is reconfigured in ATPG mode. If the BIST fault coverage in Problem 5.21 is lower than the ATPG fault coverage, insert as many test points as required in the logic BIST system to reach the ATPG fault coverage; alternatively, run top-up ATPG in both ATPG compression and ATPG modes and report the circuit's final fault coverage.

5.24 **(A Design Practice)** Write a C/C++ program to backtrace from a circuit output (primary output or scan cell input) to determine the maximum number of inputs (w, primary inputs and scan cell outputs, excluding clocks and set/reset ports) that drive the output. Assume the scan-based circuit under test (CUT) has n inputs and m outputs. The n-input, m-output CUT is defined as an (n, w) CUT, where $w \leq n$. Find out which (n, w) CUTs are for both clock domains s38417 and s38584, given in Table 5.6.

5.25 **(A Design Practice)** Use the logic BIST programs and user's manuals contained on the Web site with data provided in Problem 5.24 to design a CBILBO-based logic BIST system using staggered single-capture for the circuit given in Section 5.8. Report the circuit's BIST fault coverage every 10,000 increments up to 100,000 pseudo-exhaustive patterns. Compare the observed BIST fault coverage with the ATPG fault coverage given in Problem 5.23, and explain why both methods produce the same or different fault coverage numbers.

Acknowledgments

The author wishes to thank Prof. Nur A. Touba of University of Texas at Austin for contributing the Fault Coverage Enhancement section; Prof. Charles Stroud of Auburn University; Khader S. Abdel-Hafez and Shianling Wu of SynTest Technologies for editing the chapter; and Teresa Chang and Renay Chang of SynTest Technologies for formatting the text and drawing all figures.

References
R5.0—Books

[Abramovici 1994] M. Abramovici, M. A. Breuer, and A. D. Friedman, *Digital Systems Testing and Testable Design*, IEEE Press, Piscataway, NJ, 1994 (revised printing).

[Bardell 1987] P. H. Bardell, W. H. McAnney, and J. Savir, *Built-In Test for VLSI: Pseudo-random Techniques*, John Wiley & Sons, Somerset, NJ, 1987.

[Bushnell 2000] M. L. Bushnell and V. D. Agrawal, *Essentials of Electronic Testing for Digital, Memory and Mixed-Signal VLSI Circuits*, Springer Science, New York, 2000.

[Golomb 1982] S. W. Golomb, *Shift Register Sequence*, Aegean Park Press, Laguna Hills, CA, 1982.

[Jha 2003] N. K. Jha and S. K. Gupta, *Testing of Digital Systems*, Cambridge University Press, Cambridge, U.K., 2003.

[McCluskey 1986] E. J. McCluskey, *Logic Design Principles: With Emphasis on Testable Semicustom Circuits*, Prentice Hall, Englewood Cliffs, NJ, 1986.

[Mourad 2000] S. Mourad and Y. Zorian, *Principles of Testing Electronic Systems*, John Wiley & Sons, Somerset, NJ, 2000.

[Nadeau-Dostie 2000] B. Nadeau-Dostie, *Design for At-Speed Test, Diagnosis and Measurement*, Springer, Boston, MA, 2000.

[Peterson 1972] W. W. Peterson and E. J. Weldon, Jr., *Error-Correcting Codes*, MIT Press, Cambridge, MA, 1972.

[Rajski 1998] J. Rajski and J. Tyszer, *Arithmetic Built-In Self-Test for Embedded Systems*, Prentice Hall, Englewood Cliffs, NJ, 1998.

[Stroud 2002] C. E. Stroud, *A Designer's Guide to Built-In Self-Test*, Springer Science, Boston, MA, 2002.

[Wakerly 2000] J. Wakerly, *Digital Design Principles and Practices*, 3rd ed., Prentice Hall, Englewood Cliffs, NJ, 2000.

R5.2—BIST Design Rules

[Abdel-Hafez 2004] K. S. Abdel-Hafez, L.-T. Wang, A. Kifli, F.-S. Hsu, X. Wen, M.-C. Lin, and H.-P. Wang, Method and Apparatus for Testing Asynchronous Set/Reset Faults in a Scan-Based Integrated Circuit, U.S. Patent Application No. 20040153926, August 5, 2004.

[Al-Yamani 2002] A. A. Al-Yamani, S. Mitra, and E. J. McCluskey, *Avoiding Illegal States in Pseudorandom Testing of Digital Circuits*, Technical Report (CRC TR) No. 02-2, Center for Reliable Computing, Stanford University, December 2002.

[Cheung 1996] B. Cheung and L.-T. Wang, The seven deadly sins of scan-based designs, *Integrated Syst. Des.*, August, 1996 (www.eetimes.com/editorial/1997/test9708.html).

[Eichelberger 1977] E. B. Eichelberger and T. W. Williams, A logic design structure for LSI testability, in *Proc. Des. Automat. Conf.*, June 1977, pp. 462–468.

R5.3—Test Pattern Generation

[Akers 1985] S. B. Akers, On the use of linear sums in exhaustive testing, in *Digest of Papers, Fault-Tolerant Computing Symp.*, June 1985, pp. 148–153.

[Barzilai 1981] Z. Barzilai, J. Savir, G. Markowsky, and M. G. Smith, The weighted syndrome sums approach to VLSI testing, *IEEE Trans. Comput.*, 30(12), 996–1000, December 1981.

[Barzilai 1983] Z. Barzilai, D. Coppersmith, and A. Rosenberg, Exhaustive bit pattern generation in discontiguous positions with applications to VLSI testing, *IEEE Trans. Comput.*, 32(2), 190–194, 1983.

[Bershteyn 1993] M. Bershteyn, Calculation of multiple sets of weights for weighted random testing, in *Proc. Int. Test Conf.*, October 1993, pp. 1031–1040.

[Bozorgui-Nesbat 1980] S. Bozorgui-Nesbat and E. J. McCluskey, Structured design for testability to eliminate test pattern generation, in *Digest of Papers, Fault-Tolerant Computing Symp.*, June 1980, pp. 158–163.

[Breuer 1987] M. A. Breuer and N. K. Nanda, Simplified Delay Testing for LSI Circuit Faults, U.S. Patent No. 4,672,307, June 9, 1987.

[Bushnell 1995] M. L. Bushnell and I. Shaik, Robust Delay Fault Built-In Self-Testing Method and Apparatus, U.S. Patent No. 5,422,891, June 6, 1995.

[Chakrabarty 1997] K. Chakrabarty, B. T. Murray, J. Liu, and M. Zhu, Test width compression for built-in self-testing, in *Proc. Int. Test Conf.*, November 1997, pp. 328–337.

[Chen 1986] C. L. Chen, Linear dependencies in linear feedback shift register, *IEEE Trans. Comput.*, 35(12), 1086–1087, 1986.

[Chen 1987] C. L. Chen, Exhaustive test pattern generation using cyclic codes, *IEEE Trans. Comput.*, 37(3), 329–338, 1987.

[Chen 1988] C.-A. Chen and S. K. Gupta, Efficient BIST TPG designs and test set compaction via input reduction, *IEEE Trans. Comput.-Aided Des.*, 17(8), 692–705, 1988.

[Chin 1984] C. K. Chin and E. J. McCluskey, *Weighted Pattern Generation for Built-In Self-Test*, Technical Report (CRC TR) No. 84-7, Center for Reliable Computing, Stanford University, August 1984.

[Chin 1987] C. K. Chin and E. J. McCluskey, Test length for pseudorandom testing, *IEEE Trans. Comput.*, 36(2), 252–256, 1987.

[Das 1990] A. K. Das, M. Pandey, A. Gupta, and P. P. Chaudhuri, Built-in self-test structures around cellular automata and counters, *IEEE Proc. Comput. Digital Tech., Part E*, 37(1), 268–276, 1990.

[Furuya 1991] K. Furuya and E. J. McCluskey, Test-pattern test capabilities of autonomous TPG circuits, in *Proc. Int. Test Conf.*, October 1991, pp. 704–711.

[Girard 2002] P. Girard, Survey of low-power testing of VLSI circuits, *IEEE Des. Test Comput.*, May/June 82–92, 2002.

[Gloster 1988] C. S. Gloster, Jr., and F. Brglez, Boundary scan with cellular built-in self-test, in *Proc. Int. Test Conf.*, September 1988, pp. 138–145.

[Golan 1988] P. Golan, O. Novak, and J. Hlavicka, Pseudoexhaustive test pattern generator with enhanced fault coverage, *IEEE Trans. Comput.*, 37(4), 496–500, 1988.

[Hamzaoglu 2000] I. Hamzaoglu and J. H. Patel, Reducing test application time for built-in self-test test pattern generators, in *Proc. VLSI Test Symp.*, April 2000, pp. 369–375.

[Hortensius 1989] P. D. Hortensius, R. D. McLeod, W. Pries, D. M. Miller, and H C. Card, Cellular automata-based pseudorandom number generators for built-in self-test, *IEEE Trans. Comput.-Aided Des.*, 8(8), 842–859, 1989.

[Hsiao 1977] M. Y. Hsiao, A. M. Patel, and D. K. Pradhan, Store address generator with online fault detection capability, *IEEE Trans. Comput.*, 26(11), 1144–1147, 1977.

[Kapur 1994] R. Kapur, S. Patil, T. J. Snethen, and T. W. Williams, Design of an efficient weighted random pattern generation system, in *Proc. Int. Test Conf.*, October 1994, pp. 491–500.

[Khara 1987] M. Khara and A. Albicki, Cellular automata used for test pattern generation, *Proc. Int. Conf. on Computer Design*, October 1987, pp. 56–59.

[Lai 2005] L. Lai, J. H. Patel, T. Rinderknecht, and W.-T. Cheng, Hardware efficient LBIST with complementary weights, in *Proc. Int. Conf. on Computer Design*, October 2005, pp. 479–481.

[Lempel 1985] A. Lempel and M. Cohn, Design of universal test sequences for VLSI, *IEEE Trans. Inform. Theory*, IT-31(1), 10–17, 1985.

[Lisanke 1987] R. Lisanke, F. Brglez, A. J. de Geus, and D. Gregory, Testability-driven random test-pattern generation, *IEEE Trans. Comput.-Aided Des.*, 6(6), 1082–1087, 1987.

[McCluskey 1981] E. J. McCluskey and S. Bozorgui-Nesbat, Design for autonomous test, *IEEE Trans. Comput.*, 30(11), 860–875, 1981.

[McCluskey 1982] E. J. McCluskey, Built-in verification test, in *Proc. Int. Test Conf.*, November 1982, pp. 183–190.

[McCluskey 1984] E. J. McCluskey, Verification testing: A pseudoexhaustive test technique, *IEEE Trans. Comput.*, 33(6), 541–546, 1984.

[Patashnik 1983] O. Patashnik, *Circuit Segmentation for Pseudo-Exhaustive Testing*, Technical Report (CRC TR) No. 83–14, Center for Reliable Computing, Stanford University, 1983.

[Savir 1980] J. Savir, Syndrome-testable design of combinational circuits, *IEEE Trans. Comput.*, 29(6), 442–451, 1980.

[Savir 1984a] J. Savir, G. S. Ditlow, and P. H. Bardell, Random pattern testability, *IEEE Trans. Comput.*, 33(1), 79–90, 1984.

[Savir 1984b] J. Savir and P. H. Bardell, On random pattern test length, *IEEE Trans. Comput.*, 33(6), 467–474, 1984.

[Schnurmann 1975] H. D. Schnurmann, E. Lindbloom, and R. G. Carpenter, The weighted random test-pattern generator, *IEEE Trans. Comput.*, 24(7), 695–700, 1975.

[Serra 1990] M. Serra, T. Slater, J. C. Muzio, and D. M. Miller, The analysis of one-dimensional linear cellular automata and their aliasing properties, *IEEE Trans. Comput.-Aided Des.*, 9(7), 767–778, 1990.

[Seth 1990] S. C. Seth, V. D. Agrawal, and H. Farlet, A statistical theory of digital circuit testability, *IEEE Trans. Comput.*, 39(4), 582–586, 1990.

[Slater 1990] T. Slater and M. Serra, *Tables of Linear Hybrid 90/150 Cellular Automata*, Technical Report DCS-105-IR, Department of Computer Science, University of Victoria, British Columbia, Canada 1990.

[Tang 1983] D. T. Tang and L. S. Woo, Exhaustive test pattern generation with constant weight vectors, *IEEE Trans. Comput.*, 32(12), 1145–1150, 1983.

[Tang 1984] D. T. Tang and C. L. Chen, Logic test pattern generation using linear codes, *IEEE Trans. Comput.*, 33(9), 845–850, 1984.

[Udell 1986] J. G. Udell, Jr., Test set generation for pseudoexhaustive BIST, in *Proc. Int. Conf. on Comput.-Aided Des.*, November 1986, pp. 52–55.

[Udell 1987] J. G. Udell, Jr., *Reconfigurable Hardware for Pseudo-Exhaustive Test*, Technical Report (CRC TR) No. 87-4, Center for Reliable Computing, Stanford University, February 1987.

[van Sas 1990] J. van Sas, F. Catthoor, and H. D. Man, Cellular automata-based self-test for programmable data paths, in *Proc. Int. Test Conf.*, September 1990, pp. 769–778.

[Virazel 2002] A. Virazel and H.-J. Wunderlich, High defect coverage with low-power test sequences in a BIST environment, *IEEE Des. Test Comput.*, 19(5), 44–52, 2002.

[Vasanthavada 1985] N. Vasanthavada and P. N. Marinos, An operationally efficient scheme for exhaustive test-pattern generation using linear codes, in *Proc. Int. Test Conf.*, November 1985, pp. 476–482.

[Wagner 1987] K. D. Wagner, C. K. Chin, and E. J. McCluskey, Pseudorandom testing, *IEEE Trans. Comput.*, 36(3), 332–343, 1987.

[Waicukauski 1989] J. A. Waicukauski, E. Lindbloom, E. B. Eichelberger, and O. P. Forenza, WRP: A method for generating weighted random test patterns, *IBM J. Res. Dev.*, 33(2), 149–161, 1989.

[Wang 1986a] L.-T. Wang and E. J. McCluskey, Condensed linear feedback shift register (LFSR) testing: A pseudoexhaustive test technique, *IEEE Trans. Comput.*, 35(4), 367–370, 1986.

[Wang 1986b] L.-T. Wang and E. J. McCluskey, Complete feedback shift register design for built-in self-test, in *Proc. Int. Conf. on Comput.-Aided Des.*, November 1986, pp. 56–59.

[Wang 1987] L.-T. Wang and E. J. McCluskey, Circuits for pseudo-exhaustive test pattern generation using shortened cyclic codes, in *Proc. Int. Conf. on Computer Design: VLSI in Computers & Processors*, October 1987, pp. 450–453.

[Wang 1988a] L.-T. Wang and E. J. McCluskey, Hybrid designs generating maximum-length sequences, special issue on testable and maintainable design, *IEEE Trans. Comput.-Aided Des.*, 7(1), 91–99, 1988.

[Wang 1988b] L.-T. Wang and E. J. McCluskey, Circuits for pseudo-exhaustive test pattern generation, *IEEE Trans. Comput.-Aided Des.*, 7(10), 1068–1080, 1988.

[Wang 1988c] L.-T. Wang and E. J. McCluskey, Linear feedback shift register design using cyclic codes, *IEEE Trans. Comput.*, 37(10), 1302–1306, 1987.

[Wang 1989] L.-T. Wang, M. Marhoefer, and E. J. McCluskey, A self-test and self-diagnosis architecture for boards using boundary scan, in *Proc. of the First European Test Conf.*, April 1989, pp. 119–126.

[Wang 1999] S. Wang and S. K. Gupta, LT-RTPG: A new test-per-scan BIST TPG for low heat dissipation, in *Proc. Int. Test Conf.*, October 1999, pp. 85–94.

[Williams 1985] T. W. Williams, Test length in a self-testing environment, *IEEE Des. Test Comput.*, 2, 59–63, April 1985.

[Wolfram 1983] S. Wolfram, Statistical mechanics of cellular automata, *Rev. Mod. Phys.*, 55(3), 601–644, 1983.

[Wunderlich 1987] H.-J. Wunderlich, Self test using unequiprobable random patterns, in *Digest of Papers, Fault-Tolerant Computing Symp.*, July 1987, pp. 258–263.

[Wunderlich 1988] H.-J. Wunderlich, Multiple distributions for biased random test patterns, in *Proc. Int. Test Conf.*, September 1988, pp. 236–244.

R5.4—Output Response Analysis

[Frohwerk 1977] R. A. Frohwerk, Signature analysis: A new digital field service method, *Hewlett-Packard J.*, 28, 2–8, 1977.

[Hassan 1984] S. Z. Hassan and E. J. McCluskey, Increased fault coverage through multiple signatures, in *Digest of Papers, Fault-Tolerant Computing Symp.*, June 1984, pp. 354–359.

[Hayes 1976] J. P. Hayes, Transition count testing of combinational logic circuits, *IEEE Trans. Comput.*, 25(6), 613–620, 1976.

[McCluskey 1981] E. J. McCluskey and S. Bozorgui-Nesbat, Design for autonomous test, *IEEE Trans. Comput.*, 30(11), 860–875, 1981.

[McCluskey 1985] E. J. McCluskey, Built-in self-test structures, *IEEE Des. Test Comput.*, 2(2), 29–36, 1985.

[Savir 1985] J. Savir and W. H. McAnney, On the masking probability with ones count and transition count, in *Proc. Int. Conf. on Comput.-Aided Des.*, pp. 111–113, November 1985.

[Williams 1987] T. W. Williams, W. Daehn, M. Gruetzner, and C. W. Starke, Aliasing errors in signature analysis registers, *IEEE Des. Test Comput.*, 4(4), 39–45, 1987.

R5.5—Logic BIST Architectures

[Bardell 1982] P. H. Bardell and W. H. McAnney, Self-testing of multiple logic modules, in *Proc. Int. Test Conf.*, November 1982, pp. 200–204.

[Benowitz 1975] N. Benowitz, D. F. Calhoun, G. E. Alderson, J. E. Bauer, and C. T. Joeckel, An advanced fault isolation system for digital logic, *IEEE Trans. Comput.*, 24(5), 489–497, 1975.

[Carletta 1994] J. Carletta and C. Papachristou, Structural constraints for circular self-test paths, in *Proc. VLSI Test Symp.*, April 1994, pp. 87–92.

[Cheon 2005] B. Cheon, E. Lee, L.-T. Wang, X. Wen, P. Hsu, J. Cho, J. Park, H. Chao, and S. Wu, At-speed logic BIST for IP cores, *Proc. Design, Automation and Test in Europe*, March 2005, pp. 860–861.

[Eichelberger 1983] E. B. Eichelberger and E. Lindbloom, Random-pattern coverage enhancement and diagnosis for LSSD logic self-test," *IBM Journal of Research and Development*, 27(3), 265–272, March 1983.

[Könemann 1979] B. Könemann, J. Mucha, and G. Zwiehoff, Built-in logic block observation techniques, *Proc. Int'l Test Conf.*, October 1979, pp. 37–41.

[Könemann 1980] B. Könemann, J. Mucha, and G. Zwiehoff, Built-in test for complex digital circuits, *IEEE Journal of Solid-State Circuits*, 15(3), 315–318, June 1980.

[Krasniewski 1989] A. Krasniewski and S. Pilarski, Circular self-test path: A low cost BIST technique for VLSI circuits, *IEEE Trans. on Comput.-Aided Des.*, 8(1), 46–55, January 1989.

[Liu 1987] D. L. Liu and E. J. McCluskey, High fault coverage self-test structures for CMOS ICs, *Proc. Custom Integrated Circuits Conf.*, May 1987, pp. 68–71.

[McCluskey 1981] E. J. McCluskey and S. Bozorgui-Nesbat, Design for autonomous test, *IEEE Trans. on Computers*, 30(11), 860–875, November 1981.

[McCluskey 1985] E. J. McCluskey, Built-in self-test structures, *IEEE Des. Test Comput.*, 2(2), pp. 29–36, April 1985.

[Perkins 1980] C. C. Perkins, S. Sangani, H. Stopper, and W. Valitski, Design for in-situ chip testing with a compact tester, in *Proc. Int. Test Conf.*, November 1980, pp. 29–41.

[Pilarski 1992] S. Pilarski, A. Krasniewski and T. Kameda, Estimating test effectiveness of the circular self-test path technique, *IEEE Trans. on Comput.-Aided Des.*, 11(10), 1301–1316, October 1992.

[Sedmak 1979] R. M. Sedmak, Design for self-verification: An approach for dealing with testability problems in VLSI-based designs, in *Proc. Int. Test Conf.*, October 1979, pp. 112–124.

[Sedmak 1980] R. M. Sedmak, Implementation techniques for self-verification, in *Proc. Int. Test Conf.*, October 1980, pp. 267–278.

[Stroud 1988] C. Stroud, An automated built-in self-test approach for general sequential logic synthesis, in *Proc. Design Automation Conf.*, June 1988, pp. 3–8.

[Touba 2002] N. A. Touba, Circular BIST with state skipping, *IEEE Trans. on Very Large Scale Integration (VLSI) Systems*, 10(5), 668–672, October 2002.

[Wang 1986c] L.-T. Wang and E. J. McCluskey, Concurrent built-in logic block observer (CBILBO), in *Proc. Int. Symp. on Circuits and Systems*, 3, 1054–1057, May 1986.

R5.6—Fault Coverage Enhancement

[Aboulhamid 1983] M. E. Aboulhamid and E. Cerny, A class of test generators for built-in testing, *IEEE Trans. Comput.*, C-32(10), 957–959, 1983.

[Agarwal 1981] V. K. Agrawal. Store and generate built-in testing approach, in *Digest of Papers, Fault-Tolerant Computing Symp.*, June 1981, pp. 35–40.

[Al-Yamani 2005] A. Al-Yamini, S. Mitra, and E. J. McCluskey, Optimized reseeding by seed ordering and encoding, *IEEE Trans. Comput.-Aided Des.*, 24(2), 264–270, 2005.

[Boubezari 1999] S. Boubezari, E. Cerny, B. Kaminska, and B. Nadeau-Dostie, Testability analysis and test-point insertion in RTL VHDL specifications for scan-based BIST, *IEEE Trans. Comput.-Aided Des.*, 18(9), 1327–1340, 1999.

[Brglez 1984] F. Brglez, On testability of combinational networks, in *Proc. Int. Symp. on Circuits and Systems*, May 1984, pp. 221–225.

[Cheng 1995] K.-T. Cheng and C.-J. Lin, Timing-driven test point insertion for full-scan and partial-scan BIST, in *Proc. Int. Test Conf.*, October 1995, pp. 506–514.

[Dandapani 1984] R. Dandapani, J. Patel, and J. Abraham, Design of test pattern generators for built-in test, in *Proc. Int. Test Conf.*, October 1984, pp. 315–319.

[Das 2000] D. Das and N. A. Touba, Reducing test data volume using external/LBIST hybrid test patterns, in *Proc. Int. Test Conf.*, October 2000, pp. 115–122.

[Dorsch 2001] R. Dorsch and H.-J. Wunderlich, Tailoring ATPG for embedded testing, in *Proc. Int. Test Conf.*, October 2001, pp. 530–537.

[Edirisooriya 1992] G. Edirisooriya and J. P. Robinson, Design of low cost ROM based test generators, in *Proc. VLSI Test Symp.*, April 1992, pp. 61–66.

[Hellebrand 1995a] S. Hellebrand, J. Rajski, S. Tarnick, S. Venkataramann, and B. Courtois, Generation of vector patterns through reseeding of multiple-polynomial linear feedback shift registers, *IEEE Trans. Comput.*, 44(2), 223–233, 1995.

[Hellebrand 1995b] S. Hellebrand, B. Reeb, S. Tarnick, and H.-J. Wunderlich, Pattern generation for a deterministic BIST scheme, in *Proc. Int. Conf. on Comput.-Aided Des.*, November 1995, pp. 88–94.

[Ichino 2001] K. Ichino, T. Asakawa, S. Fukumoto, K. Iwasaki, and S. Kajihara, Hybrid BIST using partially rotational scan, in *Proc. Asian Test Symp.*, November 2001, pp. 379–384.

[Iyengar 1989] V. S. Iyengar and D. Brand, Synthesis of pseudo-random pattern testable designs, in *Proc. Int. Test Conf.*, August 1989, pp. 501–508.

[Jas 2004] A. Jas, C. V. Krishna, and N. A. Touba, Weighted pseudo-random hybrid BIST, *IEEE Trans. VLSI Syst.*, 12(12), 1277–1283, 2004.

[Jervan 2003] G. Jervan, P. Eles, Z. Peng, R. Ubar, and M. Jenihhin, Test time minimization for hybrid BIST of core-based systems, in *Proc. Asian Test Symp.*, November 2003, pp. 318–323.

[Kiefer 1998] G. Kiefer and H.-J. Wunderlich, Deterministic BIST with multiple scan chains, in *Proc. Int. Test Conf.*, October 1998, pp. 1057–1064.

[Könemann 1991] B. Könemann, LFSR-coded test patterns for scan designs, in *Proc. European Test Conf.*, April 1991, pp. 237–242.

[Krishna 2003] C. V. Krishna and N. A. Touba, Hybrid BIST using an incrementally guided LFSR, in *Proc. Symp. on Defect and Fault Tolerance*, November 2003, pp. 217–224.

[Krishnamurthy 1987] B. Krishnamurthy, A dynamic programming approach to the test point insertion problem, in *Proc. Des. Automat. Conf.*, June 1987, pp. 695–704.

[Lei 2005] L. Lei and K. Chakrabarty, Hybrid BIST based on repeating sequences and cluster analysis, in *Proc. Design, Automation and Test in Europe*, March 2005, pp. 1142–1147.

[Liang 2001] H.-G. Liang, S. Hellebrand, and H.-J. Wunderlich, Two-dimensional test data compression for scan-based deterministic BIST, in *Proc. Int. Test Conf.*, September 2001, pp. 894–902.

[Nakao 1997] M. Nakao, K. Hatayama, and I. Higashi, Accelerated test points selection method for scan-based BIST, in *Proc. Asian Test Symp.*, November 1997, pp. 359–364.

[Rajski 1998] J. Rajski, J. Tyszer, and N. Zacharia, Test data decompression for multiple scan designs with boundary scan, *IEEE Trans. Comput.*, 47(11), 1188–1200, 1998.

[Seiss 1991] B. H. Seiss, P. M. Trouborst, and M. H. Schulz, Test point insertion for scan-based BIST, in *Proc. European Test Conf.*, April 1991, pp. 253–262.

[Sugihara 1998] M. Sugihara, H. Date, and H. Yasuura, A novel test methodology for core-based system LSIs and a testing time minimization problem, in *Proc. Int. Test Conf.*, October 1998, pp. 465–472.

[Tamarapalli 1996] N. Tamarapalli and J. Rajski, Constructive multi-phase test point insertion for scan-based BIST, in *Proc. Int. Test Conf.*, October 1996, pp. 649–658.

[Touba 1995] N. A. Touba and E. J. McCluskey, Transformed pseudo-random patterns for BIST, in *Proc. VLSI Test Symp.*, April 1995, pp. 410–416.

[Touba 1996] N. A. Touba and E. J. McCluskey, Test point insertion based on path tracing, in *Proc. VLSI Test Symp.*, April 1996, pp. 2–8.

[Touba 1999] N. A. Touba and E. J. McCluskey, RP-SYN: Synthesis of random-pattern testable circuits with test point insertion, *IEEE Trans. Comput.-Aided Des.*, 18(8), 1202–1213, 1999.

[Touba 2001] N. A. Touba and E. J. McCluskey, Bit-fixing in pseudorandom sequences for scan BIST, *IEEE Trans. Comput.-Aided Des.*, 20(4), 545–555, 2001.

[Tsai 1998] H.-C. Tsai, K.-T. Cheng, C.-J. Lin, and S. Bhawmik, Efficient test point selection for scan-based BIST, *IEEE Trans. VLSI Syst.*, 6(4), 667–676, 1998.

[Wohl 2003] P. Wohl, J. Waicukauski, S. Patel, and M. Amin, X-tolerant compression and applications of scan-ATPG patterns in a BIST architecture, in *Proc. Int. Test Conf.*, October 2003, pp. 727–736.

[Xiang 2005] D. Xiang, M.-J. Chen, J.-G. Sun, and H. Fujiwara, Improving test effectiveness of scan-based BIST by scan chain partitioning, *IEEE Trans. Comput.-Aided Des.*, 24(6), 916–927, 2005.

R5.7—BIST Timing Control

[Bhawmik 1997] S. Bhawmik, Method and Apparatus for Built-In Self-Test with Multiple Clock Circuits, U.S. Patent No. 5,680,543, October 21, 1997.

[Hetherington 1999] G. Hetherington, T. Fryars, N. Tamarapalli, M. Kassab, A. Hassan, and J. Rajski, Logic BIST for large industrial designs: Real issues and case studies, in *Proc. Int. Test Conf.*, October 1999, pp. 358–367.

[Li 2002] J. C.-M. Li and E. J. McCluskey, Diagnosis of sequence-dependent chips, in *Proc. VLSI Test Symp.*, April 2002, pp. 187–192.

[Ma 1995] S. C. Ma, P. Franco, and E. J. McCluskey, An experimental chip to evaluate test techniques: Experimental results, in *Proc. Int. Test Conf.*, October 1995, pp. 663–672.

[McCluskey 2000] E. J. McCluskey and C.-W. Tseng, Stuck-fault tests vs. actual defects, in *Proc. Int. Test Conf.*, October 2000, pp. 336–343.

[McCluskey 2004] E. J. McCluskey, A. Al-Yamani, J. C.-M. Li, C.-W. Tseng, E. Volkerink, F.-F. Ferhani, E. Li, and S. Mitra, ELF-Murphy data on defects and test sets, in *Proc. VLSI Test Symp.*, April 2004, pp. 16–22.

[Nadeau-Dostie 1994] B. Nadeau-Dostie, A. Hassan, D. Burek, and S. Sunter, Multiple Clock Rate Test Apparatus for Testing Digital Systems, U.S. Patent No. 5,349,587, September 20, 1994.

[Rajski 2003] J. Rajski, A. Hassan, R. Thompson, and N. Tamarapalli, Method and Apparatus for At-Speed Testing of Digital Circuits, U.S. Patent Application No. 20030097614, May 22, 2003.

[Savir 1993] J. Savir and S. Patil, Scan-based transition test, *IEEE Trans. Comput.-Aided Des.*, 12(8), 1232–1241, 1993.

[Savir 1994] J. Savir and S. Patil, Broad-side delay test, *IEEE Trans. Comput.-Aided Des.*, 13(8), 1057–1064, 1994.

[Tseng 2001] C.-W. Tseng and E. J. McCluskey, Multiple-output propagation transition fault test, in *Proc. Int. Test Conf.*, October 2001, pp. 358–366.

[Wang 2005a] L.-T. Wang, X. Wen, P.-C. Hsu, S. Wu, and J. Guo, At-speed logic BIST architecture for multi-clock designs, in *Proc. Int. Conf. on Computer Design: VLSI in Computers and Processors*, October 2005, pp. 475–478.

[Wang 2005b] L.-T. Wang, M.-C. Lin, X. Wen, H.-P. Wang, C.-C. Hsu, S.-C. Kao, and F.-S. Hsu, Multiple-Capture DFT System for Scan-Based Integrated Circuits, U.S. Patent No. 6,954,887, October 11, 2005.

[Wang 2006] L.-T. Wang, P.-C. Hsu, S.-C. Kao, M.-C. Lin, H.-P. Wang, H.-J. Chao, and X. Wen, Multiple-Capture DFT System for Detecting or Locating Crossing Clock-Domain Faults during Self-Test or Scan-Test, U.S. Patent No. 7,007,213, February 28, 2006.

R5.8—A Design Practice

[Brglez 1989] F. Brglez, D. Bryan, and K. Kozminski, Combinational profiles of sequential benchmark circuits, in *Proc. Int. Symp. on Circuits and Systems*, August 1989, pp. 1929–1934.

[Rajski 2003] J. Rajski, A. Hassan, R. Thompson, and N. Tamarapalli, Method and Apparatus for At-Speed Testing of Digital Circuits, U.S. Patent Application No. 20030097614, May 22, 2003.

[Wang 2004] L.-T. Wang, X. Wen, K.S. Abdel-Hafez, S.-H. Lin, H.-P. Wang, M.-T. Chang, P.-C. Hsu, S.-C. Kao, M.-C. Lin, and C.-C. Hsu, Method and Apparatus for Unifying Self-Test with Scan-Test during Prototype Debug and Production Test, U.S. Patent Application No. 20040268181, December 30, 2004.

[Wang 2005a] L.-T. Wang, X. Wen, P.-C. Hsu, S. Wu, and J. Guo, At-speed logic BIST architecture for multi-clock designs, in *Proc. Int. Conf. on Computer Design: VLSI in Computers and Processors*, October 2005, pp. 475–478.

R5.9—Concluding Remarks

[Bardell 1982] P. H. Bardell and W. H. McAnney, Self-Testing of multiple logic modules, in *Proc. Int. Test Conf.*, November 1982, pp. 200–204.

[Wang 1986c] L.-T. Wang and E. J. McCluskey, Concurrent built-in logic block observer (CBILBO), in *Proc. Int. Symp. on Circuits and Systems*, 3, May 1986, pp. 1054–1057.

R5.8—A Design Practice

[Brglez 1989] F. Brglez, D. Bryan, and K. Kozminski, Combinational profiles of sequential benchmark circuits, in *Proc. Int. Symp. on Circuits and Systems*, August 1989, pp. 1929–1934.

[Rajski 2003] J. Rajski, A. Hassan, R. Thompson, and N. Tamarapalli, Method and Apparatus for At-Speed Testing of Digital Circuits, U.S. Patent Application No. 20030097614, May 22, 2003.

[Wang 2004] L.-T. Wang, X. Wen, K.S. Abdel-Hafez, S.-H. Lin, H.-P. Wang, M.-T. Chang, P.-C. Hsu, S.-C. Kao, M.-C. Lin, and C.-C. Hsu, Method and Apparatus for Unifying Self-Test with Scan-Test during Prototype Debug and Production Test, U.S. Patent Application No. 20040268181, December 30, 2004.

[Wang 2005a] L.-T. Wang, X. Wen, P.-C. Hsu, S. Wu, and J. Guo, At-speed logic BIST architecture for multi-clock designs, in *Proc. Int. Conf. on Computer Design: VLSI in Computers and Processors*, October 2005, pp. 475–478.

R5.9—Concluding Remarks

[Bardell 1982] P. H. Bardell and W. H. McAnney, Self-Testing of multiple logic modules, in *Proc. Int. Test Conf.*, November 1982, pp. 200–204.

[Wang 1986c] L.-T. Wang and E. J. McCluskey, Concurrent built-in logic block observer (CBILBO), in *Proc. Int. Symp. on Circuits and Systems*, 3, May 1986, pp. 1054–1057.

CHAPTER 6

TEST COMPRESSION

Xiaowei Li
Chinese Academy of Sciences, Beijing, China

Kuen-Jong Lee
National Cheng Kung University, Tainan, Taiwan

Nur A. Touba
University of Texas, Austin, Texas

ABOUT THIS CHAPTER

Test compression involves compressing the amount of test data (both stimulus and response) that must be stored on *automatic test equipment* (ATE) for testing with a deterministic (*automatic test pattern generation* [ATPG]-generated) test set. This is done by adding some additional on-chip hardware before the scan chains to decompress the test stimulus coming from the ATE and after the scan chains to compress the response going to the ATE. This differs from *built-in self-test* (BIST) and hybrid BIST in that the test vectors that are applied to the *circuit under test* (CUT) are exactly the same as the test vectors in the original deterministic (ATPG-generated) test set (no additional pseudo-random vectors are applied). *Test compression* can provide a $10\times$ or even $100\times$ reduction in the amount of test data stored on the ATE. This greatly reduces ATE memory requirements and even more importantly reduces test time because less data has to be transferred across the limited bandwidth between the ATE and the chip. Moreover, test compression methodologies are easy to adopt in industry because they are compatible with the conventional design rules and test generation flows used for scan testing.

This chapter begins with an introduction to the basic concepts and principles of *test compression*. Then we focus on test stimulus compression and describe three different categories of schemes: (1) using data compression codes, (2) employing linear decompression, and (3) broadcasting the same value to multiple scan chains. Next we focus on test response compaction and look at different ways for dealing with unknown (nondeterministic) values in the output response. Finally, we discuss commercial tools that are used for implementing test compression in industry.

6.1   INTRODUCTION

Automatic test equipment (ATE) has limited speed, memory, and I/O channels. The test data bandwidth between the tester and the chip, as illustrated in Figure 6.1, is relatively low and generally is a bottleneck with regard to how fast a chip can be tested [Khoche 2000]. The chip cannot be tested any faster than the amount of time required to transfer the test data which is equal to:

$$\frac{Amount\ of\ Test\ Data\ on\ Tester}{(Number\ of\ Tester\ Channels)(Tester\ Clock\ Rate)}$$

The idea in *test compression* is to compress the amount of test data (both stimulus and response) that is stored on the tester. This provides two advantages. The first is that it reduces the amount of tester memory that is required. The second and more important advantage is that it reduces test time because less test data has to be transferred across the low bandwidth link between the tester and the chip. Test compression is achieved by adding some additional on-chip hardware before the scan chains to decompress the test stimulus coming from the tester and after the scan chains to compact the response going to the tester. This is illustrated in Figure 6.2. This extra on-chip hardware allows the test data to be stored on the tester in a compressed form.

    Test data is inherently highly compressible. Test vectors have many unspecified bits that are not assigned values during ATPG (*i.e.*, they are *"don't cares"* that can be filled with any value with no impact on the fault coverage). In fact, typically only 1 to 5% of the bits have specified (*care*) values, and even the specified values tend to be highly correlated due to the fact that faults are structurally related in the circuit. Consequently, lossless compression techniques can be used to significantly reduce the amount of test stimulus data that must be stored on the tester. The on-chip **decompressor** expands the compressed test stimulus back into the original test vectors (matching in all the care bits) as they are shifted into the scan chains. Output

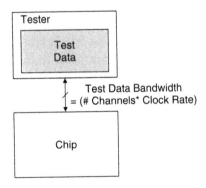

■ **FIGURE 6.1**

Block diagram illustrating test data bandwidth.

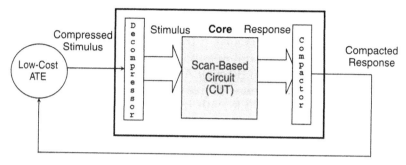

■ **FIGURE 6.2**

Architecture for test compression.

response is even more compressible than test stimulus because **lossy compression** (also known as "**compaction**") can be used. This is similar to the techniques used in BIST as described in Chapter 5. Output response compaction converts long output response sequences into short signatures. Because the compaction is lossy, some fault coverage can be lost due to aliasing when a faulty output response signature is identical to the fault-free output response signature; however, with proper design of the compaction circuitry, the probability of aliasing can be kept negligibly small. A more challenging issue for output response compaction is dealing with unknown (nondeterministic) values (commonly referred to as X's) that might appear in the output sequence, as they can corrupt compacted signatures. This can be addressed by "**X-blocking**" or "**X-bounding**," where the design is modified to eliminate any sources of X's in the output as is done in BIST; however, this adds additional design complexity. Alternatively, "**X-masking**" can be done to selectively mask off the X's in the output sequence, or an "**X-tolerant**" compaction technique can be used. If it is still impossible to prevent all X's from reaching the compactor, then an "**X-impact**" compaction technique that uses ATPG assignments to avoid propagating X's can be used. This technique is discussed further in Section 6.3.

Test compression differs from (**logic**) **BIST** and **hybrid BIST**. Traditional stand-alone BIST does all the test pattern generation and output response analysis on-chip, without requiring any tester storage. The advantage of stand-alone BIST is that it can perform self-test out in the field where there is no access to the tester; however, to achieve high fault coverage with stand-alone BIST, typically a lot of overhead is required (either test points or deterministic pattern embedding logic). Moreover, stringent BIST design rules are necessary in order to eliminate all X's in the output response. If BIST is only going to be used for manufacturing test, then hybrid BIST, where some data are stored on the tester to aid in detecting *random-pattern resistant faults*, offers a more efficient solution. Hybrid BIST and test compression are similar in that they both use on-chip hardware to reduce the amount of data stored on the tester. The difference is that test compression applies a precise deterministic (ATPG-generated) test set to the CUT whereas hybrid BIST applies a large number of pseudo-random patterns plus a smaller number of deterministic patterns for the *random-pattern resistant faults*. Hybrid BIST can reduce the amount of test data on the tester more than test compression, but it generally requires longer overall test time because more test patterns are applied to the CUT than with test compression (in essence, it trades off more test time for less tester storage). The advantage of test compression is that the exact set of patterns that are applied to the CUT is selected through ATPG and thus can be minimized with respect to the desired fault coverage. Moreover, test compression is easy to adopt in industry because it is compatible with the conventional design rules and test generation flows used for scan testing.

The amount of test data required to test **integrated circuits** (ICs) is growing rapidly in each new generation of technology. Increasing integration density results in larger designs with more scan cells and more faults. Moreover, achieving high test quality in ever smaller geometries requires more test patterns targeting delay faults and other fault models beyond stuck-at faults. As the amount of test data has increased, test compression has become very attractive as the additional hardware overhead is relatively low and significant reductions (10× or even 100×) in the amount of test data that must be stored on the tester can be achieved. One benefit of test compression is that it can extend the life of older "legacy" testers that may have limited memory by making it possible to fit all of the test data in the tester memory (note that reloading the tester memory is very time consuming and thus highly undesirable). Even for testers that have plenty of memory, test compression is still very attractive because it can reduce the test time for a given test data bandwidth.

6.2   TEST STIMULUS COMPRESSION

A **test cube** is defined as a deterministic test vector in which the bits that are not assigned values by the ATPG procedure are left as "don't cares" (X's). Normally, ATPG procedures perform *random fill*, in which all the X's in the test cubes are filled randomly with 1's and 0's to create fully specified test vectors; however, for test stimulus compression, random fill is not performed during ATPG so the resulting test set consists of incompletely specified test cubes. The X's make the test cubes much easier to compress than fully specified test vectors. As mentioned earlier, test stimulus compression should be an information lossless procedure with respect to the specified (care) bits in order to preserve the fault coverage of the original test cubes. After decompression, the resulting test patterns shifted into the scan chains should match the original test cubes in all the specified (care) bits. Many schemes for compressing test cubes have been proposed. They can be broadly classified into the three categories shown below; these schemes are described in detail in the following subsections (shown in parentheses):

1. **Code-based schemes** (6.2.1)—These schemes use data compression codes to encode the test cubes.

2. **Linear-decompression-based schemes** (6.2.2)—These schemes decompress the data using only linear operations (*e.g., linear feedback shift registers* [LFSRs] and *exclusive-OR* [XOR] networks).

3. **Broadcast-scan-based schemes** (6.2.3)—These schemes are based on broadcasting the same value to multiple scan chains.

6.2.1 Code-Based Schemes

One approach for test compression is to use data compression codes to encode the test cubes. Data compression codes partition the original data into *symbols*, and then each *symbol* is replaced with a *codeword* to form the compressed data. The decompression is performed by having a decoder that simply converts each codeword into the corresponding symbol.

Data compression codes can be classified into four categories, depending on whether the symbols have a fixed size (*i.e.*, each symbol contains exactly n bits) or a variable size (*i.e.*, different symbols have different numbers of bits) and whether the codewords have a fixed or variable size. Table 6.1 provides examples of data compression codes in each category. Note that the codes listed in Table 6.1 are only representative examples of each category and are by no means an exhaustive list. One representative example code from each category is described in detail in this subsection. Brief descriptions and references for other codes that have been proposed in each category are also provided.

6.2.1.1 Dictionary Code (Fixed-to-Fixed)

In **fixed-to-fixed coding**, the original test cubes are partitioned into n-bit blocks to form the symbols. These symbols are then encoded with codewords that each have b bits. In order to get compression, b must be less than n. One can view each symbol as an entry in a **dictionary** and each codeword as an index into the dictionary that points to the corresponding symbol. There are 2^n possible symbols and 2^b possible codewords, so not all possible symbols can be in the dictionary. If $S_{dictionary}$ is the set of symbols that are in dictionary and S_{data} is the set of symbols that occur in the original data, then if $S_{data} \subseteq S_{dictionary}$, it is a complete dictionary; otherwise, it is a partial dictionary. Compression can be achieved with a complete dictionary provided that the number of distinct symbols that occur in the original data $|S_{data}|$ is much less than 2^n, the number of all possible symbols. The compression ratio that can be achieved with a complete dictionary is equal to:

$$2^{n-\lceil \log_2 |s_{data}| \rceil} : 1$$

A test compression scheme that uses a complete dictionary was described in [Reddy 2002]. The scheme is illustrated in Figure 6.3. There are n scan chains, and the test cubes are partitioned into n-bit symbols such that each **scan slice** corresponds to a symbol. Each scan slice is comprised of the n-bits that are loaded

TABLE 6.1 ■ Four Categories of Data Compression Codes

Category	Example Data Compression Code	Ref.
Fixed-to-fixed	Dictionary code	[Reddy 2002]
Fixed-to-variable	Huffman code	[Jas 2003]
Variable-to-fixed	Run-length code	[Jas 1998]
Variable-to-variable	Golomb code	[Chandra 2001a]

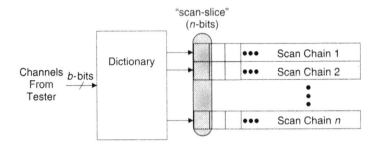

Test compression using a complete dictionary.

into the scan chains in each clock cycle as illustrated in Figure 6.3. The X's in the test cubes are filled so as to minimize the number of distinct symbols (*i.e.*, $|S_{data}|$). The size of each codeword is b bits, where $b = \lceil \log_2 |S_{data}| \rceil$. Note that, with this scheme, b channels from the tester can be used to load n scan chains. Normally, b channels from the tester can only load b scan chains. By having more scan chains, the length of each scan chain becomes shorter, thus reducing the number of clock cycles required to load each scan vector and therefore reducing the test time. This is a good illustration of how test compression reduces not only tester storage but also test time.

A drawback of using a complete dictionary is that the size of the dictionary can become very large, resulting in too much overhead for the decompressor. In [Li 2003], a partial dictionary coding scheme was proposed in which the size of the dictionary is selected by the user based on how much area the user wants to allocate for the decompressor. If the size of the dictionary is 2^b, then the 2^b symbols that occur most frequently in the test cubes are placed in the dictionary. For any symbol that is not in the dictionary, the symbol is left unencoded and the dictionary is bypassed. An extra bit is added to each codeword to indicate whether or not to use the dictionary.

In [Würtenberger 2004], a partial dictionary is used along with a "correction" network that flips bits to convert a dictionary entry into the desired scan slice. By using the correction network, the size of the dictionary can be reduced.

6.2.1.2   *Huffman Code (Fixed-to-Variable)*

In **fixed-to-variable coding**, the original test cubes are partitioned into n-bit blocks to form the symbols. These symbols are then encoded using variable-length codewords. One form of fixed-to-variable coding is **statistical coding**, where the idea is to calculate the frequency of occurrence of the different symbols in the original test cubes and make the codewords that occur most frequently have fewer bits and those that occur least frequently more bits. This minimizes the average length of a codeword. A **Huffman code** [Huffman 1952] is an optimal statistical code that is proven to provide the shortest average codeword length among all uniquely decodable fixed-to-variable length codes. A Huffman code is obtained by constructing a

Huffman tree as described in [Huffman 1952]. The path from the root to each leaf in the Huffman tree gives the codeword for the binary string corresponding to the leaf. An example of constructing a Huffman code can be seen in Table 6.2 and Figures 6.4 and 6.5. An example of a test set divided into 4-bit symbols is shown in Figure 6.4. Table 6.2 shows the frequency of occurrence of each of the possible symbols. The example shown in Figure 6.4 has a total of 60 4-bit symbols. Figure 6.5 shows the Huffman tree for this frequency distribution, and the corresponding codewords are shown in Table 6.2.

In [Jas 2003], a scheme for test compression based on Huffman coding was described. The test cubes are partitioned into symbols and then the X's in the test cubes are filled to maximally skew the frequency of occurrence of the symbols. A **selective Huffman code** in which only the k most frequently occurring symbols are encoded is used. The reason for this is that using a full Huffman code that encodes all n-bit symbols requires a decoder with $2^n - 1$ states. By only selectively encoding the k most frequently occurring symbols, the decoder requires only $n + k$

TABLE 6.2 ■ Statistical Coding Based on Symbol Frequencies for Test Set in Figure 6.4

Symbol	Frequency	Pattern	Huffman Code	Selective Code
S_0	22	0 0 1 0	1 0	1 0
S_1	13	0 1 0 0	0 0	1 1 0
S_2	7	0 1 1 0	1 1 0	1 1 1
S_3	5	0 1 1 1	0 1 0	0 0 1 1 1
S_4	3	0 0 0 0	0 1 1 0	0 0 0 0 0
S_5	2	1 0 0 0	0 1 1 1	0 1 0 0 0
S_6	2	0 1 0 1	1 1 1 0 0	0 0 1 0 1
S_7	1	1 0 1 1	1 1 1 0 1 0	0 1 0 1 1
S_8	1	1 1 0 0	1 1 1 0 1 1	0 1 1 0 0
S_9	1	0 0 0 1	1 1 1 1 0 0	0 0 0 0 1
S_{10}	1	1 1 0 1	1 1 1 1 0 1	0 1 1 0 1
S_{11}	1	1 1 1 1	1 1 1 1 1 0	0 1 1 1 1
S_{12}	1	0 0 1 1	1 1 1 1 1 1	0 0 0 1 1
S_{13}	0	1 1 1 0	—	—
S_{14}	0	1 0 1 0	—	—
S_{15}	0	1 0 0 1	—	—

```
0010 0100 0010 0110 0000 0010 1011 0100 0010 0100 0110 0010
0010 0100 0010 0110 0000 0110 0010 0100 0110 0010 0010 0000
0010 0110 0010 0010 0010 0100 0100 0110 0010 0010 1000 0101
0001 0100 0010 0111 0010 0010 0111 0111 0100 0100 1000 0101
1100 0100 0100 0111 0010 0010 0111 1101 0010 0100 1111 0011
```

■ **FIGURE 6.4**

Example of test set divided into 4-bit blocks.

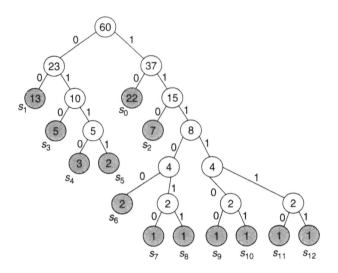

■ **FIGURE 6.5**

Huffman tree for the code shown in Table 6.4.

states. It was shown in [Jas 2003] that a selective Huffman code achieves only slightly less compression than a full Huffman code for the same symbol size while using a much smaller decoder. Because the decoder size grows only linearly with selective Huffman encoding, it is possible to use a much larger symbol size, which significantly improves the effectiveness of the code thereby achieving much more overall compression.

In selective Huffman coding, an extra bit is added at the beginning of each codeword to indicate whether or not it is coded. As an example, consider selective Huffman coding for the test set shown in Figure 6.4, where only the three most frequency occurring symbols are encoded (*i.e.*, $k = 3$). A Huffman tree is built only for the three most frequently occurring symbols, as shown in Figure 6.6. The codewords are then constructed as shown in Table 6.2. The first bit of the codewords for the three most frequently occurring symbols is "1" to indicate that they are coded (and hence must pass through the decoder). The first bit of the rest

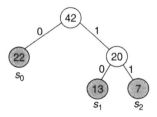

■ **FIGURE 6.6**

Huffman tree for the three highest frequency symbols in Table 6.2.

of the codewords is a "0" to indicate that they are not coded (*i.e.*, the remainder of the codeword is simply the unencoded symbol itself).

A method for improving the compression with a statistical code by modifying the test cubes without losing fault coverage is described in [Ichihara 2000]. The goal is to modify the specified bits in the test cubes in a way that maximally skews the frequency distribution.

Another type of fixed-to-variable coding, which differs from statistical coding, exploits the fact that most scan slices have a relatively small number of specified bits. If there are b channels coming from the tester, the techniques described in [Reda 2002], [Han 2005b], and [Wang 2005a] use a variable number of b-bit codewords to decode the specified bits in each scan slice. Each b-bit codeword can decode a small number of specified bits. For scan slices with very few specified bits, a single b-bit codeword may be sufficient, while for more heavily specified scan slices several b-bit codewords may be required to decode all the specified bits. [Reda 2002] and [Han 2005b] actually encode the specified bits that differ between the previous scan slice and the current scan slice to take advantage of correlation between the scan slices.

6.2.1.3 Run-Length Code (Variable-to-Fixed)

In **variable-to-fixed coding**, the original test cubes are partitioned into variable-length symbols, and the codewords are each b-bits long. In **run-length coding**, one particular variable-to-fixed coding scheme, the symbols consist of runs of consecutive 0's or 1's. An example of a 3-bit run-length code for runs of 0's is given in Table 6.3. Each codeword is 3 bits long and encodes different length runs of 0's. As an example, the sequence 001 0001 01 0000001 1 000001 can be encoded into 010 011 001 110 000 101, which is a reduction from 23 bits to 18 bits. For very long runs of 0's (longer than 7), codeword 111 can be used repeatedly as needed. Note that only data with an unbalanced number of 0's and 1's can be efficiently compressed by a run-length code.

In [Jas 1998], test compression based on a run-length code was proposed using a *cyclical scan architecture* as shown in Figure 6.7. The cyclical scan architecture XORs the data currently being shifted in with the previous test vector. Thus, instead

TABLE 6.3 ■ 3-Bit Run-Length Code

Symbol	Codeword
1	0 0 0
0 1	0 0 1
0 0 1	0 1 0
0 0 0 1	0 1 1
0 0 0 0 1	1 0 0
0 0 0 0 0 1	1 0 1
0 0 0 0 0 0 1	1 1 0
0 0 0 0 0 0 0	1 1 1

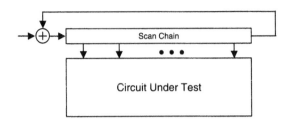

■ **FIGURE 6.7**

Cyclical scan architecture for applying difference vectors.

of applying the original test set, $T_D = \{t_1, t_2, t_3, \ldots, t_n\}$, a *difference vector set*, $T_{diff} = \{t_1, t_1 \oplus t_2, t_2 \oplus t_3, \ldots, t_{n-1} \oplus t_n\}$, is applied instead. The advantage of this is that the test vectors can be ordered so similar test vectors come after each other so the *difference vectors* have many 0's. This enhances the effectiveness of run-length coding. Other types of variable-to-fixed codes that have been proposed for test stimulus compression include LZ77 [Wolff 2002] and LZW [Knieser 2003].

6.2.1.4 *Golomb Code (Variable-to-Variable)*

In **variable-to-variable coding**, both the symbols and codewords have variable length. A **Golomb code** [Chandra 2001] is a variable-to-variable code that evolved from the run-length code. To construct a Golomb code, a specific parameter m, called the group size (usually a power of 2), is first chosen. All the run-lengths are divided into groups of size m denoted by $A_1, A_2, A_3, \ldots$. The set of run-lengths $\{0, 1, 2, m-1\}$ form the first group A_1, the set of run-lengths $\{m, m+1, \ldots, 2m\}$ form the second group A_2, and so on. Each codeword of a Golomb code consists of two parts: a *group prefix* and a *tail*. A run-length L that belongs to group A_k is assigned a group prefix $(k-1)$ of ones followed by a zero. The tail is an index of the run-length in a group. If m is chosen to be a power of 2 (*i.e.*, $m = 2^N$ for some integer N), then each group contains 2^N members, and a $\log_2 m$-bit-long sequence (tail) can uniquely identify each member within the group. Table 6.4 shows an example of Golomb code in which each group contains four run-lengths.

Example 6.1

Figure 6.8 shows an example using the Golomb code of Table 6.4. In the original test sequence T_D, the run-length of the 0's before the first 1 is 2. Based on Table 6.4, the sequence "001" is encoded as "010," in which the "0" is the prefix and "10" is the tail. Similar procedures are repeated until all the run-lengths are processed. It can be found that the length of the test sequence can be reduced from 43 to 32.

In a Golomb code, each group contains the same number of run-lengths and thus may still be inefficient in some cases. In [Chandra 2003], a *frequency-directed run-length* (FDR) **code** was proposed that has variable-length tails based on the group index. It can be constructed such that a shorter run-length can be encoded into a shorter codeword to give better compression. An even more optimized run-length

TABLE 6.4 ■ Golomb Code with Four Run-Lengths for Each Group

Group	Run-Length	Group Prefix	Tail	Codeword
A_1	0	0	0 0	0 0 0
	1		0 1	0 0 1
	2		1 0	0 1 0
	3		1 1	0 1 1
A_2	4	1 0	0 0	1 0 0 0
	5		0 1	1 0 0 1
	6		1 0	1 0 1 0
	7		1 1	1 0 1 1
A_3	8	1 1 0	0 0	1 1 0 0 0
	9		0 1	1 1 0 0 1
	10		1 0	1 1 0 1 0
	11		1 1	1 1 0 1 1
...	...	...	...	...

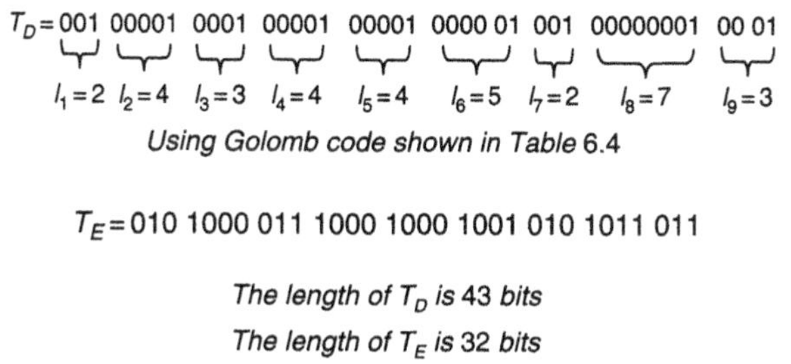

$T_D = 001\ 00001\ 0001\ 00001\ 00001\ 0000\ 01\ 001\ 00000001\ 00\ 01$

$l_1 = 2\quad l_2 = 4\quad l_3 = 3\quad l_4 = 4\quad l_5 = 4\quad l_6 = 5\quad l_7 = 2\quad l_8 = 7\quad l_9 = 3$

Using Golomb code shown in Table 6.4

$T_E = 010\ 1000\ 011\ 1000\ 1000\ 1001\ 010\ 1011\ 011$

The length of T_D is 43 bits

The length of T_E is 32 bits

■ **FIGURE 6.8**

A test sequence and its encoded test data using Golomb code.

code is the ***variable-length-input Huffman code*** (VIHC) described in [Gonciari 2003]. In [Würtenberger 2003], a hybrid approach that combines dictionary coding with run-length coding was proposed. Other variable-to-variable codes that are not based on run-length coding include packet-based codes [Volkerink 2002] and nine-coded compression [Tehranipoor 2005].

6.2.2 Linear-Decompression-Based Schemes

Another class of test stimulus compression schemes is based on using **linear decompressors** to expand the data coming from the tester to fill the scan chains. Any decompressor that consists of only XOR gates and flip-flops is a **linear decompressor**. Linear decompressors have a very useful property: Their *output space (i.e., the space of all possible test vectors that they can generate)* is a linear subspace that is spanned by a Boolean matrix. In other words, for any linear decompressor that expands an m-bit compressed stimulus from the tester into an n-bit stimulus (test vector), there exists a Boolean matrix $A_{n \times m}$ such that the set of test vectors that can be generated by the linear decompressor is spanned by A. A test vector Z can be compressed by a particular linear decompressor if and only if there exists a solution to a system of linear equations, $AX = Z$, where A is the characteristic matrix of the linear decompressor and X is a set of *free variables* stored on the tester (every bit stored on the tester can be thought of as a "free variable" that can be assigned any value, 0 or 1).

The characteristic matrix for a linear decompressor can be obtained by doing symbolic simulation where each free variable coming from the tester is represented by a symbol. An example of this is shown in Figure 6.9, where a sequential linear decompressor containing an LFSR is used. The initial state of the LFSR is represented by the free variables $X_1 - X_4$, and the free variables $X_5 - X_{10}$ are shifted in from two channels as the scan chains are loaded. After symbolic simulation, the final values in the scan chains are represented by the equations for $Z_1 - Z_{12}$. The corresponding system of linear equations for this linear decompressor is shown in Figure 6.10.

The symbolic simulation goes as follows. Assume that the initial seed $X_1 - X_4$ has been already loaded into the flip-flops. In the first clock cycle, the top flip-flop is loaded with the XOR of X_2 and X_5, the second flip-flop is loaded with X_3, the third flip-flop is loaded with the XOR of X_1 and X_4, and the bottom flip-flop is loaded with the XOR of X_1 and X_6. Thus, we obtain $Z_1 = X_2 \oplus X_5$, $Z_2 = X_3$, $Z_3 = X_1 \oplus X_4$, and $Z_4 = X_1 \oplus X_6$. In the second clock cycle, the top flip-flop is loaded with the XOR of

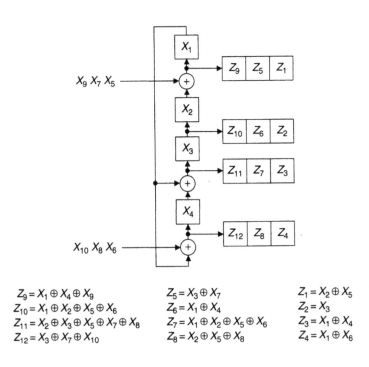

$$Z_9 = X_1 \oplus X_4 \oplus X_9$$
$$Z_{10} = X_1 \oplus X_2 \oplus X_5 \oplus X_6$$
$$Z_{11} = X_2 \oplus X_3 \oplus X_5 \oplus X_7 \oplus X_8$$
$$Z_{12} = X_3 \oplus X_7 \oplus X_{10}$$

$$Z_5 = X_3 \oplus X_7$$
$$Z_6 = X_1 \oplus X_4$$
$$Z_7 = X_1 \oplus X_2 \oplus X_5 \oplus X_6$$
$$Z_8 = X_2 \oplus X_5 \oplus X_8$$

$$Z_1 = X_2 \oplus X_5$$
$$Z_2 = X_3$$
$$Z_3 = X_1 \oplus X_4$$
$$Z_4 = X_1 \oplus X_6$$

■ **FIGURE 6.9**

Example of symbolic simulation for linear decompressor.

$$
\begin{pmatrix}
0\ 1\ 0\ 0\ 1\ 0\ 0\ 0\ 0\ 0 \\
0\ 0\ 1\ 0\ 0\ 0\ 0\ 0\ 0\ 0 \\
1\ 0\ 0\ 1\ 0\ 0\ 0\ 0\ 0\ 0 \\
1\ 0\ 0\ 0\ 0\ 1\ 0\ 0\ 0\ 0 \\
0\ 0\ 1\ 0\ 0\ 0\ 1\ 0\ 0\ 0 \\
1\ 0\ 0\ 1\ 0\ 0\ 0\ 0\ 0\ 0 \\
1\ 1\ 0\ 0\ 1\ 1\ 0\ 0\ 0\ 0 \\
0\ 1\ 0\ 0\ 1\ 0\ 0\ 1\ 0\ 0 \\
1\ 0\ 0\ 1\ 0\ 0\ 0\ 0\ 1\ 0 \\
1\ 1\ 0\ 0\ 1\ 1\ 0\ 0\ 0\ 0 \\
0\ 1\ 1\ 0\ 1\ 0\ 1\ 1\ 0\ 0 \\
0\ 0\ 1\ 0\ 0\ 0\ 1\ 0\ 0\ 1
\end{pmatrix}
\begin{bmatrix}
X_1 \\ X_2 \\ X_3 \\ X_4 \\ X_5 \\ X_6 \\ X_7 \\ X_8 \\ X_9 \\ X_{10}
\end{bmatrix}
=
\begin{bmatrix}
Z_1 \\ Z_2 \\ Z_3 \\ Z_4 \\ Z_5 \\ Z_6 \\ Z_7 \\ Z_8 \\ Z_9 \\ Z_{10} \\ Z_{11} \\ Z_{12}
\end{bmatrix}
$$

■ **FIGURE 6.10**

System of linear equations for the decompressor in Figure 6.9.

the contents of the second flip-flop (X_3) and X_7; the second flip-flop is loaded with the contents of the third flip-flop $(X_1 \oplus X_4)$; the third flip-flop is loaded with the XOR of the contents of the first flip-flop $(X_2 \oplus X_5)$ and the fourth flip-flop $(X_1 \oplus X_6)$; and the bottom flip-flop is loaded with the XOR of the contents of the first flip-flop $(X_2 \oplus X_5)$ and X_8. Thus, we obtain $Z_5 = X_3 \oplus X_7$, $Z_6 = X_1 \oplus X_4$, $Z_7 = X_1 \oplus X_2 \oplus X_5 \oplus X_6$, and $Z_8 = X_2 \oplus X_5 \oplus X_8$. In the third clock cycle, the top flip-flop is loaded with the XOR of the contents of the second flip-flop $(X_1 \oplus X_4)$ and X_9; the second flip-flop is loaded with the contents of the third flip-flop $(X_1 \oplus X_2 \oplus X_5 \oplus X_6)$; the third flip-flop is loaded with the XOR of the contents of the first flip-flop $(X_3 \oplus X_7)$ and the fourth flip-flop $(X_2 \oplus X_5 \oplus X_8)$; and the bottom flip-flop is loaded with the XOR of the contents of the first flip-flop $(X_3 \oplus X_7)$ and X_{10}. Thus, we obtain $Z_9 = X_4 \oplus X_9$, $Z_{10} = X_1 \oplus X_6$, $Z_{11} = X_2 \oplus X_5 \oplus X_8$, and $Z_{12} = X_3 \oplus X_7 \oplus X_{10}$. At this point, the scan chains are fully loaded with a test cube, so the simulation is complete.

For a linear decompressor, finding an assignment for the free variables that will encode a particular test cube can be done by solving the system of linear equations for the specified bits in the test cube. Solving the linear equations can be done with Gauss–Jordan elimination [Cullen 1997] in time complexity $O(mn^2)$, where m is the number of columns (free variables) and n is the number of rows (number of specified bits in the test cube). Note that for Boolean matrices, XOR is used in place of addition and AND is used in place of multiplication.

An example of solving the linear equations for a test cube is shown in Figure 6.11. There are five specified bits in the test cube, so the five linear equations corresponding to those bit positions must be simultaneously solved. One solution is shown, but note that there are many solutions to this system of linear equations. In Figure 6.12, an example of a test cube that does not have a solution is shown (note that row three cannot be solved). In this case, it is not possible to encode that test cube with this particular linear decompressor.

There are a few different strategies for handling unencodable test cubes. A simple approach is to just bypass the decompressor when applying those test cubes (directly shift them in unencoded form); however, that can significantly degrade the

$Z = 1\text{--}011\text{-----}0$

$$
\begin{bmatrix}
0 & 1 & 0 & 0 & 1 & 0 & 0 & 0 & 0 & 0 & | & 1 \\
1 & 0 & 0 & 0 & 0 & 1 & 0 & 0 & 0 & 0 & | & 0 \\
0 & 0 & 1 & 0 & 0 & 0 & 1 & 0 & 0 & 0 & | & 1 \\
1 & 0 & 0 & 1 & 0 & 0 & 0 & 0 & 0 & 0 & | & 1 \\
0 & 0 & 1 & 0 & 0 & 0 & 1 & 0 & 0 & 1 & | & 0
\end{bmatrix}
\xrightarrow[\text{Elimination}]{\text{Gaussian}}
$$

$X = 0111000001$

$$
\begin{bmatrix}
1 & 0 & 0 & 0 & 0 & 1 & 0 & 0 & 0 & 0 & | & 0 \\
0 & 1 & 0 & 0 & 1 & 0 & 0 & 0 & 0 & 0 & | & 1 \\
0 & 0 & 1 & 0 & 0 & 0 & 1 & 0 & 0 & 0 & | & 1 \\
0 & 0 & 0 & 1 & 0 & 1 & 0 & 0 & 0 & 0 & | & 1 \\
0 & 0 & 0 & 0 & 0 & 0 & 0 & 0 & 0 & 1 & | & 1
\end{bmatrix}
$$

■ **FIGURE 6.11**

Example of solving system for linear equations.

$Z = 1\text{-}0\text{--}1\text{------}$

$$
\begin{bmatrix}
0 & 1 & 0 & 0 & 1 & 0 & 0 & 0 & 0 & 0 & | & 1 \\
1 & 0 & 0 & 1 & 0 & 0 & 0 & 0 & 0 & 0 & | & 0 \\
1 & 0 & 0 & 1 & 0 & 0 & 0 & 0 & 0 & 0 & | & 1
\end{bmatrix}
\xrightarrow[\text{Elimination}]{\text{Gaussian}}
$$

$X = \text{No Solution}$

$$
\begin{bmatrix}
1 & 0 & 0 & 1 & 0 & 0 & 0 & 0 & 0 & 0 & | & 0 \\
0 & 1 & 0 & 0 & 1 & 0 & 0 & 0 & 0 & 0 & | & 1 \\
0 & 0 & 0 & 0 & 0 & 0 & 0 & 0 & 0 & 0 & | & 1
\end{bmatrix}
$$

■ **FIGURE 6.12**

Example of system of linear equations with no solution.

overall compression. Another approach is to rerun the ATPG for the unencodable test cubes to try to find a different test cube that is encodable. A third approach is to redesign the linear decompressor so it uses more free variables when decompressing the test cubes, thereby making it easier to solve the linear equations. Note that it is very unlikely to be able to encode a test cube that has more specified bits than the number of free variables used to encode it. On the other hand, for linear decompressors that have diverse linear equations (*e.g.*, an LFSR with a *primitive polynomial*), if the number of free variables is sufficiently larger than the number of specified bits, the probability of not being able to encode the test cube becomes negligibly small. For an LFSR with a primitive polynomial, it has been shown that if the number of free variables is 20 more than the number of specified bits, then the probability of not finding a solution is less than 10^{-6} [Chen 1986] [Könemann 1991].

A figure of merit for linear decompressors is **encoding efficiency**, which is defined as follows:

$$\frac{\textit{Specified Bits in Test Set}}{\textit{Bits Stored on Tester}}$$

How close a linear decompressor's encoding efficiency is to 1 is a measure of its optimality. In general, it is not possible to achieve higher than an encoding efficiency of 1 because the probability of solving the linear equations when there are more specified bits than free variables is extremely low and would likely only happen for very few test cubes in a test set.

Many different linear decompressor designs have been proposed. They can be categorized based on whether they use combinational logic or sequential logic and based on whether they use a fixed number of free variables when encoding each test cube or whether the number of free variables varies for different test cubes.

6.2.2.1 Combinational Linear Decompressors

The simplest linear decompressors use only combinational XOR networks. Each scan chain is fed by the XOR of some subset of the channels coming from the tester [Bayraktaroglu 2001] [Bayraktaroglu 2003] [Könemann 2003]. The advantages compared with sequential linear decompressors are simpler hardware and control. The drawback is that, in order to encode a test cube, each scan slice must be encoded using only the free variables that are shifted from the tester in a single clock cycle (which is equal to the number of channels). The worst-case most highly specified scan slices tend to limit the amount of compression that can be achieved because the number of channels from the tester has to be sufficiently large to encode the most highly specified scan slices. Consequently, it is very difficult to obtain a high encoding efficiency (typically it will be less than 0.25); for the other less specified scan slices, a lot of the free variables end up getting wasted because those scan slices could have been encoded with many fewer free variables.

One approach for improving the encoding efficiency of combinational linear decompressors that was proposed in [Krishna 2003] is to dynamically adjust the number of scan chains that are loaded in each clock cycle. So, for a highly specified scan slice, four clock cycles could be used in which 25% of the scan chains are loaded in each cycle, while for a lightly specified scan slice, only one clock cycle can be used in which 100% of the scan slices are loaded. This allows a better matching of the number of free variables with the number of specified bits to achieve a higher encoding efficiency. Note that it requires that the scan clock be divided into multiple domains.

6.2.2.2 Fixed-Length Sequential Linear Decompressors

Sequential linear decompressors are based on linear finite-state machines such as LFSRs, cellular automata, or ring generators [Mrugalski 2004]. The advantage of a sequential linear decompressor is that it allows free variables from earlier clock cycles to be used when encoding a scan slice in the current clock cycle. This provides much greater flexibility than combinational decompressors and helps avoid the problem of the worst-case most highly specified scan slices limiting the overall compression. The more flip-flops that are used in the sequential linear decompressor, the greater the flexibility that is provided. Results in [Krishna 2001] and [Rajski 2004] have shown that a well-designed sequential linear decompressor with a sufficient number of flip-flops can provide greater than 0.95 encoding efficiency.

A typical design of a sequential linear decompressor is shown in Figure 6.13. Different variants of this were described in [Krishna 2001], [Könemann 2001], and [Rajski 2004]. There are b channels from the tester that inject free variables into a linear finite-state machine (in this case, it is shown as an LFSR, but it could also be a cellular automaton or a ring generator). The LFSR is then followed by a combinational linear XOR network that expands the outputs of the LFSR to fill the scan chains. When decompressing each test cube, the state of the LFSR is first reset and then a few initial clock cycles are used to load the LFSR with some initial free variables. After that, in each clock cycle the scan chains are loaded as additional free variables are injected into the LFSR. The total number of free variables that

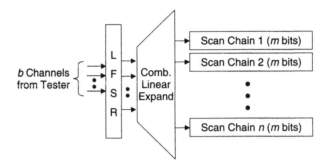

■ **FIGURE 6.13**

Typical sequential linear decompressor.

are used to generate each test cube is equal to $b(q+m)$, where b is the number of channels from the tester, q is the number of clock cycles used to initially load the LFSR, and m is the length of the longest scan chain. The reason for resetting the LFSR between test cubes is that it decouples the system of linear equations that have to be solved when encoding each test cube. If the LFSR is not reset, then the complexity of the linear equations would grow very large, as each test cube would depend on all the free variables injected up to that point. The reason why q clock cycles are used to initially load the LFSR before beginning to fill the scan chains is to create a reservoir of free variables that can be drawn upon in case the first scan slices are heavily specified.

The simplest way to perform sequential linear decompression is to just shift in the same number of free variables for every test cube. The control logic in this case is very simple because every test cube is decompressed in exactly the same way. The drawback of this approach is that the encoding efficiency is limited by the worst-case most heavily specified test cube. If s_{max} is the maximum number of specified bits in any test cube, then the number of free variables used to encode each test cube would have to be at least s_{max}. If s_{avg} is the average number of specified bits in any test cube, then the highest encoding efficiency that can be achieved is s_{avg}/s_{max} because every test cube is encoded with at least s_{max} free variables. If the difference between s_{avg} and s_{max} can be kept small, then high encoding efficiency can be achieved. One way to do this is to constrain the ATPG so s_{max} does not become too large and stays near s_{avg} (this approach was taken in [Rajski 2004]).

6.2.2.3 Variable-Length Sequential Linear Decompressors

An alternative to using a fixed number of free variables for decompressing every test cube is to use a sequential linear decompressor that can vary the number of free variables that are used for each test cube. The advantage of this is that better encoding efficiency can be achieved by using only as many free variables as are needed to encode each test cube. The cost is that more control logic and control information is needed.

One approach for implementing variable-length sequential decompression that was described in [Könemann 2001] and [Volkerink 2003] is to have an extra channel from the tester that gates the scan clock. If there is a heavily specified scan slice (or window of scan slices for [Volkerink 2003]), this extra gating channel can be used to stop the scan shifting for one or more cycles to allow the LFSR to accumulate a sufficient number of free variables to solve for the current scan slice before proceeding to the next one. With this approach it is very easy to control the number of clock cycles and hence the number of free variables that are used for decompressing each test cube. The drawback is the need for the additional gating channel, which diminishes the amount of test compression achieved in proportion with the total number of channels being used (note that, if 16 channels are used, then an additional channel would subtract around 6% from the overall compression).

An alternative approach that eliminates the additional gating channel was described in [Krishna 2004]. The idea is that, when decompressing a test cube, in the first clock cycle the first b bits coming from the tester are used to specify how many clock cycles should be used for decompressing the test cube. These first b bits are used to initialize a counter that counts down until it reaches 0, at which point the scan vector is applied to the CUT and the next test cube is decompressed.

As was mentioned earlier, the more flip-flops that are used in a sequential linear decompressor, the more flexibility it provides in solving the linear equations because more free variables from earlier clock cycles are retained and can be utilized. This improves the encoding efficiency. One idea for increasing the number of flip-flops without incurring a lot of overhead is to configure the scan chains themselves into a large LFSR as originally suggested in [Rajski 1998]. A particular architecture for this was proposed in [Krishna 2004]; it involves constructing a highly efficient three-stage linear decompressor as shown in Figure 6.14, where the parts in gray are configured from the scan cells themselves. The three stages of decompression are used to achieve high encoding efficiency for any distribution of specified bits in a test cube. The first stage is a combinational linear decompressor, the second stage is a vertical LFSR, and the third stage is a set of large horizontal LFSRs. The combinational linear decompressor is designed in a way that makes the large horizontal LFSRs linearly independent, thus allowing any test cube to be decompressed with this architecture, including fully specified ones. Encoding efficiencies greater than 0.99 can be obtained with this architecture without requiring any constraints on the ATPG. The drawback is the need to add logic in the scan chains to configure them as LFSRs.

6.2.2.4 *Combined Linear and Nonlinear Decompressors*

The amount of compression that can be achieved with linear decompression is limited by the number of specified bits in the test cubes. While linear decompressors are very efficient at exploiting "don't cares" in the test set, they cannot exploit correlations in the specified bits; hence, they cannot compress the test data to less than the total number of specified bits in the test data. The specified bits tend to be highly correlated, and one strategy that takes advantage of this is to combine

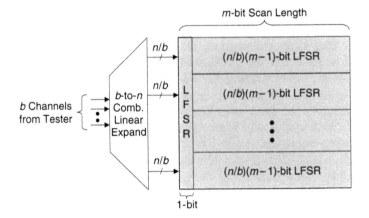

■ **FIGURE 6.14**

Three-stage sequential linear decompressor [Krishna 2004].

linear and nonlinear decompression together to achieve greater compression than either can alone.

In [Krishna 2002], the inputs to a linear decompressor are encoded using a nonlinear code. A method is described in [Krishna 2002] for selecting the solution to the system of linear equations for each test cube in such a way that they can be effectively compressed by a statistical code. The statistical code reduces the number of bits that must be stored on the tester for the linear decompressor.

In [Sun 2004], dictionary coding is combined with a linear decompressor. For each scan slice, either the dictionary is used to generate it, or, if it is not present in the dictionary, the linear decompressor is then used to generate it.

In [Li 2005] and [Ward 2005], a nonlinear decompressor is placed between the linear decompressor and the scan chains as shown in Figure 6.15. In [Li 2005], the nonlinear decompressor is constructed by identifying scan chains that are compatible with a nonlinear combination of two other scan chains. For example, if scan chain a can be driven by the AND of the scan-in of scan chains b and c (this is

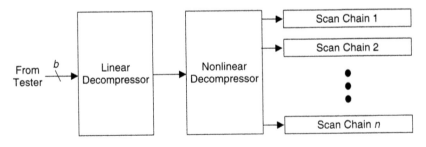

■ **FIGURE 6.15**

One approach for combining linear and nonlinear decompressors.

checked by making sure that, in any scan slice where scan chain *a* has a specified "1," neither scan chain *b* nor *c* has a specified "0"). A limitation of this approach is that as the scan length and number of test patterns increase, it becomes increasingly difficult to find such compatibility relationships among the scan chains. In [Ward 2005], a statistical code that compresses the number of specified bits is used. By reducing the number of specified bits that the linear decompressor has to produce, greater compression can be achieved because fewer free variables are required from the tester to solve the linear equations.

6.2.3 Broadcast-Scan-Based Schemes

The third class of test stimulus compression schemes is based on broadcasting the same value to multiple scan chains. This was first proposed in [Lee 1998] and [Lee 1999]. Due to its simplicity and effectiveness, this method has been used as the basis of many test compression architectures, including some commercial ***design for testability*** (DFT) tools.

6.2.3.1 Broadcast Scan

To illustrate the basic concept of **broadcast scan**, first consider two independent circuits C_1 and C_2. Assume that these two circuits have their own test sets $T_1 = <T_{11}, t_{12}, \ldots, t_{1k}>$ and $T_2 = <T_{21}, t_{22}, \ldots, t_{2l}>$, respectively. In general, a test set may consist of random patterns and deterministic patterns. In the beginning of the ATPG process, usually random patterns are initially used to detect the easy-to-detect faults. If the same random patterns are used when generating both T_1 and T_2 then we may have $t_{11} = t_{21}, t_{12} = t_{22}, \ldots$, up to some *i*th pattern. After most faults have been detected by the random patterns, deterministic patterns are generated for the remaining difficult-to-detect faults. Generally, these patterns have many "don't care" bits. For example, when generating $t_{1(i+1)}$, many "don't care" bits may still exist when no more faults in C_1 can be detected. Using a test pattern with bits assigned so far for C_1, we can further assign specific values to the "don't care" bits in the pattern to detect faults in C_2. Thus, the final pattern would be effective in detecting faults in both C_1 and C_2.

The concept of pattern sharing can be extended to multiple circuits as illustrated in Figure 6.16. The problem is how to guide the ATPG tool to generate the patterns to be shared. If the ATPG tool has an option that allows the user to place the constraint that some inputs must always have the same values then the problem is solved. Note that the inputs here may include the primary inputs as well as the pseudo-primary inputs (*i.e.*, the outputs of D flip-flops in a full-scan design). If the ATPG tool does not have this property, the concept of a *virtual circuit* presented in [Lee 1998] and [Lee 1999] can be used to deal with this problem. As shown in Figure 6.17, one may connect the inputs of the circuits that are to share the test patterns in a 1-to-1 mapping manner. This circuit is then handed to the ATPG tool as a single circuit, with the number of inputs being the maximum number of inputs among the circuits. The test compression can then be automatically done by the ATPG tool which will target faults as if it were a single circuit. This way of "cheating"

Scan_input

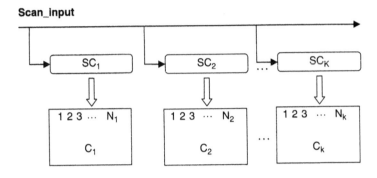

■ **FIGURE 6.16**

Broadcasting to scan chains driving independent circuits.

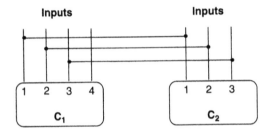

■ **FIGURE 6.17**

Forcing ATPG tool to generate patterns for broadcast scan.

the ATPG tool to generate compact tests does not require any modification to the ATPG program and hence can be applied to any ATPG tool.

One major advantage of using *broadcast scan* for independent circuits is that all faults that are detectable in all original circuits will also be detectable with the broadcast structure. This is because if one test vector can detect a fault in a stand-alone circuit then it will still be possible to apply this vector to detect the fault in the broadcast structure. Thus, the broadcast scan method will not affect the fault coverage if all circuits are independent. Note that broadcast scan can also be applied to multiple scan chains of a single circuit if all subcircuits driven by the scan chains are independent. An example of this is the pipelined circuit shown in Figure 6.18, where each scan chain is driving an independent circuit. The response data from all subchains can be compacted by a ***multiple-input signature register*** (MISR) or other space/time compactor.

6.2.3.2 Illinois Scan

If *broadcast scan* is used for multiple scan chains of a single circuit where the subcircuits driven by the scan chains are not independent, then the property of always being able to detect all faults is lost. The reason for this is that if two scan

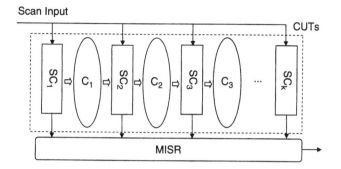

Broadcast scan for a pipelined circuit.

chains are sharing the same channel, then the ith scan cell in each of the two scan chains will always be loaded with identical values. If some fault requires two such scan cells to have opposite values in order to be detected, it will not be possible to detect this fault with broadcast scan.

To address the problem of some faults not being detected when using broadcast scan for multiple scan chains of a single circuit, the **Illinois scan architecture** was proposed in [Hamzaoglu 1999] and [Hsu 2001]. It got its name because the authors are with the University of Illinois at Urbana–Champaign. The Illinois scan architecture consists of two modes of operations, namely a *broadcast mode* and a *serial scan mode*, which are illustrated in Figure 6.19. The *broadcast mode* is first used to detect most faults in the circuit. During this mode, a scan chain is divided into multiple subchains called *segments* and the same vector can be shifted into

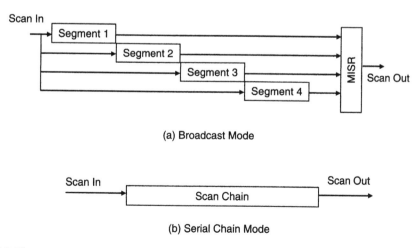

(a) Broadcast Mode

(b) Serial Chain Mode

■ **FIGURE 6.19**

Two modes of Illinois scan architecture.

all segments through a single shared scan-in input. The response data from all subchains are then compacted by a MISR or other space/time compactor. For the remaining faults that cannot be detected in broadcast mode, the *serial scan mode* is used where any possible test pattern can be applied. This ensures that complete fault coverage can be achieved. The extra logic required to implement the Illinois scan architecture consists of several multiplexers and some simple control logic to switch between the two modes. The area overhead of this logic is typically quite small compared to the overall chip area.

The main drawback of the Illinois scan architecture is that no test compression is achieved when it is run in serial scan mode. This can significantly degrade the overall **compression ratio** that is achieved depending on how many test patterns must be applied in serial scan mode. To reduce the number of patterns that have to be applied in serial scan mode, multiple-input broadcast scan or reconfigurable broadcast scan can be used. These techniques are described next.

6.2.3.3   Multiple-Input Broadcast Scan

Instead of using only one channel to drive all scan chains, a **multiple-input broadcast scan** could be used where there is more than one channel [Shah 2004]. Each channel can drive some subset of the scan chains. If two scan chains must be independently controlled to detect a fault, then they could be assigned to different channels. The more channels that are used and the shorter each scan chain is, the easier it is to detect more faults because fewer constraints are placed on the ATPG. Determining a configuration that requires the minimum number of channels to detect all detectable faults is thus highly desired with a multiple-input broadcast scan technique.

6.2.3.4   Reconfigurable Broadcast Scan

Multiple-input broadcast scan may require a large number of channels to achieve high fault coverage. To reduce the number of channels that are required, a **reconfigurable broadcast scan** method can be used. The idea is to provide the capability to reconfigure the set of scan chains that each channel drives. Two possible reconfiguration schemes have been proposed, namely **static reconfiguration** [Pandey 2002] [Samaranayake 2003], and **dynamic reconfiguration** [Li 2004] [Sitchinava 2004] [Wang 2004] [Han 2005a]. In *static reconfiguration*, the reconfiguration can only be done when a new pattern is to be applied. For this method, the target fault set can be divided into several subsets, and each subset can be tested by a single configuration. After testing one subset of faults, the configuration can be changed to test another subset of faults. In *dynamic reconfiguration*, the configuration can be changed while scanning in a pattern. This provides more reconfiguration flexibility and hence can in general lead to better results with fewer channels. This is especially important for hard cores when the test patterns provided by core vendor cannot be regenerated. The drawback of dynamic reconfiguration *versus* static reconfiguration is that more control information is needed for reconfiguring at the right time, whereas for static reconfiguration the control information is much less because the reconfiguration is done only a few times (only after all the test patterns using a particular configuration have been applied). Dynamic reconfiguration is illustrated in the following example.

Example 6.2

For the patterns shown in Figure 6.20, we only need four channels to drive the eight scan chains if two broadcast configurations are used to control the generation of the first five and last five bits for each scan chain, respectively. The first configuration is: $1 \rightarrow \{2, 3, 6\}$, $2 \rightarrow \{7\}$, $3 \rightarrow \{5, 8\}$, $4 \rightarrow \{1, 4\}$, where $A \rightarrow \{B_1, B_2, \ldots, B_n\}$ means that the Ath channel drives the B_1th, B_2th, ..., B_nth scan chains. The other configuration is: $1 \rightarrow \{1, 6\}$, $2 \rightarrow \{2, 4\}$, $3 \rightarrow \{3, 5, 7, 8\}$. The block diagram of a ***multiplexer*** (MUX) network for the above example is shown in Figure 6.21. It consists of five two-input multiplexers. Because there are two configurations, only one control line for the MUX network is required. In this example, when the control line is 0 (1), the first (second) configuration is selected.

6.2.3.5 *Virtual Scan*

Rather than using MUX networks for test stimulus compression, combinational logic networks can also be used as decompressors. The combinational logic network can consist of any combination of simple combinational gates, such as buffers, inverters, AND/OR gates, MUXs, and XOR gates. This scheme, referred to as **virtual scan**, is different from *reconfigurable broadcast scan* and *combinational linear decompression* where pure MUX and XOR networks are allowed, respectively. The combinational logic network can be specified as a set of constraints or just as an expanded circuit for ATPG. In either case, the test cubes that ATPG generates are the compressed stimuli for the decompressor itself. There is no need to solve linear equations, and *dynamic compaction* can be effectively utilized during the ATPG process.

   The virtual scan scheme was proposed in [Wang 2002] and [Wang 2004]. In these papers, the decompressor was referred to as a **broadcaster**. The authors also proposed adding additional logic, when required, through *VirtualScan inputs* to reduce or remove the constraints imposed on the decompressor (broadcaster), thereby yielding very little or no fault coverage loss caused by test stimulus compression.

Scan Chain 1	1 X 1 X X	X 0 0 X X
Scan Chain 2	X X 0 X 1	0 X 1 X 1
Scan Chain 3	X X X X 1	1 1 X X 1
Scan Chain 4	1 1 X X 0	0 0 X 0 1
Scan Chain 5	0 X 1 X X	X X X X X
Scan Chain 6	X 0 X 1 X	0 X 0 0 X
Scan Chain 7	0 X 0 X X	1 1 X X X
Scan Chain 8	X X 1 X X	X X 1 X X

■ **FIGURE 6.20**

Test patterns for example of dynamic reconfiguration.

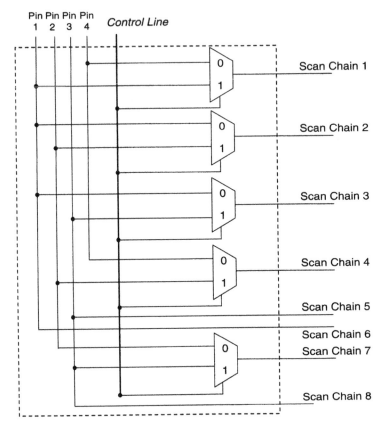

■ FIGURE 6.21

Block diagram of MUX network for Figure 6.20.

In a broad sense, virtual scan is a generalized class of broadcast scan, Illinois scan, multiple-input broadcast scan, reconfigurable broadcast scan, and combinational linear decompression. The advantage of using virtual scan is that it allows the ATPG to directly search for a test cube that can be applied by the decompressor and allows very effective dynamic compaction. Thus, virtual scan may produce shorter test sets than any test stimulus compression scheme based on solving linear equations; however, because this scheme may impose XOR constraints directly on the original circuit, it may take longer than those based on solving linear equations to generate test cubes or compressed stimuli.

6.3   TEST RESPONSE COMPACTION

Test response compaction is performed at the outputs of the scan chains. The purpose is to reduce the amount of test response that must be transferred back to the tester. While test stimulus compression must be lossless, test response compaction can be lossy. A large number of different test response compaction schemes have been presented and described to various extents in the literature. The schemes differ in the following attributes: time *versus* space, circuit-function-specific *versus* circuit-function-independent, linearity *versus* nonlinearity.

Prior to describing the distinguishing attributes, it is useful to introduce some notations. Consider the general case of compaction where an $m \times n$ matrix of test data $D = [d_{ij}]$ of $m \times n$ bits is transformed into a $p \times q$ matrix $C = [c_{ij}]$ of $p \times q$ bits, where $p < m$ and/or $q < n$. Denote the transformation operator Φ as a matrix operator such that $C = \Phi(D)$. We refer to the ratio $m{:}p$ as the space compaction ratio and the ratio $n{:}q$ as the time compaction ratio.

- **(I) Time *versus* space**—The column index of test data matrix D is referred to as the *time dimension* because it corresponds to the output bits from a single circuit output resulting from the application of different input test patterns. If Φ is such that C has its time dimension $q < n$, then time compaction occurs. The row index of the test data matrix D is referred to as the *space dimension* because it corresponds to the output bits from different circuit outputs resulting from the application of the input test pattern. Thus, if Φ is such that C has its space dimension $p < m$, then space compaction occurs.

Figure 6.22 can help explain the difference between space compaction and time compaction. A **space compactor** compacts an m-bit-wide output pattern to a p-bit-wide output pattern (where $p < m$), whereas a **time compactor** compacts n output patterns to q output patterns (where $q < n$).

It is possible to have both time and space compaction performed concurrently. The scheme combining both time and space compaction is referred to as mixed time and space compaction:

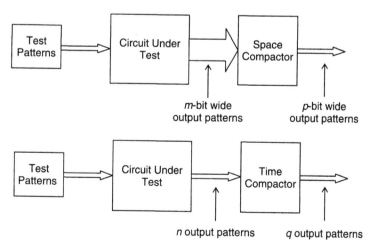

■ **FIGURE 6.22**

Time and space compaction schemes for response data.

- **(II) Circuit-function-specific *versus* circuit-function-independent—** *Circuit-function-specific* (CFS) is a characteristic referring to how a particular Φ is selected. The compaction function Φ is *circuit function independent* (CFI) (*i.e.*, not circuit function specific) if Φ is selected regardless of the test data expected to originate from either the fault-free or faulty *circuit*. A Φ that is CFI implies that it is selected independently from the circuit functionality in test mode (*i.e.*, independently from the functionality resulting from the application of a particular test set to the circuit). This is important, because frequently the compactor is designed before the actual design of the chip. If the compactor depends on the functionality of the design in the chip, then last-minute design changes may require you to modify the compactor. This may impact time-to-market.

- **(III) Linearity *versus* nonlinearity—**A (response) compactor is said to be *linear* if it consists of only XOR gates and flip-flops. For linear compactors, the compacting function Φ is such that each bit of the compacted data matrix C can be expressed as a Boolean sum (XOR sum) of any number of bits of the data matrix D. Thus, for linear compactors, Φ is a linear operator.

Using the above attributes, we classify a number of known compaction functions in Table 6.5.

The test response bits are obtained by using a logic simulator to simulate the fault-free design for the test stimulus. Unfortunately, for complex designs, logic simulators cannot always deterministically predict the logic values of all test response bits. For example, the simulator may not accurately predict values due to floating buses, uninitialized and uncontrollable storage elements, bus contention, or multiple clock domains or simply because the simulation model is inaccurate. The response bits whose logic values are not accurately predicted by the simulators are also called *unknown test response bits* or *X*'s. *X*'s significantly complicate response

TABLE 6.5 ■ Taxonomy of Various Response Compaction Schemes

Compaction Scheme	I		II		III	
	Space	Time	CFS	CFI	Linearity	Nonlinearity
Zero-aliasing compactor [Chakrabarty 1998] [Pouya 1998]	✓		✓			✓
Parity tree [Karpovsky 1987]	✓			✓	✓	
Enhanced parity tree [Sinanoglu 2003]	✓	✓	✓		✓	
X-Compact [Mitra 2004]	✓			✓	✓	
q-Compactor [Han 2003]	✓	✓		✓	✓	
Convolutional compactor [Rajski 2005]	✓	✓		✓	✓	
OPMISR [Barnhart 2002]	✓	✓		✓	✓	
Block compactor [Wang 2003]	✓	✓		✓	✓	
i-Compact [Patel 2003]	✓			✓	✓	
Compactor for SA [Wohl 2001]	✓	✓		✓	✓	
Scalable selector [Wohl 2004]	✓			✓		✓

compression. For example, one unknown test response can render a signature of a MISR unknown and unusable.

Test response compaction may induce some loss of information. Although the complete information of the original response is lost, the objective of fault detection can be still achieved. Because of the loss, response compaction techniques face two major challenges: (1) aliasing and (2) fault diagnosis.

- *Aliasing* is a problem where different uncompressed data *A* and uncompressed data *B* yield the same compressed data *C* after compaction, $C = \Phi(A) = \Phi(B)$. For example, a faulty circuit with an erroneous response can produce the same *signature* as the fault-free circuit, preventing the faulty circuit from being detected by the test. Hence, the response compaction techniques must be employed in a way to minimize aliasing.

- Another important challenge for response compaction techniques is the ability to perform fault diagnosis. In response compaction, a better diagnosis is to locate the failing scan cells in the scan chains from the outputs of the compactor without configuring the chip in a special diagnosis mode.

In the following subsections, we will discuss in detail three types of response compaction techniques: (1) space compaction, (2) time compaction, and (3) mixed space and time compaction.

6.3.1   Space Compaction

A space compactor is a combinational circuit for compacting *m* outputs of the circuit under test to *n* test outputs, where $n < m$. Space compaction can be regarded as the inverse procedure of *linear expansion* (which was described in Section 6.2.3). It can be expressed as a function of the input vector (*i.e.*, the data being scanned out) and the output vector (the data being monitored):

$$Y = \Phi(X)$$

where *X* is an *m*-bit input vector and *Y* is an *n*-bit output vector, $n < m$.

Some linear codes can be used to implement space compaction. Parity tree circuits have frequently been proposed for space compaction because of their good error propagation properties; however, while experimental results indicate that a high percentage of single stuck-at faults in typical logic circuits are detected with a parity tree space compactor, zero-aliasing compaction is rarely achieved. To provide better characteristics than a parity tree, a number of other compaction methods have been developed [Wohl 2001] [Wohl 2003a] [Das 2003] [Mitra 2004]. These include the enhanced single-error-correcting, double-error-detecting, or odd-error-detecting codes methods, which can reduce the aliasing ratio.

6.3.1.1   *Zero-Aliasing Linear Compaction*

Consider a space compaction function $Y = \Phi(X)$. A space compactor is said to be *transparent* if any two different values X_1 and X_2 that appear at its input produce different values Y_1 and Y_2 at its output, *i.e.*, $\Phi(X_1) \neq \Phi(X_2)$. If X_1 is the compactor input value in a correctly working circuit and X_2 is the input value due to a fault, then because the corresponding output vectors will always be different for a *transparent* compactor it will be **zero-aliasing** [Chakrabarty 1998a,b].

For a space compaction function to guarantee zero-aliasing for the complete input vector space, the number of output bits for the compactor will have to be equal to or greater than the number of input bits. This means that if we want to design a full zero-aliasing space compactor then we cannot obtain any benefit of compaction. Therefore a practical space compactor cannot be zero-aliasing for all possible errors. Thus, the objective is to make it zero-aliasing only for the set of errors that can occur due to some set F of faults that actually occur in the circuit.

An upper bound on the number of outputs of a compactor, given a specified test set T and a circuit C, is calculated in Theorem 6.1.

Theorem 6.1

For any test set T, for a circuit that implements function C, there exists a zero-aliasing output space compactor for C with q outputs where $q = \lceil \log_2(|T|+1) \rceil$.

In the worst case, every fault-free response will be distinct. Each faulty input X needs to be mapped to a different Y (using function C). This means that such a space compactor guarantees zero-aliasing. It must produce $|T|+1$ different output combinations, which implies that it must have at least $\lceil \log_2(|T|+1) \rceil$ output lines.

Theorem 6.1 gives an upper bound on the number of outputs of the space compactor. This does not mean that we can only design the compactor with q outputs. A more efficient space compaction circuit may be possible by taking the fault set into account. This optimization can be realized by a graph model, also called a *response graph*. The response graph $G = (V, E)$ consists of the set of vertices $V = \{v_1, v_2, \ldots, v_n\}$ corresponding to all possible responses of circuit C to test set T given fault set F, and the set edges E where $(v_j, v_k) \in E$ if and only if there exists a test pattern $t \in T$ for which the fault-free response is v_j, and a fault $f \in F$ such that the faulty response of C for test t is v_k.

As an example, consider the ISCAS C17 benchmark circuit given in Figure 6.23. Figure 6.23a shows the fault-free response $\boldsymbol{R} = (00, 11, 11, 00)$ for the minimal test set $\boldsymbol{T} = (10010, 11010, 10101, 01111)$. The response graph for this circuit is shown in Figure 6.23b. In this figure, for example, if there is a ***stuck-at-1*** (SA1) fault at X_1, the fault-free response is "00" and the erroneous response is "10," so there is an edge between the state "00" and "10." Other edges can be concluded by the definition.

Theorem 6.2

Let G be a response graph. If G is 2^q *colorable*, then there exists a q-output zero-aliasing space compactor for the circuit C.

Every vertex v of G corresponds to an input vector X to the space compactor. The color assigned to v can be associated with the output of the compactor for input X. If G is 2^q colorable, then the compactor realizes 2^q different output values and can be represented with q output bits. Moreover, every faulty input X' different from X

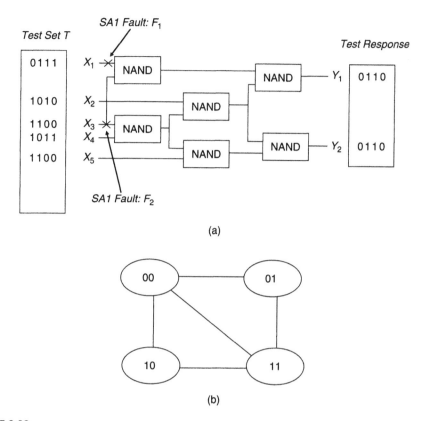

■ FIGURE 6.23

An example of response graph: (a) The C17 circuit with a complete set of test patterns and fault-free response, and (b) response graph.

produces a different output of the space compactor; therefore, all faults in C that cause X' are detected. Hence, zero aliasing is ensured with q outputs.

The problem with the above compacters is that unknown test responses may prevent error detectability. The next section presents a compaction tree that tolerates unknown test responses.

6.3.1.2 X-Compact

X-compact [Mitra 2004] is an **X-tolerant response compaction** technique that has been used in several designs. The combinational compactor circuit designed using the X-compact technique is called an **X-compactor**. Figure 6.24 shows an example of an X-compactor with 8 inputs and 5 outputs. It is composed of 4 3-input XOR gates and 11 2-input XOR gates.

The X-compactor can be represented as a binary matrix (matrix with only 0's and 1's) with n rows and k columns; this matrix is called the *X-compact matrix*. Each row of the X-compact matrix corresponds to a scan chain and each column corresponds to an X-compactor output. The entry in row i and column j of the

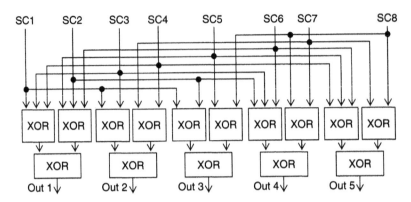

■ **FIGURE 6.24**

An X-compactor with eight inputs and five outputs.

matrix is 1 if and only if the jth X-compactor output depends on the ith scan chain output; otherwise, the matrix entry is 0. The corresponding X-compact matrix M of the X-compactor shown in Figure 6.24 is:

$$M = \begin{bmatrix} 1 & 1 & 1 & 0 & 0 \\ 1 & 0 & 1 & 1 & 0 \\ 1 & 1 & 0 & 1 & 0 \\ 1 & 1 & 0 & 0 & 1 \\ 1 & 0 & 1 & 0 & 1 \\ 1 & 0 & 0 & 1 & 1 \\ 0 & 1 & 0 & 1 & 1 \\ 0 & 0 & 1 & 1 & 1 \end{bmatrix}$$

For a conventional sequential compactor, such as a MISR, there are two sources of *aliasing*: error masking and error cancellation. **Error masking** occurs when one or more errors captured in the compactor during a single cycle propagate through the feedback path and cancel out with errors in the later cycles. **Error cancellation** occurs when an error bit captured in a shift register is shifted and eventually cancelled by another error bit. The error cancellation is a type of aliasing specific to the multiple-input sequential compactor. Because the X-compactor is a combinational compactor, it only results in error masking. To handle aliasing, the following theorems provide a basis for systematically designing X-compactors.

Theorem 6.3

If only a single scan chain produces an error at any scan-out cycle, the X-compactor is guaranteed to produce errors at the X-compactor outputs at that scan-out cycle, if and only if no row of the X-compact matrix contains all 0's.

Theorem 6.4

Errors from any one, two, or an odd number of scan chains at the same scan-out cycle are guaranteed to produce errors at the X-compactor outputs at that scan-out cycle, if every row of the X-compact matrix is nonzero, distinct, and contains an odd number of 1's.

If all rows of the X-compact matrix are distinct and contain an odd number of 1's, then a bitwise XOR of any two rows is nonzero. Also, the bitwise XOR of any odd number of rows is also nonzero. Hence, errors from any one or any two or any odd number of scan chains at the same scan-out cycle are guaranteed to produce errors at the compactor outputs at that scan-out cycle. Because all rows of the X-compact matrix of Figure 6.24 are distinct and odd, then by Theorem 6.4 simultaneous errors from any two or odd scan chains at the same scan-out cycle are guaranteed to be detected.

The X-compact technique is nonintrusive and independent of the test patterns used to test the circuit. Insertion of the X-compactor does not require any major change to the ATPG flow; however, the X-compactor cannot guarantee that errors other than those described in Theorem 6.3 and Theorem 6.4 are detectable.

6.3.1.3 X-Blocking

Instead of tolerating X's on the response compactor, X's can also be blocked before reaching the response compactor. During design, these potential X-generators (X-sources) can be identified using a scan design rule checker. When an X-generator is likely to reach the response compactor, it must be fixed [Naruse 2003] [Patel 2003]. The process is often referred to as **X-blocking** or **X-bounding**.

In X-blocking, the output of an X-source can be blocked anywhere along its propagation paths before X's reach the compactor. An example is shown in Figure 6.25. When the X-source has been blocked at a nearby location during test and will not reach the compactor, there is no need to fix further; however, care must be taken to ensure that no observation points are added between the X-source and the location at which it is blocked. For example, a non-scan flip-flop is a potential X-generator (X-source). If the non-scan flip-flop has two outputs (Q and QB), then one can add

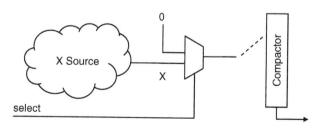

■ **FIGURE 6.25**

A simple illustration of the X-blocking scheme.

a control point to each of the outputs and activate it in test mode. Alternatively, if the flip-flip has an asynchronous set/reset pin, a control point can be added to permanently force the flip-flip to 0 or 1 during test. While a control point can be added to force the non-scan flip-flop to a constant value, it is recommended that for better fault coverage inserting a MUX control point driven by a nearby existing scan cell is preferred.

X-blocking can ensure that no X's will be observed; however, it does not provide a means for observing faults that can only propagate to an observable point through the now-blocked X-source. This can result in fault coverage loss. If the number of such faults for a given bounded X-generator justifies the cost, one or more observation points can be added before the X-source to provide an observable point to which those faults can propagate. These X-blocking or X-bounding methods have been extensively discussed in Section 5.2 (BIST Design Rules) of Chapter 5.

6.3.1.4 X-Masking

While it may not result in fault coverage loss, the X-blocking technique does add area overhead and may impact delay due to the inserted logic. It is not surprising to find that, in complex designs, more than 25% of scan cycles could contain one or more X's in the test response. It is difficult to eliminate these residual X's by DFT; thus, an encoder with high X-tolerance is very attractive. Instead of blocking the X's where they are generated, the X's can also be masked off right before the response compactor [Wohl 2004] [Han 2005c] [Volkerink 2005] [Rajski 2005]. An example X-masking circuit is shown in Figure 6.26. The mask controller applies a logic value 1 at the appropriate time to mask off any scan output that contains an X.

Mask data is required to indicate when the masking should take place. The mask data can be stored in compressed format and can be decompressed using on-chip hardware. Possible compression techniques are *weighted pseudo-random LFSR reseeding* or *run-length encoding* [Volkerink 2005].

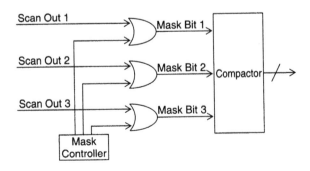

■ **FIGURE 6.26**

An example X-masking circuit.

6.3.1.5 X-Impact

While X-compact, X-blocking, and X-masking each can achieve significant reduction in fault coverage loss caused by X's present at the inputs of a space compactor, the **X-impact** technique described in [Wang 2004] is helpful in that it simply uses ATPG to algorithmically handle the impact of residual X's on the space compactor without adding any extra circuitry.

Example 6.3

An example of algorithmically handling an X-impact is shown in Figure 6.27. Here, $SC1$ to $SC4$ are scan cells connected to a space compactor composed of XOR gates $G7$ and $G8$. Lines $a, b, \dots, h$ are internal signals, and line f is assumed to be connected to an X-source (memory, non-scan storage element, etc.). Now consider the detection of the **stuck-at-0** (SA0) fault f_1. Logic value 1 should be assigned to both lines d and e in order to activate f_1. The fault effect will be captured by scan cell $SC3$. If the X on f propagates to $SC4$, then the compactor output q will become X and f_1 cannot be detected. To avoid this, ATPG can try to assign either 1 to line g or 0 to line h in order to block the X from reaching $SC4$. If it is impossible to achieve this assignment, ATPG can then try to assign 1 to line c, 0 to line b, and 0 to line a in order to propagate the fault effect to $SC2$. As a result, fault f_1 can be detected. Thus, X-impact is avoided by algorithmic assignment without adding any extra circuitry.

Example 6.4

It is also possible to use the X-impact approach to reduce aliasing. An example of algorithmically handling aliasing is shown in Figure 6.28. Here, $SC1$ to $SC4$ are scan cells connected to a compactor composed of XOR gates $G7$ and $G8$. Lines $a, b, \dots, h$ are internal signals. Now consider the detection of the stuck-at-1 fault f_2. Logic value 1 should be assigned to lines c, d, and e in order to activate f_2, and logic value 0 should be assigned to line b in order to propagate the fault effect to $SC2$. If line a is set to 1, the fault effect will also propagate to $SC1$. In this case,

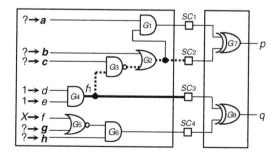

■ **FIGURE 6.27**

Handling of X-impact.

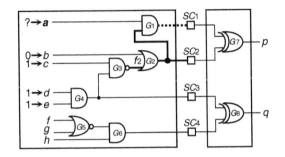

■ **FIGURE 6.28**

Handling of aliasing.

aliasing will cause the compactor output p to have a fault-free value, resulting in an undetected f_2. To avoid this, ATPG can try to assign 0 to line a in order to block the fault effect from reaching $SC1$. As a result, fault f_2 can be detected. Thus, aliasing can be avoided by algorithmic assignment without any extra circuitry.

6.3.2   Time Compaction

A time compactor uses sequential logic (whereas a space compactor uses combinational logic) to compact test responses. Because sequential logic is used, one must make sure that no unknown (X) values from the circuit under test will reach the compactor. If that happens, X-bounding or X-masking must be employed.

The most widely adopted response compactor using time compaction is the **_multiple-input signature register_** (MISR). Consider the n-stage MISR shown in Figure 6.29. The internal structure of the n-stage MISR can be described by specifying a **characteristic polynomial** of degree n, $f(x)$, in which the symbol h_i is either 1 or 0, depending on the existence or absence of the feedback path, where

$$f(x) = 1 + h_1 x + h_2 x^2 + \cdots + h_{n-1} x^{n-1} + x^n$$

The MISR uses n extra XOR gates for compacting nm-bit output sequences, M_0 to M_{m-1}, into the modular LFSR simultaneously. The final contents stored in

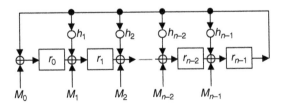

■ **FIGURE 6.29**

An n-stage multiple-input signature register (MISR).

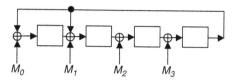

■ **FIGURE 6.30**

A four-stage MISR.

M_0	1 0 0 1 0
M_1	0 1 0 1 0
M_2	1 1 0 0 0
M_3	1 0 0 1 1
M	1 0 0 1 1 0 1 1

■ **FIGURE 6.31**

An equivalent M sequence.

the MISR after compaction are called the (final) *signature* of the MISR. For more information on signature analysis and the MISR design, the reader is referred to Section 5.4.3 (Signature Analysis) of Chapter 5.

Example 6.5

Consider the four-stage MISR shown in Figure 6.30 using $f(x) = 1 + x + x^4$. Let $M_0 = \{10010\}$, $M_1 = \{01010\}$, $M_2 = \{11000\}$, and $M_3 = \{10011\}$. From this information, the signature R of the MISR can be calculated as $\{1011\}$. Using the formula $M(x) = M_0(x) + xM_1(x) + x^2M_2(x) + x^3M_3(x)$ as discussed in Section 5.4.3 of Chapter 5, we obtain $M(x) = 1 + x^3 + x^4 + x^6 + x^7$ or $M = \{10011011\}$, as shown in Figure 6.31. The final signature is stored in the rightmost four bits of the M sequence; therefore, $R = \{1011\}$.

6.3.3 Mixed Time and Space Compaction

In the previous two sections, we introduced different kinds of compactors for space compaction and time compaction independently. In this section, we introduce mixed time and space compactors [Saluja 1983]. A mixed time and space compactor combines the advantages of a time compactor and a space compactor. Many mixed time and space compactors have been proposed in the literature, including OPMISR [Barnhart 2002], convolutional compactor [Rajski 2005], and q-compactor [Han 2003] [Han 2005c,d].

Because q-compactor is simple, this section uses it to introduce the conceptual architecture of a mixed time and space compactor. Figure 6.32 shows an example of a q-compactor assuming the inputs are coming from scan chain outputs. The spatial part of the q-compactor consists of single-output XOR networks (called

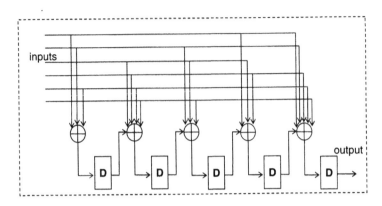

■ **FIGURE 6.32**

An example *q*-compactor with single output.

spread networks) connected to the flip-flops by means of additional two-input XOR gates interspersed between successive storage elements. As can be seen, every error in a scan cell can reach storage elements and then outputs in several possible ways. The spread network that determines this property is defined in terms of *spread polynomials* indicating how particular scan chains are connected to the register flip-flops.

Different from a conventional MISR, the *q*-compactor presented in Figure 6.32 does not have a feedback path; consequently, any error or *X* injected into the compactor is shifted out after at most five cycles. The shifted-out data will be compared with the expected data and then the error will be detected.

Example 6.6

An example of a *q*-compactor with six inputs, one output, and five storage elements—five per output—is shown in Figure 6.32. For the sake of simplicity, the injector network is shown here in a linear form rather than as a balanced tree.

6.4 INDUSTRY PRACTICES (EDITED BY LAUNG-TERNG WANG)

During the last few years, several test compression products and solutions have been introduced by some of the major DFT vendors in the CAD industry. These products differ significantly with regard to technology, design overhead, design rules, and the ease of use and implementation. A few second-generation products have also been introduced by a few of the vendors. In this section, we briefly review a few of the products introduced by companies such as Cadence [Cadence 2006], Mentor Graphics [Mentor 2006], SynTest [SynTest 2006], Synopsys [Synopsys 2006], and LogicVision [LogicVision 2006].

Current industry solutions can be grouped under two main categories for stimulus decompression. The first category uses *linear-decompression-based schemes*, and the second category employs *broadcast-scan-based schemes*. The main difference between the two categories is the manner in which the ATPG engine is used. The first category includes products such as ETCompression from LogicVision [LogicVision 2006], TestKompress from Mentor Graphics [Rajski 2004], and SOCBIST from Synopsys [Wohl 2003b]. The second category includes products such as OPMISR+ from Cadence [Cadence 2006], VirtualScan [Wang 2004] and UltraScan [Wang 2005b] from SynTest, and DFT MAX from Synopsys [Sitchinava 2004].

For designs using linear-decompression-based schemes, test compression is achieved in two distinct steps. During the first step, conventional ATPG is used to generate sparse ATPG patterns (called *test cubes*), in which *dynamic compaction* is performed in a nonaggressive manner, while leaving unspecified bit locations in each test cube as X. This is accomplished by not aggressively performing the *random fill* operation on the test cubes which is used to increase coverage of individual patterns and hence reduce the total pattern count. During the second step, a system of liner equations that describe the hardware mapping from the external scan input ports to the internal scan chain inputs is solved in order to map each test cube into a compressed stimulus that can be applied externally. If a mapping is not found, a new attempt at generating a new test cube is required.

For designs using broadcast-scan-based schemes, only a single step is required to perform test compression. This is achieved by embedding the constraints introduced by the decompressor as part of the ATPG tool such that the tool operates with much more restricted constraints. Hence, while in conventional ATPG each individual scan cell can be set to 0 or 1 independently, for broadcast-scan-based schemes the values to which related scan cells can be set are constrained. Thus, a limitation of this solution is that, in some cases, the constraints among scan cells can preclude some faults from being tested. These faults are typically tested as part of a later top-up ATPG process if required, similar to using linear-decompression-based schemes.

On the response compaction side, industry solutions have utilized either space compactors such as XOR networks or time compactors such as MISRs to compact the test responses. Currently, space compactors have a higher acceptance rate in the industry, as they do not involve the process of guaranteeing that no unknown (X) values are generated in the circuit under test.

In this section, we briefly describe a number of test compression solutions and products currently supported by the EDA DFT vendors, including OPMISR+ from Cadence, TestKompress from Mentor Graphics, VirtualScan and UltraScan from SynTest, DFT MAX from Synopsys, and ETCompression from LogicVision. A summary of the different compression architectures used in these commercial products is listed at the end of the section.

6.4.1   OPMISR+[1]

OPMISR+ is the name of the test compression methodology that is a part of Cadence Design System's Encounter Test product. It has its roots in IBM's logic

---

[1] Contributed by Brion Keller.

BIST and ATPG technology. Due to ever-increasing test data volume and test application time, the company decided to go with on-chip compression in early 1999.

OPMISR+ originally was an intermediate step toward a more sophisticated compression approach called SmartBIST [Barnhart 2000] [Könemann 2001]. SmartBIST combined the nearly complete output response compression of a *multiple-input shift register* (MISR) with a combinational or sequential "decompression" scheme based on a linear (XOR) spreader network fed from the scan inputs, optionally with a ***pseudo-random pattern generator*** (PRPG) in between. The structure of SmartBIST clearly borrows heavily from the STUMPS logic BIST architecture [Bardell 1982]. The first compression capability implemented, OPMISR [Barnhart 2001], was released in late 2000; it included just the output compression of a MISR. The enhanced version of OPMISR, called OPMISR+ [Barnhart 2002], appeared a year later, adding space compaction (also part of the eventual SmartBIST) so that when broadcast scan was used, the composite MISR signatures could be compared in a single tester cycle without requiring too many pins to do so.

The general scan architecture for OPMISR+ is shown in Figure 6.33. By fanning out each scan input to multiple internal scan chains, it is possible to support many more scan chains than scan pins. The scan chains scan out into a set of MISRs that in aggregate create a signature for the whole design. Normally this signature is checked after each test and then reset, but it is also possible to accumulate the signature from many or all tests and check just the final signature for a go/no-go assessment of each chip, similar to a logic BIST approach but with high fault coverage and no need for test points to be inserted in the functional paths.

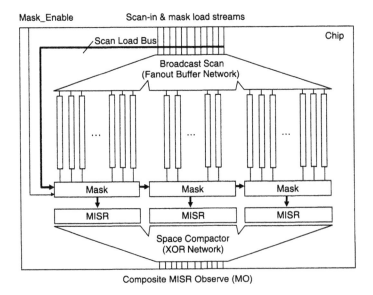

■ **FIGURE 6.33**

OPMISR+ architecture.

The architecture shown in Figure 6.33 shows the composite MISR signature being visible at **MISR Observe** (MO) pins through a space compactor. These MO pins can be shared with the **scan-input** (SI) pins to allow testing on reduced-pin testers. If the composite signatures are read out only at the end of each test, there is very little bandwidth needed for the MO pins, so they do not need to consume valuable tester scan pin resources; this allows the limited set of tester scan pins to be all dedicated to supplying input stimulus. For example, if a tester supports 32 scan pins, these would typically support up to 16 scan chains, with 2 tester scan pins attached to each chip scan chain. Because there is no need for any scan output pins, all 32 scan pins can be used to load scan data into the device, doubling the bandwidth for loading each test; with no output drivers switching during scan, power and noise during scan are also reduced. Doubling the number of scan-in pins in use helps to mitigate any potential issues associated with scan correlation because fewer chains have to share the same scan-in pin; even so, it is best to avoid fanout to scan chains in the same physical locality of the design to reduce the chance of correlations causing a problem. Also, like logic BIST, the signatures can be observed serially (not shown in Figure 6.33) instead of in parallel through the space compactor, which reduces the compression overhead at the cost of increased time to observe the signatures.

Signatures become corrupted if they ever capture an unknown or unpredictable (X) response value, so it is required that either the design be free of all such unpredictability or that some means be provided to keep these values from corrupting the MISR signatures. As Figure 6.33 shows, mask registers and associated logic between the scan chain outputs and the MISRs can be used to eliminate these unknowns. The mask registers can be loaded using the full bandwidth provided by the scan input pins, and one or more Mask_Enable signals select between no masking and use of one of the mask registers on each scan cycle. Each Mask_Enable consumes a scan pin tester resource because they may change value on each scan cycle.

One additional capability made useful by having no scan output streams is to utilize certain testers' capability to repeat when the data values on all pins (both stimulus and response) repeat on consecutive tester cycles. Filling the "don't care" bits in the scan-in stream by repeating the previous or next care bit value for each scan-in pin has shown that application of a simple run-length encoding provides an additional reduction in test data volume in the tester scan buffer.

6.4.2 Embedded Deterministic Test[2]

TestKompress® is the first commercially available on-chip test compression product and was introduced by Mentor Graphics [Mentor 2006] in 2001. It uses the **embedded deterministic test** (EDT) technology [Rajski 2002] [Rajski 2004] shown in Figure 6.34. The EDT architecture consists of an on-chip decompressor located between the external scan input ports and the internal scan chains, as well as an on-chip selective compactor inserted between the internal scan chains and the external scan output ports.

---

[2] Contributed by Janusz Rajski.

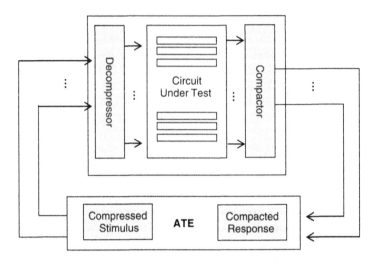

■ **FIGURE 6.34**

EDT (TestKompress®) architecture.

Because the decompressor determines the effectiveness of TestKompress stimuli compression, it was designed to achieve high compression ratio, very high speed of operation, very low silicon area, and high modularity. The decompressor, as shown in Figure 6.35, performs continuous flow decompression; that is, it has the ability to receive new information as the data are being decompressed and loaded to the scan chains. This property reduces dramatically the hardware overhead. The first silicon with EDT [Rajski 2002] used only a 20-bit ring generator with 5 injectors (external scan input ports) but was able to encode over 2000 positions in the scan chains. Conventional reseeding would require an LFSR of length 2000 and a shadow register to match that encoding capacity in this case. The sequential design of the decompressor provides a buffering function that enables sharing of

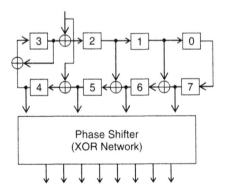

■ **FIGURE 6.35**

Ring generator.

information between shift cycles that have very different numbers of specified bits. The decompressor can operate in an overdrive mode [Rajski 2004] where the stimuli of the input channels stay constant for a number of shift cycles. In this case, the ratio of volume compression is not limited by the ratio of internal scan chains to external channels.

The compression algorithm is tightly integrated with the dynamic compaction of the ATPG engine. The linear equation solver works iteratively with ATPG to maximize compression. Every time ATPG generates a test cube for a new fault, the solver is invoked to compress it. As long as the solver can compress a test cube, the ATPG algorithm attempts to target additional faults and specify more bits. The solver operates incrementally. In every iteration, the system of linear equations gradually increases.

The TestKompress compaction scheme, shown in Figure 6.36, is designed to preserve fault coverage. It provides the ability to deterministically handle X-values propagating to scan cells, eliminate aliasing effects completely, and support scan chain and combinational logic diagnosis. It comprises a number of space compactors driven by outputs of selected scan chains. While the space compactors are essentially XOR trees, they are not necessarily combinational circuits. If the propagation delay through the XOR tree becomes unacceptable with respect to the shift frequency, the XOR tree can be pipelined to allow faster operation.

A distinct feature of the selective compactor is its ability to selectively mask some scan chains to ensure detection of the captured fault effects on other scan chains. This feature is implemented by gating logic that is capable of forcing some scan chain outputs to 0 while allowing data stored in other scan chains to pass through the compactor. The gating logic is controlled by a decoder driven by a select register loaded by the decompressor. The compactor guarantees observability of any scan cell regardless of the number and configuration of X-values. This functionality is essential in achieving very high fault coverage in designs with X-values. It is

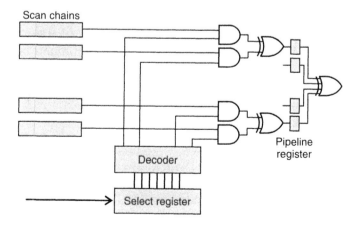

■ **FIGURE 6.36**

Selective compactor.

especially important for at-speed testing of designs with false and multiple-cycle paths.

In addition to traditional fault models, TestKompress supports bridging faults extracted from layout as well as a wide range of fault models and functionality needed for at-speed testing. That includes transition faults and path-delay faults with an on-chip *phase-locked loop* (PLL) controlled launch and capture, small delay defects with timing, and analysis of false and multiple-cycle paths defined by design constraints. TestKompress also provides support for direct combinational logic and scan chain diagnosis from fail log data for compressed patterns [Leininger 2005]. This functionality is very useful in high-volume diagnosis performed for yield learning.

6.4.3 VirtualScan and UltraScan[3]

The VirtualScan and UltraScan test data volume and test application time reduction solutions were introduced by SynTest in 2002 and 2005, respectively [SynTest 2006]. VirtualScan [Wang 2004] was the first commercial product based on the *broadcast scan scheme* using combinational logic for pattern decompression. The VirtualScan architecture consists of three major parts: (1) a full-scan circuit; (2) a broadcaster with a 1-to-n scan configuration, which is driven by the external scan input ports and which drives the internal scan chain inputs of the full-scan circuit; and (3) a space compactor located between internal scan chain outputs of the full-scan circuit and the external scan output ports. The broadcaster, comprised of a network of combinational logic gates, is used to decompress an input compressed stimulus into decompressed stimulus for driving the scan data into the scan cells of all scan chains. The space compactor, comprised of a network of XOR gates, is used to compact the captured test responses.

Figure 6.37 shows the general architecture of a VirtualScan circuit with a split ratio of four. The broadcaster has a 1-to-4 scan configuration, meaning that the broadcaster is used to split one original scan chain into four shorter balanced scan chains. The broadcaster is used to drive the shorter scan chains by broadcasting the m-bit input compressed stimulus to $4m$-bit decompressed stimulus. This transformation can be implemented using any number of combinational logic gates, including AND, OR, NAND, NOR, MUX, XOR, and XNOR gates as well as buffers and inverters. Because the longest scan chain length is reduced by four times, this places a maximum limit on the maximum test data volume and test cycle reduction that can be achieved. Due to the stronger ATPG constraints introduced by the broadcaster, the actual reduction ratio achieved for a split ratio of four would typically be less than four. However, when required, additional logic provided by extra *VirtualScan inputs* added to the broadcaster can be used to further reduce or remove any fault coverage loss caused by test compression.

UltraScan [Wang 2005b] is an extended version of VirtualScan. The UltraScan circuit consists of three major parts: (1) a VirtualScan circuit, (2) a *time-division*

---

[3] Contributed by Laung-Terng Wang.

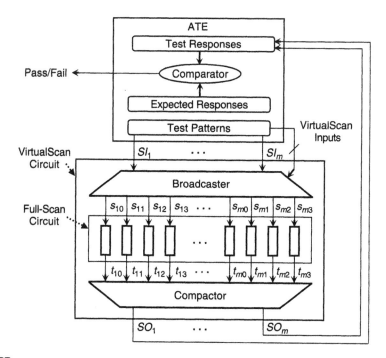

■ FIGURE 6.37

VirtualScan architecture.

demultiplexer (TDDM) placed between the external scan input ports and the internal VirtualScan inputs, and (3) a *time-division multiplexer* (TDM) placed between the internal VirtualScan outputs and the external scan output ports. It relies on the fact that the frequency at which I/O pads are operated is typically much higher than the frequency at which the scan chains are operated. By matching the bandwidth difference between the I/O pad frequency and the scan chain shift clock frequency, one can easily reduce the test application time by a factor that is determined by dividing the frequency of the I/O pads by the frequency of the scan chains [Khoche 2002]. In general, the UltraScan technology can be applied to other test compression solutions as well.

Figure 6.38 shows the general UltraScan architecture using the VirtualScan circuit with a split ratio of four. Surrounding the VirtualScan circuit, a time-division demultiplexer and a time-division multiplexer (TDDM/TDM) pair have been added, as well as a clock controller to create the UltraScan circuit. The TDDM and TDM pair can be built out of combinational circuits such as multiplexers and demultiplexers or sequential circuits such as shift-registers for bandwidth matching [Khoche 2002] [Wang 2005b].

In this UltraScan circuit, a small number of high-speed input pads, typically 16 to 32, are used as external scan input ports, which are connected to the inputs of the TDDM circuit. The TDDM circuit uses a high-speed clock, provided externally or generated internally using a phase-locked loop, to demultiplex the *high-speed*

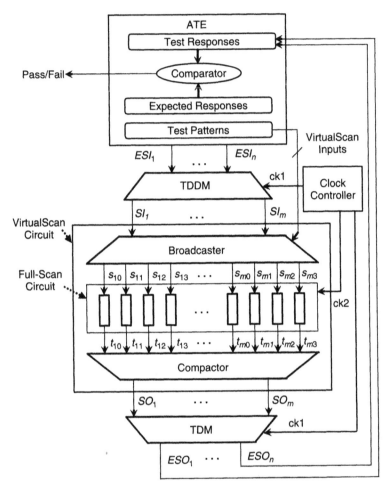

■ **FIGURE 6.38**

UltraScan architecture.

compressed stimuli into compressed stimuli operating at a slower data rate for scan shift. Similarly, the TDM circuit will use the same high-speed clock to capture and shift out the test responses to high-speed output pads for comparison. The demultiplexing ratio, the ratio between the high-speed data rate and the low-speed data rate, is typically 16, which means that designers can generate 256 to 512 internal scan chains from the external scan I/O ports. The clock controller is used to derive the scan shift clock by dividing the high-speed clock by the demultiplexing ratio. In this example, for a desired scan shift clock frequency of 10 MHz, the external I/O pads are operated at 160 MHz. Note that the TDDM/TDM circuit does not compress test data volume but only reduces test application time or test pin count. It is also possible to use UltraScan to reduce test power using a similar approach as described in [Whetsel 1998] and [Khoche 2002].

6.4.4 Adaptive Scan[4]

Adaptive scan [Sitchinava 2004] is the recent test compression architecture adopted by Synopsys as part of their DFT MAX solution [Synopsys 2006]. This compression solution is designed to be the next-generation scan architecture. To address the need for reduced test data volume and test application time, combinational logic has been added to traditional scan implementation to allow the small input–output interface of scan to be used for a large number of scan chains. ***Multiplexers*** (MUXs) are added on the input side to maintain the simple relationship between scan cells and scan-in values. This allows for a simple upgrade to the highly tuned combinational ATPG algorithms to support the needs for compression. XORs are added on the output side to maintain the high observability of scan chains.

The adaptive scan architecture of DFT MAX is shown in Figure 6.39. The combinational MUX network for stimulus decompression is controlled by select lines that allow mappings available through the data paths of the MUXs to be reconfigured on a per-shift basis. This allows a very large number of scan configurations to be implemented with very low area overhead. An X-tolerant XOR network for response compaction allows for good fault coverage in the place of X's in the test response.

The output compactor used in adaptive scan consists of a network of XOR gates. Unlike conventional compactors, adaptive scan adopts a space compactor that is capable of compacting test responses while tolerating unknown (X) values [Mitra 2004]. This can reduce fault coverage loss caused by X's in the test response. While mainstream designs have few X's, an optional masking control is available on the

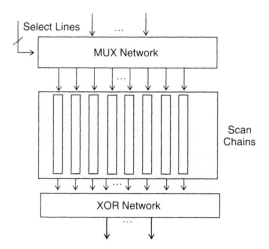

■ **FIGURE 6.39**

Adaptive scan architecture.

---

[4] Contributed by Rohit Kapur.

output side to provide a complete solution to the user when a larger number of X's exist in the test response. The XOR circuitry of adaptive scan is designed to support diagnosis of a high volume of scan pattern failures observed on the tester.

Because adaptive scan adds few combinational gates to existing scan flows, compression can be tightly integrated—and delivered—within the company's flagship products, DFT compiler for one-pass scan synthesis, and TextraMAX for ATPG [Synopsys 2006]. Every test capability that was available in conventional scan is also available in the adaptive scan implementation; for example, PLL support, ATPG compaction support, adjacent-fill, and physical integration are all inherited from the previously available traditional scan implementation.

6.4.5 ETCompression[5]

Finally, LogicVision's deterministic test compression solution, ETCompression, builds upon their *embedded logic test* (ELT) technology [LogicVision 2006]. Figure 6.40 shows the ETCompression architecture. A pseudo-random pattern generator drives the scan chains and has an autonomous (BIST) mode and a reseeding mode. ETCompression can be used with or without support of the autonomous mode. A *multiple-input signature register* (MISR) compresses the scan chain output values in both modes. A run-time programmable X-masking circuit is used to mask unknown (X) values that would corrupt the MISR signature. The input

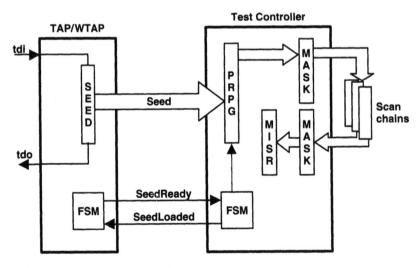

■ **FIGURE 6.40**

ETCompression architecture.

---

[5] Contributed by Benoit Nadeau-Dostie.

mask is used to load constant values in scan chains with hold-time problems. This reduces the number of X-values propagating to other scan chains.

During the reseeding mode, the seed to be used for the next pattern is shifted at low speed (typically 10 to 50 MHz) from the tester to a shadow register in the *test access port* (TAP) or *wrapper TAP* (WTAP) of an embedded block (or core). The TAP and WTAP are implemented according to IEEE 1149.1 and 1500, respectively. Both standards are the main topics in Chapter 10. In the meantime, the current pattern is being decompressed by the PRPG and loaded in scan chains at a frequency which is run time programmable and which is often higher than the tester speed. This makes the PRPG reseeding approach attractive because it does not require shifting test data in and out of the scan chains at the frequency imposed by the tester interface. Using a faster frequency to load scan chains increases throughput and allows operating the circuit at a power level that is representative of the functional mode, which has been shown to be very useful in the characterization of power grids [Nadeau-Dostie 2005].

The transfer of the seed from the shadow register to the PRPG is performed using a simple asynchronous protocol because the frequency and phase of the clocks might not be related. When the PRPG is not decompressing a pattern, the next seed is transferred if it is available (*i.e.*, SeedReady is active). If not, it waits until it is available. The TAP (or WTAP) is then informed that the next seed can be shifted in from the tester (*i.e.*, SeedLoaded is active). The two signals are then reset and the process repeats as many times as there are seeds.

The clocking used for both modes contributes a lot to the level of test compression achievable beyond the calculation of seeds from test cubes. First, a *launch-on-shift* (or *skewed-load*) approach is used which has been shown to require up to an order of magnitude fewer patterns to achieve the same transition fault coverage [Jayaram 2003]. The scan-enable signal is pipelined locally to each domain to facilitate timing closure. Second, all multiple-cycle paths and cross-domain logic are tested concurrently so there is no need to rerun patterns with different clock edge placement and masking configurations. This is done in a such a way that both fault simulation and test generation are purely combinational to minimize run time. These techniques are explained in [Nadeau-Dostie 2000].

During the capture phase, all functional clocks are enabled to produce a burst of five clock cycles. The burst is long enough to make sure that the supply has time to stabilize before the launch and capture cycles [Rearick 2005]. For each clock domain, the clock burst is configurable at run time to mimic the functional mode of operation from a timing and power point of view. This is essential to catch subtle problems related to crosstalk or IR drop, for example, as explained in [Nadeau-Dostie 2005]. The alignment of synchronous clock domains is preserved.

In order to further improve test compression efficiency, ETCompression supports test point insertion and the hierarchical test approach described in [Pateras 2003]. Test points are inserted in a nonoptimized gate-level representation of the circuit using the algorithm proposed in [Seiss 1991]. Layout tools are now capable of restructuring the logic and eliminating any timing impact. The hierarchical test approach allows the use of the deterministic mode only on a few problematic blocks while other blocks are tested in autonomous mode. The approach allows the use of

TABLE 6.6 ■ Summary of Industry Practices

Industry Practice	Stimulus Decompressor	Response Compactor
OPMISR+	Broadcast scan (Illinois scan)	MISR with XOR network
TestKompress	Ring generator	XOR network
VirtualScan	Combinational logic network	XOR network
DFT MAX	Combinational MUX network	XOR network
ETCompression	(Reseeding) PRPG	MISR
UltraScan	TDDM	TDM

Note: MISR, multiple-input signature register; MUX, multiplexer; PRPG, pseudo-random pattern generator; TDDM, time-division demultiplexer; TDM, time-division multiplexer; XOR, exclusive-OR.

functional flip-flops to provide isolation of the core. These flip-flops are then used in both internal and external testing of the core and allow at-speed testing of the interface with the rest of the circuit.

6.4.6 Summary

A summary of the different compression architectures used in the commercial products is shown in Table 6.6. It can be seen that the solutions offered by the current EDA DFT vendors are quite diverse on stimulus decompression and response compaction. For stimulus decompression, OPMISR+, VirtualScan, and DFT MAX are broadcast scan based, while TestKompress and ETCompression are linear decompression based. For response compaction, OPMISR+ and ETCompression include MISRs, while other solutions purely adopt (X-tolerant) XOR networks. For at-speed delay testing, ETCompression uses the launch-on-shift (or skewed-load) approach for ATPG, while other solutions support launch-on-capture (or double-capture). The UltraScan TDDM/TDM architecture can be implemented on top of any test compression solution to further reduce test application time and test pin count.

6.5 CONCLUDING REMARKS

Test compression is an effective method for reducing test data volume and test application time with relatively small cost. Due to these advantages, test compression is beginning to be adopted in different industrial designs. Many EDA vendors have released first- and even second-generation tools for test compression and integrated it successfully as part of the design flow. Test compression has proven to be easy to implement and capable of producing high-quality tests and has been demonstrated to be an efficient test structure for embedded hard cores. This has allowed test compression to become more widely accepted than logic BIST. While code-based test compression schemes produce good results, at present the industry seems to favor solutions based on broadcast scan and linear decompression.

One remaining issue for test compression is standardization. Currently, different vendors have proposed their own proprietary solutions, which prevent users from utilizing different ATPG compression software with different compression architectures. Fortunately, a working group is now being organized by the IEEE to address this problem.

6.6 EXERCISES

6.1 **(Dictionary Coding)** For the given test data, T_D = 0000 0110 0000 0000 0 100 0000 0001 1100 0000 0100. If it is partitioned into 4-bit symbols, how many entries would be required in a complete dictionary? What would be the compression ratio using the complete dictionary?

6.2 **(Golomb Coding)** For the given test data, T_D = 00000110000000000 1000000000001110000000100. If a Golomb code with $m = 4$ is used for compression, show the compressed test data T_E and the compression ratio.

6.3 **(Compatibility Analysis)** Given two definitions:

   a. **Incompatible**—For a scan chain segment S_i, define $S_i[q]$ as the value of the qth scan cell in S_i. Two scan chain segments S_i and S_j are said to be incompatible, if $\exists q(1 \le q \le T)$ such that $S_i[q] \oplus S_j[q] = 1$, where T is the largest number of scan cells in both scan chain segments.

   b. **CI-graph $G(V, E)$**—Assuming that each node in a graph V represents a scan chain segment, a CI-graph $G(V, E)$ is constructed by associating an edge E between any two nodes whose values, V_i and V_j, are incompatible.

If there are eight scan chains, each containing five scan cells, then for the following test pattern, construct the corresponding CI-graph:

	1st	2nd	3rd	4th	5th
Chain 1	1	X	1	X	X
Chain 2	X	X	0	X	1
Chain 3	X	X	X	X	1
Chain 4	1	1	X	X	0
Chain 5	0	X	1	X	X
Chain 6	X	0	X	1	X
Chain 7	0	X	0	X	X
Chain 8	X	X	1	X	X

(X: don't-care bit)

6.4 **(Linear Decompressor)** What is the characteristic A matrix for the sequential linear decompressor shown below such that $AX = Z$?

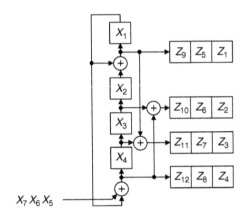

6.5 **(Linear Decompressor)** For the sequential linear decompressor shown in Figure 6.9 whose corresponding system of linear equations is shown in Figure 6.10, find the compressed stimulus $X_1 - X_{10}$ necessary to encode the following test cube: $<Z_1, \ldots, Z_{12}> = <0\text{-}\text{-}\text{-}1\text{-}0\text{-}\text{-}010>$.

6.6 **(X-Compactor)** For the X-compact matrix of the compactor shown below, design the corresponding X-compactor. What errors can the X-compactor detect?

$$\begin{bmatrix} 0 & 1 & 1 & 1 & 0 \\ 0 & 1 & 0 & 1 & 1 \\ 1 & 1 & 0 & 0 & 1 \\ 1 & 1 & 0 & 1 & 0 \\ 1 & 0 & 1 & 0 & 1 \\ 1 & 0 & 0 & 1 & 1 \\ 1 & 0 & 1 & 1 & 0 \\ 0 & 0 & 1 & 1 & 1 \end{bmatrix}$$

6.7 **(X-Compactor)** Prove the X-tolerant theorem of the X-compactor given in Section 6.3.3.

6.8 **(A Design Practice)** Use the VirtualScan program and user's manuals provided online to split the ISCAS s38584 design [ISCAS 1989] into 4, 8, and 16 scan chains. Calculate the fault coverage loss in each case. Then, perform top-up ATPG in each case and report the additional number of test patterns required to uncover the fault coverage loss. Report the actual compression ratio in each case.

Acknowledgments

The authors wish to thank Dr. Erik H. Volkerink of Agilent Technologies for contributing a portion of the Test Response Compaction section; Dr. Laung-Terng

(L.-T.) Wang of SynTest Technologies, Dr. Brion Keller of Cadence Design Systems, Dr. Janusz Rajski of Mentor Graphics, Dr. Rohit Kapur of Synopsys, and Dr. Benoit Nadeau-Dostie of LogicVision for contributing the Industry Practices section; Khader S. Abdel-Hafez of SynTest Technologies for reviewing the manuscript; and Dr. Yinhe Han of the Institute of Computing Technology, Chinese Academy of Sciences, for formatting the text and drawing the figures.

References
R6.0—Books

[Chakrabarty 2002] K. Chakrabarty, V. Iyengar, and A. Chandra, *Test Resource Partitioning for System-on-a-Chip*, Kluwer Academic, Norwell, MA, 2002.

[Nadeau-Dostie 2000] B. Nadeau-Dostie, *Design for At-Speed Test, Diagnosis, and Measurement*, Springer, Boston, MA, 2000.

R6.1—Introduction

[Khoche 2000] A. Khoche and J. Rivoir, I/O bandwidth bottleneck for test: Is it real?, in *Proc. Int. Workshop on Test Resource Partitioning*, Paper 2.3, Atlantic City, NJ, 2000, 6 pp.

R6.2—Test Stimulus Compression

[Bayraktaroglu 2001] I. Bayraktaroglu and A. Orailoglu, Test volume and application time reduction through scan chain concealment, in *Proc. Design Automation Conf.*, June 2001, pp. 151–155.

[Bayraktaroglu 2003] I. Bayraktaroglu and A. Orailoglu, Concurrent application of compaction and compression for test time and data volume reduction in scan designs, *IEEE Trans. Comput.*, 52(11), 1480–1489, 2003.

[Chandra 2001] A. Chandra and K. Chakrabarty, System-on-a-chip test-data compression and decompression architectures based on Golomb codes, *IEEE Trans. Comput.-Aided Des.*, 20(3), 355–368, 2001.

[Chandra 2003] A. Chandra and K. Chakrabarty, Test data compression and test resource partitioning for system-on-a-chip using frequency-directed run-length (FDR) codes, *IEEE Trans. Comput.*, 52(8), 1076–1088, 2003.

[Chen 1986] C. L. Chen, Linear dependencies in linear feedback shift registers, *IEEE Trans. Comput.*, C-35(12), 1086–1088, 1986.

[Cullen 1997] C. G. Cullen, *Linear Algebra with Applications*, Addison-Wesley, Boston, MA, 1997.

[Gonciari 2003] P. T. Gonciari, B. M. Al-Hashimi, and N. Nicolici, Variable-length input Huffman coding for system-on-a-chip test, *IEEE Trans. Comput.-Aided Des.*, 22(6), 783–796, 2003.

[Hamzaoglu 1999] I. Hamzaoglu and J. H. Patel, Reducing test application time for full scan embedded cores, in *Proc. Int. Symp. on Fault-Tolerant Computing*, July 1999, pp. 260–267.

[Han 2005a] Y. Han, S. Swaminathan, Y. Hu, A. Chandra, and X. Li, Scan data volume reduction using periodically alterable MUXs decompressor, in *Proc. 14th IEEE Asian Test Symp.*, November 2005, pp. 372–377.

[Han 2005b] Y. Han, Y. Hu, X. Li, H. Li, A. Chandra, and X. Wen, Wrapper scan chains design for rapid and low power testing of embedded cores, *IEICE Trans. Inform. Syst.*, E88–D(9), 2126–2134, 2005.

[Hsu 2001] F. F. Hsu, K. M. Butler, and J. H. Patel, A case study on the implementation of Illinois scan architecture, in *Proc. IEEE Int. Test Conf.*, October 2001, pp. 538–547.

[Huffman 1952] D. A. Huffman, A method for the construction of minimum redundancy codes, in *Proc. IRE*, 40(9), 1098–1101, 1952.

[Ichihara 2000] H. Ichihara, K. Kinoshita, I. Pomeranz, and S. M. Reddy, Test transformation to improve compaction by statistical encoding, in *Proc. of VLSI Design*, January 2000, pp. 294–299.

[Jas 1998] A. Jas and N. A. Touba, Test vector compression via cyclical scan chains and its application to testing core-based designs, in *Proc. IEEE Int. Test Conf.*, October 1998, pp. 458–464.

[Jas 2003] A. Jas, J. Ghosh-Dastidar, M. Ng, and N. A. Touba, An efficient test vector compression scheme using selective Huffman coding, *IEEE Trans. Comput.-Aided Des.*, 22(6), 797–806, 2003.

[Knieser 2003] M. Knieser, F. Wolff, C. Papachristou, D. Weyer, and D. McIntyre, A technique for high ratio LZW compression, in *Proc. Design, Automation, and Test in Europe*, March 2003, pp. 116–121.

[Könemann 1991] B. Könemann, LFSR-coded test patterns for scan designs, in *Proc. European Test Conf.*, April 1991, pp. 237–242.

[Könemann 2001] B. Könemann, C. Barnhart, B. Keller, T. Snethen, O. Farnsworth, and D. Wheater, A SmartBIST variant with guaranteed encoding, in *Proc. Asian Test Symp.*, November 2001, pp. 325–330.

[Könemann 2003] B. Köenemann, C. Barnhart, and B. Keller, Real-Time Decoder for Scan Test Patterns, U.S. Patent No. 6,611,933, August 26, 2003.

[Krishna 2001] C. V. Krishna, A. Jas, and N. A. Touba, Test vector encoding using partial LFSR reseeding, in *Proc. IEEE Int. Test Conf.*, October 2001, pp. 885–893.

[Krishna 2002] C. V. Krishna, A. Jas, and N. A. Touba, Reducing test data volume using LFSR reseeding with seed compression, in *Proc. IEEE Int. Test Conf.*, October 2002, pp. 321–330.

[Krishna 2003] C. V. Krishna and N. A. Touba, Adjustable width linear combinational scan vector decompression, in *Proc. Int. Conf. on Computer-Aided Design*, September 2003, pp. 863–866.

[Krishna 2004] C. V. Krishna and N. A. Touba, 3-stage variable length continuous-flow scan vector decompression scheme, in *Proc. VLSI Test Symp.*, April 2004, pp. 79–86.

[Lee 1998] K.-J. Lee, J. J. Chen, and C. H. Huang, Using a single input to support multiple scan chains, in *Proc. Int. Conf. on Computer-Aided Design*, November 1998, pp. 74–78.

[Lee 1999] K.-J. Lee, J. J. Chen, and C. H. Huang, Broadcasting test patterns to multiple circuits, *IEEE Trans. Comput.-Aided Des.*, 18(12), 1793–1802, 1999.

[Li 2003] L. Li, K. Chakrabarty, and N. A. Touba, Test data compression using dictionaries with selective entries and fixed-length indices, *ACM Trans. Design Autom. Electr. Syst.*, 8(4), 470–490, 2003.

[Li 2004] L. Li and K. Chakrabarty, Test set embedding for deterministic BIST using a reconfigurable interconnection network, *IEEE Trans. Comput.-Aided Des.*, 23(9), 1289–1305, 2004.

[Li 2005] L. Li, K. Chakrabarty, S. Kajihara, and S. Swaminathan, Efficient space/time compression to reduce test data volume and testing time for IP cores, in *Proc. VLSI Design*, January 2005, pp. 53–58.

[Mrugalski 2004] G. Mrugalski, J. Rajski, and J. Tyszer, Ring generators: New devices for embedded test applications, *IEEE Trans. Comput.-Aided Des.*, 23(9), 1306–1320, 2004.

[Pandey 2002] A. R. Pandey and J. H. Patel, Reconfiguration technique for reducing test time and test volume in Illinois scan architecture based designs, in *Proc. IEEE VLSI Test Symp.*, April 2002, pp. 9–15.

[Rajski 1998] J. Rajski, J. Tyszer, and N. Zacharia, Test data decompression for multiple scan designs with boundary scan, *IEEE Trans. Comput.*, 47(11), 1188–1200, 1998.

[Rajski 2004] J. Rajski, J. Tyszer, M. Kassab, and N. Mukherje, Embedded deterministic test, *IEEE Trans. Comput.-Aided Des.*, 23(5), 776–792, 2004.

[Reda 2002] S. Reda and A. Orailoglu, Reducing test application time through test data mutation encoding, in *Proc. Design, Automation, and Test in Europe*, March 2002, pp. 387–393.

[Reddy 2002] S. M. Reddy, K. Miyase, S. Kajihara, and I. Pomeranz, On test data volume reduction for multiple scan chain designs, in *Proc. IEEE VLSI Test Symp.*, April 2002, pp. 103–108.

[Samaranayake 2002] S. Samaranayake, N. Sitchinava, R. Kapur, M. Amin, and T. W. Williams, Dynamic scan: Driving down the cost of test, *IEEE Comput.*, 35(10), 65–70, 2002.

[Samaranayake 2003] S. Samaranayake, E. Gizdarski, N. Sitchinava, F. Neuveux, R. Kapur, and T. W. Williams, A reconfigurable shared scan-in architecture, in *Proc. IEEE VLSI Test Symp.*, April 2003, pp. 9–14.

[Shah 2004] M. A. Shah and J. H. Patel, Enhancement of the Illinois scan architecture for use with multiple scan inputs, in *Proc. Annual Symp. on VLSI*, February 2004, pp. 167–172.

[Sitchinava 2004] N. Sitchinava, S. Samaranayake, R. Kapur, E. Gizdarski, F. Neuveux, and T. W. Williams, Changing the scan enable during shift, in *Proc. IEEE VLSI Test Symp.*, April 2004, pp. 73–78.

[Sun 2004] X. Sun, L. Kinney, and B. Vinnakota, Combining dictionary coding and LFSR reseeding for test data compression, in *Proc. IEEE Design Automation Conf.*, June 2004, pp. 944–947.

[Tehranipoor 2005] M. Tehranipoor, M. Nourani, and K. Chakrabarty, Nine-coded compression technique for testing embedded cores in SoCs, *IEEE Trans. VLSI*, 13(6), 719–731, 2005.

[Volkerink 2002] E. H. Volkerink, A. Khoche, and S. Mitra, Packet-based input test data compression techniques, in *Proc. IEEE Int. Test Conf.*, October 2002, pp. 154–163.

[Volkerink 2003] E. H. Volkerink and S. Mitra, Efficient seed utilization for reseeding based compression, in *Proc. IEEE VLSI Test Symp.*, April 2003, pp. 232–237.

[Wang 2002] L.-T. Wang, H.-P. Wang, X. Wen, M.-C. Lin, S.-H. Lin, D.-C. Yeh, S.-W. Tsai, and K. S. Abdel-Hafez, Method and Apparatus for Broadcasting Scan Patterns in a Scan-Based Integrated Circuit, U.S. Patent Appl. No. 20030154433, January 16, 2002.

[Wang 2004] L.-T. Wang, X. Wen, H. Furukawa, F.-S. Hsu, S.-H. Lin, S.-W. Tsai, K. S. Abdel-Hafez, and S. Wu, VirtualScan: A new compressed scan technology for test cost reduction, in *Proc. IEEE Int. Test Conf.*, October 2004, pp. 916–925.

[Wang 2005a] Z. Wang and K. Chakrabarty, Test data compression for IP embedded cores using selective encoding of scan slices, in *Proc. IEEE Int. Test Conf.*, October 2005, pp. 581–590.

[Ward 2005] S. Ward, C. Schattauer, and N. A. Touba, Using statistical transformations to improve compression for linear decompressors, in *Proc. IEEE Int. Symp. on Defect and Fault Tolerance in VLSI Systems*, October 2005, pp. 43–50.

[Wolff 2002] F. G. Wolff and C. Papachristou, Multiscan-based test compression and hardware decompression using LZ77, in *Proc. IEEE Int. Test Conf.*, October 2002, pp. 331–339.

[Würtenberger 2003] A. Würtenberger, C. S. Tautermann, and S. Hellebrand, A hybrid coding strategy for optimized test data compression, in *Proc. IEEE Int. Test Conf.*, October 2003, pp. 451–459.

[Würtenberger 2004] A. Würtenberger, C. S. Tautermann, and S. Hellebrand, Data compression for multiple scan chains using dictionaries with corrections, in *Proc. IEEE Int. Test Conf.*, October 2004, pp. 926–935.

R6.3—Test Response Compaction

[Barnhart 2002] C. Barnhart, V. Brunkhorst, F. Distler, O. Farnsworth, A. Ferko, B. Keller, D. Scott, B. Könemann, and T. Onodera, Extending OPMISR beyond 10× scan test efficiency, *IEEE Des. Test Comput.*, 19(5), 65–73, 2002.

[Chakrabarty 1998] K. Chakrabarty, B. T. Murray, and J. P. Hayes, Optimal zero-aliasing space compaction of test responses, *IEEE Trans. Comput.*, 47(11), 1171–1187, 1998.

[Das 2003] S. R. Das, M. Sudarma, M. H. Assaf, E. M. Petriu, W.-B. Jone, K. Chakrabarty, and M. Sahinoglu, Parity bit signature in response data compaction and built-in self-testing of VLSI circuits with nonexhaustive test sets, *IEEE Trans. Instrument. Measure.*, 52(5), 1363–1380, 2003.

[Han 2003] Y. Han, Y. Xu, A. Chandra, H. Li, and X. Li, Test resource partitioning based on efficient response compaction for test time and tester channels reduction, in *Proc. Asian Test Symp.*, November 2003, pp. 440–445.

[Han 2005c] Y. Han, Y. Hu, H. Li, and X. Li, Theoretic analysis and enhanced X-tolerance of test response compact based on convolutional code, in *Proc. Asia and South Pacific Design Automation Conf.*, January 2005, pp. 53–58.

[Han 2005d] Y. Han, X. Li, H. Li, and A. Chandra, Test resource partitioning based on efficient response compaction for test time and tester channels reduction, *J. Comput. Sci. Technol.*, 20(2), 201–210, 2005.

[Ivanov 1996] A. Ivanov, B. Tsuji, and Y. Zorian, Programmable space compactors for BIST, *IEEE Trans. Comput.*, 45(12), 1393–1405, 1996.

[Karpovsky 1987] M. Karpovsky and P. Nagvajara, Optimal time and space compression of test response for VLSI devices, in *Proc. IEEE Int. Test Conf.*, October 1987, pp. 523–529.

[Mitra 2004] S. Mitra and K. S. Kim, X-compact: An efficient response compaction technique, *IEEE Trans. Comput.-Aided Des.*, 23(3), 421–432, 2004.

[Naruse 2003] M. Naruse, I. Pomeranz, S. M. Reddy, and S. Kundu, On-chip compression of output responses with unknown values using LFSR reseeding, in *Proc. IEEE Int. Test Conf.*, October 2003, pp. 1060–1068.

[Patel 2003] J. H. Patel, S. S. Lumetta, and S. M. Reddy, Application of Saluja-Karpovsky compactors to test responses with many unkowns. in *Proc. IEEE VLSI Test Symp.*, April 2003, pp. 107–112.

[Pouya 1998] B. Pouya and N. A. Touba, Synthesis of zero-aliasing space elementary-tree space compactors, in *Proc. IEEE VLSI Test Symp.*, April 1998, pp. 70–77.

[Rajski 2005] J. Rajski, J. Tyszer, C. Wang, and S. M. Reddy, Finite memory test response compactors for embedded test applications, *IEEE Trans. Comput.-Aided Des.*, 24(4), 622–634, 2005.

[Saluja 1983] K. K. Saluja and M. Karpovsky, Test compression hardware through data compression in space and time, in *Proc. IEEE Int. Test Conf.*, October 1983, pp. 83–88.

[Sinanoglu 2003] O. Sinanoglu and A. Orailoglu, Compacting test responses for deeply embedded SoC cores, *IEEE Design Test Comput.*, 20(4), 22–30, 2003.

[Volkerink 2005] E. H. Volkerink and S. Mitra, Response compaction with any number of unknowns using a new LFSR architecture, in *Proc. Design Automation Conf.*, June 2005, pp. 117–122.

[Wang 2003] C. Wang, S. M. Reddy, I. Pomeranz, J. Rajski, and J. Tyszer, On compacting test response data containing unknown values, in *Proc. Int. Conf. on Computer-Aided Design*, September 2003, pp. 855–862.

[Wang 2004] L.-T. Wang, X. Wen, H. Furukawa, F.-S. Hsu, S.-H. Lin, S.-W. Tsai, K. S. Abdel-Hafez, and S. Wu, VirtualScan: A new compressed scan technology for test cost reduction, in *Proc. IEEE Int. Test Conf.*, October 2004, pp. 916–925.

[Williams 1989] T. W. Williams and W. Daehn, Aliasing errors in multiple input signature analysis registers, in *Proc. European Test Conf.*, April 1989, pp. 338–345.

[Wohl 2001] P. Wohl, J. A. Waicukauski, and T. W. Williams, Design of compactors for signature-analyzers in built-in self-test, in *Proc. IEEE Int. Test Conf.*, October 2001, pp. 54–63.

[Wohl 2003a] P. Wohl and L. Huisman, Analysis and design of optimal combinational compactors, in *Proc. IEEE VLSI Test Symp.*, April 2003, pp. 101–106.

[Wohl 2004] P. Wohl, J. A. Waicukauski, and S. Patel, Scalable selector architecture for X-tolerant deterministic BIST, in *Proc. Design Automation Conf.*, June 2004, pp. 934–939.

R6.4—Industry Practices

[Bardell 1982] P. H. Bardell and W. H. McAnney, Self-testing of multi-chip logic modules, in *Proc. IEEE Int. Test Conf.*, October 1982, pp. 200–204.

[Barnhart 2000] C. Barnhart, B. Keller, and B. Könemann, Logic DFT and test resource partitioning for 100M gate ASICs (parts I, II and III), presented at the Test Resource Partitioning Workshop, Session 4, presentations 1, 2, and 3 (no hardcopy), 2000.

[Barnhart 2001] C. Barnhart, V. Burnkhorst, F. Distler, O. Farnsworth, B. Keller, and B. Könemann, OPMISR: The foundation for compressed ATPG vectors, in *Proc. IEEE Int. Test Conf.*, October 2001, pp. 748–757.

[Barnhart 2002] C. Barnhart, V. Brunkhorst, F. Distler, O. Farnsworth, A. Ferko, B. Keller, D. Scott, B. Könemann, and T. Onodera, Extending OPMISR beyond 10× scan test efficiency, *IEEE Design Test Comput.*, 19(5), 65–73, 2002.

[Cadence 2006] Cadence Design Systems (http://www.cadence.com), 2006.

[Chickermane 2004] V. Chickermane, B. Foutz, and B. Keller, Channel masking synthesis for efficient on-chip test compression, in *Proc. IEEE Int. Test Conf.*, October 2004, pp. 452–461.

[Jayaram 2003] V. B. Jayaram, J. Saxena, and K. M. Butler, Scan-based transition-fault test can do job, *EE Times*, October 24, 2003, pp. 60–66.

[Khoche 2002] A. Khoche, Test resource partitioning for scan architectures using bandwidth matching, *IEEE Test Resource Partitioning Workshop*, October 2002, pp. 1.4-1–1.4-8.

[Könemann 2001] B. Könemann, C. Barnhart, B. Keller, T. Snethen, O. Farnsworth, and D. Wheater, A SmartBIST variant with guaranteed encoding, in *Proc. Asian Test Symp.*, November 2001, pp. 325–330.

[Könemann 1991] B. Könemann, LFSR-coded test patterns for scan designs, in *Proc. European Test Conf.*, April 1991, pp. 237–242.

[Leininger 2005] A. Leininger, P. Muhmenthaler, W.-T. Cheng, N. Tamarapalli, W. Yang, and H. Tsai, Compression mode diagnosis enables high volume manitoring diagnosis flow, in *Proc. IEEE Int. Test Conf.*, Paper 7.3, October 2005, 10 pp.

[LogicVision 2006] LogicVision (http://www.logicvision.com), 2006.

[Mentor 2006] Mentor Graphics (http://www.mentor.com), 2006.

[Mitra 2004] S. Mitra and K. S. Kim, X-compact: An efficient response compaction technique, *IEEE Trans. Comput.-Aided Des.*, 23(3), 421–432, 2004.

[Nadeau-Dostie 2005] B. Nadeau-Dostie, J.-F. Côté, and F. Maamari, Structural test with functional characteristics, in *Proc. Int. Workshop on Current and Defect-Based Testing Workshop*, May 2005, pp. 57–60.

[Pateras 2003] S. Pateras, Achieving at-speed structural test, *IEEE Design Test Comput.*, 20(5), 26–33, 2003.

[Rajski 2002] J. Rajski, J. Tyszer, M. Kassab, N. Mukherjee, R. Thompson, Kun-Han Tsai, A. Hertwig, N. Tamarapalli, G. Mrugalski, G. Eide, and J. Qian, Embedded deterministic test for low cost manufacturing test, in *Proc. IEEE Int. Test Conf.*, October 2002, pp. 301–310.

[Rajski 2004] J. Rajski, J. Tyszer, M. Kassab, and N. Mukherjee, Embedded deterministic test, *IEEE Trans. Comput.-Aided Des.*, 23(5), 776–792, 2004.

[Rearick 2005] J. Rearick and R. Rodgers, Calibrating clock stretch during AC-scan test, in *Proc. IEEE Int. Test Conf.*, Paper 11.3, October 2005, 8 pp.

[Seiss 1991] B. H. Seiss, P. Trouborst, and M. Schulz, Test point insertion for scan-based BIST, in *Proc. European Test Conf.*, April 1991, pp. 253–262.

[Sitchinava 2004] N. Sitchinava, S. Samaranayake, R. Kapur, E. Gizdarski, F. Neuveux, and T. W. Williams, Changing the scan enable during shift, in *Proc. IEEE VLSI Test Symp.*, April 2004, pp. 73–78.

[Synopsys 2006] Synopsys (http://www.synopsys.com), 2006.

[SynTest 2006] SynTest Technologies (http://www.syntest.com), 2006.

[Wang 2004] L.-T. Wang, X. Wen, H. Furukawa, F.-S. Hsu, S.-H. Lin, S.-W. Tsai, K. S. Abdel-Hafez, and S. Wu, VirtualScan: A new compressed scan technology for test cost reduction, in *Proc. IEEE Int. Test Conf.*, October 2004, pp. 916–925.

[Wang 2005b] L.-T. Wang, K. S. Abdel-Hafez, X. Wen, B. Sheu, S. Wu, S.-H. Lin, and M.-T. Chang, UltraScan: Using time-division demultiplexing/multiplexing (TDDM/TDM) with VirtualScan for test cost reduction, in *Proc. IEEE Int. Test Conf.*, Paper 36.4, November 2005, 8 pp.

[Whetsel 1998] L. Whetsel, Core test connectivity, communication, and control, in *Proc. IEEE Int. Test Conf.*, October 1998, pp. 303–312.

[Wohl 2003b] P. Wohl, J. A. Waicukauski, S. Patel, and M. B. Amin, Efficient compression and application of deterministic patterns in a logic BIST architecture, in *Proc. Design Automation Conf.*, June 2003, pp. 566–569.

CHAPTER 7

Logic Diagnosis

Shi-Yu Huang
National Tsing Hua University, Hsinchu, Taiwan

ABOUT THIS CHAPTER

Given a logic circuit that fails a test, logic diagnosis is the process of narrowing down the possible locations of the defect. By reducing the candidate locations down to possibly only a few, subsequent physical failure analysis becomes much faster and easier when searching for the root causes of failure. For ***integrated circuit*** (IC) products, logic diagnosis is crucial in order to ramp up the manufacturing yield and in some cases to reduce the product debug time as well. This chapter begins by introducing the basic concepts of logic diagnosis. We then review the diagnosis techniques for combinational logic, scan chains, and logic ***built-in self-test*** (BIST). For combinational logic, it is assumed that a fault-free scan chain is in place to assist the diagnosis process. The two most commonly used paradigms—namely, **cause–effect analysis** and **effect–cause analysis**—along with their variants are introduced. Next, we describe three different methods for diagnosing faults within the scan chains, including hardware-assisted, modified inject-and-evaluate, and signal-profiling-based methods. Finally, we discuss the challenges of diagnosis in a logic BIST environment.

7.1 INTRODUCTION

During the IC design and manufacturing cycle, a manufacturing test screens out the bad chips. Diagnosis is used to find out why the bad chips failed, which is especially important when the yield is low or when a customer returns a failed chip. Typically, a successful IC product goes through two manufacturing stages: (1) prototype or **pilot-run** stage, and (2) high-volume manufacturing stage.

During the prototype stage, a small number of samples are produced to validate the functionality on the tester and on demo/prototype boards. During this stage, the prototype samples could fail badly due to design bugs or unstable manufacturing processes. Some of the reasons for this include the following:

- **Misunderstandings about the functionality**. A complex product is generally defined or built by multiple engineers. Because the specifications are usually written in English, there can be ambiguities, inconsistencies, and contradictions (as is true for anything that is created by humans). The actual *hardware description language* (HDL) model (*register-transfer level* [RTL] code) or gate-level netlist may not conform to the desired specification under certain scenarios or the specification may simply have been misinterpreted. Functional test generation and simulation are very time consuming. Designers may not be able to verify their designs comprehensively before the tape-out; however, a rigorous functional verification methodology should be able to reduce the probability of this type of failure.

- **Timing failure and circuit marginality issues**. Fabricated silicon may not execute as fast as what is expected based on timing simulations, or it may not operate properly at certain supply voltages or temperatures. The mismatch between simulated behavior and actual behavior can be due to the inaccuracy of the tools or *signal integrity* (SI) effects that were not considered appropriately. For example, the voltage drop due to power grid resistance, coupling effects among signals, and other effects could slow down the actual operating speed of specific circuits within a chip. Such a timing failure requires the identification of the failing segments or paths of a circuit to guide the circuit optimization before respinning the design.

- **Inappropriate layout design**. For advanced nanometer technologies, the actual geometries of the devices and interconnecting wires fabricated on silicon will deviate from the drawn layout. This is due to optical effects during the lithography process (using light that has a much longer wavelength than the geometries that have to be printed). These mismatches give rise to potential shorts or opens, thereby leading to circuit failure. In light of this, certain *design-for-manufacturability* (DFM) rules could be added to the design rule set to ensure improved manufacturability; nevertheless, this is a slow learning process for each new technology generation. Diagnosis is required to shed some light on the layout patterns that may cause these types of failures in the early stages.

During the design validation stage, the same failure may appear repeatedly for a high percentage of the prototype chips. One can usually determine whether it is a functionality failure or timing failure by turning down the clock speed. Sometimes, certain circuit marginality issues also can be exposed by changes in the supply voltage or temperature, etc. Then, the diagnosis (or debugging) process is initiated by a joint team that includes designers, layout editors, testing engineers, and perhaps process engineers.

After a design has passed the prototype stage, bugs and circuit marginality issues are mostly resolved and the product can ramp up to high-volume production. During this ramp-up stage, the yield could be low or fluctuating. Yield improvement is necessary, and it can be accomplished by tuning the fabrication process. Even when a product reaches the peak run rate, the manufacturing yield could still fluctuate from one wafer to another. Continuous yield monitoring is necessary from time to time to respond to any unexpected low-yield situations [Khare 1995]. At this stage, the chip failures are more or less due to manufacturing imperfections, some of which are catastrophic (*e.g.*, shorts and opens) and some parametric due to process variations. Published results for failure analysis have revealed a number of common defect mechanisms, including via misalignment, via/contact voiding, missing region of interconnecting metal (often called *mouse-bites*), oxide breakdown, and shorts between the drain and source of a transistor [Segura 2002].

For yield improvement, yield/failure analysis engineers must actually inspect the silicon by all means available (so they can identify the failure mechanisms and figure out ways to rectify them); such methods would include etching away certain layers, imaging the silicon surface by ***scanning electronic microscopy*** (SEM) or ***focused ion beam*** (FIB) systems.[1] The silicon de-layering and imaging process is often laborious and time consuming. In a chip with millions of transistors, such a process is doomed to fail if not guided by a good diagnosis tool.

The problem of diagnosis is illustrated in Figure 7.1, where we compare the behavior of a fault-free model (which is a gate-level circuit or a transistor schematic) with a failing chip. The fault-free model will be referred to as the ***circuit under diagnosis*** (CUD). Under the application of certain test patterns, the failing chip and CUD produce different responses at certain primary outputs. Similar to testing, we need to incorporate design-for-testability circuitry to reduce the complexity of diagnosis to a manageable level. For example, the interface signals between the logic

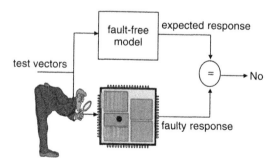

Question: Where are the defect locations?

■ **FIGURE 7.1**

The problem of logic diagnosis.

---

[1] The usage of **FIB** has been widespread in the semiconductor industry. It provides site-specific trimming, milling, and deposition. In addition to defect analysis, FIB is also used for device trimming, circuit modification, or even mask repair.

components and embedded memories can be made controllable and observable, and most flip-flops, if not all, should have been included in the scan chains. With the DFT support, the question boils down to which logic gates or interconnect wires are responsible for producing the mismatches between the circuit model and the failing chip.

Most existing diagnosis tools try to solve the problem in the logic domain. After analysis, some of them report a handful of candidates for the defect site, with each candidate referring to a gate or a signal. All of these candidates are considered to have equal probabilities of being a defect site. Some tools report a ranked list of candidates. A candidate with a higher ranking is considered more likely to be a defect site.

Throughout this chapter, we use the following terminology:

Definition 7.1 (Output Pair)

(z_1, z_2) is called an output pair if z_1 and z_2 are corresponding outputs from the CUD and the failing chip, respectively.

Definition 7.2 (Failing Test Vector)

A test vector v is called a *failing test vector* if it creates a mismatch at any output pair.

The quality of a diagnosis tool can be measured in a number of ways. The most important criterion is whether it is good at pinpointing the defect sites. The second important criterion is whether it can complete the analysis within a reasonable time period (*e.g.*, overnight for a fairly large chip). When it comes to the first criterion (*i.e.*, the ability to pinpointing the defect sites), several different quality indexes have been proposed in the literature:

- **Diagnostic resolution**—The total number of defect candidates reported by a tool is defined as the diagnostic resolution. Ideally, the diagnostic resolution is just 1. In some sense, this index shows how focused the diagnosis tool is; however, a tool could have a good resolution (produce very few candidates) but still miss the target all the time. So, good resolution does not necessarily imply good accuracy.

- **First-hit index**—Diagnostic resolution has no meaning for a tool that only reports a ranked list, rather than a small number of candidates. In this case, the accuracy can be measured by how fast one can hit a true defect site. By definition, the first-hit index refers to the index of the first candidate in the ranked list that turns out to be a true defect site. The smaller this number (the closer it is to the top of the list), the more accurate the diagnosis process.

- **Top-10 hit**—It is possible that the chip could contain multiple defects. Because one cannot afford to target too many gates or signals during each inspecting session, candidates beyond the top 10 are usually ignored in a reported ranked list. Among the top-10 candidates, it is desirable that more than one defect is hit; therefore, the top-10 hit is defined as the number of defects hit by the top-10 candidates as a quality index for multiple-defect diagnosis. The larger this number, the better the diagnosis result.

- **Success rate**—The percentage of hitting at least one defect in one chip inspection session is defined as the success rate. This reflects the ultimate goal of failure analysis; however, this index is more judgmental because the success rate depends on how much time one is willing to spend. Also, the above first-hit index or top-10 hit indexes are linked closely to the success rate. A diagnosis algorithm with a better first-hit index or top-10 hit could translate to a higher success rate.

Diagnosis has long been compared to the job of a criminal detective or medical doctor. In both cases, one would like to identify the root cause by analyzing the observed **syndrome**. Here, for logic diagnosis, the syndrome refers to when and for what output the chip produces a wrong binary response during the test application. These three types of jobs can be compared as follows:

- **Logic diagnosis**—(failing chip) → (syndrome) → (failing gates or wires)
- **Criminal detection**—(crime) → (crime scene) → (criminal)
- **Medical diagnosis**—(patient) → (syndrome) → (disease)

A chip failure can occur anywhere. It can be in the flip-flops, combinational logic, or even the ***design for testability*** (DFT) circuitry (such as scan chains or logic BIST circuitry). In the past, most diagnosis work has been focused on combinational logic; however, the amount of DFT circuitry (*e.g.*, scan chains and logic BIST) has increased, and failures there have become increasingly likely.

In the rest of this chapter, we discuss combinational logic diagnosis and then address the diagnosis of failing scan chains. Finally, we describe methods for diagnosis in a logic BIST environment.

7.2 COMBINATIONAL LOGIC DIAGNOSIS

In combinational logic diagnosis, we assume that the faults to be identified are within the combinational logic (in between the scan flip-flops) that performs the desired logic functionality. For the moment, we assume that the flip-flops and the scan chains are fault free. Two major paradigms have been proposed: *cause–effect analysis* and *effect–cause analysis* [Abramovici 1994]. After we have introduced these two paradigms, we will discuss how to apply them to a fairly large chip with multiple defects.

7.2.1 Cause–Effect Analysis

Cause–effect analysis begins by confining the *causes* of failure to a specific fault type (generally stuck-at faults). Intensive fault simulation is performed to build a **fault dictionary** for deriving and recording the test responses with respect to the applied test set and fault type. Once this dictionary is built, the *effect* or syndrome of the failing chip is analyzed using table look-up. In other words, the syndrome of the failing chip is matched up with the recorded possible syndromes in the dictionary. The closest one implies the most likely fault; therefore, it is often also referred to as the **fault-dictionary based paradigm**.

Example 7.1

Consider a circuit under diagnosis, as shown in Figure 7.2. The circuit has three inputs $\{a, b, c\}$ and one output $\{g\}$. Assume that five test vectors are generated in advance $\{v_1, v_2, v_3, v_4, v_5\}$. Based on the single stuck-at fault assumption, the fault universe will be $\{f_1, f_2, f_3, f_4, f_5\}$ after equivalent fault collapsing. Figure 7.2b shows the full-response table of output signal g obtained by complete fault simulation, including those for the fault-free circuit and the five faulty circuits. One row corresponds to a circuit (either fault-free or faulty), whereas one column corresponds to the response of one test vector. From these simulation results, we first conclude that the test set has 100% fault coverage. Next, we will build a simple fault dictionary to aid the diagnosis process.

A possible fault dictionary is shown in Figure 7.2c (this type of fault dictionary is specifically called a *diagnostic tree*). Note that this dictionary may not be the most economical in terms of size. We only use it to demonstrate the diagnosis process. The basic idea is to refine the *fault candidates* iteratively. Initially, the candidate set (as in an oval) contains all faults. After examining the response of

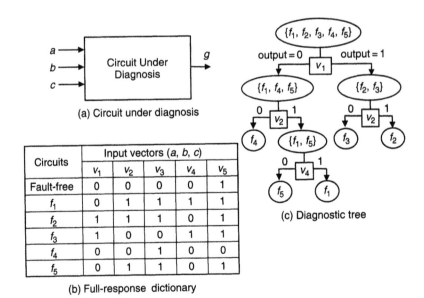

(a) Circuit under diagnosis

Circuits	Input vectors (a, b, c)				
	V_1	V_2	V_3	V_4	V_5
Fault-free	0	0	0	0	1
f_1	0	1	1	1	1
f_2	1	1	1	0	1
f_3	1	0	0	1	1
f_4	0	0	1	0	0
f_5	0	1	1	0	1

(b) Full-response dictionary

(c) Diagnostic tree

■ **FIGURE 7.2**

Example of cause–effect analysis.

the failing chip to the first test vector v_1, we are able to narrow the candidate set down to one of a number of groups. In this example, there are two candidate sets: $\{f_1, f_4, f_5\}$ and $\{f_2, f_3\}$. Because the CUD has only one output, we can only achieve binary partitioning; however, in general, the partitioning could be faster. In this example, the refinement continues until we examine the responses of $v_1, v_2,$ and v_4. We stop here, as the cardinality of each candidate set has been reduced to 1. The overall diagnosis process is simply a traversal from the root of this tree to one of its leaf nodes, representing the final fault candidates. For example, if the response of a failing chip at output signal g under the five test vectors $\{v_1, v_2, v_3, v_4, v_5\}$ is $\{0, 1, 1, 0, 1\}$, by traversing the diagnosis tree we can immediately deduce that the only faulty circuit that could produce the observed failing syndrome is f_5.

It may take time and space to construct the fault dictionary; however, once the dictionary is built, conducting *syndrome analysis* is usually fast. Because the fault dictionary is built once initially, the overall diagnosis process is computationally efficient; however, for practical applications, this approach could be limited by a number of problems:

- **Dictionary size problem**—A fault dictionary records every output response of each modeled fault at each clock cycle. Without proper compaction, the size is proportional to the product of three factors: $(F \cdot V \cdot O)$, where F is the number of modeled faults, V is the number of test vectors, and O is the number of outputs.[2] In a logic chip with one million gates, 10,000 flip-flops, and 10,000 test vectors, the size will amount to 10^{12} bits, requiring extremely large storage. The entire dictionary also has to be regenerated even if a small logic change is made. With proper compression techniques, this problem can be relieved to some extent [Richman 1985] [Pomeranz 1992] [Chess 1999]; however, the excessive storage requirement and the inability to scale to ever-larger circuits still pose a serious limitation.

- **Unmodeled-fault problem**—The dictionary is built using a single stuck-at fault assumption. If the CUD truly contains a stuck-at fault only, then the result is highly accurate. However, realistic defects may not behave as single stuck-at faults, but often exhibit themselves as bridging faults (having two or more interconnects shorted together). In addition, the defects may have resistive characteristics [Aitken 1996]. These realistic defects compound the ever-growing storage requirement and could easily lead to misleading results. Although extensions have been proposed to resolve this problem to some extent by targeting popular bridging faults [Wu 2000], in general, the diagnostic accuracy of this approach is not as good as effect–cause analysis.

7.2.1.1 *Compaction and Compression of Fault Dictionary*

We address the dictionary size problem in this subsection by showing how to compact or compress a fault dictionary. With respect to terminology, *compression*

---

[2] Here we assume that the CUD has full-scan chains. The input of a flip-flop is also considered a pseudo output.

refers to techniques that reduce the size of a fault dictionary without sacrificing the diagnostic resolution for the modeled faults, whereas *compaction* refers to techniques that could degrade the resolution.

Example 7.2 (Pass–Fail Dictionary)

The simplest way of compacting a dictionary is to replace an output response vector by a single pass-or-fail bit. The size of the resulting dictionary is then independent of the number of outputs and only proportional to the order of $(F \cdot V)$, where F is the total number of faults and V is the total number of test vectors. An example is shown in Figure 7.3b.

Example 7.3 (P&R Compression Dictionary)

The above pass–fail dictionary leads to a high compaction ratio; however, it is often a lossy technique, meaning that it may become infeasible to locate the root causes of failure. Pomeranz and Reddy (P&R) [Pomeranz 1992] solved the problem by selectively putting back a sufficient amount of output response information to restore the full diagnostic resolution, thus making it a compression technique. The

Fault ID	Output Response (z_1, z_2)			
	t_1	t_2	t_3	t_4
f_1	10	10	11	10
f_2	00	00	11	00
f_3	00	00	00	00
f_4	01	00	00	01
f_5	01	00	01	01
f_6	01	00	01	01
f_7	10	00	10	00
f_8	11	11	11	11

(a) Full-response table

Fault ID	Pass (0) or Fail (1)			
	t_1	t_2	t_3	t_4
f_1	1	1	0	1
f_2	1	0	0	1
f_3	1	0	1	1
f_4	1	0	1	0
f_5	1	0	1	0
f_6	1	0	1	0
f_7	1	0	1	1
f_8	0	1	0	1

(b) Pass-fail dictionary

Fault ID	Pass-fail + Extra outputs			
	t_1	t_2	t_3	t_4
f_1	1 1	1	0 1	1
f_2	1 0	0	0 1	1
f_3	1 0	0	1 0	1
f_4	1 0	0	1 0	0
f_5	1 0	0	1 1	0
f_6	1 0	0	1 1	0
f_7	1 1	0	1 0	1
f_8	1 1	1	0 1	1

Response of z_1 Response of z_2

(c) P&R compression dictionary

■ **FIGURE 7.3**

P&R compression fault dictionary.

restoration process begins by first putting every pass-or-fail-bit **output column** with respect to each specific test vector into the dictionary. It then incrementally takes one output column at a time until the full resolution is achieved. In the example shown in Figure 7.3c, there are originally eight output columns, two for each test vector. This algorithm chooses two of them to restore the full resolution (*i.e.*, z_1 in response to test vector t_1 and z_2 in response to test vector t_3). It can be seen that the diagnostic resolution is originally two in the full-response table (faults f_5 and f_6 are not distinguishable). In the pass–fail dictionary, the resolution rises to three, with the equivalence groups being $\{f_1, f_2, (f_3, f_7), (f_4, f_5, f_6), f_8\}$. Finally, in the compression dictionary the resolution has been reduced back to two again, with the equivalence groups being $\{f_1, f_2, f_3, f_4, (f_5, f_6), f_7, f_8\}$.

Example 7.4 (Detection Dictionary)

Another popular dictionary organization is called a *detection dictionary*. It is based on the idea that we only need to record the *failing output vectors* (*i.e.*, the output vectors that are different from their counterparts in the fault-free circuit). Statistically, many output vectors may be fault free; therefore, we can drop a lot of unnecessary information without sacrificing resolution. One drawback of a detection dictionary is that the structure becomes irregular. As shown in Figure 7.4, we need to specify a *test vector identifier* along with a failing output vector. For example, for fault f_1, the detection information is specified as (t_1:10, t_2:10, t_4:10), meaning that the output responses under test vectors t_1, t_2, and t_4 are failing in the presence of fault f_1. In other words, the detection information for each fault is now a list of failing output vectors. Sometimes, a *drop-on-k* heuristic may be used to further reduce the size at the cost of some minor resolution degradation. The basic idea is to stop the recording after k failing output vectors have been collected. The total size of such a dictionary is on the order of $(F \cdot \log(V) \cdot k \cdot O)$, where F is the total number of modeled faults, $\log(V)$ is the number of bits for encoding a test vector identifier, k is the maximum number of failing output vectors, and O is the output number.

7.2.2   Effect–Cause Analysis

Unlike the fault-dictionary-based paradigm, effect–cause analysis directly examines the syndrome (*i.e.*, the effect) of the failing chip to derive the fault candidates (*i.e.*, the cause) through Boolean reasoning on the CUD. Effect–cause analysis is superior to the fault-dictionary-based paradigm in a number of aspects:

- It does not assume an *a priori* fault model and thus is more suitable to handle non-stuck-at faults (*e.g.*, bridging faults).

- It can be adapted to cases where there are multiple faults in the failing chip, especially when these faults are structurally codependent.

- It can be adapted to partial-scan designs more easily.

The only minor drawback of effect–cause analysis is that it takes longer to complete because a unique round of analysis is required for each failing chip. This

Fault	Output Response (z_1, z_2)			
ID	t_1	t_2	t_3	t_4
f_1	10	10	11	10
f_2	00	00	11	00
f_3	00	00	00	00
f_4	01	00	00	01
f_5	01	00	01	01
f_6	01	00	01	01
f_7	10	00	10	00
f_8	11	11	11	11

(a) Full-response table

failing output vectors

Fault	Pass (1) or Fail (0)			
ID	t_1	t_2	t_3	t_4
f_1	1	1	0	1
f_2	1	0	0	1
f_3	1	0	1	1
f_4	1	0	1	0
f_5	1	0	1	0
f_6	1	0	1	0
f_7	1	0	1	1
f_8	0	1	0	1

(b) Pass-fail dictionary

Fault ID	Detection information (Test ID : Output Vector)
f_1	t_1:10 t_2:10 t_4:10;
f_2	t_1:00 t_4:00;
f_3	t_1:00 t_3:00 t_4:00;
f_4	t_1:01 t_3:00;
f_5	t_1:01 t_3:01;
f_6	t_1:01 t_3:01;
f_7	t_1:10 t_3:10 t_4:00;
f_8	t_2:10 t_4:11;

(c) Detection dictionary

■ FIGURE 7.4

Example of detection dictionary.

is a more dynamic process compared to the more static fault-dictionary-based paradigm, in which certain information is reused constantly. Because logic diagnosis is used to guide the time-consuming physical silicon inspection, the analysis time is in general not a very important factor.

In the following discussion, we assume that the CUD has been implemented with full-scan and its functionality is represented as a combinational gate-level circuit. The **primary output** (PO) signals of the CUD and the failing chip are denoted as $\{z_1^C, z_2^C, \ldots, z_m^C\}$ and $\{z_1^F, z_2^F, \ldots, z_m^F\}$, respectively, where m is the total number of primary outputs. We assume that the set of test vectors, denoted as $TV = \{v_1, v_2, \ldots, v_t\}$, has been generated in advance.

Definition 7.3 (Mismatched Output)

An output pair (z_i^F, z_i^C) is said to be *mismatched* if there exists a test vector v such that v, when applied to both CUD and the failing chip, produces different binary values at z_i^F and z_i^C. In particular, we call the primary output z_i^C in the CUD a *mismatched output*, whereas the primary output z_i^F in the failing chip is called a *failing output*. In logic diagnosis, the failing chip is like a black box that cannot be analyzed. The best we can do is to reason upon the circuit model under diagnosis.

Example 7.5

In Figure 7.5, the test vector *v*, when applied to both the CUD and the failing chip, produces mismatches at the first and the fifth output pairs.

In the rest of this subsection we first discuss a **structural pruning** technique that can narrow down the potential *fault candidate area* in the CUD. Next, we introduce an efficient **backtrace algorithm**. Finally, we present a versatile **inject-and-evaluate paradigm** that can achieve higher accuracy.

7.2.2.1 *Structural Pruning*

The **fanin cone** of an output in the CUD refers to the collection of the logic gates that can reach this output structurally. Depending on the number of faults in the failing chip, we can employ **cone intersection** or **cone union** to prune out those logic gates that could not possibly produce the faulty behavior [Waicukauski 1989].

Example 7.6

As illustrated in Figure 7.6, if there is only one fault, then we take the intersection of the fanin cones of the *mismatched outputs*. The resulting area of gates is the *fault candidate area*. On the other hand, if there is more than one fault in the failing chip, then we should perform *cone union* instead. This is because every gate in the fanin cone of any mismatched output could now be responsible for the observed syndrome partially, if not completely. Obviously, the pruning capability of *cone intersection* is much more effective than that of cone union; however, the number of faults in a failing chip is not known in advance before the diagnosis process, so cone union is a conservative and safer technique. As a matter of fact, cone intersection could lead to an empty fault candidate area if there are multiple faults, as shown in Figure 7.6b.

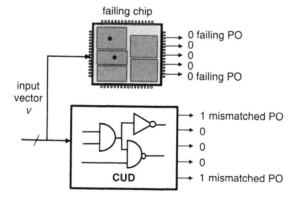

■ FIGURE 7.5

Illustration of mismatched outputs.

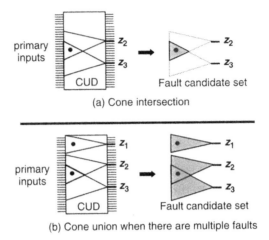

(a) Cone intersection

(b) Cone union when there are multiple faults

■ **FIGURE 7.6**

Structural pruning techniques.

7.2.2.2 Backtrace Algorithm

Structural pruning techniques are often used as the first-step process in effect–cause analysis so subsequent diagnosis can focus on a smaller area before applying a more accurate procedure to pinpoint the fault locations. Backtrace is one such **functional pruning** technique. By *functional pruning*, we mean that this method disqualifies candidates from the fault candidate area by examining the signal values inside the CUD with simulation.

Backtrace is similar to *critical path tracing*, which was originally proposed by [Abramovici 1990] for fast fault simulation and then subsequently applied to logic diagnosis [Kuehlmann 1994]. This algorithm iterates through each failing test vector. For each failing test vector, it performs fault-free simulation on the CUD first, then it checks the mismatched outputs one at a time. From each mismatched output, it traces the CUD backward to find the signals that can account for the output mismatch.

Backtrace is based on two rules: the *controlling rule* and the *noncontrolling rule*. When performing backtrace, complex cells (*e.g.*, an AOI cell) are decomposed into AND, OR, NAND, NOR, and NOT primitives first. Here, we explain the backtrace rules for a NAND gate only; the implications for other types of gates are left as exercises. Assume that we have reached a gate (say, g) with a binary value that is either 0 or 1. Based on its value, we want to make decisions about which inputs of g can be further held responsible if g has been classified responsible:

- **Controlling case** (*i.e.*, 0 for a NAND gate)—Only fanin signals of g with controlling values (*i.e.*, 0) are considered responsible. On the other hand, those fanin signals with noncontrolling values (*i.e.*, 1) should not be held responsible because the current value of signal g is not decided by these fanin signals.

- **Noncontrolling case** (*i.e.*, 1 for a NAND gate)—In this case, the current value of signal g is determined by all fanin signals; therefore, every fanin signal is held responsible.

Example 7.7

Figure 7.7 demonstrates the backtrace algorithm. Here, we make no distinction between a logic gate and its output signal. The figure shows a trace starting from signal e. At the end, the fault candidate area is marked by bold lines.

At the end of the backtrace there will be one fault candidate set for each mismatched output under a specific failing test vector. How to combine these fault candidate sets into a final set is an issue. If a single fault is assumed, then the intersection of all of these fault candidate sets will be determined to derive the final fault candidate set. As we mentioned previously, if it turns out that this final set is empty, it implies that there could be multiple faults in the failing chip.

7.2.2.3 Inject-and-Evaluate Paradigm

The backtrace algorithm is generally efficient; however, it may not be accurate enough in some cases. To address this issue, [Pomeranz 1995] pioneered an alternative method, referred to here as the **inject-and-evaluate paradigm**. Computationally, it performs the diagnosis by a sequence of fault injections and evaluations. This paradigm was further adopted and polished by a number of other researchers [Huang 1997] [Venkataraman 1997] [Veneris 1997] [Boppana 1999] to achieve an even higher accuracy or to deal with realistic defects, such as bridging fault diagnosis [Bartenstein 2001] or multiple-fault diagnosis [Huang 2001] [Wang 2003] [Liu 2005].

The basic outline of the inject-and-evaluate paradigm is shown in Figure 7.8. The structural pruning technique is first used as a preprocessing step to derive the initial candidate set. Then, the procedure enters two levels of loops. The outer loop iterates through every failing test vector, whereas the inner loop examines one

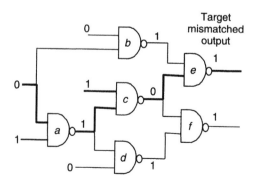

■ **FIGURE 7.7**

Demonstration of the backtrace algorithm.

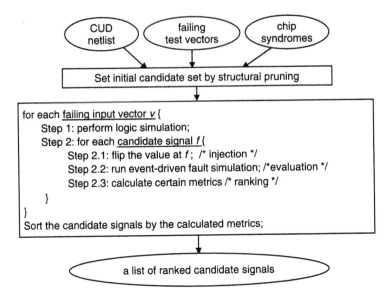

■ **FIGURE 7.8**

The outline of an inject-and-evaluate paradigm.

candidate signal at a time. The key steps in the body of these loops are described below:

- **Step 1.** For each failing test vector v, perform fault-free simulation on the CUD, and record the value of each signal in the circuit structure.

- **Step 2.** For each candidate signal f, further perform the inject-and-evaluate process in three minor steps:

  Step 2.1. *Flip* the current fault-free value at signal f and create a *value-change* event. This is the *injection* part of the process.

  Step 2.2. Perform an *event-driven simulation* to evaluate the effect due to the above injection. This is the *evaluation* part of the process.

  Step 2.3. Calculate certain *ranking metrics* by examining the *reactions* of the output signals in response to the injection made. This is the *ranking* part of the process.

For diagnosis, we need to grade the possibility of a signal being one of the defect sites. This requires a decision-making mechanism regarding which signals are more responsible for the chip failure under a specific test vector. The existing variations of the inject-and-evaluate paradigm may differ from one another in how the injection is made and how the *ranking metrics* are calculated. Most of them, however, comply with the following *reproduction principle* when it comes to the basic intuition.

Definition 7.4 (Reproduction Principle)

A signal f is regarded as a likely fault candidate if one can *reproduce* the failing syndrome on the CUD's outputs by manipulating the signal value at f.

When there are multiple faults in the chip, we may not be able to find the perfect spot that we can manipulate to reproduce the syndrome for every failing test vector. In that case, certain metrics of *partial reproduction* should be used. From this point of view, the inject-and-evaluate paradigm is similar to acupuncture:

- Inject-and-evaluate—(find a spot for injection) → (reproduce the syndrome)
- Medical acupuncture—(find a point to inject hot needle) → (cure the illness)

The only difference is that we cannot actually cure a failing chip. We can only hope to make the CUD behave like the failing chip. In other words, we only want to *resolve the mismatch* between the chip and the CUD by injections. In the following, we first formally define *resolving a mismatched output* and *curable output* before giving an example to illustrate the entire procedure.

Definition 7.5 (Resolving a Mismatched Output)

In the sequel, resolving a mismatched output z_i^C refers to a mechanism that injects binary values to certain signals in the CUD, so the response of z_i^C becomes equivalent to its counterpart in the failing chip, z_i^F.

Definition 7.6 (Curable Output)

Under a specific test vector v, a mismatched output is called a *curable output* of a signal (say, f) if the mismatch can be resolved by an *injection* at signal f. Such an injection is referred to as a **cure injection**.

In [Pomeranz 1995], a signal with more curable outputs is regarded as being more likely to be a fault location. That is, the *curable output number* is used as the major **ranking metric** in the inject-and-evaluate paradigm.

Example 7.8

Consider the CUD shown in Figure 7.9. A failing test vector under consideration is $\{(x_1, x_2, x_3, x_4)|(0, 1, 1, 1)\}$. Assume that the response of the failing chip to this vector is $\{(z_1, z_2)|(0, 1)\}$, whereas the response of the CUD is $\{(z_1, z_2)|(1, 1)\}$. As a result, there is a mismatch at the first output pair. Now we want to determine if signal f in the CUD is responsible for the mismatch. The value at this signal is changed from 1 to 0. This effect ripples through the circuit and changes both output values from 1 to 0; therefore, the mismatch at the first output is resolved (or cured). We conclude that the curable output number for signal f is 1. Similar operations can be performed for the other signals. The final candidate list is produced by sorting the signals with their curable output numbers.

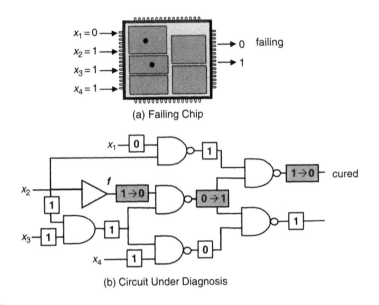

(a) Failing Chip

(b) Circuit Under Diagnosis

■ **FIGURE 7.9**

An example of the inject-and-evaluate paradigm.

Computational Details

The implementation of the above event propagation could affect the efficiency significantly. In detail, it involves the processing of one failing test vector v using event-driven fault simulation.

- **Step 1. Set the initial event queue** to have only one *signal-value pair*, (f, $f(v)'$), where f is a candidate signal under consideration and $f(v)$ is the original fault-free value at this signal. The injection has been made by flipping $f(v)$ to $f(v)'$. Each event is also associated with a *key*, given by the topological level of the signal in the signal–value pair. An event queue is an important data structure in this process which can be implemented as an array maintaining the *minimum heap* structure. In other words, the first element of the array is always the event with the smallest key. Whenever an element (*i.e.*, an event) is inserted into this queue, the array needs to be adjusted to maintain its heap property. This property makes the subsequent event retrieval more efficient. Using a heap is only one method; one can also use other data structures (*e.g.*, a two-dimensional linked list) to avoid unnecessary insertion time.

- **Step 2. Perform an event-processing loop** until there is no event left in the event queue. The body in this loop consists of a number of subtasks. First, the event of the minimum key (*i.e.*, the first element in the array implementing the event queue) is retrieved. Let the signal in this retrieved event be f_{event}. Then, the logic gates driven by signal f_{event} are reevaluated. If any of these logic gates results in a value different from its fault-free value after the reevaluation, then a new event is created. Every new event is inserted into the queue and the queue is adjusted to satisfy the heap property. Such an adjustment has a time complexity of $O(log|\text{queue-size}|)$. The reason why we always retrieve the event of minimum key first is to guarantee that the simulation is conducted following the topological order in which logic gates closer to the primary inputs are always processed before any gates in their fanout region. Without enforcing this rule, the results could be erroneous.

- **Step 3. Update the total number of *curable outputs*** of the candidate signal under consideration. When the event queue is exhausted, the computation of the final values at the outputs of the CUD due to the flipping of a signal's value is completed. As defined previously, an output is counted as a *curable output* if and only if it satisfies two conditions: (1) it is originally mismatched with its counterpart in the failing chip, and (2) it flips after the event-driven simulation.

- **Step 4. Roll back** the value of each signal of the CUD to its fault-free signal before moving on to the next candidate signal. During the execution of the loop in step 2, every new event including the initial event should be recorded in a data structure called an *event history* to support an efficient roll back; for example, we only need to roll back the fault-free values of those signals that are changed during the event-driven fault simulation. Without such support, one may need to roll back every signal's fault-free value. In that case, the time complexity of computing a signal's curable measures will become $O(n)$, where n is the total number of signals in the CUD, and the advantage of using event-driven simulation may disappear because the computation is now dominated by the roll-back process. After the roll back, the next fault-simulation run for another candidate signal can then be started.

Curable-Vector-Based Metric

It has been found that an inject-and-evaluate method that is based on the number of curable outputs is sometimes not accurate enough. This is because a signal reaching out to a larger number of outputs tends to have a larger number of curable outputs as well. These signals may overwhelm the true faulty signal. A better metric, based on what is referred to as *curable vectors* here, was incorporated in [Bartenstein 2001], [Huang 2001], and [Venkataraman 2001].

Definition 7.7 (Curable Vector)

A test vector v is called a *curable vector* of a signal f if every mismatched output in the CUD with respect to v can be resolved without creating new mismatched outputs by an injection at f. A curable vector is called a SLAT (*single location at a time*) pattern in [Bartenstein 2001].

Using the number of *curable vectors* is a better metric than using the number of *curable outputs* in three aspects. First, it takes into account the side effect of an injection (*i.e.*, the newly created mismatch) when grading the effect. Second, it checks the *reaction* of all outputs simultaneously, instead of one by one. Third, it can be proved that a signal f is not a *single-fault candidate* unless every failing test vector is also a curable vector of signal f, assuming that there is only one fault in the failing chip.

Example 7.9

Figure 7.10 illustrates a curable vector. During the fault-free simulation, the mismatched outputs of the CUD under a failing test vector v are marked by "×" and the matched outputs are marked by "o." After we flip the value of signal f from 1 to 0, the value-change event propagates to every mismatched output (*i.e.*, the first, fourth, and fifth outputs) but not to any originally matched output (*i.e.*, the second and third outputs). Therefore, the failing test vector v is curable by signal f.

For a failing chip with only one single fault, there always exists a signal in the CUD that can cure all failing test vectors. This is based on an observation that, if the failing syndrome is created by one fault, then one should be able to clean it up completely at the fault site, too. In the following, we will further generalize this idea to the cases where there are multiple faults.

Ranking Heuristic for Multiple-Fault Diagnosis

At the end of the entire inject-and-evaluate process, the number of *curable vectors* and *curable outputs* for each signal has been calculated. These two metrics should be combined to indicate how likely a signal would be a fault location. Experiments show that the following rule can yield good results: *A signal with a larger number of curable vectors is ranked higher. When there is a tie, use the number of curable*

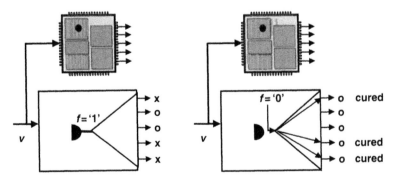

■ **FIGURE 7.10**

Illustration of a curable vector.

outputs as the tie-breaker [Huang 2001]. The reason why the priority goes to the number of curable vectors can be further explained from two points of view:

- The curable vector is a very stringent condition. If a signal is able to cure a failing vector, it is a strong indication that the signal is one of the fault sites. In some sense, one curable vector is considered stronger than many curable outputs here because it is not easy for a non-fault site signal to possess a curable vector.

- Even if there are multiple faults in the failing chip, it is very likely that certain failing test vectors only activate a single fault, or, in other words, the failing syndrome is contributed by only one fault. This kind of vector thus has a higher value in diagnosis than others that activate multiple faults, because it sends out a strong message by being a curable vector of the acting fault.

Reward-and-Penalty Principle

The ranking heuristic can be further refined with respect to the tie-breaking mechanism. Instead of simply considering the *curable output count*, we can take into account the *new mismatched output count* as well. Based on the *reward-and-penalty* principle, we can use the following rule for combining them as the second-level ranking metric for each signal:

$$rank2 = (\text{Curable Output Count} - 0.5 \times \text{New Mismatched Output Count})$$

Still, the curable vector count is used as the first-level metric without change. For the second-level metric, a *reward* is granted to a signal with more curable outputs, whereas a *penalty* is imposed on a signal with a relatively large number of new mismatched outputs.

Example 7.10

Consider a CUD and a failing chip as shown in Figure 7.11. Again, a failing test vector under consideration is $\{(x_1, x_2, x_3, x_4)|(0, 1, 1, 0)\}$ is applied to both of them. Assume that the response of the failing chip is $\{(z_1, z_2)|(0, 1)\}$, whereas the response of the CUD is $\{(z_1, z_2)|(1, 1)\}$. As a result, there is a mismatch at the first output pair. After an injection is made at signal f, the mismatch at the first output is cured; however, the originally matched output now becomes a new mismatched output. As a result, we know that this test vector is not curable by signal f. Also, *rank2* with respect to this vector is (reward $- 0.5 \cdot$ penalty) $= (1 - 0.5 \cdot 1) = 0.5$.

In summary, we have discussed the basic effect–cause analysis in this subsection, including the backtrace algorithm and the inject-and-evaluate paradigm. For the latter, we also discussed several concepts for deriving the final ranked list of candidates (*e.g.*, the curable output, the curable vector, and the reward-and-penalty principle). Unlike fault-dictionary-based methods, these approaches do not rely on any specific fault model and thus can handle the real causes better. In terms of accuracy, experimental results in the literature show that the *first-hit index* is roughly seven if only the curable output count is used as the ranking metric. This

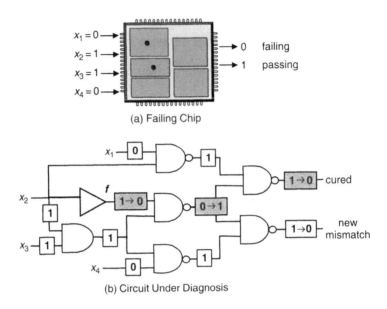

(a) Failing Chip

(b) Circuit Under Diagnosis

■ **FIGURE 7.11**

Illustration of a new mismatched output.

can be improved to around five on average for ISCAS 1985 benchmark circuits [Brglez 1985], if the curable vector concept is incorporated, and to three if the reward-and-penalty principle is further applied [Huang 2001].

An Application to Bridging Fault Diagnosis Using Multiplets

A bridging defect is very common in an IC due to the routing characteristics of today's manufacturing process. When two interconnects are shorted by a defect, the defect often affects the output signals of two logic cells. When it comes to the diagnosis of such a defect, one may wish to identify both affected cells instead of just one of them so as to pinpoint the exact location of the defect more accurately by analyzing the layout.

An approach called the *SLAT* (single location at a time) paradigm [Bartenstein 2001] was proposed to address this issue quite effectively. The overall flow is shown in Figure 7.12. It has two phases: (1) finding **SLAT vectors**, and (2) finding **valid fault multiplets**.

In the first phase, finding SLAT vectors, the paradigm uses a procedure similar to the aforementioned inject-and-evaluate paradigm to find out all possible *curable vectors* (called *SLAT vectors* here) for each signal. The results can be viewed as a two-dimensional table, as shown in Figure 7.13. This table is referred to as *SLAT table*. The horizontal axis is the fault index, whereas the vertical axis is the failing input vector index. For each row, the table shows all possible cure injection locations (*i.e.*, the signals where an injection exists to cure all output mismatches) for a particular failing input vector. For each column, the table shows all SLAT vectors

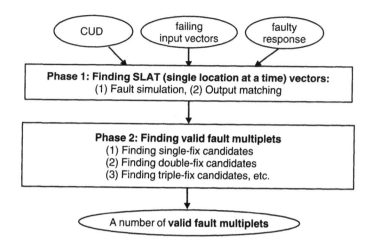

■ **FIGURE 7.12**

Flowchart of the SLAT paradigm for bridging fault diagnosis.

Failing Input Vectors	Signals in the CUD						
	f_1	f_2	f_3	f_4	f_5	f_6	f_7
v_1	*				*		
v_2	*	*	*				
v_3			*	*			*
v_4					*	*	
v_5		*			*		
v_6		*			*		
v_7	*		*				
v_8			*				*
v_9			*		*		
v_{10}					*		*

An asterisk (*) means the corresponding vector
is a SLAT vector of the corresponding signal.

■ **FIGURE 7.13**

Example of finding a valid fault multiplet.

for a particular signal. If there exists a signal in the SLAT table that is able to cure all failing input vectors, then this signal is identified as a **single-fix candidate**. There could be more than one single-fix candidate if there is only one fault in the failing chip. These candidates are then reported as the final results; otherwise, the process continues to the second phase.

In the second phase, finding fault multiplets, this algorithm aims to find a set of signals, called a **multiplet**, so these signals can take turns explaining the failing syndrome of each failing input vector. In other words, the union of the SLAT vectors of signals in a valid fault multiplet will cover the entire failing input vector set. It starts by finding a signal pair (*i.e.*, a *double-fix candidate*). If that is not possible, it further increases the cardinality of the multiplets incrementally to three and to four if necessary.

Example 7.11 (Finding Valid Fault Multiplets)

Consider the SLAT table given in Figure 7.13. We can first check to see if there is any single-fix candidate. The answer is no. Hence, we can proceed to find double-fix candidates. It turns out that the union of the SLAT vectors of signal pair (f_3, f_5) covers the entire failing input vector set. So, these two signals form a *valid fault multiplet*.

It is possible that the SLAT approach could report zero valid fault multiplet if there are multiple faults in the failing chip. The reason is that there might be a failing input vector whose failing syndrome is jointly created by multiple signals such that it cannot be perfectly explained by any single signal. However, this approach is especially effective for bridging fault diagnosis. In a bridging fault (AND-bridging, OR-bridging, or even dominant bridging), there is a **victim signal** (*i.e.*, a signal whose value flips erroneously due to the influence of a stronger **aggressor** through the bridging defect) at all times. In light of this, every failing input vector should be a SLAT vector of some signal in the CUD if there is only one bridging fault in the chip under diagnosis.

7.2.3 Chip-Level Strategy

The previous methods are mostly at the block level and primarily aimed to identify one fault only. Strategies on top of these block-level techniques are needed in order to successfully diagnose a large chip targeting multiple faults simultaneously so as to increase the success rate. In other words, we hope that more faults are included in the final top-10 candidate list. In this subsection, we present such a strategy. It proceeds in two phases. In the first phase we concentrate on the identification of the so-called structurally *independent faults* (IFs) based on a concept referred to as *word-level prime candidate*, while in the second phase we further trace the locations of the more elusive structurally *dependent faults* (DFs). Experimental results show that this strategy is able to find three to four faults within the top-10 list for three real-life designs randomly injected with five node-type or stuck-at faults.

7.2.3.1 Direct Partitioning

A faulty chip might have more than one fault. Some of them may be *structurally dependent*, while others are *structurally independent*. More precisely, a fault is referred to as a *structurally independent fault* if the fanout cone of this fault does not overlap with that of any other fault. On the other hand, a fault is referred to as a *structurally dependent fault* if it is not a structurally independent fault. As demonstrated in Figure 7.14, f_1 is an independent fault, whereas f_2 and f_3 are dependent faults.

Intuitively, an independent fault is easier to identify than a dependent fault. This is mainly because an IF is the sole cause of the syndrome at its reachable outputs.

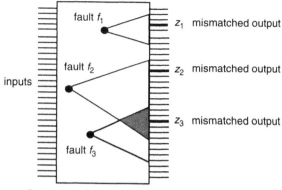

Fault f_1 is an independent fault.
Faults f_2 and f_3 are dependent faults.

■ **FIGURE 7.14**

Structurally independent and dependent faults.

For example, the mismatched output z_1 in Figure 7.14 is only reachable by fault f_1. Hence, the syndrome at z_1 is caused exclusively by fault f_1. Provided that we know the exact reachable outputs of an IF, we can trace back to the faulty location using existing techniques. Based on the **divide-and-conquer strategy** [Wang 2003], a *multiple-fault diagnosis* can be decomposed into a number of block-level single-fault problems ideally. After that, we can try to identify one fault in each block.

The decomposition can be performed by partitioning the outputs into groups. Each group of outputs and their respective fanin cones form a block for single-fault diagnosis. As demonstrated in Figure 7.15, one can first construct a **dependency graph**, with each node denoting a mismatched output and each edge (z_i, z_j) denoting

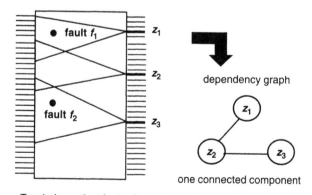

Two independent faults, f_1 and f_2, lead to one diagnosis block.

■ **FIGURE 7.15**

Limitation of direct partitioning.

the *overlapping relationship* between the fanin cones of two mismatched outputs z_i and z_j. Each *connected component* in this dependence graph will then correspond to a set of mismatched outputs, whose joint fanin cones form a **diagnosis block**.

In practice, such a direct partitioning may fail to isolate independent faults. As demonstrated in Figure 7.15, faults f_1 and f_2 are independent faults, but it turns out that they are included in the same *diagnosis block*. This phenomenon can be further explained as follows. Let the reachable mismatched outputs of fault f_1 be $\{z_1\}$ and the reachable mismatched outputs of fault f_2 be $\{z_2, z_3\}$. Although the reachable mismatched output sets of these two faults are disjointed, satisfying the condition of independent faults, the fanin cones of z_1 and z_2 are overlapped. As a result, the logic blocks containing faults f_1 and f_2 are mixed together. Due to the mixture, the subsequent block-level diagnosis may not be able to identify both of them easily.

7.2.3.2 Two-Phase Strategy

A two-phase strategy can overcome this limitation. The ultimate goal is *to identify every structurally independent fault and find dependent faults as much as possible*. In the first phase, a concept called **prime candidate** is used. Without performing partitioning, certain structurally independent faults can be identified precisely using this concept. In the second phase, the syndromes (*i.e.*, the mismatched outputs and their respective fanin cones) caused by the prime candidates are eliminated first. Then, the dependency-graph-based partitioning becomes more effective and the faults can be targeted one by one through a number of block-level diagnosis runs. On average, this methodology will result in 2.6 faults in a design with 5 randomly injected faults.

Definition 7.8 (Partially Curable Vector)

A failing test vector v is partially curable by a signal f if the mismatches at *all mismatched outputs reachable by* f can be resolved by an injection at signal f. Here, a partially curable vector differs from the conventional curable vector only in the *output range of interest*. Instead of the entire output set, we confine it to the target signal's structurally reachable outputs.

Definition 7.9 (Structurally Independent Fault Candidate [SIC] Point)

A signal f is a SIC point if every failing test vector is a *partially curable vector* of signal f.

Observation 7.1

A signal at a structurally independent fault site is a SIC point.

Observation 7.2

A SIC point is not necessarily a structurally independent fault site.

These two observations imply the possibility of aliasing. That is, certain non-faulty signals may disguise themselves as SIC points, thereby reducing the accuracy of diagnosis. The following example illustrates this problem.

Example 7.12 (Aliasing SIC Points)

As shown in Figure 7.16a, faults f_1 and f_2 are two structurally independent faults. After the inject-and-evaluate process, it may turn out that $\{f_1, f_2, f_3\}$ are three SIC points. Among them, f_3 is aliasing. Similarly in Figure 7.16b, f_1 and f_2 are assumed to be two structurally dependent faults. After diagnosis, the SIC points are $\{f_3, f_4, f_5\}$. These three SIC points are all aliasing and misleading. In this case, they are all aliasing partially because they are closer to the outputs. With a small number of reachable outputs, they have a greater chance of aliasing as a SIC point. A more stringent *filter* is appropriate in order to screen out these *false SIC points* (*i.e.*, signals that are not truly at the fault sites). We first eliminate the false SIC points in the case of Figure 7.16a through a criterion called **prime candidate**.

Definition 7.10 (Prime Candidate)

A signal f is a prime candidate for a structurally independent fault if the following two conditions are satisfied: (1) Signal f is a SIC point, and (2) the set of outputs reachable by signal f is not a proper subset of that of any other SIC point.

In the example of Figure 7.16a, the SIC points are $\{f_1, f_2, f_3\}$, while the prime candidates are $\{f_1, f_2\}$. The false SIC point f_3 has been successfully screened out from the prime candidate set; however, this condition does not eliminate the false SIC points, as in the case of Figure 7.16b, in which the false SIC points are signals closer to the outputs.

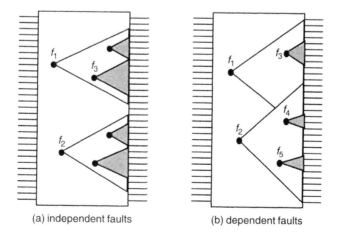

(a) independent faults   (b) dependent faults

■ **FIGURE 7.16**

The aliasing problem of SIC points.

In order to screen out the false SIC points that are closer to outputs, we can take advantage of a common practice that we often specify multiple-bit signals into groups in a design. This implies that structural correlations exist among the output signals of a CUD. Furthermore, these correlations can be easily recognized by their names.

Example 7.13 (Output Group)

Assume that output signal Z and register R are specified in a design as follows:

```
Module design( Z, ... )
output[31:0]  Z;
reg[31:0]  R;
...
Endmodule
```

Then, $\{Z[31], Z[30], \ldots, Z[0]\}$ and $\{R[31], R[30], \ldots, R[0]\}$ form two output groups in the combinational circuit under diagnosis.

With the notion of output group, we can further define reachable output group and *word-level prime candidate* to exploit the structural correlations among outputs.

Definition 7.11 (Reachable Output Group)

Let Z be an output group containing a number of outputs $\{z_1, z_2, \ldots, z_k\}$. Then Z is called a *reachable output group* of a signal f if there exists a path from f to an output in Z.

Definition 7.12 (Word-Level Prime Candidate)

A signal f is a word-level prime candidate if the following three conditions are satisfied:

- Signal f is a SIC point.
- The set of reachable outputs of f is not a proper subset of that of any other SIC point.
- Every mismatched output in the reachable output groups of signal f is also reachable from signal f.

Intuitively, the above three conditions jointly require that a word-level prime candidate should be able to cure all syndromes at its *reachable output groups* for all failing test vectors, not just the syndromes at its reachable outputs. The following example gives an illustration.

Example 7.14

Assume that the outputs have been divided into two groups, Z and R. In Figure 7.17a, the prime candidates are $\{f_1, f_2\}$. After taking into account the word-level information, both prime candidates survive. On the other hand, the prime candidates in the

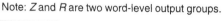

Note: Z and R are two word-level output groups.

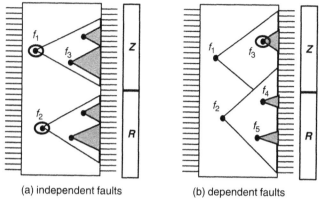

(a) independent faults   (b) dependent faults

■ **FIGURE 7.17**

Circuit reduction after partitioning.

case of Figure 7.17b are originally $\{f_3, f_4, f_5\}$. After taking into account the word-level information, signals $\{f_4, f_5\}$ are excluded from the candidate list because neither of them can assume the full responsibility of the syndromes at their common reachable output group, R. In this example, f_3 is assumed to cover all syndromes at output group Z, thus it will survive as a word-level prime candidate.

The criterion of word-level prime candidate serves as a powerful *filter*. Experiments on a crypto-processor reveal that on the average there are 72.3 prime candidates. On the other hand, there are only 3.7 word-level prime candidates. The reduction from 72.3 to 3.7 is significant. Specifically, 2.4 out of the original 72.3 prime candidates are true fault locations, while 1.2 out of 3.7 word-level prime candidates are true. It shows that certain true faults are still mistakenly screened out when word-level information is exploited; however, this minor *over-killing effect* is not catastrophic because of two considerations. First, the sharpness has been greatly improved; failure analysis based on this guidance will be able to hit one fault in less than four guesses, and this is the ultimate goal of diagnosis. Second, it is possible that the true faults *escaping* the first phase of detection can still be targeted in the second phase.

After having identified the word-level prime candidates, we perform a reduction before moving on to the divide-and-conquer process. In this reduction, some subcircuits must be removed in the CUD. The subcircuits being removed include the fanin cone of every mismatched output reachable by the identified word-level prime candidates.

Example 7.15

Consider the example shown in Figure 7.18. Assume that f is a word-level prime candidate. The mismatched outputs reachable by f are $\{y, z\}$; hence, we remove all fanin cones of $\{y, z\}$. The reason why these logic gates in the CUD are removed

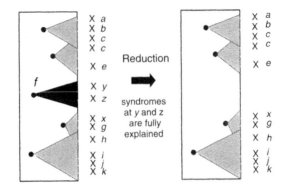

■ **FIGURE 7.18**

Circuit reduction based on word-level prime candidates.

from further consideration is because all the syndromes at $\{y, z\}$ have been perfectly explained by the identified word-level prime candidates.

Circuit reduction offers two advantages: (1) The subsequent diagnosis can be focused on those mismatched outputs that are not well explained yet, and (2) it is more likely to partition the remaining CUD into separate subcircuits because the structural correlations among them have been eliminated to some extent.

7.2.3.3 Overall Chip-Level Diagnostic Flow

The overall procedure is shown in Figure 7.19. The inputs include a netlist as the CUD, a set of failing test vectors, and the faulty response of the chip. First, we run fault-free simulation on the CUD using every failing test vector to determine the mismatched outputs. Second, we adopt the common *structural pruning* techniques to narrow the fault candidate area down to the joint fanin cones of the mismatched outputs. In other words, we use the concept of cone union instead of cone intersection. Only signals in the candidate area will be considered for further *prime candidate* checking. In practice, structural pruning could be a powerful technique that reduces the run time of the diagnosis dramatically. It allows one to focus on a small subcircuit even though the CUD is extremely large.

In the first phase, fault simulation using each failing test vector is necessary in the identification process of the word-level prime candidates. Fortunately, the number of failing test vectors is not necessarily large, so this process may not be that time consuming. Comprehensive experiments indicate that 32 failing test vectors are normally enough.

In the second phase, the CUD reduction is quickly carried out through simple structural analysis, then a dependency graph is incorporated to break down the remaining CUD to one or several *diagnosis blocks*, as described previously. Each diagnosis block requires a separate diagnosis session using block-level techniques. Finally, one ranked list of candidates will be generated for each block. These ranked lists are merged into a single one in an interleaved manner. For example, if the word-level prime candidates are denoted as $P = \{p_1, p_2, \ldots, p_k\}$ and two extra ranked

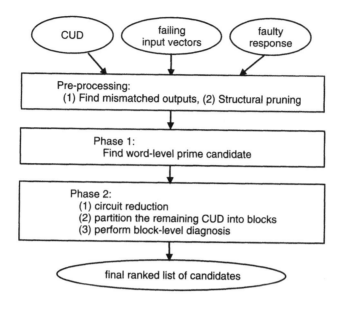

■ **FIGURE 7.19**

Overall chip-level diagnostic flow.

lists generated in the second phase are $S_1 = \{\alpha_1, \alpha_2, \ldots, \alpha_m\}$ and $S_2 = \{\beta_1, \beta_2, \ldots, \beta_n\}$, then the final list will be $F = \{p_1, p_2, \ldots, p_k | \alpha_1, \beta_1, \alpha_2, \beta_2, \ldots\}$.

In summary, the concept of *word-level prime candidate* can be used as a chip-level strategy for identifying structurally independent faults. Once such candidates are found, they are removed along with their fanin cones to unlink the structural correlations among mismatched outputs, thereby easing the subsequent partitioning process. By doing so, other independent or dependent faults can be isolated and the divide-and-conquer methodology can be more effective in finding other structurally dependent faults in the second phase.

7.2.4   Diagnostic Test Pattern Generation

The diagnostic resolution is linked not only to the diagnosis algorithm but also to the test set quality. It is often dictated by the *maximum number of faults in an equivalence class* partitioned by the test set. Here, an equivalence class of faults is a set of faults that cannot be differentiated from one another any further by the test set. To be more specific, given two stuck-at faults f_1 and f_2, a test vector v is said to differentiate these two faults if v produces at least one output mismatch when applied to the two faulty circuits with f_1 and f_2, respectively.

It is common that the high-volume manufacturing test set is used for diagnosis as well; however, such a high fault-coverage test set does not always guarantee a high diagnostic resolution or accuracy. In order to enhance the diagnostic quality, a ***diagnostic test pattern generation*** (DTPG) process might be needed. In general,

DTPG augments the original test set by differentiating vectors that can further refine the yet indistinguishable fault classes.

Definition 7.13

Given a pair of faults, f_1 and f_2, we would like to generate a differentiating vector. This vector generation can be done in a model demonstrated in Figure 7.20, where the inputs of the two faulty circuits have been merged, whereas each corresponding output pair is connected to an XOR gate and every XOR gate is further connected to an OR gate that produces the only final output denoted as z. It can be proved that if a vector v can detect $z/0$ faults, then v is a differentiating vector for f_1 and f_2.

7.2.5   Summary of Combinational Logic Diagnosis

In this section, we have discussed methods for combinational logic diagnosis. We have covered two major paradigms: cause–effect analysis and effect–cause analysis. For cause–effect analysis, we described techniques for constructing a compact fault dictionary so as to deal with large designs. For effect–cause analysis, we discussed the structural pruning techniques, the basic backtrace algorithm, the powerful inject-and-evaluate paradigm, and finally a chip-level divide-and-conquer strategy for dealing with multiple faults.

For bridging fault diagnosis, one may desire to locate the exact *signal pairs* that are shorted together as mentioned previously in the SLAT paradigm. There are

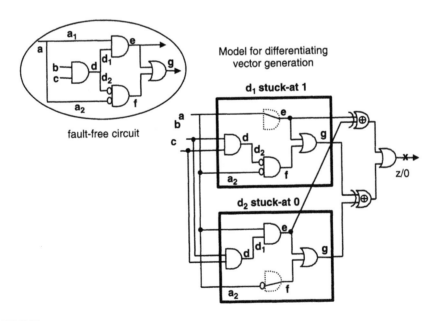

■ **FIGURE 7.20**

Differentiating vector generation for diagnosis.

many other approaches to diagnosing bridging faults [Millman 1990] [Chess 1995] [Olson 2000] [Wu 2000] [Lavo 2002]. It is worth mentioning that a bridging fault could sometimes give rise to an analog voltage level at the fault site. Such an ambiguous signal could be interpreted differently either as a logic "0" or "1" by its fanout gates with different threshold voltages, a phenomenon often referred to as the *Byzantine general's phenomenon* [Lavo 1996]. To cope with this phenomenon, *symbolic simulation* can be incorporated inside the inject-and-evaluate paradigm [Huang 2002] [Wen 2004] [Smith 2005]. The symbolic simulation techniques can be further applied to the diagnosis of partial-scan designs, by a *fading scheme* [Huang 2004]. Open faults in an interconnecting wire are another common cause of failure. An interconnecting wire physically branches out like a huge tree structure on the silicon. For fault analysis purposes, we need to locate the fault down to a specific segment. A formulation for such a segment-level open fault diagnosis can also be found in [Huang 2003].

7.3   SCAN CHAIN DIAGNOSIS

Scan chains have long been touted as an effective aid for logic circuit testing and diagnosis; however, it has been reported that scan chain failures account for almost 50% of chip failure in some cases. These failures are often revealed during the normal **flush test** for a scan chain in which a set of random patterns are shifted in and out of a flip-flop chain to ensure that scan shifting is not blocked. Even though the test is simple, identifying where the scan chain is blocked is not. Locating scan chain defects is also important for yield improvement. In this section we will introduce three major types of scan chain diagnosis methods: (1) hardware-assisted, (2) modified inject-and-evaluate paradigm, and (3) signal-profiling based.

7.3.1   Preliminaries for Scan Chain Diagnosis

For simplicity, without loss of generality, we assume that there is only one scan chain in the circuit under diagnosis. As shown in Figure 7.21, there is a **scan input** (SI) pin and a **scan output** (SO) pin. The flip-flops in the scan chain are ordered from SI to SO sequentially, denoted as $(q_1, q_2, \ldots, q_n)$, assuming that there are n flip-flops. We further define the *snapshot image* and the *observed image* of a given scan chain as follows.

Definition 7.14 (Snapshot Image)

The snapshot image of a scan chain is the logic value combination of all the scanned flip-flops at a particular time instance. Note that the snapshot image at any clock cycle of the fault-free circuit under diagnosis is available through functional simulation as long as a test sequence is given; however, the snapshot images of a scan chain in a failing chip are obviously not available due to the blockage.

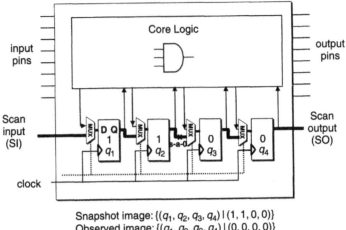

Snapshot image: $\{(q_1, q_2, q_3, q_4) \mid (1, 1, 0, 0)\}$
Observed image: $\{(q_1, q_2, q_3, q_4) \mid (0, 0, 0, 0)\}$

■ **FIGURE 7.21**

Snapshot image and observed image of a scan chain.

Definition 7.15 (Observed Image)

The observed image of a scan chain is the scanned-out version of a snapshot image. For a fault-free circuit, it is equivalent to the snapshot image. For a failing chip, it consists of the bitstream collected at the scan output pin. These two images could be different due to the presence of faults inside the scan chain or the combinational logic.

Example 7.16

In Figure 7.21, a stuck-at fault in a scan chain acts as a *signal distortion element* that produces an observed image different from its original snapshot image. In this case, the snapshot image is originally $\{(q_1, q_2, q_3, q_4) \mid (1, 1, 0, 0)\}$, whereas the observed image is $\{(q_1, q_2, q_3, q_4) \mid (0, 0, 0, 0)\}$.

In this section, we use the following test commands.

Definition 7.16 (Scan_Shift Command)

A command that serially applies a bit into the scan chain through the scan input pin and retrieves a bit out of the scan chain from the scan output pin at the same time. The traditional flush test can be regarded as a sequence of *Scan_Shift* commands.

Definition 7.17 (Capture Command)

A functional-mode command that forces each flip-flop to take its functional input, instead of the scan input, as the content. A system clock (if different from the scan clock) might be needed.

Definition 7.18 (Apply Command)

The apply command is a functional-mode command that applies a vector to the chip input pins of the circuit under diagnosis.

Like the diagnosis of faults inside the combinational logic, either cause–effect analysis or effect–cause analysis can be applied to scan chain diagnosis based on some fault models. As shown in Figure 7.22, a number of fault types have commonly been targeted [Kundu 1993] [Huang 2003]. These fault types are usually classified by two characteristics: (1) functional or timing, and (2) permanent or **intermittent**. An intermittent fault refers to a fault that occurs nondeterministically or only under a certain operating environment (*e.g.*, when the power supply is noisy or unstable). For functional faults, stuck-at and bridging faults are mostly used. For timing faults, two types of timing violations associated with the flip-flops need to be considered: **setup time violation** and **hold-time violation**. Setup time violation is mostly due to *late signal* at a flip-flop's input, while the hold-time violation is due to a *too-early signal change* at a flip-flop's output. Often, the setup time violation fault is further divided into the two subtypes of *slow-to-fall* and *slow-to-rise* faults to reflect the difference between the up-transition and down-transition driving strengths of the logic cells.

Figure 7.23 illustrates a number of faulty syndromes observed at a scan output pin under a number of common fault types. Here, we assume that we flush into the scan chain a bitstream "0011001100110011" forming a pattern. Due to a fault effect at certain flip-flop, the observed bitstream could be altered; for example, it could become "001001001001" in the presence of a slow-to-rise fault. This means that an up-transition at the output of a flip-flop cannot be finished in one clock cycle. While a flush test is not a good vehicle for locating the fault, it can be used to classify the fault type. For example, an all-0 syndrome indicates a stuck-at-0 fault,

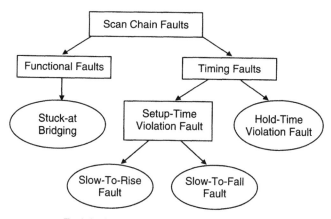

Each fault could be permanent or intermittent.

■ **FIGURE 7.22**

Common fault types in a scan chain.

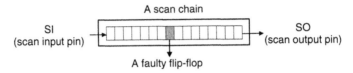

Fault Type	Scan-In Pattern	Observed Syndrome
Stuck-at-0	1100110011001100	<u>0000000000000000</u>
Stuck-at-1	1100110011001100	<u>1111111111111111</u>
Slow-to-Rise	1100110011001100	1<u>000</u>1<u>000</u>1<u>000</u>1<u>000</u>
Slow-to-Fall	1100110011001100	1101110111011100

The rightmost bit goes into the scan first
The rightmost bit gets out of the scan first
An underlined bit in the observed image is failing.

■ **FIGURE 7.23**

Example faulty syndromes of a scan chain.

and missing a few transitions indicates a transition fault. What is not modeled in this figure is the common bridging fault (*e.g.*, a flip-flop output unexpectedly shorted with a logic cell's output). Diagnosis of a bridging fault in the scan chain is usually more challenging due to its huge possible fault candidate space and its intermittent nature.

In addition to the fault models, the **diagnostic test procedures** also affect the accuracy of diagnosis. In our discussion, we assume that one is free to mix any kind of vectors, including functional vectors, scan vectors, and flush vectors, as long as the faults can be properly revealed.

7.3.2 Hardware-Assisted Method

A number of hardware-assisted approaches have been proposed in the literature [Schafer 1992] [Edirisooriva 1995] [Narayanan 1997] [Wu 1998]. The basic idea is that a scan chain is difficult to diagnose, so one may need to insert extra logic to facilitate the process. There are a number of reasons why scan chain diagnosis is seemingly more difficult than combinational logic diagnosis.

- In combinational logic diagnosis, the scan chain is assumed to have been validated, but when it comes to scan chain diagnosis we cannot assume the integrity of the combinational logic. This implies that a good scan chain diagnosis method should be robust enough to endure the harsh conditions when certain cells in the combinational logic are also faulty.

- The observability of a scan chain is limited. We can only retrieve the failing syndrome from the scan output pins. With less information, it is more difficult to trace back to the cause of failure.

- A simple fault in the scan chain could have *global effects*. For example, when we flush the scan chain with a simple stuck-at-0 fault by a random bitstream, then what we get at the scan output pin is an all-zero bitstream. This is because every bit we observed passes through the faulty flip-flop and gets distorted.

A ***design-for-diagnosis*** (DFD) scheme augments a scan flip-flop, as shown in Figure 7.24. An XOR gate is placed in the front and controlled by an extra control signal *Invert*. This type of flip-flop has three modes of operation: (1) normal operation, (2) scan-shifting operation, and (3) **inversion operation**. Under inversion operation, all flip-flops are inverted simultaneously, triggered by a global signal *Invert* that connects to each scan cell.

With the extra supporting circuitry, the *diagnostic test procedure* using a modified flush test can proceed as follows:

1. Prepare a flush pattern (*e.g.*, all-1 pattern).

2. Scan in this pattern into the scan chain. If there are *n* flip-flops in the chain, we need to apply *n* scan-shifting commands.

3. Invert the scan chain by setting signal *invert* to one.

4. Scan the image out of the scan chain. Check the observed image to see if there is a stuck-at-0 flip-flop in the scan chain.

5. Repeat steps 1 to 4 by flushing the all-0 pattern to see if there is a stuck-at-1 flip-flop.

The following example illustrates how this procedure can successfully locate a stuck-at fault.

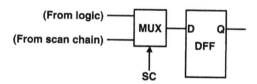

(a) A normal scan flip-flop

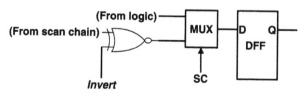

(b) A modified scan flip-flop for diagnosis

■ **FIGURE 7.24**

Augmentation of a scan flip-flop for diagnosis.

Example 7.17

Assume that we have a stuck-at-0 fault at the fifth flip-flop counting from the scan input side, and the bitstream to be flushed into the chain has an all-1 pattern. The fault location divides the scan chain into two parts: **SI-to-fault** and **fault-to-SO**, as shown in Figure 7.25. After we have scanned in the pattern, the *snapshot image* in the failing chip will become (1111000000000000). This is because the fault-to-SO part has been distorted by the stuck-at-0 fault, whereas the SI-to-fault part is not affected. Before we do the scan-out, we invert this snapshot image to (0000011111111111). Then, we scan it out and obtain an observed image as (0000011111111111). The fault location is revealed at the edge between the 0's and 1's.

7.3.3 Modified Inject-and-Evaluate Paradigm

The inject-and-evaluate paradigm can also be applied to scan chain diagnosis with a specific fault model [Stanley 2001] [Huang 2003]. Unlike the hardware-assisted method, this is a software method without any area overhead. As noted previously, the *flush test*, although ineffective for fault location, can be used for classifying the fault type first. Once the fault type is known, the subsequent fault injection process can be more realistic and thus lead to a more accurate result. A diagnostic test procedure of this type operates the same as normal scan testing, which goes through a *scan–capture–scan* scenario for each test vector.

Example 7.18

An example is shown in Figure 7.26. We assume that there is a stuck-at-0 fault at the output of the second flip-flop. After we have scanned in a bitstream (1011), the snapshot image becomes (1000); that is, the two bits nearest the scan output pin have been corrupted. Next, a system clock is applied to capture the next-state

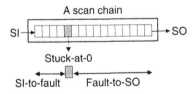

(1) Original bitstream pattern = (1111111111111111)
(2) After scan-in: snapshot image = (1111000000000000)
(3) After inversion: snapshot image = (0000011111111111)
(4) After scan-out: observed image = (0000011111111111)

The fault location is at the edge between 0's and 1's.

■ **FIGURE 7.25**

Fault location via inversion operation.

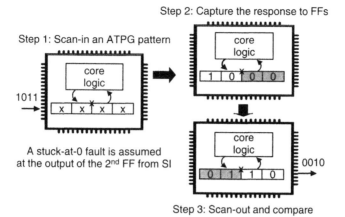

Step 1: Scan-in an ATPG pattern

Step 2: Capture the response to FFs

1011

A stuck-at-0 fault is assumed
at the output of the 2nd FF from SI

0010

Step 3: Scan-out and compare

■ **FIGURE 7.26**

The scan–capture–scan test procedure.

values to the flip-flops, thereby creating a new snapshot image, say (0110). Finally, this new snapshot image is scanned out to become the observed image (0010).

It can be seen from the above process that the distortion on the observed image is quite severe. This means that what we observed at the scan output pin could be quite different from what we have anticipated from a fault-free chip. But, still, this scan–capture–scan process maps a test vector (as a bitstream) to an output vector (as the observed image). Based on this mapping relation, we can try to inject a fault in the scan chain to see which one can faithfully *reproduce* the syndrome we observed. When there is no perfect match, some ranking heuristic as discussed in the subsection on combinational logic diagnosis can be applied here as well.

In light of the above discussion, we have the following inject-and-evaluate paradigm for scan chain diagnosis:

1. Run a *flush test* to guess the type of the faults.

2. Pick one *scan vector* as a bitstream. Simulate the scan–capture–scan process on the CUD to derive the fault-free (observed) image.

3. Pick one possible fault candidate. Inject the fault effect into the scan chain.

4. Simulate the scan–capture–scan process on the CUD and derive its failing (observed) image.

5. Compare the failing image with the fault-free image. Accumulate the *matching score* for each fault candidate.

6. Go back to step 3 if there are more candidates.

7. Go back to step 2 if there are more scan vectors.

8. Rank candidates based on their matching scores. The higher the score, the higher the rank.

In general, this paradigm is not as effective as that of the combinational logic diagnosis, due to a number of reasons. First, it is not easy to find a *universal injection model* to represent the fault effect of a scan chain fault. In combinational logic diagnosis, the fault effects of different types of faults (*e.g.*, stuck-at, bridging, or interconnect open) can be denoted as a *signal flipping* at the fault locations when they are activated. This, however, is not true for scan chain diagnosis. A stuck-at fault and a bridging fault in the scan chain could behave totally differently. The fault effect modeling for a bridging fault in the scan–capture–scan process is even more challenging. Furthermore, unlike the single-cycle process in the combinational logic diagnosis, a test session now takes multiple clock cycles to complete. It is not easy to figure out at what clock cycle the bridging fault really takes the toll. Second, the distortion is so severe that many faults could have similar effects on the final observed image. As a result, the differentiation among faults may not be very phenomenal.

7.3.4 Signal-Profiling-Based Method

Instead of using scan vectors as the vehicle, the signal-profiling-based method drives the failing chip through *functional mode* with selected **diagnostic test sequences**; thus, the fault effect can be reflected to the observed image in a more systematic way. After that, signal-processing techniques such as *filtering* and *edge detection* can be applied to reveal the location of the faulty flip-flops. The entire flow consists of two major parts: (1) the diagnostic test sequence selection, and (2) the subsequent analysis of the observed syndromes.

7.3.4.1 *Diagnostic Test Sequence Selection*

The *diagnostic test sequence* in this method is literally the functional sequence derived from the simulation testbench. It could start from a given reset state or an unknown state. The primary objective for such a sequence is to bring the failing chip through a state sequence as randomly as possible. The randomness is measured by the *signal*-1 *frequency* of the flip-flops, illustrated by an example in Figure 7.27.

In this example, a three-vector sequence is applied to the CUD. In the time-frame expansion model, the value of each flip-flop is shown. For example, the first flip-flop goes through sequences of $\{0 \rightarrow 0 \rightarrow 0 \rightarrow 0\}$ in four cycles, while the second flip-flop goes through $\{0 \rightarrow 1 \rightarrow 0 \rightarrow 1\}$. By counting the occurrences of signal-1's, their signal-1 probabilities over the time can be calculated as 0.0 and 0.5, respectively. In other words, the second flip-flop has become randomized; however, the randomness of the first one is not yet adequate. Sequences like this will be chosen one by one until every flip-flop has a fairly random value.

7.3.4.2 *Run-and-Scan Test Application*

Unlike the traditional *scan–capture–scan* test procedure, this method adopts a procedure referred to as **run-and-scan**, meaning that the procedure involves running a number of functional sequences followed by a number of scan-out operations. At the end of each test sequence application, only the final snapshot image of the scan

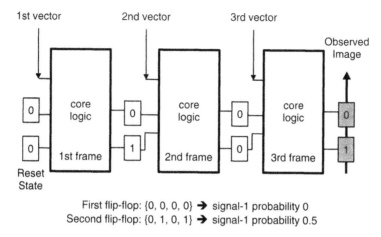

First flip-flop: {0, 0, 0, 0} ➔ signal-1 probability 0
Second flip-flop: {0, 1, 0, 1} ➔ signal-1 probability 0.5

■ **FIGURE 7.27**

Illustration of signal frequencies at flip-flops.

chain is shifted out and recorded as the observed images. In this methodology, a large number of test sequences (*e.g.*, 100) are required to achieve accurate diagnosis results.

Assume that a long functional sequence, denoted by $(v_1, v_2, \ldots)$ is provided by the designer. During the design process, this long functional sequence is assumed to have been simulated at the register-transfer level or gate level. The flip-flop values in response to this sequence at each clock cycle are therefore available instantly, in the form of a ***value–change–dump*** (VCD) file when performing scan chain diagnosis.

The test sequence selection is first done by selecting a number of clock cycles, say [1, 4, 5, 7], with an attempt to make the values of each flip-flop at these selected clock cycles as random as possible. In other words, the signal-1 frequency of each flip-flop at these selected clock cycles is as close to 0.5 as possible. Once this has been done, we can generate their corresponding test sequences. For each selected clock cycle j, we simply take the prefix of the functional sequence up to this clock cycle as the corresponding sequence, such as $(v_1, v_2, \ldots, v_j)$. As illustrated in Figure 7.28, one test sequence is generated for each selected clock cycle in [1, 4, 5, 7]. Each of them starts from a known reset state or an unknown state. The final snapshot image at the flip-flops in response to each test sequence will then be scanned out for subsequent analysis. In this case, the snapshot images to be scanned out are $\{(q_1, q_2)|(0, 0), (1, 0), (0, 1), (1, 1)\}$. The values of both q_1 and q_2 switch between 0 and 1 over the time; therefore, the goal of randomness is met to some extent.

7.3.4.3 *Why Functional Sequences?*

The reason why we use functional sequences can be explained by analyzing how the fault effect manifests itself during the test application. As discussed in the previous subsections, there are two stages during the *run-and-scan* test application,

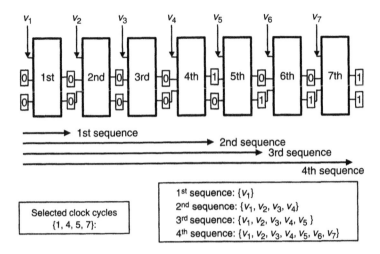

■ FIGURE 7.28

Example of test sequence selection.

i.e., (1) *functional sequence stage*, and (2) *scan-shifting stage*. In the first stage, the snapshot image of the scan chain inside the failing chip is only slightly affected by the fault. Such fault effects could be regarded as white noise that could slightly affect a number of flip-flops only. On the other hand, the fault effect is much more prominent during the scan-shifting stage, when a snapshot image is shifted out to become the failing observed image. The distortion at this stage is usually dramatically biased; therefore, it contains information about the fault location.

An image can be viewed as the composition of two parts, the *SI-to-fault* part and the *fault-to-SO* part. When shifting out a snapshot image, the *fault-to-SO* part is not affected, while the *SI-to-fault* part could be seriously distorted. As a result, it is very likely that there will be a big difference between the failing image and fault-free image for the *SI-to-fault* part but not for the *fault-to-SO* part.

Example 7.19

In the example shown in Figure 7.29, there are four flip-flops in the scan chain: $\{q_1, q_2, q_3, q_4\}$. Assume that the application of t test sequences gives rise to t corresponding snapshot images of the scan chain. The signal-1 frequencies of the flip-flop bits among these t snapshot images are assumed to be (0.4, 0.5, 0.6, 0.4) for the fault-free model and (0.41, 0.51, 0.61, 0.41) for the failing chip. This represents a likely condition that the failing snapshot image is only a perturbation of the fault-free snapshot image in terms of the signal-1 frequency; however, after we have shifted out the scan chain contents as the observed images, their difference is more significant. In this illustration, it remains (0.4, 0.5, 0.6, 0.4) for the fault-free model, but it becomes (0, 0, 0.61, 0.41) for the failing chip. Hence, we can infer that the fault occurs at the second flip-flop that distorts the signals at the first two bits dramatically.

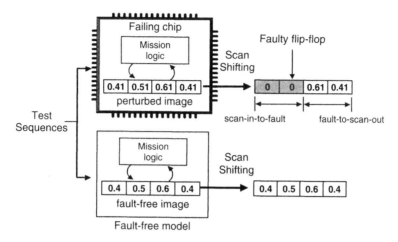

■ **FIGURE 7.29**

Global distortion due to scan shifting.

The above discussion implies that this approach could be relatively robust even when there are also certain faults in the combinational logic. The reason is that such fault is likely to cause *white-noise* type of fault effects on the flip-flops; therefore, the biased fault effects during the scan shifting stage can still dominate the overall failing syndrome at the chip's output and thereby provide diagnosis information. Because of this property of graceful degradation, our experiments even show that this approach is applicable when the faults in the scan chain are intermittent (*e.g.*, bridging faults).

The proposed approach is also applicable to designs without a completely known reset state (*e.g.*, a design without completely resettable data path registers). In that case, we can resort to the three-valued logic simulation in deriving the diagnostic test sequences and the snapshot images.

7.3.4.4 *Profiling-Based Analysis*

In this subsection we discuss how to perform signal processing on the observed images to locate the faulty flip-flop. The overall procedure is shown in Figure 7.30:

- **Step 1.** Profile the signal-1 frequency of each flip-flop bit from the fault-free observed images. The result is a *fault-free profile*, denoted as $good(i)$, where i is the flip-flop index.

- **Step 2.** Profile the signal-1 frequency of each flip-flop bit from the set of failing observed images. The result is a *failing profile*, denoted as $bad(i)$.

- **Step 3.** Compute the *difference profile* between the fault-free images and the failing images. For a flip-flop q_i, we calculate the frequency that its fault-free value is different from its failing value as the *difference frequency*. Once the difference frequency has been derived for each flip-flop, we can derive the

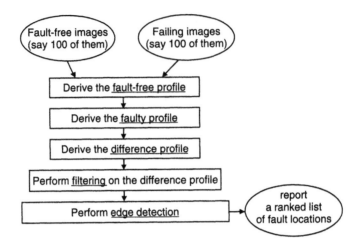

Profiling-based analysis flow.

profile over the entire scan chain to analyze the trend for revealing the fault location.

- **Step 4.** Perform *filtering* on the difference profile to eliminate the glitches.

- **Step 5.** Perform *edge detection* to derive a *suspicion profile*. In this profile, the flip-flops with higher suspicion values are considered more likely to be a fault location.

Example 7.20

Figure 7.31 shows the fault-free profile and failing profile of a real circuit. If the fault to be diagnosed is a stuck-at fault, there might be an all-0 or all-1 region in the *failing profile* closer to the scan input part. The boundary of the all-0 region shown is right at the faulty flip-flop location.

A stuck-at fault can be easily detected from the failing profile; however, more sophisticated analysis such as filtering and edge detection are required for bridging faults or transition faults (*e.g.*, slow-to-rise or slow-to-fall). *Filtering* is mainly used to smooth out the *difference profile* in such a way that the small glitches can be removed. A simple *average-sum filtering* is often adequate in this application.

Definition 7.19 (Average-Sum Filtering)

Assume that the difference profile is given and denoted as $D[i]$, where i is the index of a flip-flop. We use the following formula to compute a *smoothed difference profile*, $SD[i]$:

$$SD[i] = 0.2^*(D[i-2]+D[i-1]+D[i]+D[i+1]+D[i+2])$$

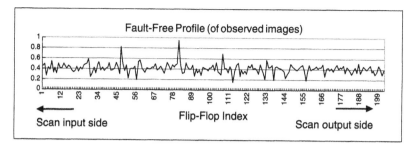

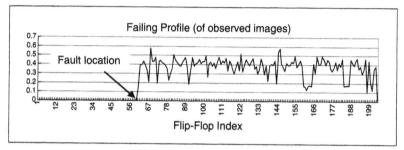

- **FIGURE 7.31**

Failing profile in the presence of a stuck-at fault.

The smoothed difference profile tends to have a trend that it is higher at the left-hand side (*i.e.*, the scan input side) and lower at the right-hand side (*i.e.*, the scan output side), with a transition region in between. The left-hand side is higher because the flip-flops closer to the scan input pin could become more distorted during the scan-shifting operations, whereas the flip-flops closer to the scan output pin are mostly unaffected. The true location of the faulty flip-flop is likely to be the *left-boundary of the transition region in the difference profile*. To detect this boundary, we can use the simple edge detection formula defined below.

Definition 7.20 (Edge Detection)

On the smoothed difference profile $SD[i]$, the following formula can be used to compute the faulty frequency of each flip-flop as a suspicious profile:

$$suspicion[i] = [-1, -1, -1, 1, 1, 1] \cdot \begin{bmatrix} |SD[i] - SD[i-3]| \\ |SD[i] - SD[i-2]| \\ |SD[i] - SD[i-1]| \\ SD[i] - SD[i+1] \\ SD[i] - SD[i+2] \\ SD[i] - SD[i+3] \end{bmatrix}$$

This is a weighted-sum formula that tends to maximize at the left boundary of the transition region in the difference profile. Intuitively, for a flip-flop at location i, we take three of its left neighbors ($SD[i-3]$, $SD[i-2]$, and $SD[i-1]$) and three of its right neighbors ($SD[i+1]$, $SD[i+2]$, and $SD[i+3]$) in the computation of *suspicion*[i]. This formula is anticipated to peak at a flip-flop location where its left-hand side neighbors are roughly the same as its *SD* value, while its right-hand side neighbors are sharply lower. The calculation is on the *reward-and-penalty* basis:

- We add points to the overall score of a flip-flop i if its right-hand-side neighbors' values are lower than the current $SD[i]$. This is a reward mechanism.

- We deduct points from the overall score of a flip-flop i if its left-hand-side neighbors' values deviate from the current $SD[i]$. This is a penalty mechanism.

- The final suspicion degrees of the flip flops are then sorted in a decreasing order as the final *ranked list* of fault candidates.

Figure 7.32 illustrates the *difference profile*, the *smoothed difference profile*, and the *suspicion profile* for a scan chain with a stuck-at-1 fault at flip-flop 23. The horizontal axis of each of these three profiles is the index of the flip-flops, ordered from the scan input pin toward the scan output pin. In other words, a flip-flop with a smaller index is a flip-flop closer to the scan input pin. It can be seen that a peak in the suspicion profile clearly indicates the location of the faulty flip-flop. Similarly, Figure 7.33 shows the profiles when there is a slow-to-rise fault in the scan chain.

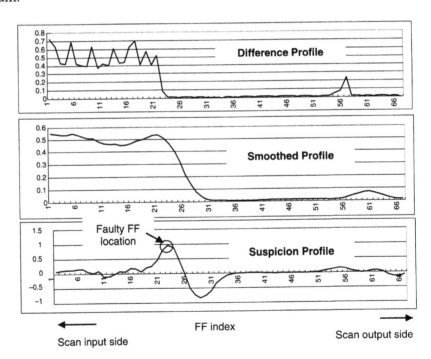

FIGURE 7.32

Profiles of a scan chain with a stuck-at fault.

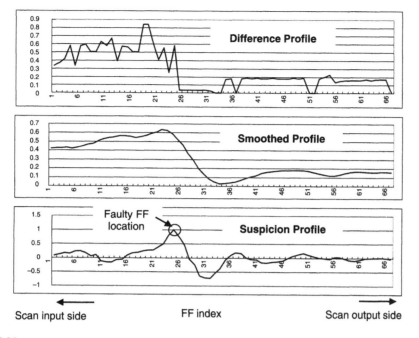

■ **FIGURE 7.33**

Profiles of a scan chain with a transition fault.

7.3.5   Summary of Scan Chain Diagnosis

Diagnosis of flip-flop faults in a scan chain is sometimes difficult because of two reasons. First, the observation is limited at the scan output pins only. Second, the fault effect could be global and overwhelming in the sense that it affects entire observed images during the scan-shifting operation. Three basic methods have been introduced in this section: hardware-assisted, modified inject-and-paradigm, and signal-profiling-based. In general, techniques for scan chain diagnosis are not as mature as those for combinational logic. New methods are still emerging. A number of historical approaches proposed in the literature are worth reading. For example, [Kundu 1993] used sequential ***automatic test pattern generation*** (ATPG) to find a test sequence that sets the flip-flops to some proper values and then scans them out for analysis. To overcome the high complexity of sequential ATPG, [Cheney 2000] proposed using random test patterns as the diagnostic test patterns. Fault simulation and response-matching heuristics were then utilized to gauge the most likely fault candidates. In general, such a process takes a long time because it has to enumerate a large number of fault candidates in the faulty scan chain. [Stanley 2001] proposed taking advantage of the fault-free scan chains as the vehicle for setting the values of the faulty chain, thereby reducing the difficulty of the diagnostic test generation process. By doing so, only combinational ATPG is required for most cases. [Guo 2001] further proposed powerful upper bounding and lower bounding techniques to narrow the fault locations down to a small region, thereby reducing the fault simulation time significantly. [Huang 2003] [Huang 2004] enhanced the bounding techniques and introduced probability for modeling the intermittent faults. More recently, [Li 2005b,c] further optimized this framework by incorporating the so-called *single-excitation patterns* and ATPG techniques to provide better diagnosis resolution.

7.4 LOGIC BIST DIAGNOSIS

Built-in self-test (BIST) involves using on-chip hardware for both test pattern generation and output response analysis (see Chapter 5 for details). The most economical forms of logic BIST involve using a ***pseudo-random pattern generator*** (PRPG) to apply a large number of test patterns and using a ***multiple-input signature register*** (MISR) to combine the output responses into a single *signature*. If the resulting signature is incorrect, then the chip fails the test. Diagnosing the cause of the failure in a logic BIST environment is very challenging because the output response is so highly compacted. This section begins with an overview of the problem of diagnosis in a logic BIST environment, and then describes practical techniques for determining which test vectors failed (time information) and which scan cells captured errors (space information).

7.4.1 Overview of Logic BIST Diagnosis

A BIST architecture that is widely used in industry is the **STUMPS** architecture [Bardell 1982] which is illustrated in Figure 7.34. The core logic contains multiple scan chains which are loaded from a PRPG. After a test vector has been shifted in, the system clock is applied and the output response is captured back into the scan chains. As the next test vector is shifted in, the output response gets shifted out and compacted into a MISR. For diagnosis, a scan-in port can be connected to

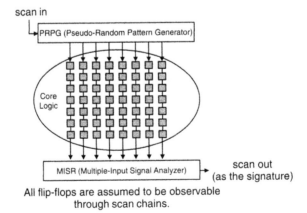

All flip-flops are assumed to be observable
through scan chains.

■ **FIGURE 7.34**

An example of the logic BIST STUMPS architecture.

the PRPG to externally load an initial seed (starting pattern), and a scan-out port can be connected to the MISR to shift out a final signature for observation.

At the end of the BIST session, the final signature is so highly compacted that very little information can be extracted from it for diagnostic purposes unless the number of bits in error is very small. In general, there is no bound on the multiplicity of errors during BIST because a single defect can cause a large number of vectors to produce faulty responses and a large number of scan cells can capture those faulty responses. Diagnosis in a BIST environment adds an extra level of difficulty compared with diagnosis in a non-BIST environment because it requires determining from compacted output responses which test vectors have produced a faulty response (*time information*) or which scan cells have captured errors (*space information*). For example, consider the case where 1 million test vectors are applied during a BIST session to a circuit with 20,000 scan cells. Time information would refer to which of the 1 million vectors failed, and space information would refer to which of the 20,000 scan cells captured a faulty response. Interval-based methods can be used to obtain time information and are described in Section 7.4.2. Masking-based methods used to obtain space information are described in Section 7.4.3.

One simple but highly inefficient way to perform BIST diagnosis is to just bypass the MISR and shift out the full output response for every test vector to an external tester. The problem with this approach is that typically a very large number of test patterns are applied to the circuit during BIST (orders of magnitude more than are applied in conventional deterministic testing); consequently, the tester may not have sufficient memory to store the full output response data for every vector. Moreover, the time required for collecting and processing all of this data is generally not as cost effective as other more efficient BIST diagnosis approaches that are described in the remainder of this section.

7.4.2 Interval-Based Methods

One general approach for diagnosis in a BIST environment is, instead of running the entire BIST session to generate a single signature, to run the BIST in shorter intervals and generate a signature for each interval [Savir 1997] [Song 1999] [Liu 2003]. If the signature for an interval is faulty, then it is known that the circuit is failing for at least one test vector in the interval. Running the BIST for smaller intervals requires being able to start the BIST from *designated seeds* (a *seed* is a starting state for the PRPG; see Chapter 5 for more details) in addition to the normal initial seed and being able to run the BIST for a *designated length*. This capability can be provided by making both the PRPG and the *pattern counter* scannable (the pattern counter counts down from its initial value to 0, at which point the BIST is stopped).

One approach for finding the first failing test vector, which was described in [Song 1999], involves performing a *binary search* using intervals. Given a circuit that has failed the entire BIST session, the following search process is used:

- ■ **Step 1.** Load the PRPG with the normal initial seed, but load the pattern counter with only half the normal BIST length. This specifies an interval equal to the first half of the normal BIST session.

- **Step 2.** Run the BIST for the specified interval.

- **Step 3.** When the BIST is complete, see if the signature is correct.

- **Step 4.** If the signature is faulty, then the first failing vector exists in the first half of the test vectors in the previous interval. Step 2 is then repeated using the same initial seed as previously but only half the previous length. This specifies a new interval equal to the first half of the previous interval.

- **Step 5.** If the signature is correct, then the first failing vector exists in the second half of the test vectors in the previous interval. Step 2 is then repeated using a seed that corresponds to the start of the second half of the patterns in the previous interval, and only half the previous length is used. This specifies a new interval equal to the second half of the previous interval.

- **Step 6.** The above procedure iterates until the interval becomes equal to only a single pattern (*i.e.*, the interval length is equal to 1). At this point, the first failing pattern has been identified.

Example 7.21

Figure 7.35 illustrates the binary search process. Assume that the full BIST session contains 28 test vectors and vector 4 is the first failing vector. In this case, five interval BIST sessions are required, where the interval lengths go through the sequence of $\{14 \rightarrow 7 \rightarrow 4 \rightarrow 2 \rightarrow 3\}$ test vectors to locate the first failing test vector (in this case, vector 4). When the first failing test vector is found, an interval BIST session is then run again to stop after vector 4. The contents of all the scan chains (which hold the output response for vector 4) are then shifted out to the tester for diagnosis. If necessary, the binary search process can be resumed to find the

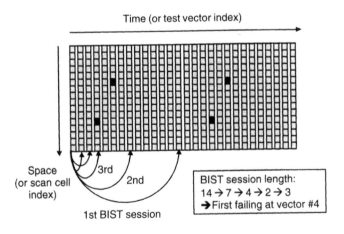

■ FIGURE 7.35

A binary search process to locate the first failing test vector.

second failing vector by initializing the PRPG with the seed of the fifth test vector, instead of the initial seed.

The binary search method requires $\log_2(BIST\ length)$ steps to identify each failing test vector. An interval-based method that requires only two steps to collect all diagnostic information (although it requires more tester memory) was described in [Wohl 2002]. It is called *interval unloading* and involves dividing the entire BIST session into small fixed-size intervals (*e.g.*, 32 patterns each). In the first step, the BIST is run for each interval and the compacted signature for each interval is scanned out to the tester for comparison with the fault-free signature to identify the failing intervals. In the second step, the BIST is run for each failing interval, and the full uncompacted output response for each failing interval is shifted out to the tester. This approach requires much less tester memory than storing the entire uncompacted output response for the whole BIST session on the tester because only the output responses for the failing intervals are shifted out. The advantage of interval unloading compared with the binary search method is that only two steps are required to obtain all failing output responses, thereby saving time by avoiding repeated tester runs. The drawback is that more tester memory and more post-processing is required compared with the binary search method, which precisely identifies the failing test vectors.

Example 7.22

Figure 7.36 shows a small example of interval unloading. It is assumed that there are 28 test vectors in a BIST session which are divided into 7 intervals containing 4 patterns each. Among them, the first and fifth intervals are failing. After checking the failing signatures in the first step, in the second step the tester runs the two failing intervals again and stores their uncompacted output response. During the

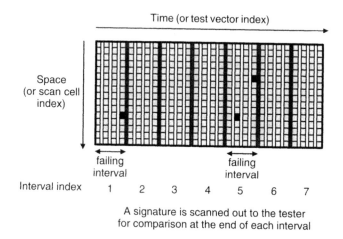

■ FIGURE 7.36

A BIST session divided into intervals.

second step, a total of $4 \times 2 = 8$ uncompacted output responses must be scanned out as the source information for combinational logic diagnosis.

7.4.3 Masking-Based Methods

Another general approach for diagnosis in a BIST environment involves running the entire BIST session multiple times while masking out the output response for different sets of scan cells each time [Wu 1999] [Rajski 1999] [Ghosh-Dastidar 2000] [Bayraktaroglu 2002b] [Liu 2004]. In this approach, a single signature is generated in each BIST run, but the set of scan cells whose output response is compacted in the signature is different each time. With this method, it is possible to obtain diagnostic information about which scan cells are capturing errors (*i.e.*, space information) using only signatures; that is, it is not necessary to store and postprocess any uncompacted output responses.

A masking based method that uses *pseudo-random masking* to obtain space information very quickly in a single tester run was described in [Rajski 1999]. The idea is to have a pseudo-random pattern generator (*e.g.*, a ***linear feedback shift register*** [LFSR]), which generates a selection signal during each BIST run. The selection signal selects some scan cells whose output response is shifted into the signature analyzer, while the output responses for the rest of the scan cells are masked out so they do not affect the final signature. If the final signature for a BIST run is correct, then all scan cells whose output responses were included in the signature are considered to be fault free. By masking different sets of scan cells in each BIST run, the process of elimination can be used to deduce which scan cells are capturing errors. The accuracy in identifying the set of scan cells capturing errors for this method depends on how many BIST runs are used, how many scan cells there are, and how many scan cells produce errors. A nice feature of this method is that all information needed for diagnosis can be obtained in a single tester run provided the tester has sufficient memory to store the final signatures for all BIST runs.

If complete accuracy of the failing scan cells is desired, then a method that uses *deterministic masking*, which was described in [Ghosh-Dastidar 2000], can be used. Rather than pseudo-randomly selecting the scan cells whose output responses are included in the signature for a BIST run, the scan cells are selected in a deterministic manner using some scan cell partitioning logic that allows a binary search to be used. The scan cells can be represented as a matrix, as shown in Figure 7.37, where each column corresponds to a scan chain and each row corresponds to a *scan slice* (*i.e.*, a set of scan cells that get shifted into the MISR in the same clock cycle). The partitioning logic selects a set of scan chains (columns) and a consecutive string of scan slices (rows). This partition is represented as a tuple (X, Y, Z) where X is the set of scan chains in the partition, Y is the bottom-most scan slice in the partition counting up from the MISR, and Z is the top-most scan slice in the partition counting up from the MISR. In Figure 7.37, an example of a scan cell partition corresponding to $(\{3, 4\}, 2, 6)$ is shaded.

The additional ***design-for-diagnosis*** (DFD) circuitry required for this method consists of the scan cell partitioning logic, which allows only the output response for the scan cells in the selected partition to be compacted in the MISR while the output

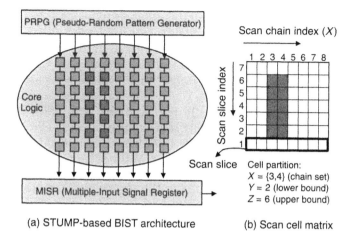

(a) STUMP-based BIST architecture   (b) Scan cell matrix

■ **FIGURE 7.37**

Matrix representation for scan cells.

responses for the rest of the scan cells are masked out. This DFD hardware shown in Figure 7.38 consists of three registers (X, Y, Z), one counter, two comparators, and some gating logic [Ghosh-Dastidar 2000]. For the example partition ({3, 4}, 2, 6), all scan chains are masked off except scan chains 3 and 4. This is accomplished by setting the third and the fourth bits in register X to logic 1, and all other bits to

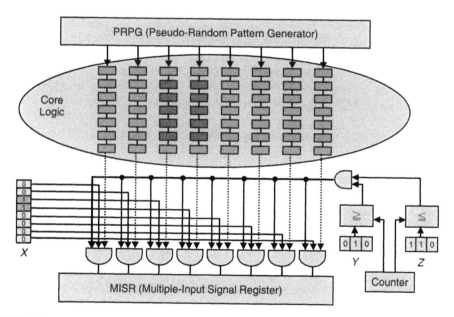

■ **FIGURE 7.38**

A DFD architecture for deterministic masking in logic BIST.

logic 0. For these two selected scan chains, only scan slices 2 to 6 are allowed to pass into the MISR. This is accomplished by setting the lower-bound register Y to 2 and the upper-bound register Z to 6. A counter is used to keep track of which scan slice is being compacted into the MISR in each clock cycle. Only the scan slices that fall between the lower-bound and upper-bound values can pass the gating logic and enter the MISR.

Given the ability to deterministically partition the scan cells, a binary search can be conducted to find the scan cells that capture errors. The binary search is performed by partitioning the scan cells in half and running the BIST session to generate a signature for the selected partition. If the selected partition has a faulty signature, then it is partitioned in half and the process is repeated. If it has a fault-free signature, then all of the scan cells in the partition are marked as fault-free, the nonselected partition is partitioned in half, and the process is continued. The worst-case number of BIST sessions that are required to identify all scan cells that capture errors is equal to $(2e)\lfloor \log_2(N/e)\rfloor + 2e - 1$, where N is the total number of scan cells, and e is the number of scan cells that capture errors.

One issue is how to determine the fault-free signature for each partition. This can be done by using the **superposition principle** to reduce the complexity. This principle states that the *fault-free signature of a flip-flop partition shifted into the MISR is simply the bitwise sum of the fault-free signature of each individual flip-flop* [Bardell 1987]. Based on this property, the fault-free signature of any selected flip-flop segment can be formed rapidly without time-consuming simulation.

Deterministic masking [Ghosh-Dastidar 2000] is an *adaptive diagnostic approach* where each step of the diagnosis depends on the results from the previous step. In this case, each step of the binary search depends on the results of the previous BIST run. The drawback is that multiple tester runs are required to obtain all diagnostic information. Pseudo-random masking [Rajski 1999] is a *nonadaptive diagnostic approach* where all diagnostic data can be obtained in a single tester run; consequently, pseudo-random masking is faster, but it may not accurately identify all scan cells that capture errors. The advantage of deterministic masking is that it is guaranteed to find the exact set of scan cells capturing errors.

Example 7.23 (Binary Search Using Deterministic Masking)

Consider the example shown in Figure 7.39a, where the third and seventh flip-flops of the fourth scan chain are assumed to capture errors. Binary search on just the scan chains can be used to identify the scan chains that capture errors. In six BIST sessions, the fourth scan chain can be identified as the only scan chain capturing errors. The value of X is set to 4, and the next step is to determine which scan cells in the fourth scan chain capture errors. As shown in Figure 7.39b, the search process can be viewed as a tree structure where each node denotes a BIST session. The associated scan cell partition is specified by its corresponding Y and Z values. In this example, the two flip-flops capturing errors are identified after nine additional BIST sessions (each corresponding to a node in the tree structure of Figure 7.39b). In addition to this basic search strategy, a more sophisticated one can also be found in [Ghosh-Dastidar 2000] in which structural information can be utilized to further reduce the total number of BIST sessions.

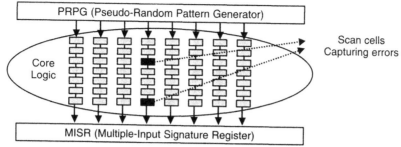

(a) Scan cells capturing errors in the fourth scan chain

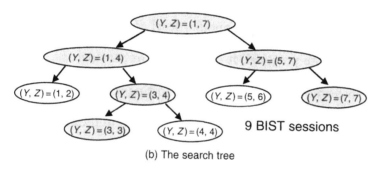

9 BIST sessions

(b) The search tree

■ **FIGURE 7.39**

Example of the search process for scan cells capturing errors.

Once the failing scan cells have been identified, the principles of combinational logic diagnosis discussed in the previous section can be applied to locate the faults in the combinational logic. Under a single stuck-at fault assumption, each candidate fault must satisfy two conditions [Bayraktaroglu 2002b]: (1) It causes a faulty response at each of the scan cells that capture errors sometime during the BIST session, and (2) it should not cause a faulty response at any of the scan cells that do not capture errors during the entire BIST session. Checking these two conditions can be done either by fault simulation or by table look-up from the fault dictionary as discussed in Section 7.2. Note that the accuracy and resolution when using only space information for diagnosis can be lower than when time information is used, because time information identifies the test patterns for which the errors are occurring, thereby adding more specific conditions for the candidate faults.

7.5   CONCLUDING REMARKS

Logic diagnosis is an important task for companies that design, test, and manufacture ICs. It provides guidance for subsequent failure analysis for design debugging and yield ramp-up. Over the years, logic diagnosis techniques have become more and more mature in dealing with faults in the combinational logic. State-of-the-art approaches can capture not just stuck-at faults but also bridging faults and other types of faults. Elegant solutions have been developed for a number of formerly nasty problems (*e.g.*, multiple fault diagnosis). Recently, major attention has been focused on diagnosing scan chain problems. As for diagnosis in a scan-based logic BIST environment, hardware support is often required to achieve satisfactory results.

Looking forward to the future challenges, performance (speed) debug is an area that is starting to attract some attention [Krstic 2003]. While speed debug is not critical for diagnosing production fallout, it is the most important issue to be resolved during early silicon debug as this impacts time-to-market greatly. Also, on-chip test compression/decompression has been a common practice for reducing the test volume on the tester. Information loss due to the use of the compression/decompression circuitry is adding another level of difficulty to diagnosis and has emerged as another challenging problem to be solved [Cheng 2004]. To make a diagnosis tool even more practical, it is necessary to consider the layout information. Based on layout analysis, one can further identify a defect down to its physical location, instead of just pointing out its hosting cell. The parametric yield loss due to lithographic-induced variations is on the rise as technology advances to the nanometer scale. Whether existing methods can adapt to these kinds of defects or not remains to be seen [Vogels 2004].

7.6 EXERCISES

7.1 **(Fault Dictionary Compaction)** Consider a fault dictionary as shown in Figure 7.40. There are 10 test vectors $\{t_1, \ldots, t_{10}\}$ and 16 faults $\{f_1, \ldots, f_{16}\}$. Here, a "1" entry means an output response is failing, whereas "0" means not failing. The detection fault dictionary attempts to represent only the failing bits in the dictionary.

   a. Calculate the compaction ratio using the detection fault dictionary.

   b. Calculate the compression ratio if a *drop-on*-2 heuristic is further used.

7.2 **(Single Fault Diagnosis)** The *backtrace* algorithm discussed in this chapter is not optimal even when the CUD has only one output and there is only one fault in the failing chip. Construct an example (with a simple netlist and an injected stuck-at fault) such that the *backtrace* algorithm exaggerates the fault candidate list unnecessarily under a particular test vector.

7.3 **(Bridging Fault Diagnosis)** A bridging defect often affects two logic cells in a chip, resulting in a multiple fault diagnosis problem. A test vector, however, only activates one fault at any given time in most cases; that is, there could be only one victim cell at a time due to the bridging defect. It is known that every failing test vector should be curable by the real faulty signal if there is only one fault in the chip. If the one-victim-at-a-time assumption is true, generalize the above statement to the case when there is only one bridging defect (that results in two cell faults, as discussed above) in the failing chip.

Fault ID	Failing-or-not flags for two outputs									
	t_1	t_2	t_3	t_4	t_5	t_6	t_7	t_8	t_9	t_{10}
f_1	10	00	00	00	01	00	10	00	00	10
f_2	00	00	00	00	00	00	00	00	00	01
f_3	00	00	00	00	00	01	00	00	00	00
f_4	00	00	00	00	00	00	00	01	00	00
f_5	00	00	00	00	00	00	00	00	00	00
f_6	00	00	01	00	00	00	00	00	00	00
f_7	10	00	00	00	00	00	00	00	00	00
f_8	00	10	00	00	00	11	00	01	10	00
f_9	00	00	00	11	10	01	00	00	00	00
f_{10}	00	00	01	00	01	00	00	00	10	00
f_{11}	00	00	00	00	00	00	10	00	00	00
f_{12}	00	00	00	00	00	11	00	00	00	00
f_{13}	01	00	00	00	00	00	00	00	00	00
f_{14}	10	00	00	00	00	00	00	00	00	00
f_{15}	00	01	10	00	00	00	01	01	00	00
f_{16}	00	11	00	10	00	00	00	10	00	01

■ **FIGURE 7.40**

A fault dictionary to be compacted.

7.4 **(Diagnosis of Byzantine Faults)** The inject-and-evaluate paradigm can be modified to deal with the so-called Byzantine faults (*i.e.*, faults that exhibit the Byzantine general's phenomenon). As shown in Figure 7.41, an injection at a suspect signal f is made by replacing each fanout of signal f by a dummy variable $\alpha_i (1 \leq i \leq n)$, where n is the total number of fanouts of signal f. The effect of the symbolic injection can be evaluated by event-driven symbolic evaluation. Assume that there are three outputs in a circuit under diagnosis (CUD): $\{z_1, z_2, z_3\}$. For a particular test vector v, the chip produces the failing response $\{(z_1, z_2, z_3)|(1, 0, 0)\}$, and the CUD produces $\{(z_1, z_2, z_3)|(1, 1, 0)\}$. After we have made the symbolic injection and evaluation, the CUD's outputs become $\{(z_1 = \alpha_1', z_2 = (\alpha_1' \cdot \alpha_2)', z_3 = 0)\}$. Note that the response of each output could become a Boolean function in terms of the injected variables, $\{\alpha_1, \alpha_2, \alpha_3\}$. These functions are called *react functions* because they represent how the responses of the output will react to the value combinations of the injected variables. Based on the above information, answer the following two questions and explain your answers. (Refer to [Boppana 1999] and [Huang 2002].)

a. Is test vector v a curable vector of signal f if the Byzantine general's phenomenon is possible?

b. Is test vector v a curable vector of signal f if there is no Byzantine general's phenomenon?

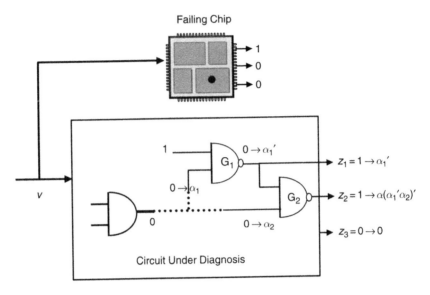

Diagnosis example for Problem 7.4.

7.5 **(Interconnect Fault Modeling)** The layout of a signal is assumed to be a binary tree structure with one source and a number of destination fanout points. An open defect in this tree structure may only affect part of it, while leaving the rest intact (as shown in Figure 7.42); for example, an open defect at the stem will be very different from a defect at a middle segment in the

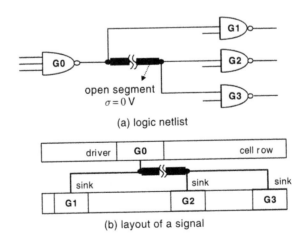

Illustration of an open segment fault in the layout of a signal.

tree, or at an ending branch. Consider the layout of a signal with k fanout points. Determine how many different faulty behaviors an open defect could cause, assuming a binary tree structure. Here, we assume that the voltage of the undriven segments due to the open defect is 0 V. (*Hint:* This problem can be formulated as a graph problem to solve it [Huang 2003].)

7.6 **(Scan Chain Fault Classification)** The flush test is often used as a classifier before the scan chain diagnosis process. Consider the observed image in response to a bitstream "0011001100110011" flushed into the scan chain in the presence of a stuck-at-0 fault, a stuck-at-1 fault, a bridging fault, or a slow-to-rise transition fault. Are these images distinct enough for fault type classification? What are the rules you may use to make the differentiation? (Note that a realistic bridging fault could act like AND-bridging at one time and an OR-bridging at another.)

7.7 **(Hold-Time Violation Fault)** Draw a master–slave edge-trigger flip-flop using only NAND gates and inverters. Mark clearly the input signal D, the clock signal *CLK*, and the output signal Q. Describe one situation when the hold-time constraint could be violated and discuss its resulting outcome in terms of its impact on the scan chain shift operation.

7.8 **(Logic BIST Diagnosis)** Consider a BIST'ed logic design with 4 flip-flops $\{q_1, q_2, q_3, q_4\}$, and 10 stuck-at faults $\{f_1, f_2, \ldots, f_{10}\}$. Fault simulation indicates that the set of faults that could ever cause flip-flop q_1 to fail during the entire BIST session is $FS(q_1) = \{f_1, f_2, f_3, f_4\}$. Similarly, $FS(q_2) = \{f_1, f_3, f_7, f_9, f_{10}\}$, $FS(q_3) = \{f_1, f_2, f_5, f_7, f_8, f_9\}$, and $FS(q_4) = \{f_8, f_{10}\}$. A logic BIST diagnosis algorithm identifies the failing flip-flops as q_1 and q_3 but not q_2 and q_4 for a failing chip. What can you conclude from the above information about the fault location in this chip?

7.9 **(A Diagnosis Practice and Design Practice)** Try to design a 32-bit ripple-carry adder in Verilog. Inject one stuck-at fault at random into the adder to mimic a failing chip. Use the diagnosis programs and user's manuals provided online to show if you can locate the fault location.

7.10 **(A Diagnosis Practice)** Inject a bridging fault at random into the adder you designed in Problem 7.9 to mimic a failing chip. Use the diagnosis programs and user's manuals provided online to show if you can locate the fault locations.

Acknowledgments

The author wishes to thank Prof. Nur A. Touba of the University of Texas at Austin for contributing a portion of the Logic BIST Diagnosis section and T. M. Mak of Intel, Prof. Chien-Mo James Li of the National Taiwan University, and Prof. Xiaoqing Wen of the Kyushu Institute of Technology for their kind help in reviewing the text.

References
R7.0—Books

[Abramovici 1994] M. Abramovici, M. A. Breuer, and A. D. Friedman, *Digital Systems Testing and Testable Design*, IEEE Press, Piscataway, NJ, 1994.

[Bardell 1987] P. H. Bardell, W. H. McAnney, and J. Savir, *Built-In Test for VLSI: Pseudo-random Techniques*, John Wiley & Sons, Somerset, NJ, 1987.

[Bushnell 2000] M. L. Bushnell and V. D. Agrawal, *Essentials of Electronic Testing for Digital, Memory and Mixed-Signal VLSI Circuits*, Springer Science, New York, 2000.

[Jha 2002] N. K. Jha and S. K. Gupta, *Testing of Digital Systems*, Cambridge University Press, Cambridge, U.K., 2002.

[Krstic 1998] A. Krstic and K.-T. Cheng, *Delay Fault Testing for VLSI Circuits*, Kluwer Academic, Boston, MA, 1998.

[Nadeau-Dostie 2000] B. Nadeau-Dostie, *Design for At-Speed Test, Diagnosis and Measurement*, Kluwer Academic, Boston, MA, 2000.

R7.1—Introduction

[Khare 1995] J. B. Khare, W. Maly, S. Griep, and D. Schmitt-Landsiedel, Yield-oriented computer-aided defect diagnosis, *IEEE Trans. Semiconductor Manufacturing*, 8-2, 195–206, 1995.

[Segura 2002] J. Segura, A. Keshavarzi, J. Soden, and C. Hawkins, Parametric failures in CMOS ICs: A defect-based analysis, in *Proc. Int. Test Conf.*, October 2002, pp. 90–99.

R7.2—Combinational Logic Diagnosis

[Abramovici 1990] M. Abramovici, P. R. Menon, and D. T. Miller, Critical path tracing: An alternative to fault simulation, *IEEE Design Test Comput.*, 1(1), 83–93, 1984.

[Acken 1992] J. M. Acken and S. D. Millman, Fault model evolution for diagnosis: Accuracy vs. precision, in *Proc. Custom Integrated Circuits Conf.*, May 1992, pp. 13.4.1–13.4.4.

[Aitken 1995] R. C. Aitken and P. C. Maxwell, Better models or better algorithms? On techniques to improve fault diagnosis, *Hewlett-Packard J.*, February 1995.

[Aitken 1996] R. C. Aitken, Modelling the unmodellable: Algorithmic fault diagnosis, in *Proc. IEEE Int. Test Conf.*, October 1996, pp. 931–940.

[Bartenstein 2001] T. Bartenstein, D. Herberlin, L. Huisman, and D. Sliwinski, Diagnosing combinational logic design using the single location at-a-time (SLAT) paradigm, in *Proc. IEEE Int. Test Conf.*, October 2001, pp. 287–296.

[Blanton 2002] R. D. Blanton, J. T. Chen, R. Desineni, K. N. Dwarakanath, W. Maly, and T. J. Vogels, Fault tuples in diagnosis of deep-submicron circuits, in *Proc. IEEE Int. Test Conf.*, October 2002, pp. 233–241.

[Boateng 1998] K. O. Boateng, H. Takahashi, and Y. Takamatsu, Multiple gate delay fault diagnosis using test-pairs for marginal delays, *IEICE Trans. Inform. Syst.*, E81-D(7), 706–714, 1998.

[Boppana 1998] V. Boppana and M. Fujita, Modeling the unknown! Toward model-independent fault and error diagnosis, in *Proc. IEEE Int. Test Conf.*, October 1998, pp. 1094–1101.

[Boppana 1999] B. Boppana, R. Mukherjee, J. Jain, and M. Fujita, Multiple error diagnosis based on Xlists, in *Proc. Design Automation Conf.*, June 1999 pp. 100–110.

[Brglez 1985] F. Brglez and H. Fujiwara, A neutral netlist of 10 combinational benchmark designs and a special translator in Fortran, in *Proc. Int. Symp. on Circuits and Systems*, June 1985, pp. 663–698.

[Chess 1998] B. Chess, D. B. Lavo, F. J. Ferguson, and T. Larrabee, Diagnosing realistic bridging faults with single stuck-at information, *IEEE Trans. Comput.-Aided Des.*, 17(3), 255–268, 1998.

[Chess 1999] B. Chess and T. Larrabee, Creating small fault dictionaries, *IEEE Trans. Comput.-Aided Des.*, 18(3), 346–356, 1999.

[Chung 1993] P. Y. Chung, Y. M. Wang, and I. N. Hajj, Diagnosis and correction of logic design errors in digital circuits, in *Proc. Design Automation Conf.*, June 1993, pp. 503–508.

[Dumas 1994] D. Dumas, P. Girard, C. Landrault, and S. Pravossoudovitch, Effectiveness of a variable sampling time strategy for delay fault diagnosis, in *Proc. European Design and Test Conf.*, February 1994, pp. 518–523.

[Girard 1992] P. Girard, C. Landrault, and S. Pravossoudovitch, Delay-fault diagnosis by critical-path tracing, *IEEE Des. Test Comput.*, 9(4), 27–32, 1992.

[Girard 1995] P. Girard, C. Landrault, S. Pravossoudovitch, and B. Rodriguez, Diagnostic of path and gate delay faults in non-scan sequential circuits, in *Proc. IEEE VLSI Test Symp.*, April 1995, pp. 380–386.

[Gong 1998] Y. Gong and S. Chakravarty, Locating bridging faults using dynamically computed stuck-at fault dictionaries, *IEEE Trans. Comput.-Aided Des.*, 17(9), 876–887, 1998.

[Hsu 1998] Y.-C. Hsu and S. K. Gupta, A new path-oriented effect–cause methodology to diagnose delay failures, in *Proc. IEEE Int. Test Conf.*, October 1998, pp. 758–767.

[Huang 1997] S.-Y. Huang, K.-T. Cheng, K.-C. Chen, and D.-T. Cheng, ErrorTracer: A fault simulation based approach to design error diagnosis, in *Proc. IEEE Int. Test Conf.*, October 1997, pp. 974–981.

[Huang 2001] S.-Y. Huang, On improving the accuracy of multiple-fault diagnosis, in *Proc. IEEE VLSI Test Symp.*, April 2001, pp. 34–39.

[Huang 2002] S.-Y. Huang, Speeding up the Byzantine fault diagnosis using symbolic simulation, in *Proc. IEEE VLSI Test Symp.*, April 2002, pp. 193–198.

[Huang 2003] S.-Y. Huang, diagnosis of Byzantine open-segment faults, in *Proc. Asian Test Symp.*, November 2002, pp. 248–253.

[Huang 2004] S.-Y. Huang, A fading algorithm for sequential fault diagnosis, in *Proc. Int. Symp. on Defect and Fault Tolerance in VLSI Systems*, October 2004, pp. 139–147.

[Huisman 2004] L. M. Huisman, Diagnosing arbitrary defects in logic designs using single location at a time (SLAT), *IEEE Trans. Comput.-Aided Des.*, 23(1), 91–101, 2004.

[Kuehlmann 1994] A. Kuehlmann, D. I. Cheng, A. Srinivasan, and D. P. Lapotin, Error diagnosis for transistor-level verification, in *Proc. Design Automation Conf.*, June 1994, pp. 218–223.

[Lavo 1996] D. B. Lavo, T. Larrabee, and B. Chess, Beyond the Byzantine generals: Unexpected behavior and bridging fault diagnosis, in *Proc. IEEE Int. Test Conf.*, October 1996, pp. 611–619.

[Lavo 1997] D. B. Lavo, B. Chess, T. Larrabee, F. J. Ferguson, J. Saxena, and K. M. Butler, Bridging fault diagnosis in the absence of physical information, in *Proc. IEEE Int. Test Conf.*, October 1997, pp. 887–893.

[Lavo 2002] D. B. Lavo, I. Hartanto, and T. Larrabee, Multiplets, models, and the search for meaning: Improving per-test fault diagnosis, in *Proc. IEEE Int. Test Conf.*, October 2002, pp. 250–259.

[Li 2005a] J. C.-M. Li and E. J. McCluskey, Diagnosis of resistive-open and stuck-open defects in digital CMOS ICs, *IEEE Trans. Comput.-Aided Des.*, 24(11), 1748–1759, 2005.

[Lin 2003] Y.-C. Lin and S.-Y. Huang, Chip-level diagnosis strategy for full-scan designs with multiple faults, in *Proc. Asian Test Symp.*, November 2003, pp. 38–44.

[Liu 2005] J. B. Liu and A. Veneris, Incremental fault diagnosis, *IEEE Trans. Comput.-Aided Des.*, 24(2), 240–251, 2005.

[Millman 1990] S. D. Millman, E. J. McCluskey, and J. M. Acken, Diagnosing CMOS bridging faults with stuck-at fault dictionaries, in *Proc. IEEE Int. Test Conf.*, October 1990, pp. 860–870.

[Olson 2000] M. Olson and X. Sun, Single bridging fault diagnosis for CMOS circuits, in *Proc. Canadian Conf. on Electrical and Computer Engineering*, May 1997, pp. 732–735.

[Pant 2001] P. Pant, Y.-C. Hsu, S. Gupta, and A. Chatterjee, Path delay fault diagnosis in combinational circuits with implicit fault enumeration, *IEEE Trans. Comput.-Aided Des.*, 20(10), 1226–1235, 2001.

[Pomeranz 1992] I. Pomeranz and S. M. Reddy, On the generation of small dictionaries for fault location, in *Proc. IEEE Int. Conf. on Comput.-Aided Des.*, November 1992, pp. 272–279.

[Pomeranz 1995] I. Pomeranz and S. M. Reddy, On correction of multiple design errors, *IEEE Trans. Comput.-Aided Des.*, 14(2), 255–264, 1995.

[Rajsuman 1991] R. Rajsuman, An analysis of feedback bridging faults in MOS VLSI, in *Proc. IEEE VLSI Test Symp.*, April 1991, pp. 53–58.

[Richman 1985] J. Richman and K. R. Bowden, The modern fault dictionary, in *Proc. IEEE Int. Test Conf.*, October 1985, pp. 696–702.

[Ryan 1993] P. G. Ryan, W. K. Fuchs, and I. Pomeranz, Fault dictionary compression and equivalence class computation for sequential circuits, in *Proc. IEEE Int. Conf. on Comput.-Aided Des.*, November 1993, pp. 508–511.

[Ryan 1998] P. G. Ryan and W. K. Fuchs, Dynamic fault dictionaries and two-stage fault isolation, *IEEE Trans. VLSI Systems*, 6(3), 176–180, 1998.

[Sikdar 2005] B. K. Sikdar, N. Ganguly, and P. P. Chaudhuri, Fault diagnosis of VLSI circuits with cellular automata based pattern classifier, *IEEE Trans. Comput.-Aided Des.*, 24(7), 1115–1131, 2005.

[Smith 2005] A. Smith, A. Veneris, M. F. Ali, and A. Viglas, Fault diagnosis and logic debugging using Boolean satisfiability, *IEEE Trans. Comput.-Aided Des.*, 24(10), 1606–1621, 2005.

[Takahashi 2002] H. Takahashi, K. O. Boateng, K. K. Saluja, and Y. Takamatsu, On diagnosing multiple stuck-at faults using multiple and single fault simulations in combinational circuits, *IEEE Trans. Comput.-Aided Des.*, 21(4), 362–368, 2002.

[Vallett 1997] D. P. Vallett, IC failure analysis, *IEEE Des. Test Comput.*, 14(3), 76–82, 1997.

[Veneris 1997] A. G. Veneris, and I. N. Hajj, A fast algorithm for locating and correcting simply design errors in VLSI digital circuits, in *Proc. Great Lakes Symp.*, March 1997, pp. 45–50.

[Veneris 2002] A. Veneris, J. B. Liu, A. Amiri, and M. S. Abadir, Incremental diagnosis and correction of multiple faults and errors, in *Proc. IEEE Design, Automation, and Test in Europe Conf.*, March 2002, pp. 716–721.

[Venkataraman 1997] S. Venkataraman and W. K. Fuchs, A Deductive Technique for Diagnosis of Bridging Faults, in *Proc. IEEE Int. Conf. on Comput.-Aided Des.*, November 1997, pp. 562–567.

[Venkataraman 2000] S. Venkataraman and S. B. Drummonds, A technique for logic fault diagnosis of interconnect open faults, in *Proc. IEEE VLSI Test Symp.*, April 2000, pp. 313–318.

[Venkataraman 2001] S. Venkataraman and S. B. Drummonds, POIROT: Applications of a logic fault diagnosis tool, *IEEE Des. Test Comput.*, 18(1), 19–30, 2001.

[Waicukauski 1988] J. A. Waicukauski and E. Lindbloom, Fault detection effectiveness of weighted random patterns, in *Proc. IEEE Int. Test Conf.*, October 1998, pp. 245–255.

[Waicukauski 1989] J. A. Waicukauski and E. Lindbloom, Failure diagnosis of structured VLSI, *IEEE Des. Test Comput.*, 6(4), 49–60, 1989.

[Wang 2002] H.-B. Wang, S.-Y. Huang, and J.-R. Huang, Gate-delay fault diagnosis using the inject-and-evaluate paradigm for full-scan designs, in *Proc. IEEE Int. Symp. on Defect and Fault Tolerance in VLSI Systems*, November 2002, pp.117–125.

[Wang 2003] Z. Wang, K. H. Tsai, M. Marek-Sadowska, and J. Rajski, An efficient and effective methodology on the multiple fault diagnosis, in *Proc. IEEE Int. Test Conf.*, October 2003, pp. 329–338.

[Wang 2005] Z. Wang, M. M. Marek-Sadowska, K.-H. Tsai, and J. Rajski, Delay-fault diagnosis using timing information, *IEEE Trans. Comput.-Aided Des.*, 24(9), 1315–1325, 2005.

[Wen 2004] X. Wen, T. Miyoshi, S. Kajihara, L.-T. Wang, K. K. Saluja, and K. Kinoshita, On per-test fault diagnosis using the X-fault model, in *Proc. IEEE Int. Conf. on Comput.-Aided Des.*, November 2004, pp. 633–640.

[Wu 2000] J. Wu and E. M. Rudnick, Bridging fault diagnosis using stuck-at fault simulation, *IEEE Trans. Comput.-Aided Des.*, 19(4), 489–495, 2000.

R7.3—Scan Chain Diagnosis

[Cheney 2000] L. Cheney and N. Sheils, A method for isolating defects in scannable sequential elements, in *Proc. Intel Design & Test Technology Conf.*, August 2000, pp.203–208.

[Edirisooriva 1995] S. Edirisooriva and G. Edirisooriva, Diagnosis of scan path failures, in *Proc. IEEE VLSI Test Symp.*, April 1995, pp. 250–255.

[Guo 2001] R. Guo and S. Venkataraman, A technique for fault diagnosis of defects in scan chains, in *Proc. IEEE Int. Test Conf.*, October 2001, pp. 268–277.

[Huang 2003] Y. Huang, W.-T. Cheng, S.-M. Reddy, C.-J. Hsieh, and Y.-T. Hung, Statistical diagnosis for intermittent scan chain hold-time fault, in *Proc. IEEE Int. Test Conf.*, October 2003, pp. 319–328.

[Huang 2004] Y. Huang, W.-T. Cheng, C.-J. Hsieh, H.-Y. Tseng, and Y.-T. Hung, Intermittent scan chain fault diagnosis based on signal probability analysis, in *Proc. IEEE Design, Automation, and Test in Europe Conf.*, February 2004.

[Huang 2005a] Y. Huang, W.-T. Cheng, and G. Growell, Using fault model relaxation to diagnose real scan chain defects, in *Proc. Asia and South Pacific Design Automation Conf.*, January 2005, pp. 1176–1179.

[Huang 2005b] Y. Huang, W.-T. Cheng, and J. Rajski, Compressed pattern diagnosis for scan-chain failures, in *Proc. IEEE Int. Test Conf.*, October 2005, pp. 744–751.

[Kundu 1993] S. Kundu, On diagnosis of faults in a scan chain, in *Proc. IEEE VLSI Test Symp.*, April 1993, pp. 303–308.

[Li 2005b] J. C.-M. Li, Diagnosis of timing faults in scan chains using single excitation patterns, *IEICE Trans. Electronics*, E88-A(4), 1024–1030, 2005.

[Li 2005c] J. C.-M. Li, Diagnosis of single stuck-at faults and multiple timing faults in scan chains, *IEEE Trans. VLSI Syst.*, 13(6), 708–718, 2005.

[Narayanan 1997] S. Narayanan and A. Das, An efficient scheme to diagnose scan chains, in *Proc. IEEE Int. Test Conf.*, October 1997, pp. 704–713.

[Schafer 1992] J. L. Schafer, Partner SRLs for improved shift register diagnosis, in *Proc. IEEE VLSI Test Symp.*, April 1992, pp. 198–201.

[Song 2004] P. Song, F. Stellari, T. Xia, and A. J. Weger, A novel scan chain diagnostics technique based on light emission from leakage current, in *Proc. IEEE Int. Test Conf.*, October 2004, pp. 140–147.

[Stanley 2001] K. Stanley, High-accuracy flush-and-scan software diagnostics, *IEEE Des. Test Comput.*, 18(6), 56–62, 2001.

[Wu 1998] Y. Wu, Diagnosis of scan chain failures, in *Proc. Int. Symp. on Defect and Fault Tolerance in VLSI Systems*, October 1998, pp. 217–222.

[Yang 2005] J.-S. Yang and S.-Y. Huang, Quick scan chain diagnosis using signal profiling, in *Proc. IEEE Int. Conf. on Computer Design*, October 2005, pp. 157–160.

R7.4—Logic BIST Diagnosis

[Aitken 1989] R. C. Aitken and V. K. Agrawal, A diagnosis method using pseudo-random vectors without intermediate signatures, in *Proc. IEEE Int. Conf. on Comput.-Aided Des.*, November 1989, pp. 574–580.

[Balakrishnan 2003] K. J. Balakrishnan and N. A. Touba, Scan-based BIST diagnosis using an embedded processor, in *Proc. Int. Symp. on Defect and Fault Tolerance in VLSI Systems*, October 2003, pp. 209–216.

[Bardell 1982] P. H. Bardell and W. H. McAnney, Self-testing of multiple logic modules, in *Proc. IEEE Int. Test Conf.*, October 1982, pp. 200–204.

[Bayraktaroglu 2000] I. Bayraktaroglu and A. Orailoglu, Improved fault diagnosis in scan-based BIST via superposition, in *Proc. Design Automation Conf.*, 55–58, 2000.

[Bayraktaroglu 2001] I. Bayraktaroglu and A. Orailoglu, Diagnosis for scan-based BIST: Reaching deep into the signatures, in *Proc. Design, Automation, and Test in Europe Conf.*, June 2001, pp. 102–109.

[Bayraktaroglu 2002a] I. Bayraktaroglu and A. Orailoglu, Cost-effective deterministic partitioning for rapid diagnosis in scan-based BIST, *IEEE Des. Test Comput.*, 19(1), 42–53, 2002.

[Bayraktaroglu 2002b] I. Bayraktaroglu and A. Orailoglu, Gate-level fault diagnosis in scan-based BIST, in *Proc. Design, Automation, and Test in Europe Conf.*, March 2002, pp. 1–6.

[Ghosh-Dastidar 1999] J. Ghosh-Dastidar, D. Das, and N. A. Touba, Fault diagnosis in scan-based BIST using both time and space information, in *Proc. IEEE Int. Test Conf.*, October 1999, pp. 95–102.

[Ghosh-Dastidar 2000] Ghosh-Dastidar and N. A. Touba, A rapid and scalable diagnosis scheme for BIST environments with a large number of scan chains, in *Proc. IEEE VLSI Test Symp.*, April 2000, pp. 79–85.

[Liu 2003] C. Liu and K. Chakrabarty, Failing vector identification based on overlapping intervals of test vectors in a scan-BIST environment, *IEEE Trans. Comput.-Aided Des.*, 22(5), 593–604, 2003.

[Liu 2004] C. Liu and K. Chakrabarty, Compact dictionaries for fault diagnosis in scan-BIST, *IEEE Trans. Comput.-Aided Des.*, 53(6), 775–780, 2004.

[McAnney 1987] W. H. McAnney and J. Savir, There is information in faulty signatures, in *Proc. IEEE Int. Test Conf.*, October 1987, pp. 630–636.

[Rajski 1997] J. Rajski and J. Tyszer, Fault diagnosis in scan-based BIST, in *Proc. IEEE Int. Test Conf.*, October 1997, pp. 894–902.

[Rajski 1999] J. Rajski and J. Tyszer, Diagnosis of scan cells in BIST environment, *IEEE Trans. Computers*, 48(7), 724–731, 1999.

[Savir 1988] J. Savir and W. H. McAnney, Identification of failing tests with cycling registers, in *Proc. IEEE Int. Test Conf.*, October 1998, pp. 322–328.

[Savir 1997] J. Savir, Salvaging test windows in BIST diagnosis, in *Proc. IEEE VLSI Test Symp.*, April 1997, pp. 416–425.

[Song 1999] P. Song, F. Motika, D. Knebel, R. Rizzolo, M. Kusko, J. Lee, and M. McManus, Diagnostic techniques for the IBM S/390 600 MHz G5 microprocessor, in *Proc. IEEE Int. Test Conf.*, October 1999, pp. 1073–1082.

[Wohl 2002] P. Wohl, J. A. Waicukauski, S. Patel, and G. Maston, Effective diagnostics through interval unloads in a BIST environment, in *Proc. IEEE Design Automation Conf.*, June 2002, pp. 249–254.

[Wu 1999] Y. Wu and S. M. I. Adham, Scan-based BIST fault diagnosis, *IEEE Trans. Comput.-Aided Des.*, 18(2), 203–211, 1999.

R7.5—Concluding Remark

[Cheng 2004] W.-T. Cheng, K.-H. Tsai, Y. Huang, N. Tamarapalli, and J. Rajski, Comparator independent direct diagnosis, in *Proc. Asian Test Symp.*, November 2004, pp. 204–209.

[Krstic 2003] A. Krstic, L.-C. Wang, K.-T. Cheng, and J.-J. Liou, Diagnosis of delay defects using statistical timing models, in *Proc. IEEE VLSI Test Symp.*, April 2003, pp. 339–344.

[Vogels 2004] T. Vogels, T. Zanon, R. Desineni, R. D. Blanton, W. Maly, J. G. Brown, J. E. Nelson, Y. Fei, X. Huang, P. Gopalakrishnan, M. Mishra, V. Rovner, and S. Tiwary, Benchmarking diagnosis algorithms with a diverse set of IC deformations, in *Proc. IEEE Int. Test Conf.*, October 2004, pp. 508–517.

CHAPTER 8

MEMORY TESTING AND BUILT-IN SELF-TEST

Cheng-Wen Wu
National Tsing Hua University, Hsinchu, Taiwan

ABOUT THIS CHAPTER

Semiconductor memory testing research dates back to the early 1960s, with a history aligned with the growth of IC industry. Although test time and test coverage have always been major concerns, the industry basically enjoys mature techniques and tools for manufacturing test of memory products. The introduction of system chips did bring forth new problems for researchers. Both the number of embedded memory cores and area occupied by memories are rapidly increasing on system chips. The yield of on-chip memories thus determines chip yield. Go/no-go testing is no longer enough for embedded memories in the *system-on-chip* (SOC) era. In addition, memories have been widely used as the technology driver; that is, they are often designed with a density that is at the extremes of the process technology. **Memory diagnosis** is quickly becoming a critical issue, as far as manufacturing yield and time-to-volume of SOC products are concerned. Effective memory diagnosis and *failure analysis* (FA) methodologies will help improve the yield of SOC products, especially with rapid evolution in new product development and advanced process technologies. These topics will be covered in this chapter and the next.

In this chapter we will first discuss memory fault models and test algorithms, then we will present a **memory fault simulator** called *random access memory simulator for error screening* (RAMSES), which consists of a simulation engine and numerous fault descriptors. The simulation engine reads the test inputs and sets the operation flags for each memory cell. Fault coverage is determined by checking the **fault descriptors** for predefined conditions. The *test algorithm generator by simulation* (TAGS) will then be presented, which is based on RAMSES and **March test algorithms**. March tests have been widely considered to be the most efficient for conventional RAM fault models. They are easy to generate and are normally short.

We will also discuss *memory built-in self-test* (BIST), which has been considered the best solution for testing embedded memories on system chips. As an example, we will present a BIST design and its implementation for embedded DRAM in detail.

It also supports ***built-in self-diagnosis*** (BISD) by feeding the errata information to the external tester. Finally, a **memory BIST compiler** called **BRAINS (Bist for RAm IN Seconds)** will be discussed, which supports major types of SRAM and DRAM by using novel BIST templates and memory specification techniques.

8.1 INTRODUCTION

With the advent of deep submicron ***very-large-scale integration*** (VLSI) technology, ***application-specific integrated circuit*** (ASIC) vendors are turning toward the SOC solution. It is a natural direction of the integration. Meanwhile, with multimillion-gate designs that are pad limited, we can see why embedded memory is such an attractive solution. Because almost all system chips contain some types of embedded memory, memories are considered one of the most universal cores. The percentage of the value for embedded memories in the overall semiconductor memory market is estimated to grow to 50% or more in the future. Because the trend is widely agreed on, the testing of embedded memories is receiving growing attention from the industry as well as the research community.

There are many challenges in merging memory (DRAM, flash memory, *etc.*) with logic [Wu 1998]. In addition to process technology issues, guaranteeing the performance, quality, and reliability of the embedded memory cores in a cost-effective way requires further research efforts. Testing embedded memory is more difficult than testing commodity memory. The first testing issue is accessibility. Accessing the DRAM core from an external memory tester is costly—in terms of pin/area overhead, performance penalty, and noise issues—when the DRAM core is embedded in a CPU or ASIC and surrounded by logic blocks. Proper ***design-for-testability*** (DFT) methodology must be provided for core isolation and tester access, and a price has to be paid for the resulting hardware overhead, performance penalty, and noise and parasitic effects. Even if these are manageable, memory testers for full qualification and testing of ***embedded DRAM*** (EDRAM) will be much more expensive due to increased speed and I/O data width, and if we also consider engineering change the overall investment will be even higher. A promising solution to this dilemma is BIST. With BIST, the tester requirement for EDRAM can be minimized, and memory tester time can be greatly reduced throughout the entire test flow of the EDRAM. Also, the total test time can be reduced because parallel testing at the memory bank and chip levels is easier. Therefore, BIST has been widely considered as a must for EDRAMs. Another advantage for BIST is that it also is a good approach to protecting ***intellectual property*** (IP); that is, the IP (EDRAM core in this case) provider needs only deliver the BIST activation and response sequences for testing and diagnosis purposes without disclosure of the design details.

Although BIST has been successfully applied to ***embedded SRAM*** (ESRAM), its success in embedded DRAM, flash memory, CAM, *etc.*, remains to be seen. Take EDRAM, for example. The need for an external memory tester cannot be removed unless redundancy analysis and repair can be done on-chip, in addition to AC testing by BIST. This obviously cannot be achieved by the BIST schemes currently used for ESRAM. Also, new failure modes or faults may have to be tested, and March algorithms such as those used in ESRAM BIST schemes are considered insufficient.

Other challenges exist. For example, memory devices normally require burn-in to reduce field failure rate, but for logic devices IddQ may be used instead. Using IddQ for memories is not trivial. What, then, should be done when we merge memory and logic to achieve the same reliability requirement? The combination of built-in current sensor and BIST logic also is an interesting topic, and the support of memory burn-in by BIST logic is another. The next challenge is design automation. Logic designers use synthesis tools while memory designers normally use the full custom design approach. Integration of the two different flows requires a lot of effort. Another challenge is the timing qualification, or AC testing, of an asynchronous memory with the synchronous BIST logic, whose timing resolution cannot compete with a typical external memory tester.

8.2 RAM FUNCTIONAL FAULT MODELS AND TEST ALGORITHMS

In memory functional testing, we normally need to characterize the device first in order to determine the most likely failure modes of the *circuit under test* (CUT). After the dominant failure modes are identified, we can select a set of tests to detect these failure modes. For easy manipulation of the failure modes or defects, we model them according to their faulty behavior. Although the functional fault models are not widely used in the industry, where defects (physical faults) and failure modes are more popular among the engineers, they are important tools for efficient methodology development in solving many memory testing issues, as will be presented in the rest of the chapter.

8.2.1 RAM Functional Fault Models

Semiconductor memories are widely considered to be one of the most important types of microelectronic components in modern digital systems [Sharma 1997]. It is reported that memories represent about 30% of the world-wide semiconductor market [Prince 1991]. The growing need for storage in computer, communications, consumer, and network applications is driving the continuous innovation of various semiconductor memory technologies. Bigger and faster memories are always desirable due to our insatiable appetites for voluminous transmission and storage of data in these applications; that is, continuing technology innovations are likely to increase the market share of commodity and embedded memories in the future. The increasing size and density of memory chips will soon make their testing the bottleneck of the entire production process. Yield loss is another issue due to the increased size and density. Memories are more vulnerable to physical defects than logic circuits because of their higher density and more complicated processing steps. Therefore, investing in memory failure analysis, fault modeling and simulation, test algorithm development and evaluation, DFT, BIST, diagnosis, *etc.*, has been considered one of the key factors in producing successful memory as well as

SOC products. Tools for fault model evaluation and test algorithm generation are fundamental for tackling the above issues efficiently.

Functional fault models are commonly used for memories. They define the functional behavior of the faulty memory. More and more fault models are being proposed to cover defects and failures in modern memory circuit and deep-submicron process technologies. Test algorithms are also being developed to detect these faults. Much work on fault models and test algorithms has been reported in the past for **random access memory** (RAM), including SRAM and DRAM [Dekker 1988a] [Nadeau-Dostie 1990] [van de Goor 1991] [Prince 1991] [Riedel 1995] [Sharma 1997] [Simonse 1998] [Huang 1999].

We consider faults that may occur in the address decoder, read/write circuitry, and memory cell array of the DRAM core. **Address decoder faults** (AFs) can be categorized as follows according to their functional behavior [Dekker 1988a] [van de Goor 1993]: (1) no cell can be accessed by a certain address; (2) multiple cells are accessed simultaneously by a certain address; (3) a certain cell is not accessible by any address; and (4) a certain cell is accessible by multiple addresses. As to the read/write circuitry (including buses, sense amplifiers, and write buffers), the typical faults are equivalent to faults in the memory cell array. For faults in the memory cell array, we follow the notation used in [van de Goor 1993]:

$\uparrow$	—	Denotes a rising transition of a cell (due to a write operation).
$\downarrow$	—	Denotes a falling transition of a cell.
$\updownarrow$	—	Denotes either a rising or a falling transition of a cell.
$\forall$	—	Denotes any operation at a cell.
$<S/F>$	—	Denotes a fault in a cell, where S is the value or operation activating the fault, F is the faulty value of the cell, $S \in \{0, 1, \uparrow, \downarrow, \updownarrow\}$, and $F \in \{0, 1\}$.
$<S_1, \ldots, S_{m-1}; S_m/F>$	—	Denotes a fault involving m cells, where $S_1, \ldots, S_{m-1}$ are the conditions of the first $m-1$ cells, respectively, that are required to activate the fault in cell m (whose state is S_m), F is the faulty value/state of cell m, and for all $0 \leq i \leq m-1$, $S_i \in \{0, 1, \uparrow, \downarrow, \updownarrow\}$.

Typical faults in the memory cell array are as follows [Dekker 1988a] [van de Goor 1993]:

1. **Stuck-at fault** (SAF)—A cell is stuck-at-1 or 0; $<\forall/1>$ denotes a stuck-at-1 and $<\forall/0>$ denotes a stuck-at-0.

2. **Stuck-open fault** (SOF)—A cell is not accessible due to, *e.g.*, a broken word-line or a permanent open switch.

3. **Transition fault** (TF)—A cell fails to transit; it can be $<\uparrow/0>$ or $<\downarrow/1>$.

4. **Data retention fault** (DRF)—A cell fails to retain its logic value after a pre-specified period of time.

5. **Coupling faults**

   a. **Inversion coupling fault** (CFin)—A transition in one cell inverts the content of another; that is, $<\uparrow / \updownarrow>$ or $<\downarrow / \updownarrow>$.

   b. **Idempotent coupling fault** (CFid)—A transition in one cell forces a constant value (1 or 0) into another; that is, $<\uparrow; 1/0>$, $<\uparrow; 0/1>$, $<\downarrow; 1/0>$, or $<\downarrow; 0/1>$.

   c. **State coupling fault** (CFst)—A coupled cell or line is forced to a certain value only if the coupling cell or line is in a given state; that is, $<0; 0/1>$, $<1; 0/1>$, $<0; 1/0>$, or $<1; 1/0>$.

6. **Read disturb fault** (RDF)—The cell value will flip when being read (repeatedly) [van de Goor 1998a].

Note that for word-oriented memories, the above single-cell fault models still apply. Also, coupling faults between cells in different words have the same behavior as in a bit-oriented memory, but coupling between cells inside the same word will virtually disappear if they can be erased by the write operation; that is, the fault can be corrected by the write operation. In that case, the coupling fault can be detected only when the coupling effect is stronger than the write operation. For example, if a 4-bit word $b_3 b_2 b_1 b_0$ has a CFst, $<0; 1/0>$, where b_3 couples b_2, then writing 0101 to $b_3 b_2 b_1 b_0$ will result in a faulty value of 0001 when CFst is stronger than the write operation; otherwise, the fault effect will be masked.

8.2.2 RAM Dynamic Faults

A **static fault** is one that has a static behavior; that is, its behavior does not change over time. A **dynamic fault**, on the other hand, has a dynamic behavior that may change over time. We give a few examples here.

A **recovery fault** occurs when some part of the memory cannot recover fast enough from a previous state. Popular recovery faults include: (1) **sense amplifier recovery** fault—the sense amplifier saturates after reading or writing a long string of 0's or 1's; and (2) **write recovery** fault—a write followed by a read or write at a different location result in reading or writing at the same location due to slow address decoder.

A **retention (refresh) fault** occurs when the memory loses its content spontaneously, not caused by the read or write operation. One example is the **sleeping sickness** of MOS DRAM that is caused by, for example, charge leakage or environment sensitivity, where the DRAM cells lose information in less than the specified hold (refresh) time—typically tens to hundreds of milliseconds. The problem usually affects a row or a column. Another example is the refresh line stuck-at fault, which also can damage the refresh mechanism of the DRAM. For SRAM, there can also be static data losses, caused by a defective pull-up device that induces excessive leakage currents which can change the state of a cell.

Another dynamic fault is the **imbalance fault**, where the bit-line voltage imbalance causes read errors.

8.2.3 Functional Test Patterns and Algorithms

In Table 8.1 we show the comparison of test time for test algorithms with varying complexity. We assume that the patterns are applied at a rate of 100M read/write operations per second. The first column shows the memory size, and other table entries are the test times with respect to different test algorithm complexities, where "s" stands for seconds, "m" for minutes, "h" for hours, "d" for days, and "y" for years. It is obvious that any test algorithm with a complexity over the linear-time complexity cannot be tolerated anymore.

The simplest (linear-time) tests that detect SAFs, TFs, and CFs are part of a family of tests called the **Marches** (*i.e.*, the **March tests**).

Definition 1

A March test consists of a finite sequence of March elements, while a **March element** is a finite sequence of operations applied to every cell in the memory array before proceeding to the next cell. An **operation** can consist of writing a 0 into a cell (w0), writing a 1 into a cell (w1), reading an expected 0 from a cell (r0), and reading an expected 1 from a cell (r1).

TABLE 8.1 ■ Test Time as a Function of Memory Size

Size	Complexity			
n	n	$n \log n$	$n^{3/2}$	n^2
1 K	0.0001 s	0.001 s	0.0033 s	0.105 s
4 K	0.0004 s	0.0048 s	0.0262 s	1.7 s
16 K	0.0016 s	0.0224 s	0.21 s	27 s
64 K	0.0064 s	0.1 s	1.678 s	7.17 m
256 K	0.0256 s	0.46 s	13.4 s	1.9 h
1 M	0.102 s	2.04 s	1.83 m	1.27 d
4 M	0.41 s	9.02 s	14.3 m	20.39 d
16 M	1.64 s	39.36 s	1.9 h	326 d
64 M	6.56 s	2.843 m	15.25 h	14.3 y
256 M	26.24 s	12.25 m	5.1 d	229 y
1 G	1.75 m	52.48 m	40.8 d	3659 y

```
Procedure ZERO-ONE
{
1: write 0 in all cells;
2: read all cells;
3: write 1 in all cells;
4: read all cells;
}
```

■ **FIGURE 8.1**

The zero-one algorithm.

Note that any March element can be done in either one of two **address orders**: the ascending order ($\Uparrow$) or descending order ($\Downarrow$). When the address order is irrelevant (*i.e.*, it can be either ascending or descending) then the symbol $\Updownarrow$ is used.

We now present some well-known conventional RAM test algorithms (patterns). The **zero-one algorithm**, also known as the *memory scan* (MSCAN) algorithm, is shown in Figure 8.1. Using the March notation for MSCAN, we can rewrite it as:

$$\{\Uparrow (w0); \Uparrow (r0); \Uparrow (w1); \Uparrow (r1)\}$$

We will discuss March tests in detail later. The MSCAN test algorithm is a minimal test, whose complexity is $O(4N)$, assuming that there are N cells in the memory (*i.e.*, the total number of read/write operations is $4N$). It is a rough estimate of the test time. It can be shown that not all $\downarrow /1$ TFs are covered by this simple test, and not all CFs are covered, either. The SAFs are covered if the address decoder is correct (however, not all AFs are covered by the test).

Theorem 1

A test detects all AFs if it contains the March elements $\Uparrow (rx, \ldots, wx')$ and $\Downarrow (rx', \ldots, wx)$, and the memory is initialized to the proper value before each March element.

The **checkerboard algorithm** (or **checkerboard pattern**) is similar to the zero-one algorithm, except that, instead of writing the all-0 and all-1 patterns (called the **solid background**), we write the 1's and 0's into alternate memory locations of the cell array in a checkerboard pattern. It is shown in Figure 8.2. The time complexity of the algorithm is also $O(4N)$. However, the checkerboard pattern is mainly used for activating failures resulting from, for example, leakage, shorts between cells, and data retention, though it also detects SAFs and half of the TFs. For that purpose, we normally wait for several seconds before reading the data after the pattern has been written into the cell array. Note that, as in MSCAN, we repeat the same operations for complementary patterns. The algorithm is also considered as the starting point for pattern sensitivity test (though it does not guarantee full coverage of pattern sensitive faults). As MSCAN, it only detects some CFs, and is not good for AFs. An important thing to note is that, when applying the checkerboard pattern, one must create the true physical checkerboard, not the logical checkerboard; that is, one

```
Procedure Checkerboard
{
while (i is odd && j is even)
  { write 0 in cell[i]; write 1 in cell[j];
  pause; read all cells;
  complement all cells;
  pause; read all cells; }
}
```

■ **FIGURE 8.2**

The checkerboard algorithm.

must obtain the design information about the actual layout and then modify the test addressing sequence accordingly.

The classical *galloping (ping-pong) pattern* (GALPAT) is shown in Figure 8.3. In the algorithm the *base cell* (BC) is read alternately with every other cell in its set—the entire cell array. The complexity is quadratic, so $O(4N^2)$. It is a very long sequence when N is large and is not a feasible test for almost all memory devices we use today, though it may still be used for characterization (not production test) of small chips. It is a strong test for most faults, though—all AFs, TFs, CFs, and SAFs are detected and located. Instead of all cells in the array, the set may also be a column, a row, or a diagonal.

An alternative to GALPAT is the *walking pattern* (WALPAT), which is similar to galloping except that the BC is read only after all others are read. For WALPAT (walking 1/0), if we consider the set as containing all cells in the RAM, then the complexity is $O(2N^2)$, which is not much better than GALPAT. Other alternatives are the **galloping diagonal, galloping row**, and **galloping column** algorithms. They are all based on GALPAT. Instead of shifting a 1 through the memory, a complete diagonal (row, or column, respectively) of 1's are shifted. The entire memory is read after each shift. The complexity is reduced from $O(4N^2)$ to $O(4N^{1.5})$. They detect all faults as GALPAT, except for some CFs. The **sliding diagonal/row/column**

```
Procedure GALPAT
{
1: write 0 in all cells;
2: for i = 0 to n−1
     {
     complement cell[i];
     for j = 0 to n−1, j != i
              { read cell[i]; read cell[j]; }
     complement cell[i];
     }
3: write 1 in all cells;
4: replay Step 2;
}
```

■ **FIGURE 8.3**

The GALPAT algorithm.

algorithm is similar to the galloping diagonal/row/column algorithm, but only those cells that are supposed to contain 1 are read after each shift. The complexity is thus further reduced to $O(4N)$. However, some CFs and TFs are not covered. More CFs and all TFs can be covered if we repeat the algorithm with a complemented data background.

The **butterfly algorithm** is shown in Figure 8.4. This test is also modified from GALPAT, with the purpose to find only AFs and SAFs. Its time complexity is $O(5N \log N)$. All SAFs and some AFs are detected.

In the **moving inversion** (MOVI) **algorithm**, the memory is initialized to contain all 0's, then this string of 0's is successively inverted to become all 1's, and vice versa. MOVI was designed as a shorter alternative for GALPAT. It has a complexity of $O(12N \log N)$. It can be used as both a functional test and an AC parametric test. As a functional test, it ensures that no cell is disturbed by a read/write operation on another unrelated cell, and it detects all AFs and SAFs. As a parametric test, it allows for the determination of the best and worst access times together with the address changes imposing these times. More details can be found in [van de Goor 1991].

Finally, the **surround disturb algorithms** attempt to examine how the cells in a particular row/column are affected when complementary data are written into adjacent cells of other rows/columns. The algorithms are designed on the premise that DRAM cells are most susceptible to interference from their nearest neighbors; thus we can eliminate global sensitivity checks to reduce complexity.

8.2.4 March Tests

A bit-oriented March C^- algorithm is given in Table 8.2 as a March test example [van de Goor 1993]. In Table 8.2 there are six **March elements**, denoted as

```
Procedure Butterfly
{
1: write 0 in all cells;
2: for i = 0 to n−1
     { complement cell i;
     dist = 1;
     while dist <= maxdist
       /* maxdist < 0.5* col/row length */
       {
       read cell at dist north from cell[i];
       read cell at dist east from cell[i];
       read cell at dist south from cell[i];
       read cell at dist west from cell[i];
       read cell[i];
       dist *= 2;   /* or dist += skip */
       }
     complement cell[i]; }
3: write 1 in all cells;
4: replay Step 2;
}
```

■ **FIGURE 8.4**

The butterfly algorithm.

TABLE 8.2 ■ The March C⁻ Algorithm

$\updownarrow (w0);$	$\Uparrow (r0w1);$	$\Uparrow (r1w0);$	$\Downarrow (r0w1);$	$\Downarrow (r1w0);$	$\updownarrow (r0)$
M_0	M_1	M_2	M_3	M_4	M_5

$M_0, M_1, \ldots, M_5$. In each March element, we first specify the address sequence: $\Uparrow$ means that the address sequence is in ascending order, $\Downarrow$ means that the address changes in descending order, and $\updownarrow$ means that either $\Uparrow$ or $\Downarrow$ is acceptable. Consider M_1, for example; the address sequence begins at the lowest address and changes in ascending order toward the highest address. For each address (memory cell), perform a read operation (with an expected 0 in the fault-free case) and write back the complemented bit immediately, then continue to the next address. The algorithm is also called the March $10N$ algorithm as it requires $10N$ read/write operations, where N is the number of memory cells (address locations).

March C⁻ is known to completely detect SAFs, unlinked AFs, unlinked TFs, and CFs (including CFins, CFids, and CFsts) [van de Goor 1993]. It also detects SOFs if M_1 is extended to $r0w1r1$, or M_2 to $r1w0r0$. The resulting algorithm is called the extended March C⁻ algorithm. In order to reduce the test cost, appropriate fault models and test algorithms should be chosen. Because the EDRAM is word oriented, the $10N$ algorithm should be modified as:

$$\updownarrow (wa); \Uparrow (rawa'); \Uparrow (ra'wa); \Downarrow (rawa'); \Downarrow (ra'wa); \updownarrow (ra)$$

where a represents a data word (*i.e.*, the background word) and a' is its complement. This word-oriented algorithm reduces to the bit-oriented one when a is a single bit.

Background words are selected based on the defined fault models and required fault coverage. Exhaustive data backgrounds normally are not affordable and not necessary. Although the word-oriented March C⁻ algorithm detects all the SAFs, unlinked AFs, TFs, and SOFs, coupling faults in the same word may not be detectable. The choice of data backgrounds determines the coverage of this kind of faults. This will be discussed later. Some other March tests are summarized below [van der Goor 1991]:

- **Modified algorithmic test sequence (MATS)**—$\{\updownarrow (w0); \updownarrow (r0, w1); \updownarrow (r1)\}$

- **MATS⁺**—$\{\updownarrow (w0); \Uparrow (r0, w1); \Downarrow (r1, w0)\}$

- **Marching 1/0**—$\{\Uparrow (w0); \Uparrow (r0, w1, r1); \Downarrow (r1, w0, r0); \Uparrow (w1); \Uparrow (r1, w0, r0); \Downarrow (r0, w1, r1)\}$

- **MATS⁺⁺**—$\{\updownarrow (w0); \Uparrow (r0, w1); \Downarrow (r1, w0, r0)\}$

- **March X**—$\{\updownarrow (w0); \Uparrow (r0, w1); \Downarrow (r1, w0); \updownarrow (r0)\}$

- **March C**—$\{\updownarrow (w0); \Uparrow (r0, w1); \Uparrow (r1, w0); \updownarrow (r0); \Downarrow (r0, w1); \Downarrow (r1, w0); \updownarrow (r0)\}$

- **March C⁻**—$\{\updownarrow (w0); \Uparrow (r0, w1); \Uparrow (r1, w0); \Downarrow (r0, w1); \Downarrow (r1, w0); \updownarrow (r0)\}$

- **March A**—{$\updownarrow$ $(w0)$; $\Uparrow$ $(r0, w1, w0, w1)$; $\Uparrow$ $(r1, w0, w1)$; $\Downarrow$ $(r1, w0, w1, w0)$; $\Downarrow$ $(r0, w1, w0)$}

- **March Y**—{$\updownarrow$ $(w0)$; $\Uparrow$ $(r0, w1, r1)$; $\Downarrow$ $(r1, w0, r0)$; $\updownarrow$ $(r0)$}

- **March B**—{$\updownarrow$ $(w0)$; $\Uparrow$ $(r0, w1, r1, w0, r0, w1)$; $\Uparrow$ $(r1, w0, w1)$; $\Downarrow$ $(r1, w0, w1, w0)$; $\Downarrow$ $(r0, w1, w0)$}

8.2.5 Comparison of RAM Test Patterns

The coverage of a March algorithm for its target faults is known by definition. However, to know its coverage of other faults will require further analysis. For example, March X [van de Goor 1993] was designed to test all AFs, SAFs, TFs, and CFins, so its coverage for these faults is 100%. If we want to know its coverage of CFids and CFsts, then analysis is required. Moreover, for a word-oriented memory, as we are discussing here, the fault coverage also depends on the selected data backgrounds. Because there are so many possible faults and test algorithms (including address sequences, read/write operations, and data patterns/backgrounds), determining the best algorithm that balances the cost and test coverage is difficult, albeit important.

We group faults into two classes: (1) single cell faults, and (2) faults involving two cells (*e.g.*, coupling faults). Class 1 faults, such as SAF, can be tested by an algorithm using any single data background, because all cells are tested in the same way as for a bit-oriented memory. Class 2 faults, however, depend on the strength of the write operation and the coupling effect. If the write operation erases the coupling effect between two cells in the same word, such faults are redundant and only coupling between two different words must be considered, so one background is sufficient. On the other hand, if the coupling effect is stronger than the write operation, coupling faults inside a word have to be considered. This is assumed in the following analysis.

The fault coverage can be derived by manual analysis, but it is tedious and sometimes impractical for complex test algorithms and fault models. Instead of manual analysis, we have implemented a novel memory fault simulator, RAMSES, for this purpose. For a word-oriented memory with 4-bit words, the data backgrounds (patterns) commonly used are 0000 (P_1), 0101 (P_2), and 0011 (P_3). To make the list complete, we also consider 0110 (P_4), 0001 (P_5), 0010 (P_6), 0100 (P_7), and 1000 (P_8). We simulated several test algorithms by RAMSES, assuming a 1-Kb word-oriented EDRAM with 4-bit words. Tables 8.3, 8.4, and 8.5 show the fault coverage simulation results of three test algorithms, where $P_{i,j}$ stands for {P_i, P_j}, $P_{i,j,k}$ for {P_i, P_j, P_k}, and P_{all} for {$P_1, P_2, \ldots, P_8$}. We show only the results for some data backgrounds, though extensive simulations have been done. We found that, in general, P_2 provides the highest fault coverage among single backgrounds, and $P_{2,3}$ is the best among double backgrounds. For triple backgrounds, $P_{1,2,3}$ provides the highest fault coverage. Intuitively, uniformity is not desirable as far as testing is concerned.

Note that for DRAMs, more faults may have to be considered, such as *neighborhood pattern-sensitive faults* (NPSFs) and linked faults. If such faults are to be targeted after failure analysis, then simulation for them should also be done in order to select the best test algorithms.

TABLE 8.3 ■ Fault Coverage of MATS^{++}

Fault	P_1	P_2	P_3	$P_{2,3}$	$P_{1,2,3}$	P_{all}
SAF	100%	100%	100%	100%	100%	100%
SOF	100%	100%	100%	100%	100%	100%
TF	100%	100%	100%	100%	100%	100%
AF	99.7%	99.9%	99.9%	100%	100%	100%
CFin	100%	100%	100%	100%	100%	100%
CFid	37.5%	37.5%	37.5%	62.6%	75.9%	89.1%
CFst	50.0%	50.0%	50.0%	75.0%	87.5%	100%

TABLE 8.4 ■ Fault Coverage of March X

Fault	P_1	P_2	P_3	$P_{2,3}$	$P_{1,2,3}$	P_{all}
SAF	100%	100%	100%	100%	100%	100%
SOF	0.8%	0.8%	0.8%	0.8%	0.8%	0.8%
TF	100%	100%	100%	100%	100%	100%
AF	99.7%	99.9%	99.9%	100%	100%	100%
CFin	100%	100%	100%	100%	100%	100%
CFid	50.0%	50.0%	50.0%	78.1%	90.7%	100%
CFst	62.5%	62.5%	62.5%	84.4%	93.0%	100%

TABLE 8.5 ■ Fault Coverage of March C$^-$

Fault	P_1	P_2	P_3	$P_{2,3}$	$P_{1,2,3}$	P_{all}
SAF	100%	100%	100%	100%	100%	100%
SOF	0.8%	0.8%	0.8%	0.8%	0.8%	0.8%
TF	100%	100%	100%	100%	100%	100%
AF	99.7%	99.9%	99.9%	100%	100%	100%
CFin	100%	100%	100%	100%	100%	100%
CFid	99.9%	99.9%	99.9%	99.95%	100%	100%
CFst	99.9%	99.9%	99.9%	99.95%	100%	100%

From the simulation results, using multiple data backgrounds significantly increases the coverage of coupling faults for MATS^{++} [van de Goor 1993] and March X as compared with single background. However, for March C$^-$, the improvement is minor; March C$^-$ with only a background P_1 covers most of the faults (the SOF fault coverage in Table 8.5 will reach 100% if M_1 is extended to *rawa'ra'*). Using an additional background will double the test time but detect only a very small percentage of additional faults (*i.e.*, intra-word coupling faults). Also, for larger DRAMs (with the same word length) the fault coverage of March C$^-$ does not decrease. Rather, the fault coverage increases because the percentage of undetected faults decreases.

8.2.6 Word-Oriented Memory

For a **word-oriented memory**, we let N represent the number of data words in the memory, each word having w bits. In this case, the read/write operations in the March tests are extended to reading and writing a word (called the **background word**, **background pattern**, or **data background**) instead of a bit at a time. For example, the word-oriented MATS^{++} is represented as $\{\updownarrow (wa); \Uparrow (ra, wa'); \Downarrow (ra', wa, ra)\}$, where a is a background word [Wu 2000]. Fault models listed above were originally developed for bit-oriented memories. Faults that occur on a single cell (*e.g.*, SAF), can still be used for word-oriented memories. Faults involving two or more cells, however, should be further classified according to whether they are within the same word or not (*i.e.*, intra-word or inter-word faults) [van de Goor 1998b]. For example, there are inter-word coupling and intra-word coupling faults, as depicted in Figure 8.5 [Wu 2000]. Traditionally, standard data backgrounds are used to test a word-oriented memory for intra-word coupling faults. However, using RAMSES we have developed a more efficient class of test algorithms, the **Cocktail–March** algorithms, and have derived the shortest one (the **March–CW** algorithm) ever reported so far for covering SAF, AF, TF, SOF, CFst, CFid, and CFin [Wu 1999].

8.2.7 Multi-Port Memory

Conventionally, a multi-port RAM is tested similarly to a single-port RAM under the same fault models, by applying the same test algorithm repeatedly to each port or each pair of ports. This approach is insufficient for detecting inter-port faults. A two-port memory array example is depicted in Figure 8.6, where the ports are denoted as port A and port B. The inter-port short fault is likely to occur on adjacent word lines or adjacent bit lines. Dedicated fault models are necessary if the detection of inter-port shorts is desired [Wu 1997] [Zhao 2000]. Some other inter-port faults have been investigated in [van de Goor 1998a], where a complete set of all possible inter-port faults is also defined. The set is very large and the corresponding test is quite long. In most cases only the shorts have to be considered. Here we consider the inter-port shorts in addition to the single-port faults as mentioned above.

The inter-port shorts can be classified as the ***bit-line short fault*** (BSF) and ***word-line short fault*** (WSF) [Wu 2001]. Figure 8.6 shows an example, in which some defects result in a BSF and a WSF. The bit lines are drawn in a simplified way for ease of presentation, though we actually consider differential pairs and all possible

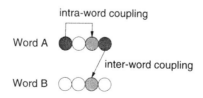

■ **FIGURE 8.5**

Inter-word and intra-word coupling faults in word-oriented memories.

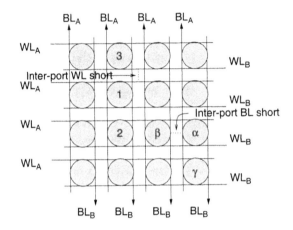

■ **FIGURE 8.6**

A two-port memory array example.

shorts during fault simulation. When there is an inter-port WSF between cells 1 and 3 as shown in Figure 8.6, a possible result is as shown in Figure 8.7, when we access cell 1 through port A and cell 2 through port B simultaneously. Due to the short fault, port B has a multiple access to cell 2 and cell 3 when port A is accessing cell 1. The resulting value of a Read to multiple cells depends on the memory design: Possible faulty results are the logic-AND or logic-OR of the two cells. Also, when the inter-port BSF (as in Figure 8.6) occurs, a possible result is as shown in Figure 8.8. The port A address of cell α can simultaneously access cells α and β as does the port B address of cell β. Again, the resulting value of a Read to multiple cells can be the logic-AND or logic-OR of the two cells, depending on the memory design. Note that an inter-port short can lead to multiple faults on the same bit line or word line; for example, cell 2 discussed above can be any cell in the column other than cells 1 and 3.

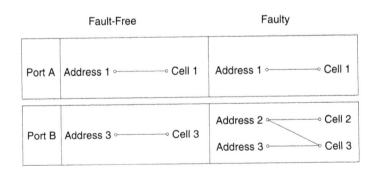

■ **FIGURE 8.7**

Behavior of an inter-port WSF.

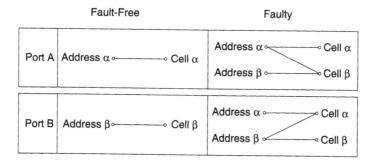

■ FIGURE 8.8

Behavior of an inter-port BSF.

The size complexity of the fault models is crucial to the time complexity of the fault simulation and test generation algorithms. Obviously, the complexity of single-cell faults is $O(N)$, and that of (two-cell) coupling faults and address decoder faults is $O(N^2)$. The availability of physical (layout) information can reduce the complexity of inter-port faults. When the address scrambling scheme is unknown, all possible effects of shorts have to be considered. When there is no physical information, from Figures 8.7 and 8.8 the complexities of inter-port word-line and bit-line shorts are $O(N^3)$ and $O(N^2)$, respectively, when they are mapped to functional faults. On the other hand, when given the address scrambling scheme and layout, a test algorithm can be developed to detect all possible shorts for the specific RAM. The complexity is reduced to the order of the number of bit lines and word lines (*i.e.*, between $O(N^{\frac{1}{2}})$ and $O(N)$) depending on the aspect ratio of the memory layout [Zhao 2000].

8.3 RAM FAULT SIMULATION AND TEST ALGORITHM GENERATION

Conventionally, fault coverage of test algorithms is proved by analysis using mathematical models such as state diagrams. Mathematical fault modeling and the *finite-state machine* (FSM) method can be used to generate the March test for full coverage of a certain set of faults, which can then be verified for completeness or irredundancy. Manual analysis is time consuming and error prone. As memory technologies are advancing very quickly, systematic approaches that can be automated are necessary to keep the test algorithms efficient and effective. For example, a systematic approach that converts tests for bit-oriented RAM to those for word-oriented RAM with complete fault coverage is proposed in [van de Goor 1998b]. In another work [Zarrineh 1998], a transition matrix model provides more detailed description of the fault models and test sequence optimization. Though elegant, these mathematical approaches cannot provide fault coverage figures during test generation and verification or data background selection. When more new and complex fault models are introduced, deriving a test algorithm becomes difficult.

Test algorithm optimization is even more difficult, because the background selection for word-oriented memories and the port selection for multi-port memories are quite complex and will vary for different memory architectures. Therefore, though not as critical as logic fault simulators and ATPG, memory fault simulators and ATPG are very helpful for memory product developers. This section is based on the pioneering work reported in [Wu 2002].

8.3.1 Fault Simulation

Memory fault simulation is different from logic fault simulation in many ways. First, unlike a logic circuit, memory has a regular structure. It consists of one or more cell arrays and the peripheral read/write circuits. However, there are many memory architectures and configurations that come with various address sizes, word lengths, and port numbers. Second, in a logic circuit we usually use only stuck-at faults, but there are multiple fault models for memories as discussed above. Furthermore, instead of the single-fault models, we have to assume multiple faults—each of the fault models can appear multiple times in the memory circuit at different locations. Finally, parallel simulation techniques developed for logic fault simulation are not suitable for simulating memory faults. Therefore, it is necessary to develop fault simulation algorithms that are dedicated for memories.

In this section, we introduce the sequential fault simulation first, which is a general and straightforward simulation algorithm. The space complexity and time complexity are evaluated. After that, we present an improved fault simulation algorithm for implementing RAMSES, the fast memory fault simulator. Fault simulation techniques for word-oriented memories and multi-port memories are also presented.

In sequential fault simulation, faults are injected into the system one by one and test patterns are applied to each and every faulty system. Outputs are then observed for evaluating the *fault coverage* (FC) of the test patterns. Consider a memory M that consists of N bits of data, with the list of target faults $f_0, f_1, \ldots, f_{k-1}$. The sequence of test patterns is $t_0, t_1, \ldots, t_{s-1}$. The sequential fault simulation procedure is shown as Figure 8.9.

```
for each i, 0 ≤ i ≤ k − 1, begin
  inject fᵢ to M;
  for each j, 0 ≤ j ≤ s − 1, begin
    apply tⱼ to M;
    if (output neq fault_free_output) begin
      set_detect(fᵢ);
      break;
    end-if
  end-for
end-for
```

■ FIGURE 8.9

The sequential fault simulation procedure.

The time complexity of the sequential fault simulation is $T = k \times s$, where k is the fault count and s is the length of the test algorithm (*i.e.*, number of test patterns). For single-cell faults, $k = O(N)$, but for two-cell coupling faults, $k = O(N^2)$. Also, for March tests, $s = O(N)$. Therefore, the time complexity for sequential fault simulation under March tests is $T = O(N^3)$, when two-cell coupling faults are the most complex faults in the target fault set. Furthermore, the space complexity of sequential fault simulation is dominated by the fault count: $O(N^2)$, in this case.

Due to the high complexity, sequential fault simulation is obviously not a practical solution for real-world applications. Nevertheless, it is easy to implement and is still useful for verifying the correctness of other simulation algorithms.

We have noted that it is feasible to simulate a smaller version of the memory under test for the purpose of FC evaluation, because of the regularity in memory structures [Wu 1999]. Simulation results of a small memory (*e.g.*, 1 Kb) are the same with a large one (*e.g.*, 16 Mb) for most fault models, though scaling requires certain calculation to avoid FC error. We also have noted that the sequential simulation algorithm has a high percentage of redundancy. We present an improved algorithm next.

8.3.2 RAMSES

RAMSES is a fast memory fault simulator that features the notion of **fault descriptor**. As illustrated in Figure 8.10, the simulator consists of the RAMSES simulation engine and numerous fault descriptors. The simulation engine executes the RAMSES simulation algorithm, which is not dedicated to a specific fault model. Target fault models are defined by their specific fault descriptors. For a user-defined memory specification and a test algorithm, RAMSES reports the FC for each fault model that is defined by a fault descriptor.

A fault descriptor consists of four primary attributes:

1. The *aggressor* (AGR) is an operation or condition that can activate the fault effect.

2. The *victim* (VTM) is the operation affected by the fault; that is, it will produce an observable faulty output.

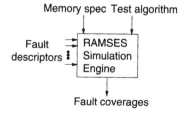

■ **FIGURE 8.10**

RAMSES I/O.

TABLE 8.6 ■ Two Fault Descriptors

RDF⟨r0/ ↑⟩	**CFin** ⟨↑; ↕⟩
AGR := R0	AGR := UTR
SPT := @	SPT := *
VTM := R0	VTM := R0, R1
RCV := W0, W1	RCV := W0, W1

3. The **suspect** (SPT) is the possible location of the aggressor, and it also indicates the possible victim location for each aggressor.

4. The **recoverer** (RCV) is the operation that can mask or recover the fault effect on the victim.

Two fault descriptor examples are listed in Table 8.6. For the read disturb fault (*i.e.*, RDF ⟨r0/ ↑⟩), a Read-0 operation activates the fault in the local cell, so the aggressor is the Read-0 operation, and the suspect is the local cell represented by @. The victim is also a Read-0 operation. Because the faulty cell's value will be flipped to 1 after the Read operation, a Read-0 operation can detect the fault. However, if the cell is written a 0 or a 1 before the victim operation takes place, then the cell value is recovered and the fault effect is erased. Therefore, the recoverer can be a Write-0 operation or a Write-1 operation.

The second example shows that CFin⟨↑; ↕⟩ is activated by an up transition of the cell. It can be observed by either a Read-0 or Read-1 operation at the victim cell and can be recovered by either a Write-0 or Write-1 operation at the victim cell. The * means that all other cells are possible coupling cells, and it also indicates that, for any aggressor cell, all other cells are possible victim cells. The core algorithm for RAMSES is summarized as Figure 8.11.

```
for each operation begin
  set_op_flags;
  if (AGR ⊂ op_flags) begin
    for each victim cell begin
      set victim flags;
      set aggressor address;
    end-for
  end-if
  if (OP eq RCV) begin
    clear victim flag;
    clear aggressor entry;
    else if (OP eq VTM) begin
      mark detected;
    end-if
  end-if
end-for
```

■ **FIGURE 8.11**

The algorithm for RAMSES.

The simulation algorithm is simple and extensible. For each test operation, various operation flags are set. For example, the Write-1 (W1) flag is set for a Write-1 operation, and the *up transition* (UTR) flag is set for a Write-1 operation only when it causes a 0 to 1 transition of a cell. There are more operation flags such as *down transition* (DTR), *last-read value* (LRV), and *last-write value* (LWV). These flags are updated for each test operation to record the current state of the cell. New flags can be added easily if it is necessary for a certain memory architecture.

After setting the operation flags, the attributes described in the fault descriptors are checked for the fault activation and fault detection conditions. If the AGR matches in the operation flags, then the cell is in the aggressor mode, and the victim flags should be set for all possible victims. The aggressor address is recorded by each victim. If the RCV matches in the operation flags, the victim flag and the aggressor entry are cleared for the memory cell. If the VTM matches in the operation flags, the fault effect is observable and it is marked as detected.

An example of the fault simulation algorithm in execution under a simple two-element test algorithm, for the fault CFin⟨↑; ↕⟩, is illustrated in Figure 8.12. There are four memory cells under simulation: cells 0, 1, 2, and 3. The initial background is all-0 after the first March element. The fault descriptor is shown in Table 8.6. In the beginning of the second March element, a Read-0 is applied to cell 0, followed immediately by a Write-1 to make an up transition in cell 0. The cell is in the aggressor mode according to the fault descriptor, so RAMSES will set the victim flags of all other cells. The aggressor address is recorded by the victim cells. Next, a Read-0 is applied to cell 1, and the cell 0 to cell 1 coupling fault is marked as detected by RAMSES. Note that the value of the cell is not changed to the faulty value, and the fault detection condition is determined only by checking the flags. After that, a Write-1 is applied to cell 1; and the up transition makes it an aggressor, and its victim flag is cleared because of the Write-1. The victim flags are set for all others and the victims record the aggressor address. In the final step shown in the figure, a Read-0 is applied to cell 2, and two coupling faults are marked as

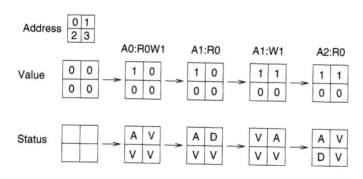

■ FIGURE 8.12

Fault simulation example for CFin⟨↑; ↕⟩.

detected: the cell 0 to cell 2 coupling and the cell 1 to cell 2 coupling. The algorithm continues like this until all cells have been visited.

For word-oriented memories, single cell faults can be simulated in the same way, but the relative strength of the Write operation and the coupling effect should be defined first in order to simulate intra-word coupling faults. If the Write operation is stronger than the coupling effect, then the coupling effect will be masked by the Write operation, and the fault is a redundant fault. Therefore, intra-word coupling faults can be detected only when the coupling effect is stronger than the Write operation. RAMSES deals with such coupling faults by disabling write recovery when the aggressor is in the same word as the victim.

For multi-port memories, additional operation flags are required for inter-port faults. Port specification also is necessary for the operation flags. For example, a Read-0 operation must explicitly specify the port from which it reads—R0(A) specifies a Read-0 from port A. Other attributes follow the same rule.

8.3.3   Test Algorithm Generation by Simulation

The *test algorithm generator by simulation* (TAGS) [Wu 2000] was developed based on the notion of a **March template**. A March template is defined as a sequence of Read/Write operations similar to a March test, but without the explicit specification of address sequences and data backgrounds. For example, $(w)(rw)(rwr)$ is a March template. The template consists of one or more template elements, such as (w) and (rw). From a March template, we can derive various March tests by applying different data and address sequence combinations. The FC of each test is calculated by RAMSES. Note that the generation of all possible March templates or March tests has an exponential complexity for both time and space. However, we have observed that most of the exhaustively generated March tests are inefficient and can be discarded. Test generation options and filtering conditions have been developed to greatly lower the complexity. We will introduce the TAGS algorithm for bit-oriented memories first, then the extended ones to handle word-oriented memories and multi-port memories.

The test generation procedure by TAGS is summarized as follows:

1. Initialize the test length as $1N$. Let the template set contain a single template $t = (w)$.

2. Increase the test length by $1N$. For each template t in the template set, add a Read/Write operation to t using one of the generation options (to be shown later). Repeat the step for all possible cases to form a new template set, except when any of the filtering conditions (to be shown later) is true.

3. A series of March tests is generated by assigning address orders in various combinations to each template in the set, together with consistent data backgrounds. When the address order for a stand-alone Read or Write can be either ⇑ or ⇓, we use ⇑ by default.

4. Simulate the resulting March tests using RAMSES.

5. Drop the tests that have no FC improvement over tests in the previous iteration, or if the improvement is completely covered by another test in the current iteration.

6. Repeat steps 2 to 5 using the new template set until the given fault set is 100% covered or the test length limit is reached.

The generation options are heuristics for generating effective March templates. Read/Write operations are inserted into an existing March template in many ways that can make it activate more memory faults or observe more errors. There are many possible combinations of March templates even with very simple generation option; therefore, some filtering conditions are necessary for dropping the ineffective templates to reduce the time complexity. We now describe the generation options and filtering conditions that we apply for the target fault models. The generation options are as follows; longer templates are derived from shorter ones by using only these options:

- Insert a stand-alone Read operation (*i.e.*, (r)) anywhere in t except in the beginning.

- Pick a template element and insert a Read operation in the beginning or append one at the end.

- Insert a stand-alone Write operation (*i.e.*, (w)) anywhere in t except in the beginning or at the end.

- Pick a template element and insert a Write operation in the beginning or append one at the end.

The filtering conditions are as follows; ineffective templates are dropped if any of the conditions is true:

- There are three consecutive Read operations, $(\cdots rrr \cdots)$, in a template element.

- There are three consecutive Read template elements, $\cdots (r)(r)(r) \cdots$, in the template.

We show a simple example to delineate the TAGS approach. Assume that, before we get to step 2, the template set contains only one template: $\{(w)(r)\}$. After we finish step 2, the new template set is $\{(w)(wr), (w)(rr), (w)(rw), (w)(r)(r), (w)(w)(r)\}$. Then, after step 3, the resulting March tests are as given in Table 8.7. After the fault simulation by RAMSES, only three of these eight algorithms are selected (see Table 8.8), and all the others are dropped. The procedure is repeated for the new template set generated for each iteration.

The following example shows the complete result generated by TAGS. Given the target faults, SAF, TF, AF, SOF, CFin, CFst, CFid, and RDF, and an unlimited test length, the March test algorithms generated by TAGS are shown in Table 8.9, where M_i^j represents the ith test algorithm with complexity jN. As shown in the table, a series of tests is generated by TAGS, with increasing complexity (test length) and FC. The test generation process stops at $12N$, when the FC reaches 100%; that is,

TABLE 8.7 ■ The 3N March Tests Generated by TAGS After Step 3

No.	Test
1	$\Uparrow (w0) \Uparrow (w1, r1)$
2	$\Uparrow (w0) \Downarrow (w1, r1)$
3	$\Uparrow (w0) \Uparrow (r0, r0)$
4	$\Uparrow (w0) \Downarrow (r0, r0)$
5	$\Uparrow (w0) \Uparrow (r0, w1)$
6	$\Uparrow (w0) \Downarrow (r0, w1)$
7	$\Uparrow (w0) \Uparrow (r0) \Uparrow (r0)$
8	$\Uparrow (w0) \Uparrow (w1) \Uparrow (r1)$

TABLE 8.8 ■ The 3N March Tests Selected by RAMSES

No.	Test
3	$\Uparrow (w0) \Uparrow (r0, r0)$
5	$\Uparrow (w0) \Uparrow (r0, w1)$
8	$\Uparrow (w0) \Uparrow (w1) \Uparrow (r1)$

it returns a complete test M_1^{12}. The test is an irredundant March test for the above faults—dropping any operation or element causes an FC loss.

RAMSES simulation results for the first test algorithm in every pass j (M_1^j) are shown in Figure 8.13. The trade-off between test length and FC can be observed—in general, the overall FC increases as the test length increases, using the presented approach. However, the FC for a particular fault may stay the same until a certain test element is added to the test; for example, SOF can be detected by $\Updownarrow (r0, w1, r1)$ or $\Updownarrow (r1, w0, r0)$, so its FC is almost 0 until the inclusion of any of the above test elements (*i.e.*, until the 12N algorithm is found). Of course, a different strategy in TAGS can generate $\Updownarrow (r0, w1, r1)$ or $\Updownarrow (r1, w0, r0)$ in an earlier stage, but then the full coverage of some other faults will be delayed during the test generation process. Note that a Read from a cell with SOF will get the data from the previous Read [Dekker 1988a], so it is clear that any of the $\Updownarrow (r0, w1, r1)$ and $\Updownarrow (r1, w0, r0)$ can fully detect SOF. However, what was not clear but can be reported by TAGS is that some March elements other than $\Updownarrow (ra, wa', ra)$ will also detect a few SOFs.

For example, the SOF of the first cell on each word line (*i.e.*, the one closest to the address decoder) can be detected by the $r1$ operation of $\{\Updownarrow (r0, w1) \Uparrow (r1, w0)\}$. Therefore, the SOF coverage is very low but not zero before $\Updownarrow (r0, w1, r1)$ or $\Updownarrow (r1, w0, r0)$ is included in the test.

Figure 8.13 shows only one test for each test length. In most cases, there are more than one test for a given test length. For example, in Figure 8.14 we depict the FC numbers for all $8N$ tests. In this figure, for example, M_2^8 detects 75% of CFid, because 3 out of the 4 March elements that are necessary for detecting CFid are part of the test. Only the March element $\Downarrow (r1, w0)$ is missing. As a second

TABLE 8.9 ■ Example Test Algorithms Generated by TAGS

1N	M_1^1	⇑ (w0)
2N	M_1^2	⇑ (w0) ⇑ (r0)
3N	M_1^3	⇑ (w0) ⇑ (w1) ⇑ (r1)
3N	M_2^3	⇑ (w0) ⇑ (r0, r0)
3N	M_3^3	⇑ (w0) ⇑ (r0, w1)
4N	M_1^4	⇑ (w0) ⇑ (r0) ⇑ (r0, w1)
4N	M_2^4	⇑ (w0) ⇑ (w1, r1) ⇑ (r1)
4N	M_3^4	⇑ (w0) ⇑ (r0, w1) ⇑ (r1)
4N	M_4^4	⇑ (w0) ⇑ (r0, w1, r1)
5N	M_1^5	⇑ (w0) ⇑ (r0, w1, r1) ⇑ (r1)
5N	M_2^5	⇑ (w0) ⇑ (r0, w1) ⇓ (r1, w0)
5N	M_3^5	⇑ (w0) ⇑ (w1) ⇑ (r1, w0) ⇑ (r0)
5N	M_4^5	⇑ (w0) ⇑ (w1) ⇑ (r1, w0, r0)
6N	M_1^6	⇑ (w0) ⇑ (r0, w1) ⇑ (r1, w0, r0)
6N	M_2^6	⇑ (w0) ⇑ (r0) ⇑ (r0, w1, r1) ⇑ (r1)
6N	M_3^6	⇑ (w0) ⇑ (r0) ⇑ (r0, w1) ⇓ (r1, w0)
6N	M_4^6	⇑ (w0) ⇑ (w1) ⇑ (r1, w0, r0) ⇑ (r0)
6N	M_5^6	⇑ (w0) ⇑ (r0, w1, r1) ⇓ (r1, w0)
6N	M_6^6	⇑ (w0) ⇑ (w1, r1) ⇑ (r1, w0, r0)
6N	M_7^6	⇑ (w0) ⇑ (r0, w1) ⇓ (r1, w0) ⇑ (r0)
6N	M_8^6	⇑ (w0) ⇑ (r0, w1, r1, w0) ⇑ (r0)
6N	M_9^6	⇑ (w0) ⇑ (r0, w1, r1) ⇑ (r1, w0)
6N	M_{10}^6	⇑ (w0) ⇑ (w1, r1) ⇑ (r1, w0) ⇑ (r0)
6N	M_{11}^6	⇑ (w0) ⇑ (r0, w1) ⇓ (r1, w0, r0)
7N	M_1^7	⇑ (w0) ⇑ (r0, w1) ⇑ (r1, w0) ⇓ (r0, w1)
7N	M_2^7	⇑ (w0) ⇑ (r0, w1, r1) ⇑ (r1, w0) ⇑ (r0)
7N	M_3^7	⇑ (w0) ⇑ (w1, r1) ⇑ (r1, w0, r0) ⇑ (r0)
7N	M_4^7	⇑ (w0) ⇑ (r0, w1) ⇓ (r1, w0, r0) ⇑ (r0)
7N	M_5^7	⇑ (w0) ⇑ (r0, w1, r1) ⇑ (r1, w0, r0)
7N	M_6^7	⇑ (w0) ⇑ (r0, w1) ⇑ (r1) ⇓ (r1, w0) ⇑ (r0)
7N	M_7^7	⇑ (w0) ⇑ (r0) ⇑ (r0, w1, r1) ⇓ (r1, w0)
7N	M_8^7	⇑ (w0) ⇑ (r0) ⇑ (r0, w1) ⇑ (r1) ⇓ (r1, w0)
7N	M_9^7	⇑ (w0) ⇑ (r0, w1) ⇑ (w0) ⇓ (r0, w1, r1)
8N	M_1^8	⇑ (w0) ⇑ (r0, w1) ⇑ (r1) ⇓ (r1, w0, r0) ⇑ (r0)
8N	M_2^8	⇑ (w0) ⇑ (r0, w1) ⇑ (r1, w0) ⇓ (r0, w1) ⇑ (r1)
8N	M_3^8	⇑ (w0) ⇑ (r0, w1, r1) ⇑ (r1, w0, r0) ⇑ (r0)
8N	M_4^8	⇑ (w0) ⇑ (r0, w1, r1) ⇑ (r1, w0) ⇓ (r0, w1)
8N	M_5^8	⇑ (w0) ⇑ (w1, r1) ⇑ (r1, w0) ⇑ (r0) ⇓ (r0, w1)
9N	M_1^9	⇑ (w0) ⇑ (r0, w1) ⇑ (r1, w0) ⇓ (r0, w1) ⇓ (r1, w0)
9N	M_2^9	⇑ (w0) ⇑ (r0, w1, r1) ⇑ (r1, w0) ⇓ (r0, w1) ⇑ (r1)
9N	M_3^9	⇑ (w0) ⇑ (r0, w1, r1) ⇑ (r1, w0) ⇑ (r0) ⇓ (r0, w1)
10N	M_1^{10}	⇑ (w0) ⇑ (r0, w1) ⇑ (r1, w0) ⇓ (r0, w1, r1) ⇓ (r1, w0)
10N	M_2^{10}	⇑ (w0) ⇑ (r0, w1) ⇑ (r1, w0) ⇓ (r0, w1) ⇓ (r1, w0) ⇑ (r0)
10N	M_3^{10}	⇑ (w0) ⇑ (r0, r0, w1, r1) ⇑ (r1, w0) ⇓ (r0, w1) ⇑ (r1)
11N	M_1^{11}	⇑ (w0) ⇑ (r0, w1, r1) ⇑ (r1, w0) ⇓ (r0, w1) ⇓ (r1, w0) ⇑ (r0)
11N	M_2^{11}	⇑ (w0) ⇑ (r0, w1) ⇑ (r1, r1, w0, r0) ⇓ (r0, w1) ⇓ (r1, w0)
12N	M_1^{12}	⇑ (w0) ⇑ (r0, w1) ⇑ (r1, w0, r0) ⇓ (r0, w1, r1) ⇓ (r1, w0) ⇑ (r0)

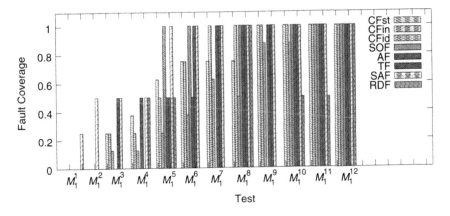

RAMSES simulation results for M_1^i.

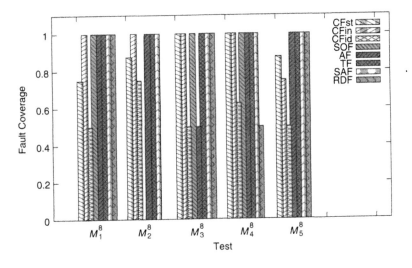

Fault coverage simulation results for the $8N$ tests.

example, 87.5% of CFst are detected by M_2^8, with a 75% FC contributed by the same March elements that detect CFid, and the other 12.5% contributed by the final $\Uparrow (r1)$ element. Again, the March element $\Downarrow (r1, w0)$ is missing in this test as far as complete detection is concerned. We can generate various tests of the same length with different fault detection capabilities. For example, if SAF, AF, SOF, and RDF are the most important fault models, then M_1^8 is the best $8N$ test. However, it is not good for CFst or CFin as compared with other $8N$ tests. This information is especially valuable when a 100% test of all faults is not affordable or necessary. Selection of a test algorithm from the set can be based on the priority of faults that are considered, which is process and product dependent. Note also that the real test time can vary even for algorithms with the same complexity, as different memories have different Read/Write-mode implementations.

A word-oriented memory has Read/Write operations that access the memory array by a word, instead of by a bit. Word-oriented memories can be tested by applying a bit-oriented test algorithm repeatedly with a set of different data backgrounds [Dekker 1988a] [van de Goor 1991] [van de Goor 1998b]. The repeating procedure multiplies the testing time. For example, three different data backgrounds (*i.e.*, 0000, 0101, and 0011) are required to test a 4-bit word-oriented memory. When using a $10N$ test algorithm, the testing time will be $30N$. We have shown in [Huang 1999] that testing word-oriented memories by repeatedly applying a single March test with different data backgrounds is not cost effective. Most faults are covered by the test even with only a single data background. Additional test runs with other data backgrounds only cover a small number of additional faults (*e.g.*, intra-word coupling faults).

The **Cocktail–March** algorithms [Wu 1999] are a class of more efficient March tests for word-oriented memories. We have extended **TAGS** for word-oriented memories based on the Cocktail–March algorithms; by mixing different test algorithms and different data backgrounds, the overall test length can be significantly reduced. For a target fault set, the steps for **TAGS** to generate the test algorithms are as follows:

1. Construct the bit-oriented memory test algorithms.

2. Generate the initial Cocktail–March test (assuming the word length is m):

    a. Generate a set of data backgrounds $P = \{p_0, p_1, \ldots, p_K\}$, where $K = \lceil \log_2 w \rceil$. For $1 \leq j \leq K$, the background word p_j is represented as $p_j = b_{m-1} \cdots b_1 b_0$, where $b_i = 1$ if $i \bmod 2^j < 2^{j-1}$, and $b_i = 0$ otherwise. Table 8.10 shows an example for 8-bit backgrounds, where the all-zero background is also called the **solid background**.

    b. Assign each and every data background, one by one, to the complete test algorithm generated in step 1 (which is the initial *candidate* test algorithm), as in the traditional method, resulting in a cascade of multiple March algorithms.

TABLE 8.10 ■ 8-Bit Data Backgrounds

p_j	$b_7 b_6 b_5 b_4 b_3 b_2 b_1 b_0$
p_0	00000000
p_1	01010101
p_2	00110011
p_3	00001111

3. Optimize the Cocktail–March test (which is now a cascade of multiple March algorithms) for each p_j, except p_0:

   a. Generate a new Cocktail–March test by replacing the March algorithm having p_j as its background with a shorter one from the set of algorithms generated in step 1.

   b. Run RAMSES for the new Cocktail–March.

   c. Repeat steps 3a and 3b until the FC drops and cannot be recovered by any other test algorithm of the same length.

   d. Store the candidate test algorithms used in the previous step.

4. Optimize the Cocktail–March test from the previous step for p_0 (the solid background):

   a. Generate a new Cocktail–March test by replacing the March algorithm having p_0 as its background with a shorter one from the set of test algorithms generated in step 1. Repeat with every test candidate for other backgrounds.

   b. Run RAMSES for the new Cocktail–March.

   c. Repeat steps 4a and 4b for all candidate test algorithms from step 3d until the FC drops and cannot be recovered by any other test algorithm of the same length or by selecting other candidates.

We now give an example to illustrate the test generation procedure. Again, the target fault models are SAF, TF, AF, SOF, CFin, CFst, CFid, and RDF. The memory under test is an 8-bit word-oriented memory. After step 1, a set of test algorithms is generated as in Table 8.9. Step 2 generates a set of data backgrounds as shown in Table 8.10, then initializes the Cocktail–March by assigning the bit-oriented complete test (M_1^{12}) to all data backgrounds. The initial $48\,N$ test algorithm is as follows:

Background	p_0	p_1	p_2	p_3
Candidates	M_1^{12}	M_1^{12}	M_1^{12}	M_1^{12}

We optimize it for p_1 by replacing M_1^{12} with $M_1^{11}, M_2^{11}, M_1^{10}$, and M_2^{10}, one by one, and calculate the FC values by RAMSES. This optimization procedure stops when the test length drops to $5\,N$, when we observe the occurrence of an FC drop. We repeat the optimization step for p_2 and p_3, respectively. For each of p_1, p_2, and p_3, usable $5\,N$ candidate tests are the same (M_3^5 and M_4^5), as shown below:

Background	p_0	p_1	p_2	p_3
Candidates	M_1^{12}	$M_3^5 M_4^5$	$M_3^5 M_4^5$	$M_3^5 M_4^5$

In step 4, the test algorithm for the solid background is optimized by the selection of candidate tests for p_1, p_2, and p_3. Finally, the optimized Cocktail–March test is generated, reducing the test length from $48N$ to $27N$ in this case (a 43.7% test time reduction). The final Cocktail–March algorithm is as follows:

Background	Test
$p_0(00000000)$	$\Uparrow (wa) \Uparrow (ra, wa', ra') \Uparrow (ra', wa, ra) \Downarrow (ra, wa') \Downarrow (ra', wa) \Uparrow (ra)$
$p_1(01010101)$	$\Uparrow (wa) \Uparrow (wa') \Uparrow (ra', wa, ra)$
$p_2(00110011)$	$\Uparrow (wa) \Uparrow (wa') \Uparrow (ra', wa, ra)$
$p_3(00001111)$	$\Uparrow (wa) \Uparrow (wa') \Uparrow (ra', wa, ra)$

In this case, single-cell faults and inter-word-related faults are already covered by the bit-oriented test using just the solid background. Additional backgrounds only cover a small number of extra intra-word faults. Figure 8.15 shows the FC improvement of intra-word coupling faults with respect to the test length, while the FC for other faults is already 100%. Note that a similar observation can be derived from [van de Goor 1998b]. However, the test of intra-word related faults other than the conventional faults is difficult to perform manually. The fault simulator makes automatic test generation easy. With RAMSES, the coverage of even untargeted faults also can be evaluated, and instead of "detected or not" the FC figures are accurately reported.

For comparison, we now present two more cases for 8-bit memories. In case 1, the target faults are SAF, AF, SOF, RDF, and CFst, and the TAGS results are listed in Table 8.11. Note that with TAGS, test generation for different target faults is fast. For example, in case 2 the target faults are SAF, TF, AF, SOF, and CFid, and

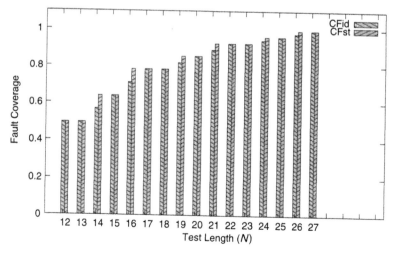

■ **FIGURE 8.15**

Intra-word fault coverage improvement.

TABLE 8.11 ■ Case 1 Test for SAF, AF, SOF, RDF, and CFst

Background	Test
p_0 (00000000)	$\Uparrow (wa) \Uparrow (ra) \Downarrow (ra, wa') \Uparrow (wa) \Uparrow (ra, wa', ra') \Uparrow (ra', wa)$
p_1 (01010101)	$\Uparrow (wa) \Uparrow (ra, wa', ra')$
p_2 (00110011)	$\Uparrow (wa) \Uparrow (ra, wa', ra')$
p_3 (00001111)	$\Uparrow (wa) \Uparrow (ra, wa', ra')$

TABLE 8.12 ■ Case 2 Test for Case 2 for SAF, TF, AF, SOF, and CFid

Background	Test
p_0 (00000000)	$\Uparrow (wa) \Uparrow (ra, wa') \Uparrow (ra', wa) \Downarrow (ra, wa') \Downarrow (ra', wa) \Uparrow (ra)$
p_1 (01010101)	$\Uparrow (wa) \Uparrow (wa') \Uparrow (ra', wa, ra)$
p_2 (00110011)	$\Uparrow (wa) \Uparrow (wa') \Uparrow (ra', wa, ra)$
p_3 (00001111)	$\Uparrow (wa) \Uparrow (wa') \Uparrow (ra', wa, ra)$

TABLE 8.13 ■ FC Comparison Between Case 1 and Case 2 Tests for a 1-K RAM

	SAF	TF	SOF	RDF	CFin	CFst	CFid	AF
Case 1	1	1	1	1	1	1	0.797	1
Case 2	1	1	1	0.0	1	1	1	1

the generated test algorithm is shown in Table 8.12. Moreover, the test coverage can easily be evaluated by RAMSES (see Table 8.13). When several test algorithms are generated by TAGS, the coverage of untargeted faults can be considered for algorithm selection.

The test length of word-oriented Cocktail–March is $(12 + 5 \log_2 B)N$ when the target faults include SAF, TF, AF, SOF, CFin, CFst, CFid, and RDF, where B is the word length and N is the address count (address space) for word-oriented memories. As a comparison, in [van de Goor 1998b] the length of the test algorithm based on March C$^-$ for intra-word CFid, CFst, and CFdst is $(10 + 6 \log_2 B)N$. CFdst is the disturb CF [van de Goor 1998b]. The $10N$ base algorithm they use is different simply because of the different set of target faults. Note that, with TAGS, the length of the additional test for covering the intra-word faults is $(5 \log_2 B)N$, while the length of the extra test for covering the intra-word faults is $(6 \log_2 B)N$ as reported in [van de Goor 1998b]. Note also that, in our case, though the specified intra-word faults do not include CFdst, the generated test covers CFdst as well.

8.4   MEMORY BUILT-IN SELF-TEST

Many *built-in self-test* (BIST) schemes have been presented in the past for embedded memories [Dekker 1988b] [Nadeau-Dostie 1990] [Treuer 1993] [Camurati 1995] [Tanoi 1997] [Dreibelbis 1998]. As an example, we present a BIST design and its implementation for EDRAM [Huang 1999] [Cheng 2000]. It supports BISD by feeding the errata information to the external tester. Moreover, using a specific test sequence the BIST scheme can test some critical timing faults of the EDRAM, reducing the tester time for AC parametric test. Our BIST design supports wafer test, pre-burn-in test, burn-in, and final test. Furthermore, it is field-programmable (*i.e.*, test algorithms using predetermined test elements such as various March elements, surround test elements, refresh modes, *etc.*) and can be programmed by the user. The hardware can be optimized for any specific EDRAM with a set of predetermined test elements. This is different from the microprogram-controlled BIST as shown in [Dreibelbis 1998] which has a higher flexibility as well as overhead. Note that the BIST design begins at the ***register-transfer level*** (RTL), and test element insertion (for higher test coverage) and deletion (for lower hardware overhead) can be done relatively easily. We will also discuss briefly how to test the timing faults of the DRAM core. Several March-based test algorithms are simulated for various fault models as well as different word-oriented data backgrounds using RAMSES. As discussed previously, using the March C^- algorithm with only the solid 0 and 1 background patterns covers almost all the *stuck-at faults* (SAFs), *transition faults* (TFs), *address faults* (AFs), and *coupling faults* (CFs). We will show that the area overhead of the BIST circuit is very low — about 1.3% for a 1-Mb DRAM and below 0.3% for a 16-Mb one. With RAMSES and the BIST design approach, designing and implementing appropriate BIST circuits for various EDRAMs can be done in a systematic way with little effort.

8.4.1   RAM Specification and BIST Design Strategy

We use a $1 M \times 4$ ***extended-data-out*** (EDO) DRAM as our example for explaining the BIST design. Of course the BIST scheme can easily be applied to other EDRAM architectures. The EDRAM is assumed to have four memory banks. Each bank is organized as a 1-Mb array; that is, it has 256 K addressable locations each containing 4 bits. The block diagram of the EDRAM with the BIST scheme is shown in Figure 8.16, for which the BIST scheme will be explained later. The timing controller controls the address buffers, data I/O buffers, and refresh mechanism via the xRAS, xCAS, and xWE signals, which represent row address strobe, column address strobe, and write enable, respectively. As shown in the figure, EDRAMs normally use separate I/O channels instead of multiplexed pins as in commodity DRAMs. Consequently, row and column addresses and ***data input*** (D) and ***output*** (Q) channels are all separated.

One of the challenges in memory BIST is that the asynchronous memory core (traditional RAMs are asynchronous) is to be tested by the synchronous BIST logic. This is especially difficult in EDRAM BIST. To illustrate our strategy for coping with this, typical EDO DRAM timing specifications are used as an example. Note that the strategy is not limited to the given EDO DRAM architecture. We show the typical EDO page-mode Read/Write cycle in Figure 8.17. Although in this case D and Q are sharing the same I/O channel, our strategy still works (the timing control of separate I/O channels is in fact easier). The values of the timing parameters

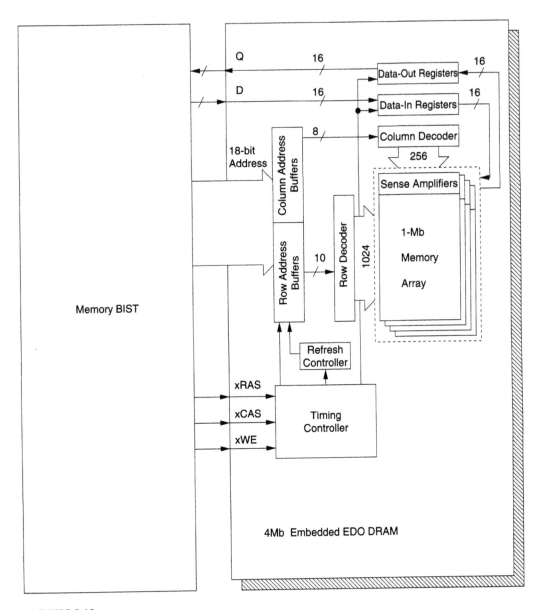

■ FIGURE 8.16

Block diagram of the embedded EDO DRAM.

shown in Figure 8.17 are listed in Table 8.14. The timing of the EDO page mode is mainly dependent on the edges of the four signals, namely xRAS, xCAS, xWE, and xOE. They determine the time to latch the row address, column address, and input data for the memory core, as well as the output data for use by other cores.

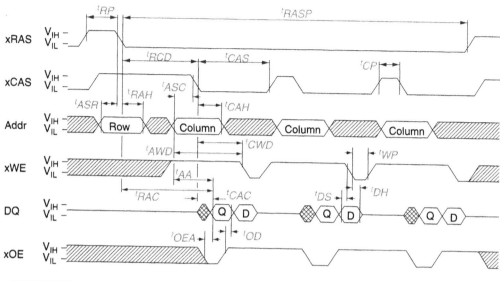

■ **FIGURE 8.17**

Timing diagram of the EDO page-mode Read/Write cycle.

TABLE 8.14 ■ Timing Parameter Values of the EDO Page-Mode Read/Write Cycle

Parameter	Min (ns)	Max (ns)	Description
t_{AA}		25	Access time from column address
t_{ASC}	0		Setup time for column address
t_{ASR}	0		Setup time for row address
t_{AWD}	42		Column address to xWE delay
t_{CAC}		13	Access time from xCAS
t_{CAH}	10		Hold time for column address
t_{CAS}	10	10000	Pulse width for xCAS active
t_{CP}	10		Pulse width for xCAS precharge
t_{CWD}	28		xCAS to xWE delay
t_{DH}	10		Hold time for D
t_{DS}	0		Setup time for D
t_{OD}	0	12	Output disable
t_{OEA}		12	Access time from xOE
t_{RAC}		50	Access time from xRAS
t_{RASP}	55	125000	Pulse width for xRAS (EDO page mode)
t_{RCD}	12		Delay time from xRAS to xCAS
t_{RAH}	10		Hold time for row address
t_{RP}	30		Pulse width for xRAS precharge
t_{WP}	5		Pulse width for write

For the case of EDRAM, which has no pin count limitation, D and Q can be separated to simplify the control, so the output enable signal (xOE) can be removed without affecting the EDRAM functionality. The BIST logic, however, still requires xOE to indicate the arrival of the output data.

We have to determine an appropriate clock period for BIST based on the xCAS cycle time (period) of the EDO page mode (*i.e.*, t_{CAS} in Table 8.14). In our example, the minimum t_{CAS} is 10 ns, which can be used as the basis of the test clock period. A test clock no faster than 100 MHz can be selected. With BIST, we need only a simple logic tester instead of an expensive memory tester—a slower and less expensive logic tester can be used to activate the BIST logic and receive the test result. The BIST sequencer (*i.e.*, timing sequence generator, which will be presented later) generates timing signals based on the clock period; that is, a high or low duration of a timing signal will be converted to a certain number of clock periods. Therefore, once the conversion is done and fixed in a BIST design, the clock period should be determined with care to avoid violation of the timing specifications of the EDRAM. In our example, the clock period (and the xCAS period) is assumed to be 20 ns (though it can be reduced to close to 10 ns). Once the clock period is fixed, the other two related timing parameters (*i.e.*, xRAS and xWE) can be determined accordingly. Also, the address, D, and Q signals in the original timing diagram (Figure 8.17) can be shifted and stretched according to the following rules. Before we give the rules, note that, for example, t_{RAC} is specified as no more than 50 ns (see Table 8.14). This means that the EDRAM design guarantees that Q is available 50 ns after the falling transition of xRAS (see also Figure 8.17), so the sample of Q should be done at least 50 ns after xRAS:

1. The row address must be ready before xRAS is pulled down to low and the column address must be ready before xCAS is pulled down to low. The time for the address to be stable before the address strobe is usually more than 1 clock cycle, so it meets the 0-ns setup time requirement in our design. Also, the address will be kept stable for more than one clock cycle to meet the 10-ns hold time requirement.

2. The timing requirement for the input data, D, is the same as the column address.

3. The major parameters related to the output data, Q, are t_{AA} (access time from column address), t_{RAC} (access time from xRAS), and t_{CAC} (access time from xCAS), which are 25 ns (max), 50 ns (max), and 13 ns (max), respectively. The xCAS low period (t_{CAS}) should span at least two clock cycles because Q will settle at the beginning of the second cycle, and the clock cycle (20 ns) is longer than 13 ns. Because the column address is ready one clock cycle before the transition of xCAS (see the first rule), we let the time from xCAS to Q to be two clock cycles to satisfy the t_{RAC} constraint. Finally, the first falling transition of xCAS in page mode is delayed for one more clock cycle, so there are at least three clock cycles from xRAS to Q, and the t_{RAC} specification also is satisfied.

4. If the write operation is considered (as in the page-mode Read/Write cycle), the key parameters are t_{AWD} and t_{CWD}, which represent the column address to xWE delay time and the xCAS to xWE delay time, respectively. Because Q is sampled at the second clock after xCAS goes low, one more clock cycle has to be inserted into the low period of xCAS, making it at least three.

Based on the above rules, the waveforms of those critical timing parameters can be generated by the sequencer to meet the specification. Slight adjustments may have to be made for other timing parameters to be considered. The waveform diagram of *rawa'* generated by the sequencer based on the above discussion is shown in Figure 8.18, which is plotted by a timing simulator. Waveforms for other March elements and the retention test and refresh test elements, for example, can also be generated using similar rules.

8.4.2   BIST Architectures and Functions

Figure 8.19 shows the block diagram of the BIST design and the interface between the BIST logic and the EDRAM. The BIST logic is activated by the **BIST activation control** (BAC) input; that is, the EDRAM is in normal mode when BAC = 0, and it is in BIST mode when BAC = 1. The BIST controller is a *finite-state machine* (FSM), whose state transition is controlled by the **BIST control selection** (BCS) input. The BIST controller also controls the scan chains—test patterns and commands can be shifted in from the **BIST scan-in** (BSI) input and results can be shifted out from the **BIST scan-out** (BSO) output. As shown in Figure 8.19, it has multiple chains internally. The Decode Logic and Test Mode Selection modules determine the proper data register to scan in the test commands and subsequently activate the sequencer. The sequencer generates the timing sequence for the EDRAM, with the help of some built-in counters and the timing generator. The output data (Q) from

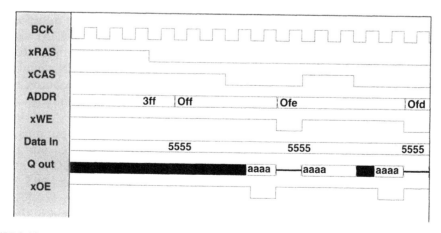

■ FIGURE 8.18

A timing diagram generated by the sequencer.

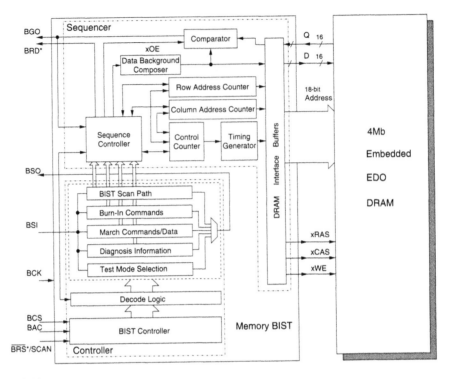

Block diagram of the BIST scheme connected to the embedded EDO DRAM.

the EDRAM will be compared with the original input data (D) generated by the sequence controller. The comparison is done by the comparator, which will report any discrepancy.

Apart from BAC, BCS, BSI, and BSO, the BIST logic has three additional I/O signals. The **BIST ready flag** (BRD*) is used to indicate when the BIST sequence is finished, so the **Go/No-Go indicator signal** (BGO) can be sampled to check whether the EDRAM is functioning correctly or not. The $\overline{\text{BRS}^*}$/SCAN signal is used as both the reset and scan test control—all registers in the BIST controller FSM are scanned, and before we use the BIST logic to test the EDRAM the logic itself is scan tested. Finally, we need a **BIST clock** (BCK) input.

Note that BCK and BAC have to be dedicated; that is, these two input pins cannot be shared (*e.g.*, using multiplexers) by others, but BRD is optional and may be removed if pin count is a concern (in that case, we can encode BGO to signal the completion of the BIST sequence and show the test result). The **reset** ($\overline{\text{BRS}^*}$) also is optional, as a short synchronizing sequence for the BIST controller (an FSM) can be used as the reset sequence. However, the SCAN pin is still required in that case. Apart from BCK and BAC, all other I/O signals of the BIST logic can share pins with signals outside the DRAM core (*i.e.*, multiplexed pins can be used to reduce pin overhead).

The BIST supports the following test modes:

1. *Scan test mode* is used for testing the BIST logic. We use scan test to verify BIST logic, except for the BIST control FSM. This test mode will be executed at the beginning of the BIST sequence to ensure the correct functionality of the BIST circuit. In addition, all the registers in the DRAM core can be tested in this mode.

2. *Memory BIST mode* is used for functional testing of the DRAM using March-based algorithms. Various operation modes of the DRAM are exercised (*e.g.,* non-page mode test, page mode test, refresh test, retention test). Diagnosis may also be supported in this mode. In that case, the BIST logic can shift out the address of any faulty cell, column, or row to the external tester by the scan mechanism. Retention faults can be tested in this mode or in a separate test mode.

3. *Burn-in (BI) mode* is used for stress testing of the DRAM to screen out unreliable parts that may fail in their infancy. In this mode we use the BIST logic to exercise the entire memory cell array in a more efficient way than the normal Read/Write access. The default BI test is to use a March algorithm supported in the memory BIST mode.

4. *Timing fault test mode* is used for testing some critical timing faults by running the BIST clock at an appropriate speed. The timing faults to be tested include incorrect setup time, hold time, data arrival time, *etc.*, of various control and data signals. Note that some of the timing faults, such as incorrect setup time and hold time, can be detected simultaneously when we perform the functional test (in the memory BIST mode). Some others can be tested by using different BIST clock periods or by an external memory tester.

Other test modes can be included in the design if necessary, as the control scheme is flexible. Of course, DC parameter testing still has to be done by the ***precision measurement unit*** (PMU) of the external tester.

8.4.3 BIST Implementation

As shown in Figure 8.19, the BIST logic is divided into two parts: the controller and the sequencer. The ***controller*** takes charge of the overall BIST flow, while the ***sequencer*** generates the address, data, and timing sequences for the EDRAM. At the ASIC level, logic BIST and memory BIST can share the same controller, and the on-chip processor can function as the sequencer during the memory BIST mode. However, for the DRAM core that is delivered as an IP to be embedded in various chips, a complete BIST circuit has to be integrated with the DRAM core. We consider the latter case.

The controller contains an FSM (labeled as the ***BIST controller*** in Figure 8.19). After the SCAN test mode has been completed successfully, we enter the memory BIST mode. The FSM actually controls the scan test and BIST flow to test the rest of the BIST circuitry and the EDRAM. We show the state diagram of the

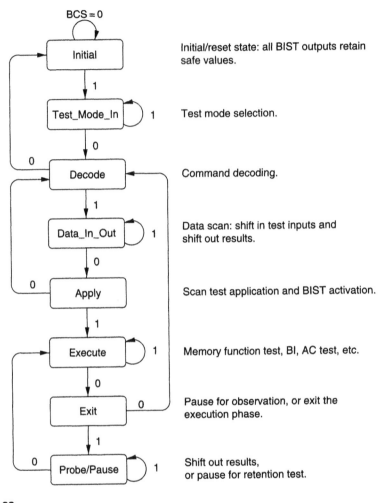

BCS = 0

Initial	Initial/reset state: all BIST outputs retain safe values.

1

Test_Mode_In) 1	Test mode selection.

0

0

Decode	Command decoding.

1

Data_In_Out) 1	Data scan: shift in test inputs and shift out results.

0

0

Apply	Scan test application and BIST activation.

1

Execute) 1	Memory function test, BI, AC test, etc.

0

0

Exit	Pause for observation, or exit the execution phase.

1

0

Probe/Pause) 1	Shift out results, or pause for retention test.

■ **FIGURE 8.20**

State diagram of the BIST controller.

FSM in Figure 8.20. Each arrow in Figure 8.20 represents a state transition that is controlled by BCS. We first enter the Initial state by asserting $\overline{\text{BRS}}^*$/SCAN low or applying a *synchronizing sequence*—note that by applying four continuous 0's on BCS we can return to the Initial state from any other state. From the Initial state, we can enter the Test_Mode_In state if BCS=1. In this state, the intended test mode can be selected. All the internal control signals will be generated in the Decode state, including those for the selection of the proper scan chain for the data sequence to be shifted in. User-specified parameters and the test algorithm are shifted in during the Data_In_Out state. Note that the Decode, Data_In_Out, and Apply states form a loop for running the scan test. Other test modes are performed in the Execute state. For memory core testing and diagnosis, we enter the bottom loop, which contains the Execute, Exit, and Probe/Pause states, and collect the error information in the Probe/Pause state. We can also run retention test in the Probe/Pause state, which allows pausing for a time interval determined by the user. An alternative approach is to add an extra mode in the sequencer, using a counter for measuring the time interval from, for example, xCAS to xWE. Appropriate timing sequences can be derived using similar rules as for March tests. When diagnosis is required, the sequencer will test the entire memory core; that is, the process will not stop immediately when an error is detected. It is not necessary to continue the testing process when an error is found if we perform testing but not diagnosis—the sequencer will simply halt and indicate that an error is found, and the controller can go back to the Decode state through the Exit state. From there either the Initial state can be reached or we can re-enter the Data_In_Out state. The Apply, Execute, Exit, and Probe/Pause states can be merged if diagnosis is not required.

In Figure 8.21 we show the timing diagram (the entire control sequence) for the BIST circuit. As discussed above, when BAC = 1, the EDRAM enters the memory BIST mode, in which every signal is synchronized to the BIST clock, BCK. The BRS/SCAN signal is pulled high at the beginning of the memory BIST mode to perform the scan test to verify the correctness of BIST controller FSM, as depicted in the figure. A scan chain is formed between BSI and BSO for applying patterns and collecting responses in this phase. After the scan, BRS/SCAN is pulled low to reset the BIST controller (BCS remains low to generate the reset sequence if necessary). The BIST controller then performs scan test for the rest of the BIST circuit. The test algorithm is subsequently applied according to the control flow discussed above and the FSM shown in Figure 8.20. Finally, we let BAC = 0 to return the EDRAM to normal mode after BRD is asserted high and BGO is sampled.

In the controller we have implemented several default Read/Write commands, address orders, data backgrounds, and EDO DRAM access modes. The built-in

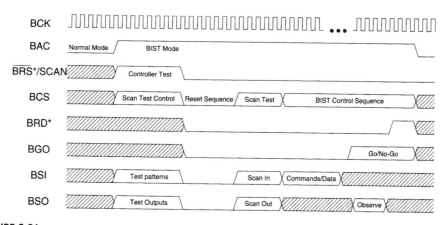

■ **FIGURE 8.21**

Control sequence of the BIST circuit.

Read/Write commands are *ra* (read the expected word *a*), *wa* (write word *a*), *rawa'* (read word *a*, complement and then write back immediately), and *rawa'ra'*. The default address orders include ⇑ and ⇓, which can be implemented by an up–down counter. The built-in access modes to be used in conjunction with the address orders are row scan, column scan, page-mode column scan, and refresh. The data background word (*a*) is to be supplied online. Each March element is a combination of the appropriate Read/Write command, address order, access mode, and data background. In addition to the March commands, the BIST design also supports diagnosis, BI, and retention test. Other test commands can be integrated easily. In our BIST scheme, a test algorithm is a sequence of commands entered from the BSI pin to the scan chains, and decoded and executed (see Figure 8.19). The end of a test algorithm is detected when the controller encounters a special **end-of-algorithm** (EOA) command. In the default implementation most of the March algorithms can be programmed, such as the extended March C⁻, March X, March Y, MATS⁺⁺, etc.

In designing the sequencer, our major goal has been flexibility. Our sequencer design can be used for a wide range of DRAM cores—it is appropriate for various operation modes, memory dimensions, and timing specifications. Figure 8.22 shows the state diagram of the **sequence controller** (*i.e.*, the FSM used as the controller in the sequencer) (see Figure 8.19). As shown in the state diagram, we have implemented the timing sequence generation modules for the single Read/Write commands and the page-mode Read/Write commands for the March elements defined in the controller. Also, a refresh timing generation module is implemented

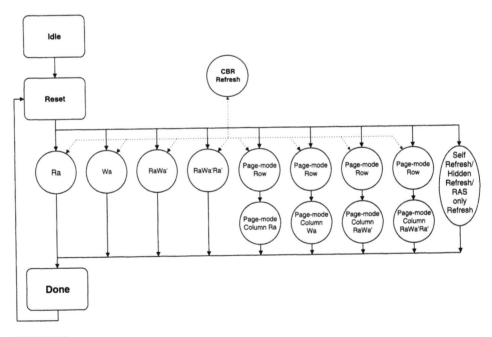

■ **FIGURE 8.22**

State diagram of the sequence controller for March tests.

for refresh tests. We show the default implementation of the sequence controller for March tests. Extending it to other test algorithms can be done easily as will be discussed below. An important concern for the DRAM core is that the outputs of the sequencer have to be glitch free, and they have to be in high-impedance when BIST is not in use (*i.e.*, when the EDRAM is in normal operation mode). This has been considered in the implementation. In Figure 8.22, the state transition is on the rising edge of BCK while the control (timing) signals for the DRAM core are applied on the falling edge of BCK. Consequently, the outputs of the sequencer are guaranteed to be glitch free.

We now briefly describe the state diagram shown in Figure 8.22. When the EDRAM is in normal mode, the sequence controller stays in the Idle state. It waits in the Idle state until the BIST controller enters the Execute state, where the sequence controller will fetch the March commands first, then enter the Reset state, followed by the sequence for the specified memory access mode. For different memory access modes, such as *ra*, *wa*, *rawa'*, and *rawa'ra'*, the timing waveform is periodic, and the period depends on the row access cycle. Proper CAS-before-RAS (CBR) refresh cycles are inserted to meet the refresh timing. The page-mode access cycle is divided into the row access cycle and column access cycle. The row address is latched first, then the column addresses of the whole page are latched one by one. In the Self-Refresh/Hidden-Refresh/RAS-only-Refresh state, the EDRAM is tested for its refresh mechanism.

As shown in Figure 8.19, the sequencer design is counter based. If the memory size increases, only the lengths of the row address counter and column address counter and the size of the comparator will increase. Note that only one additional bit is required for an address counter when the memory size is doubled, so the hardware overhead is low. The control counter is designed to meet the refresh time specification (*i.e.*, it is used for retention/refresh test). The refresh time specifications for different DRAMs currently in use do not differ much, regardless of their sizes, so the area overhead of the sequencer actually drops when the size of the DRAM core increases. In our example, a 21-bit counter suffices if the refresh cycle does not exceed 32 ms. The size of the entire BIST logic for the EDO DRAM core, without supporting BI and redundancy analysis, is about 2 to 3 thousand gates.

The sequencer generates the required output signals for the DRAM core based on the command decoded by the controller. Signal generation is done by a small ***look-up table*** (LUT). The LUT-based design reduces the design effort and hardware cost because new test commands can be added easily. When the timing specifications change, the LUT content is generated automatically by a simple program. This is an important step toward a BIST compiler for EDRAMs. It is configurable at the RT level but does not modify the architecture. For non-March algorithms, such as pseudo-random and surround-disturb tests, specific address counters or counter configurations have to be designed and included in the sequencer, and new commands have to be added to the state diagram shown in Figure 8.22.

We now discuss the area overhead of the BIST core. A commercial synthesis tool and a single-poly triple-metal logic cell library is used to estimate the area of the BIST circuit. Figure 8.23 shows the BIST area overhead percentage plot with respect to various DRAM core sizes (from 1 to 64 Mb). The numbers for the DRAM

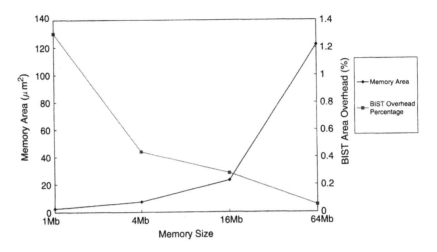

Area overhead figures of the BIST core.

area are based on existing $0.25\,\mu$m and $0.35\,\mu$m EDO DRAM chips reported by major DRAM vendors. Note that the comparison is based on the area estimated by the synthesis tool, so it is not exact. Also, it is impossible (and not necessary) for us to project the precise size of the BIST circuits on all these DRAM chips. Because the overhead is very low, as can be seen from the figure, we expect the exact area overhead numbers to be close to those shown in the figure; that is, the area overhead for the default BIST design is about 1.3% for a 1-Mb EDRAM, and it is negligible for a 64-Mb EDRAM. Even for the 16-Mb DRAM, which is a popular EDRAM candidate currently, the area overhead is less than 0.3%. Clearly, the larger the DRAM core is, the smaller the BIST area overhead will be. Because the area overhead is low, as illustrated in Figure 8.23, one can include more test modes and algorithms to increase the test coverage if necessary, as long as test time does not become a problem. The test time by non-page-mode March C⁻, for example, is about 0.4 s for the 4-Mb DRAM core (assuming a 50-MHz clock). It increases approximately in proportion to the address space. To reduce test time, parallel testing of multiple banks or even multiple words by separate BIST sequencers can be explored, but it requires modification of the memory core and should be done very carefully. Note again that an important benefit of BIST is that, after dicing, an external memory tester is not required until after burn-in.

8.4.4   BRAINS: A RAM BIST Compiler

Embedded memories, unlike commodity ones, are usually customized for different ASIC or SOC applications. The BIST circuits also have to be customized in such a case. An automatic BIST circuit generation tool thus is required to increase productivity when embedded memory cores are frequently used.

There are many commercial RAM BIST compilers (generators) in the market, such as those from Mentor Graphics, SynTest, and VirageLogic. We present here a memory BIST compiler called **BRAINS** (Bist for RAm IN Seconds), which supports SRAM and DRAM by using novel BIST templates and memory specification techniques [Huang 1999] [Huang 2000]. As BRAINS is an academic work, details can be discussed here. The compiler generates BIST design in a synthesizable HDL (*i.e.*, Verilog) upon receiving the memory specifications and test requirements provided by the user. The synthesizable BIST core can then be optimized for different fabrication processes. It can be shared among multiple memory cores. The BIST generated by BRAINS is programmable for various March tests. It is optimized automatically for user-specified March elements. The March-based programmability provides easy application of various March tests and is efficient in terms of area and performance. Furthermore, BRAINS uses unified specification techniques to generate BIST circuits for different types of embedded memory architectures and configurations. In comparison, the microcode-based design offers higher flexibility for programming non-March tests (at a lower speed), and the FSM-based design achieves smaller area overhead (for some fixed test algorithms) [Zarrineh 1999a,b].

BRAINS is a BIST compiler for both SRAM and DRAM. It consists of four components: (1) the **BIST templates**, (2) the **memory library**, (3) the **BIST intermediate description** (BID) data format, and (4) the **compiler engine**. When BRAINS receives the memory specifications and the test requirements as inputs, it translates them into the BID format, and then the compiler engine generates the BIST design according to the inputs and the memory library by creating a custom BIST module or by using an existing module from the set of BIST templates.

BRAINS generates the BIST circuit for March-based testing. It does so by using various BIST templates as building blocks. In practice, design migration cannot be done by simply changing the parameters. It usually requires more tedious adjustments, especially for DRAM cores. The adjustments for the BIST compiler have to be done within the BIST templates. Three different kinds of template are defined: (1) the **controller**, (2) the **sequencer**, and (3) the **test pattern generator** (TPG). We use the templates to construct the BIST architecture, as shown in Figure 8.24.

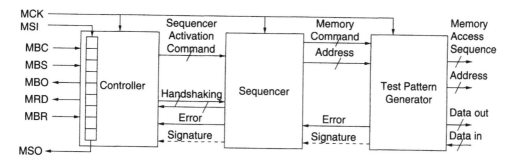

■ **FIGURE 8.24**

BIST architecture using the three templates.

The controller manages the overall operation of the BIST circuit. The test activation sequence (from tester) is received and handled by the controller, and the test result is also produced by the controller. Different test algorithms, test modes (such as BIST, burn-in, diagnosis, and repair analysis modes), data backgrounds, test ports (for multi-port memory), and memory IDs (for multiple memory cores) can be programmed into the controller. The control state machine is improved from that presented in [Huang 1999], with a larger configuration space.

During a March test, the sequencer generates the address sequence (either ⇑ or ⇓ [van de Goor 1991]) and various memory access commands based on the specifications of the memory under test. For example, the sequencer generates the *read, write, refresh, precharge, load_mode_register, active,* and *nop* (no operation) commands for an SDRAM, but only the *read, write,* and *nop* commands for a simple single-port SRAM. Some standard memory access commands for typical memory types are defined in the memory library, but customized commands specified by the user can be included as well. The sequencer generates the high-level memory access commands rather than the low-level memory access sequence (physical waveform).

The sequencer architecture (template) is shown in Figure 8.25. The control module receives the March commands from the controller. It controls the address generator, sequence generator, and memory command generator. The address generator generates the ascending (⇑) and descending (⇓) address sequences during the March test as specified by the March elements. The sequence generator generates the Read/Write sequence in the March elements (such as *wa, rawa', rawa'ra',* etc.). It also generates the initialization sequence for the memory core. The optional error handling module in the sequencer is used to scan out the error address, error signature, and the corresponding March operation that activated the fault to the external tester for diagnosis and analysis.

The TPG (see Figure 8.26) converts the high-level memory access commands provided by the sequencer to the low-level (physical) timing, address, and data sequences that can be sent directly to the memory core. The timing, address, and

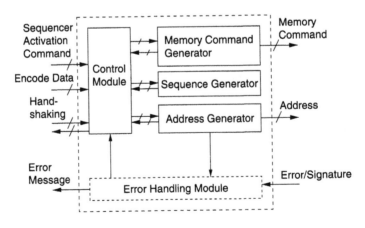

■ **FIGURE 8.25**

Block diagram of the sequencer template.

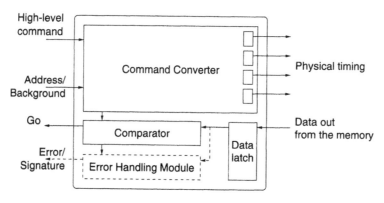

■ **FIGURE 8.26**

Block diagram of the test pattern generator.

data sequences can be high speed, double edge triggered, packetized, or even of different signal levels. The TPG also compares the data output (Q) from the memory with the original data pattern (D) to determine whether an error exists. In the diagnosis mode, the error handling modules of the sequencer and TPG are used to scan out the error information. Both the sequencer and the TPG are highly modularized.

The I/O interface of advanced memories, such as *double-data-rate* (DDR) DRAM and Rambus DRAM, are becoming more and more complex due to the need for a wider application range and higher bandwidth. In BRAINS, the characteristics of the memory interface can be modeled by the *memory specification techniques*. The BIST circuit generated by BRAINS thus tests common SRAMs and DRAMs via the specified memory interface, since testing the memory cell array directly usually cannot be done. Note that the BIST circuit and the memory controller can therefore share the memory test pattern generator (the TPG module), which further reduces the overhead for the whole system. Using the memory specification techniques, the access of different memory architectures can be done systematically and automatically, even for customized and advanced memory architectures.

In the *system-on-chip* (SOC) design environment, integrating many memory cores into a chip is common, especially for **SRAM** cores. BRAINS provides the configuration that multiple memory cores can share the same Controller and/or the same Sequencer according to the floorplan. Each of the memory cores has a unique ID number in this configuration. The memory ID is scanned in with the test commands and the BIST will activate the designated memory core to apply the March test and block other cores. The overall area overhead and the control complexity of BIST reduce because of the reduction of the number of Controllers and/or Sequencers.

The *BIST intermediate description* (BID) format is used for easy configuration and composition of the BIST templates into the synthesizable design. The BID format is the interface between the user data and the compiler engine. The BID file can be generated by the user-data capture module (*i.e.*, the user-interface software module) after parsing the input data. It also can be converted from the user data by an experienced designer. Two input categories are defined inside the BID file: the *memory specifications* and *test requirements*.

- The memory specifications define the memory under test by the memory specification technique, including memory architecture, configuration, IO labels and description, timing parameters, and memory access commands. For a built-in memory architecture (that exists in the library), there is no need to specify the details of the memory access commands. For a customized memory, new memory access commands can be defined, and existing ones can be modified.

- The test requirements specify the architecture and the functional capability of the BIST circuit. First, the desired test modes are specified. The default test algorithm can be given to generate the BIST activation sequence and the test bench. The compiler engine reads the test algorithm and determines which March element will be provided. The March-based algorithm is used for its simplicity and linear time complexity. The test coverage is guaranteed by generating an appropriate test algorithm using TAGS (Test Algorithm Generation by Simulation) that we have developed previously [Wu 2000]. During field test, a different March algorithm can be applied that consists of the specified March elements. The user also can define a March element not in the specified algorithm for future use. Primitive March elements are predefined in the compiler engine, however, new March elements can be created by simply assigning their Read/Write sequences. The compiler engine will generate the memory access commands for the new March elements with timing optimization.

In Figure 8.27 we give the BID file example for a customized memory architecture. Memory architectures such as SRAM, EDO DRAM, SDRAM, DDR DRAM, or a customized memory can be chosen. The access mode, diagnosis support, and the shared configuration can also be specified. In the file, the `parameter` operator specifies the memory size and configuration, the `timing` operator specifies the timing parameter limits, and the `hold_time` operator specifies the AC characteristics of the RAM. The compiler engine will create the BIST circuit synthesis script for the synthesis tool according to the specifications. Finally, the `March` operator specifies the March test algorithm, in which the > and < symbols stand for ascending (⇑) and descending (⇓) address sequences, respectively, and the notation a represents the data background and b is its complement. In this sample BID file, the customized memory access commands are defined by the `@custom_memory_command` operator. The memory IO ports are defined by the `@custom_memory_pin` operator, which specifies the port name (label), type, width, and latency. The `@custom_memory_task` operator defines the memory task using a combination of memory access commands. The compiler optimizes the March element by arranging the different memory tasks with the timing requirement. New March elements and initialization sequences thus can easily be created.

Built-in memory types include synchronous SRAM, asynchronous SRAM, dual-port SRAM, two-port register file, EDO DRAM, SDRAM, and DDR DRAM. Each of them is described by its own BID file with predefined `@custom_memory_pin`, `@custom_memory_command`, and `@custom_memory_task` operators and other

```
### Memory Architecture
@config memory_arch       = MYSDRAM;
@config diag              = full;        # Diagnosis support?
@config access_mode       = interleave;  # Linear or interleaving access
@config share_sequencer   = No;          # Shared Sequencer

### Memory Configuration
@parameter clock_period   = 10;
@parameter row_number     = 4096;
@parameter col_number     = 1024;
@parameter bank_number    = 4;
@parameter r_latency      = 3;           # Read latency (CAS la-
tency for SDRAM)
@parameter word_length    = 16;
@parameter t_REF          = 15625;

### Timing parameters
@timing t_RP        precharge refresh 20;  # Precharge to Ac-
tive/Refresh delay
@timing t_RCD       active    read    30;  # Active to R/W delay
@timing t_RCD       active    write   30;  #
@timing t_RAS_min   active    active  60;  # Active to Precharge Maxi-
mum delay
...

@custom_memory_pin   WE,   output, 1,  1, 1;
@custom_memory_pin   RAS,  output, 1,  1, 1;
@custom_memory_pin   CAS,  output, 1,  1, 1;
@custom_memory_pin   DQ,   output, 12, 1, 1;
@custom_memory_pin   ADDR, output, 10, 1, 1;
@custom_memory_pin   BA,   output, 10, 1, 1;
...

@custom_memory_command write   =
    {WE=0,RAS=1,CAS=0,BA=SEQ_BA,DQ=SEQ_PARITY?data_out:~data_out,ADDR=SEQ_ADDR};
@custom_memory_command read    =
    {WE=1,RAS=1,CAS=0,BA=SEQ_BA,DQ=SEQ_PARITY?data_out:~data_out,ADDR=SEQ_ADDR};
@custom_memory_command active  =
    {WE=1,RAS=0,CAS=1,BA=SEQ_BA,DQ=SEQ_PARITY?data_out:~data_out,ADDR=SEQ_ADDR};
@custom_memory_command precharge =
    {WE=0,RAS=1,CAS=0,BA=SEQ_BA,DQ=SEQ_PARITY?data_out:~data_out,ADDR=SEQ_ADDR};
@custom_memory_command refresh =
    {WE=0,RAS=0,CAS=0,BA=SEQ_BA,DQ=SEQ_PARITY?data_out:~data_out,ADDR=SEQ_ADDR};
@custom_memory_command nop     =
    {WE=1,RAS=1,CAS=1,BA=SEQ_BA,DQ=SEQ_PARITY?data_out:~data_out,ADDR=SEQ_ADDR};
@custom_memory_command lmr     =
    {WE=1,RAS=0,CAS=0,BA=0,DQ=SEQ_PARITY?data_out:~data_out,ADDR=32};
...
@custom_memory_task read          = {read};
@custom_memory_task write         = {write};
@custom_memory_task nop           = {nop};
@custom_memory_task refresh       = {refresh};
@custom_memory_task precharge     = {precharge};
@custom_memory_task load_mode_reg = {lmr};
@custom_memory_task active        = {active};
...
@hold_time ADDR     = 1;
@hold_time WE       = 1;
@hold_time RAS      = 1;
@hold_time CAS      = 1;
...

### Test requirement
@March     >(wa), >(rawb), >(rbwa), <(rawb), <(rbwara), <(ra);
```

■ **FIGURE 8.27**

The BID file example for a predefined memory architecture.

default settings. The compiler engine loads the generic templates for the predefined control and data paths of the BIST circuit, and configures it based on the BID file. The user can make minor modifications for different configurations or customized memory architectures. To customize the access timing for the new memory architecture, the BID file can be edited by inserting additional commands at the memory-access-command level and/or modifying the custom timing sequence at the physical-timing level. Consequently, only a small portion of the details have to be dealt with by the user.

The BIST circuit compilation flow using BRAINS is given in Figure 8.28. The memory specifications and test requirements are provided via the user interface. The memory specifications include the timing parameters, memory architecture (synchronous/asynchronous SRAM, single-port/multi-port SRAM, EDO DRAM, SDRAM, DDR DRAM, etc.), memory configuration (data width, address width), *etc.* The test requirements include the test algorithm requirements (which affect the choice of the March elements and the programmability), address ordering (counting or pseudo-random, interleaved or non-interleaved), supported test modes (go/no-go test, burn-in test, diagnosis test), etc.

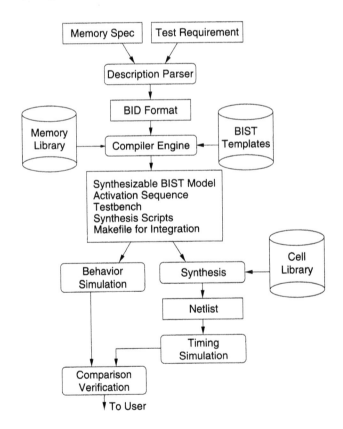

■ **FIGURE 8.28**

The BIST circuit compilation flow using BRAINS.

After the memory specifications and test requirements are entered, the description parser (a preprocessor) generates a BID file which defines the BIST templates. The user also can modify the BID file to customize the BIST circuit. The compiler engine then parses the BID file and loads the BIST templates and memory library to generate the controller, sequencer, and test pattern generator. The compiler engine configures the programmability of the BIST circuit and refines the memory access timing according to the timing specifications and test requirements, as discussed above. It generates the synthesizable RTL model for the BIST circuit, BIST activation sequence, test bench, synthesis scripts, and the UNIX Makefile for integrated command-level operations. The synthesizable BIST model is in the Verilog format. The BIST activation sequence can be used to control the BIST from a simple tester interface. Different test algorithms can be applied during field test. The test bench contains stimuli that can be used for behavior-level and gate-level simulations. Automatic synthesis can be done by a synthesis tool using the synthesis scripts. The generated logic circuit (in the net-list level) is then simulated and compared with the behavior-level result for design verification.

Table 8.15 shows the comparison among different memory architectures and configurations from the experimental results of BRAINS. The first column shows the four different memory architectures used in the experiments (*i.e.*, single-port

TABLE 8.15 ■ Comparison Among Different Memory Architectures and Configurations

Architecture	Configuration	Diag.	Bank Access	Shared DQ	# Gates	Overhead
Single-port SRAM	$8K \times 16$	No	–	No	1438	2.60%
Single-port SRAM	$8K \times 16$	Yes	–	No	1940	3.70%
Single-port SRAM	$16K \times 16$	No	–	No	1474	1.41%
Single-port SRAM	$16K \times 16$	Yes	–	No	1988	1.88%
Two-port Register File	$2K \times 32$	No	–	No	1876	3.8%
Two-port Register File	$2K \times 32$	Yes	–	No	2590	5.25%
Two-port Register File	$4K \times 32$	No	–	No	1908	1.94%
Two-port Register File	$4K \times 32$	Yes	–	No	2628	2.66%
Asyn single-port SRAM	$8K \times 16$	No	–	No	1444	2.75%
Asyn single-port SRAM	$8K \times 16$	Yes	–	No	1989	3.79%
Asyn single-port SRAM	$16K \times 16$	No	–	No	1476	1.41%
Asyn single-port SRAM	$16K \times 16$	Yes	–	No	2039	1.95%
SDRAM	$16M \times 4$	No	Non-interleaved	Yes	1587	0.033%
SDRAM	$16M \times 4$	No	Interleaved	Yes	1693	0.036%
SDRAM	$16M \times 4$	Yes	Non-interleaved	Yes	2003	0.042%
SDRAM	$16M \times 4$	Yes	Interleaved	Yes	2175	0.046%
SDRAM	$8M \times 8$	No	Non-interleaved	Yes	1683	0.036%
SDRAM	$8M \times 8$	No	Interleaved	Yes	1766	0.038%
SDRAM	$8M \times 8$	Yes	Non-interleaved	Yes	2264	0.048%
SDRAM	$8M \times 8$	Yes	Interleaved	Yes	2375	0.051%
SDRAM	$16M \times 8$	No	Non-interleaved	Yes	1679	0.018%
SDRAM	$16M \times 8$	No	Interleaved	Yes	1813	0.020%
SDRAM	$16M \times 8$	Yes	Non-interleaved	Yes	2309	0.025%
SDRAM	$16M \times 8$	Yes	Interleaved	Yes	2421	0.026%

TABLE 8.16 ■ Area Comparison Between Shared BIST and Non-Shared BIST

	Controller	Sequencer	TPG	Total area (# gates)
Non-shared	473×4	343×4	588×4	$1404 \times 4 = 5616$
Shared	496	423	606×4	3356

SRAM, two-port register file, asynchronous single-port SRAM, and SDRAM). The second column lists the memory configurations. The third column shows whether diagnosis is supported. The fourth column gives the bank access method: non-interleaved or interleaved. The fifth column indicates whether or not the input data (D) and output data (Q) are shared. Basically, shared DQ requires more complicated implementation because of the tristate bidirectional data bus and more complex timing. The last two columns list the area overhead of the BIST circuit in terms of gate counts and the percentages, respectively. The results were obtained by a popular synthesis tool using a $0.35 \mu m$ CMOS standard cell library.

From Table 8.15, the area of the BIST circuit increases slightly when the size of the memory doubles. In the BIST circuit, the area percentages of the controller, sequencer, and TPG are approximately 34%, 25%, and 41%, respectively, for a typical ($8 K \times 16$) SRAM configuration. When diagnosis support is specified, the error handling module requires about 33% more of the original area for SRAM BIST.

For a $16 M \times 4$ DRAM, the area percentages of the controller, sequencer, and TPG are approximately 25%, 36%, and 39%, respectively. The percentage of the sequencer increases due to more complex timing. From Table 8.15, doubling the address space introduces 2.5% more area, while doubling the data width introduces 6% more area. Interleaved access costs around 7% more. In general, the area overhead of BIST grows roughly in the log scale with respect to the memory size, so it is relatively small for large memory cores.

Table 8.16 illustrates the case of testing multiple memory cores, assuming the design of BIST for four identical $8 K \times 16$ synchronous single-port SRAM cores. Two implementations are applied: four identical BISTs for the four SRAMs and shared controller and sequencer with four dedicated TPGs (see Figure 8.29). The area of the shared controller and sequencer increases slightly, however, the overall area overhead reduces greatly to about 60%.

8.5 CONCLUDING REMARKS

In this chapter we have discussed important aspects of semiconductor memory testing, including fault models, test algorithms, fault simulation, automatic test algorithm generation, and BIST. The BIST architecture presented in this chapter supports March-based tests and diagnosis. By selecting an appropriate clock period, it also tests the timing specifications. The approach is flexible because additional test commands (other than March elements) can be included with little effort. It is cost effective as the test time is short, the hardware overhead is low, and the test coverage is high. Burn-in also can be supported if the design of the DRAM core can be modified for that purpose.

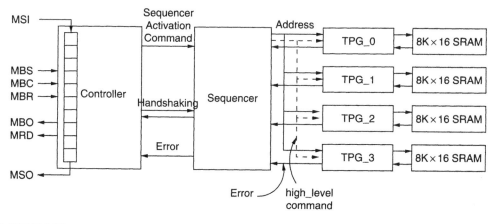

■ **FIGURE 8.29**

Architecture of the shared BIST for multiple memory cores.

A BIST compiler for embedded memories called BRAINS also has been presented, which can be used for common SRAM and DRAM cores, such as synchronous SRAM, asynchronous SRAM, dual-port SRAM, two-port register file, ZBT SRAM, EDO DRAM, SDRAM, DDR DRAM, etc. Given the memory specifications and test requirements, BRAINS generates the synthesizable RTL code for the BIST circuit in Verilog, as well as its activation sequence, test bench, and synthesis scripts. It performs at-speed testing and diagnosis of the RAM under test, and the March tests are programmable. Such a tool can be used for a wide range of RAM architectures and configurations. Also, BIST circuits can be shared among multiple memories to reduce overall area overhead. Therefore, it is critical for SOC design.

8.6 EXERCISES

8.1 (Fault Models)

a. Show that a test detects all AFs if it contains the March elements $\Uparrow$ $(rx, \ldots, wx')$ and $\Downarrow (rx', \ldots, wx)$, and the memory is initialized to the proper value before each March element. Show that MATS ($\{\Updownarrow (w0); \Updownarrow (r0, w1); \Updownarrow (r1)\}$) detects all AFs.

b. What are the requirements for a March test to detect the TFs? Explain.

c. What are the requirements for a March test to detect the AFs linked with TFs? Explain.

d. Design an irredundant March test to detect AFs, SAFs, and $\downarrow/1$ TFs. Prove that it is irredundant and complete.

8.2 **(Marching 1/0)** Prove that Marching 1/0 is a redundant test for AFs, SAFs, and TFs.

8.3 **(MATS^{++})** Prove that MATS^{++} is complete and irredundant for AFs, SAFs, and TFs.

8.4 **(March Element)** Determine the March element type in the following procedure. What faults can it detect?

```
Procedure My-March
{ for (i=0; i<n; i++) write 0 in cell[i];
pause; /* detects retention of 0 */
for (i=0; i<n; i++) read cell[i];
for (i=0 && j=n-1; i<n/2 && j>(n/2-1); i++ && j--)
  { write 1 in cell[i];
        read cell[i];
        write 1 in cell[j];
        read cell[j]; }
pause; /* detects retention of 1 */
for (i=0; i<n; i++) read cell[i];
for (i = n/2-1 && j=n/2; i>=0 && j<n; i-- && j++)
  { write 0 in cell[i];
        read cell[i];
        write 0 in cell[j];
        read cell[j]; } }
```

8.5 **(Fault Coverage)** Do the following four March tests have identical fault coverage?

  a. $\{\Uparrow (w0); \Uparrow (r0, w1); \Downarrow (r1, w0, r0)\}$

  b. $\{\Uparrow (w1); \Uparrow (r1, w0); \Downarrow (r0, w1, r1)\}$

  c. $\{\Downarrow (w0); \Downarrow (r0, w1); \Uparrow (r1, w0, r0)\}$

  d. $\{\Downarrow (w1); \Downarrow (r1, w0); \Uparrow (r0, w1, r1)\}$

8.6 **(Test Time)** Can we save test time by testing a bit-oriented memory as a word-oriented memory? Explain.

8.7 **(Time Complexity)**

  a. What are the time complexities for MOVI, GALPAT, and Butterfly, respectively?

  b. What is the time for the Butterfly algorithm to test a 1-Gb bit-oriented RAM, assuming the Read/Write access rate is 100M operations per second?

c. What is the time for the MOVI algorithm to test a 1-Gb word-oriented RAM, assuming the Read/Write access rate is 100M operations per second and the word length is 16? Standard backgrounds are used.

8.8 **(Address Line Fault)** A register file has 16 registers with 4 address lines $(A_0, A_1, A_2,$ and $A_3)$. Design a simple March test to detect the SA0 fault on address line A_1.

8.9 **(RAMSES and TAGS)** Briefly describe how RAMSES and TAGS work, and give their respective worst-case time complexity.

8.10 **(Fault Descriptors)** Give the fault descriptors for all the coupling faults defined in this chapter.

8.11 **(BIST Circuit)** Design a BIST circuit at the behavioral RT level. Use the March–CW test as the default test algorithm for the BIST circuit but allow the user to program the March test algorithm from at least eight different March elements:

- Memory size: 4K × 16:

- Pin description:

Pin	Description
A[11:0]	Address (A[11] = MSB)
D[15:0]	Data Inputs (D[15] = MSB)
CLK	Clock Input
CEN	Chip Enable
WEN	Write Enable
Q[15:0]	Data Outputs (Q[15] = MSB)

- Function table:

CEN	WEN	Data Out	Mode
H	X	Previous Data	Standby
L	L	Data In	Write
L	H	SRAM Data	Read

- Read-cycle timing diagram:

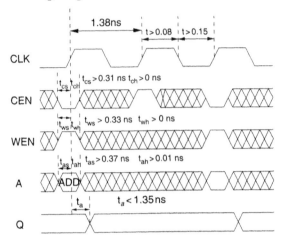

- Write-cycle timing diagram:

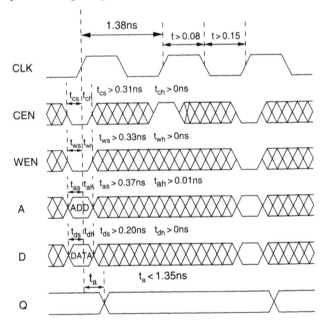

8.12 **(A Design Practice)** Repeat Problem 8.11 and use TurboBIST-Memory provided online for the same SRAM, and compare the results with respect to test algorithm, BIST circuit overhead, timing penalty, fault coverage, design effort, etc.

8.13   **(FIFO Test)** A FIFO has separate Read and Write ports. Each port has its own address register that automatically increments upon completion of a Read (Write) operation; that is, the Read (Write) address increments upon completion of a Read (Write) operation. Note that the Read address and Write address function independently, and a Reset operation resets both the Read address and Write address to 0.

    a.   What are the restrictions for March tests when applied to FIFOs?

    b.   Repeat the previous problem for a FIFO with similar specifications.

8.14   (Scrambling)

    a.   What is data scrambling? Why do we perform data scrambling?

    b.   What is address scrambling? Why do we perform address scrambling?

    c.   What are the faults for which scrambling must be considered during testing? Explain.

8.15   **(ROM BIST)** Suppose we have a ROM of size $1\,K \times 35$, and would like to test it using signature analyzer-based BIST. Design the BIST circuit, where a 24-bit signature is to be used for go/no-go comparison.

Acknowledgments

Many of my students have contributed to the work presented in this chapter, especially C.-T. Huang, J.-R. Huang, C.-F. Wu, K.-L Cheng, C.-W. Wang, C. Cheng, etc. It has been my great pleasure working with so many talented students.

References

R8.1—Introduction

[Wu 1998] C.-W. Wu, Testing embedded memories: is BIST the ultimate solution?, in *Proc. Asian Test Symp.*, November 1998, pp. 516–517.

R8.2—RAM Functional Fault Models and Test Algorithms

[Dekker 1988a] R. Dekker, F. Beenker, and L. Thijssen, Fault modeling and test algorithm development for static random access memories, in *Proc. IEEE Int. Test Conf.*, October 1988, pp. 343–352.

[Huang 1999] C.-T. Huang, J.-R. Huang, C.-F. Wu, C.-W. Wu, and T.-Y. Chang, A programmable BIST core for embedded DRAM, *IEEE Des. Test Comput.*, 16(1), 59–70, 1999.

[Nadeau-Dostie 1990] B. Nadeau-Dostie, A. Silburt, and V. K. Agarwal, Serial interface for embedded-memory testing, *IEEE Des. Test Comput.*, 7(2), 52–63, 1990.

[Prince 1991] B. Prince, *Semiconductor Memories: A Handbook of Design, Manufacture and Application*, 2nd ed., John Wiley & Sons, Chichester, 1991.

[Riedel 1995] M. Riedel and J. Rajski, Fault coverage analysis of RAM test algorithms, *Proc. IEEE VLSI Test Symp.*, April 1995, pp. 227–234.

[Sharma 1997] A. K. Sharma, *Semiconductor Memories: Technology, Testing, and Reliability*, IEEE Press, Piscataway, NJ, 1997.

[Simonse 1998] E. Simonse, Circuit Structures, Design Requirements, and Fault Simulations for CMOS SRAMs, Master's thesis, Faculty of Information Technology and Systems, Delft University of Technology, The Netherlands.

[van de Goor 1991] A. J. van de Goor, *Testing Semiconductor Memories: Theory and Practice*, John Wiley & Sons, Chichester, 1991.

[van de Goor 1993] A. J. van de Goor, Using march tests to test SRAMs, *IEEE Des. Test Comput.*, 10(1), 8–14, 1993.

[van de Goor 1998a] A. J. van de Goor and S. Hamdioui, Fault models and tests for two-port memories, in *Proc. IEEE VLSI Test Symp.*, April 1998, pp. 401–410.

[van de Goor 1998b] A. J. van de Goor and I. B. S. Tlili, March tests for word-oriented memories, in *Proc. IEEE Design, Automation, and Test in Europe Conf.*, March 1998, pp. 501–508.

[Wu 1997] Y. Wu and S. Gupta, Built-in self-test for multi-port RAMs, in *Proc. Asian Test Symp.*, November 1997, pp. 398–403.

[Wu 1999] C.-F. Wu, C.-T. Huang, and C.-W. Wu, RAMSES: a fast memory fault simulator, in *Proc. IEEE Int. Symp. on Defect and Fault Tolerance in VLSI Systems*, October 1999, pp. 165–173.

[Wu 2000] C.-F. Wu, C.-T. Huang, K.-L. Cheng, and C.-W. Wu, Simulation-based test algorithm generation for random access memories, in *Proc. IEEE VLSI Test Symp.*, April 2000, pp. 291–296.

[Wu 2001] C.-F. Wu, C.-T. Huang, K.-L. Cheng, C.-W. Wang, and C.-W. Wu, Simulation-based test algorithm generation and port scheduling for multi-port memories, in *Proc. IEEE/ACM Design Automation Conf.*, June 2001, pp. 301–306.

[Zhao 2000] J. Zhao, S. Irrinki, M. Puri, and F. Lombardi, Detection of inter-port faults in multi-port static RAMs, in *Proc. IEEE VLSI Test Symp.*, April 2000, pp. 297–302.

R8.3—RAM Fault Simulation and Test Algorithm Generation

[Huang 1999] C.-T. Huang, J.-R. Huang, C.-F. Wu, C.-W. Wu, and T.-Y. Chang, A programmable BIST core for embedded DRAM, *IEEE Des. Test Comput.*, 16(1), 59–70, 1999.

[van de Goor 1991] A. J. van de Goor, *Testing Semiconductor Memories: Theory and Practice*, John Wiley & Sons, Chichester, 1991.

[van de Goor 1998b] A. J. van de Goor and I. B. S. Tlili, March tests for word-oriented memories, in *Proc. IEEE Design, Automation, and Test in Europe Conf.*, March 1998, pp. 501–508.

[Wu 1999] C.-F. Wu, C.-T. Huang, and C.-W. Wu, RAMSES: a fast memory fault simulator, in *Proc. IEEE Int. Symp. on Defect and Fault Tolerance in VLSI Systems*, October 1999, pp. 165–173.

[Wu 2002] C.-F. Wu, C.-T. Huang, K.-L. Cheng, and C.-W. Wu, Fault simulation and test algorithm generation for random access memories, *IEEE Trans. Comput.-Aided Des. Integrated Circuits Syst.*, 21(4), 480–490, 2002.

[Zarrineh 1998] K. Zarrineh, S. J. Upadhyaya, and S. Chakravarty, A new framework for generating optimal march tests for memory arrays, in *Proc. IEEE Int. Test Conf.*, October 1998, pp. 73–82.

R8.4—Memory Built-In Self-Test

[Camurati 1995] P. Camurati, P. Prinetto, M. S. Reorda, S. Barbagallo, A. Burri, and D. Medina, Industrial BIST of embedded RAMs, *IEEE Des. Test Comput.*, 12(3), 86–95, 1995.

[Cheng 2000] C. Cheng, C.-T. Huang, J.-R. Huang, C.-W. Wu, C.-J. Wey, and M.-C. Tsai, BRAINS: a BIST complier for embedded memories, in *Proc. IEEE Int. Symp. on Defect and Fault Tolerance in VLSI Systems*, October 2000, pp. 299–307.

[Dekker 1988a] R. Dekker, F. Beenker, and L. Thijssen, Fault modeling and test algorithm development for static random access memories, in *Proc. IEEE Int. Test Conf.*, October 1988, pp. 343–352.

[Dekker 1988b] R. Dekker, F. Beenker, and L. Thijssen, A realistic self-test machine for static random access memories, in *Proc. IEEE Int. Test Conf.*, October 1988, pp. 353–361.

[Dreibelbis 1998] J. Dreibelbis, J. Barth, Jr., R. Kho, and H. Kalter, An ASIC library granular DRAM macro with built-in self test, in *Proc. IEEE Int. Solid-State Circuits Conf.*, February 1998, pp. 74–75.

[Huang 1999] C.-T. Huang, J.-R. Huang, C.-F. Wu, C.-W. Wu, and T.-Y. Chang, A programmable BIST core for embedded DRAM, *IEEE Des. Test Comput.*, 16(1), 59–70, 1999.

[Huang 2000] C.-T. Huang, J.-R. Huang, and C.-W. Wu, A programmable built-in self-test core for embedded memories, in *Proc. Asia and South Pacific Design Automation Conf.*, January 2000, pp. 11–12.

[Nadeau-Dostie 1990] B. Nadeau-Dostie, A. Silburt, and V. K. Agarwal, Serial interface for embedded-memory testing, *IEEE Des. Test Comput.*, 7(2), 52–63, 1990.

[Tanoi 1997] S. Tanoi, Y. Tokunaga, T. Tanabe, K. Takahashi, A. Okada, M. Itoh, Y. Nagatomo, Y. Ohtsuki, and M. Uesugi, On-wafer BIST of a 200-Gb/s failed bit search for 1-Gb DRAM, *IEEE J. Solid-State Circuits*, 32(11), 1735–1742, 1997.

[Treuer 1993] R. P. Treuer and V. K. Agarwal, Built-in self-diagnosis for repairable embedded RAMs, *IEEE Des. Test Comput.*, 10(2), 24–33, 1993.

[van de Goor 1991] A. J. van de Goor, *Testing Semiconductor Memories: Theory and Practice*, John Wiley & Sons, Chichester, 1991.

[Wu 2000] C.-F. Wu, C.-T. Huang, K.-L. Cheng, and C.-W. Wu, Simulation-based test algorithm generation for random access memories, in *Proc. IEEE VLSI Test Symp.*, April 2000, pp. 291–296.

[Zarrineh 1999a] K. Zarrineh and S. J. Upadhyaya, On programmable memory built-in self test architectures, in *Proc. IEEE Design, Automation, and Test in Europe Conf.*, March 1999, pp. 708–713.

[Zarrineh 1999b] K. Zarrineh and S. J. Upadhyaya, Programmable memory BIST and a new synthesis framework, in *Proc. Fault Tolerant Computing Symp.*, June 1999, pp. 352–355.

CHAPTER 9

MEMORY DIAGNOSIS AND BUILT-IN SELF-REPAIR

Cheng-Wen Wu
National Tsing Hua University, Hsinchu, Taiwan

ABOUT THIS CHAPTER

The purpose of **memory diagnosis** is twofold: (1) locating failures and subsequently repairing them, and (2) analyzing failures and defects and subsequently improving design and process. Both are important for enhancing manufacturing yield.

In this chapter we first present a hybrid BIST design—with diagnosis support—for embedded RAM. In association with the BIST design, we also will show a diagnosis system (called **MECA**) for automatic identification of the fault site and fault type. The BIST design has a test mode that supports fault location for subsequent laser repair or self-repair and an online programming mode for custom diagnostic test commands. An efficient test algorithm was built in to cover all stuck-at, transition, state coupling, idempotent coupling, inversion coupling, address decoder, and stuck-open faults of the word-oriented memory cores. The default algorithm is March–CW, one of the Cocktail–March algorithms that we described previously for efficient testing of word-oriented memories. The online programming mode makes it possible for the user to apply more sophisticated diagnosis algorithms. In addition to the fault locations necessary for repair, the syndromes of the detected faults can also be exported by the BIST circuit. By recording the fault locations and syndromes, the diagnosis system can identify the fault type of each faulty cell.

Redundancy analysis (RA) algorithms are presented next, including a conventional algorithm and a greedy algorithm that can be efficiently implemented on chip, which is called the *essential spare pivoting* (ESP) algorithm. We will also discuss a simulator for evaluating repair efficiency for different RA algorithms.

Finally, we present a *built-in self-repair* (BISR) scheme for memories with two-dimensional redundancy structures. The BISR design is composed of a BIST module and a *built-in redundancy analysis* (BIRA) module. It supports three test modes: (1) main memory testing, (2) spare memory testing, and (3) repair. The BIRA module also serves as the reconfiguration (address remapping) unit in normal mode.

9.1 INTRODUCTION

9.1.1 Why Memory Diagnosis?

Embedding predesigned and preverified cores into a system chip is currently a popular methodology for *system-on-chip* (SOC). This *reuse* methodology is believed to be indispensable for maintaining an affordable product development cycle. Memories are among the most frequently used cores in SOC. Embedded memories are occupying a major portion of the silicon area and consuming most of the transistors of a typical SOC. Therefore, the yield of such a system chip is mainly controlled by the embedded memories. Test and diagnosis of embedded memories thus are important issues in SOC development.

We discussed BIST in the previous chapter. There are some important issues that a pure BIST scheme does not solve, such as diagnosis and repair. High density, high operating clock rate, and deep submicron technology are giving us more new failures and faults in memory cores. Conventional memory *automatic test equipment* (ATE) designed for mass production test provides only limited information for failure analysis that usually is insufficient for fast debugging. Designers need a diagnosis-supporting mechanism within the BIST circuit and sometimes a *built-in self-repair* (BISR) scheme to increase product quality, reliability, and yield.

9.1.2 Why Memory Repair?

To avoid yield loss, **redundant elements** or **spare elements** (*i.e.*, spare rows and columns of storage cells) are often added so most faulty cells can be repaired (*i.e.*, replaced by spare cells) [Cenker 1979] [Smith 1981] [Benevit 1982]. *Redundancy*, however, adds to cost in another form. Analysis of redundancies to maximize yield (after repair) and minimize cost is an important process during manufacturing. *Redundancy analysis* (RA) using expensive memory testers is becoming inefficient (and therefore not cost-effective) as chip density continues to grow. The use of embedded memories creates yet another problem—embedded memories are even more difficult to deal with using external testers [Huang 1999]. Although BIST is a promising solution, if BIST schemes are only for functional testing, they cannot replace external memory testers entirely. BIST with diagnosis support is still not enough because of the large amount of diagnosis data that must be transferred through the channel with limited bandwidth to external tester. Therefore, *built-in redundancy analysis* (BIRA) and *built-in self-repair* (BISR) are now among the top items to be incorporated with memory cores.

9.2 REFINED FAULT MODELS AND DIAGNOSTIC TEST ALGORITHMS

The functional fault models we use for the memory under test are the same as those discussed in Chapter 8, including *stuck-at fault* (SAF), *transition fault* (TF), *stuck-open fault* (SOF), *address decoder fault* (AF), *inversion coupling fault* (CFin), *idempotent coupling fault* (CFid), *state coupling fault* (CFst), etc. However, in diagnosis we need to know not only the fault site but also the root cause of the failure. It is helpful to refine fault models so more detailed behavior can be identified.

A coupling fault can be expressed by the state or operations of the coupling cell (called the *aggressor*), the state or operations of the coupled cell (called the *victim*), and the fault content. For example, $\langle \downarrow ; \updownarrow \rangle$ is a CFin where the victim is inverted when the aggressor goes through a falling transition; $\langle 1; 1/0 \rangle$ is a CFst where the aggressor with state 1 forces the victim to transit from 1 to 0; and $\langle \uparrow ; 1/0 \rangle$ is a CFid where the victim is forced to transit from 1 to 0 when the aggressor goes through a rising transition. Table 9.1 lists all the fault models considered here, with the notation for each specific fault type. In the table, *Agr* is the state of the aggressor, *Vtm* the state of the victim (in the form of fault-free/faulty state or an inversion), and *Addr* the relation between the aggressor address and victim address. Note that for diagnosis, we need to specify the relation between the aggressor address and victim address (*i.e.*, which one is higher in the address space). In the *Addr* column of the table, A > V means that the aggressor address is greater than the victim address, and A < V means the other way around.

It has been shown that all the above faults can be completely and efficiently tested by the *March–CW* algorithm [Wu 1999], which is a fast Cocktail–March test algorithm extended from March C$^-$. It consists of several conventional March tests, each of them running on a specific data background. The March–CW algorithm for 8-bit word-oriented memories is shown in Figure 9.1.

Here, we focus on *synchronous SRAM. Asynchronous SRAM* is easier to deal with as far as functional test is concerned, because they are conventional RAMs without pipelining or sophisticated interface. By inspecting the timing specifications of the asynchronous SRAM, an appropriate clock cycle can be determined and the SRAM is tested as a synchronous one [Huang 1999]. This is quite similar to what a memory controller does. The Read/Write timing between the BIST circuit and the synchronous SRAM has to be synchronized by the system clock. During the Read or Write cycle of the target SRAM (see Figure 9.2), the address and *Output-Enable* (OE) signals should be asserted before the positive clock edge, and the data should be ready at the next positive clock edge. In this particular case the *input data* (DI) and *output data* (DO) use separate IO channels; therefore, the Write operation can write to both the cell array (from the data input channel) and data output channel if output is enabled. This is called the *write-through* operation.

The embedded synchronous SRAM may also have a pipelined access mode to raise clock rate. In the pipelined architecture, Read/Write latency can be more than one clock cycle. The corresponding data are ready after several cycles, when the address and OE signals are asserted. Similarly, in burst access mode we can use a single address and the Read/Write command to access a contiguous run of data. For functional test and diagnosis, we run March tests using single Read/Write modes (*i.e.*, without pipelining or burst access), as it is effective and easy to implement. The pipeline and burst-mode logic is not our concern here—it can be tested easily (*e.g.*, by a pipelined/burst-mode Read/Write sequence issued from an embedded processor on the same SOC).

TABLE 9.1 ■ Fault Models and Notation

Name	Agr	Vtm	Addr
SAF$_0$	—	1/0	—
SAF$_1$	—	0/1	—
TF$_0$	—	$\downarrow$/1	—
TF$_1$	—	$\uparrow$/1	—
CFin$_0$	$\downarrow$	$\updownarrow$	A < V
CFin$_1$	$\downarrow$	$\updownarrow$	A > V
CFin$_2$	$\uparrow$	$\updownarrow$	A < V
CFin$_3$	$\uparrow$	$\updownarrow$	A > V
CFst$_0$	0	1/0	A < V
CFst$_1$	0	1/0	A > V
CFst$_2$	0	0/1	A < V
CFst$_3$	0	0/1	A > V
CFst$_4$	1	1/0	A < V
CFst$_5$	1	1/0	A > V
CFst$_6$	1	0/1	A < V
CFst$_7$	1	0/1	A > V
CFid$_0$	$\downarrow$	1/0	A < V
CFid$_1$	$\downarrow$	1/0	A > V
CFid$_2$	$\downarrow$	0/1	A < V
CFid$_3$	$\downarrow$	0/1	A > V
CFid$_4$	$\uparrow$	1/0	A < V
CFid$_5$	$\uparrow$	1/0	A > V
CFid$_6$	$\uparrow$	0/1	A < V
CFid$_7$	$\uparrow$	0/1	A > V
AF$_0$	—	—	A < V
AF$_1$	—	—	A > V
SOF$_0$	—	—	addr = 0
SOF$_1$	—	—	0 < addr < N − 1
SOF$_2$	—	—	addr = N − 1

1. $\{\updownarrow (wa_1); \Uparrow (ra_1, wa_1'); \Uparrow (ra_1', wa_1); \Downarrow (ra_1, wa_1'); \Downarrow (ra_1', wa_1, ra_1); \updownarrow (ra_1)\}$

2. $\{\updownarrow (wa_2, wa_2', ra_2', wa_2, ra_2)\}$

3. $\{\updownarrow (wa_3, wa_3', ra_3', wa_3, ra_3)\}$

4. $\{\updownarrow (wa_4, wa_4', ra_4', wa_4, ra_4)\}$

■ **FIGURE 9.1**

March–CW for 8-bit word-oriented memories, where $a_1 = 00000000$, $a_2 = 00001111$, $a_3 = 00110011$, and $a_4 = 01010101$.

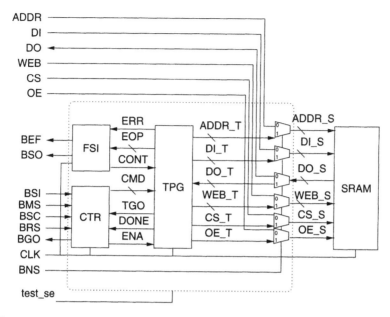

■ FIGURE 9.2

Block diagram of BIST and SRAM core.

9.3 BIST WITH DIAGNOSTIC SUPPORT

Figure 9.2 shows the architecture of the BIST design. It also shows connections between BIST and the embedded SRAM. The clock of the synchronous SRAM can be shared by the BIST circuit, so there is no need for an additional test clock. The BIST core consists of three blocks: ***controller*** (CTR), ***test pattern generator*** (TPG), and ***fault site indicator*** (FSI). It also has a set of multiplexers to form the test collar for the SRAM, switching between the BIST and Normal operation modes under the control of the BNS (BIST/Normal Select) signal. The test_se signal enables scan test for the BIST circuit itself. In Scan-Test mode, the ***BIST Serial-In*** (BSI) terminal takes the scan input data, and the ***BIST Serial-Out*** (BSO) terminal is the output data port of the scan chain. The BIST scan chain can be linked to other scan chains on the chip.

9.3.1 Controller

The CTR controls the overall test procedure and issues test commands for the ***test pattern generator*** (TPG) that generates test patterns for the targeted SRAM. It has two operation modes: Test and Analysis (or Diagnosis). The modes are selected by the ***BIST Mode Select*** (BMS) signal. In Test mode, the CTR sends a set of built-in commands to the TPG, based on the default test algorithm. In Analysis mode, the test algorithm is user programmable, and test commands are shifted in from the

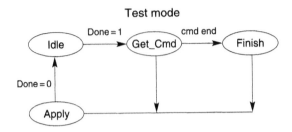

■ **FIGURE 9.3**

The CTR state diagram in Test mode.

BSI input. Note that the built-in commands are also programmable at the RT level; that is, they can be modified by the circuit designer before synthesis of the BIST circuit, so, new fault models can be covered. State diagrams of the CTR in Test and Analysis modes are shown in Figures 9.3 and 9.4, respectively.

In Test mode, the **BIST State Control** (BSC) input should stay at 1, and state transition follows the default sequence automatically. By applying a 0 to BSC, we force CTR to the Idle state (also known as the Reset state). The **BIST Reset** (BRS) signal also drives CTR to the Idle state. From the Idle state, CTR goes to the Get_Cmd state to fetch the first command from a look-up table and send it by the CMD channel to TPG. It then goes to the Apply state to generate the handshaking signal (ENA) to enable TPG. When TPG is executing the test command, CTR returns to the Idle state. When TPG completes the test command, the DONE signal goes high, and CTR enters the Get_Cmd state again to fetch the next command. The process is repeated as long as there are commands to be executed. In the Get_Cmd state, if all the commands have been executed, CTR goes to the Finish state and sends the Null command to terminate Test mode. Also, the **BIST Go/No-Go** (BGO) signal reports the testing result. The built-in test algorithm (*i.e.*, March–CW) is stored in a look-up table as a set of test commands.

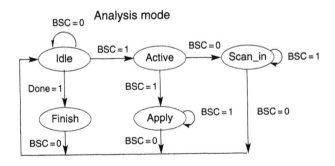

■ **FIGURE 9.4**

The CTR state diagram in Analysis mode.

In Analysis mode, the state transition of CTR is controlled by an external host (tester) through the BSC signal. After the Idle state, CTR goes to the Active state and then the Scan_In state if we apply 10 to BSC. In this state, CTR gets the command serially from the BSI input, then goes to the Apply state to enable TPG in a similar way as Test Mode. Note that CTR will go through the Idle and Active states again before entering the Apply state. This allows one to pause at the Idle state before applying test patterns. After the command is executed, BGO output will indicate the completion of the test command. The host can then apply the next command.

A test command consists of a 1-bit March direction (ascending or descending), a 4-bit operation code (Opcode), and a data background word whose width depends on the data width of SRAM, as shown in Figure 9.5. The 4-bit Opcode allows at most 16 different memory operations, such as *wa, rawa', rawa'ra', warawa'ra',* etc. Note that the Opcode set is configurable at the register transfer level like the built-in test algorithm mentioned above. We can use the test commands to generate commonly used March-based test/diagnosis algorithms.

9.3.2 Test Pattern Generator

The test pattern generator executes the test commands issued by CTR and generates the corresponding SRAM input signals, including the data, address, and control signals. A TPG state diagram is illustrated in Figure 9.6. Again, the Idle state is the default state after resetting TPG. When it is enabled by CTR, TPG goes to the Init state to initialize Address Counter and Session Counter. Session Counter keeps track of the SRAM Read/Write operation (session) that is being executed on the current address. For example, if the test command is *rawa'*, then the session value is 0 for *ra* and 1 for *wa'*, respectively. The session value is used to select the current operation being applied to the SRAM under test. Depending on the selected operation, either the **Write-Enable** (WE) or **Output-Enable** (OE) signal is asserted.

After decoding the Opcode, we obtain the SRAM operation sequence. If the Opcode is not a terminal command, TPG goes through the Ifetch, Exec, Dfetch, and Compare states in sequence; otherwise, it goes to the Go state and sets the TGO (see Figure 9.2) value to 1 (for Go) or 0 (for No-Go) depending on whether faults have been detected. In the Ifetch-Exec-Dfetch-Compare state sequence, TPG fetches the timing control data from the lookup table according to the current session value, then waits in the Exec state for a period of time equal to the SRAM access latency. In the experimental case the latency is one clock cycle. In the Dfetch state we get the output data from SRAM, then compare it with the fault-free data to obtain the error syndrome. If the syndrome value is non-zero (*i.e.*, errors are found), TPG goes to the Wait state and stores the information of the current operation in the **Error**

bits	1	4	W
	U/D	OP	Data

■ **FIGURE 9.5**

The test command format.

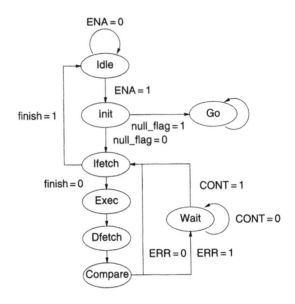

The test pattern generator state diagram.

Operation Protocol (EOP) registers. The error operation information includes the error address, the fault-detecting operation indicated by the session number, and the output data from memory. The format of EOP registers is shown in Figure 9.7. TPG resumes its execution cycle upon receiving the ***Continue*** (CONT) signal from FSI. If the syndrome value is zero (*i.e.*, no errors are found), the session number is incremented by 1 and TPG repeats the Ifetch-Exec-Dfetch-Dfetch state loop until we reach the last session for the current address. When the last session is reached, the session counter is reset to 0 and we advance to the next address. After finishing all addresses, TPG goes to the Idle state and sets the DONE signal to 1. Figure 9.8 shows the timing diagram of the waveform generated by TPG when executing the *rawa'* (write through) test element at address 000c, with the all-0 (*i.e.*, solid) data background. Note that TPG is easily configurable at the register transfer level to accommodate SRAMs with different latencies.

9.3.3 Fault Site Indicator (FSI)

The fault site indicator receives error information from TPG and sends it to the BSO output serially using a scan chain. When a fault is detected, TPG enters the Wait

bits	logN	3	W
	Addr	Session #	Syndrome

The EOP register format.

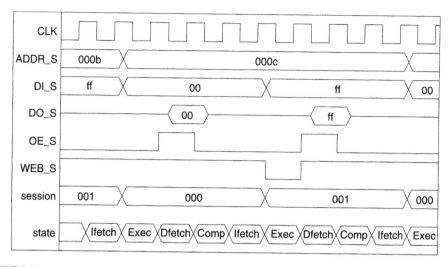

■ **FIGURE 9.8**

Timing diagram generated by TPG.

state, issues the **Error** (ERR) signal, and sets the *Error Operation Protocol* (EOP) data for the FSI. FSI then sets the **BIST Error Flag** (BEF) signal and sends the error information to the BSO output. After the EOP content is completely shifted out, FSI sets the CONT signal to allow TPG to continue the execution of the current test command. The timing diagram for FSI is shown in Figure 9.9.

When more area overhead is acceptable, the data compression function for error syndromes [Li 2000a,b] can be implemented in FSI. Data compression can reduce the size of the diagnosis data transmitted from the BIST circuit to the external tester. Testing time and ATE memory requirement thus can be reduced significantly.

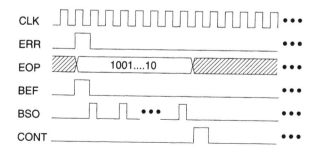

■ **FIGURE 9.9**

FSI timing diagram.

9.4 RAM DEFECT DIAGNOSIS AND FAILURE ANALYSIS

After collecting the EOP data from FSI, the error **bitmap** of the faulty SRAM for every Read operation can be obtained. If the SRAM has redundant resources (such as spare rows/columns), it can be repaired according to these error addresses. Moreover, the designers and process engineers can identify the fault type and failure mode by offline analysis using the EOP data. Arranged properly, the error data can be represented in the form of March syndromes [Wu 2000b]. A **March syndrome** shows the comparison results of all Read operations, which are either correct (represented by a 0) or incorrect (represented by a 1), during the testing process. For example, if a particular address has a syndrome $\langle 1100 \rangle$, then it means the first and second Read operations for this address return incorrect values. With the help of RAMSES [Wu 1999], a fault dictionary can be generated from these syndromes. The **fault dictionary** is constructed from the simulated responses under the given test algorithm and fault models [Abramovici 1990]. The fault dictionary approach for diagnosis is suited to memories, because the dictionary is just a small table—there is no need to create a different entry for each different memory cell. Table 9.2 shows the fault dictionary of the following March test called IFA9N:

$$E_0 \quad E_1 \, E_2 \quad E_3 \, E_4 \quad E_5 \, E_6 \quad E_7 \, E_8$$

$$\Uparrow (w0); \Uparrow (r0, w1); \Uparrow (r1, w0); \Downarrow (r0, w1); \Downarrow (r1, w0)$$

In the table, E_0 to E_8 represent the nine Read/Write operations in the algorithm. The value for a Write operation is always 0, because a March test will not detect faults during the Write operation. In a normal situation only Read operations can detect faults, and the Write operations can only activate faults; however, if the SRAM has a write-through mechanism, then the Write operation may detect fault in the write-through logic. The *March signature* for each fault model is the error-domain response of the March test when the fault exists and is represented by the corresponding row in the table. That is, the fault dictionary is the table of March signatures for all target fault models. In a March signature for fault f, there is a 1 in the table entry E_i if and only if fault f is detected by the (Read) operation E_i. For example, $CFin_3$ is detected by E_3 and E_5, so its March signature is $\langle 000101000 \rangle$.

To measure the quality of a diagnosis test algorithm, we use the **diagnosability ratio** (DR), defined as the ratio of the number of distinguishable fault types among the number of total detectable fault types. For example, the DR of IFA9N is $10/27 = 37\%$, which is quite low. All the equivalent classes of fault types are listed as follows:

$\{SAF_1\}$
$\{CFin_2\}$
$\{CFin_3\}$
$\{TF_1, SAF_0\}$
$\{SOF_0, AF_0\}$
$\{SOF_2, AF_1, CFin_1\}$
$\{CFid_6, CFst_3, CFst_6\}$

TABLE 9.2 ■ Fault Dictionary for IFA9N

Name/Operation	E_0	E_1	E_2	E_3	E_4	E_5	E_6	E_7	E_8
SAF_0	0	0	0	1	0	0	0	1	0
SAF_1	0	1	0	0	0	1	0	0	0
TF_0	0	0	0	0	0	1	0	0	0
TF_1	0	0	0	1	0	0	0	1	0
$CFin_0$	0	0	0	1	0	0	0	0	0
$CFin_1$	0	0	0	0	0	1	0	1	0
$CFin_2$	0	1	0	0	0	0	0	1	0
$CFin_3$	0	0	0	1	0	1	0	0	0
$CFst_0$	0	0	0	1	0	0	0	0	0
$CFst_1$	0	0	0	0	0	0	0	1	0
$CFst_2$	0	0	0	0	0	1	0	0	0
$CFst_3$	0	1	0	0	0	0	0	0	0
$CFst_4$	0	0	0	0	0	0	0	1	0
$CFst_5$	0	0	0	1	0	0	0	0	0
$CFst_6$	0	1	0	0	0	0	0	0	0
$CFst_7$	0	0	0	0	0	1	0	0	0
$CFid_0$	0	0	0	1	0	0	0	0	0
$CFid_1$	0	0	0	0	0	0	0	1	0
$CFid_2$	0	0	0	0	0	0	0	0	0
$CFid_3$	0	0	0	0	0	1	0	0	0
$CFid_4$	0	0	0	0	0	0	0	1	0
$CFid_5$	0	0	0	1	0	0	0	0	0
$CFid_6$	0	1	0	0	0	0	0	0	0
$CFid_7$	0	0	0	0	0	1	0	0	0
SOF_0	0	1	0	1	0	0	0	0	0
SOF_1	0	0	0	0	0	0	0	0	0
SOF_2	0	0	0	0	0	1	0	1	0
AF_0	0	1	0	1	0	0	0	0	0
AF_1	0	0	0	0	0	1	0	1	0

$\{CFst_1, CFst_4, CFid_1 CFid_4\}$
$\{TF_0, CFid_3, CFid_7, CFst_2, CFst_7\}$
$\{CFst_0, CFst_5, CFid_0, CFid_5, CFin_0\}$

The fault types in each group are *indistinguishable under the simple IFA9N test algorithm*, as they have identical fault signatures. However, they may be *distinguishable under other test algorithms*. To distinguish them, a longer and more complex algorithm is required. The IFA9N algorithm apparently is not a good diagnosis algorithm. In [Li 2001c], a $17N$ algorithm is shown to be effective for the target faults discussed here. Because the BIST circuit is programmable, the $17N$ diagnosis algorithm can be applied easily.

We also have developed an automatic diagnosis system—the **Memory Error Catch and Analysis** (MECA) system—that can identify fault types by comparing the syndrome of each faulty cell with all March signatures in the fault dictionary

[Wu 2000b]. When the syndrome of a faulty cell matches a March signature in the dictionary, the system will report the corresponding fault type. For example, if a faulty cell has a syndrome $\langle 000101000 \rangle$, then the system compares it with the fault dictionary, and reports that the cell has a $CFin_3$ fault.

From the ATE data log the **error bitmaps** can be obtained. Error analysis is a procedure that takes the error bitmaps and the fault dictionary as input and generates **fault bitmaps** that contain fault locations and fault types. The MECA system is shown in Figure 9.10. The main components are RAMSES (the memory fault simulator [Wu 1999]), TAGS (the test algorithm generator [Wu 2000a]), and the Analysis Engine. For any RAM under test, we have user-specified test requirements (*i.e.*, the target fault models, fault coverage, diagnosability ratio, and test length). RAMSES evaluates the fault coverage and diagnosability ratio and generates the fault dictionary for the March test. TAGS generates a March test based on RAMSES results to meet the test requirements. After applying the test, the results are sent to the Analysis Engine, which in turn generates error bitmaps and subsequently fault bitmaps.

With the BIST circuit and MECA system, we can easily construct fault bitmaps for RAM, one for each fault model. In each fault bitmap, the distribution of the faulty cells is detailed by a visual diagram.

As an example, we use the IFA9N test algorithm for an industrial $16K \times 9$ SRAM. Table 9.3 shows the summary of diagnosis result of this memory chip. According to the diagnosis result, we can see that more than 95% of the faults are SAF_1, SAF_0, and TF_1. The number of coupling faults is small compared with stuck-at faults for this particular case. The address scrambling table was provided by the SRAM designer, so we were able to generate fault bitmaps with the correct floorplan and physical location of each cell.

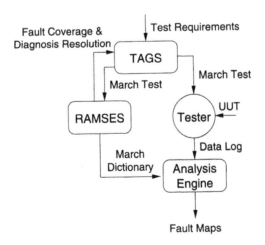

■ FIGURE 9.10

The MECA system.

TABLE 9.3 ■ Summary of Diagnosis Results Using IFA9N

Equivalent Fault Class	Instances
$\{SAF_1\}$	51.7%
$\{TF_1, SAF_0\}$	43.9%
$\{SOF_2, AF_1, CFin_1\}$	0.2%
$\{CFin_2\}$	0.1%
$\{CFin_3\}$	0.1%
$\{SOF_0, AF_0\}$	0.1%
$\{CFid_6, CFst_3, CFst_6\}$	0.1%
$\{CFst_1, CFst_4, CFid_1 CFid_4\}$	0.1%
$\{TF_0, CFid_3, CFid_7, CFst_2, CFst_7\}$	0.1%
$\{CFst_0, CFst_5, CFid_0, CFid_5, CFin_0\}$	0.1%

9.5 RAM REDUNDANCY ANALYSIS ALGORITHMS

9.5.1 Conventional Redundancy Analysis Algorithms

Conventionally, *redundancy analysis* (RA) is performed on a memory tester using software. The tester stores the bitmap (a map of the faulty cells) after a diagnostic test and performs redundancy analysis based on the bitmap. Software analysis is slow, so normally only simple heuristic algorithms are used. Most such algorithms consist of two phases: **must-repair** followed by **final-repair**. In the must-repair phase, all faulty lines that must be repaired are identified first, limiting the number of remaining faulty cells. In the final-repair phase, simple algorithms, such as the *fault-driven, row-first,* and *column-first* algorithms, are used. For example, a greedy redundancy analysis algorithm called the *repair-most* (RM) algorithm was presented in [Tarr 1984]. The RM algorithm also consists of the must-repair phase and final-repair phase. Error counters for the respective faulty rows and columns are required in the RM analysis. The *fault-driven* algorithm (based on exhaustive search) generates all possible spare allocations to find the optimal one [Day 1985]. The exhaustive search approach is slow. Also, finding that the optimal solution is *NP* complete has been proven by transforming it in polynomial time to the bipartite-graph clique problem [Kuo 1987]. Branch-and-bound and approximation algorithms can be used to reduce search time. An improved approach called the *faulty-line covering* technique and a heuristic criterion allowing fast repair were later presented in [Huang 1990]. In addition, there are other redundancy analysis techniques which are mainly for fast repairability decision [Wey 1987] [Haddad 1991]. Most of the conventional redundancy analysis approaches, however, assume the availability of a memory tester with high computing power, memory capacity, and flexibility. Such memory testers would be very expensive. It is clear that analysis time (or tester throughput) is critical as far as cost is concerned [Haddad 1991]. The use of BIRA (with BIST) will greatly increase tester throughput.

Before discussing more efficient approaches, we must first define the terminology and notation. A memory *block* consists of M rows and N columns of storage cells (*i.e.*, an $M \times N$ array of cells). The origin of the cell array is the upper left corner. There are r spare rows and c spare columns. We follow the definitions given in [Huang 1990].

Definition 9.1

A **faulty line** l is either a row or a column in which one or more faulty cells exist. The number of faulty cells in l is F_l. A faulty line is either a **faulty row** or a **faulty column**.

Definition 9.2

A faulty line is said to be *covered* if all faulty cells on the line have been scheduled to be repaired by specific spare rows and/or spare columns.

Definition 9.3

A faulty cell that does not share any row or column with any other faulty cell is referred to as an **orthogonal faulty cell**.

Let the numbers of available spare rows and available spare columns during the analysis process be denoted as r_a and c_a, respectively. Also, F denotes the total number of faulty cells in the memory block, and $F_\perp$ represents the number of orthogonal faulty cells. Two important early termination conditions are as follows:

Condition 1. After must-repair phase, $F > 2r_a c_a$.
Condition 2. $F_\perp > r_a + c_a$.

If either condition is true, then the analysis process stops. Early termination conditions help identify memories that cannot be repaired by available spares. Also, during analysis, any faulty line that consists of k faulty cells requires either 1 spare line in the same direction or k perpendicular spare lines. Therefore, any faulty row (column) with k faulty cells, where $k > c_a (k > r_a)$, is a *must-repair faulty line*. We thus have the following two additional early termination conditions:

Condition 3. $n_r > r_a$ if $c_a = 0$, where n_r is the number of faulty rows not covered so far.
Condition 4. $n_c > c_a$ if $r_a = 0$, where n_c is the number of faulty columns not covered so far.

If any of the above four conditions is true, then the memory block is unrepairable.

Unlike the fault-driven approach, the repair-most approach is straightforward and simple. The F_l values of all faulty lines are calculated during bitmap construction. The faulty lines are ordered according to their F_l values—the first line to be repaired is the one with the largest F_l value. Row and column counters are

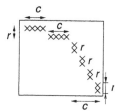

■ **FIGURE 9.11**

A defective memory block of the worst case.

still required for keeping fault counts of the faulty lines. Apparently bitmap is a necessary tool for repair-most analysis.

A defective memory block of the worst case is shown in Figure 9.11, where $F = 2r_a c_a$, and all available spare rows and columns have to be used for repairing the faulty lines. Specifically, available spare rows cover r_a faulty rows, where each faulty row has c_a faulty cells, and among these faulty rows no cells share the same column address. Therefore, the faulty cells in these rows are located in $c_a \times r_a$ different columns. Similarly, the c_a faulty lines covered by available spare columns have $r_a \times c_a$ row addresses for the faulty cells. In this special case, the width of the bitmap is $c_a r_a + c_a$ and the height is $r_a c_a + r_a$. Therefore, in general, the size of the bitmap can be limited to $(r_a c_a + r_a) \times (c_a r_a + c_a)$ instead of $M \times N$ after the must-repair phase. Note that we assume $F_r \leq c_a$ and $F_c \leq r_a$ after must-repair phase, where $F_r(F_c)$ is the maximum number of faulty cells in a row (column).

In addition to the bitmap, a total of $(r_a c_a + r_a)$ (each has $\log(c_a + 1)$ bits) row counters and $(c_a r_a + c_a)$ (each has $\log(r_a + 1)$ bits) column counters are required. The row counter for a faulty row l stores F_l, which is tagged by the $\log N$-bit row address. Similarly, the column counter for a faulty column l stores F_l, which is tagged by the $\log M$-bit column address. Note that, thanks to the must-repair phase, the maximum F_l value of any faulty row (column) can be limited to $c + 1(r + 1)$ without affecting the result. Therefore, even if the must-repair phase is skipped, the maximum size of the bitmap can still be limited to $(r(c + 1) + r) \times (c(r + 1) + c)$. However, in that case the complexity of on-chip analysis circuitry could be unaffordable due to the high storage requirement of the repair-most algorithms.

For the fault-driven algorithms, including those using exhaustive search and those using heuristics, the control (*e.g.*, tree expansion) usually is complicated, so its hardware is difficult to realize. Also, it is slow. Consequently, redundancy analysis algorithms are based on the repair-most approach.

9.5.2 The Essential Spare Pivoting Algorithm

Storing full bitmap on-chip for the purpose of redundancy analysis obviously is not feasible. The goal is for the BIRA circuit to properly allocate redundancies in parallel with the BIST operation. The area overhead should be low and repair efficiency should be high. The *repair efficiency* is defined as the *repair rate* with respect to unit area overhead. Because repair-most algorithms rely on a bitmap as the basic tool for redundancy analysis, it seems that a bitmap is inevitable. This is not true. We now present the **essential spare pivoting** (ESP) algorithm, which does not use a bitmap [Huang 2003].

We observe the following general guidelines for redundancy analysis.

1. A faulty row is more suitable for row repair when there are more faults in the row than in any of the columns of the corresponding faulty cells in the row. Likewise, a faulty column is more suitable for column repair when there are more faults in the column than in any of the rows of the corresponding faulty cells in the column.

2. An orthogonal fault can be repaired by either a spare row or a spare column. Orthogonal faults should be processed after all others.

The first guideline leads to the repair-most-based algorithms. A full bitmap or various local bitmaps are required to perform analysis. Because our goal is to repair without bitmaps, we revise the first guideline as follows:

- For any faulty row (column), if the number of faulty cells is greater than or equivalent to a threshold number (E_{th}), repair it by a spare row (column).

This guideline is similar to the must-repair rule, except that the decision is based on a customized threshold number, E_{th}, instead of r_a and c_a. In the analysis procedure, we maintain a counter for the number of faults in each faulty line. When the number reaches E_{th}, it is marked as an *essential line*. The second guideline shown above states that an orthogonal fault should be recognized early but processed after all others. The reason is that, for example, while $c_a > 0$ and $r_a > 0$, if we repair an orthogonal fault by a spare row before repairing other nonorthogonal faults, we may lose the chance to repair more faults with this spare row, because that orthogonal fault can also be repaired by a spare column. Based on the discussion, the *greedy algorithm* ESP is presented in Figure 9.12. Here, we assume the spare memory is fault-free.

The function ESP_FC() collects the faulty-cell addresses and stores them in the P_R *(row pivot)* and P_C *(column pivot)* register files. Both have $r + c$ registers, and all registers are initially empty. During the FC phase, if the number of faults exceeds $r + c$, this memory is unrepairable and the process terminates. An incoming faulty-cell address ($\hat{R}, \hat{C}$) is compared with the existing row pivots and column pivots in the register file. If there is a row-address match or column-address match, the matched pivot is marked as an **essential pivot** (EP). If there is no match, the row address and column address of the current faulty cell are stored in the P_R and P_C registers, respectively. In ESP_FC() we apply the revised first guideline. Note that in the algorithm shown in Figure 9.12, E_{th} is assumed to be 2, so threshold comparison is greatly simplified. We will show in the next section that the repair rate is high in this case. We need only a flag along with each pivot to indicate whether it is an EP.

The function ESP_SA() allocates spares to repair faults according to the contents of the P_R and P_C registers. It consists of two stages. In the first stage we allocate spare rows for the essential row pivots and spare columns for the essential column

```
ESP_FC() {
    p = 0; r_a = r; c_a = c;
    for each (faulty_cell_detected()) {
        (R̂, Ĉ) = faulty_address();
        if (∃ 0 ≤ i < p such that P_{R_i}==R̂) { E_{R_i} = 1; }
        else if (∃ 0 ≤ j < p such that P_{C_j}==Ĉ) { E_{C_j} = 1; }
        else {
            P_{R_p} = R̂;
            P_{C_p} = Ĉ;
            p++;
            if (p > r + c) { repair_termination("unrepairable"); }
        }
    }
}

ESP_SA() {
    for all k, 0 ≤ k < p {
        if (E_{R_k} == 1) { allocate_spare_row(P_{R_k}); }
        if (E_{C_k} == 1) { allocate_spare_column(P_{C_k}); }
    }
    for all k, 0 ≤ k < p {
        if (E_{R_k} == E_{C_k} == 0)
            if (r_a > 0) { allocate_spare_row(P_{R_k}); }
            else if (c_a > 0) { allocate_spare_column(P_{C_k}); }
            else { repair_termination("unrepairable"); }
    }
}

allocate_spare_row(R̂)
    { R_{R_{r_a}} = R̂; r_a = r_a - 1; }
allocate_spare_column(Ĉ)
    { R_{C_{c_a}} = Ĉ; c_a = c_a - 1; }
```

■ **FIGURE 9.12**

The essential spare pivoting algorithm.

pivots. After the first stage, the pivot registers contain all and only the addresses of orthogonal faults, because they have never matched other faulty-cell addresses. We can repair these faults by either spare rows or spare columns. In ESP_SA(), we simply allocate available spare rows before spare columns.

Example 9.1

Let the memory block under test be shown in Figure 9.13. The faulty cells detected are, in sequence, cell(1,0), cell(1,6), cell(2,4), cell(3,4), cell(5,1), cell(5,2), cell(5,4), cell(5,6), cell(5,7), and cell(7,3). The FC procedure is illustrated in Figure 9.14. In the figure, the row_pivot and column_pivot registers are shown as the left and right columns of the register array, respectively. For each faulty-cell address, the row_pivot is stored in the left column and the column_pivot is stored in the right column. There is a circle on a pivot if it is an essential pivot. In the beginning, the register array is empty, so the first address (1,0) is stored in the first row

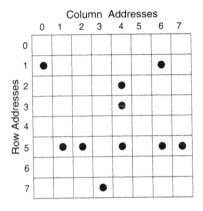

■ FIGURE 9.13

A memory block with defective cells.

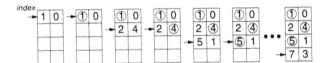

■ FIGURE 9.14

Fault collection example in ESP.

of the array directly. The second address (1,6) matches (1,0) in the row address, so the row_pivot of cell(1,0) is marked as an EP. Similarly, the address (2,4) is inserted directly, while address (3,4) matches (2,4) in its column address, thus the column_pivot of cell(2,4) is marked as an EP. This procedure continues until the address (7,3) is recorded. The SA procedure is simple: first we allocate spares for the EPs—row 1, row 5, and column 4, then we use a spare column to repair the orthogonal fault on cell(7,3).

The major advantage of ESP is mainly its simplicity in implementation which results in smaller area overhead than other algorithms. The revised first guideline provides a simple search method for orthogonal faults. In the SA stage of the ESP algorithm, orthogonal faults and nonorthogonal faults can be easily separated by checking their EP flags. The automatic recognition for orthogonal faults greatly increases the repair efficiency. These features make the ESP algorithm small, fast, and easily implementable.

The BIRA algorithms are presented for localized redundancy architecture; however, the cost function in spare allocation can be easily adapted for the redundancy architecture with global or shared spare resources. With the help of an evaluation tool for repair efficiency, the most effective BIRA algorithm with an optimized spare architecture can be found for specific manufacturing processes systematically.

9.5.3   Repair Rate and Overhead

A simulator such as BRAVES (Built-in Redundancy Analysis Verification and Evaluation System) [Huang 2003] can be used for analyzing the efficiency of the redundancy analysis and repair algorithms. The distribution of defect sizes on memory chips usually is modeled by mixed Poisson statistics using the Gamma distribution, resulting in a Polya–Eggenberger distribution [Stapper 1980]. In BRAVES, a mixed Poisson and exponential distribution is assumed, as the mixed Poisson and exponential model is accurate enough for the said purpose, and different conditions can be applied in the simulator for different redundancy analysis algorithms. There are two types of faults that we can inject into the memory to be simulated: *cell fault* and *line fault*. The cell fault represents an independent individual fault, while a line fault occurs when multiple faults exist in the same line, such as the case of a faulty wordline or bitline. Figure 9.15 shows a size distribution of cell faults using a mixed Poisson and exponential distribution model. The size distribution for line faults looks similar, except the probabilities are lower. With the simulator, we can simulate a random collection of memory instances for a specified range of spare count and failure patterns, given a test algorithm. A high repair rate implies a high yield after repair, if the area overhead is roughly the same.

Figure 9.16 shows a particular simulation result from an example, where 1552 memory blocks with a core size of 1024×64 bits are simulated, assuming $r = 10$ and c ranges from 2 to 6. The ESP result is compared with some other BIRA algorithms reported in [Huang 2003], including the optimal and repair-most algorithms. The ESP algorithm is close to optimal if most of the faults are independent cell faults,

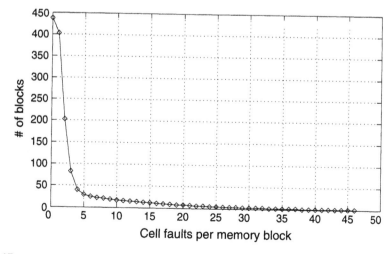

■ **FIGURE 9.15**

Size distribution for cell faults.

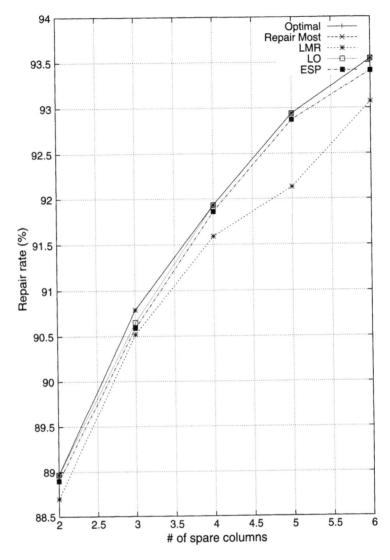

Simulation result for $r = 10$.

as is assumed in this case. Note that the relative efficiency of these algorithms will vary in different memory configurations and spare architectures.

As to hardware (area) overhead, the storage requirement is calculated for estimating the hardware overhead because all the algorithms require storage cells with matching capabilities, and the storage cells dominate the silicon area of the BIRA circuit. A row address tag and a column address tag require $\lceil \log M \rceil + 1$ bits and $\lceil \log N \rceil + 1$ bits, respectively. The orthogonal-fault heuristics require $(r + c) \cdot$

$(\lceil \log M \rceil + \lceil \log N \rceil + 1)$ storage cells. Therefore, the area overhead of the ESP algorithm is estimated as follows:

$$A_{ESP} = (r+c) \cdot [(\lceil \log M \rceil + 1) + (\lceil \log N \rceil + 1)] \qquad (9.1)$$

$$A_{spare\_register} = (\lceil \log M \rceil + 1) \cdot r + (\lceil \log N \rceil + 1) \cdot c \qquad (9.2)$$

where $A_{spare\_register}$ denotes the area for the spare row and column registers, which are required for all the three algorithms.

9.6 BUILT-IN SELF-REPAIR

9.6.1 Redundancy Organization

The redundancy organization of a RAM affects not only the repair rate but also the area cost of the BIRA circuitry. Figure 9.17 shows a RAM cell array with redundancy rows and columns. In the figure, a 512-bit RAM has two spare rows at the bottom and four spare columns on the right. If there is a faulty row, any of the spare rows (SR0 and SR1) can be used to replace it; however, spare columns are used differently: We partition them into several ***spare column groups*** (SCGs). In the figure, two spare columns are grouped into an SCG; the **group size** is 2—group SCG0 (resp. SCG1) contains columns SC0 and SC1 (resp. SC2 and SC3). Also, a word is divided into multiple subwords, where a subword contains consecutive bits of the word, whose length is the same as the group size. For example, assume that the number of words of a row in the RAM as shown in Figure 9.17 is four and the word length is eight. Then a word can be divided into four subwords as shown in the figure. Moreover, each SCG is logically divided into segments for better utilization;

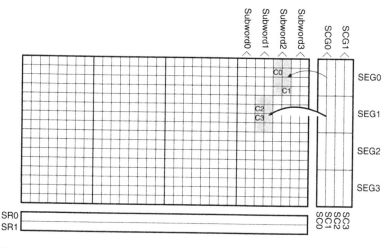

■ **FIGURE 9.17**

An example RAM with two spare rows and two spare column groups.

that is, the segments of a spare column are not physically divided by local sense amplifiers. The switching from main memory to spare column groups is controlled by the BISR circuit, so only the multiplexers induce additional access time and area cost. In Figure 9.17, one SCG has four segments (SEG0, SEG1, SEG2, and SEG3). The segments are identified by the first two ***most significant bits*** (MSBs) of the row address. Let the row addresses be $a_3a_2a_1a_0$, then a_3a_2 specifies the segment where the addressed row sits. Different segments of an SCG can be used to repair defective cells in different columns of the main memory. For example, if cells C0, C1, C2, and C3 (see Figure 9.17) are faulty, then SEG0 and SEG1 of SCG0 can be used to replace them.

If the number of segments of each SCG is increased, the utilization of SCGs is also increased. Thus, better repair rates usually can be achieved. However, the hardware complexity of BISR is increased, as more storage elements are required to store redundancy information. Note that spare rows also can be logically divided into segments, though this is not shown in the example. We have developed a simulator for evaluating the RA algorithms of redundancy-repairable memories [Huang 2002] [Huang 2003]. The simulator can help the user to determine a good redundancy organization for the applied redundancy analysis algorithm.

9.6.2   BISR Architecture and Procedure

Figure 9.18 depicts the block diagram of the presented BISR scheme, including the BIST module, BIRA module, and test wrapper for the memory. The BIST circuit detects the faults in the main memory and spare memory. It is programmable at the March element level [Huang 1999]. The BIRA circuit performs redundancy allocation using a novel RA algorithm (to be discussed later). The test wrapper switches the memory between test/repair mode and normal mode. In test/repair mode the memory is accessed by the BIST module, while in normal mode the wrapper selects the data outputs either from the main memory or the spare memory

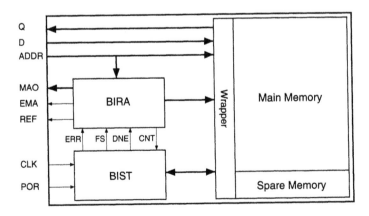

■ **FIGURE 9.18**

Block diagram of the presented BISR scheme.

(replacing the faulty memory cells) depending on the control signals from the BIRA module.

This BISR is a *soft repair* scheme; therefore, the BISR module will perform testing, analysis, and repair upon every power-up. As Figure 9.18 shows, the BIST circuit is activated by the ***Power-On Reset*** (POR) signal. More details of the BIST design are provided in Section 9.6.3. The BISR procedure is shown in Figure 9.19. When we turn on the power, the BIST module starts to test the spare memory. Once a fault is detected, it informs the BIRA module to mark the defective spare row or column as faulty through the ***Error*** (ERR) and ***Fault Syndrome*** (FS) signals. After finishing the spare memory test, it tests the main memory. If a fault is detected (ERR outputs a pulse), the test process pauses and the BIST module exports FS to the BIRA module, which then performs the RA procedure. When the procedure is completed and the memory testing is not yet finished, the BIRA module issues a ***Continue*** (CNT) signal to resume the test process. During the RA procedure, if a spare row is requested but there are no more spare rows, the BIRA module exports the faulty row address through the ***Export Mask Address*** (EMA) and ***Mask Address Output*** (MAO) signals. The memory will then be operated at a downgraded mode (*i.e.*, with a smaller usable capacity) by software-based address remapping. For example, assume that a memory with multiple blocks is used for buffering, and the blocks are chained by pointers. If some block is faulty and should be masked, then the pointers are updated to invalidate the block. The size of the memory is reduced, as one block is removed. The system still works if a smaller buffer is allowed, though performance may be affected. This approach effectively increases the product yield. The number of blocks that can be invalidated normally depends on the performance penalty that can be tolerated. If downgrade mode is not allowed, MAO is removed and EMA indicates whether the memory is repairable.

When the main memory test and RA are finished, the ***Repair End Flag*** (REF) signal goes high and the BIRA module switches to the normal mode. The BIRA

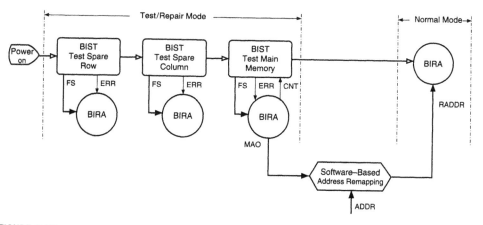

■ **FIGURE 9.19**

Power-on BISR procedure.

module then serves as the address remapper, and the memory can be accessed using the original address (ADDR). When the memory is accessed, ADDR is compared with the fault addresses stored in the BIRA module. If ADDR is the same as any of the fault addresses, the BIRA module controls the wrapper to remap the access to spare memory.

Subsequently, we will describe the redundancy analysis procedure. Before proceeding to the presentation of redundancy analysis, however, we must first define *subword*. Let a *subword* be consecutive bits of a word, whose length is the same as the group size. For example, in Figure 9.17 we assume that there are four words in each row and each word has eight bits. There are two bits in each subword as the group size is two, so a word has four two-bit subwords. To reduce complexity, we use two row-repair rules:

1. The first row-repair rule is that if a row has multiple faulty subwords, we repair the faulty row by a spare row if available. Take Figure 9.20a as an example. If a word has two faulty subwords (marked "X" in the figure), we actually could repair them using two spare column group segments. However, for multiple spare column group segments, there can be many possible ways to repair the faulty subwords, resulting in complex output multiplexing and RA; thus, the memory performance is degraded and the cost of the BIRA circuit increases.

2. The other row-repair rule is that, if there are multiple faulty subwords with the same column address and different row addresses within a segment, the last detected faulty subword should be repaired with an available spare row. In Figure 9.20b, a faulty subword is detected and repaired with a spare column segment first, and a faulty subword is then detected, where its column address is the same as the column address of the repaired faulty subword but their row addresses are different. The faulty subword will be repaired with an available spare row. In this case, when the memory accesses cells in the overlapped region in the functional mode, the address remapper gives priority to the spare row.

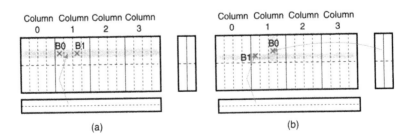

(a)   (b)

■ **FIGURE 9.20**

Faulty memory examples.

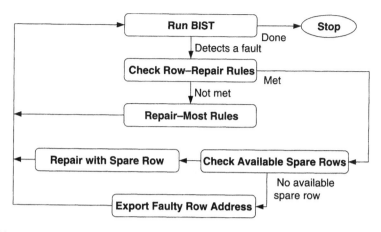

Flow diagram of the BIRA procedure.

After the *row-repair* phase, spare allocation for the remaining faulty elements is performed according to the *repair-most* rules [Tarr 1984]. Figure 9.21 provides a flow diagram of the presented BIRA algorithm. The BIRA procedure consists of the following major steps:

1. Run BIST; pause and jump to step 2 when it detects a fault. Stop when BIST is done.

2. Check whether both row-repair rules can be applied. If so, go to step 4.

3. Allocate a spare row or spare column group segment to repair corresponding faulty cells according to the repair-most rules. Resume step 1.

4. Check if there are available spare rows. If so, repair by a spare row and resume step 1.

5. Export the corresponding faulty row address; resume step 1.

9.6.3 BIST Module

The BIST module block diagram is shown in Figure 9.22, which consists of a *controller* (CTR) and a *test pattern generator* (TPG) for handling test operations and generating test stimuli, respectively. In addition to **Clock** (CLK), the BIST only requires the **Power-On Reset** (POR) signal to initiate the test procedure. Thus, POR generates a pulse signal when power is turned on. The pulse triggers the BIST circuit to initiate the test procedure. The **BIST Done** (BDN), ERR (error indicator), FS (fault syndrome), and CNT (continue) are signals between the BIST and BIRA modules. TPG output signals are connected to the memory under test. The **BIST Normal Selection** (BNS) signal is used to switch the memory between test/repair mode and normal mode.

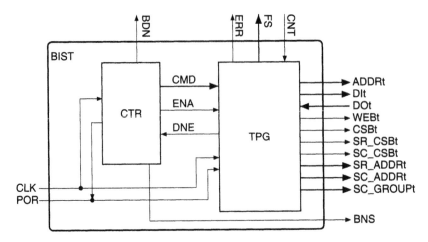

■ **FIGURE 9.22**

BIST module block diagram.

CTR is a typical finite state machine [Huang 1999] [Cheng 2000]. TPG executes the **test command** (CMD) provided by CTR. When a fault is detected, it pauses and sends the ERR and FS signals to inform the BIRA module to perform RA. When RA finishes, the BIRA module sends the CNT signal to resume the TPG process. The BIST implementation is typical, similar to our previous design [Huang 1999] that is presented in chapter 8.

9.6.4 BIRA Module

The BIRA module has three components—**multiple faulty subwords detector** (MFSD), **process element** (PE), and **address remapping unit** (ARU)—as shown in Figure 9.23. When power is on, all flip-flops are reset to the initial state. Signal normal is 0 and FS is connected to ARU. The **Input Address** (ADDR) is sent to ARU when it is in normal mode (normal = 1). Initially, signals solid_flag, faulty_flag, repaired_flag, row_match, and col_match are all reset to 0. The PE evaluates the status of these signals and issues control signals solid_en, repair_en, update, and export_mask_addr to ARU, which then updates the status of its registers. Signals REF (repair end flag), EMA (export mask address), and MAO (mask address output) are connected to ATE.

The MFSD detects whether the number of faulty subwords on a row is larger than one. The PE is implemented by a *finite state machine* (FSM), whose state transition diagram is shown in Figure 9.24. The initial state is MONITOR, which monitors the ERR signal from the BIST circuit. If a fault is detected, PE goes to the DFETCH state to load the status data into the flip-flops. In the COMPARE state, PE compares the faulty address with the previously stored addresses. If there is a match, PE goes back to the MONITOR state through the CONTINUE state; otherwise, it goes to the CHECK_RMR state. If the status is must-repair (by row),

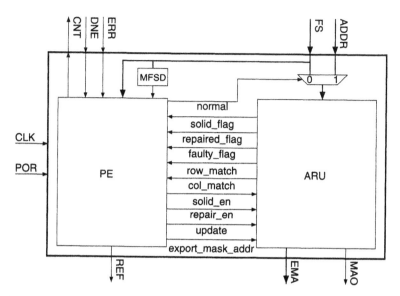

■ **FIGURE 9.23**

BIRA module block diagram.

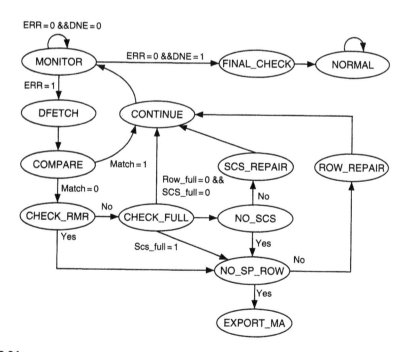

■ **FIGURE 9.24**

State diagram of PE.

then it goes to the NO_SP_ROW state and checks whether there are available spare rows. If no spare row is available, it sends a signal to ARU that will then export the faulty row address for software-based repair later in the downgraded operation mode. If, on the other hand, a spare row is available, then the faulty row is replaced by the spare row in the ROW_REPAIR state, and PE goes back to the MONITOR state through the CONTINUE state, where a continue signal is issued to the BIST circuit. If, in the CHECK_RMR state, the must-repair conditions are not satisfied, PE will go to the CHECK_FULL state to see if the solid flags (explained below) in ARU are on—if all spare rows (resp. column segments) are full, spare column segments (resp. rows) are used for repair (unless both are full). It then goes to either the NO_SP_SCS state or NO_SP_ROW state. Note that if both are full, either spare rows or spare column segments can be used for repair. In this case, spare column segments are used for repair and then the FSM goes to NO_SP_SCS state. Finally, in the MONITOR state, if the BIST module issues a done signal to PE, the FSM will go to the NORMAL state through the FINAL_CHECK state. In the FINAL_CHECK state, the FSM checks and sets the repair flags (explained below) of the remaining faulty addresses which are not repaired.

In the test/repair mode, ARU stores the addresses of the faulty cells detected so far and compares the current faulty-cell address with the stored ones. Figure 9.25

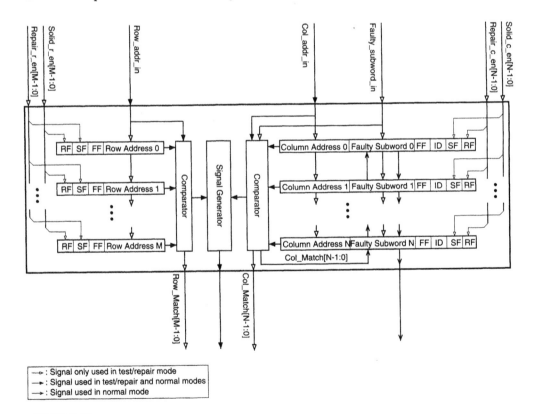

■ **FIGURE 9.25**

Block diagram of ARU.

shows an ARU block diagram, which mainly contains the storage elements (registers), comparators, and a signal generator. Each storage element stores faulty cell information. Assume that there are M spare rows and N spare column group segments. Then M row storage elements and N column storage elements are implemented. Each storage element has three status flags: (1) the **Fault Flag** (FF) denotes whether the corresponding spare element is defective (FF = 1) or fault-free (FF = 0); (2) the **Repair Flag** (RF) indicates whether the spare element is used to repair the defective main memory (RF = 1) or not (RF = 0); and (3) the **Solid Flag** (SF) shows whether the storage element has loaded the faulty cell information (SF = 1) or not (SF = 0). Each column storage element has an **identification** (ID) field to store its segment number.

When the system is operated in test/repair mode, the Row_addr_in (row address input), Col_addr_in (column address input), and Faulty_subword_in (faulty subword input) signals come from the BIST circuit, and the Row_r_en (row repair enable), Row_s_en (row solid enable), Col_r_en (column repair enable), and Col_solid_en (column solid enable) signals are from PE. The spare rows and columns are first tested. If a fault is detected, the FF of its corresponding storage element is set to 1. The main memory is tested next. If a fault is detected, the row address, column address, and faulty subwords are compared with the data stored in the storage elements with SF = 1 and FF = 0. The results are exported from the Row_match and Col_match terminals to PE. If there is no match, fault information is written into an empty storage element, and PE sets its SF (SF = 1) through the Solid_r_en or Solid_c_en input.

The Row_addr_in and Col_addr_in inputs are the address inputs in normal mode. When the memory is accessed, the address also is compared with those stored in the storage elements. If it is the same as one of the stored addresses, the signal generator triggers the control signals to reconfigure the I/Os between the main memory and spare memory. Compared with the conventional laser repair scheme, the presented BISR scheme must execute the address remapping operation, however, the address remapping operation and the main memory access operation are executed in parallel. Moreover, the delay time of the access path through ARU and the small spare memory usually is shorter than that through the main memory. A slight performance degradation results from the wrapper, however. For example, the performance degradation of the wrapper is about 0.3 ns for a typical 0.25 μm CMOS standard-cell design. This delay penalty can be minimized by implementing the wrapper with a full-custom design.

9.6.5 An Industrial Case

We now show a repairable SRAM core with the presented BISR methodology. Figure 9.26 shows the repairable SRAM, which is composed of two blocks with $8K \times 32$ bits each. Four spare rows and two spare column groups—with a group size of four—are implemented using two separate memory blocks. In the figure, data input DI is broadcast to main memory and spare memory. The chip select signals determine which memory the data should be written into. When the memory is in normal mode, the data output may come from main memory or spare memory, depending on the control signals sc_group and sr_csb.

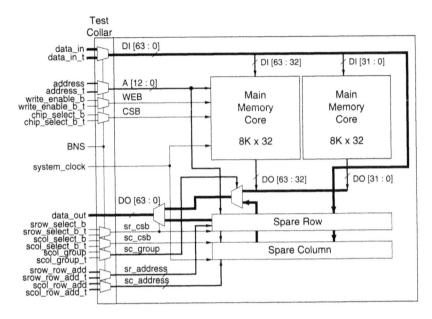

■ **FIGURE 9.26**

An 8K × 64 repairable SRAM with 4 spare rows and 2 spare column groups.

The gate count of the BISR (including BIST) design is about 5.6K using a typical synthesis procedure and a standard 0.25 μm CMOS cell library. Figure 9.27 shows the layout of the 8K × 64 repairable SRAM. The areas of the SRAM (BANK 1 and BANK 2), BISR module, and spare elements (SPARE ROW and SPARE COL) are $653836 \mu m^2$, $301040 \mu m^2$, and $298856 \mu m^2$, respectively. The hardware overhead of

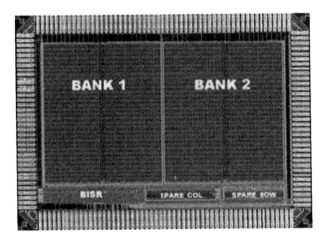

■ **FIGURE 9.27**

Layout of the 8K × 64 repairable SRAM.

the spare elements (HO_{spare}) and that of the BISR module (HO_{bisr}) are calculated and shown below:

$$HO_{spare} = \frac{298856}{6538362} \times 100\% = 4.57\%$$

$$HO_{bisr} = \frac{301040}{6538362} \times 100\% = 4.6\%$$

The total hardware overhead for this repairable SRAM is about 9.17%. We guarantee 100% repair rate if the number of random faults is no more than 10 (it will be analyzed in the next section).

Figure 9.28 shows part of the timing diagram for MAO and EMA from post-layout simulation. It shows that if spare rows are exhausted but a spare row is still required to repair the defective memory, the address of the defective row is exported to the ATE through the MAO output. When the BIRA circuit wants to export a mask address, the EMA signal becomes 1 such that the ATE can correctly receive the valid mask address. Figure 9.29 shows a waveform sample of the data inputs/outputs and some control signals of the spare memories during the normal-mode memory access. If sc_csb = 0, the 4-bit data out is from the Spare Column, controlled by

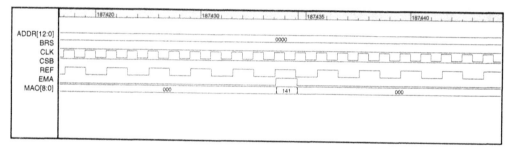

■ **FIGURE 9.28**

A waveform sample of the EMA and MAO signals.

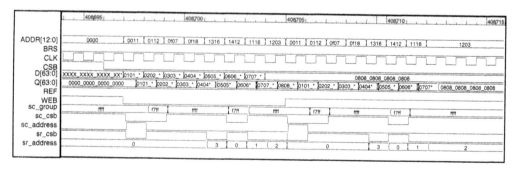

■ **FIGURE 9.29**

A waveform sample showing normal-mode memory access.

sc_group. In this example, sc_group = f7ff; that is, the 12th 4-bit data output is from Qc[3:0] (data outputs of the Spare Column). If both sr_csb and sc_csb are 0, the data output is from the Spare Row, controlled by sr_csb; that is, the data output is from Qr[63:0] (data output of the Spare Row). This avoids data access conflict.

9.6.6 Repair Rate and Yield

Table 9.4 summarizes the ***repair rate*** (RR) for various redundancy configurations based on the presented and exhaustive redundancy analysis algorithms. The number

TABLE 9.4 ■ Simulation Results

N_{SR}	N_{SC}	N_{SCG}	RR	1MA	2MA	3MA	4MA	5MA	>5 MA	RR (Best)
0	4	1	44.90%	89	70	17	60	36	25	72.28%
0	8	2	81.45%	26	22	16	17	10	9	98.69%
0	12	3	95.73%	8	4	1	3	3	4	100%
1	0	0	18.37%	99	191	4	69	45	32	18.54%
1	4	1	73.10%	38	40	35	16	9	7	86.14%
1	8	2	94.43%	5	7	12	1	3	2	99.81%
1	12	3	99.26%	1	1	1	1	0	0	100%
2	0	0	36.55%	192	2	71	46	18	13	37.08%
2	4	1	86.09%	36	16	12	3	8	0	94.01%
2	8	2	99.26%	3	1	0	0	0	0	100%
2	12	3	100%	0	0	0	0	0	0	100%
3	0	0	72.17%	0	75	43	18	7	7	55.06%
3	4	1	96.10%	7	5	4	3	2	0	97.38%
3	8	2	99.81%	1	0	0	0	0	0	100%
3	12	3	100%	0	0	0	0	0	0	100%
4	0	0	72.36%	73	44	18	8	5	1	71.91%
4	4	1	98.52%	4	3	0	0	0	0	98.69%
4	8	2	100%	0	0	0	0	0	0	100%
4	12	3	100%	0	0	0	0	0	0	100%
5	0	0	85.90%	44	18	7	6	1	0	85.77%
5	4	1	99.81%	1	0	0	0	0	0	99.81%
5	8	2	100%	0	0	0	0	0	0	100%
5	12	3	100%	0	0	0	0	0	0	100%
6	0	0	94.06%	18	7	6	1	0	0	94.01%
6	4	1	100%	0	0	0	0	0	0	100%
6	8	2	100%	0	0	0	0	0	0	100%
6	12	3	100%	0	0	0	0	0	0	100%
7	0	0	97.40%	8	5	1	0	0	0	97.57%
7	4	1	100%	0	0	0	0	0	0	100%
7	8	2	100%	0	0	0	0	0	0	100%
7	12	3	100%	0	0	0	0	0	0	100%
8	0	0	98.70%	6	1	0	0	0	0	98.69%
8	4	1	100%	0	0	0	0	0	0	100%
8	8	2	100%	0	0	0	0	0	0	100%
8	12	3	100%	0	0	0	0	0	0	100%

of injected random faults is from 1 to 10, and the number of memory samples is 534. The defect distribution assumed is pessimistic, which is used for evaluating the presented scheme. Mature products have a far lower defect density. Note that the exhaustive RA algorithm is simulated based on the assumption that a single spare row and spare column can be used to repair any defective row and column, respectively. It guarantees 100% RR under such type of redundancy organization [Kawagoe 2000]. In the table, N_{SR}, N_{SC}, and N_{SCG} denote the numbers of spare rows, spare columns, and spare column groups. The RR column reports RR of the presented approach. The results show that the RR difference between the presented approach and the best (exhaustive search without grouping) is very small for most of the redundancy configurations. In the xMA columns of Table 9.4, the values represent the numbers of unrepairable memories for the respective spare configurations that can still be used in the downgraded mode if we mask out x faulty-cell row addresses. For example, the 1MA column shows the numbers of unrepairable memories that can still be used if one masked address is allowed—the memory thus has one less usable address. According to the table, if $N_{SR} = 2$ and $N_{SCG} = 2$, the number of unrepairable memories is four. Among them, three can work in the downgraded mode if one masked address is allowed, and one memory can work in the downgraded mode if two masked addresses are allowed. In the industrial chip design, $N_{SR} = 4$ and $N_{SCG} = 2$. Therefore, the BISR design can achieve 100% RR with low area cost. However, if the CRESTA in [Kawagoe 2000] is implemented, it requires $C_2^{12} = 66$ subanalyzers to try 66 possible solutions, resulting in very high hardware cost.

We now discuss the relation between column group size and RR. Figure 9.30 plots the RR for different spare configurations with a group size of two. The number

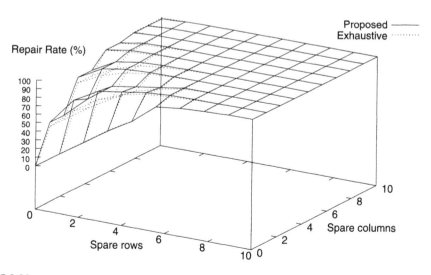

■ **FIGURE 9.30**

RR comparison when the group size is two.

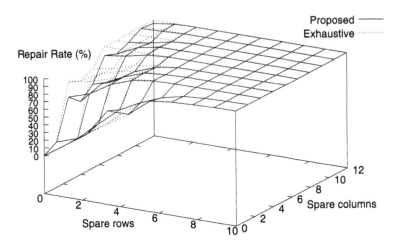

Repair Rate (%)

Proposed ——
Exhaustive ········

Spare columns

Spare rows

■ **FIGURE 9.31**

RR comparison when the group size is four.

of injected random faults varies from 1 to 10 and the total number of memory samples is 534. Note that the conventional algorithm being compared here is the one without spare row or column grouping or segmentation, but an exhaustive search is assumed. The figure shows that the RR of the presented approach is even better than the conventional one in many instances. Figure 9.31 shows a similar comparison, but now the group size is four. The result of our approach in this case is not as good as the previous one; however, the area cost of the BIRA circuit is lower. For example, if there are four spare columns, then the number of required column storage elements for the case where group size is two (two spare column groups) is larger than the one whose group size is four (only one spare column group). Note also that the RR is in fact related to how the defects are distributed. An analysis of the fail bit patterns of the target memory design and process technology is required to achieve the most cost-effective solution.

According to the Poisson model, $Y = e^{-AD}$, the chip yield Y decreases exponentially with increasing area (A) and manufacturing defect density (D). Let the defect density of the memory and random logic be D_m and D_l, respectively, and $D_m = 2D_l$. Thus, the yield (Y_m) of the memory without BISR circuit can be estimated as:

$$Y_m = e^{-A_m D_m}$$

where A_m is the memory area. We use the simplest model (the yield of a chip is the product of the yield of all the constituent modules) to estimate the yield (Y_{mbisr}) of the memory with BISR circuit. Therefore, Y_{mbisr} can be expressed as:

$$Y_{mbisr} = Y'_m \cdot Y_{bisr}$$

where Y_{bisr} is the yield of the BISR circuit. The Y_{bisr} can be calculated as $Y_{bisr} = e^{-A_l D_l}$, where A_l is the area of the BISR circuit. Also, Y'_m is the yield of the memory after executing the BISR process. It is associated with the *repair rate* (RR) and can be expressed as $Y'_m = Y_m + (1 - Y_m) \times RR$, as RR is the ratio of the number of repaired memories to the number of defective memories. Thus, Y_{mbisr} can be estimated as follows:

$$Y_{mbisr} = (Y_m + (1 - Y_m) \times RR) \cdot Y_{bisr}$$

For example, consider the yield improvement of the case described in Section 9.6.5. The ratio of the area of BISR circuit to the area of memory is about 4.6%. The yield of the BISR circuit is shown as follows:

$$Y_{bisr} = e^{-0.046 A_m \times 0.5 D_m}$$

$$= (e^{-A_m D_m})^{0.023}$$

$$= Y_m^{0.023}$$

as $A_l = 0.046 A_m$ and $D_m = 2D_l$. Therefore, the Y_{mbisr} of the design is:

$$Y_{mbisr} = (Y_m + (1 - Y_m) \times RR) \cdot Y_m^{0.023}$$

Note that the yield of spare elements is neglected. The reason is that the spare elements of the presented design are tested first. Thus, a memory with defective spare elements may be repairable or not depending on the number of defective spare elements and the error pattern of the memory. For simplicity of analysis, therefore, we do not consider the yield of spare elements. Figure 9.32 shows the yields of the memory with the presented BISR scheme with respect to the

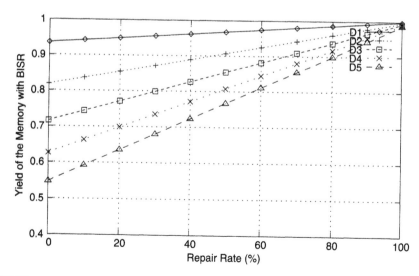

■ **FIGURE 9.32**

Yields of repairable memories with respect to repair rates.

repair rates, where the D1, D2, D3, D4, and D5 represent the defect densities 0.01, 0.03, 0.05, 0.07, and $0.09\,\text{defect}/\mu m^2$, respectively. As the figure shows, if high RR can be achieved, the yield of the memories with BISR scheme can be greatly enhanced.

9.7 CONCLUDING REMARKS

In this chapter, we have discussed BIST with diagnosis support, BIRA, and BISR for RAM. Fault type identification is done by an offline diagnosis process using the MECA system. It is useful for RAM designers and process engineers, as it helps debug the RAM design and process for yield enhancement during both the development and mass production stages. Note that diagnosis algorithms are usually more complex than ordinary test algorithms, and they are used only during the product development phase, so hard-wiring the diagnosis algorithms does not make sense. Thus, a programmable TPG scheme is desirable.

The ESP algorithm for BIRA is shown to be cost effective. It is a greedy-algorithm-based approach that greatly simplifies the control circuit and results in relatively low time and area overhead among the known BIRA schemes. It also achieves a high repair rate for a mature fabrication process with small area overhead.

The BISR circuit discussed is composed of a BIST module and a BIRA module. The BIST circuit supports three operation modes—main memory testing, spare memory testing, and repair. The BIRA circuit executes the presented RA algorithm for two-dimensional redundancy—spare rows and spare columns. The spare columns are grouped and segmented. Compared with the conventional approach (without grouping and segmentation) using exhaustive search, the discussed scheme outperforms in many instances and can be implemented with low area cost.

9.8 EXERCISES

9.1 (March Dictionary) Construct the March dictionary based on MSCAN, considering the same test set (without TF) that we have used for the March 17N algorithm.

9.2 (Repair)

    a. Define the must-repair phase in redundancy allocation (RA) of a RAM with two-dimensional redundant (spare) elements (*i.e.*, rows and columns).

    b. Assume that the must-repair phase has been finished for a RAM with R rows and C columns. Let there be n faulty cells, r available spare rows, and c available spare columns. Derive the upper bound of n such that the RAM is reparable.

c. Derive the size of the smallest bitmap that we need to perform the bitmap-based final-repair phase of the RA procedure to guarantee a 100% repair rate for reparable RAMs.

d. Give any efficient final repair algorithm described in this chapter or any other literature (except the row-first and column-first algorithms), assuming that the must-repair phase has been finished.

9.3 **(Repair Rate and Yield)** The repair rate of a memory is denoted as R and defined as the ratio of the number of repaired memories to the number of defective memories in a product run. Note that the spare memory and BISR circuit increases the area, thus reducing the yield if we do not consider the repair mechanism. Let $Y = e^{-AD}$, where Y, A, and D are the yield, area, and defect density of the chip, respectively. Assume memory and logic circuits have the same defect density, and A_m, A_r, and A_b denote the main memory area, redundant memory area, and logic (BIST/BIRA) circuit area, respectively. Derive the yield improvement (ΔY) of the memory chip after repair.

9.4 **(BISR Design)** Based on the $4K \times 16$ SRAM BIST circuit defined in Chapter 8, design a BISR scheme assuming two spare rows and two spare columns. Provide the block diagrams and algorithms, and explain how the scheme works. You can use any existing BIRA approach or develop your own.

Acknowledgments

Many of my students have contributed to the work presented in this chapter, especially C.-T. Huang, C.-F. Wu, J.-F. Li, K.-L. Cheng, C.-W. Wang, R.-S. Tzeng, R.-F. Huang, J.-C. Yeh, etc. It has been my great pleasure working with so many talented students.

References

R9.1—Introduction

[Benevit 1982] C. A. Benevit, J. M. Cassard, K. J. Dimmler, A. C. Dumbri, M. G. Mound, F. J. Procyk, W. Rosenzweig, and A. W. Yanof, A 256K dynamic random-access memory, *IEEE J. Solid-State Circuits*, 17(5), 857–862, 1982.

[Cenker 1979] R. P. Cenker, D. G. Clemons, W. P. Huber, J. P. Petrizzi, F. J. Procyk, and G. M. Trout, A fault-tolerant 64K dynamic random-access memory, *IEEE Trans. Electron. Devices*, 26(6), 853–860, 1979.

[Huang 1999] C.-T. Huang, J.-R. Huang, C.-F. Wu, C.-W. Wu, and T.-Y. Chang, A programmable BIST core for embedded DRAM, *IEEE Des. Test Comput.*, 16(1), 59–70, 1999.

[Smith 1981] R. T. Smith, J. D. Chipala, J. F. M. Bindels, R. G. Nelson, F. H. Fischer, and T. F. Mantz, Laser programmable redundancy and yield improvement in a 64K DRAM, *IEEE J. Solid-State Circuits*, 16(5), 506–514, 1981.

R9.2—Refined Fault Models and Diagnostic Test Algorithms

[Huang 1999] C.-T. Huang, J.-R. Huang, C.-F. Wu, C.-W. Wu, and T.-Y. Chang, A programmable BIST core for embedded DRAM, *IEEE Des. Test Comput.*, 16(1), 59–70, 1999.

[Wu 1999] C.-F. Wu, C.-T. Huang, and C.-W. Wu, RAMSES: a fast memory fault simulator, in *Proc. Int. Symp. on Defect and Fault Tolerance in VLSI Systems*, October 1999, pp. 165–173.

R9.3—BIST with Diagnostic Support

[Li 2001a] J.-F. Li, R.-S. Tzeng, and C.-W. Wu, Using syndrome compression for memory built-in self-diagnosis, in *Proc. Int. Symp. on VLSI Technology, Systems, and Applications (VLSI-TSA)*, April 2001, pp. 303–306.

[Li 2001b] J.-F. Li and C.-W. Wu, Memory fault diagnosis by syndrome compression, in *Proc. IEEE Design, Automation, and Test in Europe Conf.*, March 2001, pp. 97–101.

R9.4—RAM Defect Diagnosis and Failure Analysis

[Abramovici 1990] M. Abramovici, M. A. Breuer, and A. D. Friedman, *Digital Systems Testing and Testable Design*, Computer Science Press, New York, 1990.

[Li 2001c] J.-F. Li, K.-L. Cheng, C.-T. Huang, and C.-W. Wu, March-based RAM diagnosis algorithms for stuck-at and coupling faults, in *Proc. IEEE Int. Test Conf.*, October 2001, pp. 758–767.

[Wu 1999] C.-F. Wu, C.-T. Huang, and C.-W. Wu, RAMSES: a fast memory fault simulator, in *Proc. Int. Symp. on Defect and Fault Tolerance in VLSI Systems*, October 1999, pp. 165–173.

[Wu 2000a] C.-F. Wu, C.-T. Huang, K.-L. Cheng, and C.-W. Wu, Simulation-based test algorithm generation for random access memories, in *Proc. IEEE VLSI Test Symp.*, April 2000, pp. 291–296.

[Wu 2000b] C.-F. Wu, C.-T. Huang, C.-W. Wang, K.-L. Cheng, and C.-W. Wu, Error catch and analysis for semiconductor memories using March tests, in *Proc. IEEE/ACM Int. Conf. on Comput.-Aided Des.*, November 2000, pp. 468–471.

R9.5—RAM Redundancy Analysis Algorithms

[Day 1985] J. R. Day, A fault-driven, comprehensive redundancy algorithm, *IEEE Des. Test Comput.*, 2, 35–44, 1985.

[Haddad 1991] R. W. Haddad, A. T. Dahbura, and A. B. Sharma, Increased throughput for the testing and repair of RAMs with redundancy, *IEEE Trans. Comput.*, 40(2), 154–166, 1991.

[Huang 1990] W.-K. Huang, Y.-N. Shen, and F. Lombardi, New approaches for the repair of memories with redundancy by row/column deletion for yield enhancement, *IEEE Trans. Comput.-Aided Des. Integrated Circuits Syst.*, 9(3), 323–328, 1990.

[Huang 2003] C.-T. Huang, C.-F. Wu, J.-F. Li, and C.-W. Wu, Built-in redundancy analysis for memory yield improvement, *IEEE Trans. Reliability*, 52(4), 386–399, 2003.

[Kuo 1987] S.-Y. Kuo and W. K. Fuchs, Efficient spare allocation in reconfigurable arrays, *IEEE Des. Test Comput.*, 4(1), 24–31, 1987.

[Stapper 1980] C. H. Stapper, A. N. McLaren, and M. Dreckmann, Yield model for productivity optimization of VLSI memory chips with redundancy and partially good product, *IBM J. R&D*, 24(3), 398–409, 1980.

[Tarr 1984] M. Tarr, D. Boudreau, and R. Murphy, Defect analysis system speeds test and repair of redundant memories, *Electronics*, 57, 175–179, 1984.

[Wey 1987] C.-L. Wey and F. Lombardi, On the repair of redundant RAMs, *IEEE Trans. Comput.-Aided Des. Integrated Circuits Syst.*, 6(3), 222–231, 1987.

R9.6—Built-In Self-Repair

[Cheng 2000] C. Cheng, C.-T. Huang, J.-R. Huang, C.-W. Wu, C.-J. Wey, and M.-C. Tsai, BRAINS: a BIST complier for embedded memories, in *Proc. Int. Symp. on Defect and Fault Tolerance in VLSI Systems*, October 2000, pp. 299–307.

[Huang 1999] C.-T. Huang, J.-R. Huang, C.-F. Wu, C.-W. Wu, and T.-Y. Chang, A programmable BIST core for embedded DRAM, *IEEE Des. Test Comput.*, 16(1), 59–70, 1999.

[Huang 2002] R.-F. Huang, J.-F. Li, J.-C. Yeh, and C.-W. Wu, A simulator for evaluating redundancy analysis algorithms of repairable embedded memories, in *Proc. IEEE Int. Workshop on Memory Technology, Design, and Testing*, July 2002, pp. 68–73.

[Huang 2003] C.-T. Huang, C.-F. Wu, J.-F. Li, and C.-W. Wu, Built-in redundancy analysis for memory yield improvement, *IEEE Trans. Reliability*, 52(4), 386–399, 2003.

[Kawagoe 2000] T. Kawagoe, J. Ohtani, M. Niiro, T. Ooishi, M. Hamada, and H. Hidaka, A built-in self-repair analyzer (CRESTA) for embedded DRAMs, *Proc. IEEE Int. Test Conf.*, October 2000, pp. 567–574.

[Tarr 1984] M. Tarr, D. Boudreau, and R. Murphy, Defect analysis system speeds test and repair of redundant memories, *Electronics*, 57, 175–179, 1984.

CHAPTER 10

BOUNDARY SCAN AND CORE-BASED TESTING

Kuen-Jong Lee
National Cheng Kung University, Tainan, Taiwan

ABOUT THIS CHAPTER

Boundary scan, also known as the IEEE 1149.1 or JTAG standard, appears to be the most successful test standard ever approved by the IEEE. Initially targeting board-level testing for digital circuits, this standard has now been adopted by industry for use in most large IC chips and has been used to access many other applications, including power management, clock control, debugging, verification, and chip reconfiguration. An extended boundary-scan standard for the I/O protocol of high-speed networks (namely, 1149.6) has recently been established, and it further enhances the applicability of boundary scan.

Core-based test problems arise when IC design shifts to the ***system-on-chip*** (SOC) paradigm where cores or ***intellectual properties*** (IPs) become the building blocks of a design. Because the relationships of chips to boards and cores to SOC are analogous, a test standard similar to 1149.1 (namely, 1500) was approved by the IEEE in 2005. This embedded-core-based test standard inherited most of the properties of 1149.1 and additionally solves many new test problems related to SOC design. It can be expected that in the near future an increasing number of SOC designs will incorporate this standard.

This chapter begins with an introduction to the boundary-scan family of standards and their current status. The 1149.1 standard is then described in detail. On-chip design to support scan and BIST by 1149.1 and board/system-level controller design for 1149.1 are also covered. The IEEE 1149.6 extension is then discussed. With regard to core-based testing, new test problems that have arisen during the SOC era are examined. The kernel of the 1500 standard (*i.e.*, the 1500 core wrapper) is then detailed. The Core Test Language, which standardizes the description of core test information, is also presented. Test control architectures to support 1500 design with the plug-and-play feature and hierarchical test structures are then discussed. Finally, a comparison between 1149.1 and 1500 is made.

10.1   INTRODUCTION

Testing a stand-alone chip is relatively easy because all I/O pins are controllable and observable with external test equipment. Once a chip is mounted on a ***printed-circuit board*** (PCB), the problem becomes much more complex. The conventional "bed-of-nails" board-level test method by which testing relies on probing on-board test pins and vias has already encountered difficulties in dealing with multiple-layer boards. With the advent of surface mount packages and ***multiple chip modules*** (MCMs), this method becomes infeasible as no or few through-hole pins are available for probing [Parker 2001].

In the mid-1980s, a group of test engineers from several European electronics system companies began to get together to search for possible solutions to this problem. This group, known as the Joint European Test Action Group (JETAG), finally concluded that the best way to address this problem is to chain all the boundary I/O pins of a chip into a shift register and use a concept similar to scan design to gain back the I/O accessibility of the chip. In 1988, JETAG was joined by representatives from North American companies who had also been working on this problem and had come to a similar conclusion. The combined group was renamed the Joint Test Action Group (JTAG). Through the efforts of JTAG, the idea of "boundary scan" was formally converted into a test architecture and a set of associated design rules, which were quickly approved by the IEEE as a test standard (Std. 1149.1) in 1990. Since then, the standard has been employed by most electronics companies when building large chips. Today, almost all general-purpose CPU, DSP, and FPGA and many application-specific designs comply with the 1149.1 standard. Because boundary scan provides a simple and efficient protocol for data communication, this standard has also been employed in many other applications, including power management, embedded instrumentation control, clock/PLL control, debugging/diagnosis, verification, and chip reconfiguration [Rearick 2005].

Standard 1149.1, however, defines only a general-purpose boundary-scan implementation for digital chips. Several other boundary-scan standards for different, more specific test objectives have also been established, as described next.

10.1.1   IEEE 1149 Standard Family

Boundary scan is in fact a family of test methodologies aimed at resolving a wide range of test problems: from chip level to system level, from logic cores to interconnects between cores, from digital circuits to analog or mixed-mode circuits, and from ordinary digital designs to very high-speed designs. Table 10.1 provides an overview of the boundary-scan family, now known as the IEEE 1149.x standards, and their standard setup status.

Standard 1149.1, usually referred to as the *digital boundary-scan standard*, was approved by the IEEE in 1990. Following approval of the standard, increasing demand for a standard hardware description language to describe this standard has motivated the development of the ***Boundary-Scan Description Language*** (BSDL). Thus, soon after the first revision of the digital boundary-scan standard in 1993

TABLE 10.1 ■ IEEE 1149 Standard Family

Number	Main Objectives	Status
1149.1	Testing of digital chips and interconnects between chips	Std. 1149.1-1990 Std. 1149.1a-1993 Std. 1149.1b-1994 (BSDL) Std. 1149.1-2001
1149.2	Extended digital serial interface	Discontinued
1149.3	Direct access testability interface	Discontinued
1149.4	Mixed-signal test bus	Std. 1149.4-1999
1149.5	Standard module test and maintenance (MTM) bus	Std. 1149.5-1995 (not endorsed by IEEE since 2003)
1149.6	High-speed network interface protocol	Std. 1149.6-2003

(1149.1a), the BSDL also became an IEEE standard (1149.1b) in 1994. These two standards, however, have now been merged back to 1149.1 [IEEE 1149.1-2001] [Parker 2001].

The 1149.2 (Extended Digital Serial Subset) standard was aimed primarily at *application-specific integrated circuits* (ASICs) and tried to add high-speed boundary-scan test capability, while 1149.3 targeted the direct access interface of a chip, emphasizing system testability specifications. It was argued that some features of 1149.2 could be covered by scan design [Dervisoglu 1992] [Petersen 1992] and other features by 1149.1 [Ungar 2001], so it was discontinued in 1993. Standard 1149.3 started out as a system test bus but was also defunct shortly, due to its overlap with 1149.5 [Petersen 1992] [Ungar 2001]. Standard 1149.4 [IEEE 1149.4-1999] defines the chip-level test architecture for circuits with analog I/O, now referred to as *analog boundary scan*. This standard is discussed in Chapter 11 of this book. Standard 1149.5, approved in 1995, defines the test and maintenance bus protocol at the module level. This standard, however, is no longer endorsed by the IEEE (since 2003) due to the lack of industry support [Treuren 2005]. Standard 1149.6 [IEEE 1149.6-2003], approved by the IEEE in 2003, is an extension of 1149.1 designed to standardize boundary scan for high-speed (1+ Gbps) I/O designs [Eklow 2003a] [Eklow 2003b]. The objective of this standard is to ensure simple, robust, and minimally intrusive boundary-scan testing of advanced digital networks not adequately addressed by 1149.1, especially for those networks that are AC-coupled, differential, or both. As currently high-speed I/O pins have reached multiple-gigabit-per-second rates, this standard is gaining more and more popularity in industry. Thus, in this chapter we focus on 1149.1 and 1149.6 with regard to boundary-scan standards.

10.1.2 Core-Based Design and Test Considerations

In the SOC era, conventional gate-based or cell-based design methodologies are no longer sufficient. The core-based design methodology, in which cores or intellectual properties form the basic building blocks of a system, has become the main design

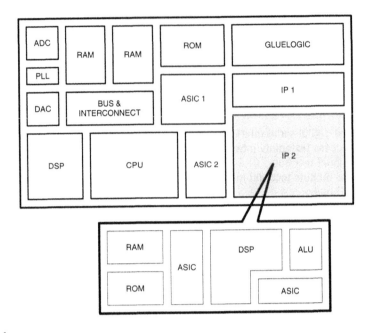

■ **FIGURE 10.1**

Core-based SOC design.

methodology for complex SOC. As shown in Figure 10.1, a typical SOC design may contain CPU, DSP, memory modules (RAM and ROM), mixed-signal devices (ADC/DAC and PLL), some buses/interconnect modules, glue logic, ASIC cores, and IPs. An IP or ASIC core may itself be a complex core containing processors, ASICs, and local memories.

Many test problems are encountered in such a complex system that are not seen in simpler designs. A SOC test developer (or integrator) has to consider how to develop a complete test for cores provided by different vendors, delivered in different formats (*e.g.*, soft or hard cores), implemented with different technologies, operating at different speeds, etc. The developer must also consider test application issues, such as accessible test resources, allowable test time, tolerable test power, and available automation tools.

In this chapter, we examine the test problems existing in a core-based design and describe how to deal with these problems efficiently. We introduce a new IEEE test standard: 1500 [IEEE 1500-2005]. This standard is similar to 1149.1 in that its main objective is to standardize boundary test circuitry (called *wrappers*) for cores. Standard 1500, however, further provides parallel access capability for a core so test efficiency for an SOC can be significantly improved. Furthermore, unlike 1149.1, where control signals are mainly generated by a finite state machine that is controlled by a single input, in 1500 the control signals can be directly applied to a core, thus providing more test flexibility. We also describe the ***Core Test Language*** (CTL) [IEEE P1450.6-2001], which is a language for capturing and expressing test-related information for cores complying with 1500. By using CTL, the SOC integrator or automation tools can successfully generate all information and circuitry required to test the SOC. We also discuss several test control architectures that support the 1500 wrappers. Examples of implementing hierarchical test control with plug-and-play features are provided. Finally, we provide a comparison between 1149.1 and 1500.

10.2   DIGITAL BOUNDARY SCAN (IEEE Std. 1149.1)

In this section we describe the digital boundary-scan standard based on the IEEE Std. 1149.1-2001 version [IEEE 1149.1-2001].

10.2.1   Basic Concept

Standard 1149.1 defines a test access protocol and a boundary-scan architecture for digital integrated circuits and the digital portions of mixed analog/digital integrated circuits. As shown in Figure 10.2, the name *boundary scan* is due to the insertion of a *boundary-scan cell* to each I/O pin of the original circuit and the chaining of these cells into a shift register called the *boundary-scan register*. Chips complying with this standard can be readily integrated into a PCB with their I/O accessible through the boundary-scan registers. Figure 10.3 shows a board containing four ICs for which the boundary-scan registers are interconnected into a single boundary-scan chain. Through this chain the I/Os of each chip are controllable and observable via serial scan and Capture/Update operations, thereby enabling the testing of internal logic of each chip as well as interconnects among the chips. In addition, 1149.1 also provides the important feature where the data capturing and shifting can be done on the boundary-scan logic without interfering with the normal circuit operations. This feature can greatly enhance the capabilities of design debugging and fault diagnosis for the chips.

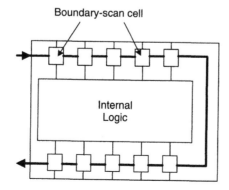

■ **FIGURE 10.2**

Basic idea of boundary scan.

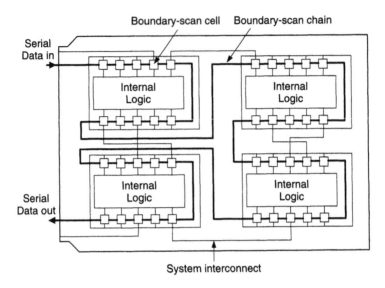

Boundary-scan cell   Boundary-scan chain

Serial Data in

Internal Logic

Internal Logic

Serial Data out

Internal Logic

Internal Logic

System interconnect

■ **FIGURE 10.3**

A board containing four ICs with boundary scan.

10.2.2 Overall 1149.1 Test Architecture and Operations

In addition to the boundary-scan register described in the previous section, extra control circuitry and data storage are required for each chip. Figure 10.4 shows a chip with the boundary-scan architecture. The internal logic represents the original circuit of the chip. It may contain some internal registers that can be supported by boundary scan, such as scan chains, **built-in self-test** (BIST) generators or compressors, or any other storage that will make use of the boundary-scan functionality. The boundary-scan circuitry can be divided into four main hardware components:

- A **test access port** (TAP), which consists of four mandatory terminals—**test data input** (TDI), **test data output** (TDO), **test mode select** (TMS), and **test clock** (TCK)—and one optional terminal, **test reset** (TRST*)

- A **TAP controller** (TAPC)

- An **instruction register** (IR) and its associated *decoder*

- Several test data registers, including the mandatory *boundary-scan register* and *bypass register*, and some optional miscellaneous registers, such as the *device-ID register*, and some design-specific test data registers

The test access port, which defines the bus protocol of the boundary scan, consists of additional I/O pins necessary for each chip employing the standard. The TAP controller is a 16-state, finite-state machine that controls each step of the

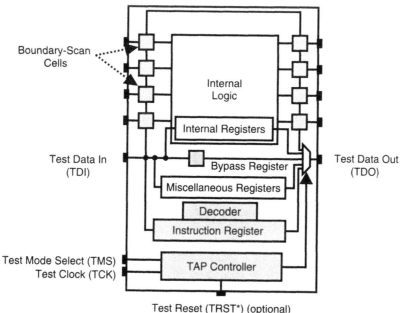

Boundary-Scan Cells

Internal Logic

Internal Registers

Test Data In (TDI)

Bypass Register

Miscellaneous Registers

Decoder

Instruction Register

Test Mode Select (TMS)
Test Clock (TCK)

TAP Controller

Test Data Out (TDO)

Test Reset (TRST*) (optional)

■ **FIGURE 10.4**

Boundary-scan architecture.

boundary-scan operations. Each instruction to be carried out by the boundary-scan architecture must be serially loaded into the *instruction register* through the **test data input** (TDI) pin. The test signals to configure the boundary-scan-related test hardware for the current test instruction are provided by the associated decoder. The *test data registers* are used to store test data or some system-related information (such as the chip ID, company name, etc.).

In addition to the hardware components, IEEE Std. 1149.1 also defines a set of test instructions, including four mandatory ones (BYPASS, SAMPLE, PRELOAD, and EXTEST) and several optional ones, including INTEST, RUNBIST, CLAMP, IDCODE, USERCODE, and HIGHZ. It also allows the users to define their own instructions. An outline of a typical test procedure using boundary scan, which will be detailed in the following sections, is as follows:

1. A boundary-scan test instruction is shifted into the IR through the TDI.

2. The instruction is decoded by the decoder associated with the IR to generate the required control signals so as to properly configure the test logic.

3. A test pattern is shifted into the selected data register through the TDI and then applied to the logic to be tested.

4. The test response is captured into some data register.

5. The captured response is shifted out through the TDO for observation and, at the same time, a new test pattern can be scanned in through the TDI.

6. Steps 3 to 5 are repeated until all test patterns are shifted in and applied, and all test responses are shifted out.

Detailed structures and functions of the boundary-scan components are described next. Figure 10.5 shows an example of the boundary-scan circuitry extracted from Figure 10.4 which provides more detailed information and is used in the following discussion.

10.2.3 Test Access Port and Bus Protocols

The TAP of 1149.1 contains four mandatory pins and one optional pin, as described below:

- **Test clock input** (TCK) is a clock input to synchronize the test operations between the various parts of a chip or between different chips on a PCB. This input must be independent of the system clocks so the serial test data path between components of a chip or different chips can be used independently of the system clocks, which may vary significantly in frequency from one component to another; so the board interconnect testing can be properly carried out; and so the shifting and capturing of test data can be executed concurrently with normal system operation, thereby facilitating online system monitoring for a design without changing the state of the on-chip system logic.

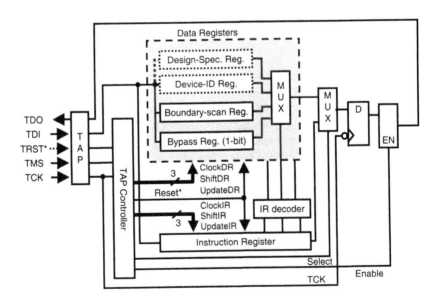

■ **FIGURE 10.5**

Boundary-scan circuitry in a chip.

- **Test data input** (TDI) is an input to allow test instructions and test data to be serially loaded into the *instruction register* and the various test data registers, respectively. Values presented at TDI are clocked into the selected register on a rising edge of TCK. It is expected that the bus master (automatic test equipment, on-board bus controller, etc.) will change the signal driven to the TDI input on the falling edge of TCK. The design of the circuitry fed from TDI shall be such that an undriven input produces a logical response equivalent to a logic 1.

- **Test data output** (TDO) is an output to allow various test data to be driven out. As shown in Figure 10.5, changes in the state of the signal driven through TDO should occur only on the falling edge of TCK. Also, the TDO driver must be set to its inactive driving state except when the scanning of data through this terminal is in progress. The EN block controlled by the Enable signal provides this capability. Note that data should be propagated from TDI to TDO without inversion.

- **Test mode select** (TMS) is the sole test control input to the TAP controller. All boundary-scan test operations such as shifting, capturing, and updating of test data are controlled by the test sequence applied to this input. Signals presented at TMS are sampled by the TAP controller on the rising edge of TCK. It is also expected that the bus master will change the signal driven to the TMS input on the falling edge of TCK. This input should also be driven to logic 1 when it is inactivated.

- **Test reset** (TRST*) is an optional pin used to reset the TAP controller. If the TRST* pin is implemented, the TAP controller can be asynchronously reset to the Test–Logic–Reset controller state (to be discussed later) when a logic "0" is applied at TRST*. This in turn will reset other boundary-scan logic to the state required by the Test–Logic–Reset state. This pin should not be used to reset the system logic so the test logic can be reset independently of the on-chip system logic. If this input is omitted, the system must have some circuitry that can reset the TAP controller during power-on. In Section 10.2.5, we show that the TAP controller can also be synchronously reset.

10.2.4 Data Registers and Boundary-Scan Cells

Standard 1149.1 specifies several test data registers, as shown in Figure 10.5. Two mandatory test data registers—the *boundary-scan register* and the *bypass register*—must be included in any boundary-scan architecture. Other registers, such as the *device identification register* and design-specific test data registers, can be added optionally.

The **boundary-scan register** (BSR) is the collection of the **boundary-scan cells** (BSCs) inserted at the I/O pins of the original circuit, as shown in Figure 10.4. Various designs for the boundary-scan cells exist. A typical BSC is shown in Figure 10.6. This cell can be used as either an input or output cell. As an input BSC, the IN signal line corresponds to a chip input pad, and the OUT signal line is tied to an input of the internal logic. As an output BSC, IN corresponds to the

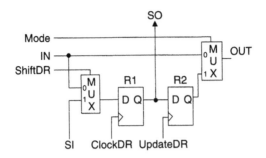

A typical boundary-scan cell (BSC).

output of the internal logic, and OUT is tied to an output pad. Data driven on the OUT signal are controlled by the Mode signal. During the **normal** mode operation (Mode = 0), data passes from IN directly to OUT and the cell is transparent to the functional logic. In **Test** mode (Mode = 1), test data driven by the R2 flip-flop pass through the multiplexer to the OUT signal. The test operations of a BSC are controlled by three output signals of the TAP controller: ClockDR, ShiftDR, and UpdateDR (see Figures 10.5 and 10.6). Three main test operations—*Capture, Shift,* and *Update*—are defined. In the Capture operation, ShiftDR is set to 0, one clock pulse is applied to ClockDR, and the test data at IN will be captured into the D-FF R1 (known as the capture flip-flop). In the Shift operation, ShiftDR is set to 1 and clock pulses are applied to ClockDR such that test data can be shifted in from SI and the test response can be scanned out through SO. The boundary-scan register is formed by connecting the SO of the previous cell to the SI of the next cell. In the Update operation, the data stored in R1 are propagated to R2 (known as the update flip-flop) by applying a clock pulse to UpdateDR. If the Mode is set to 1 at this time, then the output of R2 is connected to OUT. Note that the Capture and Shift operations can also be executed when the cell is in the normal mode operation, as mentioned in Section 10.2.1. One can also latch test data in R2 (and at the OUT terminal if Mode = 1) while other test data are shifted in/out.

Several other scan cell designs exist. Standard 1149.1-2001 defines ten boundary-scan cells (BC1–BC10), which include observation-only and bidirectional cells [IEEE 1149.1-2001].

The *bypass register* is a single-bit register that is used to bypass a chip when it is not involved in the current test operation. This can significantly reduce test time required to shift in/out test data through the long TDI–TDO path.

Standard 1149.1 also defines an optional data register called the *device-ID register,* which can be used to load information about the product (the manufacturer, part number, and version number) or the configuration of the chip. The loaded information can be shifted out for observation after the chip is mounted onto a board. One application of this register is to identify during the debugging and diagnosis process the manufacturers and revisions of chips on a board that come from multiple sources.

10.2.5 TAP Controller

The **TAP controller** (TAPC) is a 16-state, finite-state machine that operates according to the state diagram shown in Figure 10.7. The TAPC can change state only on the rising edge of TCK. The next state is determined by the logic level of TMS. The output signals of the TAPC determine the test operation to be carried out. As shown in Figure 10.5, nine control signals—ClockDR, ShiftDR, UpdateDR, ClockIR, ShiftIR, UpdateIR, Select, TCK, and Enable—as well as Reset* (optional) are produced by the TAPC. The main functions of the TAPC include:

- Resetting the boundary-scan architecture

- Providing control signals to load instructions into the instruction register

- Providing signals to perform test functions such as Capture and Update (application) of test data

- Providing control signals to shift test data from TDI to TDO

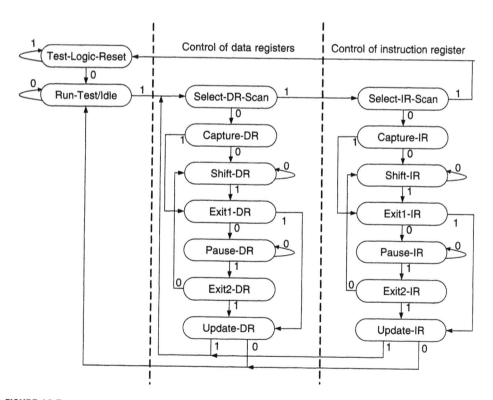

■ **FIGURE 10.7**

State diagram of TAP controller.

The 16 states can be divided into three parts. The first part (the 2 states at left) contains the reset and the "Run-Test/Idle" states, the second (the 7 states in the middle) and third (the 7 states at right) parts control the operations of the data and instruction registers, respectively. Because the only difference between the second and the third parts is the registers they deal with, only the states in the first and second parts are described in the following text. Details about the third part can be inferred from the descriptions about the second part.

- **Test–Logic–Reset**—In this state the boundary-scan circuitry is disabled and the system operates in its normal mode. Whenever a 0 signal is applied to the TRST* port, the TAPC enters this state. One should also notice that the TAPC can also be synchronously reset; whatever state the TAP controller is in, it will return to this state if a logic 1 is applied to TMS for five consecutive TCK cycles (*i.e.*, five rising edges of TCK). It should also be noted that during this state if a glitch occurs at TMS that forces the TAP controller to enter the Run-Test/Idle state (discussed next), the TAP controller can still return to this state if TMS is kept stable at 1 for the next three TCK cycles.

- **Run-Test/Idle**—In this state, the boundary-scan circuitry is waiting for some test operations synchronized with the TCK (such as BIST) to complete. It is different from the Test–Logic–Reset in that during this state activities on selected test logic may still be in progress.

- **Select-DR-Scan**—This is a temporary state in preparation for entering the data register manipulation column (the middle part) of Figure 10.7.

- **Capture-DR**—In this state, data can be loaded in parallel to the data registers selected by the current instruction. It is in this state where the current test results and normal operation status are captured.

- **Shift-DR**—In this state, test data are scanned in series through the data registers selected by the current instruction. Upon entering this state, the TAP controller will stay in this state as long as TMS = 0. For each clock cycle, one bit of test data will be shifted into (out of) the selected data register through TDI (TDO).

- **Exit-DR**—This is also a temporary state. All parallel-loaded (from the Capture-DR state) or shifted (from the Shift-DR state) data are held in the selected data register in this state in preparation to enter the update or pause state.

- **Pause-DR**—The boundary-scan logic pauses its function here to wait for some external operations. For example, when a long sequence of test data is to be loaded to the chips under test, the external tester may have to reload the data from time to time. The Pause-DR is a state that allows the boundary-scan architecture to wait for more data to shift in/out.

- **Exit2-DR**—This state either indicates completion of the current capturing/shifting operation and allows the TAPC to enter the update state or represents the end of the Pause-DR operation, allowing the TAP controller to go back to the Shift-DR state for more data to shift in/out.

- **Update-DR**—In this state, data are latched onto the parallel output of the selected test data registers from the shift register path on the falling edge of TCK; for example, the test data stored in the first stage of boundary-scan cells (R1 in Figure 10.6) are loaded to the second stage (R2) in this state. Note that, with a two-stage register design, test data can be held at the parallel output of the selected register while other data are shifted in the associated shift register path.

10.2.6 Instruction Register and Instruction Set

The *instruction register* is used to store the instruction to be executed. By the standard, this register must be a two-stage design such that when a new instruction is being shifted in the current instruction can be latched at the parallel output of the IR so as to prevent the possibility of having an indeterminate state at the output of IR. Four mandatory boundary-scan test instructions (SAMPLE, PRELOAD, BYPASS, and EXTEST) are defined in 1149.1. In addition, a commonly used instruction (namely, the INTEST instruction) is recommended. We will describe these instructions in detail. There are other useful instructions, including RUNBIST, CLAMP, IDCODE, USRCODE, and HIGHZ, which will also be described.

- **BYPASS**—When dealing with board-level testing, it is often required to send test data to or receive test results from only one or two specific chips. The BYPASS instruction is used to "bypass" the boundary-scan registers on unused chips so as to prevent long Shift operations, as shown in Figure 10.8. The BYPASS register must capture a default bit of 0 at the Capture-DR state when this instruction is executed. Furthermore, the instruction register must be forced to contain the BYPASS instruction whenever the TAP is reset, unless an IDCODE instruction (to be discussed later) is implemented, in which case IDCODE is the default instruction to be loaded to the BYPASS register.

- **SAMPLE**—Figure 10.9 illustrates execution of the SAMPLE instruction. The SAMPLE operation can be completed by simply executing the Capture operation (on the rising edge of TCK in the Capture-DR state) such that the required test data can be loaded in parallel to the selected data registers. This means that a snapshot of the normal operation of the chip can be taken and examined. This instruction also allows the capture of the signals applied to the primary inputs of a chip and capture of the responses appearing at the output of the internal logic. It is required that when the SAMPLE instruction is executed, the operation of the boundary-scan test logic must have no effect on the internal logic or on the flow of signals between the internal logic and the I/O pins of the chip.

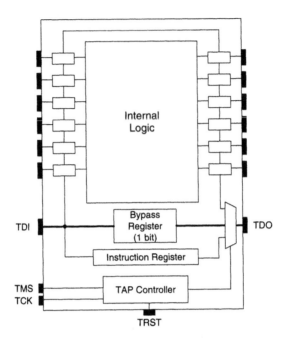

■ FIGURE 10.8

Execution of the BYPASS instruction.

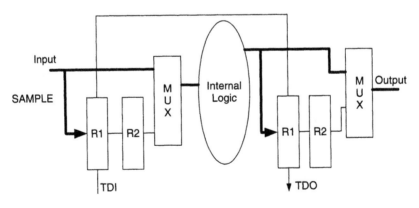

■ FIGURE 10.9

Execution of SAMPLE instruction.

- **PRELOAD**—The PRELOAD instruction allows test data to be shifted into or out of the selected data register during the Shift-DR state without causing interference to the normal operation of the internal logic, as shown in Figure 10.10. The shifted data are then latched to the parallel output (R2) of

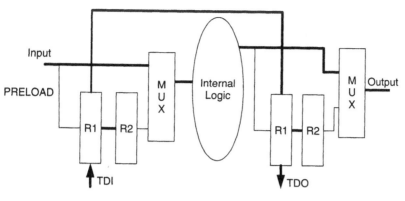

Execution of PRELOAD instruction.

the selected data registers (on the falling edge of TCK in the Update-DR controller state) for immediate or later use. This allows an initial data pattern to be placed at the latched parallel outputs of boundary-scan register cells using the PRELOAD instruction before the selection of another boundary-scan test operation (say, EXTEST). Without the PRELOAD instruction, indeterminate data would be driven until the following operation, such as a sequence of Shift operations, has been completed.

■ **EXTEST**—The EXTEST instruction is used to test the circuitry external to the chips, typically the interconnects between chips and between boards. Its execution is illustrated in Figure 10.11. Assume that an interconnect line from Chip1 to Chip2 as shown is to be tested. First, the test pattern is shifted into the "driving terminals" of Chip1 through its TDI pin during the ShiftDR state of the TAP controller. Second, an Update operation in Chip1 is executed on the falling edge of the TCK such that the shifted test data bit is loaded to the corresponding output pin of Chip1 (similar to the PRELOAD instruction). A Capture operation is then executed in Chip2 on the rising edge of the TCK and the test data bit is captured at the driven terminal of Chip2 (similar to the SAMPLE instruction); therefore, there will be two and a half cycles of latency between the Update operation of Chip1 and the Capture operation of Chip2 if they are controlled by the same TMS. Finally, the ShiftDR operation is executed in Chip2 and the received test response can be scanned out through the TDO of Chip2 for examination. New test patterns can also be shifted in at this time and the cycle can be repeated.

Based on the above procedure, it can be found that the first two steps can also be accomplished by using the PRELOAD instruction. In fact, before the selection of the EXTEST instruction, data can be loaded onto the latched parallel outputs using PRELOAD. Then, as soon as the EXTEST instruction has been transferred to the parallel output of the instruction register, the preloaded data are driven through the chip output pins. This ensures that the

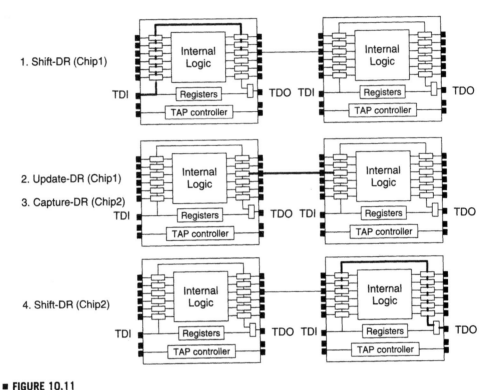

Execution of EXTEST instruction.

required test data are driven immediately when the change to the EXTEST instruction takes place in the Update-IR controller state, hence preventing the possible appearance of indeterminate data at the chip outputs during the shifting operation if the EXTEST instruction was executed without first executing the PRELOAD instruction.

Note that in the above description we only examine one interconnect. Clearly, several interconnects can be examined simultaneously. The only requirement is to shift and load the required data bits to the appropriate driving positions and then capture these data at the driven positions. The test responses can then be scanned out through the TDO for further examination. Methods to minimize test time for all interconnects can be found in [Jarwala 1989], [Chan 1992], and [Kim 2004].

■ **INTEST**—Figure 10.12 shows the steps of the INTEST instruction. During the first step, test data are shifted into the boundary-scan cells that drive internal logic. In the second step, an Update operation is executed and the shifted data are loaded to the second stage of the boundary-scan cell. At the same time the data are applied to the internal logic. The TAP controller then goes back to the Capture-DR state to capture the test result at the boundary-scan cells,

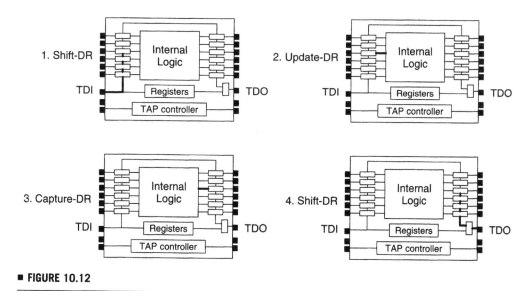

■ **FIGURE 10.12**

Execution of INTEST instruction.

which receive data from internal logic. Finally, the Shift-DR operation shifts the test results out for observation. This operation may be repeated for several cycles.

■ **RUNBIST**—The RUNBIST instruction provides a means of running a user-accessible self-test function within the chip using a single instruction. This permits all chips on a board that support the RUNBIST instruction to execute their own self-test process concurrently, thereby greatly reducing the total test time. It is also permitted to include further private or public instructions which give access to individual self-test functions one at a time. Because this is an optional instruction, the signals to control the BIST circuitry in each chip have to be generated by user-designed logic. One way to accomplish this is illustrated in Section 10.2.8.

■ **CLAMP**—When the CLAMP instruction is executed, the state of all signals driven from the output pins of the chip should be completely defined by the data held in the boundary-scan register, preferably in a "safe" state. The data may be shifted into the boundary-scan register by a previous PRELOAD instruction. It should be pointed out that, similar to the BYPASS instruction, the CLAMP instruction also places only the BYPASS register between TDI and TDO, thus greatly reducing the shift length.

■ **IDCODE**—This instruction shall be provided when the optional 32-bit device identification register (device-ID register) is included in the chip. When the IDCODE instruction is used, the vendor's identification code (containing the manufacturer's identity, the part number, and the version number of the chip) prestored somewhere in the chip shall be loaded into the device-ID register, which can then be shifted out for examination through the TDI–TDO shift path. This instruction, if implemented, should be the default instruction in the reset state.

- **USERCODE**—The USERCODE instruction allows a user-programmable 32-bit identification code to be loaded into the user-defined device-ID register and then shifted out for examination. This instruction is useful when the chip can be programmed in various ways and it is necessary to determine the way in which the chip is programmed. This instruction shall be used in a programmable device such as an FPGA chip if programming the chip via boundary-scan test logic is not allowed. When this instruction is included, it is the responsibility of the chip provider to implement the code somewhere in the chip.

- **HIGHZ**—When the HIGHZ instruction is selected for a chip, all output pins of the chip shall be placed in an inactive-drive state. This will allow an in-circuit system tester to drive the chip outputs to some desired state (*e.g.*, to test other chips) without damaging the chip. This instruction also requires that the BYPASS register be the only register on the TDI–TDO path.

10.2.7 Boundary-Scan Description Language

The *boundary-scan description language* (BSDL) has been included as part of IEEE Std. 1149.1-2001. This *VHSIC hardware description language* (VHDL)-compatible language provides information about how a boundary-scan IC is implemented, which can be used by ATPG software or system integrators to develop the test for the chip. Descriptions for mandatory logic, such as the TAP and BYPASS registers, do not have to be provided. These are already provided in a standard way. The designer only has to describe the design-specific attributes, such as the length of boundary-scan register, the user-defined boundary-scan instructions, the decoder for his or her own instructions, and the I/O pins assignment (*e.g.*, which pin is to be used as the TDI pin). In general these descriptions are quite easy to prepare. Currently, many *computer-aided design* (CAD) tools already support automatic generation of the boundary-scan design, thus it may not even be necessary for a designer to write the BSDL file; the tools can automatically generate the necessary boundary-scan circuitry as long as the specific boundary-scan information for the chip is provided (in a setup file, for example).

10.2.8 On-Chip Test Support with Boundary Scan

Standard 1149.1 defines the test circuitry on the boundary of a chip but does not describe how to use boundary-scan circuitry to support the DFT circuitry built inside a chip, such as scan or BIST design. In Figure 10.13 we show a possible

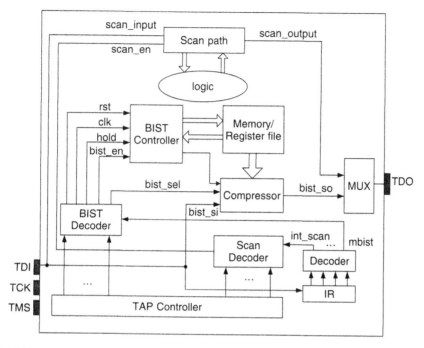

■ **FIGURE 10.13**

Scan and BIST support with boundary scan.

way to support these DFT operations. First, we may add new test instructions for these operations such as RUNBIST and RUNSCAN. Second, the decoder for the IR can be designed in such a way that one control line will be added for each new instruction and each time a new instruction is loaded into the IR the corresponding control line will be activated (set to 1, for example). Third, the control line can then be used to enable a decoder specifically designed for the instruction such that the control signals generated by the TAPC can be converted into the appropriate control signals required by the associated DFT circuitry.

We can use the sample BIST circuitry shown in Figure 10.13 to illustrate this design. When the RUNBIST instruction is loaded into the IR, the *mbist* signal will be activated and the BIST decoder can be enabled. Initially, the TAP controller is in the Test–Logic–Reset state which will provide a reset signal via the *rst* signal to reset the BIST controller. If the compressor has to be initialized, we can either reset it or place the TAPC into the shift mode and activate the *bist-sel* signal such that an initial value can be sent to the Compressor through TDI and *bist_si*. Then, we may put the TAPC in the Run-Idle/Test state and enable the BIST controller via the *bist_en* signal to start the BIST operation. After the BIST operations are completed, the TAPC is again put into the shift mode so as to shift out the compressed results for evaluation. The *hold* signal shown in Figure 10.13 can be used to hold the BIST controller when necessary (*e.g.*, waiting for the TAP to reach some required state or waiting for latency testing).

The above design can be extended to execute BIST for multiple components—concurrently, one by one, or in a mixed manner. Because 1149.1 allows the user to define any number of test instructions, one can define different test instructions if multiple test sessions are required.

For the scan support, similar mechanisms can be used as shown in Figure 10.13; however, because only one test input (TDI) is available, the test time can be quite long if a single scan chain is used. The test compaction techniques described in Chapter 6, such as broadcast scan, can be used to reduce the test time.

10.2.9   Board and System-Level Boundary-Scan Control Architectures

When the 1149.1 test logic has been successfully integrated into the device, the next problem the designer faces is how to provide test data and control signals to chips on a board or system. There exist several test architectures for this purpose, as shown in Figure 10.14 [Zak 1992] [O'Donnell 1994] [Gibbs 2003] [Treuren 2005]. These are described next.

- **Single-ring architecture with shared TMS** (Figure 10.14a)—In this architecture, all boundary-scan registers of chips are daisy-chained together and the TMS signal is broadcast to all chips. All chips will always execute the same Capture, Shift, or Update operation under the control of the TAPC. Note that different chips may still receive different test instructions through the long TDI–TDO chain; for example, some chips may receive the BYPASS instruction while others receive the EXTEST instruction.

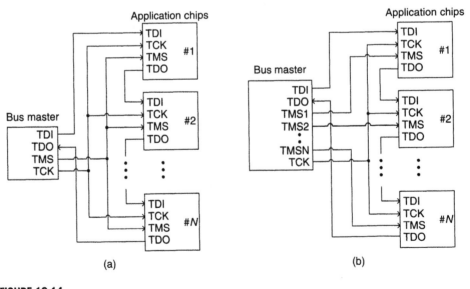

(a)   (b)

■ **FIGURE 10.14**

Bus master for chips with boundary scan: (a) ring architecture with shared TMS; (b) ring architecture with separate TMS; (c) star architecture; (d) multidrop architecture; (e) hierarchical architecture.

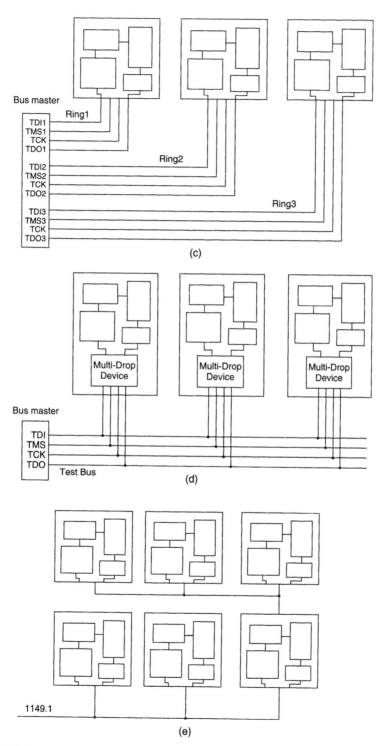

■ **FIGURE 10.14** (*Continued*)

- **Single-ring architecture with separate TMS** (Figure 10.14b)—In this architecture, each chip receives its own TMS signal, hence one can provide different instructions to different chips as well as operate the chips with different control signals. With this architecture, virtually all chips can be tested independently. The single-ring architecture is easy to implement. It is generally sufficient to test chips on a single board; however, for a system or backplane that contains a number of boards where each board contains ICs with boundary-scan architectures, the single-ring configuration may become inefficient due to the long and often cumbersome scan chain that has to pass through all chips in all boards. This architecture also runs into problems when boards are removed or added, as some type of jumper or bridge is required when a board is removed; otherwise, the chain will be broken.

- **Star (multi-ring) architecture** (Figure 10.14c)—In the star-architecture, every board in the system gets a dedicated set of boundary-scan data and control signals. Though test for chips in each board can be efficiently carried out as if only a single board exists, such an approach requires a larger number of (connection) traces in the backplane. The advantage of the ring and star architectures is that they do not require any additional components or new protocols beyond what is required by the boundary-scan specification. This makes them straightforward to implement, but in a large multiple-board system they are often too cumbersome to use.

- **Multidrop architecture** (Figure 10.14d)—A multidrop architecture uses only one set of 1149.1 data and control signals which is wired in parallel to each board in the system. To ensure that boundary-scan operations are applied to only one board at a time, an addressable gateway device must be implemented on each board. A special selection protocol is applied to the boundary-scan bus by the system-level boundary-scan controller to connect the chosen board's scan chain to the bus. This solves the problems mentioned above at the expense of extra selection logic in each board. The multidrop architecture is the one used most in industry; for example, the Addressable Shadow Port developed by Texas Instruments (TI) [Whetsel 1992] [Joshi 2003] [TI 1999] [TI 2003] and the Scan-Bridge developed by National Semiconductor (NS) [NS 2004a] [NS 2004b] belong to this category.

- **Hierarchical architecture** (Figure 10.14e)—A hierarchical connection is essentially a nested test structure, as shown in Figure 10.14e. This architecture has the following advantages: For a complex system the bit length of the serial scan path can become quite large. By breaking the single chains into multiple hierarchical chains, chip access can become much more efficient. The hierarchical structure is also helpful when performing system-level test integration because a complex system is often designed in a top-down fashion which naturally bears a hierarchical structure. The hierarchical architecture can be implemented together with a multidrop system such that the scan path of one board can be used to drive another entire level of multidrop test, as shown in Figure 10.14e.

All of the above-mentioned architectures use the boundary-scan protocol for board or system testing with at least four dedicated lines required. Methods to test chips by using a simplified bus and protocol not compliant to the 1149.1 standard have also been proposed [Bäckström 2005]. The motivation behind such designs lies in the fact that usually embedded go/no-go and efficient local diagnosis are crucial in order to quickly identify and diagnose problems in a system; therefore the information to be transferred at the backplane level can be as simple as the initialization of a test procedure and the go/no-go report of the test results. The solution presented in [Bäckström 2005] belongs to this category and requires only two wires in the backplane.

10.3 BOUNDARY SCAN FOR ADVANCED NETWORKS (IEEE 1149.6)

10.3.1 Rationale for 1149.6

As serial data I/O have increased to multiple-gigabit-per-second rates, advanced signaling techniques such as differential and AC-coupled networks have begun to emerge. These signaling techniques present significant challenges for the IEEE 1149.1 standard [IEEE 1149.1-2001]. The presence of the coupling capacitor in AC-coupled networks blocks DC signals. As a result, the DC level that is applied to the net during a boundary-scan EXTEST instruction decays over time to an undefined logic level. This is shown in Figure 10.15 [IEEE 1149.6-2003], where "C" and "U" refer to the 1149.1 Capture logic (R1 flip-flop in Figure 10.6) and Update logic (R2 flip-flop in Figure 10.6), respectively. In this case, the Capture-DR state must occur within a minimum time after driving the signal in the Update-DR state. This in turn places a minimum frequency requirement on TCK that cannot be supported by the 1149.1 standard. The IEEE 1149.6 standard addresses this issue by capturing the edges of data transitions instead of capturing data levels. By capturing edges instead of levels, the minimum TCK frequency requirement is removed.

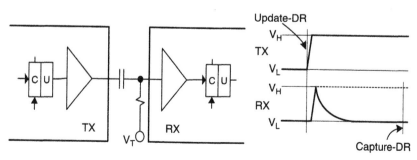

■ **FIGURE 10.15**

Capturing an AC-coupled signal with IEEE 1149.1.

In addition to AC-coupling, differential signaling presents problems for the 1149.1 standard because of its fault tolerance. Standard 1149.1 specifies two ways to address differential networks:

- Placing boundary cells on both outputs of the driver and both inputs of the receiver

- Placing a single boundary cell attached internally to the driver and another boundary cell internally to the receiver (see Figure 10.16)

Only the second choice is practical for high-speed networks, as the boundary cells would present an unreasonable load to the high-speed driver. Herein lies the problem. With only a single boundary-scan cell connected to the output of the differential receiving buffer, opens on either of the input pins may not be detected due to the fault-tolerant nature of differential receivers. An open on one of the two input pins will most likely result in that input to the receiver buffer being terminated to a bias or threshold voltage. Assuming that the other input pin is connected, the boundary-scan cell will still detect the correct state of the network. While these faults are undetectable during a boundary-scan test, they will certainly cause failures during operation at functional speeds.

A better boundary-scan circuit would consist of a single cell internal to the driver and a boundary-scan cell on each input of the receiver. There is little performance penalty on the driver side and better coverage on the receiver side. This circuit can be implemented without violating the 1149.1 standard; however, the standard does not discuss this implementation, and most 1149.1 tools do not recognize such a configuration. This is the implementation that IEEE 1149.6 uses for differential networks.

The following sections describe the four key components of the IEEE 1149.6 circuit: (1) analog test receiver, (2) digital driver logic, (3) digital receiver logic, and (4) 1149.6 TAP. The analog test receiver and digital receiver logic replace the Rx side boundary-scan logic shown in Figure 10.15. The digital driver logic replaces the Tx side boundary-scan logic shown in Figure 10.15.

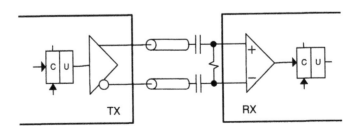

■ **FIGURE 10.16**

The IEEE 1149.1 configuration for differential signaling.

10.3.2 1149.6 Analog Test Receiver

The analog test receiver is the most critical part of the 1149.6 implementation. It is the test receiver that is able to capture transition edges (as described above). The challenge for the test receiver is to capture the edges without the noise immunity that is built into the differential receiver (as there is a test receiver on each input pin). The test receiver uses a "self-referenced" comparator, along with voltage and delay hysteresis, to capture a valid edge and filter any unwanted noise. The test receiver uses a low-pass filter to create a delayed reference signal. An example of the test receiver and its response to AC- and DC-coupled signals is shown in Figure 10.17.

10.3.3 1149.6 Digital Driver Logic

The 1149.6 digital driver logic is a simple extension to the IEEE 1149.1 driver. Unlike the 1149.1 driver, the 1149.6 driver is required to drive a pulse (or a sequence of pulses) when it is executing the 1149.6 EXTEST_PULSE (or EXTEST_TRAIN) instruction. The EXTEST_PULSE instruction is used to drive the output signal to the opposite state, wait for the signal to fully decay, and then drive the signal to the correct value (this is the value that gets captured). By allowing the signal to fully decay, the maximum voltage swing is generated on the next driven edge, allowing

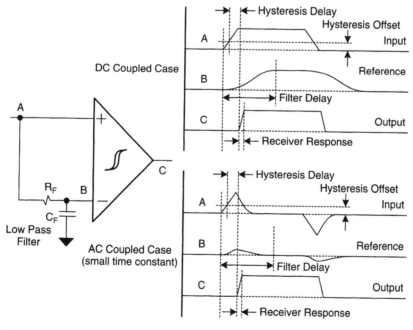

■ **FIGURE 10.17**

Analog test receiver response to AC- and DC-coupled signals.

for better capture by the analog test receiver. In rare cases, a continuous wave-form may be required for some high-speed logic. In this case, the EXTEST_TRAIN instruction is used instead of the EXTEST_PULSE. EXTEST_TRAIN will generate a continuous waveform based on TCK. The TCK frequency must be adjusted to allow for maximum decay without affecting the receiver side logic. The digital driver logic must also support the 1149.1 EXTEST instruction. It simply extends the 1149.1 logic by multiplexing the 1149.6 signal into the 1149.1 shift/update circuit, after the update flip-flop (see Figure 10.18). The 1149.6 driver logic is selected by setting the AC Mode signal to a logic 1. The AC Mode signal is a new signal generated by the TAPC in response to an EXTEST_PULSE or EXTEST_TRAIN instruction. RTI state is also a new signal required for 1149.6. This signal is driven by the TAPC when the controller is in the Run-Test/Idle state while executing an EXTEST_PULSE or EXTEST_TRAIN instruction. The additional logic required for 1149.6 is circled in the figure.

10.3.4 1149.6 Digital Receiver Logic

The digital receiver logic (with the analog test receiver) is shown in Figure 10.19. The digital receiver logic takes the output of the analog test receiver and sets a capture flip-flop to a corresponding logical zero or one. The digital test receiver logic also ensures that a valid transition has been captured on every test vector by initializing the state of the capture memory prior to the transition being driven onto the net (this element is shown as the "hysteresis memory," or Hyst Mem). Without this initialization, it would be impossible to determine if two sequential

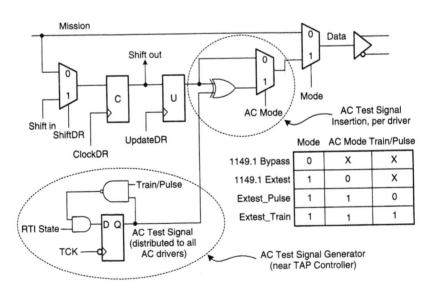

	Mode	AC Mode	Train/Pulse
1149.1 Bypass	0	X	X
1149.1 Extest	1	0	X
Extest_Pulse	1	1	0
Extest_Train	1	1	1

■ **FIGURE 10.18**

Digital driver logic.

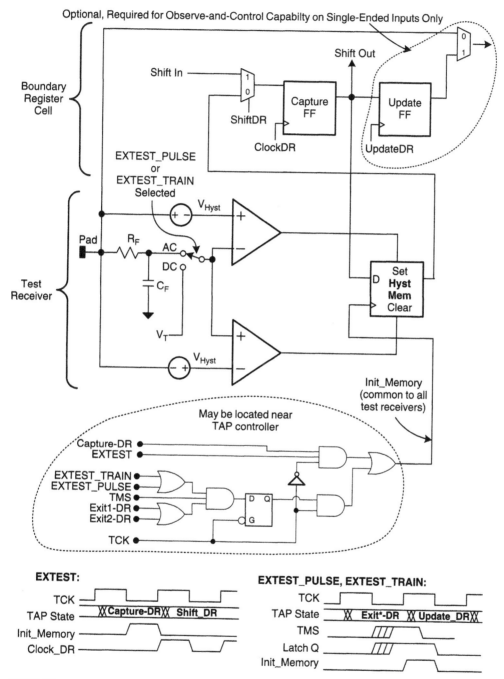

NOTE: The generated clock (Init_Memory) shown is suitable for rising edge-sensitive behavior only.

■ **FIGURE 10.19**

Digital receiver logic.

transitions in the same direction (positive or negative) occurred or if only one transition occurred; that is, if a positive transition occurs and is captured in the memory, and the subsequent test vector also generates a positive transition, then there is no way to determine if the second transition occurred without clearing the contents of the hysteresis memory before the second transition occurs. The capture memory is initialized with the contents of data shifted into the capture flip-flop. The memory is set or cleared based on the transition detected by the analog test receiver. The capture memory is then loaded into the capture flip-flop corresponding to a rising TCK during the Capture-DR state.

10.3.5   1149.6 Test Access Port (TAP)

Changes were made to the 1149.1 TAP to allow the 1149.6 driver logic to generate pulses. It was determined that the 1149.6 TAP would require an excursion through the Run-Test/Idle state to allow for the generation of the pulse or pulses required by the EXTEST_PULSE and EXTEST_TRAIN instructions. Entry into the Run-Test/Idle state during the execution of either EXTEST_PULSE or EXTEST_TRAIN would generate the AC Test signal, which was shown in Figure 10.18 (1149.6 Driver). This would in turn cause the data that were driven onto the net during the Update-DR state to be inverted upon entry into the Run-Test/Idle state (on the first falling edge of TCK) and to be inverted again upon exiting Run-Test/Idle (on the first falling edge of TCK in the Select-DR state). As mentioned previously, the data signal is inverted and then allowed to fully decay in order to guarantee the maximum transition from the driver. This behavior is shown in Figure 10.20.

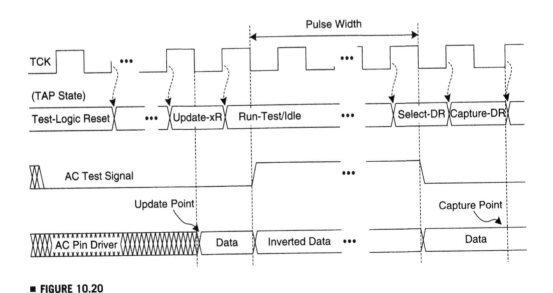

■ **FIGURE 10.20**

Driver behavior during EXTEST_PULSE instruction.

10.3.6 Summary

The IEEE 1149.6 standard is an extension of the IEEE 1149.1 standard. The 1149.6 standard must comply with all 1149.1 rules. The 1149.6 logic allows for testing of AC-coupled networks by capturing edges of pulses that are generated by 1149.6 drivers. A special analog test receiver is used to capture these edges. The 1149.6 receiver logic is placed on both inputs of the differential receiver logic. Special hysteresis logic filters out noise and captures only valid transitions. These extensions allow for an equivalent level of testing (to 1149.1) for high-speed digital interconnects.

10.4 EMBEDDED CORE TEST STANDARD (IEEE Std. 1500)

10.4.1 SOC (System-on-Chip) Test Problems

As shown in Figure 10.1, a typical SOC design may contain a large number of cores. The relationship between cores and SOCs appears to be analogous to that between chips and boards/systems; hence, a test architecture similar to a boundary scan should also be highly desirable for SOC testing. Many concepts developed for boundary scan have indeed been applied to SOC testing; however, a major difference exists between core-based SOC testing and chip-based board/system testing. The components in the latter are chips that have been manufactured and tested before they are put on a board; thus, the main test problem for the system integrator is the interconnects between chips on a board and between boards in a system. For SOC testing, all cores are not manufactured before they are integrated into an SOC. It is the responsibility of the system integrator to test all of the cores, as well as the interconnects, after the chips are manufactured. Many test problems arise, as discussed below [Zorian 1997] [Wu 2001].

- **Mixing technologies**—An SOC may contain cores with logic, processor, memory, and analog design/manufacturing technologies. All of these cores must be tested after the SOC is manufactured. It is almost impossible for a system integrator to develop all the tests alone. Assistance from the core providers must be enforced, and a standard way to communicate between the core providers and the system integrator must be established.

- **Deeply embedded cores**—A core may be deeply embedded in a chip. This requires some kind of ***test access mechanism*** (TAM) through which the core can be efficiently accessed and tested. It is preferred that the cores to be integrated have a *plug-and-play* feature under the TAM so as to make system integration manageable.

- **Hierarchical cores**—A core itself may contain some cores in a hierarchical manner. A TAM only for cores at the top level of a hierarchy is insufficient. An efficient and effective hierarchical test structure is needed to test the cores at the lower level of a hierarchy. It is also desired that the plug-and-play feature can be carried out for cores at any hierarchical level to simplify the integration work.

- **Different core providers**—Cores may come from different vendors. The cores available to the system integrator may be soft cores, hard cores, or firm cores. The integrator needs to know how to test these cores after putting them together. Some kind of design and test standard is again essential for the integration purpose.

- **IP protection/test reuse**—Detailed internal structural information of a core is usually unavailable due to IP protection considerations. It is thus desirable to be able to reuse the test provided by the core developer with very limited or no modifications. A standard core test interface and protocol are also essential to address this problem.

- **Higher performance core I/Os than SOC pins**—The clock rate inside a core can be significantly higher than what can be provided from SOC pins; for example, many contemporary CPUs are running multiple-gigahertz clocks while their chip I/Os are still limited to the range of hundreds of megahertz. Test clocks provided from external testers usually cannot support at-speed testing even if the core can be isolated and well accessed through some TAM. Raising the test clock rate using a dedicated phase-lock loop would significantly complicate the design, resulting in unacceptable test costs. In this case, employing normal functional units to create the required at-speed test environment seems to be the only way to achieve efficient, effective, and economic testing.

- **Expensive and inefficient external *automatic test equipment* (ATE)**—The specifications of ATE for testing digital, analog, and memory devices are significantly different. If the SOC testing relies solely on external ATE, then the ATE must be capable of generating/examining all types of test signals and hence will be extremely expensive. Moving some test control or test data generation mechanism into the chip can potentially reduce the use of external ATE and reduce the test cost.

- **Long test application time**—Clearly, long test application time is required if the cores in an SOC are tested sequentially. This will worsen the already acute problem of the extremely high cost of ATE. Moreover, this may seriously affect the time-to-market, resulting in disastrous market loss. Parallel testing or test scheduling is necessary to reduce test time and hence the adverse effects described above.

- **Large test power**—While great efforts have been devoted to low power design, test power issues have only recently gained attention. While parallel testing of as many cores as possible is desired to reduce test time, excess test power resulting in incorrect test results or even damaging the devices under test is possible. A test schedule must be carefully planned so as not to violate any constraint or limit of test power.

- **Testable design automation**—As SOC testing involves many new complicated problems, new DFT insertion tools that solve these problems are highly desirable. These tools should include the automatic generation of standard test circuitry, test architecture, and test plan or schedule. Test pattern generation and formatting are also necessary for test reuse.

Many of the above problems strongly suggest that a standard be established such that core developers and system integrators can communicate efficiently to address issues of SOC testing. Other problems require a good test plan or schedule to carry out actual test procedures as efficiently as possible while maintaining the integrity of the test. The inefficiency and ineffectiveness of the external ATE also have to be dealt with to prevent testing from becoming the bottleneck of SOC industry. In this section, we focus on the standard for core-based testing.

Standard 1500 is a test standard approved by IEEE in 2005 which inherits most of the properties of 1149.1 and can further address many of the SOC test problems described above. The main goal of 1500 is to standardize a core test architecture by defining a core test interface between an embedded core and the system so as to facilitate test reuse for embedded cores through core access mechanisms, provide testability for system interconnect and **user-defined logic** (UDL), and enable core test with **plug-and-play** (PnP) protocols. The IEEE 1500 standard supports both serial and parallel test access mechanisms and provides a rich set of instructions for core and interconnect testing. It also defines features that enable core isolation and protection. A system chip complying with 1500 should contain the following hardware components:

- One standard test wrapper for each core
- Signal sources and sinks for test pattern provision and reception
- On-chip TAMs to connect the wrapper to the sources/sinks

In addition to the hardware components, the 1500 uses a test-specific language called *Core Test Language* (CTL) to communicate information between core providers and core users. This language is now considered part of the 1500 standard, and describing the 1500 wrapper as well as the test data for a reusable core is required.

It should be noted that, similar to boundary scan, the 1500 standard itself only standardizes the core test mechanism for core access/isolation protocols and test mode control. The system-level test access mechanism still must be defined by the system integrator. Also, any test methods inside each core must be defined and implemented by the core providers.

10.4.2 Overall Architecture

The most important feature of the 1500 standard is the provision of a *wrapper* on the boundary (I/O terminals) of each core, thereby allowing the test interface of the core to be standardized and the test commands to be executed. An overall architecture of IEEE 1500 is shown in Figure 10.21, where a system with N cores, each wrapped

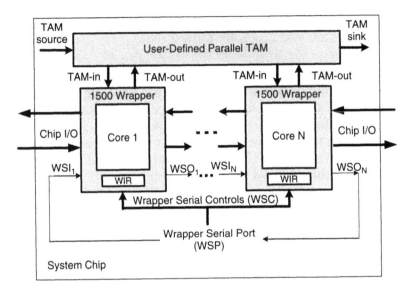

A system overview of the IEEE 1500 standard.

by an IEEE 1500 wrapper, is illustrated. The **wrapper serial port** (WSP) is a set of I/O terminals of the wrapper for serial operations. It consists of the **wrapper serial input** (WSI), the **wrapper serial output** (WSO), and several **wrapper serial control** (WSC) terminals. Each wrapper has a **wrapper instruction register** (WIR), which is used to store the instruction to be executed in the corresponding core. The WSP supports the serial test mode similar to that in a boundary-scan architecture, but without using a TAP controller (*i.e.*, the serial control signals of 1500 are directly applied to the core without the conversion of the TAP controller). This is discussed further in Section 10.4.3.

In addition to the serial test mode, the 1500 standard also supports a parallel test mode by incorporating a user-defined, parallel **test access mechanism** (TAM). Each core can have its own TAM-in and TAM-out ports consisting of a number of data or control lines for parallel test operations. The user-defined, parallel TAM can transport test signals from the TAM-source to the cores through TAM-in and from the cores to the TAM-sink through TAM-out. In Figure 10.22 the interface of a core is highlighted and both parallel and serial data/control signals are indicated; the **wrapper parallel control** (WPC) and **wrapper parallel input** (WPI) signals correspond to the TAM-in port, and the **wrapper parallel output** (WPO) signals correspond to the TAM-out port in Figure 10.21. It should be pointed out that for the 1500 standard the serial ports are mandatory while the parallel ports are optional; however, the parallel interface represents one main difference between 1500 and 1149.1 that leads to a significant time reduction for SOC testing.

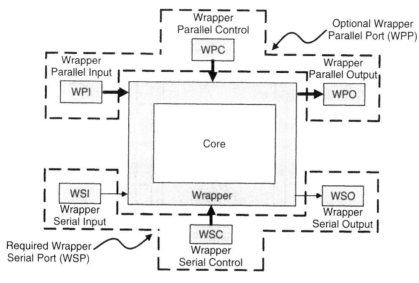

FIGURE 10.22

Test interface of a core wrapper.

10.4.3 Wrapper Components and Functions

Figure 10.23 shows the detailed hardware structure of the IEEE 1500 standard core wrapper, which is comprised of five components, as described below.

1. A *wrapper serial port* (WSP), which consists of a *wrapper serial input* (WSI), a *wrapper serial output* (WSO), and several *wrapper serial control* (WSC) terminals. Similar to the TDI and TDO of 1149.1, the WSI and WSO terminals are used to scan in and scan out wrapper test instructions and data. Both WSI and WSO are mandatory for 1500. The WSC contains six mandatory control terminals (WRSTN, WRCK, SelectWIR, CaptureWR, ShiftWR, and UpdateWR), one optional control terminal (TransferDR), and some optional clock terminals (AUXCKn). These terminals are described next. Some operations enabled by these terminals will be described later.

   - **WRCK**—This mandatory wrapper clock terminal is dedicated to the operation of the 1500 standard functions.

   - **AUXCK*n***—These optional auxiliary 1500 clocks can be used for some implementations of wrapper boundary registers. The "n" indicates the number of these auxiliary clocks that shall be indexed by $1, 2, \ldots, n$. These clocks may be shared with the system clocks. When they are employed, the user must clearly define the timing relation between these signals and WRCK.

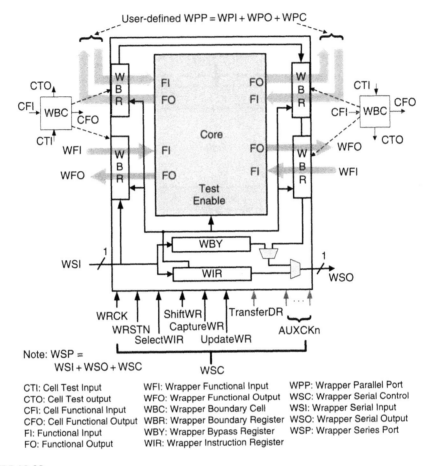

User-defined WPP = WPI + WPO + WPC

CTI: Cell Test Input
CTO: Cell Test output
CFI: Cell Functional Input
CFO: Cell Functional Output
FI: Functional Input
FO: Functional Output

WFI: Wrapper Functional Input
WFO: Wrapper Functional Output
WBC: Wrapper Boundary Cell
WBR: Wrapper Boundary Register
WBY: Wrapper Bypass Register
WIR: Wrapper Instruction Register

WPP: Wrapper Parallel Port
WSC: Wrapper Serial Control
WSI: Wrapper Serial Input
WSO: Wrapper Serial Output
WSP: Wrapper Series Port

■ **FIGURE 10.23**

Serial test circuitry of the 1500 standard for a core.

- **WRSTN**—This mandatory wrapper reset terminal resets the wrapper circuitry and puts the wrapper into the normal system mode when asserted. The wrapper bypass instruction, which is similar to the bypass instruction in 1149.1, shall be automatically loaded into the wrapper instruction register whenever WRSTN is asserted.

- **SelectWIR**—This mandatory terminal is used to determine whether an instruction or data type of operation is to be performed. When SelectWIR = 1, WIR is selected and connected between WSI and WSO; otherwise, some data register(s) is connected between WSI and WSO.

- **CaptureWR**—This mandatory terminal is used to enable the capture operation for the selected register(s).

- **ShiftWR**—This mandatory terminal is used to enable the Shift operation for the selected register(s).

- **UpdateWR**—This mandatory terminal is used to enable the Update operation for the selected register(s).

- **TransferDR**—This optional terminal enables the Transfer operation for selected register(s) that implement the Transfer function.

Similar to 1149.1, correct timing is also critical for proper operation of the 1500 circuitry. In the 1500 standard, it is required that the SelectWIR, ShiftWR, CaptureWR, and TransferDR be sampled at the rising edge of WRCK and the UpdateWR be sampled at the falling edge of the WRCK.

2. A **wrapper parallel port** (WPP), which consists of user-defined *wrapper parallel input* (WPI) terminals, *wrapper parallel output* (WPO) terminals, and *wrapper parallel control* (WPC) terminals. All of these terminals are optional. A WPP may include the clock terminals of WSC (the WRCK and AUXCK terminals) but may not replace other WSC terminals.

3. A **wrapper instruction register** (WIR), which is used to store the instruction to be executed, similar to the instruction register in 1149.1. The WIR is unconditionally selected whenever the SelectWIR of WSC is set to 1, regardless of the current wrapper instruction and selected wrapper/core data registers. It is implemented using a two-stage design such that the shifting of a new instruction will not interfere with the current instruction. The two main differences between the instruction registers of 1149.1 and 1500 are described below.

   - First, because there is no finite-state machine in 1500, the control signals provided by the WIR are derived from both the current wrapper instruction and the current states of the signals connected to the WSC terminals. Figure 10.24 shows the circuitry design of a WIR. It consists of a shift stage and a decode/update stage. Three sets of essential control signals (DR_Select, WBY_Cntrl, and WBR_Cntrl) are used to select the appropriate register(s), control the operations of the wrapper bypass register, and control the operation of the wrapper boundary register, respectively. Other control signals, such as those for other **wrapper data registers** (WDR), those for **core data registers** (CDR), and those for the core itself, can also be generated. Note that these control signals are defined in a quite flexible way and the user may use different names for these signals.

   - The second difference is that, in addition to the serial shift-in operation of a new instruction through the WSI–WSO chain, the 1500 standard optionally provides a parallel load mode, as shown in Figure 10.24. This permits the WIR to capture test control information directly (remember that the WSC terminals are also inputs to WIR) or to capture data that can be used to test the WIR or other 1500 circuitry.

4. A **wrapper bypass register** (WBY), which bypasses test signals similar to the bypass register in 1149.1. The WBY should be selected and connected between WSI and WSO when WRSTN is asserted or when the current instruction is the

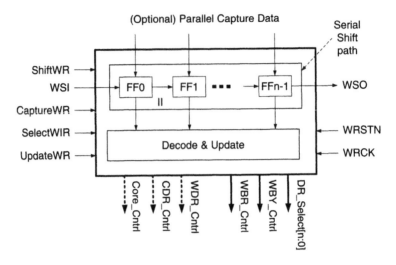

■ **FIGURE 10.24**

WIR circuitry design.

wrapper bypass instruction. It is also the default register to be put between WSI and WSO when no other wrapper data register is selected or when an unused wrapper instruction opcode appears in the WIR.

5. A ***wrapper boundary register*** (WBR), which consists of ***wrapper boundary cells*** (WBCs) similar to the boundary-scan register in 1149.1. Each WBC has four data terminals: ***cell functional input*** (CFI), ***cell functional output*** (CFO), ***cell test input*** (CTI), and ***cell test output*** (CTO), as shown in Figure 10.23. The functional modes and operation events of WBR are described next.

Functional modes of WBR—Four modes are defined as follows:

- *Normal mode*—The WBR is transparent to the system and the core executes its normal functions.

- *Inward facing mode*—The test access is for the core itself; that is, the functional inputs of the core are controlled and the functional outputs of the core are observed by the WBR (see Figure 10.23).

- *Outward facing mode*—The test access is for external circuitry; that is, the wrapper functional outputs and the wrapper parallel outputs (of WPP) are controlled by the WBR, and the wrapper functional inputs and wrapper parallel inputs are observed (or captured) by the WBR (see Figure 10.23).

- *Nonhazardous (safe) mode*—The functional inputs of the core and wrapper functional outputs are controlled by WBR to a safe state.

Operation events of WBR and WBC—Five events (Shift, Capture, Update, Transfer, and Apply) are supported by the WBR (or WBC) in 1500. We will use "bubble" diagrams to help illustrate these events. The four symbols shown in Figure 10.25 are used to represent the structures of all WBCs, where a circle represents a storage element, an arrow represents a data path, a vertical line together with two or more input arrows and one output arrow represents a decision point, and two arrows emerging from a single point represent two data paths from the same signal source. Each circle may be characterized by one or more characters from the set of S, C, U, T, and F to indicate that the corresponding storage element is responsible for the Shift, Capture, Update, Transfer, and any Functional event, respectively.

Several bubble diagrams representing different types of WBCs provided in the IEEE 1500-2005 standard document are provided in Figure 10.26, where their names are shown under each WBC. The events supported by each WBC are indicated by the characters in the storage elements; for example, Figure 10.26a shows a simple WBC that contains only one storage element and supports only the Shift and Capture events, while Figure 10.26e shows a WBC that contains three storage elements, two of which are used for the Shift and Transfer events and the third one for the Update, Capture, and Functional events. Note that Figure 10.26d corresponds to a boundary-scan cell shown in Figure 10.6. Also note that Figure 10.26g does not support the Capture event. This may represent a WBC for a core terminal that can be exempted from being wrapped, such as a clock or a reset terminal of the core. These five events are described below:

- *Shift event*—A mandatory event whereby the data stored in the WBR shift path are advanced one storage position closer to the WBR's serial test output. The data present at the WBR's serial test input are loaded into the shift path storage element closest to the WBR's input.

- *Capture event*—An event whereby the data present on the CFI or CFO of a WBC are captured and stored in a storage element within the WBC; for example, Figures 10.26a shows a WBC that captures data from CFI, while Figure 10.26h shows a WBC that can capture data from both the CFI and CFO. Unless the WBC is used for a terminal that can be exempted from being wrapped such as a clock or reset terminal, this event is mandatory for all WBCs.

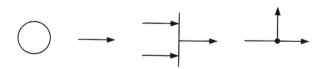

■ FIGURE 10.25

The four symbols used in bubble diagrams.

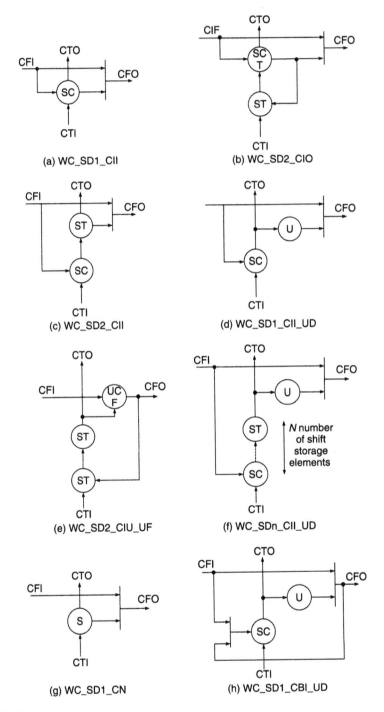

■ FIGURE 10.26

Some typical WBCs represented by bubble diagrams.

- *Update event*—An optional event whereby data stored in a WBC's shift path storage element closest to CTO are loaded into an off-shift-path storage element of the WBC. Note that this event is optional in 1500; for example, the WBCs in parts a, b, c, and g of Figure 10.26 do not support this event.

- *Transfer event*—An optional event that either moves data to the storage element closest to the CTI of a WBC, if the data stored by the Capture event are not on this storage element (Figure 10.26b,e), or moves the data one storage position closer to the CTO (Figure 10.26c,e,f). There are two purposes for the Transfer event. First, in order to preserve as many capture values as there are storage elements in the shift path, captured data should enter a WBC's shift path via the storage element closest to CTI. Second, to provide sequential stimuli data such as those required for delay testing, Transfer moves data through the shift path so that each bit may be sequentially loaded into the update storage element and then applied to the CFO of the WBC.

- *Apply event*—A derivative event inferred from the operation of the other four events (Shift, Update, Transfer, and Capture) whereby test data become active and effective as test stimuli. While the wrapper is in inward (outward) facing mode, the Apply event causes test data to be applied from input (output) cells onto the functional inputs of the core (WBR's functional outputs). The test data are the data stored in the shift path storage element closest to CTO unless the Update event is supported, in which case the test data shall be the data stored in the off-shift-path storage element by the Update event. It should be noted that the Apply event is a virtual event inferred from other events and hence is not specifically represented in the bubble diagrams shown in Figure 10.26.

In the following we give two examples to illustrate how the signals generated by the WIR control the operations of the WBR and WBCs.

Example 10.1

In Figure 10.27, we provide an example of a WIR interface to the WBY, WBR, WDRs, and CDRs of a core. The WSO of the core is connected to the serial output of WIR (*i.e.*, WIR_WSO) or the serial output of one of the data registers—WBY, WBR, CDRs, or other WDRs (*i.e.*, DR_WSO) under the direct control of the **SelectWIR** signal. The serial output (DR_WSO) of the data registers is in turn controlled by the **Dr_Select[$n...0$]** generated by the WIR. All other signals are applied to the data registers as described in the next example.

Example 10.2

In Figure 10.28, we show the schematic diagram of the WBC WC_SD2_CIO, the bubble diagram for which is given in Figure 10.26b. In this design, the **Mode** signal is generated by the WIR and is used to determine whether the cell is in the normal or test mode operation. The **IO_FACE** signal, also generated by the WIR, is used

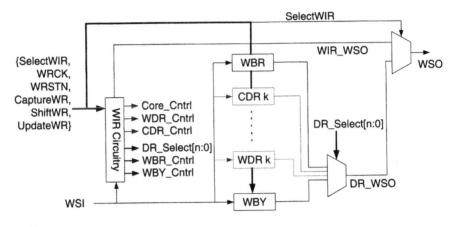

■ **FIGURE 10.27**

Example of WIR interface with WBY, WBR, WDRs, and CDRs.

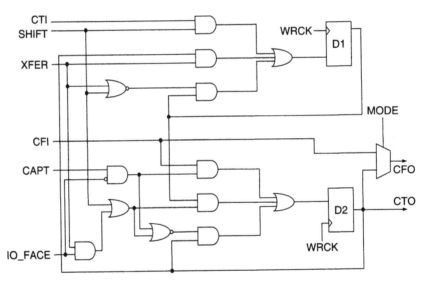

■ **FIGURE 10.28**

Schematic diagram of WBC WC_SD2_CIO.

to determine whether the core is in the inward or outward facing configuration. Both of these two signals are **WBR_Cntrl** signals, as they are generated for the WBCs. The **SHIFT**, **CAPTURE**, and **XFER** signals, which are derived directly from the ShiftWR, CaptureWR, and TransferDR signals, determine which event is being executed. D1 and D2 flip-flops correspond to the storage elements closest to CTI and CTO, respectively, as shown in Figure 10.26b. As D1 supports two events (S and T in the bubble diagram), its input comes from three sources. When **SHIFT** is asserted, it receives data from CTI; when **XFER** is asserted, it receives data from the output of D2; and when neither of these two signals is asserted, it holds its current state. The D2 flip-flop supports three events (S, C, and T). It participates in the Shift event whenever **SHIFT** is asserted; it captures the data at CFO when **CAPT** is asserted and **IO_FACT** = 0 (the instruction is an inward facing one); and it receives data from the D1 flip-flop when **XFER** is asserted and **IO_FACT** = 1. When neither of these events is executed, it holds its state.

10.4.4 Instruction Set

The IEEE 1500 standard has a richer instruction set than 1149.1, as many instructions have additional parallel options. In this section, all instructions defined in 1500-2005 are described. These instructions follow the following naming convention:

W<S/P/H>_<Command>{_<Configuration>}—(*e.g.*, WS_INTEST_RING), where the first "**W**" is the preface for all 1500 instructions. **S**, **P**, and **H** represent the serial, parallel, and hybrid test modes, respectively. A serial instruction uses WSI, WSO, and WSC only. A parallel instruction mainly uses WPI, WPO, and WPC and must have only the WBY between WSI and WSO. A hybrid instruction involves the use of both serial and parallel ports. An instruction simultaneously using parallel ports and configuring registers other than WBY between WSI and WSO is considered to be a hybrid one. In the following description, an "x" may be used for some instructions. In these cases, the "x" should be replaced by S, P, or H. **Command** represents the operation of the instruction (*e.g.*, EXTEST or BYPASS). **Configuration** describes the test configuration selected by the instruction. In 1500-2005, this field only appears in the WS_INTEST_SCAN and WS_INTEST_RING instructions where the "SCAN" in WS_INTEST_SCAN indicates that the internal scan chain is included in the WSI–WSO chain in addition to the WBR, while the "RING" in WS_INTEST_RING indicates that only the WBR is between WSI and WSO. Next, we describe these instructions in more detail with the help of Figure 10.29.

- **WS_BYPASS:** (Figure 10.29a)—This mandatory instruction is used to bypass the test information and enables the normal functional configuration of the wrapper. The WSI–WSO connection only passes through the WBY to allow rapid data movement.

- **WS_EXTEST** (Figure 10.29b)—This mandatory instruction allows testing of core-to-core interconnects and the off-chip *user-defined logic* (UDL) using a single scan chain configuration. Only the WBR is connected for serial access between WSI and WSO during the Shift operation. The test data can be loaded into the WBR using a Wx_PRELOAD instruction prior to the WS_EXTEST instruction. The WBR shall be operated in the output facing mode, meaning that the values present at the CFOs of the WBCs can be applied to the external UDL or interconnects and the test results from these external circuits can be

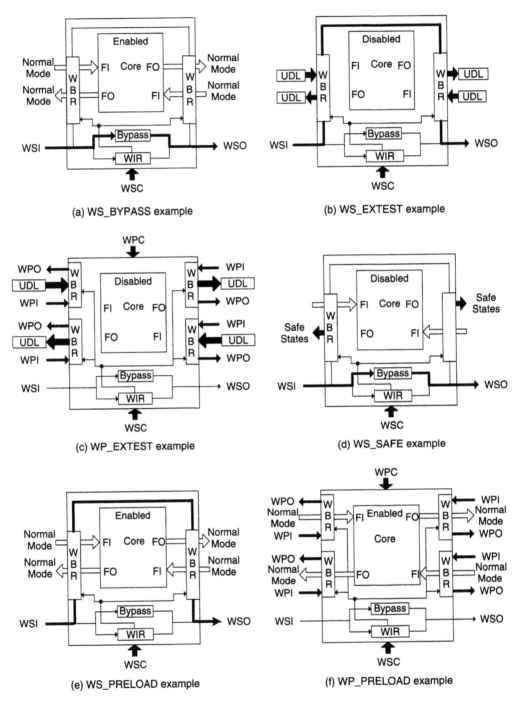

(a) WS_BYPASS example

(b) WS_EXTEST example

(c) WP_EXTEST example

(d) WS_SAFE example

(e) WS_PRELOAD example

(f) WP_PRELOAD example

■ **FIGURE 10.29**

Test instructions for the 1500 standard: (a) WS_BYPASS, (b) WS_EXTEST, (c) WP_EXTEST, (d) WS_SAFE, (e) WS_PRELOAD, (f) WP_PRELOAD, (g) WS_CLAMP, (h) WS_INTEST_RING, (i) WS_INTEST_SCAN.

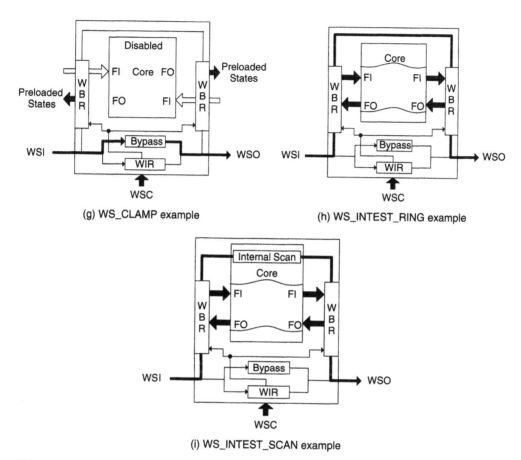

(g) WS_CLAMP example

(h) WS_INTEST_RING example

(i) WS_INTEST_SCAN example

■ **FIGURE 10.29** *(Continued)*

captured to the WBR through the CFIs of the WBCs. The execution of this instruction is similar to that of the EXTEST instruction of 1149.1.

- **WP_EXTEST** (Figure 10.29c)—This optional instruction allows testing of core-to-core interconnects and the off-chip UDL using a multiple scan chain configuration. The WBR can be divided into multiple segments (multiple scan chains) and the test data can be loaded into WBR using multiple WPI–WBR–WPO paths under the control of a user-defined WPC. Other than the parallel data transmission, this instruction mimics the WS_EXTEST instruction.

- **Wx_EXTEST**—When the "x" in this user-specified instruction is replaced by S (P), this instruction is the same as the WS_EXTEST (WP_EXTEST) instruction and follows all rules defined in the 1500 standard. When the "x" is replaced by H, the instruction becomes a hybrid one, which may allow more flexible capabilities to carry out the EXTEST operation using a mixed series and parallel configuration.

- **WS_SAFE** (Figure 10.29d)—This optional instruction provides a straight-forward way for the core integrator to put a wrapper into a static and safe state. If this instruction is present, the wrapper functional outputs of the WBR shall be hardwired to constant values that have been predetermined when the cores are wrapped. Note that the safe state can also be achieved by using a preload instruction followed by a WS_CLAMP instruction as described below.

- **WS_PRELOAD** (Figure 10.29e)—This conditionally required instruction allows test data to be serially loaded into the WBR without interfering with the operation of cores or UDL attached to the WBR. It is mandatory when the WBR is composed entirely of cells with a shift path supporting the Shift operation that keeps the WFO terminals static. Similar to the PRELOAD instruction in 1149.1, this instruction can be used to load test data for external testing prior to the WS_EXTEST instruction to prevent possible indeterminate states at the wrapper functional outputs when loading the WS_EXTEST instruction (see Section 10.2.6).

- **WP_PRELOAD** (Figure 10.29f)—This optional instruction allows the WBR to be divided into multiple segments, and all segments can be loaded with test data simultaneously. This instruction is typically utilized before some defined instructions such as WP_EXTEST. It is usually preferred over the WS_PRELOAD due to better data transfer efficiency. One can expect that with the WP_PRELOAD and WP_EXTEST instructions, plus the parallel loading capability of the WIR, interconnect and UDL testing can be carried out quite efficiently.

- **WS_CLAMP** (Figure 10.29g)—This optional instruction allows the state of signals driven by the wrapper functional outputs to be determined by the data prestored in the WBR. The test control for this instruction shall be provided by the WSC, and the WBY shall be connected for serial access between WSI and WSO. The core with this instruction shall be put into a quiet mode (*e.g.*, reset or clock off). It is the responsibility of the user of this instruction to make sure the wrapper functional outputs of the WBR are in the desired state (*e.g.*, via Wx_PRELOAD) when this instruction is used.

- **WS_INTEST_RING** (Figure 10.29h)—This optional instruction allows single-step testing of the core circuitry with each test pattern and response being shifted through the WBR, where the single-step testing of the core means that the core will move one step forward in its operation each time shifting of the WBR is completed and applied. One such example is the single-step testing for a CPU. The WBR shall be the only register connected between WSI and WSO and shall be in the inward facing mode during this instruction. The test pattern can be applied to the core and the response can be captured by the WBR by enabling the Apply and the Capture events, respectively.

- **WS_INTEST_SCAN** (Figure 10.29i)—This instruction is the same as the WS_INTEST_RING instruction except that an internal scan chain of the core can be concatenated with the WBR to form a single scan chain. This allows more inside access to the core and can potentially provide better fault coverage than the WS_INTEST_RING instruction.

- **Wx_INTEST**—For 1500, at least one INTEST instruction is required. This can be a WS_INTEST_RING, a WS_INTEST_SCAN, or a user-defined instruction with the name of Wx_TEST, where "x" can also be S, P, and H. Note that in the 1500-2005 standard, parallel INTEST instructions are not defined; however, similar to the WP_EXTEST instruction, a WP_INTEST instruction would be highly beneficial in reducing test time if a large number of test access lines are available for the WPP.

10.4.5 Core Test Language (CTL)

The *core test language* (CTL) is a language for capturing and expressing test-related information for reusable cores. It standardizes the description of all the information that the core providers must give to system integrators or design automation tools so a complete test for the embedded cores, interconnects among cores, and any user-defined logic around the cores can be created. CTL is an extension of the IEEE 1450 *Standard Test Interface Language* (STIL) standard [IEEE 1450-1999] and is now defined in the IEEE draft standard P1450.6 (where "P" indicates proposed). While STIL is a language mainly for representing IC test patterns and waveforms, CTL provides additional information for core-specific controls to configure a core and its surrounding logic, as well as the requirements and constraints on the implementation of chip-level test interfaces for the core. Within CTL, one can create enough information at the boundary of the core to allow for successful (1) instantiation of a wrapper, (2) mapping of the core terminals to wrapper terminals, (3) reuse of the core test data, and (4) testing of the user-defined logic and wiring external to the core.

In addition to the basic information described by standard 1450 constructs, such as *Signals*, *Patterns*, *Macros*, and *Timing*, the CTL description of a design can be constructed by an environment statement consisting of various blocks that describe global information or different operating modes of the design. Following is what a CTL description of a design having two operation modes would look like:

```
STIL 1.0 {
}
Signals { // defines each of the signal names of the core
}
Patterns { // contains the parallel and scan data for testing the core
}
Timing { // defines the waveform and corresponding timing on each
         // signal for each parallel or scan pattern
}
MacroDefs { // contains the protocol for applying test data to the core or
            // chip
}
...
```

```
Environment {
    CTL {
        // Common information of all CTL blocks within the Environment
    }
    CTL mode1 {
        // Information about mode 1
    }
    CTL mode2 {
        // Information about mode 2
    }
}
```

The blocks before the Environment block are standard STIL descriptions of the core I/O signals, test patterns, timing, and protocol to apply the patterns. Note that the patterns and the protocol to apply the patterns are separated in CTL. The core integrator only has to modify the latter without requiring significant changes to the test patterns when integrating the core into a system. The Environment block in this example contains one unnamed and two named blocks. The unnamed block is used to describe global information that will be used by other blocks. Each named block describes a test mode by specifying static attributes on the boundary signals of the core and/or the sequence and pattern information for the test mode. These attributes and information are described in the following format:

```
CTL  mode_name {
    TestMode  test_mode_name;
    Internal {
        // Describe information about the core itself
    }
    PatternInformation {
        // Describe pattern information for the mode
    }
    External {
        // Describe the outside environment expected for test integration
    }
    ScanInternal chain_name {
        // Describe information of scan chain within the design
    }
    Relation {
        // Describe the relation between signals or/and signal groups
    }
    TestResourceConstraints {
        // Describe the constraints on the testability of the design
    }
    CoreInstance {
        // Describe the internal level of hierarchy of the design
    }
}
```

Each named block in the Environment block shall specify its test mode name with the **TestMode** parameter. The **Internal**, **PatternInformation**, and **External** blocks are three essential blocks to specify the test mode as described below:

- **Internal**—This block is used to describe the internal characteristics of the core signals so the core integrator can determine the pertinent test information for each terminal of the core without having to access the detailed design information. Examples of these characteristics include wrapper type, signal names, time accuracy requirements, and electrical characteristics such as analog or digital.

- **PatternInformation**—This block is used to specify the purpose of each of the test patterns provided and the test mode necessary for the execution of each pattern. The fault model used, the number of faults considered, and the fault coverage achieved can also be given in this block.

- **External**—This block describes the external characteristics that are expected from the perspective of the core boundary. Examples include connections to chip pins (input, output, or bidirectional), connections to another named core, connections to TAM, and connections to UDL.

Other information specified in the CTL blocks includes the following:

- Information about the scan chains inside the core, which is described within the **ScanInternal** block

- Relations between signals and signal groups within the scope of the CTL, which are defined in the **Relation** block

- Internal level of reference to a hierarchy design, which is provided in the **CoreInstance** block

- Test resource constraints, such as maximum scan length, maximum run time, and maximum power consumption during testing, which are specified in the **TestResourceConstraint** block

The IEEE 1500-2005 standard requires that the CTL description must be used for any design with the 1500 wrappers. Several examples of CTL to describe the various test modes using 1500 test instructions are provided in the standard document [IEEE 1500-2005]. It is recommended that the reader refer to [Marinissen 1999], [Kapur 2001], and [IEEE P1450.6-2001] for more comprehensive descriptions of CTL.

10.4.6 Core Test Supporting and System Test Configurations

As mentioned in Section 10.4.1, unlike the chip-to-board relation in 1149.1, where chips are tested before being mounted onto a board, it is the responsibility of the test integrator to test all cores in an SOC. Fortunately, the 1500 standard does provide an efficient solution for testing the cores by allowing parallel test access with the user-defined WPP. Thus, the development of core testing in 1500 is easier than in 1149.1. For example, one can use the WPP to support the testing of a core with multiple scan chains by directly connecting some WPI, WPO, and WPC to the scan inputs, scan outputs, and scan enables of the core, respectively. Similarly, one can also provide necessary interface signals directly to a core to carry out the BIST operation of the core. The main problem here for 1500 is how to transfer test data between system test sinks/sources and the wrappers (see Figure 10.21).

Because the serial mode of 1500 resembles that of 1149.1, the test configurations described in Section 10.2.9 also apply to a 1500-based system; however, because parallel access is allowed, it is beneficial to make use of the WPP to execute the core testing more efficiently. Figure 10.30 shows a general test structure with both WSP and WPP indicated. Because the WSP is mandatory and user-defined test instructions can be loaded into the WIR to configure the core into any desired test structure, it is advantageous for the user to use the WSI and WSC to enable the cores to be tested through an ENA (enable) signal for simple test control of the whole chip. This, of course, is not mandatory, and users may design their own test enable protocols.

There are several user-defined parallel TAM configurations. Figure 10.31 shows four such configurations: *multiplexed, daisy-chained, direct-access* (or distribution-based), and *locally controlled* [Aerts 1998] [Waayers 2005]. In the multiplexed architecture, only one core can get access to the available SOC TAM wires at a time, hence interconnect testing between cores is not easy to achieve. In the daisy-chain architecture, all cores can access all TAM wires during a test session and each core can be tested sequentially; however, if only a subset of cores is to be accessed simultaneously, a mechanism to bypass parallel test data is required (which may not be easy to design). Even if a parallel bypass mechanism is provided, a significant increase in test time per test pattern will be inevitable, as analyzed in [Aerts 1998]. In the direct-access architecture, the available TAM wires are distributed

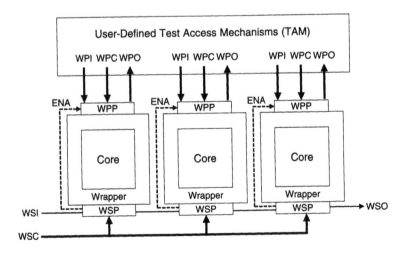

■ **FIGURE 10.30**

General parallel TAM structure.

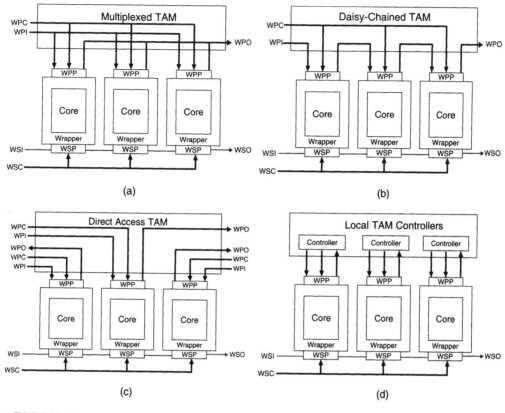

■ **FIGURE 10.31**

Various architectures for parallel TAM.

over the wrapped cores. The optimal number of TAM wires to be assigned to a core depends on test requirements of the core (*e.g.*, ATPG results and scan chain structure), and a test plan considering the requirements of all cores may be necessary in order to minimize the total test time. In the locally controlled TAM, a dedicated controller for each core is used, which allows the test procedures for all cores to be carried out simultaneously. This, however, requires tremendous hardware overhead.

Clearly, each of the above configurations has its own pros and cons. As a result, a combination of these configurations and TAM reconfigurations during testing is usually employed in a real system. For example, the Test Bus architecture presented in [Varma 1998] uses both the multiplexing and the distribution configurations. The TestRail proposed in [Marinissen 1998] can be connected in several ways following the daisy-chain configuration, the distribution-based architecture, or a combination of the two. Many test architecture design algorithms have been proposed to minimize the overall SOC test time for a given number of TAM wires by determining the number of distinct TAMs, their widths, and the assignment of these

TAMs to cores. For example, optimization for Test Bus architectures can be found in [Ivengar 2002], [Koranna 2002], and [Larsson 2002], whereas optimization of TestRail architectures is described in [Goel 2002] and [Goel 2003a]. Further discussion on the optimization of the wrapper and TAM design can be found in [Goel 2003b], [Sehgal 2004], and [Waayers 2005].

10.4.7 Hierarchical Test Control and Plug-and-Play

Hierarchical design is a natural way to design a complex system starting from the behavioral or architectural level of description. Testing for a hierarchical core containing a number of internal cores may be achieved by flatting all cores during testing. This, however, may require significantly more routing area because the routing for testing is quite different from that for normal operation. On the other hand, the 1500 standard has been designed to accommodate the easy *plug-and-play* (PnP) of cores. Unless some test clocks (AUXCKn) and WBR are to be shared by some system functional operation, PnP of a core into a SOC should be readily applicable with the 1500 wrapper implemented in a flat design [IEEE 1500-2005]. To retain the PnP feature for a hierarchical core, however, is a nontrivial problem, as one has to enable the cores to be tested in addition to transferring test data between the SOC and the cores, preferably through the hierarchy. In this section,we describe three test architectures that deal with this problem [Benabdenbi 2000] [Lee 2000] [Li 2002].

In [Benabdenbi 2000], a test architecture called **Core Access Switch** (CAS) is proposed, where each core is wrapped by a 1500 wrapper and a CAS block is allocated to each core as shown in Figure 10.32. These CAS blocks are connected in series with N test signals, and a test controller is used to provide the N test signals to all CAS blocks (CAS 1, CAS 2, ...). Each CAS block can be operated in the configuration, bypass, or test mode, as shown in Figure 10.33. In the beginning of a test session, all CAS blocks are in the configuration mode, in which test instructions are sent to all CAS blocks and each CAS is configured to the required configuration

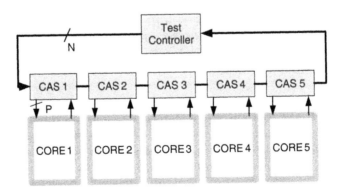

■ **FIGURE 10.32**

The Core Access Switch (CAS) architecture.

according to the instruction it receives. When testing is carried out, each core is either in the bypass mode or in the test mode, as shown in Figure 10.33.

For the test mode, CAS can support four test types according to the test requirements of the cores, as shown in Figure 10.34. Each CAS receives N test signals and converts them into P signals in one of the following ways:

- For a core with k multiple scan chains, $P = k$ and test data are provided to the core through the k chains simultaneously.

- For a BISTed core, P can be as small as 1.

- For a core to be tested using external source and sink, P depends on the nature of the source and sink.

- For a hierarchical core, the CAS technique allows the internal cores to be CASed, and in this configuration P is equal to the width of the internal test bus.

In the last option, the CAS structure allows hierarchical cores to be tested with a single test controller. This poses a problem when the PnP feature in a hierarchical structure is considered. Referring to Figure 10.34, if the entire circuit is to be used

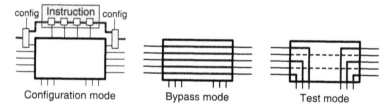

Configuration mode   Bypass mode   Test mode

■ FIGURE 10.33

Different functional modes of CAS.

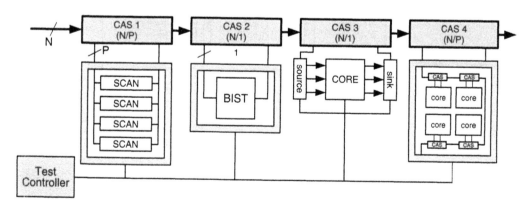

■ FIGURE 10.34

Various types of test supporting using the CAS structure.

as a core in a plug-and-play manner, the test controller in the circuit must also be included in the inner core; thus, one has to define how two test controllers at different hierarchical levels can cooperate in order to complete all test procedures. This appears to be quite difficult in the CAS structure. For the CAS design to achieve PnP in a hierarchical manner, no cores should be equipped with a test controller except for the core at the highest level.

In [Lee 2000] a hierarchical test architecture that does allow PnP of a hierarchical core is proposed. As shown in Figure 10.35, this architecture supports cores wrapped by the 1149.1 (JTAG) and 1500 (CTAG) standards. In addition, a hierarchical core can be plugged and played if it is wrapped by a proposed hierarchical core (H-core) wrapper. The H-core wrapper contains a *center test controller* (CTC), which consists of an 1149.1 TAP controller; a *hierarchical test controller* (HTC); and a programmable switch, as shown in Figure 10.36. The I/O of a hierarchical core contains the 5 required 1149.1 signals and 6 other signals (a total of 11 extra signals). The 6 extra signals are mainly used to enable cores through the hierarchy and provide test data and control signals similar to 1149.1 to the cores at the lower level.

The CTC is the mechanism used to select the cores to be tested and to distribute test signals to the selected cores. If a JTAG or CTAG core inside a hierarchical core is to be tested, then the upper-level controller will send the TAP control signals (TCS, including TCK, TRST, and TMS) to the hierarchical core. The CTC of the core will either distribute the signals directly to the JTAG core or convert the signals to *wrapper control signals* (WCSs) that are compatible with the CTAG core via the

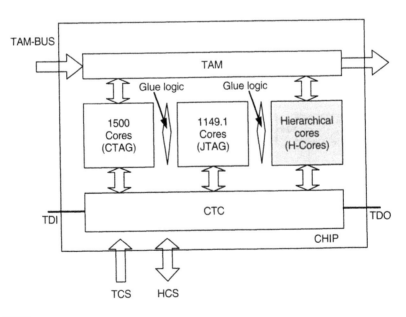

■ **FIGURE 10.35**

A hierarchical test architecture supporting the plug-and-play feature.

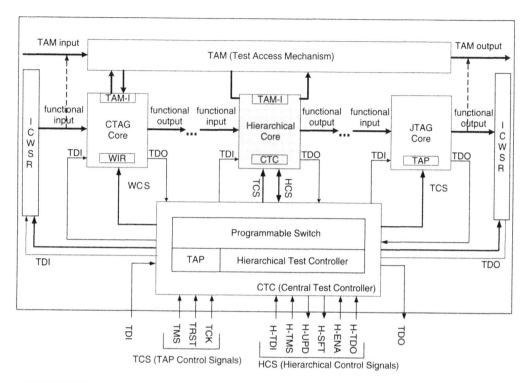

■ **FIGURE 10.36**

Detailed I/O and CTC of the hierarchical test architecture.

TAP controller of the CTC. If a hierarchical core inside a core is to be tested, then the upper-level controller will send the *hierarchical control signals* (HCSs) to the core and the CTC of the core will distribute the signals directly to the hierarchical core. The programmable switch block contains switches controlled by the HTC that will determine which cores are to be tested next and then set up the appropriate connections. The extra signals provided to the hierarchical cores allow the enabling of a core in a hierarchy directly from another core in an upper or lower level. This can save a lot of setup time when a new core in a hierarchy is to be tested as it is not necessary to send the new setup signals through cores outside the hierarchy. If a direct enabling of a core in a hierarchy is not considered, then these extra signals may not be necessary and fewer control signals can be used for implementing the PnP feature for a hierarchical core.

In Figure 10.37, a hierarchical test architecture with only five extra I/Os is proposed [Li 2002]. A *hierarchical test manager* (HTM) is used to generate the test signals required by a hierarchical core so as to handle the test operations at a lower level. Its upstream I/Os are compatible with the IEEE 1149.1 standard; that is, the I/Os consist of the serial test control signals (denoted as TCS_UP and which include TCK_UP, TMS_UP, and TRST_UP) and the serial test data signals (TDI_UP and TDO_UP). Its downstream I/Os consist of the 1500 control signals (PCS), TCS_DN

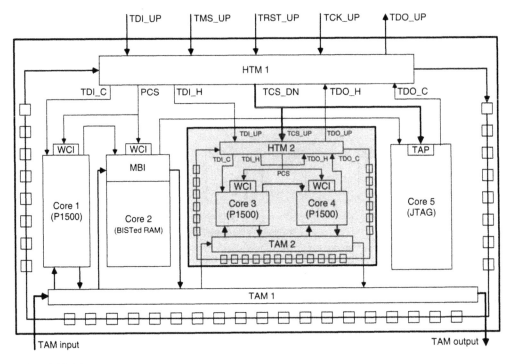

TDI_UP　TMS_UP　TRST_UP　TCK_UP　TDO_UP

■ **FIGURE 10.37**

A hierarchical test architecture with I/Os compatible with 1149.1.

(including TCK_DN, TMS_DN, and TRST_DN), the serial data I/Os (TDI_H and TDO_H) for the HTMs at the next level, and the serial data I/Os (TDI_C and TDO_C) for the cores at the same level. The TAM provides parallel test data transport capacity for the 1500 and BISTed memory cores.

When a hierarchical core is to be tested, the TCS_DN of the upper-level HTM is encoded into the various required test signals (PCS and TCS_DN) of the lower-level HTM. Also, the TDI_H (TDO_H) of the upper-level HTM is connected to TDI_C (TDO_C) or TDI_H (TDO_H) of the lower-level HTM, according to the types of cores to be tested. With this architecture, the hierarchical core test problem can be solved with a core I/O that is compatible with the 1149.1 standard.

10.5　COMPARISONS BETWEEN THE 1500 AND 1149.1 STANDARDS

While 1500 is primarily targeted at core testing and 1149.1 at board-level testing, the 1500 architecture was designed to allow interface compatibility with the IEEE 1149.1 *test access port* (TAP) controller from the beginning. The wrapper's WSC interface signals can indeed be generated by the IEEE 1149.1 TAP controller except for the optional TransferDR signal. Thus, while the IEEE 1500 standard does not require or suggest it, the WSC interface may be controlled by an IEEE

TABLE 10.2 ■ Comparison between 1500 and 1149.1

	1149.1	**1500**
Main objective	Board-level testing	Core-based testing
Parallel mode	No	Yes
Extra data/control I/Os	Four mandatory (TMS, TCK, TDI, TDO) + one optional (TRST)	Two mandatory (WSI, WSO) + six mandatory WSC + one optional (TransferDR) + optional AUXCKn(s) + optional WPP
FSM (TAP controller)	Yes	No
Transfer mode	No	Optional
Latency between Shift, Capture, and Update operations	Yes (*e.g.*, two and a half cycles between Update and Capture)	No (due to the direct control of WSC)
Mandatory instructions	Four (EXTEST, BYPASS, PRELOAD, SAMPLE)	Two (WS_EXTEST, WS_BYPASS) + one Wx_INTEST + one conditionally required WS_PRELOAD

1149.1 TAP controller if the system integrator of an SOC wishes to do so in order to allow access to IEEE 1500 wrappers via the dedicated TAP pins on the SOC; however, there does exist a major difference between 1149.1 and 1500. With the direct application capability of the 1500 WSP protocol, delay testing can be achieved via executing a Capture event immediately after an Update or Transfer event, while the two and a half cycles of IEEE 1149.1 *test clock* (TCK) latency between the TAP's UpdateDR and Capture-DR states limit the ability of the TAP protocol to execute some delay tests. Table 10.2 lists the main differences between the 1500 and 1149.1 standards.

10.6   CONCLUDING REMARKS

This chapter focused on the 1149.1, 1149.6, and 1500 test standards. Test architectures to support these standards were also discussed. Currently, boundary scan is widely used throughout the industry; most commercial computer-aided test tools now provide automatic synthesis capability for boundary-scan design; the 1500 standard is new but is becoming popular in the SOC paradigm. Given the pervasiveness of the 1149.1 TAP, the expected popularity of 1500, and the ubiquity of internal scan and BIST test features, it seems natural to build upon the foundation of a boundary scan and 1500 to support TAP or wrapper-based access to internal chip/core test features in a standardized manner. A framework for the extension of the boundary scan standards has been launched by a working group called the Internal JTAG (IJTAG-IEEE P1687) [Rearick 2005].

Issues that were discussed but not fully addressed in this chapter include long test time, high test power, and the inefficiency of ATE. Many test scheduling algorithms that aim at minimizing test time for an SOC under various constraints such as limited TAM width, maximum allowable test power, test execution precedence, etc., have been proposed [Larsson 2002] [Ivengar 2002] [Zou 2003] [Rosinger 2005]. Recently, the concept of a test platform that makes use of the resources of an SOC such as embedded processors, memories, and bus structures to carry out on-chip, at-speed testing, has also been proposed to deal with the inefficiency problem of ATE [Huang 2001] [Tsai 2001] [Krstic 2002] [Tehranipour 2003] [Lee 2005].

10.7 EXERCISES

10.1 **(1149.1 Boundary Scan)** Given a printed-circuit board that has four chips built with a boundary scan, such as the one shown in Figure 10.3, describe a test procedure via the boundary scan to test each chip and the interconnects between chips. Also, describe the instruction(s) used in each step of your procedure for each chip. Assume external ATE is used to provide and receive test data.

10.2 **(1149.1 Boundary Scan)** How many test cycles are needed to shift a 4-bit test instruction into the instruction register of a boundary-scan architecture? Assume that you start from the reset state and that after the instruction is shifted in the TAP will be in the Select-DR-Scan state.

10.3 **(1149.1 Boundary-Scan Instructions)** Continue on Problem 10.2. Assume that the loaded instruction is an INTEST instruction. Now you are going to apply 100 patterns to the internal logic and observe the test results. If the length of the boundary-scan register is 30, then how many test cycles will be required to carry out the entire test procedure? Assume that the internal logic is a combinational circuit and that after the test procedure the circuit will return to the Test–Logic–Reset state.

10.4 **(1149.1 Boundary-Scan Instructions)** Give the timing diagrams for executing the SAMPLE and the PRELOAD instructions. Can these two instructions be executed in one iteration of the seven states shown in the middle of Figure 10.7?

10.5 **(A Design Practice)** Use the boundary-scan programs and user's manuals provided online to insert the boundary-scan circuit to the ISCAS 1985 benchmark circuit, c499. Create its BSDL file and generate a Verilog verification testbench. Use any commercially available Verilog simulator to verify if the generated verification testbench passes Verilog simulation. Repeat the same exercise for the ISCAS 1989 circuit, s38417.

10.6 **(A Design Practice)** Repeat Problem 10.5. What is the area overhead for the c499 boundary-scan circuit? Compare its area overhead with that in s38417.

10.7 **(1149.6 Digital Receiver Logic)** The drawing of the digital receiver logic shown in Figure 10.38 was taken directly from the IEEE 1149.6-2003 standard

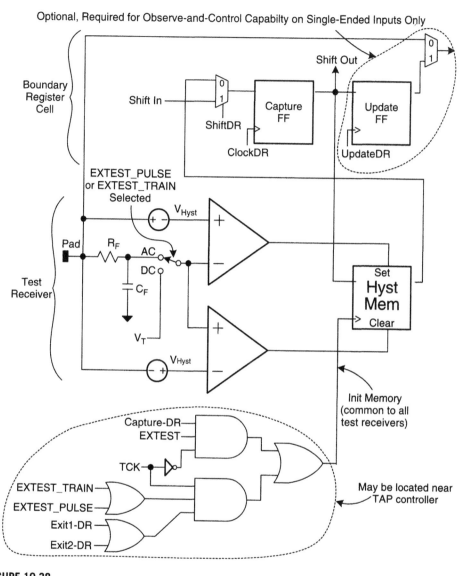

■ **FIGURE 10.38**

Digital receiver logic.

(Figure 48 in the standard). An error in this figure was discovered by the working group after the standard had been published. Try to determine the error in the drawing and what can be done to fix it.

10.8 **(1149.6 Test Access Port)** Try to derive the AC Test signal of the timing diagram shown in Figure 10.6 from the logic in Figure 10.4 and the transitions through the TAP state machine. Also, see what happens if the EXTEST_TRAIN instruction is used instead of EXTEST_PULSE.

10.9 **(1500 Wrapper Cells)** Give gate-level designs for the WBCs shown in Figures 10.26a and 10.26e. Compare their functionalities, gate counts, and the control signals required.

10.10 **(Delay Test with 1500 and 1149.1)** Describe how to execute a delay test on a core wrapped by a 1500 wrapper with a WBR consisting of WBCs shown in Figure 10.28. Use a simple core that contains only a two-input OR gate to illustrate the procedure. Give the timing diagram to show the waveforms of related signals. After completing this work, discuss whether a similar procedure can be applied to a circuit in compliance with 1149.1.

10.11 **(1500 in Serial Mode)** Similar to Problems 10.3 and 10.4, assume now that the internal logic is wrapped by a 1500 wrapper. How many test cycles are required to load the instruction and execute the scan operation using WSP only?

10.12 **(1500 in Parallel Mode)** Similar to Problem 10.11, assume that there are ten parallel TAM wires available for testing. How many test cycles will be required? Draw your test configuration and clearly state any assumptions that you made.

10.13 **(Comparison between 1149.1 and 1500)** Table 10.2 lists the main differences between the 1149.1 boundary-scan and 1500 core-based test standards. Except for the first difference (Objective), discuss the effects of each difference.

Acknowledgments

The author wishes to thank William Eklow, Cisco Systems and Chair of the IEEE 1149.6 Standard Committee, for writing the IEEE 1149.6 Boundary-Scan Extension section and proofreading the manuscript; Prof. Wen-Ben Jone of the University of Cincinnati and Dr. Kenneth P. Parker of Agilent Technologies for their valuable comments and suggestions; and Grady L. Giles of AMD and Eric Jain Marinissen of Philips Research Laboratories for their clarification of the IEEE 1500 standard.

References

R10.0—Books

[IEEE 1149.1-2001] IEEE Std. 1149.1-2001, *IEEE Standard Test Access Port and Boundary-Scan Architecture*, IEEE Press, New York, 2001.

[IEEE 1149.4-1999] IEEE Std. 1149.4-1999, *IEEE Standard for a Mixed Signal Test Bus*, IEEE Press, New York, 1999.

[IEEE 1149.6-2003] IEEE Std. 1149.6-2003, *IEEE Standard for Boundary-Scan Testing of Advance Digital Networks*, IEEE Press, New York, 2003.

[IEEE 1450-1999] IEEE Std. 1450-1999, *IEEE Standard Test Interface Language (STIL) for Digital Test Vector Data*, IEEE Press, New York, 1999.

[IEEE 1500-2005] IEEE Std. 1500-2005, *IEEE Standard for Embedded Core Test*, IEEE Press, New York, 2005.

[IEEE P1450.6-2001] CTL Web site (http://grouper.ieee/groups/ctl/ctl.pdf).

[Parker 2001] K. P. Parker, *The Boundary-Scan Handbook*, Kluwer Academic Publishers, 2001.

R10.1—Introduction

[Dervisoglu 1992] B. Dervisoglu, Boundary-scan update IEEE P1149.2: Description and status report, *IEEE Des. Test Comput.*, 9(3), 79–81, 1992.

[Eklow 2003a] B. Eklow, K. P. Parker, and C. F. Barnhart, IEEE 1149.6: A boundary-scan standard for advanced digital networks, *IEEE Des. Test Comput.*, 20(5), 76–83, 2003.

[Eklow 2003b] B. Eklow, C. Barnhart, M. Ricchetti, and T. Borroz, IEEE 1149.6: A practical perspective, in *Proc. IEEE Int. Test Conf.*, October 2003, pp. 494–502.

[Petersen 1992] K. Petersen, Boundary scan: A promising method for testing complex electronic systems. *ABB Rev.*, July 8, 1992, pp. 31–38.

[Treuren 2005] B. G. Van Treuren, B. E. Peterson, and J. M. Miranda, JTAG-based vector and chain management for system test, in *Proc. IEEE Int. Test Conf.*, October 2005, Paper 32.1, 10 pp.

[Ungar 2001] L. Y. Ungar, H. Bleeker, J. E. McDermid, and H. Hulvershorn, IEEE-1149.x standards: Achievements vs. expectations, in *Proc. AUTOTESTCON*, August 2001, pp. 188–205.

R10.2—Digital Boundary Scan (IEEE Std. 1149.1)

[Bäckström 2005] D. Bäckström, G. Carlsson, and E. Larsson, Remote boundary-scan system test control for the ATCA standard, in *Proc. IEEE Int. Test Conf.*, October 2005, Paper 32.2.

[Bhattacharya 1998] D. Bhattacharya, Hierarchical test access architecture for embedded cores in an integrated circuit, in *Proc. IEEE VLSI Test Symp.*, April 1998, pp. 10–14.

[Bhavsar 1991] D. Bhavsar, An architecture for extending the IEEE Standard 1149.1 test access port to system backplanes, in *Proc. IEEE Int. Test Conf.*, October 1991, pp. 768–776.

[Chan 1992] J. C. Chan, Boundary walking test: An accelerated scan method for greater system reliability, *IEEE Trans. on Reliability*, 41(4), 496–503, 1992.

[Gibbs 2003] C. Gibbs, Backplane test bus applications for IEEE Std. 1149.1, in *Proc. IEEE Int. Test Conf.*, October 2003, pp. 167–180.

[Jarwala 1989] N. Jarwala and C. W. Yau, A new framework for analyzing test generation and diagnosis algorithms for wiring interconnects, in *Proc. IEEE Int. Test Conf.*, October 1989, pp. 63–70.

[Joshi 2003] R. N. Joshi, K. L. Williams, and L. Whetsel, Evolution of IEEE 1149.1 addressable shadow protocol devices, in *Proc. IEEE Int. Test Conf.*, October 2003, pp. 981–987.

[Kim 2004] Y. Kim, H.-D. Kim, and S. Kang, A new maximal diagnosis algorithm for interconnect test, *IEEE Trans. Very Large Scale Integration Syst.*, 12(5), 532–537, 2004.

[NS 2004a] National Semiconductor, SCANSTA111 Enhanced SCAN Bridge Multidrop Addressable IEEE 1149.1 (JTAG) Port [data sheet], National Semiconductor, Santa Clara, CA, April 2004 (http://cache.national.com/ds/SC/SCANSTA111.pdf).

[NS 2004b] National Semiconductor, SCANSTA112 7-port Multidrop Addressable IEEE 1149.1 (JTAG) Multiplexer [data sheet], National Semiconductor, Santa Clara, CA, May 2004 (http://cache.national.com/ds/SC/SCANSTA112.pdf).

[O'Donnell 1994] G. O'Donnell, Using 1149.1 for multi-drop and hierarchical system testing, in *Proc. WESCON*, September 1994, pp. 538–541.

[Rearick 2005] J. Rearick, B. Eklow, K. Posse, A. Crouch, and B. Bennetts, IJTAG (internal JTAG): A step toward a DFT standard, in *Proc. IEEE Int. Test Conf.*, October 2005, Paper 32.3, 10 pp.

[TI 1999] Texas Instruments, SN54LVT8996, SN74LVT8996: 3.3-V 10-bit addressable scan ports multidrop-addressable IEEE Std. 1149.1 (JTAG) TAP transceiver [data sheet], Texas Instruments, Dallas, TX, December 1999 (http://focus.ti.com/lit/ds/symlink/sn74lvt8996.pdf).

[TI 2003] Texas Instruments, SN54LVT8986, SN74LVT8986: 3.3-V LINKING addressable scan ports multidrop-addressable IEEE Std. 1149.1 (JTAG) TAP transceiver [data sheet], Texas Instruments, Dallas, TX, April 2003 (http://focus.ti.com/lit/ds/symlink/sn74lvt8986.pdf).

[Whetsel 1992] L. Whetsel, A proposed method of accessing 1149.1 in a backplane environment, in *Proc. IEEE Int. Test Conf.*, October 1992, pp. 206–216.

[Whetsel 1997] L. Whetsel, An IEEE 1149.1-based test access architecture for ICs with embedded cores, in *Proc. IEEE Int. Test Conf.*, October 1997, pp. 69–78.

[Whetsel 1999] L. Whetsel, Addressable test ports: An approach to testing embedded cores, in *Proc. IEEE Int. Test Conf.*, October 1999, pp. 1055–1065.

[Zak 1992] R. C. Zak and J. V. Hill, An IEEE 1149.1 compliant testability architecture with internal scan, in *Proc. Int. Conf. on Computer Design*, 436–442, 1992.

R10.3—Boundary-Scan Extension (IEEE Std. 1149.6)

[Duzevik 2003] I. Duzevik, Design and implementation of IEEE 1149.6, in *Proc. IEEE Int. Test Conf.*, October 2003, pp. 87–95.

[Eklow 2003a] B. Eklow, K. P. Parker, and C. F. Barnhart, IEEE 1149.6: A boundary-scan standard for advanced digital networks, *IEEE Des. Test Comput.*, 20(5), 76–83, 2003.

[Eklow 2003b] B. Eklow, C. Barnhart, M. Ricchetti, and T. Borroz, IEEE 1149.6: A practical perspective, in *Proc. IEEE Int. Test Conf.*, October 2003, pp. 494–502.

[Rearick 2004] J. Rearick, J. S. Patterson, and K. Dorner, Integrating boundary scan into multi-GHz I/O circuitry, in *Proc. IEEE Int. Test Conf.*, October 2004, pp. 560–566.

[Shaikh 2004] S. A. Shaikh, IEEE Standard 1149.6 implementation for a XAUI-to-serial 10-Gbps transceiver, in *Proc. IEEE Int. Test Conf.*, October 2004, pp. 543–550.

[Sunter 2004] S. Sunter, A. Roy, and J.-F. Cote, An automated, complete, structural test solution for SERDES, in *Proc. IEEE Int. Test Conf.*, October 2004, pp. 95–104.

[Vandivier 2003] S. Vandivier, M. Wahl, and J. Rearick, First IC validation of IEEE Std. 1149.6, in *Proc. IEEE Int. Test Conf.*, October 2003, pp. 632–639.

R10.4—Embedded Core Test Standard (IEEE Std. 1500)

[Aerts 1998] J. Aerts and E. J. Marinissen, Scan chain design for test time reduction in core-based ICs, in *Proc. IEEE Int. Test Conf.*, October 1998, pp. 448–457.

[Benabdenbi 2000] M. Benabdenbi, W. Maroufi, and M. Marzouki, CAS-BUS: A scalable and reconfigurable test access mechanism for systems on a chip, in *Proc. IEEE Design, Automation, and Test in Europe Conf.*, March 2000, pp. 141–145.

[Chakrabarty 2000] K. Chakrabarty, Design of system-on-a-chip test access architectures under place-and-route and power constraints, in *Proc. IEEE Design Automation Conf.*, June 2000, pp. 432–437.

[Wu 2001] C.-W. Wu, J.-F. Li, and C.-T. Huang, Core-based system-on-chip testing: Challenges and opportunities, *J. Chinese Inst. Electr. Eng.*, 8(4), 335–353, 2001.

[Goel 2002] S. K. Goel and E. J. Marinissen, Effective and efficient test architecture design for SOCs, in *Proc. IEEE Int. Test Conf.*, October 2002, pp. 529–538.

[Goel 2003a] S. K. Goel and E. J. Marinissen, Layout-driven SOC test architecture design for test time and wire length minimization, in *Proc. IEEE Design, Automation, and Test in Europe Conf.*, March 2003, pp. 738–743.

[Goel 2003b] S. K. Goel and E. J. Marinissen, Control-aware test architecture design for modular SOC testing, in *Proc. European Test Workshop*, May 2003, pp. 57–62.

[Ivengar 2002] V. Ivengar, K. Chakrabarty, and E. J. Marinissen, Integrated wrapper/TAM co-optimization, constraint-driven test scheduling, and tester data volume reduction for SOCs, in *Proc. Design Automation Conf.*, June 2002, pp. 685–690.

[Kapur 1999] R. Kapur, B. Keller, B. Koenemann, M. Lousberg, P. Reuter, T. Taylor, and P. Varma, 1500-CTL: Towards a standard core test language, in *Proc. IEEE VLSI Test Symp.*, April 1999, pp. 489–490.

[Kapur 2001] R. Kapur, M. Lousberg, T. Taylor, B. Keller, P. Reuter, and D. Kay, CTL: The language for describing core-based test, in *Proc. IEEE Int. Test Conf.*, October 2001, pp. 131–139.

[Koranna 2002] S. Koranna and V. Iyengar, On the use of k-tuples for SOC test schedule representation, in *Proc. IEEE Int. Test Conf.*, October 2002, pp. 539–548.

[Larsson 2002] E. Larsson and Z. Peng, An integrated framework for the design and optimization of SOC test solutions, *J. Electron. Test. Theory Appl.*, 18(4/5), 385–400, 2002.

[Lee 2000] K.-J. Lee and C.-I. Huang, A hierarchical test control architecture for core-based design, in *Proc. Asian Test Symp.*, December 2000, pp. 248–253.

[Li 2002] J.-F. Li, H.-J. Huang, J.-B. Chen, C.-P. Su, C.-W. Wu, C. Cheng, S.-I Chen, C.-Y. Hwang, and H.-P. Lin, A hierarchical test scheme for system-on-chip designs, in *Proc. IEEE Design, Automation and Test in Europe Conf.*, March 2002, pp. 486–490.

[Marinissen 1998] E. J. Marinissen, R. Arendsen, G. Bos, H. Dingemanse, M. Lousberg, and C. E. Wouters, A structured and scalable mechanism for test access to embedded reusable cores, in *Proc. IEEE Int. Test Conf.*, October 1998, pp. 284–293.

[Marinissen 1999] E. J. Marinissen, Y. Zorian, R. Kapur, T. Taylor, and L. Whetsel, Towards a standard for embedded core test: An example, in *Proc. IEEE Int. Test Conf.*, October 1999, pp. 616–627.

[Marinissen 2000] E. J. Marinissen, R. Kapur, and Y. Zorian, On using IEEE P1500 SECT for test plug-n-play, in *Proc. IEEE Int. Test Conf.*, October 2000, pp. 770–777.

[Marinissen 2002] E. J. Marinissen, R. Kapur, M. Lousberg, T. McLaurin, M. Ricchetti, and Y. Zorian, On IEEE P1500's standard for embedded core test, *J. Electron. Test. Theory Appl.*, 18(4/5), 365–383, 2002.

[Sehgal 2004] A. Sehgal, S. K Goel, E. J. Marinissen, and K. Chakrabarty, IEEE P1500-compliant test wrapper design for hierarchical cores, in *Proc. IEEE Int. Test Conf.*, October 2004, pp. 1203–1212.

[Varma 1998] P. Varma and S. Bhatia, A structured test re-use methodology for core-based system chips, in *Proc. IEEE Int. Test Conf.*, October 1998, pp. 294–302.

[Waayers 2005] T. Waayers, R. Morren, and R. Grandi, Definition of a robust modular SOC test architecture: Resurrection of the single TAM daisy-chain, in *Proc. IEEE Int. Test Conf.*, October 2005, Paper 25.3, 10 pp.

[Zorian 1997] Y. Zorian, Test requirements for embedded core-based systems and IEEE P1500, in *Proc. IEEE Int. Test Conf.*, October 1997, pp. 191–199.

R10.6—Concluding Remarks

[Bäckström 2005] D. Bäckström, G. Carlsson, and E. Larsson, Remote boundary-scan system test control for the ATCA standard, in *Proc. IEEE Int. Test Conf.*, October 2005, Paper 32.2, 10 pp.

[Huang 2001] J.-R. Huang, M. K. Iyer, and K.-T. Cheng, A self-test methodology for IP cores in bus-based programmable SoCs, in *Proc. IEEE VLSI Test Symp.*, April 2001, pp. 198–203.

[Ivengar 2002] V. Iyengar and K. Chakrabarty, System-on-a-chip test scheduling with precedence relationships, preemption, and power constraints, *IEEE Trans. Comput.-Aided Des.*, 21(9), 1088–1094, 2002.

[Krstic 2002] A. Krstic, W.-C. Lai, K.-T. Cheng, L. Chen, and S. Dey, Embedded software-based self-test for programmable core-based designs, *IEEE Des. Test Comput.*, 19(4), 18–27, 2002.

[Larsson 2002] E. Larsson and Z. Peng, Integrated test scheduling, test parallelization and TAM design, in *Proc. Asian Test Symp.*, November 2002, pp. 397–404.

[Lee 2005] K.-J. Lee, C.-Y. Chu, and Y.-T. Hong, A test platform for SOC testing, in *Proc. IEEE Int. Symp. on Circuits and Systems*, May 2005, pp. 2983–2986.

[Rearick 2005] J. Rearick, B. Eklow, K. Posse, A. Crouch, and B. Bennetts, IJTAG (internal JTAG): A step toward a DFT standard, in *Proc. IEEE Int. Test Conf.*, October 2005, Paper 32.3, 10 pp.

[Rosinger 2005] P. Rosinger, B. Al-Hashimi, and K. Chakrabarty, Rapid generation of thermal-safe test schedules, in *Proc. IEEE Design, Automation and Test in Europe Conf.*, March 2005, pp. 840–845.

[Tehranipour 2003] M. H. Tehranipour, M. Nourani, S. M. Fakhraie, and A. Afzali-Kusha, Systematic test program generation for SoC testing using embedded processor, in *Proc. IEEE Int. Symp. on Circuits and Systems*, May 2003, pp. 541–544.

[Tsai 2001] C.-H. Tsai and C.-W. Wu, Processor-programmable memory BIST for bus-connected embedded memory, in *Proc. Asia and South Pacific Design Automation Conf.*, January 2001, pp. 325–330.

[Zou 2003] W. Zou, S. M. Reddy, I. Pomeranz, and Y. Huang, SOC test scheduling using simulated annealing, in *Proc. IEEE VLSI Test Symp.*, April 2003, pp. 325–330.

CHAPTER 11

ANALOG AND MIXED-SIGNAL TESTING

Chauchin Su
National Chiao Tung University, Hsinchu, Taiwan

ABOUT THIS CHAPTER

Analog and mixed-signal (AMS) circuits are becoming more critical in the *system-on-chip* (SOC) era, although they are occupying less silicon area. AMS circuits are designed using specialized techniques because a wide range of circuit structures are possible. Dedicated customization is required for various process technologies to satisfy performance requirements. Similarly, AMS testing depends strongly on the circuit and so depends on specialized approaches. This chapter introduces AMS circuits, failure modes, and fault models. It then addresses analog testing, including DC and AC parametric testing. Waveform-oriented testing and specification-oriented testing are considered. Then, mixed-signal circuits, *analog-to-digital converters* (ADCs), *digital-to-analog converters* (DACs), and their testing approaches, are discussed. Terminology and test approaches are consistent with the IEEE 1057 standard. Finally, the IEEE 1149.4 standard for mixed-signal test buses is studied. Two analog test buses are employed to deliver test stimuli and test responses in board-level analog interconnect testing and passive component measurement.

11.1 INTRODUCTION

Continuance and discreteness fundamentally distinguish analog from digital signals. Analog signals are continuous in time and amplitude, while digital signals are discrete in both domains. Additionally, digital signals are mostly binary, with V_{DD} for logical high and GND for logical low. Mixed signals are quantizations of analog signals. As digital signals, they are discrete in time and amplitude; however, they have a much higher amplitude resolution than digital signals. Figure 11.1 presents the waveforms of the three types of signals.

Due to the continuance property, analog circuits are commonly required to behave uniformly across their operational range; for example, the gain should be constant for various signal amplitudes within a particular range of frequencies. This property is referred to as **linearity**. Theoretically, as a real number, a continuous signal has an infinite resolution. Practically, such a requirement is excessive

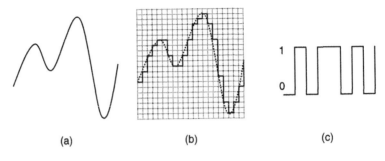

(a)   (b)   (c)

Signals: (a) analog, (b) mixed-signal, and (c) digital.

because a human cannot perceive infinitesimal variations; therefore, distortion and noise are introduced into the specification to allow some imperfection. The imperfection represents a gray area. It supports a tradeoff between cost and performance. Tradeoffs reduce the complexity of a particular design but enlarge the design space, meaning that more designs can be employed for a particular application; therefore, an analog design is specific to fine-grain applications so analog testing is used.

Most digital function can be composed from a set of primitive gates such as NAND and NOR gates. Even though different circuit structures can be used to implement these gates, their inputs and outputs are the same in all cases. Analog circuits are designed using a totally different philosophy. Special circuit structures are required for particular applications; for example, an audio amplifier has high resolution and low bandwidth, and a video amplifier has low resolution and high bandwidth. A *radiofrequency* (RF) amplifier must satisfy totally different requirements related to impedance matching, noise figure, and power efficiency. Analog and mixed-signal designs and testing involve particular approaches; in other words, different design and test methodologies are required for different analog circuits.

11.1.1 Analog Circuit Properties

The fundamental difference between analog and digital signals has now been outlined. Now, let us consider the detailed properties of circuits that influence their design and testing. Important analog properties are listed below.

- Continuous signals
- Large range of circuits
- Nonlinear characteristics
- Feedback ambiguity
- Complicated cause–effect relationship
- Absence of suitable fault model
- Accurate measurements required

11.1.1.1 *Continuous Signals*

Analog signals are continuous. The interpretations of a particular waveform plotted in Figure 11.2 differ with the domain: digital or analog. For a digital waveform, the useful specifications are logical high and low voltages (V_H and V_L) and rise and fall times (t_{LH} and t_{HL}). For an analog waveform, they are amplitude (V_A), slew rate (SR), overshoot (V_{ov}), settling time (t_{Settle}), bandwidth, phase, and others; therefore, more parameters must be considered in the analog domain.

11.1.1.2 *Large Range of Circuits*

Analog circuits include generic modules such as operational amplifiers, filters, comparators, regulators, mixers, low noise amplifiers, power amplifiers, and switches. Specialized modules include line drivers, variable gain amplifiers, oscillators, sensors, and RF transceivers, among many others. Their functions and specifications differ significantly. For each circuit, their particular characteristics must be considered; therefore, analog testing is knowledge intensive.

11.1.1.3 *Nonlinear Characteristics*

Analog active devices are nonlinear; for example, **large signal models** of diodes, bipolar transistors, and MOS transistors are all nonlinear:

$$I_D = I_s \cdot e^{V_D/n \times V_T} \tag{11.1}$$

$$I_C = I_s \cdot e^{V_{BE}/V_T} \tag{11.2}$$

$$I_D = \frac{1}{2}\mu C_{ox}\frac{W}{L}(V_{gs} - V_t)^2 \tag{11.3}$$

A desired biasing current can be obtained by solving nonlinear equations to determine a suitable biasing voltage. Although they are globally nonlinear, they are approximately linear within a small operating range. Such a model is referred to as a **small signal model**. Notably, the parameters of the linear model depend on the

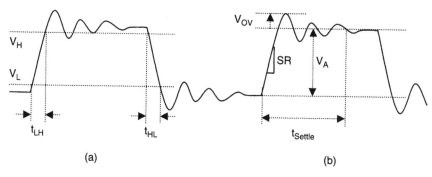

(a)   (b)

■ **FIGURE 11.2**

Different interpretations of a step response: (a) digital, and (b) analog.

bias. For example, the transconductance of a MOS transistor is $g_m = \sqrt{2\mu C_{ox}\frac{W}{L}I_D}$. Additionally, the second-order effects, including the body effect, channel length modulation, mobility variation, subthreshold, hot electrons, and others, further complicate the situation.

11.1.1.4 Feedback Ambiguity

Analog circuits use feedback extensively to increase linearity; for example, for an operational amplifier in the inverting configuration, as presented in Figure 11.3, the closed-loop gain is $A_f = \dfrac{-A}{1 + \frac{R1}{R2}A}$, where A is the open-loop gain. If A is sufficiently large, then the closed-loop gain approximates $A_f = -\frac{R2}{R1}$, a constant. However, when the closed-loop gain is less than the specified value, determining whether $R1$ is too large, $R2$ is too small, or A is too small, is difficult. A similar feedback scheme applies to a *phase-locked loop* (PLL), *automatic gain control* (AGC), power regulators, and others. More tests must be conducted to identify faulty components.

11.1.1.5 Complicated Cause–Effect Relationship

Nonlinearity and feedback complicate the cause–effect relationship. Many circuit parameters associated with a signal parameter also complicate the cause–effect relationship. Any single device failure will influence, if not all, most of the circuit and signal parameters. So, any circuit or signal parameter failure can be caused by the failure of any device; therefore, fault-specific test generation is more difficult than for digital counterparts. Specification-oriented testing is still more popular than other forms of testing, despite its being costly and time consuming.

11.1.1.6 Absence of Suitable Fault Model

In digital testing, the single-fault assumption has attracted consensus in the test community because it is simple and effective. It might not be sufficiently accurate for deep-submicron circuits, but it is satisfactory, considering the complexity of other, more accurate fault models. No simple and effective nor generic and generally acceptable analog fault model is available, because the range of circuits is large.

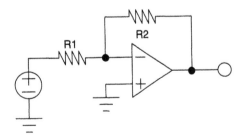

■ FIGURE 11.3

Amplifier in inverting configuration.

Establishing a model that fits all circuits and applications is difficult. Additionally, even for the same type of circuits, the essential parameters vary by case.

11.1.1.7 Requirement for Accurate Instruments for Measuring Analog Signals

Due to their continuous characteristic, analog signals can be measured only using highly accurate instruments. Additionally, different parameters may be measured using different instruments; for example, oscilloscopes and digitizers are used to make measurements in the time domain, and spectrum analyzers and network analyzers are used to make measurements in the frequency domain, thus increasing the cost and barrier for analog testing. Given their characteristics, analog testing and mixed-signal testing are not as extensively developed as digital testing. Analog testing can only be conducted by knowledgeable and experienced engineers who can handle such issues as nonlinearity, feedback ambiguity, complex cause–effect relationships, and noise. Their experience is very important, especially in diagnosis.

11.1.2 Analog Defect Mechanisms and Fault Models

Defects in *integrated circuits* (ICs) are caused by imperfections in the manufacturing process. Possible imperfections include the following:

- **Material defects**
  - Cracks
  - Crystal imperfection
  - Surface impurities
  - Ion migration
- **Processing faults**
  - Oxide thickness
  - Mobility change
  - Impurity density
  - Diffusion depth
  - Dielectric constants
  - Metal sheet resistance
  - Missing contacts
  - Dust

- **Time-dependent failures**
    - Dielectric breakdown
    - Electron migration
- **Packaging failures**
    - Contact degradation
    - Seal leakage

Most of the defects listed above are more likely to have global effects than local ones. Precise modeling at favorable cost is difficult. Some defects, such as dust and surface impurities, have localized effects. Two categories of faults models, *hard faults* and *soft faults*, are defined according to the degree of faulty effects, to simplify fault modeling and fault simulation efforts. Before the fault model is considered, defects caused by dust are used as examples in the discussion.

Figure 11.4a presents the layout of a CMOS inverter and possible defects caused by dust. Depending on whether a negative or positive emulsion is used, dust that blocks the light during lithography causes extra spots or missing spots in the layer that is being processed. Figure 11.4b shows that the dust may cause missing spots for poly and extra spots for diffusion and metal; these spots are referred to as **etching defects** and **extra defects**, respectively. An *etching defect* is evidenced by shrinking of the active region caused by the dust. Similarly, and *extra defect* occurs when the active region is growing. The significance of the dust defects depends on the sizes and locations of the dust. Hard faults and soft faults are categorized based on size, as follows.

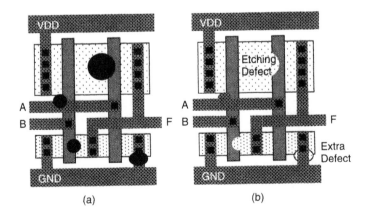

(a)   (b)

■ **FIGURE 11.4**

Layout of NOR-2 gate: (a) random dust on the layout, and (b) extra and etching defects associated with dust.

11.1.2.1 Hard Faults

If defects are to sufficiently change the circuit schematics, then they are classified as **hard faults**. If the dust is too large, it may cause an opening or a short during the metallization process for fabricating metal wires. During the implantation or diffusion process, in which transistors are fabricated, transistors may disappear or be improperly formed because of the dust; therefore, as in the example of the two-stage OPAmp in Figure 11.5, four hard fault models are defined. They are **open**, **short**, **extra device**, and **missing device**. These faults are classified as hard faults because they alter the schematic circuit diagram. The circuit diagrams must be modified to simulate the effects of the faults. Analog circuits are designed concisely with a small margin, and hard faults are more likely to cause *catastrophic errors* and *system failure*; therefore, they are quite easy to test.

11.1.2.2 Soft Faults

If defects are too minor to cause hard faults, they may change device parameters. They are then classified as **soft faults**. For example, the largest dust presented in Figure 11.4b does not entirely block the polysilicon line but shortens the effective channel length; therefore, the device parameters are changed. Another example is the opening of one of a set of parallel transistors. In advanced technology, a large transistor is commonly implemented as a set of small transistors in parallel. Although "open" is a hard fault, it does not completely alter the schematic. It only changes the effective W/L ratio of the transistor.

Soft faults are further classified into **parametric faults** and **deviation faults**. *Parametric faults* are used to model the variation in the parameter that governs a device in the circuit of interest. For example, in Figure 11.5, a change of the biasing current from $100\,\mu A$ to $80\,\mu A$, caused by an open transistor in the current mirror

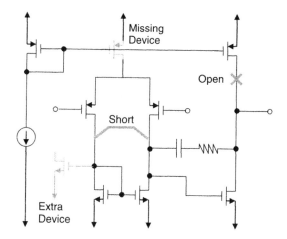

■ **FIGURE 11.5**

Hard faults: open, short, extra device, missing device.

that is comprised of five parallel devices, is a parametric fault. Another parametric fault is a change in the effective channel length from 130 nm to 100 nm caused by the dust presented in Figure 11.4b. *Deviation faults* refer to changes in the overall performance of the entire circuit of interest. For example, the above parametric faults and their simulations may indicate that the unit-gain bandwidth is reduced from 100 MHz to 70 MHz and the DC gain has decreased from 80 dB to 70 dB. These faults are referred to as deviation faults. Deviation faults and parametric faults constitute an inference mechanism for hierarchical fault modeling. A deviation fault of a child module is a parametric fault of the parent module. Higher level faults can be derived from the faults at lower levels.

Hard faults and soft faults are not mutually exclusive. One fault may be classified as both a hard fault and a soft fault. **Resistive short** or **resistive open** are popular examples. Instead of 100% short or open, the fault site may produce resistive behavior—for example, equivalent to being connected by a 10K-Ω resistor. Causes include water on the wafer or the mask during the lithography process and an excess of etching defects. The rule of thumb is that if the resistance is less than one-tenth of that of the fault-free node then it is regarded as a *resistive short*. If it is more than ten times the resistance of the fault-free node, it is regarded as a *resistive open*. If the resistance is between that associated with a resistive short and a resistive open, then it is a *parametric fault*.

[Wang 1997] studied hierarchical fault modeling approaches. He injected a certain number of dust particles at random locations on a **circuit under test** (CUT) to cause realistic faults. The size of the dust was a random variable whose distribution function was obtained by measuring dust collected from a fab. Random dusting enabled the probability of hard faults and the distribution of parametric faults to be determined.

Larger devices are less likely to have hard faults than are smaller ones; therefore, tests focus on the more probable faults to lower the computational complexity. Similarly, random dusting yields the distribution function of a parametric fault, which helps to improve the accuracy and reduce the complexity of fault modeling and test generation. Additionally, the distribution function of the referred deviation faults can also be determined from the Monte Carlo simulation in SPICE™; therefore, a comprehensive hierarchical fault model can be built based on random dusting and the inference engine outlined below.

Figure 11.6 presents an example of hierarchical fault derivation. For simplicity, an inverter amplifier is used as an example. After the random dusting of the layout in Figure 11.6a, the distribution functions of the parametric faults of the K value ($\frac{1}{2}\mu C_{ox}\frac{W}{L}$) of both transistors can be derived. Changes in K_n and K_p alter the gain of the inverter amplifier. The distributions of these values yield the distribution function of the gain via a Monte Carlo simulation. In summary, the parametric faults are represented by changes in K_n and K_p, and the deviation fault is represented by a change in the gain. They are both derived from the random dusting of the layout.

This section has provided some background information on analog circuit characteristics, defect mechanisms, and fault models. The following section addresses analog circuit testing based on the information gathered herein.

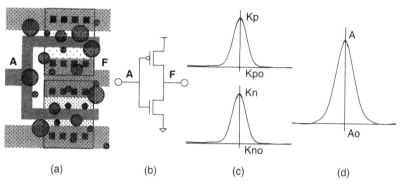

■ **FIGURE 11.6**

Hierarchical fault model derivation: (a) layout with dusting, (b) circuit schematics, (c) effect of parametric faults on the K values, and (d) effect of deviation fault on the gain.

11.2 ANALOG CIRCUIT TESTING

As stated in the preceding section, analog circuits exhibit very large variations in function, structure, and performance, so methods of testing them vary significantly. No simple, effective, or generally accepted analog fault model exists, so implementing a fault-model-based analog test generation algorithm for general classes of analog circuits is difficult. In practice, much manual engineering is still involved. In this section, test approaches, test waveforms, and AC and DC parametric testing are outlined.

11.2.1 Analog Test Approaches

Analog testing can be divided into two categories: **specification-oriented testing** and **waveform-oriented testing**. *Specification-oriented testing* tests every specification presented in the data sheet to determine the pass/failure of the circuit. Figure 11.7 presents the data sheet of OP777 from Analog Devices (Norwood, MA). Most of the test and measurement approaches can be found in the application notes published by the chip vendors [ADI 1982]. The process is tedious and may depend on accurate and extensive instrumentation setup; therefore, it is more suitable for the **bench test** than the **final test**. Notably, the *bench test* is conducted in a laboratory for the characterization purposes, and the *final test* is conducted in a test house before the chips are shipped to customers.

Waveform-oriented testing measures particular parameters of response waveforms to determine the pass/failure of the CUT. Figure 11.8 presents an example of a waveform-oriented test. The CUT is a Sallen–Key second-order low-pass filter [TI 1999] and the test stimulus is a square waveform. Four test points are sampled: A, B, C, and D. If all are within the predefined margin (marked in gray), then the CUT is regarded to be fault free and to pass; otherwise, it is faulty and has failed.

OP777/OP727/OP747–SPECIFICATIONS

ELECTRICAL CHARACTERISTICS (@ V_S = 5.0 V, V_{CM} = 2.5 V, T_A = 25°C unless otherwise noted.)

Parameter	Symbol	Conditions	Min	Typ	Max	Unit
INPUT CHARACTERISTICS						
Offset Voltage OP777	V_{OS}	+25°C < T_A < +85 °C		20	100	μV
		–40°C < T_A < +85 °C		50	200	μV
Offset Voltage OP727/OP747		+25°C < T_A < +85 °C		30	160	μV
		–40°C < T_A < +85 °C		60	300	μV
Input Bias Current	I_B	–40°C < T_A < +85 °C		5.5	11	nA
Input Offset Current	I_{OS}	–40°C < T_A < +85 °C		0.1	2	nA
Input Voltage Range			0		4	V
Common-Mode Rejection Ratio	CMRR	V_{CM} = 0 V to 4 V	104	110		dB
Large Signal Voltage Gain	A_{VO}	R_L = 10 kΩ, V_O = 0.5 V to 4.5 V	300	500		V/mV
Offset Voltage Drift OP777	$\Delta V_{OS}/\Delta T$	–40°C < T_A < +85 °C		0.3	1.3	μV/°C
Offset Voltage Drift OP727/OP747	$\Delta V_{OS}/\Delta T$	–40°C < T_A < +85 °C		0.4	1.5	μV/°C
OUTPUT CHARACTERISTICS						
Output Voltage High	V_{OH}	I_L = 1 mA, –40 °C to +85 °C	4.88	4.91		V
Output Voltage Low	V_{OL}	I_L = 1 mA, –40 °C to +85 °C		126	140	mV
Output Circuit	I_{OUT}	$V_{DROPOUT}$ < 1 V		±10		mA
POWER SUPPLY						
Power Supply Rejection Ratio	PSRR	V_S = 3 V to 30 V	120	130		dB
Supply Current/Amplifier OP777	I_{SY}	V_O = 0 V		220	270	μA
		–40°C < T_A < +85 °C		270	320	μA
Supply Current/Amplifier OP727/OP747		V_O = 0 V		235	290	μA
		–40°C < T_A < +85 °C		290	350	μA
DYNAMIC PERFORMANCE						
Slew Rate	SR	R_L = 2 kΩ		0.2		V/μs
Gain Bandwidth Product	GBP			0.7		MHz
NOISE PERFORMANCE						
Voltage Noise	e_np-p	0.1 Hz to 10 Hz		0.4		μV p-p
Voltage Noise Density	e_n	f = 1 kHz		15		nV/√Hz
Current Noise Density	i_n	f = 1 kHz		0.13		pA/√Hz

NOTES
Typical specifications: >50% of units perform equal to or better than the "typical" value.
Specifications subject to change without notice.

■ **FIGURE 11.7**

Data sheet of Analog Devices' OP777.

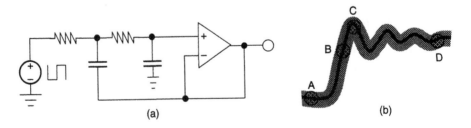

(a)   (b)

■ **FIGURE 11.8**

Waveform-oriented testing: (a) second-order Sallen–Key low-pass filter, and (b) response waveform and sample points.

The test points in waveform-oriented testing must be carefully chosen. Correlation of the test points with specifications is important to improving the *test yield* and reducing the *defect level*. Restated, each test point must be correlated with one or more specifications. If they are so correlated, then those specifications are also tested. In practice, critical specifications of the waveform-oriented testing are considered. For example, Table 11.1 presents the relationships between the four test points and the circuit specifications shown in Figure 11.8b.

Waveform-oriented testing items are not limited to the sampled points of a response waveform in the time domain. The amplitude and phase responses in the frequency domain can also be measured. For example, the tested items for a 1-MHz Sallen–Key second-order low-pass filter can include the amplitude responses at 0.1 MHz, 1 MHz, and 10 MHz, from which, the DC gain, the 3-dB frequency, and the stop-band attenuation rate can be derived and tested.

The margins of waveform-oriented testing are commonly obtained by correlating the results of specification-oriented testing with those of waveform-oriented testing. A limited number of chips are tested using both methods. Those that pass the specification-oriented test determine the margins for the waveform-oriented testing. Such a method is referred to as **correlation** in testing. The margin can also be obtained by fault simulation; however, great care must be taken because the variation in the parasitic effects of the test environment can be large. These variations include tester to tester, load board to load board, and pin to pin.

11.2.2 Analog Test Waveforms

Commonly used analog test waveforms include sine, ramp, step, triangular, chirp, arbitrary, and synthesized waveforms, which are presented in Figure 11.9.

Sinusoidal waveforms are basic frequency-domain test waveforms. They are the easiest to generate with high quality. Noise and harmonic distortion can be filtered out using a band-pass filter with a high Q value. 1-KHz, 10-KHz, 100-KHz and 1-MHz sinusoidal waveforms are regarded as standard test waveforms and used extensively. **Step waveforms** are basic time-domain test waveforms. Many time-domain specifications are defined by the step responses. They are used to measure the step response of filters and amplifiers. They are often square waveforms. A step is not necessarily defined as a step change in voltage. For *phase-locked loop* (PLL) or *automatic gain control* (AGC) circuits, step stimuli can be step changes in the frequency from f_1 to f_2 or step changes in amplitude from V_{A1} to V_{A2}. Accordingly,

TABLE 11.1 ■ Correlations between Test Points and Specifications in Figure 11.8b

Test Point	Specifications
A	DC bias, input offset
B	Slew rate, damping factor
C	Overshoot, damping factor, bandwidth
D	Settling time, DC gain

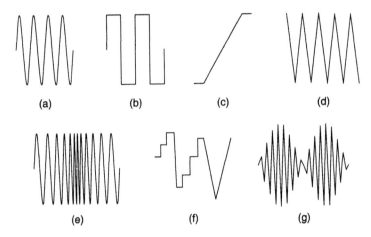

■ **FIGURE 11.9**

Analog test waveforms: (a) sinusoidal, (b) square (step), (c) ramp, (d) triangular, (e) chirp (sweep sine), (f) arbitrary, and (g) synthesized.

the **transient responses** of PLLs or AGCs can be tested to determine the behavior-level parameters, including the acquisition time and damping factors. Figure 11.10 presents these waveforms.

One important issue regarding a step waveform is the **edge rate** of the input step waveform. The *edge rate* is around $f = \dfrac{1}{3.5T_r}$. Herein, T_r is the rise time. For example, for a rise time of 1 μs, the edge rate is 286 KHz; hence, the circuit must have a bandwidth of at least 286 KHz to allow a step with a rise time of 1 μs to pass through. The edge rate is derived as presented in Figure 11.11. The rising edge is approximated as a sinusoidal waveform. For a sinusoidal wave, the gradient between ±45° and ±60° is quite linear, and the period of the sine wave is three to four times the rise time; therefore, the frequency of the edge can be approximated by $f = \dfrac{1}{3.5T_r}$.

Ramp waveforms are commonly employed in analog–digital converter testing. The sampling histograms can be used to determine the linearity of the conversion

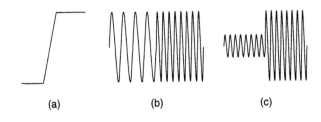

■ **FIGURE 11.10**

Step waveforms: (a) step in voltage, (b) step in frequency, and (c) step in amplitude.

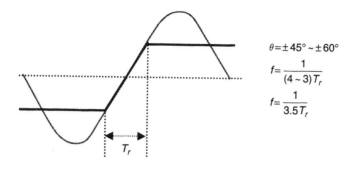

$$\theta = \pm 45^\circ \sim \pm 60^\circ$$

$$f = \frac{1}{(4 \sim 3)T_r}$$

$$f = \frac{1}{3.5T_r}$$

■ **FIGURE 11.11**

Edge rate of a step.

curve. The details are discussed in the following section. Ramps are commonly used as slow **triangular waveforms**, which have a much higher frequency than ramps. The fundamental difference between a ramp and a triangular waveform is that the ramp provides time for the circuit to reach its steady state whereas a triangular waveform tests the dynamic response. A triangular waveform has a frequency that is close to the circuit's normal operating frequency. The frequency of a ramp waveform is less than one-hundredth of the normal operating frequency. Notably, the second moment of a triangular waveform is discontinuous (a square waveform). These waveforms are of special use in testing systems of high order.

A **chirp** is also called a **sweep sine**. It is a sinusoidal waveform with a changing frequency. It is often generated using a **voltage-controlled oscillator** (VCO). The control signal can be a triangular or a sine wave. In most cases, a sine wave is preferred because all of the high-order moments are continuous. Chirps can be used to determine the frequency response of a filter.

Arbitrary waveforms are generated by **arbitrary waveform generators** (AWGs) in which ADCs are the core modules. AWGs can also be used to generate the waveforms described above. **Synthesized waveforms** are especially useful. They are commonly referred to as *RF-modulated signals*, such as **amplitude modulation** (AM), **frequency modulation** (FM), **frequency shift keying** (FSK), **binary phase shift keying** (BPSK), and **quadrature phase shift keying** (QPSK), among others. They are used in communication circuit testing.

11.2.3   DC Parametric Testing

DC parametric testing measures the DC characteristics of the CUT presented in the data sheet. Common test items are associated with various circuits. Most are I/O electric parameters, such as output rated voltage and current, input offset voltage and current, input and output impedance, and others. Specific test items include open-loop gain and unit gain bandwidth for operational amplifiers, line and load regulations for voltage regulators, and differential and integral nonlinearity for ADCs, among many others.

TABLE 11.2 ■ Operational DC Parametric Test Items

Rated output current	Rated output voltage
Open-loop gain	Slewing rate
Unity gain full power response	Unity gain small signal response
Overload recovery	Input bias current
Input offset voltage	Input offset current
Input noise	Input impedance
Supply voltage sensitivity	Common mode rejection
Maximum voltage between inputs	Maximum common mode voltage
Temperature drift	

Source: [Stata 1967].

The test items and test approaches are included in the application notes provided by the IC vendors. Table 11.2 presents the test items for operational amplifiers from Analog Devices as an example. AMS circuits cover a wide range of circuits and applications, so they cannot all be studied herein; therefore, this section focuses on some common and popular test items.

11.2.3.1   Open-Loop Gain Measurement

Figure 11.12 presents an open-loop gain measurement setup [Stata 1967]. An operational amplifier is typically a first-order system that is modeled by an open-loop gain (DC gain) and a pole at its 3-dB frequency as shown in Figure 11.12a. The open-loop gain A_o is 80 dB or 10,000, and the 3-dB frequency is 100 Hz. Given the test setup presented in Figure 11.12b, the voltage divider at the negative input (10K Ω/100 Ω) enhances the sensitivity of the input differential voltage. The inputs are low-frequency

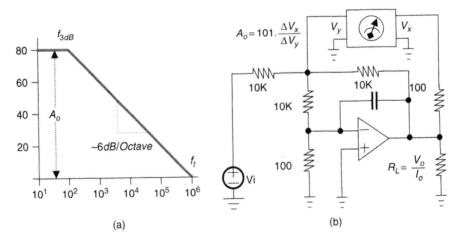

(a)   (b)

■ **FIGURE 11.12**

Open-loop gain measurement: (a) Bode plot of an operational amplifier, and (b) test setup.

AC signals. The frequency is often lower than one-tenth of the 3-dB frequency of the amplifier. Two inputs with different amplitude are applied and two measurements are made. The open-loop gain is derived using the following equation:

$$A_o = 101 \cdot \frac{\Delta V_x}{\Delta V_y} \tag{11.4}$$

As presented in Figure 11.12b, the load resistance is $R_L = \dfrac{V_o}{I_o}$. Here, V_o is the maximal output voltage and I_o is the maximal output current. Restated, the amplifier is operated at its rated output voltage and current conditions.

11.2.3.2 Unit Gain Bandwidth Measurement

Unit gain bandwidth f_t is the frequency at which the open-loop gain is unity. It is also called the **gain-bandwidth product**. If the amplifier is a first-order system with 6-dB/Octave roll-off as presented in Figure 11.12a, then:

$$f_t = A_o \cdot f_{3dB} \tag{11.5}$$

The term f_t is also referred to as the **small signal unit gain bandwidth**, where "small signal" relates to the fact that the signal is sufficiently small so that it exhibits no distortion. Many nonideal characteristics may cause distortion. Herein, the *slew rate limitation* is the most important factor. The **slew rate** (SR) is the maximal rate at which the output waveform rises. Under the slew rate limitation, the input amplitude is constrained as follows:

$$V_i \leq \frac{SR}{2\pi f_t} \tag{11.6}$$

Figure 11.13 presents two setups, in inverting and noninverting configurations, for measuring the unit gain bandwidth. Stray capacitance is associated with the

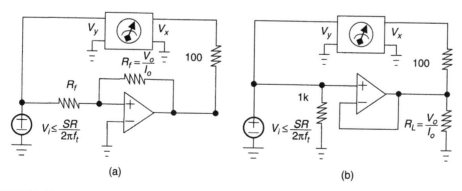

(a)   (b)

■ **FIGURE 11.13**

Test setup for measuring unit gain bandwidth: (a) inverting configuration, and (b) noninverting configuration.

feedback resistor in the inverting configuration and affects the accuracy of measurement; therefore, the noninverting configuration is preferred because its parasitic effects are smaller.

Full power unit gain bandwidth f_p is the maximal frequency measured at a closed-loop gain of unity for which the rated output voltage and current are obtained without distortion, because the slew rate is limited. In addition to testing the unit gain bandwidth, the setups in Figure 11.13 can also test the slew rate and maximal common mode voltage.

11.2.3.3 Common Mode Rejection Ratio Measurement

If two differential inputs are connected to the same input, as presented in Figure 11.14, then the output voltage should be zero for an ideal differential amplifier. The shift in the biasing condition caused by the common mode input and the slightly different gains of the positive and negative inputs cause the output to vary slightly from zero.

For a change ΔV_{CM} in the common mode, the output voltage changes by ΔV_o. Dividing ΔV_o by the DC gain ($A_o = R2/R1$) yields the **input referred common mode voltage**:

$$V_{CM\cdot i} = \frac{\Delta V_o}{A_o} = \Delta V_o \Big/ \frac{R2}{R1} \tag{11.7}$$

Restated, an input common mode change of ΔV_{CM} has the same effect as a differential mode signal of $\Delta V_o/A_o$; therefore, the **common mode rejection ratio** (CMRR)

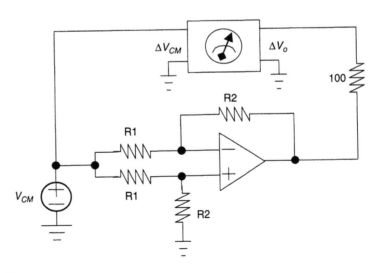

■ **FIGURE 11.14**

Test setup for measuring common mode rejection ratio.

is defined as the ratio of the common mode voltage to the input referred common mode voltage (in dB):

$$\text{CMRR} = 20\log\left(A_o \middle/ \frac{\Delta V_o}{\Delta V_{CM}}\right) \tag{11.8}$$

CMRR specifies the extent to which a circuit is immune from common mode variation.

11.2.3.4 Power Supply Rejection Ratio Measurement

The *power supply rejection ratio* (PSRR) is similar to CMRR. It shows the extent to which a circuit is immune to supply voltage variation. Changing the supply voltage changes the biasing current and voltage; therefore, a fluctuation in the supply voltage occurs at the output. As presented in Figure 11.15, for a supply voltage change of ΔV_{DD} and an output voltage change of ΔV_o, the PSRR is defined as:

$$PSRR = 20\log\left(A_o \middle/ \frac{\Delta V_o}{\Delta V_{DD}}\right) \tag{11.9}$$

Restated, a fluctuation in the power supply of ΔV_{DD} can be considered to be a differential input with an amplitude of $\Delta V_{DD}/PSRR$.

11.2.4 AC Parametric Testing

AC parametric testing refers to the testing of the AC characteristics of a circuit. AC parametric testing relates to frequency and timing parameters, including bandwidth, phase, distortion, noise, and other factors. Conventional AC parametric testing uses dedicated instruments to make measurements. Today, most mixed-signal testers have a built-in *digital signal processing* (DSP) module. The details of DSP algorithms can be found elsewhere [Oppenheim 1989]. DSP-based test techniques have also been described elsewhere [Mahoney 1987]. AC parametric tests are commonly conducted using the setup presented in Figure 11.16.

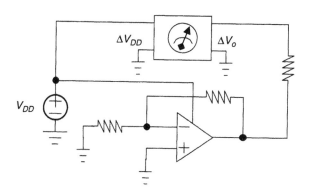

■ **FIGURE 11.15**

Test setup for measuring power supply rejection ratio.

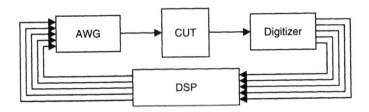

Test setup for DSP based testing.

Test stimuli are synthesized using the DSP and generated by an *arbitrary wave-form generator* (AWG). After the test waveforms have been applied, the test response is captured and transformed into digital form using a digitizer. Then, they are analyzed using a DSP processor. Many parameters can be evaluated by analyzing a single-response waveform using various DSP programs. This concept is also referred to as **virtual instrumentation**. The core modules of AWGs and digitizers are DACs and ADCs.

11.2.4.1 *Maximal Output Amplitude Measurement*

Maximal output amplitude or **maximal output swing** is the maximal amplitude of the distortion-free sinusoidal output waveform. Conventionally, the input amplitude is increased slowly until the output is distorted. The approach is time consuming because the search for the maximal input amplitude is iterative. Additionally, it requires a distortion meter to make the measurements. DSP-based testing is performed in the following steps (Figure 11.17 presents the block diagram):

1. A large-input sinusoidal waveform, usually at 1 KHz, is applied to cause a slight clipping of the output waveform.

2. The output waveform is digitized by the digitizer.

3. *Fast Fourier transformation* (FFT) transforms the time-domain response waveform into a frequency-domain series.

The amplitude of the fundamental component is the maximal output amplitude.

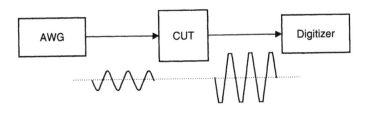

Maximal output amplitude measurement.

The maximal output amplitude can detect the first-order defects. Any changes in biasing current and/or voltage influence the output signal swing: For example, a drop in the output biasing current increases the biasing voltage; therefore, the headroom, which is the difference between V_{DD} and DC bias, is reduced. The maximal output swing is also reduced.

11.2.4.2 Frequency Response Measurement

Frequency response is conventionally measured using network and spectrum analyzers. The cost of this instrument is an issue, and the test time overhead is another because of the frequency sweeping mechanism. A spectrum analyzer uses one single-frequency test signal at a time to measure the amplitude response at a particular frequency. The response of the entire spectrum is determined after all of the frequencies have been swept.

Frequency analysis is very important for a filter. Filters are of four generalized types: low-pass, high-pass, band-pass, and band-reject or notch filters. Figure 11.18 plots their frequency-domain transfer functions. Consider the band-pass filter presented in Figure 11.19 as an example of how frequency-domain testing can be conducted using the setup in Figure 11.16. For a filter, **pass bands** and **stop bands** are first defined. Between them are the **transition bands**. The pass band is the frequency range in which a signal can pass through with little attenuation. **Pass-band ripple** models the variation of the amplitude response in the band. Signals in the stop band are heavily attenuated. The attenuation from the pass band to the stop band is defined as the **stop-band rejection ratio** (in dB). The gradient in the transition band is determined by the order of the filter. It is 6 dB/Octave per order.

In frequency-domain testing, two masks, the *upper limit mask* and the *lower limit mask*, are defined, as presented in Figure 11.20. The frequency response must fall

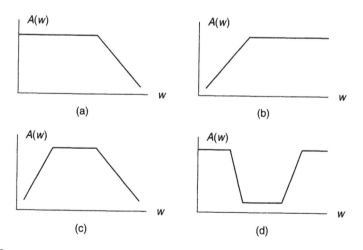

■ **FIGURE 11.18**

Frequency-domain transfer function of (a) low-pass filter, (b) high-pass filter, (c) band-pass filter, and (d) notch or band-reject filter.

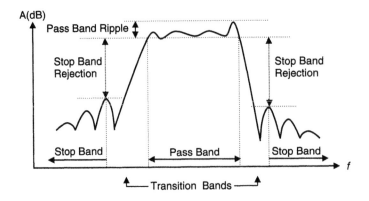

Frequency-domain transfer function of a band-pass filter.

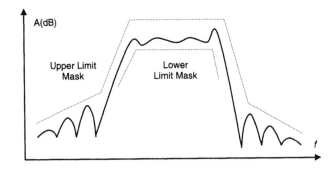

Upper and lower limit masks.

between these two limits to pass the test. A multitone signal, Eq. (11.10), is applied to test a filter:

$$v(t) = \sum A_i \sin(2\pi f_i t + \phi_i) \qquad (11.10)$$

Each tone has an amplitude of A_i, a frequency of f_i, and a phase of ϕ_i. The frequencies are carefully chosen such that those of greater interest have a higher tone density. The corners of the transitions from the pass band to the stop band, or *vice versa*, contain more information than pass and stop bands; therefore, their tone densities are higher as well, as shown in Figure 11.21. The multitone waveform can be precalculated or calculated online, then it can be generated using an AWG, as presented in Figure 11.16.

Notably, the sampling rate of AWGs must be at least ten times the highest frequency component to obtain a waveform of sufficient quality. Additionally, the number of tones must not be too large to cause AWG to saturate. Here, the highest

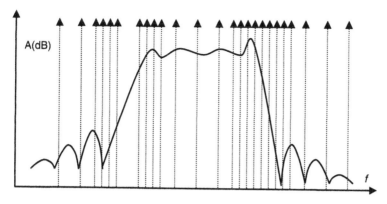

■ **FIGURE 11.21**

Multitone test signals.

allowed voltage level is the sum of all A_i; therefore, a larger number of tones corresponds to a smaller A_i. As a result, the signal-to-noise ratio of a tone is lower and the measurement is less accurate; thus, a tradeoff exists between frequency resolution (number of tones) and measurement accuracy.

After the response waveform has been captured, digitized, and transformed into the frequency domain, the resulting spectrum may look like that presented in Figure 11.22. If all of the frequency components are between upper and lower limits, then the filter is passed. If any tone is outside of these limits, then the test is failed.

11.2.4.3 SNR and Distortion Measurement

The **signal-to-noise ratio** (SNR) is the ratio of the signal power to the noise power (represented in dB), indicating the purity of the signal. An electronic system

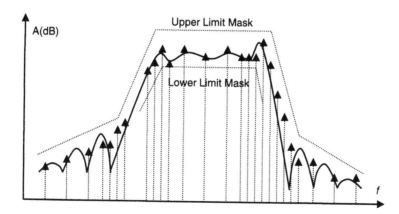

■ **FIGURE 11.22**

Multitone test response.

has many noise sources. Common noise sources include thermal noise, flicker noise, and shot noise. The integrated circuits may have noise due to power supply noise, circuit switching noise, substrate noise, and other types. Additionally, distortions, including harmonic distortion and intermodulation distortion, also reduce the SNR. Distortions are commonly caused by crossover distortion, clip, saturation, and mismatches between differential signal paths. SNR and distortion can be measured using the DSP technique as follows.

The setup of the AWG presented in Figure 11.16 generates a pure sinusoidal waveform that is applied to the CUT. The response waveform is captured, digitized, and transformed into the frequency domain. The resulting frequency-domain response may look like that presented in Figure 11.23.

The frequency components of a sinusoidal waveform are classified into three categories. F is the fundamental component, which is also the signal of interest; H_i is the ith harmonic; and N_i represents the ith noise term. SNR and distortion can be determined from the following equations:

$$\text{SNR} = 10 \log \frac{F^2}{\sum N_i^2} \tag{11.11}$$

$$\text{THD} = 10 \log \frac{F^2}{\sum H_i^2} = 100 \times \frac{F^2}{\sum H_i^2}\% \tag{11.12}$$

$$\text{SNDR} = 10 \log \frac{F^2}{\sum H_i^2 + \sum N_i^2} \tag{11.13}$$

F is either a voltage or a current, and F^2 is the corresponding power, given by $P = I^2 R = V^2/R$. Therefore, only noise components N_i are considered in the calculation of SNR. THD is the **total harmonic distortion**. It is the ratio of signal power to total harmonic power; therefore, harmonic terms H_i are included. THD can be represented in either dB or as a percentage. SNDR is the **signal-to-noise and distortion ratio**; both noise and distortion are included.

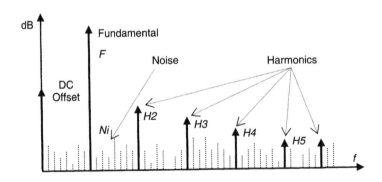

■ **FIGURE 11.23**

Spectrum with noise and distortion.

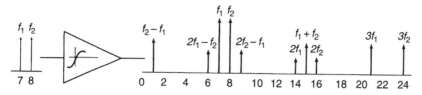

Spectrum with noise and distortion.

In addition to SNR, THD, and SNDR, the *peak harmonic* is also an important test item. Harmonics commonly have a peak at either the second or the third harmonic, from which the source of the distortion can be determined. If the peak corresponds to the second harmonic, then a crossover distortion or a symmetric nonlinear distortion is more likely. If the peak corresponds to the third harmonic, then the source of the distortion may be clipping or saturation.

11.2.4.4 *Intermodulation Distortion Measurement*

The nonlinear characteristics of a circuit cause the intermodulation of various components of a signal, in what is called ***intermodulation distortion*** (IMD). IMD is of special interest in relation to communication circuits because it may modulate the tone into adjacent bands and cause interference. IMD testing uses multitone signals; however, only two or three tones are used. Figure 11.24 presents the spectrum of the response waveform with two tones. The test signal is:

$$v(t) = A_1 \sin 2\pi f_1 t + A_2 \sin 2\pi f_2 t \qquad (11.14)$$

Any linear combinations of f_1 and f_2 may appear due to the intermodulation distortion. As presented in the figure, the closest intermodulation terms to the fundamental are the IM3 terms with frequencies at $2f_1 - f_2$ and $2f_2 - f_1$.

11.3 MIXED-SIGNAL TESTING

Mixed-signal circuits contain both analog and digital signals. *Analog-to-digital converters* (ADCs) and *digital-to-analog converters* (DACs) are two prominent examples. They allow digital circuits to interface directly with the real, analog world; however, many other circuits also contain mixed signals, such as phase-locked loops, delay-locked loops, automatic gain control circuits, switched capacitor filters, and frequency synthesizers, among others. They cannot all be discussed here; this work focuses on the testing of ADCs and DACs, because these circuits are commonly used and can be specified in the same way across various application domains and a wide performance spectrum.

Both ADCs and DACs have a wide range of applications and performance requirements. A high bit length and a low data rate are two distinguishing features in audio applications. The bit length commonly exceeds 12 bits and can be as high as 22 bits, and the conversion rate is around 1 Mbps. For video applications, the corresponding values are 8~12 bits and 10 to 100 Msps (*mega samples per second*). For wireless communication, it is 8~12 bits and 10 to 100 Msps; for high-speed data communication, it is 4~6 bits and over Gsps (*giga samples per second*). Even though they have wide application domains and a wide performance spectrum, they are specified in the same way.

11.3.1 Introduction to Analog–Digital Conversion

The purpose of an ADC is to covert an analog signal to a digital one. An ADC partitions a **conversion range** (V_{PP}) into ($2^n - 1$) **quantization steps** (q). It is also referred to as a *least significant bit* (LSB); therefore, an LSB is:

$$\text{LSB} = \frac{V_{pp}}{2^n - 1} \tag{11.15}$$

where n is the bit length. For a V_{PP} of 1 V and an n of 10, 1 LSB is around 1 mV. The **resolution** of an n-bit ADC is 2^n. Figure 11.25 plots the transfer characteristic curve of an ADC. Quantization proceeds as follows. For a voltage $(K - 0.5)\text{LSB} \le V_x < (K + 0.5)\text{LSB}$, it is regarded as K LSB; therefore, the maximal error is 0.5 LSB.

The error generated in the quantization process is called the **quantization error**. It is also called the **quantization noise**. Figure 11.26 presents the quantized and quantization error waveforms of a sine wave. The characteristics of the quantization errors include the following:

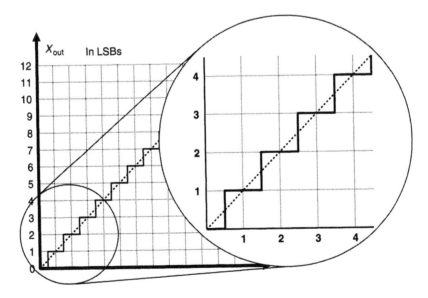

■ **FIGURE 11.25**

Transfer characteristic curve of an ADC.

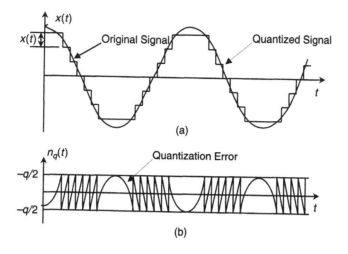

■ **FIGURE 11.26**

Quantization: (a) quantized waveform, and (b) quantization error.

- The quantization error ε is $-0.5\text{LSB} \leq \varepsilon < 0.5\text{LSB}$.
- The quantization error waveform contains many jumps.
- The error spectrum is much wider than the original signal.
- The bandwidth of the quantization error is proportional to the gradient of the signal and inversely proportional to the quantization step q.

Following quantization using an n-bit ADC, the quantization noise can be derived and transformed into the SNR. Suppose that the input sinusoidal waveform of amplitude V_A is quantized using an n-bit ADC with a conversion range of $2V_A$ at the maximal resolution. The quantization step size q is:

$$q = \frac{2V_A}{2^n} \tag{11.16}$$

The signal power is:

$$P_S = \frac{V_A^2}{2} \tag{11.17}$$

The quantization error is a sawtooth-like waveform, so its power can be approximated as a triangular waveform with an amplitude of $0.5q$; therefore, the integrated noise power is:

$$P_N = \frac{(q/2)^2}{3} = \frac{q^2}{12} = \frac{V_A^2}{3 \times 2^{2n}} \tag{11.18}$$

The signal power in Eq. (11.17) and the noise power in Eq. (11.18) yield the SNR:

$$\text{SNR} = 10\log\frac{P_S}{P_N} = 10\log(1.5 \times 2^{2n}) = (1.76 + 6.02 \times n)\,\text{dB} \tag{11.19}$$

For example, for a 10-bit ADC, the theoretical SNR is 61.96 dB after quantization. Conventional wisdom is that each added bit improves the SNR by 6 dB. With reference to Eq. (11.19), a suitable bit length can be selected based on the analog signal noise level. The rule of thumb is that the SNR should be chosen to be at least 3~6 dB more than that of the analog signal to ensure that the signal does not deteriorate. For example, if the analog signal has a SNR of 52 dB, then the bit length should be at least 9 bits to yield a SNR of 55.94 dB.

11.3.2 ADC and DAC Circuit Structure

Figure 11.27 presents an ADC topology, which is comprised of a gain stage, a filter, a multiplexer, a sample and hold circuit, and an ADC. The function of each module is detailed as follows:

- **Gain stage**—The gain stage has two functions. First, it provides an offset to shift the signal to the center of the conversion range. Second, it provides a gain that enlarges the signal to the full conversion range. These two functions allow the full range of ADC to be used and maximize the resolution. Issues that require special consideration include the noise, the nonlinearity, and the drift of the amplifier; therefore, calibration is required to minimize the error.

- **Filter**—The filter suppresses off-band noise. It is also referred to as an anti-aliasing filter because it prevents the off-band noise from aliasing into the signal band after sampling. The *oversampling ratio* (OSR) is defined as the ratio of the sampling frequency to twice the signal bandwidth. The higher the OSR is, the smaller order the filter can be. Figure 11.28 presents

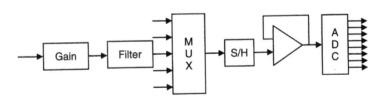

■ **FIGURE 11.27**

An ADC architecture.

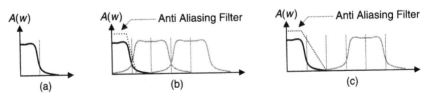

■ **FIGURE 11.28**

Sampling spectrum and anti-aliasing filter: (a) signal spectrum, (b) Nyquist rate sampling, and (c) 4× oversampling.

the signal spectrum following sampling and the meeting of the anti-aliasing filtering requirement. The **Nyquist rate** is twice of the maximal frequency of the signal, according to the Nyquist theorem. As shown in Figure 11.28b, in Nyquist rate sampling the off-band noise is aliased into the band even though a very good filter is used. 4× oversampling in Figure 11.28c does not require as high a stop-band attenuation rate as that in Figure 11.28b.

- **MUX stage**—The *multiplexer* (MUX) provides multiple channel access. The main issue associated with MUX is signal coupling caused by stray capacitance in the switches, especially CMOS switches. In CMOS switches, C_{gs} and C_{gd} are the two main coupling capacitors. To reduce the channel resistance, the transistors must be enlarged. Unfortunately, C_{gs} and C_{gd} are also increased.

- **S/H Stage**—The *sample/hold* (S/H) stage is used to sample the signal and hold it steady for one sampling period. S/H can be placed before or after the MUX. If it is placed before the MUX, then the coupling effect is minimized because the signal at input of the MUX is held steady by the S/H, and no coupling occurs. Figure 11.29 presents a S/H circuit and its operating mode.

   During the sampling phase ($t1 \sim t3$), the switch in Figure 11.29a is closed. The hold capacitor C_H is charged. At the **aperture time** $t3$, C_H is charged to its steady-state voltage within an acceptable tolerance. Normally, the tolerance is under 0.5 LSB. The switch is then opened. ($t1 \sim t3$) is called the **acquisition time**. The aperture time is uncertain and may vary from $t2$ to $t4$. Such a timing uncertainty is referred to as **jitter**. Another important specification of S/H is the **droop rate**. It is caused by the leakage current I_{Leak}. The rate is defined as:

$$V_{droop} = \frac{I_{Leak}}{C_H} \tag{11.20}$$

The leakage in a CMOS IC includes the subthreshold leakage of the channels, the reverse bias saturation leakage of the PN junctions, and the dielectric leakage of the gate oxide layer.

- **ADC**—ADC performs the actual *analog-to-digital conversion*. The important specifications of an ADC include bit length, conversion range, conversion rate, and signal bandwidth. A detailed discussion of ADC follows.

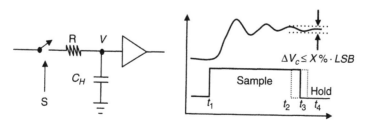

■ **FIGURE 11.29**

Sample and hold (S/H) circuit: (a) circuit structure, and (b) operational mode.

Different ADC and DAC architectures are used for different conversion rates and bit-length requirements. An example in each category is presented as a reference for further discussion. Detailed information is available in ADC and DAC design-related reference books and papers.

11.3.2.1 DAC Circuit Structure

Digital-to-analog converters convert a series of digital data to an analog waveform to drive real-world devices with audio, video, and data communication applications. DACs are of three types: resistor network, capacitor network, and transistor network. Most DACs implemented by resistor and capacitor networks are in voltage mode, while DACs implemented by transistor networks are in current mode.

Here, a conventional resistor network is considered as an example. A resistor network can be a binary weighted or resistive ladder network. Figure 11.30 presents an **R-2R ladder DAC**. The reference voltage (V_{ref}) is divided into $(1/2)V_{ref}$, $(1/4)V_{ref}$, $(1/8)V_{ref}$, and others by the R-2R ladder. They are summed by the summation circuit according to the digital data ($S_0 \sim S_5$). The output voltage is given by the following equation:

$$V_o = S_5 \cdot \frac{V_{ref}}{2^1} + S_4 \cdot \frac{V_{ref}}{2^2} + S_3 \cdot \frac{V_{ref}}{2^3} + S_2 \cdot \frac{V_{ref}}{2^4} + S_1 \cdot \frac{V_{ref}}{2^5} + S_0 \cdot \frac{V_{ref}}{2^6}$$

$$= (S_5 \cdot 2^5 + S_4 \cdot 2^4 + S_3 \cdot 2^3 + S_2 \cdot 2^2 + S_1 \cdot 2^1 + S_0 \cdot 2^0) \cdot \frac{V_{ref}}{2^6} \qquad (11.21)$$

From the above equation, digital data ($S_0 \sim S_5$) are converted into an analog voltage.

11.3.2.2 ADC Circuit Structure

An ADC converts an analog signal into digital data. Three popular ADC architectures are **flash ADC**, **sigma-delta ADC**, and **pipelined ADC**. A *flash ADC* compares the input voltage to a set of reference voltages and generates a digital output based on the results. It is also called a *parallel ADC*. The most significant feature of a flash ADC is its high speed. The speed can be as high as 1 Gsps (*giga samples per second*). It performs parallel comparison, so the hardware complexity is exponentially proportional to the bit length; therefore, the bit length is unlikely to be very

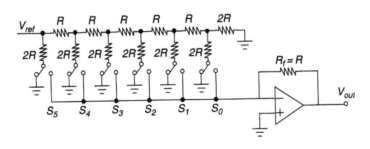

■ **FIGURE 11.30**

6-bit R-2R ladder DAC.

large. The bit length is commonly under 6 bits. *Sigma-delta ADCs* use sigma-delta modulation with high oversampling rates to achieve high resolution. They are also referred to as **oversampling ADCs**. The most important feature of a sigma-delta ADC is its high resolution. The resolution exceeds 14 bits and can be as high as 22 to 24 bits. *Pipelined ADCs* exhibit performance between that of flash and sigma-delta ADCs. It has a medium bit length and speed. Table 11.3 compares the performances of the three types of ADCs.

Resolution and throughput are two conflicting requirements. Achieving high throughput and high resolution simultaneously is difficult. Table 11.4 presents the resolution and throughput selection matrix from the data sheet of Analog Devices.

Consider an 8-bit *pipelined ADC* as an example. Figure 11.31 presents the corresponding block diagram. In each stage, the input is sampled by a *sample/hold* (S/H) module, then it is coarsely quantized by a 3-bit ADC. The outcome is transformed into a precise voltage by a 3-bit DAC. The DAC output is subtracted from the sampled valued. The result is called a **residue**. The residue is amplified eight times and sent to the next stage for finer quantization. The subtract-and-amplify module is referred to as a **residue amplifier**. A (3, 3, 3, 2) architecture with digital correction is used to minimize errors.

11.3.3 ADC/DAC Specification and Fault Models

As discussed above, ADCs and DACs are very complex mixed-signal circuits. Different circuit structures yield different resolutions and satisfy different throughput

TABLE 11.3 ■ Comparison of ADCs

ADC	Bit Length	Throughput
Flash	~6 bits	100+ M
Pipelined	8~16 bits	10~100 MHz
Sigma-delta	14+ bits	~10 M

TABLE 11.4 ■ ADC Selection Matrix from Analog Devices

Resolution/Throughput Selection Matrix

	<10 Kbps	10 Kbps to 100 Kbps	100 Kbps to 1 Mbps	1 Mbps to 10 Mbps	10 Mbps to 100 Mbps	100+ Mbps
17+	•	•	•	•		
14–16	•	•	•	•	•	•
12–13		•	•	•	•	•
10–11		•	•	•	•	•
8–9			•	•	•	•
<8			•	•	•	•

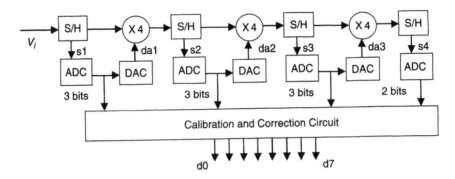

■ **FIGURE 11.31**

Eight-bit pipelined ADC.

requirements. Deriving a circuit-level fault model for all of them is difficult; therefore, a parametric fault model is defined based on the transfer function of ADCs.

Before the ADC fault model is outlined, consider AD775 as an example. Figure 11.32 presents the data sheet for AD775, which is an 8-bit 20-Msps ADC. The main specifications include resolution (8-bit), throughput (20 Msps), integral nonlinearity (1.3 bits), differential nonlinearity (0.5 bits), missing code (no), offset ($-60 \sim 45$ mV), gain error (1%), phase error (0.5°), SNDR ($41 \sim 47$ dB), THD ($-42 \sim -51$ dB), and others. These specifications are also used in fault models for testing.

- **Offset Error**—The offset error is the input voltage that generates the first code. It is represented in terms of mV or LSB. In the example presented in Figure 11.33, the transfer curve is offset by -3 LSB, so every code that is produced is 3 LSB less than the correct code.

- **Gain error**—Gain error is the deviation of the slope of the transfer curve from the ideal one. It is also called the **calibration error** because the gain can be calibrated. It is represented as the deviation of the code from the full scale code when a full scale input is applied. In Figure 11.33, the gain error is 4 LSB and is sometimes represented as a percentage.

- **Nonlinearity error**—An ideal transfer characteristic curve of an ADC is a straight line; however, the nonlinearity of the devices that form the ADC may cause the transfer curve to not be straight. As presented in Figure 11.33, the transfer curve is nonlinear. Differential nonlinearity and integral nonlinearity are defined to classify the nonlinear effects.

- *Differential nonlinearity error* (DNL)—Ideally, the code is increased by one for every increase by a quantization step q (LSB) in the input voltage. Differential nonlinearity is the maximal deviation of the input voltage from 1 LSB when the output code increases by 1, as presented in Figure 11.34. Herein, the DNL is 0.8 LSB because the input voltage must increase by 1.8 LSB to change the code from 1 to 2.

AD775—SPECIFICATIONS $(T_A = +25°C$ with AV_{DD}, $DV_{DD} = +5$ V, AV_{SS}, $DV_{SS} = 0$ V, $V_{RT} = 2.6$ V, $V_{RB} = +0.6$ V, CLOCK = 20 MHz unless otherwise noted)

Parameter	Min	AD775J Typ	Max	Units
RESOLUTION	8			Bits
DC ACCURACY				
Integral Nonlinearity (INL)		+0.5	1.3	LSB
Differential Nonlinearity (DNL)		±0.3	±0.5	LSB
No Missing Codes		GUARANTEED		
Offset				
To Top of Ladder V_{RT}	−10	−35	−60	mV
To Bottom of Ladder V_{RB}	0	+15	+45	mV
VIDEO ACCURACY[1]				
Differential Gain Error		1.0		%
Differential Phase Error		0.5		Degrees
ANALOG INPUT				
Input Range (V_{RT}–V_{RB})		2.0		V p-p
Input Capacitance		11		pF
AC SPECIFICATIONS[2]				
Signal-to-Noise and Distortion (S/(N + D))				
$f_{IN} = 1$ MHz		47		dB
$f_{IN} = 5$ MHz		41		dB
Total Harmonic Distortion (THD)				
$f_{IN} = 1$ MHz		−51		dB
$f_{IN} = 5$ MHz		−42		dB
REFERENCE INPUT				
Reference Input Resistance (R_{REF})	230	300	450	Ω
Case 1: $V_{RT} = V_{RTS}$, $V_{RB} = V_{RBS}$				
Reference Bottom Voltage (V_{RB})	0.60	0.64	0.68	V
Reference Span (V_{RT}–V_{RB})	1.96	2.09	2.21	V
Reference Ladder Current (I_{REF})	4.4	7.0	9.6	mA
Case 2: $V_{RT} = V_{RTS}$, $V_{RB} = AV_{SS}$				
Reference Span (V_{RT}–V_{RB})	2.25	2.39	2.53	V
Reference Ladder Current (I_{REF})	5	8	11	mA
POWER SUPPLIES				
Operating Voltages				
AV_{DD}	+4.75		+5.25	Volts
DV_{DD}	+4.75		+5.25	Volts
Operating Current				
IAV_{DD}		9.5		mA
IDV_{DD}		2.5		mA
$IAV_{DD} + IDV_{DD}$		12	17	mA
POWER CONSUMPTION		60	85	mW
TEMPERATURE RANGE				
Operating	−20		+75	°C

NOTES
[1]NSTC 40 IRE modulation ramp. CLOCK = 14.3 MSPS.
[2]f_{IN} amplitude = 0.3 dB full scale.

Specifications subject to change without notice. See Definition of Specifications for additional information.

■ **FIGURE 11.32**

Data sheet for Analog Devices' AD775.

- ***Integral nonlinearity error*** (INL)—Integral nonlinearity error is the maximal deviation from the ideal transfer curve in LSB. In Figure 11.34, the INL is 2.0 LSB because when the input is 6 LSB the output code is 4 rather than 6.

- **Temperature-dependent error**—Temperature-dependent error models the *temperature sensitivity* of solid-state devices. In general, the performance

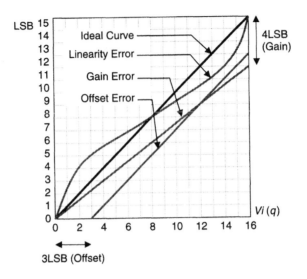

ADC transfer curves: (a) ideal, (b) linearity error, (c) gain error, and (d) offset error.

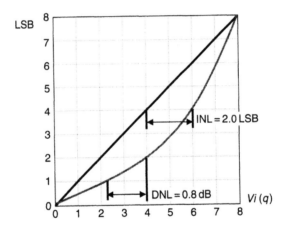

Nonlinearity errors: (a) differential nonlinearity, and (b) integral nonlinearity.

decreases as the temperature increases; therefore, the gain depends on the temperature. It exhibits the same behavior as the gain error except that the magnitude of the error depends on temperature. For the example presented in Figure 11.35, the temperature-dependent gain error is 4 LSB at 75°C. It is also sometimes represented as a percentage.

■ **Load-dependent error**—The load-dependent error models the improper output impedance of the output drivers. For a voltage mode DAC, if the

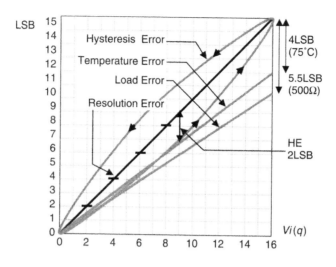

■ **FIGURE 11.35**

ADC faults: (a) hysteresis error, (b) temperature-dependent error, (c) load error, and (d) resolution error.

output impedance is too large, the gain decreases. The small load resistance is regarded as a large load herein. Similarly, for a current mode DAC, the gain decreases as the output impedance decreases. As presented in Figure 11.35, the load error is 5.5 LSB at a load resistance of 500 Ω, which is half of the rated 1-KΩ load. Again, the load error can also be expressed as a percentage.

- **Hysteresis error**—The transfer curve may exhibit hysteresis behavior, as presented in Figure 11.35. For a push–pull class-AB amplifier, hysteresis phenomena are common because the circuit that pulls up the signal differs from the circuit that pulls down the signal. The model is similar to that of the nonlinear error discussed above. The difference is that as the input signal ascends it follows one path but follows another as it descends. The model is similar to the INL, which models the maximal deviation from the ideal curve in LSB. In the example, the hysteresis error is 2 LSB.

- **Resolution error**—The resolution error captures the inability to resolve a small variation in the signal. For an n-bit ADC, the resolution is 2^n; however, noise and distortion reduce the resolution. As presented in Figure 11.35, the resolution is halved. The resolution error is represented as an effective number of bits, as discussed below.

- **Missing code**—Missing code refers to cases in which some codes are never generated. It differs from resolution error, which refers to the random appearance of codes below the resolution, and can be regarded as a random noise. Missing code generates errors at a particular voltage level. For a sinusoidal input waveform, the error is periodic, as presented in Figure 11.36; therefore, the error syndrome is the harmonic distortion in frequency domain.

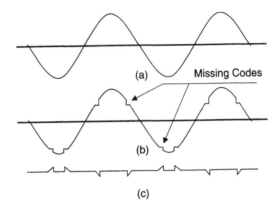

Missing code error: (a) ideal waveform, (b) quantized waveform with missing codes, and (c) error waveform.

- **Signal-to-noise ratio, signal-to-noise and distortion, total harmonic distortion, intermodulation distortion**—These noise- and distortion-related faults have the same definitions as used in analog AC parametric testing. They were discussed in Section 11.2.4 and will not be repeated here. However, for ADCs, the *effective number of bits* (ENOB) is defined based on the SNDR. As derived in Eq. (11.19), SNR = $(1.76 + 6.02n)$ dB, and the ENOB is defined as follows:

$$\text{ENOB} = \frac{\text{SNDR} - 1.76}{6.02} \qquad (11.22)$$

For example, if the SNDR of a 14-bit ADC is 70 dB, then the ENOB is 11.34 bits. Therefore, 2.66 bits of resolution error pertain, as defined above.

11.3.4   IEEE 1057 Standard

The **IEEE 1057-1994 standard** (*IEEE Standard for Digitizing Waveform Recorders*) is a very important standard for understanding ADCs and methods of testing them. It is described briefly as follows [IEEE 1057-1994].

- **Scope**—The instruments covered by the standard include electronic digitizing waveform recorders, waveform analyzers, and digitizing oscilloscopes with digital outputs. This standard applies to, but is not restricted to, general-purpose waveform recorders and analyzers.

- **Purpose**—The purpose of this standard is to provide common methods for testing and terminology for describing the performance of waveform recorders, for the benefit of users and manufacturers of such devices. The main body presents many performance features, sources of error, and test methods.

Manufacturer-Supplied Information

General Information

Model number
Dimensions and weight
Power requirement
Environmental conditions (temperature, humidity, EMC/EMI, ...)
Any special or peculiar characteristics
Available options and accessories
Exceptions to the above parameters, where applicable
Calibration interval

Minimum Specifications

Number of digitizing bits
Sample rates
Memory length

Input impedance
Analog bandwidth
Input signal ranges

Additional Specifications

Gain
Offset
Differential nonlinearity
Integral nonlinearity
Harmonic distortion
Spurious response
Maximal static error
Signal to noise ratio
Effective bits
Peak error
Random noise
Frequency response
Settling time
Slew limit
Overshoot and precursors
Aperture uncertainty
Long-term stability
Maximum common mode signal level

Fixed error in sample time
Trigger delay and jitter
Trigger sensitivity
Trigger minimum rate of change
Trigger hysteresis band
Trigger coupling to signal
Crosstalk
Monotonicity
Hysteresis
Overvoltage recovery
Word error rate
Cycle time
Common mode rejection ratio
Differential input impedance
Maximum operating common
Mode signal level
Transition duration of step response

17 Test Methods

General methods
Input impedance
Gain and offset
Noise
Analog bandwidth
Frequency response
Step response parameters
Time base errors
Linearity, harmonic distortion,
and spurious response

Triggering
Crosstalk
Monotonicity
Hysteresis
Overvoltage recovery
Word error rate
Cycle time
Differential input specification

Source: [IEEE 1057-1994].

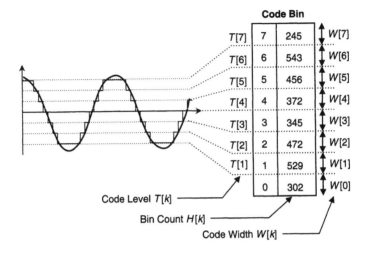

■ **FIGURE 11.37**

Code bins: (a) code level, (b) bin count, and (c) code width.

The IEEE 1057 standard is a very wide-ranging standard. It covers a total of 39 specifications and 17 test methods. For reasons of limited space, only selected test methods are described here.

11.3.5   Time-Domain ADC Testing

Time-domain ADC testing analyzes sampled data in the time domain to evaluate circuit parameters. Histograms are commonly used in time-domain testing. A periodic test waveform is applied here and the codes recorded. Commonly used waveforms include ramp and sine waves. The test items include gain, offset, and linearity errors. In IEEE 1057, code bins are defined as follows.

11.3.5.1   Code Bins

Figure 11.37 presents terms related to code bins. A **code bin** is a digital output that corresponds to a particular set of input values. Consider **code bin k** as an example. An input with value $T[k] \leq V_i < T[K+1]$ generates a digital output code k. Here, $T[k]$ is the **code level k**. It is also called the **code transition level**. $T[k]$ distinguishes code k from $k-1$. The code width k is defined as:

$$W[k] = T[k+1] - T[k] \tag{11.23}$$

After an input waveform has been sampled n times, the number of occurrences of code k is recorded in code bin k. Figure 11.37 presents the code bins of the samples on the input sine wave.

11.3.5.2 Code Transition Level Test (Static)

The code transition level test involves a programmable source, such as a DAC, whose range and output parameters are compatible with the waveform recorders and whose resolution is at least four times that of the recorder. The test step for code level $T[k]$ is as follows:

1. Apply an input voltage that is slightly lower than the expected code transition level.

2. Record N data.

3. If, according to the recorded data, over 50% of the codes are under k, then increase the level by $0.25q$ (0.25 LSB).

4. Repeat step 3 until the percentage drops to 50% or below.

5. The code level $T[k]$ is linearly interpolation from the recorded percentages at this level and the preceding level.

The number of samples determines the precision of the estimate. Table 11.5 presents the relationship between the number of samples and the precision as an RMS noise percentage.

11.3.5.3 Code Transition Level Test (Dynamic)

The code transition level is dynamically tested by applying a sinusoidal waveform across the full scale, as presented in Figure 11.37. After M records have been obtained, the code level k can be derived using the following equation:

$$T[k] = C - A \cos\left[\frac{\pi \cdot H_c[k-1]}{M}\right] \tag{11.24}$$

Here, M is the number of records, C is the offset, and H_c is the cumulative bin count:

$$H_c[k] = \sum_{j=1}^{k} H(j) \tag{11.25}$$

Notably, the M samples must be uniformly distributed into Mc integer periods of the sine wave. M and M_c must not have a common factor.

TABLE 11.5 ■ Precision of Estimates of Code Transition Level

Record length	64	256	1024	4096
Precision	45%	23%	12%	6%

Source: [IEEE 1057-1994].

11.3.5.4 Gain and Offset Test

Instead of a sine wave, a slow ramp with the full range can be applied to test the gain and offset, as presented in Figure 11.38. The transfer characteristic is given by:

$$G \cdot T[k] + V_{os} + \varepsilon(k) = Q \cdot (k-1) + T_1 \qquad (11.26)$$

where G = gain, V_{OS} = offset, $\varepsilon(k)$ = residual error, Q = quantization step, and T_1 = ideal code level for code 1.

Gain and offset errors can be derived using the following equations:

$$G = Q\frac{(2^N-1)\sum\limits_{k=1}^{2^N-1} kT[k]}{(2^N-1)\sum\limits_{k=1}^{2^N-1} T^2[k] - \left(\sum\limits_{k=1}^{2^N-1} T[k]\right)^2} - Q\frac{(2^N-1)(2^{N-1})\sum\limits_{k=1}^{2^N-1} T[k]}{(2^N-1)\sum\limits_{k=1}^{2^N-1} T^2[k] - \left(\sum\limits_{k=1}^{2^N-1} T[k]\right)^2} \qquad (11.27)$$

$$V_{os} = T_1 + Q\left(2^N-1\right) - \frac{G}{2^N-1}\sum\limits_{k=1}^{2^N-1} T[k] \qquad (11.28)$$

For a ramp signal, the histogram is a horizontal line without linearity errors. Figure 11.39 presents the histogram for a ramp signal under various offset and gain error conditions.

Figure 11.39a presents an ideal case of a flat histogram. For a 3-bit ADC, if 1024 samples are taken, then every bin has a count of 128 (1024/8) because the codes are

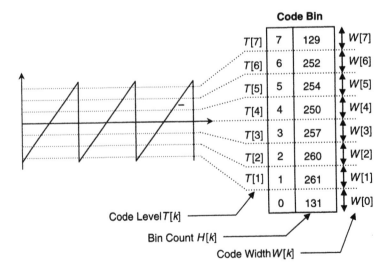

■ **FIGURE 11.38**

The code bins for a ramp signal.

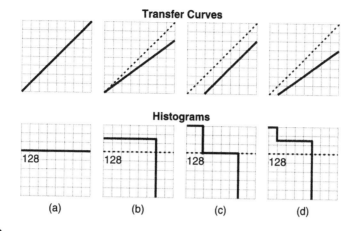

■ **FIGURE 11.39**

Ramp wave histogram: (a) ideal, (b) gain error, (c) offset error, and (d) gain + offset error.

evenly distributed to these 8 code bins. Figure 11.39b presents the case with a gain error. If the gain error is 2 LSB, then the bin count is 0 for codes 6 and 7. A total of 1024 records are evenly distributed to 6 bins with 171 or 170 records in each, so the bin count is inversely proportional to the gradient of the transfer curve. Given the offset error shown in Figure 11.39c, the bin counts for codes 1~5 remain the same: 128. Bin 0 has a count of 384 because the ADC fails to respond to signals below 2 LSB. Similarly, the histogram in case Figure 11.39d, with gain and offset errors, can be derived.

11.3.5.5 Linearity Error and Maximal Static Error

The test method presented in Figure 11.38 can also be used to test linearity errors and maximal static error. The **maximal static error** (MSE) is the maximal difference between any code transition level and its ideal value, represented as a percentage. The DNL, INL, and MSE errors are:

$$\text{DNL}(k) = \frac{G \cdot W(k) - Q'}{Q'} \tag{11.29}$$

$$\text{DNL} = \max \left| \frac{G \cdot W(k) - Q'}{Q'} \right| \tag{11.30}$$

$$\text{INL} = 100 \frac{\max |\varepsilon(k)|}{Q \cdot 2^N} \tag{11.31}$$

$$\text{MSE} = 100 \frac{\max |T(k) - Q \cdot (k-1) - T_1|}{Q \cdot 2^N} \tag{11.32}$$

where Q' is the average width of a code bin.

11.3.5.6 Sine Wave Curve-Fit Test

The digitized data in Figure 11.37 can be fitted to a three-parameter sine function. The input analog sine wave has an amplitude of A_o, frequency of ϖ_o, and offset of C_o and is described by the function:

$$y(t) = A_o \sin(\varpi_o t) + C_o \qquad (11.33)$$

The digitized sine wave can be curve-fitted using:

$$y'(t) = A \sin(\varpi t) + B \cos(\varpi t) + C \qquad (11.34)$$

The curve-fitting criterion is to minimize the mean square error of $y(t)$ and $y'(t)$. The IEEE 1057 standard offers a comprehensive list of sine-fitting algorithms. After the curve has been fit, offset, gain, phase, and frequency error can be derived as:

$$\text{Gain error:} \quad \frac{\sqrt{A^2 + B^2} - A_o}{A_o} \qquad (11.35)$$

$$\text{Offset error:} \, C - C_o \qquad (11.36)$$

$$\text{Phase error:} \, \theta = \tan^{-1}\left(-\frac{B}{A}\right) \qquad (11.37)$$

$$\text{Frequency error:} \, \frac{(\varpi - \varpi_o)}{\varpi_o} \qquad (11.38)$$

Here, the frequency error is the sampling frequency error. The sampling frequency error can be equivalent to the signal frequency error being the same percentage. For example, a 1% increase in the sampling frequency is equivalent to a 1% drop in the signal frequency.

11.3.6 Frequency-Domain ADC Testing

The frequency-domain testing of ADCs is the same as the AC parametric testing of analog circuits, described in Section 11.2.4. Similarly, the test items include SNR, SNDR, THD, and others. The ENOB can be derived from SNDR. The main difference is the item under test. In AC parametric testing, the response waveform of the CUT is tested and analyzed. The digitizer or ADC is assumed to be ideal, while in ADC testing the input waveform is assumed ideal. All the nonideal phenomena are originated from the ADC under test. Given the similarity, the frequency-domain test method will not be repeated again.

11.4   IEEE 1149.4 STANDARD FOR A MIXED-SIGNAL TEST BUS

The development of IEEE Std. 1149.4 began with a preliminary meeting in the summer of 1991, when the need was recognized for a standardized structure to be incorporated into mixed-signal circuits to combat the testability problems posed by such circuits. This meeting adopted as its mission the following [IEEE 1149.4-1999]:

To define, document, and promote the use of a standard mixed-signal test bus that can be used at the device and assembly levels to improve the controllability and observability of mixed-signal designs and to support mixed-signal built-in test structures in order to reduce both test development time and testing cost, and to improve test quality.

The architecture and means of controlling and accessing both analog and digital test data are described elsewhere [Osseiran 1999].

11.4.1   IEEE 1149.4 Overview

Figure 11.40 presents the context in which the IEEE 1149.4 standard is intended to be applied. The figure presents an electrical circuit constructed as a ***printed circuit assembly*** (PCA). The component that is subject to the standard is the shaded one at the center of Figure 11.40.

The pins of a typical mixed-signal IC are connected to:

- Other mixed-signal components (M) which may or may not conform to this standard

- Digital components (D) which may or may not conform to this standard

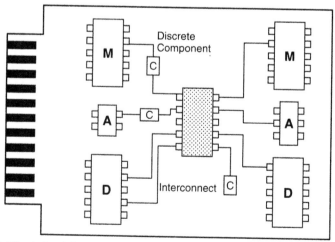

M: Mixed-signal Component   A: Analog Component   D: Digital Component

■ **FIGURE 11.40**

Mixed-signal printed circuit assembly [IEEE 1149.4-1999].

■ Analog components (A) which are unlikely to contain any associated testability features

■ Discrete components (C) such as resistors and capacitors, which do not have any associated testability features

The standard can be used in production tests and in the field service. The goal is to supply a test signal to, and to collect test responses from, edge connects without making direct physical contact with the component.

11.4.1.1 Scope of the Standard

This standard defines test features to provide standardized approaches to interconnect testing, parametric testing, and internal testing of mixed-signal PCAs.

■ **Interconnect test**—The primary goal of this standard is to support interconnect testing for PCAs comprised of analog, digital, and mixed-signal components. Any form of open and short, as presented in Figure 11.41, can be detected and diagnosed.

■ **Parametric test**—The second purpose is to characterize, measure, and test the discrete components. Discrete components perform such functions as level shifting, passive filtering, and AC coupling. They are regarded as *extended interconnects*, in contrast to the *simple interconnects* of wires only. As presented in Figure 11.42, simple, extended, and differential interconnects are included.

■ **Internal test**—The third objective, internal testing, relates to the capacity to perform comprehensive tests on the components either in isolation or while mounted on a substrate. Figure 11.43 presents a board-level connection configuration. The AT1 and AT2 ports of all of the analog chips are connected to the bus. A signal source is connected to AT1 and a response analyzer is connected to AT2. Notably, this is also a typical connection configuration for 1149.4. In internal testing, the test waveform is sent to the CUT via AT1 and AB1. The response waveform is returned to the analyzer via AB2 and AT2. The

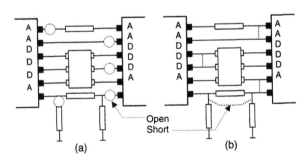

(a)   (b)

■ **FIGURE 11.41**

Interconnect testing: (a) open, and (b) short [IEEE 1149.4-1999].

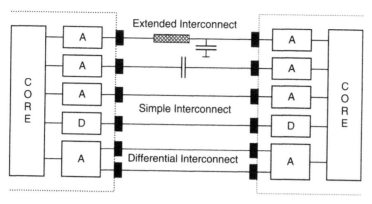

■ **FIGURE 11.42**

Interconnects: (a) simple, (b) extended, and (c) differential [IEEE 1149.4-1999].

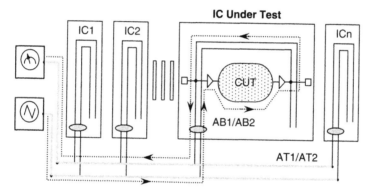

■ **FIGURE 11.43**

Internal test configuration.

internal test raises two major issues. First, the stray capacitance associated with the bus (AT1/AT2) may be very large; therefore, a high-quality signal is unlikely to be sent through the bus. Second, doing so may require the incorporation of internal test structures whose impact on the cost and performance of the circuit may be prohibitive; therefore, this aspect of the standard is not mandatory, but the addition of a designer-defined test function is unlimited.

11.4.2 IEEE 1149.4 Circuit Structures

Figure 11.44 presents the structure of an 1149.4-compliant chip. It presents all of the main mandatory components in the standard. An 1149.1 *test access port* (TAP) that supports all 1149.1 functions for digital testing [IEEE 1149.1-1990]. IEEE 1149.4 extensions include analog boundary modules on every analog function pin,

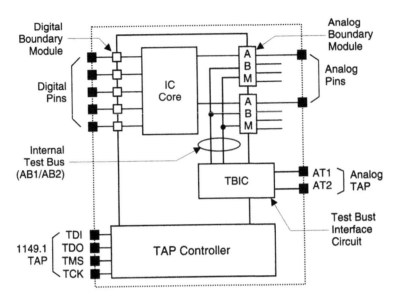

■ **FIGURE 11.44**

Structure of a basic 1149.4 chip [IEEE 1149.4-1999].

analog test access ports (AT1 and AT2), a test bus interface circuit, and a pair of internal analog test buses (AB1 and AB2):

- **Analog test access port** (ATAP)—The ATAP is an analog port that enables the test bus interface circuit to access an external analog test bus. It is comprised of a minimum of one analog input connection and one analog output connection (AT1/AT2). AT1 and AT2 carry signals to and from the **automatic test equipment** (ATE) and the CUT.

- **Analog test buses** (AB1/AB2)—AB1 and AB2 are two internal test buses. They are connected to all of the analog boundary modules. They have a function similar to that of ATAP except that they are delivering internal analog test signals. AB1/AB2 carries the signals from ABMs to the test bus interface circuit and then to AT1/AT2.

- **Test bus interface circuit** (TBIC)—The TBIC controls the connections between ATAP and AB1/AB2. It provides a link between the external test bus (AT1/AT2) and the internal test bus (AB1/AB2). Figure 11.45 presents the circuit diagram. The mandatory part, on the right, provides cross-bar connections for AT1/AT2 and AB1/AB2. It also provides switches for clamping the buses to V_C. The optional part, on the left, is provided for interconnect testing. It can send high (V_H) or low (V_L) voltages to AT1/AT2 and compare the voltage level on the bus with a threshold voltage (V_{TH}).

- **Analog boundary module** (ABM)—The ABM is the heart of the standard framework for mixed-signal testing. Figure 11.46 presents the circuit diagram

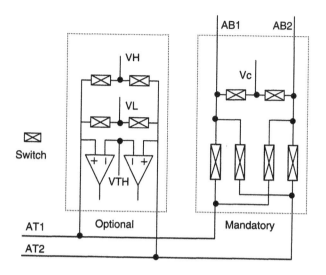

Test bus interface circuit (TBIC) [IEEE 1149.4-1999].

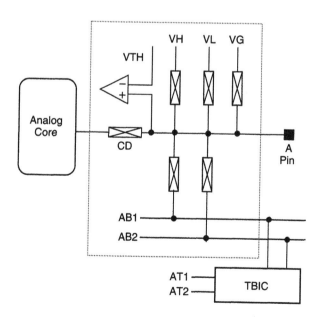

■ **FIGURE 11.46**

Analog boundary module (ABM) [IEEE 1149.4-1999].

of an ABM. Six conceptual switches in an ABM allow the pin to be connected to AB1, AB2, a high voltage (V_H), a low voltage (V_L), the reference quality voltage (V_G), and the analog core. The switch that connects to the analog core is the ***core disconnect*** (CD). It is responsible for isolating the internal core from the pin in interconnect testing mode. This switch is the most critical, because it is on the signal path. Performance degradation is the main concern. The pin is also connected to a comparator to compare to a threshold voltage (V_{TH}). Herein, V_{TH} lies in the range:

$$\frac{V_H + V_L}{2} - \frac{V_H - V_L}{4} < V_{TH} < \frac{V_H + V_L}{2} + \frac{V_H - V_L}{4} \tag{11.39}$$

In general, $V_{TH} = \frac{V_H + V_L}{2}$ is preferred. The switches to V_H/V_L and the comparator enable open/short testing without AB1/AB2 or an external instrument. This feature is discussed in detail below.

For a mixed-signal IC with a digital core, an analog core, and an ADC/DAC circuitry, the 1149.4 DFT structure presented in Figure 11.47 is recommended. A DBM chain separates ADC/DAC from the digital core, and the DAC inputs and ADC outputs can be accessed via the digital boundary scan chain. This control is equivalent to the direct control and observation of analog signals at the input of ADC and the output of DAC; therefore, the mixed-signal test effort is reduced.

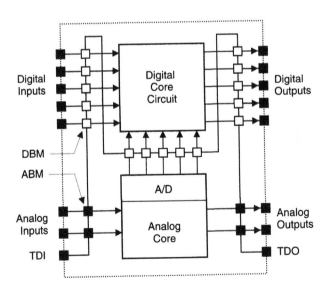

■ **FIGURE 11.47**

Circuit structure for mixed-signal ICs [IEEE 1149.4-1999].

11.4.3   IEEE 1149.4 Instructions

The IEEE 1149.4 standard is a super-set of IEEE 1149.1, so each component responds to the mandatory instructions defined in 1149.1 [IEEE 1149.1-1990]. Herein, three types of instruction are defined: mandatory instructions, optional instructions, and user-defined instructions. The contents of these instructions have already been defined in 1149.1. Their special functions, related to 1149.4, are described below.

11.4.3.1   *Mandatory Instructions*

- **BYPASS**—When the *BYPASS* instruction is selected, all ATAP pins are isolated from the internal analog test buses and from all test voltages. All analog pins are connected to the core circuit and isolated from internal and external test buses. The test logic and circuit have no effect on the operation of the core circuit.

- **SAMPLE/PRELOAD**—As with *BYPASS*, ATAP and analog pins are isolated from the analog DFT circuitry; however, this instruction has two functions. The first function, *SAMPLE*, allows the comparator (1-bit digitizer) of the ABM to capture a digitized snapshot of the analog signal. The second function, *PRELOAD*, loads a digital data pattern to specify the operation of the ABM.

- **EXTEST**—When the *EXTEST* instruction is selected, the analog pin is disconnected from the core. Here, the core disconnect switch (CD) is opened. This instruction allows the open/short of simple interconnects to be tested using the voltage sources and comparator in the ABM. Additionally, the parameters of the extended interconnects can be measured using ATE. The following section details the test modes.

- **PROBE**—When the *PROBE* instruction is chosen, the analog pin is connected to the core and the analog test buses (AB1/AB2) based on the control pattern scanned in during *RELOAD*. The *PROBE* instruction allows analog pins to be stimulated or monitored using AB1 and AB2 while the component is operated in its normal mode.

11.4.3.2   *Optional Instructions*

- **INTEST**—When the *INTEST* instruction is chosen, the analog pin is connected to the core and the analog test buses, according to the *PROBE* instruction. The stimulus can be supplied via AB1 and/or the response monitored via AB2. Figure 11.43 presents the configuration, which is used for testing the internal analog core.

- **Device identification register**—The device identification register instructions include the *IDCODE* and the *USERCODE* instructions, both of which are identical to those of IEEE 1149.1.

- **RUNBIST**—As in 1149.1, the *RUNBIST* is self-contained and leaves a single test result signature in the test data register identified by the instruction. In response to the *RUNBIST* instruction, all analog output signals are defined by the data held in the boundary-scan register or placed in inactive drive states.

- **CLAMP**—When the *CLAMP* instruction is chosen, the signals of all analog output pins are defined by the data held in the corresponding ABMs. Restated, the possible output states are high impedance, V_H, V_L, or V_G. No such state should be changed during the *CLAMP* period.

- **HIGHZ**—When the *HIGHZ* instruction is chosen, all analog function pins are disconnected from the core and from all test circuitry, such that all switches in the ABM are opened. Similarly, the AT1/AT2 pins enter the high impedance state, independently of the data held in the TBIC control register. The *HIGHZ* instruction in 1149.4 is similar to the *BYPASS* instruction. They both isolate the chip from the rest of the circuits on the board. For example, in Figure 11.43, only the CUT receives the *INTEST* instruction. The remaining chips are placed under the *HIGHZ* instruction.

Users can define their own instructions. The user-defined instruction can be treated as an extension of optional instructions. Users must follow the guidelines specified in the IEEE 1149.1 and 1149.4 standards. These guidelines are defined after, and do not conflict with, the mandatory instructions. Given the above mandatory and optional instructions, many test modes can be defined as follows.

11.4.4   IEEE 1149.4 Test Modes

Following the instructions described above, several modes for testing various items in various configurations are described below.

11.4.4.1   Open/Short Interconnect Testing

Without an external instrument, the ABM can perform interconnect testing in general and open/short in particular, as presented in Figure 11.48. The tested wire

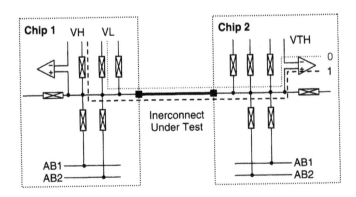

■ FIGURE 11.48

Open/short test mode.

is the bold wire that connects analog Pin1 of Chip1 to Pin2 of Chip2. ABMs are associated with both pins. The three-step test procedure is as follows:

1. Switch V_H to Pin1 and detect a logical 1 at the comparator of Pin2.

2. Switch V_L to Pin1 and detect a logical 0 at the comparator of Pin2.

3. Switch V_H to Pin1 and detect a logical 1 at the comparator of Pin2.

Notably, step 1 is exactly the same as step 3 because interconnect testing depends on going through a 0-to-1 transition and a 1-to-0 transition to detect the open faults that exhibit sequential behavior. An opened node is in a high-impedance state. At high impedance, the charge is held in the stray capacitance and can be treated as a dynamic latch; therefore, it must undergo a 0-to1 and a 1-to-0 transition to guarantee that the driver is connected and functioning.

The open/short test mode is the most important test mode for 1149.4. It is also the main feature of the ABM. V_H and V_L can be regarded as logical 1 and 0 in a digital circuit. They are commonly set to V_{DD} and GND. Additionally, if V_{TH} is set to $0.5V_{DD}$, then it can be regarded as a logical gate with a threshold voltage of $0.5V_{DD}$ and thus becomes compatible with digital logic and 1149.1. Restated, if connected to a DBM, it can output a logical 0 (V_L) and a logical 1 (V_H) and receive an input digital signal with the comparator. In summary, with the ABM, the open/short of analog interconnects can be tested using the digital interconnect testing approach.

11.4.4.2 *Extended Interconnect Measurement*

Extended interconnect measurement measures passive components, such as resistors and capacitors, used in level shifting, passive filtering, and AC coupling. It requires an external instrument. Figure 11.49 presents the test setup, and the procedure is presented as follows:

1. Condition TBIC and ABM such that AT1-AB1-APin and AT2-AB2-APin are connected.

2. Supply test current $I(t)$ from the signal source via AT1-AB1-APin to the DUT (***Device Under Test***).

3. Measure response voltage $V(t)$ from the DUT via APin-AB2-AT2 to the signal analyzer.

4. Calculate the impedance from $Z_{DUT}(s) = \dfrac{V(s)}{I(s)}$.

The preferred method of measuring impedance is to apply a current and measure voltage. Given sufficiently high output impedance, the test current can reach the DUT with little attenuation. With sufficiently high input impedance, a voltage meter can measure the voltage very accurately. Figure 11.50 presents an abstracted schematic of the setup in Figure 11.49. Herein, Z_p is the total parasitic impedance of the buses (AT and AB) and switches.

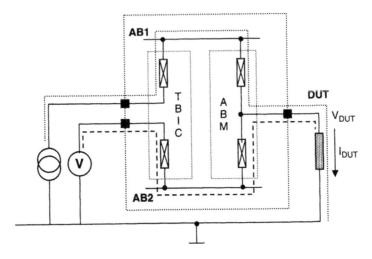

■ **FIGURE 11.49**

Parametric testing of grounded discrete component.

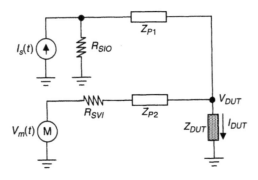

■ **FIGURE 11.50**

Equivalent circuit diagram for Figure 11.49.

Based on the assumption that R_{SVI} and R_{SVO} do not interfere with current application and voltage measurement, the current to the DUT and voltage measured are:

$$I_{DUT}(t) = I_s(t) \cdot \frac{R_{SIO}}{R_{SIO} + Z_{P1} + Z_{DUT}} \quad (11.40)$$

$$V_m(t) = V_{DUT}(t) \cdot \frac{R_{SVI}}{R_{SVI} + Z_{P2} + Z_{DUT}} \quad (11.41)$$

The above equation shows that if R_{SVO} and R_{SVI} are sufficiently large, then I_{DUT} equals I_s and V_m equals V_{DUT}. Notably, an ideal current source has infinite output impedance, and an ideal voltage meter has infinite input impedance.

In some cases, an extended interconnect is not grounded; rather, it is connected between two pins as presented in Figure 11.51. Here, in this situation, Z_{DUT} is

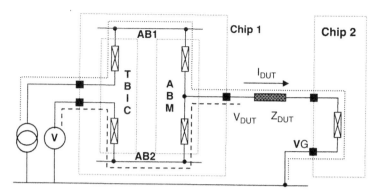

■ FIGURE 11.51

Parametric testing of floating component with zero V_G.

considered a floating component. For example, a coupling capacitor is a floating component because it is placed between two pins to pass the AC part of the signal and block the DC part.

Figure 11.51 presents the setup of testing a floating component. Here, V_G is connected to GND. The method and procedure for testing a floating component using the setup in Figure 11.51 are the same as those for testing the grounded component in Figure 11.49; however, if the floating component is active, then a DC bias voltage must be applied to ensure its normal functioning. In that case, V_G must be connected to a voltage source to supply the desired DC bias, as presented in Figure 11.52.

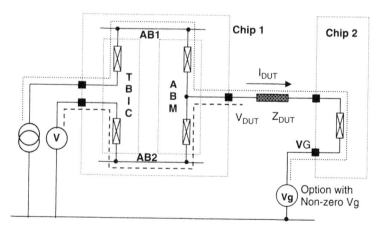

■ FIGURE 11.52

Parametric testing of floating component with non-zero V_G.

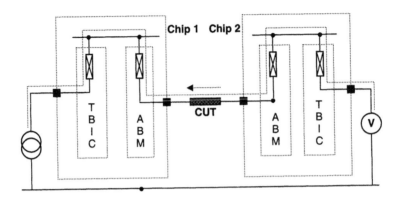

FIGURE 11.53

Parametric testing by applying voltage and measuring current.

An alternative means of measuring a floating component is to apply a voltage and measure current. Figure 11.53 presents the measurement.

Care must be taken in using this configuration. Figure 11.54 presents the circuit model that is equivalent to Figure 11.53. The voltage across the DUT and the current are derived as follows:

$$V_{DUT}(t) = V_s(t) \cdot \frac{Z_{DUT}}{R_{SVO} + Z_{P1} + Z_{DUT} + Z_{P2} + R_{SII}} \tag{11.42}$$

$$I_m(t) = \frac{V_s(t)}{R_{SVO} + Z_{P1} + Z_{DUT} + Z_{P2} + R_{SII}} \tag{11.43}$$

Although R_{SVO} and R_{SII} can both be very small, the parasitic impedance of the bus and switches, Z_{P1} and Z_{P2}, still affects the voltage across the DUT, V_{DUT}, and

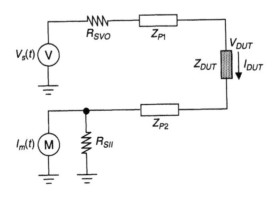

FIGURE 11.54

Circuit diagram equivalent to Figure 11.53.

the measured current. Notably, an ideal voltage source and an ideal current meter have an output impedance of $0\,\Omega$. The degenerate equations are:

$$V_{DUT}(t) = V_s(t) \cdot \frac{Z_{DUT}}{Z_{P1} + Z_{DUT} + Z_{P2}} \tag{11.44}$$

$$I_m(t) = \frac{V_s(t)}{Z_{P1} + Z_{DUT} + Z_{P2}} \tag{11.45}$$

Notably, a CMOS switch has a resistance of from $100\,\Omega$ to $10\,\mathrm{K}\,\Omega$. A 100-Ω switch is considered to have very small resistance, even smaller than most pin drivers. A typical pin driver has a maximal driving current of 1 mA to 8 mA, which is equivalent to $1.8\,\mathrm{K}\Omega$ and $225\,\Omega$ with a 1.8-V power supply; therefore, the test channel must be precisely calibrated before such a measurement can be made. The calibrated impedance is then considered when the measurement is made.

11.4.4.3 Complex Network Measurement

More general networks connected between pins can be measured similarly. Figure 11.55 presents a typical two-port network connected between P1/P2 on one chip and P3/P4 on another. All pins are assumed to be equipped with ABMs; therefore, different setups can be employed to test various transfer characteristics. Consider as an example the hybrid (H) parameters, which are defined as follows:

$$
\begin{aligned}
h_{11} &= \left. \frac{V_1}{I_1} \right|_{V_2=0} & h_{21} &= \left. \frac{I_2}{I_1} \right|_{V_2=0} \\
h_{12} &= \left. \frac{V_1}{V_2} \right|_{I_1=0} & h_{22} &= \left. \frac{I_2}{V_2} \right|_{I_1=0}
\end{aligned}
\tag{11.46}
$$

Table 11.6 presents a measurement configuration for measuring the H parameter based on Eq. (11.46). The measurements of other two-port networks, such as

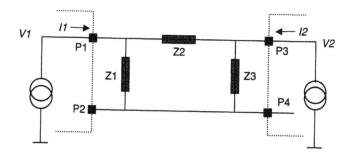

■ **FIGURE 11.55**

Testing two-port network.

TABLE 11.6 ■ Configuration for Measuring *H* Parameters

H	P1	P2	P3	P4
h_{11}	Is/Vm	GND	GND	GND
h_{12}	Vm	GND	Vs	GND
h_{21}	Is	GND	Im	GND
h_{22}	Open	GND	Vs/Im	GND

Note: Im, measure current; Is, apply current; Vm, measure voltage; Vs, apply voltage.

short-circuit admittance (*Y*), *open-circuit impedance* (*Z*), and *inverse-hybrid* (*G*), can be made in a similar way.

11.4.4.4 High-Performance Configuration

As discussed above, CMOS switches have much higher impedance than metal wires. Additionally, the stray capacitance of the test bus is a major concern. As presented in Figure 11.43, the bus is routed all over the board and connected to every 1149.4 chip. The overall stray capacitance associated with the bus can be in the range of tens of pF. A typical pin has a stray capacitance of 2~4 pF, a via has 0.5~1 pF, and a 1-cm wire has 0.25~0.5 pF; therefore, the bandwidth is severely limited by the stray capacitance of the bus and the stray resistance of the switches. Accordingly, a high-frequency signal cannot be delivered to test high-speed circuitry. For particular circuitry, high-frequency test stimuli are required to trigger frequency-dependent faults. In such cases, the standard recommends the replacement of passive switches with active **current buffers** and **voltage buffers**. Figure 11.56 presents an example of a buffered ABM and TBIC.

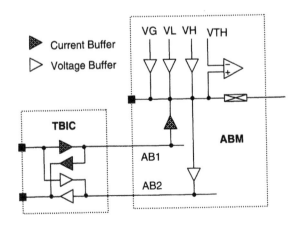

■ **FIGURE 11.56**

High-performance setup with current and voltage buffer.

11.5   CONCLUDING REMARKS

Analog and mixed-signal testing requires specialized approaches and experienced engineers, because of the large varieties of signals, the functions, and the ranges of circuits. Generic and general module-like operational amplifiers provide a good entry point for understanding the properties of analog circuits, their specifications, and the test methods. Learning that DSP approaches are so pervasive that even basic analog test items can be accomplished by the DSP-based virtual instrumentation techniques is important. Mixed-signal circuits, mainly ADCs and DACs, are key components of SOCs. IEEE 1057, with its formal terminologies and standardized test methods, provides a solid theoretical background for ADC/DAC testing. IEEE 1149.4 is one solution to extending and incorporating the digital counterpart. Overall, in analog and mixed-signal testing, experience is as important as textbook knowledge. More territory is undiscovered than has been discovered; therefore, long-term commitment, in-depth investigation, and innovative thinking will be the keys to success.

11.6   EXERCISES

11.1   (Analog Circuit Testing) For the circuit in Figure 11.12, if the amplifier under test had the following circuit parameters, what would the measured open-loop gain be?

Input impedance	$R_I = 1\,K\Omega$
DC gain	100
Output impedance	$R_O = 50\,\Omega$
Load resistance	$R_L = 100\,\Omega$

11.2   (Analog Circuit Testing) For the unit gain bandwidth test construct presented in Figure 11.13a, suppose that the amplifier has an open-loop gain of 1000, a 3-dB bandwidth of 1 KHz, and a slew rate of $1\,V/\mu s$ (other parameters assumed ideal).

   a.   What is the unit gain bandwidth and the maximal input amplitude?

   b.   What is f_{MAX} (the 3-dB frequency at which power output is maximal)?

11.3   (Analog Circuit Testing) For a regulator with an input of 6 to 9 V, an output of 5 V ± 5%, a maximal load of 1 A, and a minimum load of 1 mA, what is:

   a.   The load regulation?

   b.   The line regulation?

   c.   The drop-out voltage?

Vin (V)	Load (A)	Vout (V)	Vin (V)	Load	Vout (V)
5.4	1	4.6	6.0	1 mA	5.2
5.6	1	4.7	6.0	5 mA	5.15
5.8	1	4.8	6.0	10 mA	5.1
6.0	1	4.9	6.0	50 mA	5.05
7.0	1	5.0	6.0	100 mA	5.0
8.0	1	5.1	6.0	500 mA	4.95
9.0	1	5.15	6.0	1 A	4.9

11.4 (Analog Circuit Testing) For the power supply circuit in Figure 11.57 and device parameter shown below:

a. What are the maximal and minimal output voltages?

b. What are the line regulation and load regulation?

c. What is the drop-out voltage?

Input	5~10 V (RMS), AC, 60 Hz
Diode	$V_D = 0.7\,\text{V}, R_D = 0\,\Omega$
Zener diode	$V_Z = 5.7\,\text{V}, R_Z = 10\,\Omega$
BJT transistor	$V_{BE} = 0.7\,\text{V}, \beta = 100$
Regulator	$R = 90\,\Omega, C = 10,000\,\mu\text{F}$

11.5 (Analog Circuit Testing) For the limiter circuit presented in Figure 11.58, let $V_D = 0.7\,\text{V}, R_D = 100\,\Omega$ and $R_S = 1\,\text{K}\Omega$ (the output impedance of the signal source). (You may need SPICE simulation to solve this problem.)

a. For a sinusoidal waveform, what is the maximal distortion free amplitude?

b. For a realistic diode with $V_D = 0.7\,\text{V}$ at $I_D = 1\,\text{mA}$, redo (a).

c. If the limiter is placed as the protection circuit before an ideal ADC, what is the SNR and the maximal achievable bit length for a 2.7-V sinusoidal input?

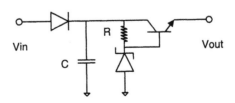

■ **FIGURE 11.57**

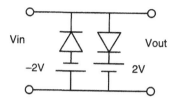

11.6 (Mixed-Signal Testing) For the sample and hold circuit presented in Figure 11.29, let $R = 100\,\Omega$ and $C_H = 10\,nF$ and $I_{Leak} = 1\,\mu A$. The bit length of the ADC is 10 and the error tolerance is 0.5 LSB. The conversion range is 1 V and the conversion time is $1\,\mu s$.

a. What is the droop rate and what is the maximal hold period such that the voltage change is less than the tolerance?

b. What is the minimum acquisition time of the S/H?

c. What is the maximal conversion throughput of the S/H and ADC?

d. Suppose that the ADC has an unlimited bit length; for such an S/H, what is the maximal achievable bit length with the ideal ADC?

e. What is the conversion rate in (d)?

11.7 (Mixed-Signal Testing) For a 10-bit ADC, the conversion range is 1 V. Suppose that the sampling clock has a jitter of 100 ps. The input is a sinusoidal waveform with an amplitude of 0.5 V.

a. What are the errors in terms of LSB caused by the jitter for inputs of 1 KHz and 1 MHz?

b. What is the maximal input signal frequency so the sampling error is less than 0.5 LSB?

c. Suppose that the ADC has an unlimited bit length. What are the maximal achievable bit lengths for 1-KHz and 1-MHz sinusoidal waveforms?

11.8 (Mixed-Signal Testing) For a 3-bit ADC, the test input is a ramp signal, and a total of 1024 samples are taken uniformly. If the histogram is as follows, what are the DNL and the INL (in terms of LSB)?

Code	000	001	010	011	100	101	110	111
Frequency	80	163	159	149	140	135	130	68

11.9 **(Mixed-Signal Testing)** For a 4-bit ADC and a ramp input, draw the histogram in the following cases:

a. An offset error of 2 LSB.

b. A gain error of 2 LSB.

c. An offset error of 2 LSB and a gain error of 2 LSB.

d. Code i has a DNL error of (in LSB):

$$\begin{cases} \mathrm{DNL}_i = 0.5 \times i/7 & i \leq 7 \\ \mathrm{DNL}_i = 0.5 - 0.5 \times (i-8)/7 & i > 7 \end{cases}$$

e. Code i has a DNL error of (in LSB):

$$\mathrm{DNL}_i = 0.5 \times \sin\left(\frac{i}{16} \cdot \pi\right)$$

f. What are the INL values of (d) and (e)?

11.10 **(Mixed-Signal Testing)** For a 4-bit ADC, the ideal transfer characteristic curve is $Y = X$ for $-0.5 \leq X \leq 0.5$. Assume that the ADC under test has a characteristic transfer curve of $Y = X + 0.1\sin(2\pi X)$.

a. Draw the 256-sample histogram of a ramp input with an amplitude of ± 0.5. The samples are obtained at $X_i = -0.5 + i/256$ for $0 \leq i \leq 256$. What are the INL and DNL derived from the histogram?

b. For $Y = X + 0.1\cos(2\pi X)$, what are the gain and offset errors?

c. After the gain and offset errors in (b) have been calibrated, what are the INL and the DNL?

11.11 **(IEEE 1149.4 Standard for a Mixed-Signal Test Bus)** Equation (11.46) and Table 11.6 provide the definitions and test configurations for the H parameters for a two-port network. For the following parameters, find the definitions and derive the test configuration:

a. Y parameters.

b. Z parameters.

c. G parameters

Acknowledgments

The author wishes to thank Prof. Fa Foster Dai of Auburn University, Prof. Jiun-Lang Huang of National Taiwan University, and Prof. Adam Osseiran of Edith Cowan University for reviewing the text and providing helpful comments.

References

R11.0—Books

[Burns 2001] M. Burns and G. W. Roberts, *An Introduction to Mixed Signal IC Test and Measurement*, Oxford University Press, London, 2001.

[IEEE 1057-1994] *IEEE Standard for Digitizing Waveform Recorders (IEEE 1057-1994)*, IEEE Press, New York, 1977.

[IEEE 1149.1-1990] *IEEE Standard Test Access Port and Boundary-Scan Architecture (IEEE 1149.1-1990)*, IEEE Press, New York, 1990.

[IEEE 1149.4-1999] *IEEE Standard for a Mixed-Signal Test Bus (IEEE 1149.4-1999)*, IEEE Press, New York, 1999.

[IEEE 181-1977] *IEEE Standard on Pulse Measurement and Analysis by Objective Techniques (IEEE 181-1977)*, IEEE Press, New York, 1994.

[Mahoney 1987] M. Mahoney, *DSP-Based Testing of Analog and Mixed-Signal Circuits*, IEEE Computer Society, New York, 1987.

[Maunder 1990] C. M. Maunder and R. E. Tulloss, *The Test Access Port and Boundary-Scan Architecture*, IEEE Computer Society Press, 1990.

[Oppenheim 1989] A. V. Oppenheim, and R. W. Schafer, *Discrete-Time Signal Processing*, Prentice Hall, Upper Saddle River, NJ, 1989.

[Osseiran 1999] A. Osseiran, *Analog and Mixed-Signal Boundary-Scan*, Kluwer Academic, Norwell, MA, 1999.

[Parker 1992] K. P. Parker, *The Boundary-Scan Handbook*, Kluwer Academic, Norwell, MA, 1992.

R11.1—Introduction

[Wang 1997] C.-P. Wang and C.-L. Wey, Development of hierarchical testability design methodologies for analog/mixed-signal integrated circuits, in *Proc. IEEE Int. Conf. on Computer Design (ICCD)*, October 1997, pp. 468–473.

R11.2—Analog Circuit Testing

[ADI 1982] Analog Devices, *How to Test Basic Operational Amplifier Parameters* [application note], Analog Devices, Norwood, MA, July 1982.

[Stata 1967] R. Stata, User's Guide to Applying and Measuring Operational Amplifier Specifications, *Analog Dialogue*, 1-3, 13-3–13-10, 1967.

[TI 1999] *Analysis of the Sallen-Key Architecture* [application report], Texas Instruments, Dallas, TX, July 1999.

CHAPTER 12

TEST TECHNOLOGY TRENDS IN THE NANOMETER AGE

Kwang-Ting (Tim) Cheng
University of California, Santa Barbara, California

Wen-Ben Jone
University of Cincinnati, Cincinnati, Ohio

Laung-Terng (L.-T.) Wang
SynTest Technologies, Inc., Sunnyvale, California

ABOUT THIS CHAPTER

Over the past three decades, we have seen semiconductor manufacturing technology advance from 4 microns to 65 nanometers. The shrinkage of feature size has made a dramatic impact on design and test. Now, we can see **system-on-chip** (SOC) designs embed 100 million transistors running in the gigahertz range. Within the next decade, there will be designs containing a billion transistors. These designs can include all varieties of *digital, analog, mixed-signal, memory, optical,* **microelectromechanical system** (MEMS), **field programmable gate array** (FPGA), and **radiofrequency** (RF) circuits. We can anticipate that testing for designs of this complexity will be a significant challenge, if not a serious problem. Data have shown it is beginning to require more than 20% of the development time and over a month to generate production test patterns of sufficient fault coverage for detecting manufacturing defects.

Today, the semiconductor industry relies heavily on two test technologies: *scan* and **built-in self-test** (BIST). *Scan* will no longer be sufficient because small feature size could cause physical failures that are difficult to detect by its *single-fault model* assumption. *BIST* will begin to cause problems if its low-fault-coverage problem is not soon solved. Faced with mountains of testing problems in the nanometer design era, it is imperative that we seek viable test solutions now. In this chapter, we discuss these test technologies by first reviewing the *International Test Technology Roadmap* published by the Semiconductor Industry Association in 2004. Then, promising test techniques will be presented to deal with highly complex nanometer designs.

12.1   TEST TECHNOLOGY ROADMAP

In 1965, Gordon Moore, Intel cofounder, predicted that the number of transistors integrated per square inch on a die would double every year [Moore 1965]. In subsequent years, the pace slowed down but the number of transistors has doubled approximately every 18 months for the past two decades. This has become the current definition of Moore's law. Most experts expect that Moore's law will hold for at least two more decades. Die size will continue growing larger and larger, but minimum feature size will shrink smaller and smaller. Although smaller transistor size could result in smaller circuit delay, a smaller feature size for interconnects does not reduce the signal propagation delay; thus, the signal propagation delay in interconnects has been the dominant factor in determining the delay of a circuit. To alleviate this problem, interconnects are made taller and taller to reduce the sheet resistance. Unfortunately, this induces **crosstalk noises** between adjacent interconnects due to capacitive and inductive coupling. This is referred to as a **signal integrity** problem, and it is extremely difficult to detect [Chen 2002]. As the clock frequency has been pushed up into the gigahertz range and supply voltage has also been scaled down along with device scaling, the power supply voltage drop caused by $L(di/dt)$ can no longer be ignored. This has caused another **power integrity** problem that again is extremely difficult to solve because finding test patterns with maximum current changes is quite difficult [Saxena 2003].

As the manufacturing technology continues to advance, precise control of the silicon process is becoming more challenging; for example, it is difficult to control the effective channel length of a transistor, and the circuit performance, such as power and delay, exhibits much larger variability. This is the **process variation** problem, and it can make delay testing extremely complex [Wang 2004]. To reduce the leakage power dissipation, many low-power design technologies have been widely used. Unfortunately, low-power circuits might result in new fault models that increase the difficulty of fault detection; for example, a *drowsy cache* that can be supplied by low voltage (*e.g.*, 0.36 V) when it is idle has been proposed recently to reduce the leakage current [Kim 2004]. Though the leakage current can be reduced by several orders of magnitude, a new fault model called *drowsy fault* that causes a memory cell to fall asleep forever can occur. Unfortunately, testing drowsy faults requires excessively long test application times, as it is necessary to drive the memory cells to sleep and then wake them up. As we move into the nanometer age, in order to keep up with Moore's law many new nanotechnologies and circuit design techniques must be invented and adopted, all of which pose severe test challenges that must be addressed concurrently. If these test issues are not solved at the same time, the cost of test would eventually surpass the cost of silicon manufacturing, as illustrated in Figure 12.1 [SIA 1997] [SIA 2004].

In 2004, the Semiconductor Industry Association (SIA) published an *International Technology Roadmap for Semiconductors* (ITRS), which includes an update to the test and test equipment trends for nanometer designs through 2010 and beyond [SIA 2004]. The ITRS is an assessment of the semiconductor technology requirements. The objective of the ITRS is to ensure advancements in the

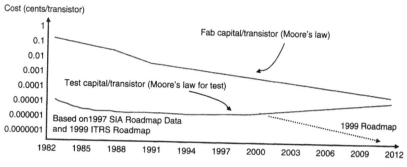

■ **FIGURE 12.1**

Fabrication capital *versus* test capital.

performance of integrated circuits. This assessment, also known as **roadmapping**, is a cooperative effort of the global industry manufacturers and suppliers, government organizations, consortia, and universities.

The ITRS identifies the technological challenges and needs facing the semiconductor industry over the next 15 years. Difficult near-term and long-term test and test equipment challenges were reported in [SIA 2004] and are listed in Tables 12.1 and 12.2. The near-term challenges for nanometer designs with feature size $\geq 45\,nm$ through 2010 include high-speed device interfaces, highly integrated designs, reliability screens, manufacturing test cost, and modeling and simulation. The long-term challenges for nanometer designs with feature size $<45\,nm$ beyond 2010 include the *device under test* (DUT) to *automatic test equipment* (ATE) interface, test methodologies, defect analysis, failure analysis, and disruptive device technologies. These difficult challenges encompass a full spectrum of test technology trends imperative for nanometer designs, including: (1) developing new *design for testability* (DFT) and *design for manufacturability* (DFM) methods for digital circuits, analog

TABLE 12.1 ■ Test and Test Equipment Difficult Challenges – Near-Term

Five Difficult Challenges ≥ 45 nm/Through 2010	Summary of Issues
High-speed Device Interfaces	A major roadblock will be the need for high frequency, high pin count probes and test sockets; research and development is urgently required to enable cost-effective solutions with reduced parasitic impedance.
High-speed Device Interfaces	High-speed serial interface speed and port count trends will continue to drive high-speed analog source/capture and jitter analysis instrument capability for characterization. DFT/DFM techniques must be developed for manufacturing.
	Device interface circuitry must not degrade equipment bandwidth and accuracy, or introduce noise, especially for high-frequency differential I/O and analog circuits.

TABLE 12.1 ■ *Continued*

Five Difficult Challenges ≥ 45 nm/Through 2010	Summary of Issues
Highly Integrated Designs	Highly structured DFT approaches are required to enable test access to embedded cores. Individual cores require special attention when using DFT and BIST to enable test.
	Analog DFT and BIST techniques must mature to simplify test interface requirements and slow ever-increasing instrument capability trends.
	Testing chips containing RF and audio circuits will be a major challenge if they also contain large numbers of noisy digital circuits.
	DFT must enable test reuse for reusable design cores to reduce test development time for highly complex designs.
Reliability Screens	Existing methodologies are limited (burn-in versus thermal runaway, IDDQ versus background current increases).
	Research is required to identify novel infant mortality defect acceleration stress conditions.
Manufacturing Test Cost	Test cell throughput enhancements are needed to reduce manufacturing test cost. Opportunities include massively parallel test, wafer-level test, wafer-level burn-in, and others. Challenges include device interfacing/contacting, power and thermal management.
	Device test needs must be managed through DFT to enable low-cost manufacturing test solutions; including reduced pin cost test, equipment reuse, and reduced test time.
	Automatic test program generators are needed to reduce test development time. Test standards are required to enable test content reuse and manufacturing agility.
Modeling and Simulation	Logic and timing accurate simulation of the ATE, device interface, and DUT is needed to enable pre-silicon test development and minimize costly post-silicon test content development/debug on expensive ATE.
	High-performance digital and analog I/O and power requirements require significant improvements to test environment simulation capability to ensure signal accuracy and power quality at the die.
	Equipment suppliers must provide accurate simulation models for pin electronics, power supplies, and device interfaces to enable interface design.

TABLE 12.2 ■ Test and Test Equipment Difficult Challenges – Long-Term

Five Difficult Challenges < 45 nm/Beyond 2010	Summary of Issues
DUT to ATE Interface	Probing capability for optical and other disruptive technologies. Support for massively parallel test—including full wafer contacting. Decreasing die size and increasing circuit density are driving dramatic increases in die thermal density. This problem is further magnified by the desire to enable parallel test to maximize manufacturing throughput. New thermal control techniques will be needed for wafer probe and component test. DFT to enable test of device pins not contacted by the interface and test equipment.
Test Methodologies	New DFT techniques (SCAN and BIST have been the mainstay for over 20 years). New test methods for control and observation are needed. Tests will need to be developed utilizing the design hierarchy. Analog DFT and BIST techniques must mature to simplify test interface requirements and slow ever increasing instrument capability trends. Logic BIST techniques must evolve to support new fault models, failure analysis, and deterministic test. EDA tools for DFT insertion must support DFT selection with considerations for functionality, coverage, cost, circuit performance and ATPG performance.
Defect Analysis	Defect types and behavior will continue to evolve with advances in fabrication process technology. Fundamental research in existing and novel fault models to address emerging defects will be required. Significant advances in EDA tools for ATPG capacity and performance for advanced fault models and DFT insertion are required to improve efficiency and reduce design complexities associated with test.
Failure Analysis	Real-time analysis of defects in multi-layer metal processes is needed. Failure analysis methods for analog devices must be developed and automated. Transition from a destructive physical inspection process to a primarily non-destructive diagnostic capability. Characterization capabilities must identify, locate, and distinguish individual defect types.
Disruptive Device Technologies	Develop new test methods for MEMS and sensors. Develop new fault models for advanced/disruptive transistor structures.

circuits (including RF and audio circuits as well as high-speed serial interfaces), MEMS, and sensors, (2) developing the means to reduce manufacturing test costs as well as enhance device reliability and yield, and (3) developing techniques to facilitate defect analysis and failure analysis. The ITRS [SIA 2004] further summarizes the design test challenges, as shown in Table 12.3.

In the following sections, we briefly present several promising test solutions to address some of the DFT/DFM needs and the difficult challenges identified by the ITRS. For more information, the reader should refer to the key references cited in each section.

TABLE 12.3 ■ Design Test Challenges

Challenges < 50 nm/Through 2009	Summary of Issues
Effective Speed Test with Increasing Core Frequencies and Widespread Proliferation of Multi-GHz Serial I/O Protocols	P, S—Continuation (avoidance) of at-speed functional test with increased clock frequencies. P, S—At-speed structure test with increased clock frequencies. P, S, A—DFT, test and on-chip measurement techniques for multi-gigahertz serial I/Os and non-deterministic interfaces.
Capacity Gap Between DFT/Test Generation/Fault Grading Tools and Design Complexity	P, S—Better EDA tools for advanced (open, delay, etc.) fault models. P, S—DFT to enable low-cost ATE. P, S—Non-intrusive logic BIST (including advanced fault models). A—AMS DFT/BIST, especially at beyond-baseband frequencies.
Quality and Yield Impact due to Test Process Diagnostic Limitations	P, S—Power and thermal management during test. P, S—Fault diagnosis and design for diagnosability. S—Yield improvement and failure analysis tools and methods. All—Increasing difficulty to fault isolate and root cause yield limiting defects.
Signal Integrity Testability and New Fault Models	P, S—Signal integrity (noise, interference, capacitive/inductive coupling, etc.) testability. A—Fault models for analog (parametric) failures.
SOC and SIP Test	S—Integration of SOC test methods into chip-level DFT. S—Integration of multiple fabric-specific test methodologies in cost-effective manufacturing flows. A—DFT, BIST and test methods compatible with core-based SOC environment and constraints. M—Embedded memory (DRAM, SRAM, Flash) built-in self-diagnosis and self-repair. All—Test reuse in context of higher integration.

TABLE 12.3 ■ *Continued*

Additional Challenges <50 nm/Beyond 2009	Summary of Issues
Integrated Self-Testing for Heterogeneous SOCs and SIPs	A—Test of multi-gigahertz RF front ends on chip. S—Use of on-chip programmable resources for SOC and SIP self-test. S, A—Dependence on self-test solutions for SOC (including RF and analog). A—(Analog) signal integrity test issues caused by interference from digital to analog circuitry. S—Test methods for heterogeneous SOC and SIP including MEMS and EO components.
Diagnosis, Reliability Screens, and Yield Improvement	A—Diagnosis and failure analysis for AMS parts. P, S—Electrical automated fault isolation techniques below gate level. P, S—Design for efficient and effective burn-in to screen out latent defects. P, S—Quality and yield impact due to test equipment limits. P, S—New timing-related fault models for defects/noise.
Fault Tolerance and Online Testing	P, S—DFT and fault tolerant design for logic soft errors. S—Logic self-repair using on-chip reconfigurability. S—System-level online testing.

Note: This table summarizes challenges to the design process advances implied by the above four trends. Each challenge is labeled with a list of the most relevant system drivers: S, system-on-chip; P, microprocessor; A, analog/mixed-signal; M, memory.

12.2 DELAY TESTING

The objective of delay testing is to detect timing defects and ensure that the design meets the desired performance specifications. The need for delay testing has evolved from a common problem faced by the semiconductor industry: Designs that function properly at low clock frequencies might fail at the desired operational speed. Traditionally, functional tests created for design verification are applied at system operational speed to screen out parts with delay defects; however, applying functional tests is becoming very expensive, given the need for a high-speed tester to apply such tests. This approach is still used extensively for high performance parts, such as microprocessors and *digital signal processors* (DSPs) for which the functional tests can be loaded into on-chip caches and then applied with a low-cost tester. Another problem with using functional tests is the lack of assurance for high test quality. Several industrial experiments (*e.g.*, [Maxwell 1991]) have shown that tests not specifically targeting delay faults have limited success in detecting timing defects. The above-mentioned problems can be alleviated by using structurally based *automatic test pattern generation* (ATPG) tests that target specific delay fault models and which can be applied through *design for testability* (DFT) structures using lower-cost testers. For the rest of the section, our discussion is focused on such structurally based delay testing approaches.

The growing need for delay testing is a result of advances in *very-large-scale integration* (VLSI) technology and an increase in the design speed. These factors are also changing the target objectives of delay tests. In the early days, most defects affecting the performance could be detected using tests for gross delay defects [Waicukauski 1987]. Aggressive timing requirements of high-speed designs have introduced the need to test smaller timing defects and distributed faults caused by statistical process variations. The increase of the circuit size has resulted in fault models that can detect distributed defects localized to a certain area of the chip [Smith 1985] [Lin 1987]. With the introduction of deep submicron technology, noise effects are becoming significant contributors to timing failures and they call for further adaptations of the fault models and testing strategies.

12.2.1   Test Application Schemes for Testing Delay Defects

To observe delay defects, it is necessary to create and propagate transitions in the circuit running *at-speed* (at its specified operating frequency). Creating transitions requires application of a vector pair, $V = <v_1, v_2>$, at the inputs of the combinational part of the circuit. The first vector initializes the relevant internal signals to desired initial logic values, while the second vector causes the desired transitions and sensitizes the transition from the target fault site to an output. The test application scheme for combinational circuits is shown in Figure 12.2. In normal operation, only one clock (system clock) is used to control the input and output latches (in a broader sense, storage elements), and its period is T_c. In this illustration, the input and output latches are controlled by two different clocks in the test mode: the input and output clocks, respectively. The period of these clocks, T_s, is assumed to be larger than T_c. The input and output clocks are skewed by an amount equal to T_c. The first vector, v_1, is applied to primary inputs at time t_0. The second vector, v_2,

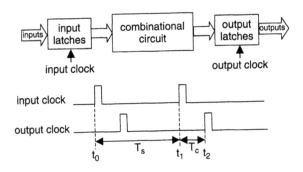

■ **FIGURE 12.2**

Testing scheme for combinational circuits.

is applied at time t_1. Time $T_s = t_1 - t_0$ is assumed to be sufficient for all signals in the circuit to stabilize under the first vector. After the second vector is applied, the circuit is allowed to settle down only until time t_2, where $t_2 - t_1 = T_c$. At time t_2, the primary output values are observed and compared to a prestored response of a fault-free circuit to determine if there is a defect.

Requiring separate clocks for input and output latches in the test mode may not be feasible for modern designs, if both clocks are not already available during normal operation; it might be too costly to resynthesize the clock trees just for the test purpose. If the input and output latches use the same clock source in the test mode, the scheme illustrated in Figure 12.2 still applies. There will be no skew between the input and the output clocks, and T_s could be equal to or larger than T_c. Times t_0, t_1, and t_2 would be at the rising edges of three consecutive clock pulses. Testing delay faults in sequential circuits is more difficult than testing delay faults in combinational circuits. Even for circuits with scan, application of an arbitrary vector pair at inputs to the combinational part of the circuit is not possible.

Generating tests for delay faults for scan designs corresponds to a two-time-frame sequential circuit test generation. In the first time frame, all primary inputs and present state lines are fully controllable. In the second time frame, only the primary inputs are fully controllable. Testing schemes for scan have been proposed in the literature [Savir 1992a,b] [Cheng 1993] [Savir 1994]. These techniques use **launch-on-capture** (also called *broad-side* test [Savir 1994]) or **launch-on-shift** (also called *skewed-load* test [Savir 1992a,b]) to obtain the second vector. In *launch-on-capture* which was referred to as *double-capture* in Chapter 5, the second vector is derived using the capture mode and represents the set of next state values obtained after the application of the first vector. In *launch-on-shift*, the second vector is obtained by shifting the contents of the scan chain by one bit after the application of the first vector using the scan-shift mode [Cheng 1993].

12.2.2 Delay Fault Models

Three delay fault models are considered: transition fault model, gate-delay fault model, and path-delay fault model. It is assumed that in the nominal design each gate has a given fall (rise) delay from each input to the output pin. Also, the interconnects are assumed to have given rise (fall) delays. Because the gate pin-to-pin delays and the interconnect delays can be combined together, the term "gate delay" will be used to denote this sum. Transition and gate-delay models are used for representing delay defects lumped at gates, while the path-delay model addresses defects that are distributed over several gates. The advantages and disadvantages of each model are discussed.

The **transition fault model** [Levendel 1986] [Waicukauski 1987] [Cheng 1993] assumes that the delay fault affects only one gate in the circuit. There are two transition faults associated with each gate: a **slow-to-rise fault** and a **slow-to-fall fault**. It is assumed that in the fault-free circuit, each gate has some nominal delay. Delay faults result in an increase of this delay. Under the transition fault model, the extra delay caused by the fault is assumed to be large enough to prevent the transition from reaching any primary output at the time of observation. In other words, the delay fault can be observed independent of whether the transition propagates through a long or a short path to any primary output; therefore, this model is also referred to as the *gross-delay fault model*.

To detect a transition fault in a combinational circuit it is necessary to apply two input vectors, $V = <v_1, v_2>$. The first vector, v_1, initializes the circuit, while the second vector, v_2, activates the fault and propagates its effect to some primary output. Vector v_2 can be found using stuck-at fault test generation tools. For example, for testing a slow-to-rise transition, the first vector initializes the fault site to 0, and the second vector is a test for a stuck-at-0 fault at the fault site. A transition fault is considered detected if a transition occurs at the fault site and a sensitized path extends from the fault site to some primary output.

The main advantage of the transition fault model is that the number of faults in the circuit is relatively small (linear in terms of the number of gates). Also, the stuck-at fault test generation and fault simulation tools can be easily modified for handling transition faults. On the other hand, the expectation that the delay fault is large enough for the effect to propagate through any path passing through the fault site might not be realistic because short paths may have a large **slack** (*slack* is defined as the difference between the clock period and the nominal delay of the path for the fault-free circuit). The assumption that the delay fault only affects one gate in the circuit might not be realistic either. A delay defect can affect more than one gate, and even though none of the individual delay faults is large enough to affect the performance of the circuit, several faults can together result in performance degradation. For practical simplicity, the transition fault model is frequently used as a qualitative delay model, and circuit delays are not considered in deriving tests.

The **gate-delay fault model** [Iyengar 1988a,b] assumes that the delay fault is lumped at one gate in the circuit; however, unlike the transition model, the gate-delay fault model does not assume that the increased delay will affect the performance independent of the propagation path through the fault site. It is assumed that only long paths through the fault site might cause performance degradation. The gate-delay fault model is a quantitative model, as it takes into account the circuit delays. To determine the ability of a test to detect a gate-delay defect, it is necessary to specify the delay size of the fault. Methods for computing the smallest delay fault size (detection threshold) guaranteed to be detected by some test have been reported in the literature [Iyengar 1988a,b] [Pramanick 1997].

The limitations of the gate-delay fault model are similar to those for the transition fault model. Because of the single gate-delay fault assumption, a test may fail to detect delay faults that are a result of the sum of several small delay defects. The main advantage of this model is that the number of faults is linear in the number of gates in the circuit.

Under the **path-delay fault model** [Smith 1985], a circuit is considered faulty if the delay of any of its paths exceeds a specified limit. A path is defined as an ordered set of gates $\{g_0, g_1, \ldots, g_n\}$, where g_0 is either a primary input or output of a flip-flop, and g_n is either a primary output or an input of a flip-flop. Also, gate g_i is an input to gate $g_{i+1} (0 \leq i \leq n-1)$. A delay defect on a path can be observed by propagating a transition through the path; therefore, a path-delay fault specification consists of a physical path and a transition that will be applied at the beginning of the path.

The delay or length of the path represents the sum of the delays of the gates and interconnections on that path.

Tests for the path-delay fault model can detect small distributed delay defects caused by *statistical process variations*. A major limitation of this fault model is that the number of paths in the circuit can be very large (possibly exponential in the number of gates). For this reason testing all path-delay faults in the circuit is impractical. Two strategies are commonly used for selecting the set of path-delay faults for testing. One is to select a minimal set of paths such that for each signal *s* in the circuit the longest path containing *s* is selected for testing [Li 1989] [Yang 2004]. The other is to select all paths with expected delays greater than a specified threshold [Sato 2005]. The reason behind selecting the longest paths is that the delay defects on shorter paths might not be large enough to affect the circuit performance. Also, if the defects on short paths are large and could affect the performance, one expects that such defects would be detected by other tests (*e.g.*, transition or gate-delay tests) that precede the path-delay fault testing. This strategy might work for circuits whose paths have very different delays so there is only a small percentage of long paths; however, often in performance-optimized designs almost all paths have long delays, and in these circuits not even all of the longest paths can be tested [Park 1991].

Various experimental results reported from industry have strongly indicated that stuck-at fault testing is not sufficient to guarantee high product quality requirements. Industrial data have shown that a large portion of the defects not detected by stuck-at fault testing represents timing failures. Transition fault tests have been shown to be effective for detecting gross-delay defects. For high-performance circuits with aggressive timing requirements, small process variations can lead to failures at the system clock rate. These defects can be detected using tests for path-delay faults. While the need for detecting delay defects is clear, the high cost for detecting them remains a problem. A possible cost-effective strategy for delay testing would include:

- Use of functional vectors that could be applied at the system's operational speed and should catch some delay defects (functional vectors should be evaluated for transition fault coverage)

- Application of ATPG tests for undetected transition faults

- Application of ATPG tests for long path-delay faults

There are more sophisticated strategies for integrating different types of delay tests. One of the drawbacks to transition fault testing is that the breadth-first search algorithm typically used in ATPG tends to select short paths through each fault site. As a result, when tested at-speed, many paths have considerable timing slack, so only relatively large delay defects can be detected. One solution is to generate one or more longest paths through each fault site [Sharma 2002] [Qiu 2003]; however, the increased path length increases test data volume [Qiu 2004]. Rather than maximizing the length of the tested paths, the alternative is to shrink the capture clock timing to minimize the slack for each pattern [Mao 1990], but the use of separate timing for each pattern drastically increases test data volume.

An alternative is to group patterns into sets of almost equal-length paths [Kruseman 2004]. The user must trade off between the number of groups and test data volume. Because the chip is being tested at faster than its operating speed, logic transitions may still be occurring when the capture clock is applied. Those storage elements fed by paths that exceed the cycle time or contain hazards must be masked off to avoid mismatch. The faults that are not detected due to masking are targeted by patterns run at the next slower clock speed [Barnhart 2004]. Applying transition fault patterns at faster than the operating speed has been shown to catch small delay defects that escape traditional transition fault tests [Kruseman 2004] [Amodeo 2005].

12.2.3 Summary

Delay testing is becoming an increasingly important part of the VLSI design testing process. Continuously increasing circuit operating frequencies results in designs in which performance specifications can be violated by very small defects. The use of traditional fault models and test strategies becomes even more inadequate as the current design trends move towards deeper submicron designs. The deep submicron process introduces new failure modes and a new set of design and test problems. Process variations are now more likely to cause marginal violations of the performance specifications. The continuous shrinking of device feature size, increased number of interconnect layers and gate density, increased current density, and higher voltage drop along the power nets give rise to noise faults, such as distributed delay defects, power supply noise, ground bounce, substrate noise, and crosstalk. Analysis shows that most of the excessive noise leads to delay faults; for example, studies have shown that the increased coupling effects produce interference between signals and may increase or decrease signal delays [Breuer 1997].

Testing delay defects continues to be a complex problem. Difficulties are related to the fault modeling, the test generation, and the test application process. Solutions currently in use still cannot satisfactorily address some of the new failure modes in deep submicron designs. Especially, most of the existing techniques are based on simplified, logic-level delay fault models and cannot be directly used to model and test timing defects in high-speed designs based on deep submicron technologies. Some of the main delay test challenges for multi-gigahertz devices are outlined in [Mak 2004]. For interested readers, a comprehensive coverage of various topics related to delay testing is available in [Krstic 1998]. The following is a subset of research topics that must be further explored to address these challenges:

- Testing delay defects in high-speed circuits requires the availability of high-speed testers; however, due to the high cost, testers in the test facilities are usually slower than the new designs that need to be tested on them. Therefore, there is a pressing need for developing practical solutions to testing fast chips on slow testers. At the current rate of design performance increase and the high cost of fast testers, the gap between the speed of the new designs and that of the testers is not likely to disappear. One emerging solution is to include the circuit's internal *phase-locked loop* (PLL) for at-speed delay testing to alleviate the dependency on high-performance external testers.

- Small distributed delay defects can best be modeled using the path-delay fault model; however, practical designs have a very large number of paths, and only a small fraction of them can be tested. The selection of paths for testing is especially difficult in performance-optimized designs because they often have a large number of paths with long propagation delays.

- Selection of critical paths for testing requires accurate timing information which is not easily available. Deep submicron process introduces new difficulties into the critical path selection because noise factors, such as power supply noise, ground bounce, and crosstalk, can significantly affect the signal delays and some paths can be more sensitive to these effects than others.

- In delay testing it also becomes important to take into account signal speedups or slowdowns resulting from various noise sources in deep submicron devices. Recent research results [Liou 2003] indicate that incorporation of statistical principles for delay testing is an effective way of addressing the issues caused by process variations and noise. A new delay test paradigm under such a principle, called **statistical delay testing and diagnosis**, has been investigated in [Liou 2003] and [Krstic 2003]. Under this new delay test paradigm, timing analysis, target-fault selection, and ATPG must be enhanced and built upon statistical models.

In **statistical timing analysis**, the delays of basic circuit elements are modeled as *correlated* random variables with presumably known ***probability density functions*** (PDFs). The modeling structure for these random variables is general enough to accommodate statistical information resulting from the manufacturing process, noise, and new defect models resulting from nanometer technologies. Timing analysis is carried out to estimate the signal arrival time at each internal signal and each primary output as a random variable, rather than as a single, worst-case value.

Under the statistical timing model, the definition of a *critical path* becomes probabilistic. If N chips are manufactured, the sets of critical paths on different chip instances could differ. In a recently developed framework [Liou 2002] [Liou 2003], the task of **statistical path selection** is divided into two phases. In the *path filtering* phase, most of the short or unsensitizable paths are quickly filtered out. The result is a set of long paths that may affect the circuit timing outcome. During the phase of *true critical path selection*, the tool attempts to select a minimal number of paths from among those *statistically long* paths for high-quality test or diagnosis applications.

Due to the statistical timing involved, delay along a target path is highly pattern dependent. In order to produce high-quality patterns, we therefore need to generate test patterns that not only sensitize the given set of statistical critical paths but also exercise the worst-case delays along these paths; however, considering timing during ATPG can significantly increase the complexity of the process. Also, with statistical timing and statistical delay defect models, the notion of *path sensitization* becomes probabilistic; thus, the key challenge is to develop a feasible **statistically constrained ATPG** method where statistical timing-sensitization constraints can be employed to guide the ATPG justification process.

12.3 COPING WITH PHYSICAL FAILURES, SOFT ERRORS, AND RELIABILITY ISSUES

Recall from Chapter 1 that *defect level* is a function of *failure rate* and *manufacturing yield*. *Failure rate* in turn is a function of *fault coverage*. Therefore, to cope with vast likely **physical failures** in nanometer designs, we need to seriously reduce the defect level to meet the ***defects per million*** (DPM) goal. This can be done by improving the *fault coverage* of the chips (devices) under test, the *manufacturing yield*, or both; however, not all chips passing manufacturing tests would function correctly in the field. Reports have shown that chips could be exposed to *alpha-particle radiation*, and *nonrecurring transient errors* caused by single or multiple event upsets, called **soft errors**, could occur [May 1979]. For nanometer *system-on-chip* (SOC) designs, there is also a growing concern whether one can find defect-free or error-free dies [Breuer 2004]. Advanced test technologies are important now in order to meet yield and DFM goals and ensure that defective chips will function correctly in the field.

There are two fundamentally complementary test technologies that can be taken to meet our goals, similar to those approaches used to improve the **reliability** of computer systems: *design for testability* (DFT) and *fault tolerance* [Lala 2001]. The fault tolerance approach aims at preventing the chip (computer system) from malfunctioning despite the presence of physical failures (errors), while design for testability uses design techniques to reduce defect level or the probability of chip (system) failures during manufacturing.

In the following subsections, we first discuss promising test technologies to deal with signal integrity and yield issues induced by physical failures. We then describe promising schemes to cope with soft errors. Finally, we present fault tolerance techniques as well as promising schemes for defect and error tolerance to ensure that defective chips can still function in nanometer designs.

12.3.1 Signal Integrity and Power Supply Noise

Signal integrity is the ability of a signal to generate correct responses in a circuit. Informally speaking, signal integrity indicates how clean or distorted a signal is. A signal with good integrity stays within *safe* (acceptable) margins for its voltage amplitude and transition delay. For example, an input signal to a flip-flop with good integrity arrives early enough to satisfy the setup/hold time requirements and does not have large undershoots that may cause erroneous logic readout or large overshoots that affect the transistor's life time.

Leaving the safe margins may not only cause failure in a system (*e.g.*, unexpected ringing) but also shorten the system's life time. The latter is due to ***time-dependent dielectric breakdown*** (TDDB) [Hunter 1999] or injection of high-energy electrons and holes (also called *hot-carriers*) into the gate oxide. Such phenomena ultimately cause permanent degradation of *metal oxide semiconductor* (MOS) transistors' performance and reliability. To quantify these, systematic methods can be employed to perform the life-time analysis and measure performance degradation of logic gates under stress (*e.g.*, repeated overshoots) [Fang 1998].

Signal integrity depends on many internal (*e.g.*, interconnects, data, characteristics of transistors) and external (*e.g.*, environmental noise, power supply, interactions with other systems) factors. By using accurate simulation in the design phase, one can apply conservative techniques (*e.g.*, stretched sizing, shielding) to minimize the effect of integrity loss; however, such vast dependencies, especially when there is a huge demand for faster chips, make it impossible (with our current state of knowledge) to have a guaranteed remedy at the design phase. Thus, testing future VLSI chips for signal integrity seems to be inevitable.

12.3.1.1 *Integrity Loss Fault Model*

True characteristics of a signal are reflected in its waveform. In practice, digital electronic components can tolerate certain levels of voltage swing and transition/propagation delay. Any portion of signal that exceeds these levels represents *integrity loss* (IL). This concept has graphically been shown in Figure 12.3, in which the horizontal and vertical shaded strips correspond to the amplitude- and delay-safe margins, respectively. The black areas illustrate the time frames in which the signal has left the safe margin and thus integrity loss has occurred.

Any portion of a signal $f(t)$ that exits the safe margins contributes to the integrity loss metric. So, conceptually we can define:

$$IL = \sum_i \left(\int_{b_i}^{e_i} |V_i - f(t)| \cdot dt \right)$$

where V_i is one of the acceptable amplitude levels (*i.e.*, a border of safe margin) and $[b_i, e_i]$ is a time frame during which integrity loss occurs.

With the existing tools and computing devices, the analysis/simulation recommended by this model would not be practical for real-world circuits; yet, it implies

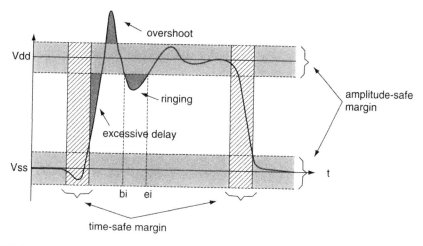

■ **FIGURE 12.3**

The concept of signal integrity loss.

three main requirements in testing VLSI chips for signal integrity: (1) determine the locations to sample and monitor IL, (2) carry out pattern generation to stimulate extreme integrity loss, and (3) design integrity loss sensors/detectors and readout circuitry. Almost all solutions presented in the literature so far point to the necessity of a combination of the above three requirements. In what follows we elaborate on these requirements.

12.3.1.2 *Location*

To have a practical evaluation of integrity loss we need to decide where to look and what to look at. Various sources of signal integrity loss in VLSI chips have been identified. The most important ones are:

- *Interconnects* that contribute to crosstalk (signal distortion due to cross-coupling effects among signals), *overshoot* (signal rising momentarily above the power supply voltage), and *electromagnetic interference* (resulting from the antenna properties). At-speed testing of crosstalk in chip interconnects [Bai 2000], testing interconnect crosstalk defects using on-chip processors [Chen 2001], a BIST method to test long interconnects for signal integrity [Nourani 2002], and using boundary scan and I_{DDT} for testing bus [Yang 2001] are some of the proposed methods. The experiments show that short interconnects as well as long interconnects are susceptible to the integrity problem. This will be a major challenge in future ultra-high-speed deep submicron chips.

- *Power supply noise*, whose large fluctuations, mainly due to simultaneous switchings, affect the functionality of some gates and eventually may lead to failure [Senthinatharr 1994] [Zhao 2000a]. Various ways of estimating **power-supply noise** (PSN) in limited forms and accuracy have been presented in the literature. To name a few, [Chang 1997] uses a scaling model, [Chen 1997a] employs a simulated switching model of power bus, and [Zheng 2000] focuses on a distributed power network model. [Lee 2001] presents the generation and characterization of three different types of noise induced by electrostatic discharge in power-supply systems.

- *Process variations*, which are deviations of parameters from their desired values due to imperfect nature of fabrication process. Sources of process variations include random dopant fluctuation, annealing effects, lithographic limitations, etc. [Borkar 2004a]. Researchers have studied the effect of process variations on reliability [Borkar 2004b], clock skew/distribution [Bowman 2001] [Zanella 2000] [Zarkesh-Ha 1999], leakage current [Keshavarzi 2002], performance [Murthy 1997], delay test [Lu 2004], and yield prediction [Jess 2003], among others.

12.3.1.3 *Pattern Generation*

Due to the nature of signal integrity loss (fault) and its intermittent occurrence, integrity fault testing must be done at speed. The pin and probing limitations further restrict the accurate observation of signal integrity losses. Therefore, an on-chip *built-in self-test* (BIST)–style mechanism is one possible choice. Conventional **pseudo-random pattern generators** (PRPGs) to stimulate maximum integrity loss on long interconnects have been tried [Nourani 2002]. In spite of the good test quality that pseudo-random patterns can achieve, the random nature of process prevents any test session from having a bound on the length.

The **maximum aggressor** (MA) fault model [Cuviello 1999] is one of the fault models proposed for crosstalk. This model assumes a signal traveling on a victim line may be affected by signals or transitions on other lines (aggressors) in its neighborhood. The traditional MA model takes only coupling capacitors into account. All aggressors are assumed to make simultaneous transitions in the same direction, while the victim line is kept quiescent (for maximal ringing) or makes an opposite transition (for maximal delay). Various approaches to analyze the crosstalk noise are described in [Chang 1997], [Zhao 2000b], and [Nagaraj 2001]. An interconnect design for integrated circuits operated in the gigahertz range is discussed in [Naffziger 1999], where the author observed that chips could fail when a specific test pattern (not included in the MA model) is applied to interconnects, due to the overall effect of coupling capacitances and mutual inductances. Similarly, according to [Cao 2002], the worst-case switching pattern to handle inductive effects for multiple signal lines may not be included in the MA fault model. Several researchers have worked on test pattern generation for crosstalk noise/delay and signal integrity [Chen 1998] [Chen 1999] [Attarha 2002].

The MA model has been extended to the **multiple transition** (MT) model in [Tehranipour 2004]. The MT pattern set is a superset of the MA set and is much more capable of testing the capacitive and inductive coupling among interconnects. The modified driving end boundary scan cells (PGBSCs) receive a few seeds and generate MT patterns at-speed to stimulate integrity violations. These cells along with new instructions extend the JTAG standard to include testing interconnects for signal integrity.

To speed up the PSN analysis and test generation process, some works exploited the concept of random search. For example, [Jiang 1997] and [Bai 2001] use a genetic algorithm (with random basis) to stimulate the worst case PSN. Another group of researchers, such as [Zhao 2000c], precharacterize cells using transistor-level simulators and annotate the information into the PSN analysis phase. A technique for vector generation for power-supply noise estimation and verification is offered in [Jiang 2001]. The authors used a genetic algorithm to derive a set of patterns producing high power-supply noise. A pattern generation method to minimize the PSN effects during test is presented in [Krstic 2001]. In [Nourani 2005], the authors identified three design metrics (*i.e.*, level, fanin, and fanout) that capture realistic PSN. Then, they employed a greedy algorithm and conventional fault simulator to quickly construct pattern pairs that simulate the worst-case PSN based on circuit topology, regardless of its functional or testing mode.

12.3.1.4 *Sensing and Readout*

Because the integrity loss is a waveform-related metric, it must be captured (sampled) right after creation. This will be practical by limiting the observation sites and by designing reasonable-cost sensors/readout circuits. Various types of sensors, potentially useful for IL detection, are reported in the literature. A BIST-style structure using D flip-flops has been proposed to detect the propagation delay deviation of operational amplifiers [Rayane 1999]. A test methodology targeting bus interconnect defects using IDDT and boundary scan has been presented in [Yang 2001]. In this work, a built-in sensor is integrated within the system; the sensor is an on-chip current mirror converting the dissipated charges into the associated test time.

The work presented in [Nourani 2002] offers inexpensive cells, called **noise detector** (ND) and **skew detector** (SD) cells, based on a modified cross-coupled *P-channel metal oxide semiconductor* (PMOS) differential sense amplifier. The authors in [Tabatabaei 2002] presented a more expensive but more accurate circuit to measure jitter and skew in the range of few picoseconds. This circuit, called an **embedded time to digital converter** (EDTC), samples signals in non-intrusive way and sends out the test information through its low speed serial information. When cost is not a concern, the concept of accurate signal monitoring has been followed up by researchers even through the idea of on-chip oscilloscope [Caignet 2001].

The authors in [Tehranipour 2004] showed that each of such sensors can be integrated within a boundary scan cell to form an **observation boundary scan cell** (OBSC). In the signal integrity test mode, the OBSC collects and sends the information on IL through the scan chain. While in this work the focus was on interconnects, any non-modeled fault (inside or outside cores) that manifests itself as integrity loss on interconnects will also be detected by that method.

A **PSN** monitor circuit presented in [Vazquez 2004] is claimed to catch high-resolution (100-ps) PSN at the power/ground lines. A power supply distribution model to control PSN has been also reported in the literature. The model presented in [Chen 1997b] identifies the hot-spots on the chip and optimizes power-supply distribution to minimize the noise. A cascaded power/ground ring for on-chip power distribution is proposed in [Cao 1997]. Another methodology for multiple power-supply distribution systems is presented in [Pham 2004]. The authors in [Zhao 2002] argue that the peak PSN can be significantly reduced based on the physical correlation of modules. They have proposed a power-supply noise-aware floor planning methodology. An analytical way of combining the PSN of nonembedded cores to estimate the PSN of a SOC (proportional to IL) is discussed in [Nourani 2005]. In addition to providing an estimate of PSN, this method can be used to group cores and design a power-supply distribution network while keeping PSN (and thus the corresponding IL) under control.

12.3.2 Parametric Defects, Process Variations, and Yield

Defects are **physical defects** that occur during manufacturing and can cause *static* or *timing physical failures*. Examples of defects include partial or spongy via and the presence of extra material between a signal line and the voltage line. Broadly speaking, defects can be *random* or *systematic*, and they can be *functional* or *parametric*. Traditional treatment of defects is more on *functional random* (*spot*) defects, which lead to existing yield models. Growing process variations and other uncertainty issues require that we look into the other three types of defects. In a narrow sense, defects are caused by *process variations* or *random localized manufacturing imperfection* [Sengupta 1999].

Process variations, such as transistor channel length variation, transistor threshold voltage variation, metal interconnect thickness variation, and intermetal layer dielectric thickness variation, have a big impact on device speed characteristics. In general, the effect of process variation shows up first in the most critical paths in the design, those with maximum and minimum delays.

Random imperfection, such as resistive bridging defects between metal lines, resistive opens on metal lines, improper via formations, and shallow trench isolation defects, are yet another source of defects and are referred to as **parametric defects**. Based on the parameters of the defect and *neighboring parasitic*, the defect may result in a *static* or *at-speed failure*.

12.3.2.1 *Defect-Based Test*

In order to detect physical failures caused by both process variations and parametric defects, one common approach is to generate multiple test sets, each targeting a fault model. A promising technique is to generate **defect-based tests** by enumerating likely defect sites (failures) from the layout [Sengupta 1999]. In either case, *at-speed tests*, consisting of *path-delay tests* and *transition tests*, must be used [Tendolkar 2000] [Lin 2003]. The *at-speed tests* can come from scan and/or BIST [Wang 2005]. One study on a 733-MHz PowerPC microprocessor design showed that if *at-speed tests* were removed from the test program, the *escape rate* went up nearly 3% [Gatej 2002].

On the other hand, it is also critical to supplement the conventional *stuck-at tests* with *bridging tests* [Sengupta 1999]. One of the most common defect types in nanometer designs is the interconnect bridge. As the number of bridges is astronomical, it is more realistic to enumerate likely bridging fault sites (*physical bridging faults*) from layout and map them to *logical bridging faults* for fault simulation or scan ATPG.

Moreover, it has been reported in [Ma 1995] that *N-detect stuck-at tests* that detect every stuck-at fault multiple (*N*) times are better at closing DPM holes than tests that detect each fault only once. This approach, called **N-detect**, works because each fault is generally targeted in several different ways, increasing the probability that the conditions necessary to activate a particular defect will exist when the observation path to the fault site opens up. *N-detect at-speed tests* can also be used [Pomeranz 1999], but a promising study shows that by generating *transition tests*, one for each reachable output for a given transition fault, **transition fault propagation to all reachable outputs** (TARO) was able to detect all defective cores that other tests could not on a test chip [Tseng 2001] [McCluskey 2004] [Park 2005]. TARO can be a good candidate for tests that require excessive thoroughness, such as sample-based quality assurance tests. For logic diagnosis, TARO can offer a much better resolution than other tests.

It is also possible to further improve yield by supplementing these tests with *IDDQ tests*, which detect many types of defects, including some timing-related defects

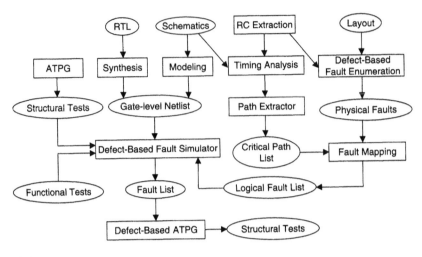

A defect-based test system architecture.

[Nigh 1998] [Nigh 2000]; however, the small geometry sizes of today's devices have caused many companies to abandon or rely less on IDDQ tests because a defective device's current will be difficult to distinguish from the normal quiescent current.

Functional testing, once the sole test method that allows for testing actual functional paths at-speed, has begun to regain acceptance in the industry. In order to meet aggressive yield and DPM goals, *functional tests* must be added to supplement *structural tests* (*at-speed tests*, *stuck-at tests*, and *bridging tests*). Figure 12.4 shows such a defect-based test system architecture [Sengupta 1999].

The key issue is how to generate these *defect-based tests* (*structural tests and functional tests*) in a timely manner in order to meet time-to-market, DFM, yield, and test budget goals all at the same time. In the nanometer age, we anticipate that BIST and *test compression* no longer will be options. Active research will be more directed toward coverage enhancement of *logic BIST*, reduction in *scan ATPG* time and test power, *physical fault modeling*, and the speed-up of *concurrent fault simulation*.

12.3.3 Soft Errors

Soft errors are the result of *transients* that are induced at the circuit when a radiation particle strikes. This radiation can range from cosmic origin (when stars are formed and die) or from every day material (*e.g.*, lead isotopes) [Ziegler 2004]. When high-energy cosmic rays reach into our atmosphere, they collide and strip off air molecules and send off neutrons. These neutrons continue their journey and penetrate through most types of matter (so shielding is largely out of the question). As these neutrons transverse through silicon, they ionize the silicon lattice and leave a trail of holes and electrons behind. These will then be moved by the electric field of surrounding diffusion and wells. As holes and electrons recombine, they charge or discharge the node appropriately. From a circuit standpoint, we just have a glitch.

Radioactive isotopes emit alpha particles as the radioactive decay process occurs. These alpha particles are larger and heavier so they will not have deep penetration; however, due to the fact that they may exist (in a stray amount) in packaging material (*e.g.*, ceramic or solder), they are located very close to the die and can lead to a relatively high error rate (in fact, early soft errors were first discovered in radioactive elements in packaging material [May 1979]). A similar event happens when alpha particles penetrate silicon. Due to their larger size, these alpha particles will lose energy rapidly and be trapped inside the material, forming more stable compounds. If such a glitch is induced in a memory element, its state can be reversed. As an example, let us examine a SRAM cell that has two back-to-back inverter pairs, as shown in Figure 12.5.

When the select transistors are off, this configuration will hold the state in a stable configuration. If a glitch is introduced at the drain of the PMOS or the source of the NMOS, the glitch can be picked up by the other inverter and the state of the cell is reversed.

A similar problem can occur for all state-holding elements (storage elements), such as D latches and D flip-flops (see Figure 12.6). If such a glitch strikes the combinational logic elements, the resulting glitch will be evaluated and passed on by the succeeding logic. Glitches are common for CMOS combinational circuits as logic inputs arrive at different times. Depending on the strength of the originating glitch, the glitch can be magnified (and passed along) or it can be filtered. The glitch can also be blocked by all other fan-in logic. Only if the glitch arrives at the time when the latch is closing is an erroneous state captured; otherwise, the glitch hitting the combinational logic will not cause any harm (an exception is a

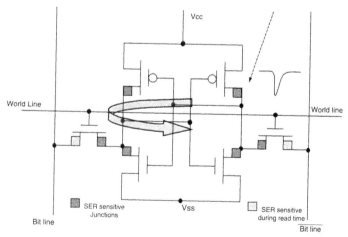

SRAM Cell flip with a radiation strike

■ **FIGURE 12.5**

Induced soft error on a SRAM cell.

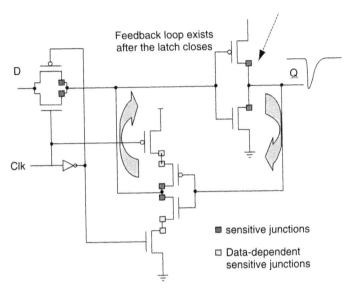

Feedback loop exists
after the latch closes

D

Q

Clk

■ sensitive junctions

□ Data-dependent
   sensitive junctions

Latch upset with radiation strike

■ **FIGURE 12.6**

Induced soft error on a D latch.

domino-type circuit, where its logic states are held at every gate by a feedback element).

Soft errors can happen to all memory and storage elements. Sometimes, they can be benign (*e.g.*, the memory elements are not used in the application); other times, they can cause a system crash or even worse—a *silent data corruption* (SDC)—if they are undetected. That is why we have to devise online detection (or fault tolerant) mechanisms to protect against such transients. Such kinds of detection and tolerant mechanisms are more fully discussed in the following section. Unlike defects or other fault types, a soft error is a transient induced at one time at one location; they are not repeatable, thus the term "soft error." This property is well utilized in the solution space.

Because the physics of soft errors involves node charging and discharging, the amount of stored charge at a given node determines how sensitive it is to a particle strike. The charge (Q) is represented by the following relationship with voltage (V) and capacitance (C):

$$Q = CV$$

As processing technology scales, capacitance for a given node decreases. This is good for performance but bad for soft error. Due to the *hot electron* type of degradation, reliability requirements also force Vcc to be lowered. This compounded effect will cause decrease in stored charge and will increase the ***soft error rate*** (SER). With scaling, we also get more transistors (roughly 2×) per chip, resulting in increase in the soft error rate [Baumann 2005]. The only saving grace is that, because the transistor junction area is also scaled, the ability for the node to collect stray charge is also reduced; however, this is not sufficient to slow the increase in soft error vulnerability. The **Moore's law** prediction of doubling transistors with every process generation [Moore 1965] will effectively double the soft error rate, so, not only should SRAM cells (such as caches and registers) be protected, but protection on storage elements as well as against glitches creeping through the combinational logic is also important. These areas are now all hot research topics.

The implication of soft errors with regard to chip testing varies. From the surface, soft errors really cannot be tested. Even good circuits are susceptible to soft errors so there is nothing to screen for. Soft errors are also not easily exercisable with electrical test stimulus. The natural occurrence of radiation also does not usually happen during the short test time of component testing. What really requires attention is an online detection scheme or a fault-tolerant scheme.

Very often, the three types of redundancy—information, time, and spatial—involve extra circuit elements. At a minimum, there is a *self-checking checker*, which tells you whether there is any error. With *information redundancy*, there are extra check bits (or code bits). With *spatial redundancy*, there is even duplicate circuitry. Each redundancy circuitry has to be tested to make sure that the redundancy scheme can detect soft errors and, when found, can signal that there is an error or simply correct the error.

As it is difficult to test every redundancy circuitry and they are probably not testable without appropriate DFT means, special attention must be given to such redundancy circuitry so it is accounted for in the overall test strategy. If a redundancy scheme is capable of correcting errors by itself, then even manufacturing defects can hide behind the redundancy scheme, as output results are always correct. The undiscovered defects will consume the correction capability, and any subsequent soft error hit to that functional circuitry or redundancy circuitry may cause an unrecoverable error.

12.3.4 Fault Tolerance

One approach to improve the reliability of a chip is to remove the source of soft errors. Because the early discovery of soft error was due to contaminated packaging material, the solution was simple: Eliminate the radiation contaminant from packaging material [May 1979]. However, trace amounts of radioactive isotopes do exist in common processing and packaging materials, such as the boron in ***borophosphosilicate glass*** (BPSG) and ceramics and the lead in solder, and their removal is very costly. Their radioactive decay still leads to some level of soft error rate. Because alpha-particle radiation (see previous section) can be stopped on the surface of the die, a die coating (epoxy resins that are deposited on the surface of the die) was introduced and used for some period of time. Die coating is only effective if the radiation comes from the outside and is of very limited value if the alpha emitter is among the materials that transistors or interconnects are made of. As we move from wire bonding packaging to flip-chip solder (tin/lead) joint (or C4) packaging, solders are never far away from the surface of the die and die coatings would have no effect at all. Today, the primary alpha-particle source is solders (lead radiation isotopes). Careful selection of raw material has resulted in low-alpha solder. Due to environmental and health concerns over lead, tin/lead solder is being phased out in packaging material which will help to reduce alpha-induced soft errors. However, the other source of radiation, *high energy neutrons*, cannot be stopped by anything associated with packaging.

All of these preventive measures combined with recent advances in manufacturing process technology have improved the system-level reliability substantially. In the past, soft errors were not critical for most computer systems for terrestrial applications. Thus, traditionally, only high-reliability applications, especially those deployed in the financial transaction, transportation, and aerospace/defense industries, have required fault tolerance to prevent the systems from crashes and silent data corruption errors.

There are three fundamental fault-tolerant schemes that can be used to protect such systems from hard errors or soft errors: (1) spatial redundancy, (2) temporal redundancy, and (3) information redundancy [Pradhan 1996] [Siewiorek 1998] [Lala 2001].

- **Spatial redundancy** relies on the assumption that defects and radiation particles will only hit on a specific device and not another device (at least not simultaneously). So, having a duplicate circuitry of the functional circuitry and with their outputs compared using a *self-checking checker* (checking circuitry), mismatches will point to an error (hard or soft error) (see Figure 5.45 in Chapter 5). This can happen at the circuit level (*e.g.*, one adder is compared with another adder while both are fed the same data) or at the system level (*e.g.*, a processor's front side bus is compared with another one on the same bus while executing the same codes). Because the computation occurs in parallel, there is little or no penalty on the overall system performance, but there has to be hardware duplication and a *self-checking checker*, resulting in higher hardware costs and a higher level of power consumption.

- **Temporal (time) redundancy** relies on the assumption that even if a functional circuitry receives a radiation strike it is very unlikely that the strike will happen on the same circuitry again at a slightly later time, so the scheme does not require another duplicate circuitry. In this case, the same computation is repeated on the same functional circuitry for a second time, and the results of the first computation are not committed without comparison to the second computation. This obviously has the benefit of not requiring additional hardware, but the software must be coded to execute the program twice, which means saving the results of the first computation on memory or disk. One serious problem with this scheme is that it cannot detect hard errors; the same erroneous result will happen when recomputed. Therefore, temporal redundancy may give a false sense of security with regard to any hard error that develops due to physical failures.

- **Information redundancy** uses ***error detecting code*** (EDC) or ***error correction code*** (ECC) to represent information contents [Peterson 1972]. Some of these coding properties are maintained even after computation, so by checking these codes before and after one can determine if a hard or soft error has occurred. Parity is one such code. Parity represents whether the number of ones in a computer word is odd or even. Normally, this parity is computed when the information is generated and stored in the memory system. Upon reading the word, parity again is recalculated and compared against the earlier stored parity bit. A mismatch identifies that an error has occurred. One major benefit of using parity code is that a single parity bit can detect any odd number of bit errors (caused by soft errors and hard errors) in each computer word; however, there is always the danger that a single radiation strike could affect more than a single bit, and when an even number of bits get flipped the errors escape detection (because parity only counts odd or even). In this case, more sophisticated codes (such as Hamming codes) can be used [Peterson 1972]. This code will allow detection of 2-bit errors as well as correction of single-bit errors. This, in general, is referred to as the *error correction code*. It requires the storage of more check bits (codes), and a computation unit that does the check code generation. It is important to note that properties such as parity or ECC are often embedded for arithmetic operations (such as add/subtract/multiply) during normal operation. Thus, they are also used for arithmetic computation protection. Because additional information is stored, it can protect against both hard and soft errors.

Having detection capability is only half the story. After an error is detected, some recovery actions have to be taken. The most common action is for the operating system to stop the application, generate the necessary error message/log, and close the application. This has varying degrees of system integrity implications, as partially computed results may have been written to disk already. Of course, this is better than simply letting the system crash, but we need better schemes that can provide more integrity. **Checkpointing** and **rollback** comprise one such scheme. Checkpointing is essentially taking a snapshot of the system states, which when revoked (rollback) will cause the system to restart from that point without reboot or terminating the application. What one has to ensure is that no error has occurred (and the system is intact) before the checkpoint. One also has to consider where (or how regular) checkpoints are done to optimize for performance (due to the checkpointing process) as well as making sure that system states are not contaminated before the checkpoint.

Thus, having explained the basic principle, what are some of the common fault-tolerant schemes used in high reliability systems? As mentioned before, duplicate and compare is one such method that is commonly used in mainframes and high-end servers [Spainhower 1999] [Bartlett 2004]. For systems that cannot fail, a more secure system is ***triple modular redundancy*** (TMR) [Sklaroff 1976] [Siewiorek 1998] [Lala 2001]. Consider the TMR example shown in Figure 12.7. Here we have three pieces of compute units, and their results are constantly compared among each other. Because we have three results, the two matched results will outvote the

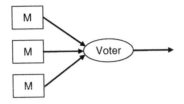

■ **FIGURE 12.7**

A triple modular redundancy (TMR) example.

mismatched result and the deemed correct result will be sent on. The aerospace industry particularly favors this approach (for obvious reasons). More recently, because a **central processing unit** (CPU) is capable of running multiple threads (different streams of instructions that can be run in parallel on a CPU), one can also send a redundant thread through another path (virtual compute units) and check the results before retiring the thread/instructions. This is called **redundant multithreading** (RMT) [Mukherjee 2002].

The incorporation of fault tolerance in a compute system did not begin with the processor. It began with the circuit that had the highest transistor density and the part of the system that had the most transistors—the main memory. The protection scheme is the use of ECC, which is quickly followed by **redundant array of inexpensive disk** (RAID). Even though the disk system does not necessarily have the highest number of transistors, disk drives are susceptible to mechanical failures; hence, it is essential to protect the data that it holds. Due to the need to have signals routed over long wires or traces, buses and backplanes that interconnect various subsystems are protected with parity. The networking communication protocols also contain error codes, such as **cyclic redundancy check** (CRC) and checksum (the sum of all the binary numbers in a particular packet of data). In the early 1990s, the importance of fault tolerance to CPUs became apparent, as on-chip cache memories have become large enough to warrant their own ECC protection. On some high-end server CPUs, register files are protected with parity, and duplicated execution blocks help to identify errors [Spainhower 1999]. So, the last holdout seems to be flip-flops and combinational logic. Researchers have come up with hardened latch/flops [Calin 1997], where circuit design has decreased the internal nodes' vulnerability to a radiation strike. More recently, enhancing the scan cells to fill the role of duplication (as the states are already duplicated) has been suggested [Mitra 2005].

As we move toward the nanotechnology era, more and more system-level functions (in the form of IP cores) will be integrated on a single piece of silicon (or package). This has substantially exposed the nanometer SOC design to ever more manufacturing defects and soft errors; therefore, it is becoming more and more important to embed online detection or fault-tolerant schemes in these chips as the distinction between computer systems and SOCs increasingly narrows in nanometer designs.

12.3.5 Defect and Error Tolerance

A couple of tolerance terminologies have surfaced recently: **defect tolerance** [Koren 1998] and **error tolerance** [Breuer 2004]. Defect tolerance is not new and used to be referred to as *redundancy repair*. Defect tolerance allows increased product yield (the percentage of good manufactured parts). Way back in the late 1980s, redundancy techniques were used in the manufacture of DRAMs. By using spare rows, columns, or blocks, defective elements can be identified during the manufacturing test process and fuses are blown to map the spare resources to replace those that are defective. The use of these techniques becomes mandatory as DRAMs scale to the gigabit level. We would not be able to buy and sell DRAMs at the price level we enjoy today without the redundancy repair process.

A similar technique is also used in the manufacture of hard disk drives. During the drive test process, defective sectors are identified and a map containing those defective sectors is stored permanently on the drive control electronics. These defective sectors are mapped so the drive will not use them to store data. In both situations, spare elements replace the defective elements. Other circuits that have regular structures also can benefit from these defect-tolerant techniques, such as FPGAs, cache memories, and processors. For field programmable gate arrays, testing can identify bad cells or routing resources; thus, the mapping and routing tools can work around those obstacles during the mapping process. For processors, it is possible to sell the product with the fewer features (*e.g.*, minus the *floating-point unit*, or FPU) upon detection of a fault during manufacturing test; however, defect tolerance has its limit. Not only can the regular circuit elements become faulty due to defects, but the spare elements themselves can also be defective. As the percentage of spare elements increases, therefore, they will occupy more area of the die and the larger die area will result in even more defects, affecting both the normal circuit elements and the spare elements. Therefore, there is a point where the law of diminishing returns begins to set in [Koren 1990] [Hirase 2001].

Error tolerance is a different concept. Conventional wisdom would seem to suggest that if an error is injected and trapped in the logic it will not perform to its intended functionality; however, some logic functionality would defy that conventional wisdom. An example would be the processing or storage of any kind of multimedia data (*e.g.*, video, pictures, or music). Compression techniques (*e.g.*, JPEG, MPEG, MP3) are generally used for these types of data, and these compression algorithms are *lossy* in nature; that is, some details of the raw data are lost in the compression process. A stuck bit in the least significant portion of the data word may or may not be distinguishable from artifacts with the compression process [Breuer 2004]. Also, these kinds of data appeal to our senses, and our senses are usually not keen enough to spot minute variances (unless one looks for them with an expert's eyes or ears). This sort of error tolerance is very application-specific, and general-purpose machines that are tolerant of all kinds of errors have not yet been designed. For example, if an error occurs at the most significant bits of the compressed data or within the control logic instead of data, the data processing can still lead to incorrect processing and may yield unacceptable picture or sound.

12.4   FPGA TESTING

Field programmable gate arrays (FPGAs) are generally composed of a two-dimensional array of **programmable logic blocks** (PLBs) interconnected by a programmable routing network with programmable I/O cells at the periphery of the device, as illustrated in Figure 12.8. Typical array sizes in terms of the number of PLBs range from 100 to over 22,000. A trend in most recent FPGAs is the addition of cores for specialized functions such as single- and dual-port RAMs, **first-in first-out** (FIFO) memories, multipliers, DSPs, and microprocessors. Memory cores vary in sizes from 128 bit to 18 Kbit, depending on the series of FPGAs where all memory cores are the same size in a given series. The number of these specialized cores is greater than 800 in the largest FPGAs currently available. As a result, the larger FPGAs, with over 22,000 PLBs and 800 specialized cores, could easily reach the 100-million transistor mark and pose a testing challenge in terms of their size and diversity of functions.

12.4.1   Impact of Programmability

The system function performed by the FPGA is controlled by an underlying configuration memory. In most of the current FPGAs, the configuration memory is a RAM ranging in size from 32 Kbits to 50 Mbits. The system function can be changed at any time by simply rewriting the configuration memory with new data, referred to as *reconfiguration*. Another trend is FPGAs that support *dynamic partial reconfiguration*, where a portion of the FPGA can be reconfigured while the remainder of the FPGA is performing normal system operation, also referred to as *runtime reconfiguration*. The size of the configuration memory is an important factor in testing FPGAs as the total test time is usually dominated by the time required to download configuration data; however, dynamic partial reconfiguration and partial configuration memory readback capabilities can help reduce testing time as will be discussed later.

Each PLB consists of one or more **look-up tables** (LUTs) and flip-flops. The LUT typically has three or four inputs and is used to implement combinational logic. In some FPGAs, the LUT can also be programmed to function as a small

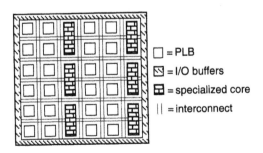

□ = PLB
⊠ = I/O buffers
⊞ = specialized core
|| = interconnect

■ **FIGURE 12.8**

Typical FPGA architecture.

RAM or shift register. The flip-flops, programmable as level-sensitive latches in some FPGAs, are used to implement sequential logic. Typical PLB sizes range from two three-input LUTs with one flip-flop to eight four-input LUTs with eight flip-flop/latches. A considerable amount of additional logic is incorporated in the PLB for implementing functions such as array multipliers, fast carry logic for adders, combining LUTs to construct larger combinational logic functions, etc. In addition to classical stuck-at faults in the PLB logic, the configuration memory bits that control logic function performed by the PLB must also be tested for stuck-at-0 and stuck-at-1 faults [Abramovici 2001]. For complete testing, the PLBs must be tested in all of their modes of operation.

The programmable interconnect network consists of wire segments of various lengths and programmable switches that connect or disconnect the wire segments to form the signal nets required by the system function. Each programmable switch is controlled by a bit in the configuration memory. The typical number of wire segments associated with each PLB ranges from 50 to over 400, while the number of programmable switches ranges from 80 to over 1000 per PLB. The number of configuration memory bits associated with the programmable routing resources is typically three to four times the number of configuration memory bits associated with the PLBs. As a result, the programmable interconnect network poses a bigger testing challenge than the programmable logic resources. The fault models used for testing the routing resources include shorts (bridging faults) and opens in the wire segments, stuck-at-1 and stuck-at-0 wire segments, and stuck-on and stuck-off programmable switches, which include the controlling configuration memory bits stuck-at-1 and stuck-at-0. While the programmable switch stuck-off fault can be detected by a simple continuity test, stuck-on faults are similar to bridging faults and require opposite logic values to be applied to the wire segments on both sides of the switch while monitoring both wire segments in order to detect the stuck-on fault [Stroud 2002b].

The programmability of FPGAs facilitates the implementation of a wide range of applications and, as a result, presents a number of testing solutions as well as a number of testing challenges. For example, FPGAs can be reprogrammed during system-level offline testing to test other components and functions on a printed-circuit board [Stroud 2002a]. Similarly, the PLBs and routing resources can be reprogrammed to test the other embedded cores within the FPGA such as memory and DSP cores [Stroud 2005b]. On the other hand, the programmability of the FPGA poses a number of challenges when it comes to complete and comprehensive testing of the FPGA itself. First, a large number of configurations must be downloaded into the FPGA to test the various programmable resources. Dynamic partial reconfiguration can reduce the total time associated with downloading these test configurations by writing only the portions of configuration memory that change from one test configuration to the next. The FPGA testing problem is further complicated by the growing size of FPGAs in terms of the PLB array, frequently changing architectures, as well as the introduction of specialized embedded cores such as RAMs and DSPs. If the FPGA can be completely tested and determined to be fault-free, the intended system function can be programmed onto the FPGA with a high probability of proper operation. When faults are detected, the system function can be reconfigured to avoid the faulty resources if the faults can be diagnosed (identified and located); therefore, diagnosis of the faulty resources is an important aspect of FPGA testing in order to take full advantage of the fault and/or defect tolerant potential of these devices.

12.4.2 Testing Approaches

Two types of testing approaches have been developed for FPGAs: *external testing* and *built-in self-test* (BIST). In external testing approaches, the FPGA is programmed for a given test configuration with the application of input test stimuli and the monitoring of output responses performed by external sources such as a test machine [Huang 1998] [Renovell 1998]. As a result, external test techniques are typically used for manufacture testing only. For FPGAs with *boundary scan* that support INTEST capabilities, the input test stimuli can be applied and output responses can be monitored via the boundary scan interface; otherwise, the FPGA I/O pins must be used, resulting in package-dependent testing. Most external test approaches seek to test all programmable resources in the FPGA independent of the system application to be programmed onto the FPGA, referred to as *application-independent testing*. *Application-dependent test* approaches, on the other hand, seek to test only those resources that will be used by the intended system function [Tahoori 2004]. This reduces the number of test configurations that must be applied as well as the total test time.

The basic idea in BIST for FPGAs is to configure some of the PLBs as **test pattern generators** (TPGs) and **output response analyzers** (ORAs). These BIST resources are then used to detect faults in PLBs [Abramovici 2001], routing resources [Harris 2002] [Stroud 2002b] [Sun 2000], and special cores such as RAMs and DSPs [Stroud 2005b]. Once the programmable resources have been tested, the FPGA is reconfigured for the intended system function without any overhead or performance penalties due to the BIST circuitry. This facilitates system-level use of the BIST configurations. Different BIST architectures are used for testing PLBs (often referred to as *logic BIST*), routing resources (often referred to as *routing BIST*), and embedded cores. It is important to note that the processes used in these BIST approaches for reconfiguring and testing the specific target resources are very similar to those used in application-independent external testing of FPGAs.

12.4.3 Built-In Self-Test of Logic Resources

The most frequently used logic BIST architecture is illustrated in Figure 12.9, where the programmable logic **blocks under test** (BUTs) and ORAs are arranged in alternating columns (or rows), and multiple identical TPGs are used to drive the alternating columns (or rows) of BUTs [Abramovici 2001]. The output responses of the identically programmed BUTs are monitored by comparison-based ORAs in neighboring columns (or rows). During a given test session, the BUTs are repeatedly reconfigured in their various modes of operation until they are completely tested. Dynamic partial reconfiguration can be used because only the BUTs must be reconfigured while the TPGs, ORAs, and interconnections remain constant for

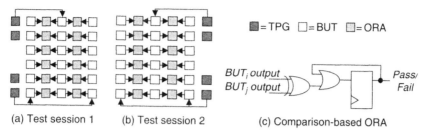

(a) Test session 1   (b) Test session 2   (c) Comparison-based ORA

■ **FIGURE 12.9**

FPGA logic BIST architecture.

the test session. During the next test session, the logic BIST architecture is flipped and the roles of the PLBs are reversed such that those PLBs previously configured as TPGs and ORAs become BUTs and *vice versa*. All PLBs can be tested in only two test sessions when at least half the PLBs are configured as BUTs during a given test session. The total number of test configurations in each test session typically ranges from 5 to 15 depending on the complexity of the PLB. After the completion of each BIST sequence the Pass/Fail contents of the ORAs can be read via either partial configuration memory readback or a scan chain constructed by incorporating a multiplexer at the input to the ORA flip-flop shown in Figure 12.9c [Stroud 2002a]. Alternatively, as a result of dynamic partial reconfiguration, the ORA contents can be read at the end of each test session with a slight loss of diagnostic resolution; faulty PLBs can still be identified but faulty modes of operation cannot. Faulty PLBs can be identified based on the BIST results using a diagnostic procedure developed for this logic BIST architecture [Abramovici 2001]. A similar architecture can be used to test and diagnose other embedded cores in the FPGA such as memories and DSPs [Stroud 2005b].

12.4.4 Built-In Self-Test of Routing Resources

Two routing BIST approaches have proven to be effective in testing the programmable interconnect resources in FPGAs including the wire segments, programmable switches, and configuration memory bits that control the switches. One is a comparison-based approach, illustrated in Figure 12.10a, in which the TPG

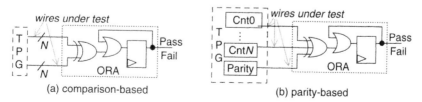

(a) comparison-based   (b) parity-based

■ **FIGURE 12.10**

FPGA routing BIST architectures.

drives exhaustive test patterns over two sets of N wires under test that are compared at the other end by comparison-based ORAs [Stroud 2002b]. The other approach is parity based, as illustrated in Figure 12.10b, where the TPG sources exhaustive test patterns over a set of N wires under test and produces a parity bit that is sent to the ORA [Sun 2000]. The ORA generates parity over the data observed on the wires under test and compares the generated parity with the parity bit sent by the TPG. This approach was later modified to send the parity over a wire under test for a total of $N + 1$ wires under test during a given BIST configuration. As in the case of logic BIST, the sets of wires under test are repeatedly reconfigured to test the various routing resources (wire segments and programmable switches) in the FPGA. The total number of test configurations required to completely test the routing resources typically ranges from 50 to 300 depending on the complexity of interconnect network and the PLB architecture used for constructing the TPGs and ORAs. Dynamic partial reconfiguration can be used to reduce the time to download test configurations. While both routing BIST approaches have been proven to be effective in detecting faults, the comparison-based approach has been extended to the diagnosis of faults in the programmable interconnect network for fault-tolerant applications [Harris 2002]. By constructing many small routing BIST circuits consisting of independent TPGs, ORAs, and sets of wires under test in the FPGA, diagnostic resolution is improved because an ORA indicating the presence of a fault also identifies the self-test area containing the fault [Stroud 2002b].

12.4.5 Recent Trends

More recent trends in FPGA testing include delay fault testing and the use of embedded processor cores for on-chip test configuration generation and application. Testing for delay faults in FPGAs is important because the transmission gates used to construct the programmable switches in the interconnect network are particularly susceptible to defects that affect the delay though the switches. External test techniques [Chmelar 2003] and BIST approaches [Abramovici 2003] have been developed to detect delay faults in FPGAs. The incorporation of embedded microprocessor cores that can write and read the FPGA configuration memory has facilitated the algorithmic generation of test configurations from within the FPGA instead of downloading test configurations. A relatively small program is stored in the program memory of the embedded processor core which is then used to reconfigure and test the programmable logic and routing resources as well as other embedded cores such as memories and DSPs. The embedded processor can then retrieve the test results and perform diagnosis [Stroud 2005a].

Recent *complex programmable logic devices* (CPLDs) are similar to FPGAs in that they contain programmable logic and routing resources as well as embedded cores such as RAMs and FIFOs. The only noticeable difference is that CPLDs use *programmable logic arrays* (PLAs) for implementing combinational logic functions instead of the LUTs typically found in FPGAs. In addition, the PLBs in CPLDs tend to be larger in terms of the size of the PLAs and the number of flip-flops. Slightly different test techniques are used to test the reprogrammable PLAs [Stroud 2002a], with the remainder of CPLD testing being the same as that for FPGAs.

Now that FPGAs are incorporating embedded cores such as memories, DSPs, and microprocessors, FPGAs are more closely resembling *system-on-chip* (SOC) implementations. At the same time, SOCs are incorporating more embedded FPGA cores. As a result, FPGA testing techniques are becoming increasingly important for a broader range of system applications.

12.5 MEMS TESTING

MEMS is the acronym for a *microelectromechanical system* [Hsu 2002]. The prefix "micro" indicates the most important feature of MEMS: its extremely small size. The typical size of MEMS components is in the range of between 1 micron (μm) and 1 millimeter (mm). This means that the key feature size of a MEMS device is usually smaller than the diameter of human hair. For feature size below 1 μm, the quantum effect cannot be ignored. It belongs to the recently emerged concept of a ***nanoelectromechanical system*** (NEMS). Thus, MEMS devices primarily concentrate on the feature sizes from $1 \sim 1000\,\mu$m. Further, the electronic and mechanical parts of a MEMS device interact with each other, so it can be called a "system." For example, in a MEMS system, the signals in a mechanical sensor can be sensed by an electronic circuit, while the actuation instructions from the electronic circuit can be implemented by a mechanical actuator. Thus, MEMS can incorporate the environment data collection, signal processing, and actuation in the same "smart" system. When compared with conventional electromechanical products, MEMS has the following specific features and corresponding advantages: (1) small volume, low weight, and high resolution; (2) high reliability; (3) low energy consumption and high efficiency; (4) multifunction capabilities and intelligentization; and (5) low cost. Typical examples of commercial MEMS devices are the ADXL series accelerometers [Chau 1998] which have been widely used in the world's automobile market.

12.5.1 Basic Concepts for Capacitive MEMS Devices

A typical MEMS differential capacitance structure is shown in Figure 12.11 where M represents the movable plate, F1 and F2 denote fixed plates, and B1 and B2 are both beams of the MEMS device. As shown in Figure 12.11, movable plate M is

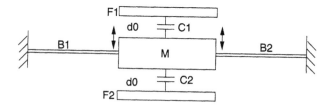

■ FIGURE 12.11

Schematic diagram of a capacitive MEMS device.

anchored to the substrate through two flexible beams, B1 and B2. It constitutes differential capacitances C_1 and C_2 with the top and bottom fixed plates (F1 and F2). In the static mode, the movable plate M is located in the center between F1 and F2, thus:

$$C_1 = C_2 = \frac{\varepsilon_0 S}{d_0}$$

where ε_0 is the dielectric constant of air, S is the overlap area between M and F1/F2, and d_0 represents the static capacitance gap between M and F1/F2. A vertical stimulus (such as acceleration) will result in the deflection of beams and a certain displacement of movable plate M along the vertical direction. Assume that the central movable mass moves upward with a displacement of x. Given $x \ll d_0$, C_1 and C_2 under the test stimulus can be derived by:

$$C_1 = \frac{\varepsilon_0 S}{(d_0 - x)} \approx \frac{\varepsilon_0 S}{d_0}\left(1 + \frac{x}{d_0}\right)$$
$$C_2 = \frac{\varepsilon_0 S}{(d_0 + x)} \approx \frac{\varepsilon_0 S}{d_0}\left(1 - \frac{x}{d_0}\right)$$

In order to sense the displacement x of movable plate M, modulation voltages V_{mp} and V_{mn} are applied to F1 and F2 separately, and we have:

$$V_{F1} = V_{mp} = V_0 sqr(\omega t),$$
$$V_{F2} = V_{mn} = -V_0 sqr(\omega t)$$

where V_0 represents the modulation voltage amplitude, ω denotes the frequency of the modulation voltage, and t gives the time for operation. According to the charge conservation law, the charges in capacitances C_1 and C_2 must be equal, so we have:

$$C_1(V_{F1} - V_M) = C_2(V_M - V_{F2})$$

where V_M is the voltage level sensed by movable plate M. Solving the above equations, we have:

$$V_M = (x/d_0)V_0 sqr(\omega t)$$

It can be observed from this result that, under the above modulation voltage biasing, the central movable plate, M, acts just as a voltage divider between the top and bottom fixed plates, F1 and F2, respectively. By measuring voltage level V_M at the central movable electrode, we can find the displacement, x, of the central movable plate, M, which in turn is directly proportional to the physical stimulus. Thus, we can derive the value of the applied physical stimulus. This is the working principle for most differential capacitive MEMS devices.

12.5.2   MEMS Built-In Self-Test
12.5.2.1   Sensitivity BIST Scheme

In the **sensitivity BIST** mode, a certain amount of driving voltage V_d can be applied to the driving plate to mimic the action of a physical stimulus (*i.e.*, test pattern) with electrostatic force. In Figure 12.11, if voltage V_d is applied to fixed plate F1, and nominal voltage V_{nom} is applied to M, an electrostatic attractive force F_d will be experienced by the central movable mass:

$$F_d = \frac{\varepsilon_0 S V_d^2}{2d^2}$$

The electrostatic force is used to apply the input stimulus during the BIST mode, and the device response to the electrostatic force is measured and compared with the good device response to check whether the device is faulty. This is the basic idea for the sensitivity test mode of a capacitive MEMS device. Note that the device in Figure 12.11 cannot implement sensitivity BIST, as the device cannot work as a sensor and actuator simultaneously. It is only used to show the basic idea about how to sense acceleration and how to generate an input stimulus by electrostatic force. To implement the sensitivity BIST technique, the device under test must contain at least an actuator plate (to generate test patterns) and a sensor plate (to sense test responses). This will be thoroughly explained in the discussion for the dual-mode BIST scheme. For vertical electrostatic driving, the driving voltage cannot exceed a threshold value by which the deflection exceeds 1/3 of the capacitance gap, d_0; otherwise, the movable plate will be stuck to the fixed plate through a positive feedback, and a short-circuit will occur. More details for sensitivity BIST can be found in [Charlot 2001].

12.5.2.2   Symmetry BIST Scheme

[Deb 2002] proposed a **symmetry BIST** scheme for capacitive MEMS devices utilizing central mass partitioning. The following presentation is mainly based on the idea of fixed-plate partitioning. A simplified MEMS capacitance structure for symmetry BIST is given in Figure 12.12, where S1 to S4 are fixed plates. As shown, each of the top and bottom fixed capacitance plates is divided into two equal portions. For simplification, the capacitance for electrostatic actuation that is necessary for BIST implementations is omitted here. The basic idea of the symmetry test scheme is to determine whether the two symmetric capacitances (*e.g.*, C_1 and C_2 in Figure 12.12) on the same side of the movable microstructure remain equal all the time, after activation.

In Figure 12.12, fixed plates S1 and S2 lie on the same side of movable plate M. The capacitance between M and S1 (S2) is defined as $C_1(C_2)$. The modulation voltages V_{mp} and V_{mn} are applied to S1 and S2 separately. If the device is fault-free, regardless of whether the movable plate is in rest or moving a certain displacement along the vertical direction, the values of C_1 and C_2 should always remain equal.

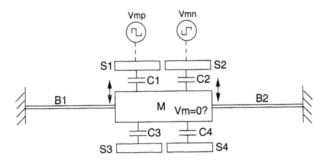

■ **FIGURE 12.12**

MEMS capacitance structure for our symmetry test scheme.

Take the voltage level on central movable plate M as V_M; according to the charge conservation law, charge Q_1 and Q_2 in capacitances C_1 and C_2 must remain equal:

$$C_1(V_{mp} - V_M) = C_2(V_M - V_{mn})$$

Because $V_{mp} = -V_{mn}$, from the above equation we have:

$$V_M = V_{mp}(C_1 - C_2)/(C_1 + C_2)$$

If C_1 equals C_2, then we have $V_M = 0$. Under the above voltage biasing scheme, the voltage level on the central movable plate is always zero for good devices; however, if any local defect alters the symmetry of the device, the movable plate will tilt and C_1 will not be equal to C_2, and output voltage V_M will not be zero anymore. Thus, by checking the voltage output on the movable plate, we can find any defect that alters the symmetry of the device. Furthermore, according to the phase polarity of V_M, we can know whether the defect lies at the left or right side of the device. For example, if a stiction defect in the right side (which introduces C_2 in Figure 12.12) of the mass causes C_2 to be smaller than C_1, V_M will have the same phase polarity as V_{mp}, and vice versa.

The above analysis is for checking both capacitances on the top side of the device; however, verification of both bottom capacitances (C_3 and C_4) can be easily performed in a similar way, and they should have the same result.

12.5.2.3 A Dual-Mode BIST Technique

A dual-mode BIST technique [Xiong 2004] for capacitive MEMS devices can be implemented by dividing the fixed capacitance plates at each side of the movable microstructure into three portions: one for electrostatic activation and the other two equal portions for capacitance sensing, as shown in Figure 12.13. Note that M is the movable plate, D1 and D2 are the fixed driving plates, and {S1, S2, S3, S4} are the fixed sensing plates. As shown in Figure 12.13, after capacitance partitioning, two BIST modes (sensitivity test and symmetry test) can be easily implemented on the device. During normal operation, we have Test Enable (TE) signal = 0. If the

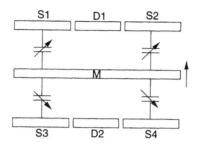

■ **FIGURE 12.13**

Fixed capacitance plates partition for MEMS device.

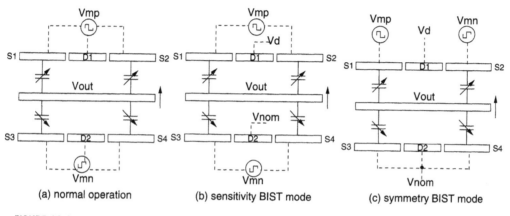

■ **FIGURE 12.14**

Voltage biasing schemes for the three modes of MEMS device.

device is a sensor, the modulation voltage V_{mp} is applied to {S1, D1, S2}, and the complementary modulation voltage V_{mn} is applied to {S3, D2, S4} (Figure 12.14a). The voltage on central mass M is sensed as the output voltage V_{out} indicating the device sensitivity. If the device is an actuator (*e.g.*, microresonator), the driving voltage V_{dp} is applied to {S1, D1, S2}, and the complementary driving voltage V_{dn} is applied to {S3, D2, S4} separately to implement the electrostatic actuation in normal operation. In short, the driving capacitance plates (D1 and D2) for BIST will also participate in the normal operation, so there is no loss of capacitance area due to the BIST implementation.

In the BIST mode (TE = 1), the Test Selection (TS) signal can select one of the two BIST modes. When TS = 0, the device is in the sensitivity test mode. Test driving voltage V_d is applied to D1 to activate the device, modulation voltage V_{mp} is applied to {S1, S2}, and V_{mn} is applied to {S3, S4} (Figure 12.14b). The voltage level on movable electrode (*i.e.*, plate) M is measured for the device sensitivity analysis. Voltage V_M is compared with the expected value (calibrated) within a tolerance level to find whether the device is faulty. When TS = 1, the device is in the symmetry test mode. In this case, test driving voltage V_d is applied to D1 to activate the device. Modulation voltage V_{mp} is applied to S1, and V_{mn} is applied to S2 separately (Figure 12.14c). The voltage level, V_M, of the movable electrode is measured to see whether it is a constant zero. A non-zero voltage output on movable electrode M indicates that a local defect is causing the asymmetry of the device. Based on the value and polarity of V_M, we can also get some idea about the approximate location of the local defect. The above discussion is for the case where the movable electrode is driven upward (V_d is applied to D1); however, when the movable electrode is driven downward (V_d is applied to D2), the implementation can be easily extended. Note that, in the BIST mode, the device should be driven in both directions for a thorough test. Because the sensitivity test and symmetry test each has its own defect coverage, by combining them together a more robust testing result can be ensured [Xiong 2004]. The defect on driving electrodes D1 and D2 can also be detected if it causes sensitivity change or left–right asymmetry in the MEMS device. For example, if the left part of D1 is missing due to improper photoetching, the mass will experience a larger electrostatic force in its right part than its left part in BIST; hence, movable mass M will tilt, and a symmetry test can detect the defect. To implement the BIST technique, a control circuit is required to switch the device among the normal operation mode and both BIST modes. Such a control circuit is not complex and only contains some switches made of analog multiplexers.

12.5.3   A BIST Example for MEMS Comb Accelerometers

A typical **surface-micromachined comb accelerometer** [Kuehnel 1994] is shown in Figure 12.15. The comb accelerometer is made of a thin layer of polysilicon on the top of a silicon substrate. The thickness of the polysilicon structure layer is about $2\,\mu m$. The fixed portion of the device includes four anchors and many left and right fixed fingers. The movable portion of the MEMS device includes four tether beams, the central movable mass, and all movable fingers extruding out of the mass. The entire movable portion floats about $1.5\,\mu m$ above the substrate. As shown in Figure 12.15, the central movable mass is connected to the four anchors through four flexible beams. The movable fingers extrude from both sides of the central mass and can move together with it. There is a pair of fixed fingers around the left and right sides of each movable finger. Each movable finger and its left and right fixed fingers constitute a differential capacitance pair (c_1 and c_2), as shown in Figure 12.16. In the static state, each movable finger stays in the middle position between the left and right fixed fingers, and the capacitance gaps of both c_1 and c_2 are equal to d_0. Let $C_1(C_2)$ represent the sum of all $c_1(c_2)$ capacitances. We have:

$$C_1 = C_2 = \frac{n_f \varepsilon_0 (L_f - \Delta)h}{d_0}$$

where n_f is the total number of differential capacitance groups, ε_0 is the dielectric constant of air, L_f is the length of each movable finger, Δ is the nonoverlapped length at the root of each movable finger, and h is the thickness of the device.

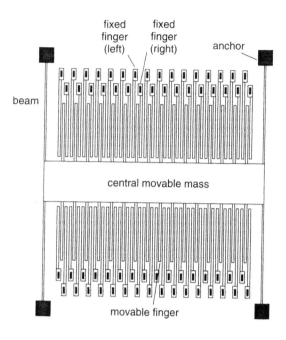

■ **FIGURE 12.15**

General design of a MEMS comb accelerometer.

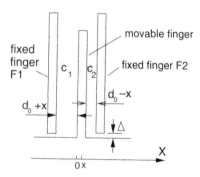

■ **FIGURE 12.16**

The schematic diagram of differential capacitance.

Assume that the mass of both the central movable mass and all the movable fingers is M. If there is an acceleration a perpendicular to the beams and parallel to the device plane, the central mass will experience an inertial force of $-M \cdot a$. This will result in a certain amount of beam deflection along the direction of the inertial force and an equivalent amount of displacement of the central mass and movable fingers. Thus, each capacitance gap will be changed accordingly, which leads to the change of corresponding capacitances (Figure 12.16).

As shown in Figure 12.16, the inertial force results in a deflection of the beams and a certain displacement x of movable fingers along the X direction. Given $x \ll d_0$, C_1 and C_2 change as follows:

$$C_1 = \frac{n_f \varepsilon_0 (L_f - \Delta) h}{(d_0 + x)} \approx \frac{n_f \varepsilon_0 (L_f - \Delta) h}{d_0} \left(1 - \frac{x}{d_0}\right)$$

$$C_2 = \frac{n_f \varepsilon_0 (L_f - \Delta) h}{(d_0 - x)} \approx \frac{n_f \varepsilon_0 (L_f - \Delta) h}{d_0} \left(1 + \frac{x}{d_0}\right)$$

By sensing the capacitance change of C_1 and C_2, we know displacement x and the acceleration experienced. This is the working principle of a MEMS comb accelerometer.

A comb accelerometer structure that can implement dual-mode BIST functions is shown in Figure 12.17. Here, M1 to M8 are movable fingers, Ms is the central mass, D1 to D8 are driving fingers, and S1 to S8 are sensing fingers. All beams are connected to the substrate through four anchors. For simplicity, only four groups of driving/sensing fingers are given here. The fixed portion of the device includes driving fingers D1 to D8 and sensing fingers S1 to S8.

During normal operation (TE = 0), modulation voltage V_{mp} is applied to {S1, S3, S5, S7, D1, D3, D5, D7}, and V_{mn} is applied to {S2, S4, S6, S8, D2, D4, D6, D8}. The voltage level in the movable fingers (V_{Ms}) is measured as the output voltage to determine the acceleration. When TE = 1 and TS = 0, the device works in the sensitivity test mode. A certain test driving voltage V_d is applied to {D1, D3, D5, D7} to activate the device with electrostatic force. The modulation voltage (V_{mp}) is

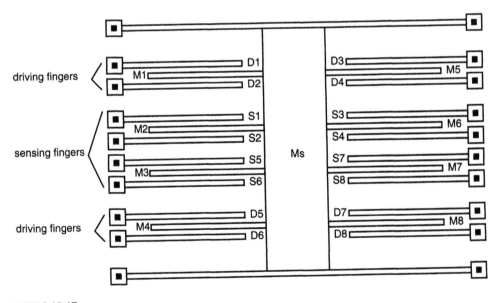

Structural diagram of a comb accelerometer.

applied to {S1, S3, S5, S7}, while V_{mn} is applied to {S2, S4, S6, S8}. The output voltage on movable mass Ms is measured for the device sensitivity. This value is compared with the expected good device value within a certain tolerance level to find whether the device is faulty. When TE = 1 and TS = 1, the device is in the symmetry test mode. Test driving voltage V_d is applied to {D1, D3, D5, D7}, modulation voltage V_{mp} is applied to {S1, S5}, and V_{mn} is applied to {S3, S7}. The sensing circuit checks whether the output voltage on the central mass is a constant zero to detect any asymmetry caused by local defects. A non-zero voltage on movable electrode Ms indicates the presence of local defects that are altering the symmetry of the device. Defects on driving electrodes can also be detected, if they cause sensitivity change or left–right asymmetry in the MEMS device. For example, if part of D1 in Figure 12.17 is missing, the movable mass will experience a smaller electrostatic force in its left part than its right part during BIST, and movable mass Ms will tilt. Symmetry test can detect this defect.

12.5.4 Conclusions

Microelectromechanical systems have achieved tremendous progress in recent decades. Various MEMS devices based upon different working principles have been developed [Hsu 2002]. MEMS has also found broad applications in various areas. With the rapid development of MEMS technology and its integration into *system-on-chip* (SOC) designs, MEMS testing (especially BIST) is becoming an even more important issue. An efficient and robust test solution is urgently needed for MEMS; however, due to the great diversity of MEMS structures and their working principles, various defect sources, multiple field coupling, and its essential analog features, MEMS testing remains a very challenging work [Tewksbury 2001]. Various efforts have been made in this area [Rosing 1999] [Rosing 2000] [Aikele 2001] [Charlot 2001] [Deb 2002] [Puers 2002] [Xiong 2004], and more research in this field must be invested in the near future.

12.6 HIGH-SPEED I/O TESTING

Even though Moore's law [Moore 1965] has dictated neither performance nor I/O channel bandwidth, system-level performance has been improving steadily with faster transistors and integration. Integration brings together circuitry that used to be on different chips, so the signals travel a shorter distance and encounter smaller loads (within a chip), rather than going through the die pad, bond wires and solder bumps, package, sockets, and printed-circuit board traces to the other components. Along the way, it also has to interact with neighboring signals (consider a typical PC motherboard), an effect called *crosstalk*. As a matter of fact, I/O bandwidth has been the limiting factor in system-level performance for some time, at least the last 10 to 15 years. Due to the need to control the cost of the PCB, many physical effects, such as those that cause energy loss and coupling as well as contribute to **deterministic jitter** (DJ) and **random jitter** (RJ), have been limiting the signaling rate. In spite of this, signaling rates at the board level have improved, although it still lags far behind what is possible within the chip; for example, a current mainstream processor runs at 3 to 4 GHz internally, but the *front-side bus* (FSB) only runs up to 800 Mbps. This has actually influenced the processor architecture to a great degree (*e.g.*, increasing the size of the cache to mitigate the FSB bandwidth requirement).

12.6.1   I/O Interface Technology and Trend

The most common signaling protocol is that of the ***common clock*** (CC) type (see Figure 12.18). In the CC scheme, the signal is launched off one chip with the system clock and received at another chip at the following clock edge. At the sending end, there is a clock to signal delay specification, and at the receiving end there is setup and hold time on either side of the following clock edge.

When the signaling rate goes up, a problem arises. The clock skew between the sending component and the receiving component (board trace delay A–B) can cut into the cycle time (see Figure 12.19). To compensate for this problem, I/O designers have come up with ***source synchronous*** (SS) and ***clock forwarding*** (CF) schemes [Ilkbahar 2001]. With such a scheme, not only will the sending component send the signal, but another strobe (similar to a clock signal) also goes along with the signal. The receiving component uses this strobe to clock the signal; hence, system-level clock skew is out of the picture. The designer is concerned only with the differential skew between the strobe and the signal. With careful design, (*e.g.*, identically sized drivers and matched layout), such signaling schemes have allowed the signaling rate to rise gradually from below 10 to 50 Mbps to today's 800 Mbps.

Although the new SS signaling scheme has improved system-level performance, another problem has begun to appear. The parallel interface that we use to transfer

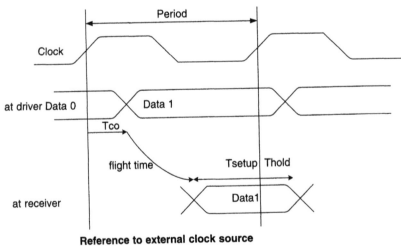

Reference to external clock source

Tester is programmed to provide clock

■ **FIGURE 12.18**

Common clock (CC) signaling scheme.

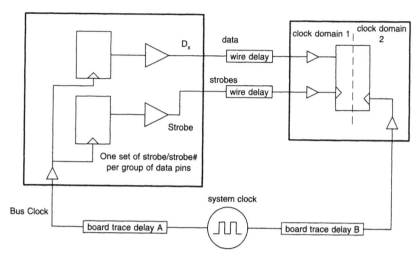

■ **FIGURE 12.19**

Source synchronous (SS) signaling scheme.

lots of data becomes a bottleneck itself. The parallel bits of data from the send-ing/receiving component have to center around the strobes, and the skews among these data bits due to uneven driving speed and propagation delay between parallel channels become the limiter to increasing data rate (see Figure 12.20). The multiple

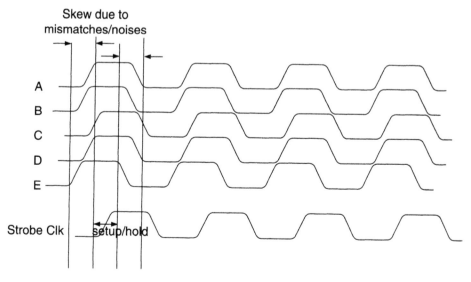

■ **FIGURE 12.20**

Skew among data bits that limits the data rate of the SS interface.

load nature of a parallel bus also creates noises that affect *signal integrity* (SI). It is generally believed that beyond 1 Gbps we will require new signaling technology.

Some signaling technology from the data communication industry was adopted for computer I/Os, with enhancements made to consider the relative short distance (<10m), high volume, and low cost characteristics of those I/O devices. For a long time, the data communication industry has used a signaling protocol that requires sending only the encoded data bits, with the clock signal embedded in the data bitstream, and recovering them at the receiver through a *clock recovery* (CR) circuit. This **serial signaling** [Athavale 2005] also minimizes noises by adhering to a clean transmission line model with only the driver on one end and the receiver on the other end (and matched impedance from the driver to the medium [wires or cables] to the receiver). To improve the *signal-to-noise ratio* (SNR) further, a differential signaling protocol over the twisted pair can cancel out *common mode noises*. For use on short distance, the voltage swing can be made small to further improve the edge rate and power consumption if the receiver sensitivity is ensured. Data encoding (*e.g.*, 8-bit to 10-bit encoding) is usually required to ensure that there are enough signaling edge transitions so the circuitry at the other end can synchronize to this incoming signal stream (see Figure 12.21). With this signaling technology, the data rate can scale from 1 ~ 10 Gbps. Inherently, noises (in the form of jitters) constitute a larger part of the cycle time, and it is possible that the data are interpreted incorrectly. To ensure that we have correct data, error checking circuits, with *cyclic redundancy check* (CRC) or *error correcting code* (ECC) capabilities, are usually necessary for error detection, and flow control/data request/resend protocol circuits are required for data recovery. These flow control and data recovery protocol circuits comprise what is referred to as the **link layer**. The transmitter and receiver circuits, as well as the medium, comprise the *physical layer* (PHY). As long as the *bit error rate* (BER) is low enough, the system will still deliver a higher level of performance as compared with previous technologies. A good CR circuit at the receiver can also track the low-frequency jitter on the data signal and will lower the receiver's BER. The CR circuit and the data-sampling flip-flop constitute a high-pass jitter receiver transfer function.

Phase-locked loop (PLL) is a commonly used CR circuit, especially for earlier network-centric I/O links, such as *fiber channel* (FC), *gigabit ethernet* (GBE), *synchronized optical network* (SONET), and *optical internetworking forum* (OIF).

To further improve the data rate and lower the BER, a phase tracking type of circuit is employed (*e.g.*, in SATA or PCI Express interfaces). Figure 12.22 shows the link architecture of modern computer-centric Gbps I/O links, such as PCI Express and FB DIMM, where *phase interpolator* (PI) is used to recover the clock [Li 2004] [Lin 2005]. In this architecture, a reference clock (*e.g.*, 100 MHz) is sent to both transmitter and receiver multiplication PLLs (*e.g.*, 25× for PCIe I). The multiplied clock (in this case, 2.5 GHz) for the transmitter will be used to drive the data signal. The multiplied 2.5-GHz clock at the receiver will first go through a PI circuit to generate a clock that is phase aligned with the data and then used subsequently to retime (strobe) the data at the data-sampling flip-flop. If the propagation delays from the reference clock to both transmitter and receiver are matched and if the PLLs for the transmitter and receiver are also matched, then the jitter on

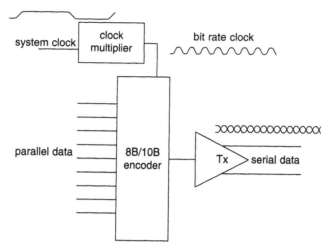

(a) Serial transmitter

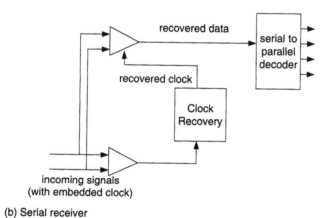

(b) Serial receiver

■ **FIGURE 12.21**

Serial receiver and clock recovery: (a) serial transmitter, and (b) serial receiver.

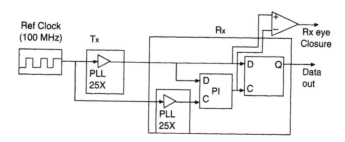

■ **FIGURE 12.22**

Link architecture diagram for a digital PI-based receiver commonly used in PCI Express and FB DIMM.

the reference clock will be completely cancelled through this "phase differential" signaling technique. Moreover, the PI circuit provides a first-order high-pass jitter tracking capability so the low-frequency jitter will be reduced. With the adoption of this advanced link architecture, combining inexpensive reference clocks, PLLs, and digital PI-based receivers, we can reduce the cost of those links, as well as maintain a high overall BER performance at 10^{-12} or lower.

12.6.2 I/O Testing and Challenges

Historically, I/O testing is handled by *automatic test equipment* (ATE) that mimics the other side of the I/O interfaces. Because the ATE is a stored stimulus and stored response system, it works well for the common clock interface. Clock is generated from a tester channel and so are the test stimuli. As the *device under test* (DUT) responds with its generated signals, the output responses are strobed by the ATE with its own timing system, so this approach works perfectly with common clock devices. With this scheme, ATE performance rises with the device I/O performance, and this methodology has worked well for decades. Of course, to get faster performance, the ATE requires more sophisticated devices (*e.g.*, GaAs or SiGe) that have driven up the cost of the ATE, but this is not a technology barrier.

The advent of the SS technology has turned the whole ATE test methodology upside down. For the ATE to send SS signals, the scenario is the same as that of the CC, and the ATE generates the strobes and the data. However, when it is time for the DUT to send data, it sends out strobes and the ATE is supposed to use those strobes to strobe the data. Here, the fixed timing system of the tester fails. Most ATEs today cannot make use of incoming signals from any channel (be it strobes or not) to strobe another channel. The ATE has to first measure where the strobe is (a trial-and-error search process) and then set the ATE internal strobes to strobe the data. Without this search methodology, the ATE cannot guess where the data are located, and erroneous results will lead to failures and yield loss. Also, even if one uses this search and timing remapping methodology, the metrology (measurement errors involved with each measurement) will cost users precious timing margins that are inherent with the SS timing signaling protocol [Ilkbahar 2001].

Although SS testing presents its own problems, the higher speed serial signaling link approach is even more challenging. Consider again the test practices of the data communication industry. Due to the complex signaling requirements, the test process involves characterizing the output signals from the DUT and also generating worst-case signals to test the receiver and its associated CR subsystem. This usually involves a bench setup consisting of an oscilloscope (real-time digitizer or equivalent sampling), *timing interval analyzers* (TIAs), or *bit error rate tester* (BERT) and signal generators that are capable of deterministic and random pattern generation as well as amplitude control and jitter injection. To aid in volume production, the ATE industry must essentially reproduce the functionalities of this necessary bench setup with a high-performance interface between the ATE driver/receiver and DUT. Furthermore, serial I/O ATE has to handle the asynchronous nature of this interface, which is different from conventional synchronous ATE. Such ATE could be very expensive and misaligned with the end-user cost expectation.

While this approach has worked for the data communication industry in the past, certain economic and technical difficulties are associated with applying this methodology in the general consumer electronic industry, where most computing devices require more data bandwidth than a single data channel (*e.g.*, PCI Express has a maximum of 32 data channels, and a microprocessor may have anywhere from 16 to 64 data channels). Because these high-performance serial channels are point to point, to establish a multiple agent system (*e.g.*, a multiple processor system), we would need more than just one set of serial channels (possibly running into the hundreds). This would increase the cost many times over and make it difficult to come up with the relays/cabling requirements in the test head of the ATE. The test time required for measuring jitters to predict the BER has further pushed designers and test engineers to look for alternative solutions.

12.6.3 High-Performance I/O Test Solutions

In the early 1990s, IBM put forth an I/O structural test methodology called **I/O wrap** [Gillis 1998]. Essentially this involves applying the transition fault test methodology to the I/O circuitry. By tying an output to an input, the output data is launched and latched back into the input buffer on the following clock. Because most signal pads are I/O in nature, the I/O wrap methodology is very convenient. Input-only or output-only pads may be connected with the DFT or on the *test interface unit* (TIU). Designers have even incorporated the DFT insertion into their synthesis tools and supported test pattern generation with their ATPG tools. The limitation is that, because this delay path is tested with the clock, one cannot characterize the delay without overstressing the rest of the peripheral circuits. So, this approach is limited to testing for gross delay defects, and timing specifications cannot be measured.

By the early 2000s, a SS test methodology (dubbed **AC I/O loopback testing**) had been proposed that uses the same loopback principle but with a twist [Tripp 2004]. Rather than just using the clock to launch and capture, the launch can be carried out by a delayed version of the clock (or the capture be done with an early version of the clock). By controlling the delay, one can actually measure the relative delay between the strobes and data, all without the need for precision timing measurement from ATE.

Essentially, this is transition fault testing of the I/O pair with a tighter clock cycle (only for the I/O circuits, thus preventing false fails from other circuits as in the case of speeding up the I/O wrap). Furthermore, this works very well with the SS scheme, where the strobes are generated by the transmitting side. In SS signaling protocol, the absolute delay of the I/O is not critical; instead, the relative delay of the strobes and its associated data bits are important. These delay timings, denoted as *time valid before* (Tvb) and *time valid after* (Tva), describe the relationship (see Figure 12.23). So, by moving the strobes from their central position to the trailing edges of the data, we are stressing Tva and the setup time of the receiver latch. If we move the strobes toward the leading edge of the data, we are stressing Tvb and the receiving latch's hold time. By stressing this combined timing, we know how much margin there is with the combined pair. If the induced delay to the clock/strobes is calibrated, we can even have more accurate measurement of this combined loop

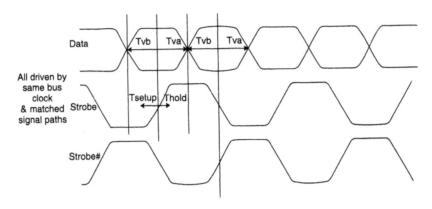

■ FIGURE 12.23

SS timing definition (Strobe# is the complement of Strobe and is sometimes added to define where the timing reference point is).

time than external instrumentation. Because the failure mechanisms for signal delay and input setup/hold time are different, the probability of aliasing is very low.

Furthermore, because we are not measuring each data bit independently (this is a bus nonetheless), the delays of all of these data bits should be close to one another unless there is a defect or local process variations. If a particular data bit is substantially different from the other data bits, we can also conclude that a defect or local process variation exists with that particular bit and declare that to be a failure. So, this can also be viewed as a **defect-based test method**, especially if no calibration is done to the induced delay (see Figure 12.24).

The authors who proposed AC I/O loopback testing also suggest that they can extend the concept to serial signals (looping outputs back to inputs), but details have not been published yet [Mak 2004a]. It is worth pointing out that those system-level serial link tests using direct loopback do not offer the worst-case fault coverage. Recent loopback methodology that has jitter injection capabilities offers better fault coverage [Laquai 2001] [Cai 2002] [Cai 2005] [Lin 2005]; however, currently available coverage is still far from what is needed. Most of the jitter injection solutions today consider only one type of jitter or noise injection rather than all types of jitter and noise injections at the same time to emulate worst-case signaling in a real system.

12.6.4 Future Challenges

Because we need to match up the data bandwidth of the chip-to-chip connection to that of the core operating speed, increasing use of serial signaling is expected. Very soon, serial signaling will replace most buses and maybe even many of the control signals as well. The methodologies for testing all of this high-speed serial signaling must consider cost and quality. Increasingly, this points to more and more self-test with DFT support [Kundu 2004], although its accuracy and fault coverage are not at the desired level [Lin 2003].

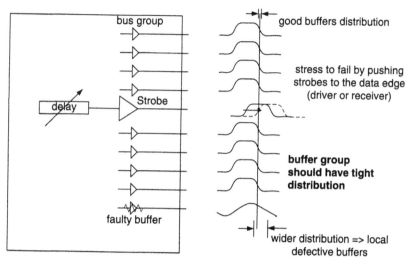

■ **FIGURE 12.24**

AC I/O loopback testing using a defect-based test method.

In order to maintain a low *bit error rate* (BER), lower cost structures, higher channel counts, and more advanced architectures and silicon technologies are expected to be developed for multiple-Gbps I/Os in the near future. In particular, more aggressive equalizations—transmitter-based, receiver-based, or hybrid—are expected to occur when the data rate increases. Furthermore, in order to reduce crosstalk, reflection, and lossy-medium-induced jitter, the ***decision feedback equalizer*** (DFE) will be widely used in the receiver to reduce the BER; consequently, test methodologies will have to advance to match and mimic future link and silicon architectures and technologies. In order to achieve an optimized test solution with acceptable accuracy, fault coverage, throughput, and cost, it is anticipated that both on-chip and off-chip test solutions will be necessary; for example, simple logical capabilities such as pattern generation and error detection should all be made internal via DFT/BIST. Other functionalities, such as jitter and noise generation and calibration, picosecond signal and jitter output measurement, and component separation, may likely remain external, before better on-chip solutions are found.

Another complication is that serial signaling comes with a layered communication protocol stack. The layer that connects to the pins is the *physical layer* (PHY). Not only does this PHY layer drive the data and perform clock/data recovery during receiving, but it also has to initialize, detect, and train/retrain between the sending and receiving ends. There is also a need for a link layer where error detection and correction, data flow control, *etc.*, are handled. In addition, there is a need for a **protocol layer**, which will turn the internal data transfer into data packets that can then be handled by the link and PHY layers. A massive increase in logic contents will result from advances in I/O subsystems. To make matter worse, I/O subsystems run on their own clocks, which are synchronized to the recovered clocks. This creates multiple clock domains on a given chip with multiple I/O channels. Cross-domain asynchronous clocking will result in nondeterministic chip responses and further lead to mismatches and yield loss [Mak 2004b].

12.7 RF TESTING

In the last decade, we have witnessed major developments in the field of personal mobile communication [Kasten 1998]. This has been accelerated by the significant research and development in the field of *radiofrequency* (RF) devices. These devices operate at very high frequencies (300 MHz and beyond) and are ubiquitous in the form of cellular phones, laptops with integrated wireless access, mobile PDAs, and various other wireless devices. Apart from the uses described above, RF circuits are used for numerous other applications (*e.g.*, medical care, air traffic control in airports, radar applications, and satellite and deep space communications). The convenience of **radiofrequency identification** (RFID) is contributing to its increasing popularity for applications such as highway tollbooths, supermarkets, and warehouses.

Production testing is an integral part of semiconductor manufacturing. To maintain an accurate and reliable environment for production testing of RF devices, having a proper measurement setup and "right" RF test instrumentation is an important factor. This involves maintaining proper impedance matching and shielding during the measurement procedure. With the extra overhead involved, it is estimated that as much as 40% of the manufacturing costs can be attributed to the test of RF-integrated circuits and systems. The production testing problem is made more difficult by the fact that most of the bench tests designed to characterize an RF part must be performed in fractions of a second (if possible) on the production floor and repeated reliably across the thousands of devices manufactured each day. In addition, the tests must be performed on a "least-cost" commercial tester in order to minimize tester maintenance and handling costs and to minimize capital outlay for test. To ensure the performance of the manufactured ICs, a predesigned test stimulus is applied to the *device under test* (DUT), and the test response data are captured. Next, by analyzing the captured test response data, the specification of the DUT is computed. This can be performed "on the fly" or as a postprocessing step after the test is performed. The computed specification value is compared to the standard specification value and, accordingly, the device is classified as good, marginal, or bad. This procedure is repeated for *every* specification of the DUT. In a production test environment, the quality of the test measurement system is evaluated according to its *accuracy* and *repeatability*, each of which is described next.

To ensure that all good devices are classified as good and *vice versa*, the performance metrics of the DUT must be measured with a high degree of accuracy. This means that the amount of measurement noise added to the test response captured by the ATE must be minimal to ensure high resolution of the captured test data and accuracy of the response analysis procedures. In addition to the need for a very accurate test stimulus application and test response capture mechanism, the test procedure must be highly repeatable. This means that, if the same test is applied several times to the DUT and the same test response analysis is performed, then the variance of the measured specification must be as small as possible. Of course, the minimum variance is bounded by the variance (power) of the measurement noise inherent to the test measurement system.

The challenges associated with RF testing have made it a pressing issue for the manufacturing industry. As a result, RF testing has received considerable attention from industry and academia in the recent past [Ferrario 2002] [Akbay 2004]. This is because production testing of RF devices faces some of the toughest challenges within the testing industry due to rapid upward scaling of frequencies of operation and the large disparity between the ability to design high-frequency devices and the lack of availability of "low-cost" testers to test them [Ozev 2001]. Currently, rapid progress is being made by tester manufacturing companies to bridge this gap. In addition, the packaging-driven integration of mixed technologies poses very difficult test access issues. Specifically, testing embedded RF, MEMS and optics subsystems is very difficult and challenging. Testing can, of course, be performed for system-level performance metrics from the externally observable pins of an integrated package; however, system-level performance metrics are usually very complicated, incur relatively large test time (*e.g.*, BER testing of a wireless transceiver system), and generate little diagnostic information for debugging problems with new designs.

In the following sections, first a brief overview of a wireless transreceiver system is presented. Next, various tests performed for different building blocks of a wireless system as well as the complete system are discussed. Finally, current test practices in the semiconductor industry are presented, followed by future trends.

12.7.1 Core RF Building Blocks

This section gives a brief overview of wireless transceiver architectures and the different modules associated with generic transceivers. Next, the methods employed during production test for measuring various specifications of the individual modules as well as the complete system are described. Figure 12.25 shows a typical direct conversion transceiver [Razavi 1997]. The basic building blocks of the transmitter shown in Figure 12.25 are the *power amplifier* (PA) and the **up-conversion mixer**. For the receiver, the key components consist of the *low noise amplifier* (LNA) and the **down-conversion mixer**. During production testing, the specifications of each individual module mentioned above, as well as those of the complete RF front-end are measured. In addition to the specifications of the RF front-end, numerous baseband specifications are also measured in order to ensure the quality of the modulated signal and the transmission link. Various specifications of a transceiver system are listed in Table 12.4. A brief description of some of the key specifications and standard test procedures for measuring the corresponding specification values are described next [Razavi 1997].

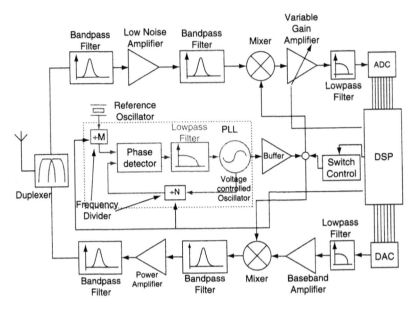

■ **FIGURE 12.25**

Block diagram of a direct down-conversion wireless transceiver.

TABLE 12.4 ■ System Components and Related Specifications

System Component	Specifications of Interest
Receiver [Agilent Rx-2002]	Gain, third-order intercept (TOI), error vector magnitude (EVM), magnitude, and phase error
LNA	Gain, TOI, noise figure
Mixer	Conversion gain, TOI, noise figure
Transmitter [Agilent Tx-2002]	Gain, TOI, EVM, adjacent channel power ratio (ACPR) [RS 1EF40-1998]
PA	Gain, output power [RS 1MA40-2002]
Mixer	Conversion gain, TOI
System	Bit error rate (BER), signal-to-noise ratio (SNR), sensitivity, dynamic range

12.7.2   RF Test Specifications and Measurement Procedures

12.7.2.1   Gain

The gain measurement is probably the easiest for any device or a system. To measure the gain of a device, the DUT is stimulated with a single tone input, with the power of the applied tone well within its linear region of operation. The ratio of the output power and the input power is specified as the gain of the DUT. Because gain is dependent on the frequency of operation, the gain measurement is specified for a particular frequency (*e.g.*, gain of LNA = 8 dB @ 900 MHz).

A note on dB and dBm: As the reader might note, the unit used to specify gain is decibel (dB). Numbers are converted to decibels using the following formula:

$$N_{dB}(\text{in dB}) = 20\log_{10}(N)$$

Therefore, a gain of 8 dB essentially means a gain of approximately 2.5. The avid reader might note that 20 dB is 10, and 40 dB is 100. A similar notation, called dBm is also used to specify power in logarithmic units, using the following formula:

$$N_{dBm}(\text{in dBm}) = 10\log_{10}(N \times 1000)$$

where N is the power in watts; therefore, 1 mW is 0 dBm and 1 W is 30 dBm (readers are encouraged to verify the numbers by themselves). The two units introduced above are used very frequently in RF design and test, and it is important for the reader to be familiar with these units for easier understanding of the subject [Razavi 1997].

12.7.2.2 Conversion Gain

The *conversion gain* (CG) is measured for mixers (both up-conversion and down-conversion) to specify the gain in signal power while frequency translation of the signal is performed via the use of a *local oscillator* (LO) signal. Thus, conversion gain is defined as the ratio of the output power of the translated frequency tone to the power of the input tone. For example, if an up-conversion mixer is supplied an input tone of 100 KHz @ −10 dBm and a LO signal at 1.575 GHz @ 0 dBm and it generates an output at 1.5751 GHz @ −6 dBm, then the conversion gain of the mixer is 4 dB (= [−6 dBm] − [−10 dBm]). This analysis is applicable to down-conversion mixers as well. Conversion gain depends not only on the frequency of the input tone but also on the frequency and power level of the LO signal. To completely specify the conversion gain test, all the above parameters must be defined (*e.g.*, CG = 4 dB, input = 100 KHz @ −10 dBm, LO = 1.575 GHz @ 0 dBm).

12.7.2.3 Third-Order Intercept

In any communication system, precise linearity of the front-end is a key requirement for ensuring high quality of transmission and reception. Effects of device nonlinearities are generally observable in the form of harmonics and distortion components at the output mode; however, these effects can be mitigated by filtering unwanted tones. Third-order intermodulation distortion products are difficult to get rid of due to the closeness of distortion frequency tones to the fundamental. To understand the origin of nonlinearity in a system, we first describe a linear system in terms of input/output responses as:

$$y(t) = A_0 + A \times u(t)$$

where $y(t)$ is the output, A_0 is the DC offset from input to output, A denotes the gain/loss of the system, and $u(t)$ is the input.

For a nonlinear system, in addition to the linear term (A), nonlinear terms of higher order are also present. Such a system can be denoted as:

$$y(t) = A_0 + A_1 \times u(t) + A_2 \times u(t)^2 + A_3 \times u(t)^3 + \ldots \text{ higher order terms}$$

Usually, fourth and higher order terms are ignored as they have very little impact on the overall system nonlinearity. Note that, of the three terms shown in the above equation, A_1 has the largest absolute magnitude, followed by A_2, A_3, and so on [Cho 2005].

If an input stimulus $u(t) = a_1 \times \sin(\omega_1 t) + a_2 \times \sin(\omega_2 t)$ is applied to the DUT (ω_1 and ω_2 are close to each other), then one can find out from the above equation that the output tones present after filtering will be ω_1, ω_2, $2\omega_1 - \omega_2$, and $2\omega_2 - \omega_1$ (the rest of the tones are filtered out). To measure the **third-order intercept** (TOI), two tones with the same amplitude and small difference in frequency are applied to the DUT, the output of which can be very easily viewed using a spectrum analyzer [Agilent SA-2005]. The output response of the DUT looks similar to Figure 12.26a. The larger tones are the applied fundamental tones (i.e., ω_1, ω_2), and the smaller tones (intermodulation) appear due to the nonlinearity of the DUT (i.e., $2\omega_1 - \omega_2$, $2\omega_2 - \omega_1$). An intuitive understanding of TOI can be seen in Figure 12.26b, where as the input power is increased, the power of the intermodulation tones increases faster (three times, hence the name *third-order intercept*) than the fundamental.

As opposed to gain measurement, during TOI test, the applied power of the input tones is in the nonlinear region of operation of the DUT. From the output spectrum, the TOI is calculated as:

$$\text{TOI} = P_{out} + |(P_{out} - P_{IMD})/2|$$

where P_{out} denotes the output power of the fundamental tones, and P_{IMD} denotes the power of the intermodulation tones.

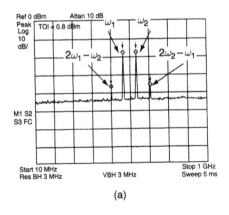

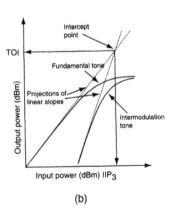

(a) (b)

■ **FIGURE 12.26**

(a) Output spectrum during TOI measurement, and (b) origin of TOI.

The TOI depends on the frequency of interest. Although two tones are used to determine the TOI, only the fundamental is specified, as the frequencies are very closely spaced to the fundamental (10–100 KHz apart). As an example, to test the TOI of a device at 1880 MHz, the tones used are 1879.9 MHz and 1880.1 MHz, and the TOI of the device is specified as TOI = 6.8 dBm @ 1880 MHz.

12.7.2.4 Noise Figure

The **noise figure** (NF), also known as *noise factor*, characterizes the noise performance of a device or a system. It is defined as:

$$NF = \frac{SNR_{in}}{SNR_{out}} = \frac{N_{out}}{N_{in}}$$

where, *SNR* denotes the signal-to-noise ratio, N_{out} is the total output noise power, and N_{in} is the amount of output noise due to input only (*i.e.*, noise from the source).

The most common method for measuring noise figure relies on the use of a **noise figure analyzer** (NFA). With the device/system biased using DC signal sources, a band-limited noise signal is supplied as input to the DUT by a calibrated noise source. The output of the DUT is measured by the NFA. As the input noise and the signal-to-noise ratio of the input noise signal are known to the NFA, the noise figure of the DUT can be calculated [Maxim 2003]. This method is capable of measuring very low NF values. For measuring high NF values, other methods such as the gain method [Maxim 2003] and the Y-factor [Agilent Y-2004] method can be used. NF is also dependent on the operating frequency of the DUT and is defined at a specific frequency (*e.g.*, NF = 1.4 dB 900 MHz).

12.7.3 Tests for System-Level Specifications

12.7.3.1 Adjacent Channel Power Ratio

Wireless transmission is performed through communication channels created by dividing the entire frequency band of communication into smaller sub-bands, also known as channels. Usually, channel spacing is on the order of few hundreds of kilohertz to few megahertz. The channels are allocated to different wireless users, both spatially and temporally, depending on the wireless protocol; thus, it is imperative for all transmitters to restrict the information within the channel allocated to each (*i.e.*, in-band). However, because of the inherent nonlinearity of RF devices, the intermodulation tones created may fall into adjacent channels, thereby creating interference. The **adjacent channel power ratio** (ACPR) specifies the amount of power leakage into adjacent bands of communication [RS 1EF40-1998]. To test for ACPR, a pseudo-random bitstream is transmitted from the baseband DSP (Figure 12.25) and the spectrum of the transmitter output is captured. From this captured output spectral response, the in-band channel power and the out-of-band channel power are measured. The ratio of these two quantities is specified as the ACPR of the transmitter. ACPR depends on the communication protocol (*i.e.*, the frequency and output power of the transmitter, the baseband modulation scheme, and the filter characteristics at the output of the PA).

12.7.3.2   *Error Vector Magnitude, Magnitude Error, and Phase Error*

The *error vector magnitude* (EVM) is a system-level specification measured at
the baseband. It describes the quality of modulation and easily identifies any
nonidealities within the system [Agilent EVMa-2005] [Agilent EVMb-2000]. To
measure the EVM, a set of known bits is modulated in the transmitter baseband to
create constellation symbols. The symbols are indicated in Figure 12.27a as asterisks
(*) for QPSK modulation [Razavi 1997]. The black astericks show the ideal symbol locations, and the grey astericks show the received and demodulated symbols.
Here, each symbol represents a pair of bits; the phase relationship between orthogonal sine and cosine waves is used to encode the two bits into four different phase
combinations. These are represented as four points in the constellation diagram
(Figure 12.27a). This information signal is up-converted through the transmitter,
received and down-converted by the receiver, and demodulated by the receiver
baseband. During the transmission process, the nonlinearities of the system, jitter,
phase noise of oscillators, and various other effects cause the modulated symbols
to deviate from the expected constellation points. EVM quantifies this amount of
deviation as [Halder 2005a]:

$$EVM_{RMS} = \sqrt{\left(\frac{1}{N}\sum_{i=1}^{N}\left|V_{ideal,i} - V_{measured,i}\right|^2\right) \bigg/ \left(\frac{1}{N}\sum_{i=1}^{N}\left|V_{ideal,i}\right|^2\right)}$$

where, $V_{ideal,i}$ and $V_{measured,i}$ are marked in Figure 12.27b. The gray dots in
Figure 12.27a show the received symbols.

Typical EVM values range from 3 to 15%. EVM also depends on the magnitude
and phase error. In cases when the system cannot deliver sufficient gain, the entire
constellation moves closer to the origin, whereas phase errors cause the constellation to rotate on the same magnitude circle (see Figure 12.27b). These effects are
usually corrected by the DSP before computing the EVM; thus, EVM is specified
with the amount of magnitude error and phase error correction applied.

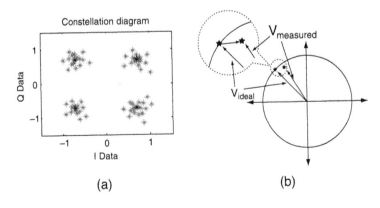

(a)   (b)

■ **FIGURE 12.27**

(a) EVM constellation diagram, and (b) origin of EVM.

12.7.4 Current and Future Trends

Various methods have been proposed in the past to address the issues related to RF testing and simplify conventional specification measurement techniques in a production test environment. Next, we discuss a few unique test methods that have emerged in the last decade to address the various issues related to RF testing.

One of the methods, known as *alternate test* [Variyam 2002], relies on strong correlation between the response of the DUT to an applied stimulus and its performance metrics (*i.e.*, specifications). To do so, the method finds a test input stimulus using an optimization algorithm such that the sensitivity of the test response to the specifications is maximized. It builds a regression-based mapping function to directly compute the specifications from the observed test response. Numerous applications of the proposed method have proven that this method is extremely powerful in a production test environment.

In [Voorakaranam 2002], problems related to high-speed signal generation and capture in RF test were mitigated by designing the test stimulus in the baseband (*i.e.*, low frequency) using an optimization algorithm. The stimulus was up-converted using a mixer, applied to the DUT (in this case, an LNA), and the response of the DUT was down-converted and captured (digitized). Using the captured response, the specifications of the DUT were estimated using the alternate test method. Another method suggested by [Cherubal 2004] generates multitone tests directly in the RF domain by controlling the frequency and amplitude values. The underlying assumption is that high-quality signal generation is possible at high frequencies using RF signal generators. It demonstrated that up to 60% of the total test time could be reduced by using this method.

The problem related to test generation for RF circuits is due to the long simulation times required for RF netlist-level transient simulations. Although various simulation techniques have been proposed (*e.g.*, periodic steady state, harmonic balance), test generation is still a bottleneck for RF DUTs. In [Halder 2003], a frequency-domain behavioral simulation framework was developed. It has been demonstrated that by using this method, test generation time can be significantly reduced without loss of accuracy. Currently, similar methods are in use for test generation purposes.

Testing system-level specifications is another difficult task to perform in RF testing. A loopback test method was proposed for testing RF transceiver systems in [Dabrowski 2003], but the method relied on spot defects only. Spot defects manifest as constant high or low signals at specific circuit nodes. As the reader knows from earlier sections, the notion of spot defects (*i.e.*, stuck-at-1 and stuck-at-0) is well understood in the realm of digital testing. RF circuits deal with signals that are analog in nature but at a much higher frequency. Parametric variations, which are a major cause of variations in RF devices, do not necessarily show up as spot defects. A more detailed study based on the effects of parametric variations on the RF system was performed in [Halder 2005b]. It was shown that by using an alternative test-based approach, both system-level and module-level specifications could be predicted with a high degree of accuracy.

In [Ferrario 2003], the authors used detectors to estimate various specifications of the DUT. In [Yin 2005] and [Valdes-Garcia 2005], the authors designed and used sensors for on-chip self testing. A combination of the above two approaches and the alternative test method has shown that test accuracy can be increased significantly [Bhattacharya 2004]. The detectors were placed at specific circuit nodes using an optimization algorithm, and the alternate test-based method was used for test generation.

In addition, significant work is underway to minimize the cost of instrumentations involved in the generation of RF test signals and signal capture/analysis. In [Sylla 2003], a novel method has been presented where RF test signals were generated using DSPs. Also, researchers are trying to minimize RF testing cost by making wafer-probe testing more efficient so bad ICs are not packaged, thus saving packaging costs, which may be significant for RF SOCs [Lau 2002]. In [Ozev 2004], a loopback delay-insertion-based DFT method was proposed to minimize the overall test cost.

12.7.4.1 *Future Trends*

The domain of RF testing is becoming increasingly difficult due to the rapid introduction of new communication protocols with complex baseband modulation techniques. In the last few years, various new protocols, such as Zigbee, *ultrawideband* (UWB) [Nekoogar 2005], *radiofrequency identification* (RFID), and WiMAX [Agilent WiMax-2005], have either been newly introduced or gained renewed interest in the design community. However, in contrast to the past, the present implementations are all monolithic and hence the complexity of test and characterization has elevated manifold. For system-level implementation of a communications protocol, observability is limited and very little diagnosis information can be obtained from test responses. A key issue related to testing newer standards stems from the higher speeds of operation. At-speed testing and characterization of the manufactured devices requires expensive test instrumentation and equipment during production testing. For example, UWB devices operate up to 10.6 GHz. At present, there are very few automatic test equipments in the industry can work at such high speeds. Finally, as devices and protocols improve, test times also become a major concern for the manufacturing industry due to the elevated complexity and the tighter margins on the test limits. For example, high-speed *serializer–deserializer* (SERDES) devices have a BER in the range of 10^{-12}; therefore, typical test times for these devices are in the range of minutes, even hours. The test industry needs to rethink the way these devices are tested and new solutions are needed to keep up with continued advancements in the RF domain.

Within this context, *built-in-self-test* (BIST) is gaining increased attention from test and design engineers alike [Veillette 1995]. BIST can potentially address all the above issues with a small overhead in terms of calibration. The sensor-based test methods described earlier have the potential to address the issues related to diagnosis and at-speed testing. In addition, the test time for high-speed devices can be reduced by applying the alternative test-based approach. [RS 7BM03-2002] elaborates on the BER test, and [Bhattacharya 2005] describes an alternative test-based method to reduce BER test time for UWB devices. Interested readers are encouraged to look at various application notes and white papers from relevant industries to keep abreast of the topic.

12.8 CONCLUDING REMARKS

The Semiconductor Industry Association (SIA) published an *International Technology Roadmap for Semiconductors* (ITRS) in [SIA 2004] which includes an update to the test and test equipment needs and difficult challenges (see Tables 12.1 and 12.2) for nanometer designs through 2010 and beyond. The ITRS calls for: (1) new *design-for-testability* (DFT) and *design-for-manufacturability* (DFM) methods for digital circuits and analog circuits (including *radiofrequency* (RF) and audio circuits), as well as high-speed serial interface, MEMS, and sensors; (2) the means to reduce manufacturing test costs as well as enhance device reliability and yield; and (3) techniques to facilitate defect analysis and failure analysis.

The SIA also published an ITRS design roadmap in [SIA 2004] which includes an update (see Table 12.3) to the design test challenges given in [SIA 2003]. Nanometer process technology, increasing clock rate, and increasing *system-on-chip* (SOC) and *system-in-package* (SIP) integrations were identified to present severe challenges to design for test.

Today, almost all $0.13\,\mu m$ designs have implemented memory *built-in self-test* (BIST), but not necessarily operated at-speed. For 90 nm designs, we have found that *at-speed memory BIST, at-speed scan testing*, and *test compression* have been extensively adopted and are becoming a must. Unfortunately, at-speed logic BIST and at-speed memory **built-in self-repair** (BISR) have not yet grown in popularity. Testing continues to be a big challenge for multi-gigahertz serial I/O protocols and **analog and mixed-signal** (AMS) circuits. As test time for AMS circuits is currently reaching 80% of the total test time spent on an ATE, more research on AMS DFT/BIST must be pursued immediately along with innovations on digital and AMS diagnosis. In the near term, logic BIST and memory BIST/BISR must also be extended to cover more realistic fault types, such as delay faults in logic BIST and drowsy faults in memory cells. Noise faults caused by parametric variations must be modeled and tested. Integration of SOC test methods (*e.g.*, test reuse, core-based testing) into chip-level DFT continues to be a challenge seeking an efficient solution.

In the long term, by 2018, according to the ITRS [SIA 2004], 90% of silicon will be embedded with BIST, so efforts should also be directed toward *embedded software-based self-testing* utilizing on-chip programmable resources. Because SOC and SIP integrations will be desperately needed, new DFT/BIST methods for *field programmable gate array* (FPGA) testing, *microelectromechanical system* (MEMS) testing, RF testing, and high-speed serial interface must be researched. As clock rates of digital circuits are aggressively driven to the limit, signal integrity test issues, especially timing-related faults and digital interference in analog circuits, must be further investigated. Continued innovation of logic BIST/BISR and emerging needs for AMS DFT/BIST/diagnosis are required. Fundamentally new long-term solutions must be developed for reliability screen and may include significant on-die hardware for stressing or special reliability measurements. Finally, as circuits must coexist with defects in nanotechnology, fault tolerance and online testing must be embedded into circuits, particularly for *logic soft errors*. In summary, the test industry must cope with an enormous spectrum of problems ranging from high-level test synthesis to noise/interface and power dissipation, as listed in Table 12.3 [SIA 2003] [SIA 2004].

In this chapter, we have presented several promising techniques to address some of the critical ITRS needs and challenges for testing nanometer designs. Due to space limitations, we can only briefly cover techniques for delay testing; coping with physical failures, soft errors, and reliability issues; FPGA testing; MEMS testing; high-speed I/O testing; and RF testing. Other important test techniques, such as *software-based self-testing* [Cheng 2006], *design for manufacturability* (DFM) [Gupta 2003], **design for yield enhancement** (DFY) [Director 1994] [Zorian 2004], and **design for debug and diagnosis** (DFD) for AMS circuits [Vinnakota 1998], unfortunately, have had to be left out of the discussion. As the test needs and challenges facing the semiconductor industry in the nanometer age are so broad and difficult in nature, conducting further research is imperative, and over the next 5 to 10 years better solutions have to be found for all the subjects mentioned here.

Acknowledgments

The authors wish to thank Prof. Duncan M. (Hank) Walker of Texas A&M University for contributing a portion of the Delay Testing section; Prof. Mehrdad Nourani of the University of Texas at Dallas for contributing to the Signal Integrity and Power Supply Noise section; T.M. Mak of Intel Corp. for contributing to the Soft Errors, Fault Tolerance, and Defect and Error Tolerance sections; Prof. Charles Stroud of Auburn University for contributing to the FPGA Testing section; T.M. Mak of Intel Corp. and Dr. Mike Peng Li of Wavecrest Corp. for contributing to the High-Speed I/O Testing section; and Dr. Soumendu Bhattacharya and Prof. Abhijit Chatterjee of the Georgia Institute of Technology for contributing to the RF Testing section; as well as Teresa Chang and her lovely daughter Alice Yu of SynTest Technologies for typing the text and drawing the figures.

References

R12.0—Books

[Athavale 2005] A. Athavale and C. Christensen, *High-Speed Serial I/O Made Simple: A Designers' Guide, with FPGA Applications*, Xilinx Connectivity Solutions, San Jose, CA, 2005.

[Cheng 2006] K.-T. Cheng, Embedded software-based self-testing for SoC design, in *The Embedded Systems Handbook*, R. Zurawski, Ed., CRC Press, Boca Raton, FL, 2006, chap. 28.

[Dally 1998] W. J. Dally and J. W. Poulton, *Digital Systems Engineering*, Cambridge University Press, Cambridge, U.K., 1998.

[Director 1994] S. W. Director and W. Maly, Eds., *Advances in CAD for VLSI*, Vol. 8, *Statistical Approaches to VLSI Design*, North Holland, Amsterdam, 1994.

[Hsu 2002] T. R. Hsu, *MEMS: Microsystems Design and Manufacture*, McGraw-Hill, New York, 2002.

[Krstic 1998] A. Krstic and K.-T. Cheng, *Delay Fault Testing for VLSI Circuits*, Kluwer Academic, Norwell, MA, 1998.

[Lala 2001] P. K. Lala, *Self-Checking and Fault-Tolerant Digital Design*, Morgan Kaufmann, San Francisco, CA, 2001.

[Nekoogar 2005] F. Nekoogar, *Ultra-Wideband Communications: Fundamentals and Applications*, First ed., Pearson Education, Upper Saddle River, NJ, 2005.

[Pradhan 1996] D. K. Pradhan, *Fault-Tolerant Computer System Design*, Prentice Hall, Upper Saddle River, NJ, 1996.

[Razavi 1997] B. Razavi, *RF Microelectronics*, First ed., Prentice Hall, Upper Saddle River, NJ, 1997.

[Senthinatharr 1994] R. Senthinatharr and J. L. Prince, *Simultaneous Switching Noise of CMOS Devices and Systems*, Kluwer Academic, Norwell, MA, 1994.

[Siewiorek 1998] D. Siewiorek and R. S. Swarz, *Reliable Computer Systems: Design and Evaluation, Third Edition*, AK Peters, MA, 1998.

[Stroud 2002a] C. Stroud, *A Designer's Guide to Built-In Self-Test*, Springer, Boston, MA, 2002.

[Vinnakota 1998] B. Vinnakota, Ed., *Analog and Mixed-Signal Test*, Prentice Hall, Upper Saddle River, NJ, 1998.

R12.1—Test Technology Roadmap

[Chen 2002] W. Y. Chen, S. K. Gupta, and M. A. Breuer, Analytical models for crosstalk excitation and propagation in VLSI Circuits, *IEEE Trans. Comput.-Aided Des.*, 21(10), 1117–1131, 2002.

[Kim 2004] N. S. Kim, K. Flautner, D. Blaauw, and T. Mudge, Circuit and microarchitectural techniques for reducing cache leakage power, *IEEE Trans. Very Large Scale Integration (VLSI) Syst.*, 12(2), 167–184, 2004.

[Moore 1965] G. Moore, Cramming more components onto integrated circuits, *Electronics*, 38, 114–117, 1965.

[Saxena 2003] J. Saxena, K. M. Butler, V. B. Jayaram, S. Kundu, N. V. Arvind, P. Sreeprakash, and M. Hachinger, A case study of IR-drop in structured at-speed testing, in *Proc. IEEE Int. Test Conf.*, October 2003, pp. 1098–1104.

[SIA 1997] SIA, *The National Technology Roadmap for Semiconductors*, Semiconductor Industry Association, San Jose, CA (http://public.itrs.net), 1997.

[SIA 2003] SIA, *The International Technology Roadmap for Semiconductors: 2003 Edition—Design*, Semiconductor Industry Association, San Jose, CA (http://public.itrs.net), 2003, pp. 30–36.

[SIA 2004] SIA, *The International Technology Roadmap for Semiconductors: 2004 Update*, Semiconductor Industry Association, San Jose, CA, (http://public.itrs.net), 2004.

[Wang 2004] L.-C. Wang, J. J. Liou, and K.-T. Cheng, Critical Path selection for delay fault testing based upon a statistical timing model, *IEEE Trans. Comput.-Aided Des.*, 23(11), 1550–1565, 2004.

R12.2—Delay Testing

[Amodeo 2005] M. Amodeo and B. Cory, Defining faster-than-at-speed delay tests, *Cadence Nanometer Test Quarterly*, 2(2), 2005 (http://www.cadence.com/newsletters/nanometer_test).

[Barnhart 2004] C. F. Barnhart, What is true-time delay test?, *Cadence Nanometer Test Quarterly*, 1(2), 2–7, 2004.

[Breuer 1997] M. A. Breuer, C. Gleason, and S. Gupta, New validation and test problems for high performance deep sub-micron VLSI circuits [tutorial notes], in *Proc. IEEE VLSI Test Symp.*, April 1997.

[Cheng 1993] K.-T. Cheng, S. Devadas, and K. Keutzer, Delay-fault test generation and synthesis for testability under a standard scan design methodology, *IEEE Trans. Comput.-Aided Des.*, 12(8), 1217–1231, 1993.

[Iyengar 1988a] V. S. Iyengar, B. Rosen, and I. Spillinger, Delay test generation 1. Concepts and coverage metrics, in *Proc. IEEE Int. Test Conf.*, October 1988, pp. 857–866.

[Iyengar 1988b] V. S. Iyengar, B. Rosen, and I. Spillinger, Delay test generation 2. Algebra and algorithms, in *Proc. IEEE Int. Test Conf.*, October 1988, pp. 867–876, 1988.

[Krstic 2003] A. Krstic, L.-C. Wang, K.-T. Cheng, J. Liou, and M. Abadir, Delay defect diagnosis based upon statistical timing models: the first step, in *Proc. IEEE Design, Automation and Test in Europe Conf.*, March 2003, pp. 328–333.

[Kruseman 2004] B. Kruseman, A. K. Majhi, G. Gronthoud, and S. Eichenberger, On hazard-free patterns for fine-delay fault testing, in *Proc. IEEE Int. Test Conf.*, October 2004, pp. 213–222.

[Levendel 1986] Y. Levendel and P. Menon, Transition faults in combinational circuits: input transition test generation and fault simulation, in *Proc. Fault Tolerant Computing Symp.*, July 1986, pp. 278–283.

[Li 1989] W. Li, S. M. Reddy, and S. Sahni, On path selection in combinational logic circuits, *IEEE Trans. Comput.-Aided Des.*, 8(1), 56–63, 1989.

[Lin 1987] C. Lin and S. M. Reddy, On delay fault testing in logic circuits, *IEEE Trans. Comput.-Aided Des.*, CAD-6(5), 694–703, 1987.

[Liou 2002] J.-J. Liou, A. Krstic, L.-C. Wang, and K.-T. Cheng, False-path-aware statistical timing analysis and efficient path selection for delay testing and timing validation, in *Proc. of 39th Design Automation Conf.*, June 2002, pp. 566–569.

[Liou 2003] J. Liou, A. Krstic, Y. Jiang, and K.-T. Cheng, Modeling, testing, and analysis for delay defects and noise effects in deep submicron devices, *IEEE Trans. Comput.-Aided Des.*, 22(6), 756–769, 2003.

[Mak 2004] T. M. Mak, A. Krstic, K.-T. Cheng, and L.-C. Wang, New challenges in delay testing of nanometer, multigigahertz designs, *IEEE Des. Test Comput.*, 21(3), 241–247, 2004.

[Mao 1990] W. W. Mao and M. D. Ciletti, A variable observation time method for testing delay faults, in *Proc. IEEE Design Automation Conf.*, June 1990, pp. 728–731.

[Maxwell 1991] P. Maxwell, R. Aitken, V. Johansen, and I. Chiang, The effect of different test sets on quality level prediction: when is 80% better than 90%, in *Proc. IEEE Int. Test Conf.*, October 1991, pp. 358–364.

[Park 1991] E. Park, B. Underwood, T. W. Williams, and M. Mercer, Delay testing quality in timing-optimized designs, in *Proc. IEEE Int. Test Conf.*, October 1991, pp. 897–905.

[Pramanick 1997] A. Pramanick and S. M. Reddy, On the fault coverage of gate delay fault detecting tests, *IEEE Trans. Comput.-Aided Des.*, 16(1), 78–94, 1997.

[Qiu 2003] W. Qiu and D. M. H. Walker, An efficient algorithm for finding the K longest testable paths through each gate in a combinational circuit, in *Proc. IEEE Int. Test Conf.*, October 2003, pp. 592–601.

[Qiu 2004] W. Qiu, J. Wang, D. M. H. Walker, D. Reddy, X. Lu, Z. Li, W. Shi, and H. Balachandran, K longest paths per gate (KLPG) test generation for scan-based sequential circuits, in *Proc. IEEE Int. Test Conf.*, October 2004, pp. 223–231.

[Sato 2005] Y. Sato, S. Hamada, T. Maeda, A. Takatori, Y. Nozuyama, and S. Kajihara, Invisible delay quality: SDQM model lights up what could not be seen, in *Proc. IEEE Int. Test Conf.*, October 2005, Paper 47.1, 9 pp.

[Savir 1992a] J. Savir, Skewed-load transition test, Part I: Calculus, in *Proc. IEEE Int. Test Conf.*, October 1992, pp. 705–713.

[Savir 1992b] J. Savir, Skewed-load transition test. Part II: Coverage, in *Proc. IEEE Int. Test Conf.*, October 1992, pp. 714–722.

[Savir 1994] J. Savir, On broad-side delay testing, in *Proc. IEEE VLSI Test Symp.*, April 1994, pp. 284–290.

[Sharma 2002] M. Sharma and J. H. Patel, Finding a small set of longest testable paths that cover every gate, in *Proc. IEEE Int. Test Conf.*, October 2002, pp. 974–982.

[Smith 1985] G. Smith, Model for delay faults based upon paths, in *Proc. IEEE Int. Test Conf.*, October 1985, pp. 342–349.

[Waicukauski 1987] J. A. Waicukauski, E. Lindbloom, B. K. Rosen, and V. S. Iyengar, Transition fault simulation, *IEEE Des. Test Comput.*, 4(2), 32–38, 1987.

[Yang 2004] K. Yang, K.-T. Cheng, and L.-C. Wang, TranGen: a SAT-based ATPG for path-oriented transition fault, in *Proc. Asia and South Pacific Design Automation Conf.*, January 2004, pp. 92–97.

R12.3—Coping with Physical Failures, Soft Errors, and Reliability Issues

[Attarha 2002] A. Attarha and M. Nourani, Test pattern generation for signal integrity faults on long interconnects, in *Proc. IEEE VLSI Test Symp.*, April 2002, pp. 336–341.

[Bai 2000] X. Bai, S. Dey, and J. Rajski, Self-test methodology for at-speed test of crosstalk in chip interconnects, in *Proc. IEEE Design Automation Conf.*, June 2000, pp. 619–624.

[Bai 2001] G. Bai, S. Bobba, and I. Haji, Maximum power supply noise estimation in VLSI Circuits using multimodal genetic algorithms, *Proc. Int. Conf. Electronics, Circuits Syst.*, 3, 1437–1440, 2001.

[Bartlett 2004] W. Bartlett and L. Spainhower, Commercial fault tolerance: a tale of two systems, *IEEE Trans. Dependable Secure Comput.*, 1(1), 87–96, 2004.

[Baumann 2005] R. Baumann, Soft errors in advanced computer systems, *IEEE Des. Test Comput.*, 22(3), 258–266, 2005.

[Borkar 2004a] S. Borkar, Microarchitecture and Design Challenges for Gigascale Integration, keynote speech for *Micro 2004 Conf.*, December 2004.

[Borkar 2004b] S. Borkar, T. Karnik, and V. De, Design and reliability challenges in nanometer technologies, in *Proc. IEEE Design Automation Conf.*, June 2004, p. 75.

[Bowman 2001] K. Bowman and J. Meindl, Impact of within-die parameter fluctuations on future maximum clock frequency distributions, in *Proc. Custom Integrated Circuits Conf.*, May 2001, pp. 229–232.

[Breuer 2004] M. Breuer, S. Gupta, and T. M. Mak, Defect and error tolerance in the presence of massive numbers of defects, *IEEE Des. Test Comput.*, 21(3), 216–227, 2004.

[Caignet 2001] F. Caignet, S. Delmas-Bendhia, and E. Sicard, The challenge of signal integrity in deep-submicrometer CMOS technology, in *Proc. IEEE*, 89(4), 556–573, 2001.

[Calin 1997] T. Calin, R. Velazco, M. Nicolaidis, S. Moss, S. D. LaLumondiere, V. T. Tran, R. Koga, and K. Clark, Topology-related upset mechanisms in design hardened storage cells, in *Proc. 4th European Conference on Radiation and Its Effects on Components and Systems (RADECS)*, September 1997, pp. 484–488.

[Cao 1997] L. Cao and J. Krusius, A new power distribution strategy for area array bonded ICs and packages of future deep sub-micron ULSI, in *Proc. Electronic Components and Technology Conf.*, June 1997, pp. 1138–1145.

[Cao 2002] Y. Cao, X. Huang, N. Chang, S. Lin, S. Nakagawa, W. Xie, D. Sylvester, and C. Hu, Effective on-chip inductance modeling for multiple signal lines and application to repeater insertion, *IEEE Trans. Very Large Scale Integration (VLSI) Syst.*, 10(6), 799–805, 2002.

[Chang 1997] Y. Chang, S. Gupta, and M. Breuer, Analysis of ground bounce in deep submicron circuits, in *Proc. IEEE VLSI Test Symp.*, April 1997, pp. 110–116.

[Chen 1997a] H. Chen and D. Ling, Power supply noise analysis methodology for deep-submicron VLSI design, in *Proc. IEEE Design Automation Conf.*, June 1997, pp. 638–643.

[Chen 1997b] H. Chen, A hierarchical power supply distribution model for full-chip switching noise analysis, *IEEE Meeting on Electrical Performance of Electronic Packaging*, October 1997, pp. 60–63.

[Chen 1998] W. Chen, S. Gupta, and M. Breuer, Test generation in VLSI circuits for crosstalk noise, in *Proc. IEEE Int. Test Conf.*, October 1998, pp. 641–650.

[Chen 1999] W. Chen, S. Gupta, and M. Breuer, Test generation for crosstalk-induced delay in integrated circuits, in *Proc. IEEE Int. Test Conf.*, October 1999, pp. 191–200.

[Chen 2001] L. Chen, X. Bai, and S. Dey, Testing for interconnect crosstalk defects using on-chip embedded processor cores, in *Proc. IEEE Design Automation Conf.*, June 2001, pp. 317–322.

[Cuviello 1999] M. Cuviello, S. Dey, X. Bai, and Y. Zhao, Fault modeling and simulation for crosstalk in system-on-chip interconnects, in *Proc. IEEE Int. Conf. on Comput.-Aided Des.*, November 1999, pp. 297–303.

[Fang 1998] P. Fang, J. Tao, J. Chen, and C. Hu, Design in hot carrier reliability for high performance logic applications, in *Proc. Custom Integrated Circuits Conf.*, October 1998, pp. 25.1.1–25.1.7.

[Gatej 2002] J. Gatej, L. Song, C. Pyron, R. Raina, and T. Munns, Evaluating ATE features in terms of test escape rates and other cost of test culprits, in *Proc. IEEE Int. Test Conf.*, October 2002, pp. 1040–1048.

[Hirase 2001] J. Hirase, Yield increase of VLSI after redundancy-repairing, in *Proc. Asian Test Symp.*, November 2001, pp. 353–358.

[Hunter 1999] W. Hunter, The statistical dependence of oxide failure rates on Vdd and Tox variations with applications to process design, circuit design and end use, in *Proc. IEEE Int. Reliability Physics Symposium*, March 1999, pp. 72–81.

[Jess 2003] J. Jess, K. Kalafala, S. R. Naidu, R. Otten, and C. Visweswariah, Statistical timing for parametric yield prediction of digital integrated circuits, in *Proc. IEEE Design Automation Conf.*, June 2003, pp. 343–347.

[Jiang 1997] Y.-M. Jiang, K.-T. Cheng, and A. Krstic, Estimation of maximum power and instantaneous current using a genetic algorithm, in *Proc. Custom Integrated Circuits Conf.*, October 1997, pp. 135–138.

[Jiang 2001] Y. Jiang and K.-T. Cheng, Vector generation for power supply noise estimation and verification of deep submicron designs, *IEEE Trans. VLSI*, 9(2), 329–340, 2001.

[Keshavarzi 2002] A. Keshavarzi, J. Tschanz, K. Roy, C. Hawkins, R. Daasch, and M. Sachdev, Leakage and process variation effects in current testing on future CMOS circuits, *IEEE Des. Test Comput.*, 19(5), 36–43, 2002.

[Koren 1990] I. Koren and A. D. Singh, Fault tolerance in VLSI Circuits, *IEEE Comput.*, 23(7), 73–83, 1990.

[Koren 1998] I. Koren and Z. Koren, Defect tolerance in VLSI circuits: techniques and yield analysis, *Proc. IEEE*, 86(9), 1819–1838, 1998.

[Krstic 2001] A. Krstic, Y.-M. Jiang, and K.-T. Cheng, Pattern generation for delay testing and dynamic timing analysis considering power-supply noise effects, *IEEE Trans. Comput.-Aided Des.*, 20(3), 416–425, 2001.

[Lee 2001] J. Lee, Y. Huh, P. Bendix, and S. Kang, Understanding and addressing the noise induced by electrostatic discharge in multiple power supply systems, in *Proc. IEEE Int. Conf. on Computer Design*, September 2001, pp. 406–411.

[Lin 2003] X. Lin, R. Press, J. Rajski, P. Reuter, T. Rinderknecht, B. Swanson, and N. Tamarapalli, High-frequency, at-speed scan testing, *IEEE Des. Test Comput.*, 20(5), 17–25, 2003.

[Lu 2004] X. Lu, Z. Li, W Qui, D. Walker, and W. Shi, Longest path selection for delay test under process variation, in *Proc. Asia and South Pacific Design Automation Conf.*, January 2004, pp. 98–103.

[Ma 1995] S. C. Ma, P. Franco, and E. J. McCluskey, An experimental chip to evaluate test techniques: experimental results, in *Proc. IEEE Int. Test Conf.*, October 1995, pp. 663–672.

[May 1979] T. C. May and M. H. Woods, Alpha-particle-induced soft errors in dynamic memories, *IEEE Trans. Electron. Devices*, ED-26(1), 2–9, 1979.

[McCluskey 2004] E. J. McCluskey, A. Al-Yamani, J. C.-M. Li, C.-W. Tseng, E. Volkerink, F.-F. Ferhani, E. Li, and S. Mitra, ELF-Murphy data on defects and test sets, in *Proc. IEEE VLSI Test Symp.*, April 2004, pp. 16–22.

[Mitra 2005] S. Mitra, N. Seifert, M. Zhang, Q. Shi, and K. S. Kim, Robust system design with built-in soft-error resilience, *IEEE Comput.*, 38(2), 43–52, 2005.

[Moore 1965] G. Moore, Cramming more components onto integrated circuits, *Electronics*, 38, 114–117, 1965.

[Mukherjee 2002] S. S. Mukherjee, M. Kontz, and S. Reinhardt, Detailed Design and Evaluation of Redundant Multithreading Alternatives, in *Proc. Int. Symp. on Computer Architecture*, pp. 99–110, 2002.

[Murthy 1997] C. Murthy and M. Gall, Process variation effects on circuit performance: TCAD simulation of 256-Mbit technology, *IEEE Trans. Comput.-Aided Des.*, 16(11), 1383–1389, 1997.

[Naffziger 1999] S. Naffziger, Design methodologies for interconnects in GHz+ ICs [tutorial lecture], in *Proc. IEEE Int. Solid-State Circuits Conf.*, February 1999.

[Nagaraj 2001] N. S. Nagaraj, P. Balsara, and C. Cantrell, Crosstalk noise verification in digital designs with interconnect process variations, in *Proc. IEEE Int. Conf. on VLSI Design*, January 2001, pp. 365–370.

[Nigh 1998] P. Nigh, D. Vallett, A. Patel, J. Wright, F. Motika, D. Forlenza, R. Kurtulik, and W. Chong, Failure analysis of timing and IDDq-only failures from the SEMATECH test methods experiment, in *Proc. IEEE Int. Test Conf.*, October 1998, pp. 43–52.

[Nigh 2000] P. Nigh and A. Gattiker, Test method evaluation experiments and data, in *Proc. IEEE Int. Test Conf.*, October 2000, pp. 454–463.

[Nourani 2002] M. Nourani and A. Attarha, Signal integrity: fault modeling and testing in high-speed SoCs, *J. Electron. Testing: Theory Applications (JETTA)*, 18(4–5), 539–554, 2002.

[Nourani 2005] M. Nourani and A. Radhakrishnan, Power-supply noise in SoCs: ATPG, estimation and control, in *Proc. IEEE Int. Test Conf.*, November 2005, pp. 22.1.1–22.1.10.

[Park 2005] I. Park, A. Al-Yamani, and E. J. McCluskey, Effective TARO pattern generation, in *Proc. IEEE VLSI Test Symp.*, May 2005, pp. 161–166.

[Pham 2004] N. Pham, M. Cases, D. Araujo, and E. Matoglu, Design methodology for multiple domain power distribution systems, in *Proc. Electronic Components and Technology Conf.*, June 2004, pp. 542–549.

[Pomeranz 1999] I. Pomeranz and S. M. Reddy, On *N*-detection test sets and variable *N*-detection test sets for transition faults, in *Proc. IEEE VLSI Test Symp.*, April 1999, pp. 173–179.

[Rayane 1999] I. Rayane, J. Velasco-Medina, and M. Nicolaidis, A digital BIST for operational amplifiers embedded in mixed-signal circuits, in *Proc. IEEE VLSI Test Symp.*, April 1999, pp. 304–310.

[Sengupta 1999] S. Sengupta, S. Kundu, S. Chakravarty, P. Parvathala, R. Galivanche, G. Kosonocky, M. Rodgers, and T. M. Mak, Defect-based test: a key enabler for successful migration to structural test, *Intel Technol. J.*, Q1, 1–14, 1999.

[Sklaroff 1976] J. R. Sklaroff, Redundancy management technique for space shuttle computers, *IBM J. R&D*, 20(1), 5–19, 1976.

[Spainhower 1999] L. Spainhower and T. A. Gregg, IBM S/390 parallel enterprise server G5 fault tolerance: a historical perspective, *IBM J. R&D*, 43(5/6), 863–873, 1999.

[Tabatabaei 2002] S. Tabatabaei and A. Ivanov, An embedded core for sub-picosecond timing measurements, in *Proc. IEEE Int. Test Conf.*, October 2002, pp. 129–137.

[Tehranipour 2004] M. Tehranipour, N. Ahmed, and M. Nourani, Testing SoC interconnects for signal integrity using extended JTAG architecture, *IEEE Trans. Comput.-Aided Des.*, 23(5), 800–811, 2004.

[Tendolkar 2000] N. Tendolkar, R. Molyneaux, C. Pyron, and R. Raina, At-speed testing of delay faults for Motorola's MPC7400, a PowerPC microprocessor, in *Proc. IEEE VLSI Test Symp.*, April 2000, pp. 3–8.

[Tseng 2001] C.-W. Tseng and E. J. McCluskey, Multiple-output propagation transition fault test, in *Proc. IEEE Int. Test Conf.*, October 2001, pp. 358–366.

[Vazquez 2004] J. Vazquez and J. Gyvez, Power supply noise monitor for signal integrity faults, in *Proc. IEEE Design, Automation and Test in Europe Conf.*, February 2004, pp. 1406–1407.

[Wang 2005] L.-T. Wang, X. Wen, P.-C. Hsu, S. Wu, and J. Guo, At-speed logic BIST architecture for multi-clock designs, in *Proc. IEEE Int. Conf. on Computer Design*, October 2005, pp. 475–478.

[Yang 2001] S. Yang, C. Papachristou, and M. Tabib-Azar, Improving bus test via I_{DDT} and boundary scan, in *Proc. IEEE Design Automation Conf.*, June 2001, pp. 307–312.

[Zanella 2000] S. Zanella, A. Nardi, A. Neviani, M. Quarantelli, S. Saxena, and C. Guardiani, Analysis of the impact of process variation on clock skew, *IEEE Trans. Semiconductor Manuf.*, 13(4), 401–407, 2000.

[Zarkesh-Ha 1999] P. Zarkesh-Ha, T. Mule, and J. Meindl, Characterization and modeling of clock skew with process variation, in *Proc. Custom Integrated Circuits Conf.*, May 1999, pp. 441–444.

[Zhao 2000a] S. Zhao and K. Roy, Estimation of switching noise on power supply lines in deep sub-micron CMOS circuits, in *Proc. IEEE Int. Conf. on VLSI Design*, January 2000, pp. 168–173.

[Zhao 2000b] Y. Zhao and S. Dey, Analysis of interconnect crosstalk defect coverage of test sets, in *Proc. IEEE Int. Test Conf.*, October 2000, pp. 492–501.

[Zhao 2000c] S. Zhao, K. Roy, and C. Koh, Estimation of inductive and resistive switching noise on power-supply network in deep submicron CMOS circuits, in *Proc. IEEE Int. Conf. on Computer Design*, September 2000, pp. 65–72.

[Zhao 2002] S. Zhao, K. Roy, and C. Koh, Power supply noise aware floorplanning and decoupling capacitance placement, in *Proc. Asia and South Pacific Design Automation Conf.*, January 2002, pp. 489–495.

[Zheng 2000] L. Zheng, B. Li, and H. Tenhunen, Efficient and accurate modeling of power supply noise on distributed on-chip power networks, in *Proc. Int. Symp. on Circuits and Systems*, May 2000, pp. 513–516.

[Ziegler 2004] J.F. Ziegler and H. Puncher, SER - History, Trends and Challenges, A Guide for Designing with Memory ICs, Cypress Semiconductors, San Jose, CA, 2004.

R12.4—FPGA Testing

[Abramovici 2001] M. Abramovici and C. Stroud, BIST-based test and diagnosis of FPGA logic blocks, *IEEE Trans. VLSI Syst.*, 9(1), 159–172, 2001.

[Abramovici 2003] M. Abramovici and C. Stroud, BIST-based delay fault testing in field programmable gate arrays, *J. Electron. Testing: Theory Appl.*, 19(5), 549–558, 2003.

[Chmelar 2003] E. Chmelar, FPGA interconnect delay fault testing, in *Proc. IEEE Int. Test Conf.*, September 2003, pp. 1239–1247.

[Harris 2002] I. Harris and R. Tessier, Testing and diagnosis of interconnect faults in cluster-based FPGA architectures, *IEEE Trans. Comput.-Aided Des.*, 21(11), 1337–1343, 2002.

[Huang 1998] W. Huang, F. Meyer, X. Chen, and F. Lombardi, Testing configurable LUT-based FPGAs, *IEEE Trans. VLSI Syst.*, 6(2), 276–283, 1998.

[Renovell 1998] M. Renovell, J. Portal, J. Figueras, and Y. Zorian, Testing the interconnect of RAM-based FPGAs, *IEEE Des. Test Comput.*, 15(1), 45–50, 1998.

[Stroud 2002b] C. Stroud, J. Nall, M. Lashinsky, and M. Abramovici, BIST-based diagnosis of FPGA interconnect, in *Proc. IEEE Int. Test Conf.*, October 2002, pp. 618–627.

[Stroud 2005a] C. Stroud, S. Garimella, and J. Sunwoo, On-chip BIST-based diagnosis of embedded programmable logic cores in system-on-chip devices, in *Proc. Int. Conf. on Computers and Their Applications*, March 2005, pp. 308–313.

[Stroud 2005b] C. Stroud and S. Garimella, Built-in self-test and diagnosis of multiple embedded cores in SoCs, in *Proc. Int. Conf. on Embedded Systems and Applications*, June 2005, pp. 130–136.

[Sun 2000] X. Sun, J. Xu, B. Chan, and P. Trouborst, Novel technique for BIST of FPGA interconnects, in *Proc. IEEE Int. Test Conf.*, October 2000, pp. 795–803.

[Tahoori 2004] M. Tahoori, Application-dependent diagnosis of FPGAs, in *Proc. IEEE Int. Test Conf.*, October 2004, pp. 645–649.

R12.5—MEMS Testing

[Aikele 2001] M. Aikele, K. Bauer, W. Ficker, F. Neubauer, U. Prechtel, J. Schalk, and H. Seidel, Resonant accelerometer with self-test, *Sensors Actuators*, A(92), 161–167, 2001.

[Charlot 2001] B. Charlot, S. Mir, F. Parrain, and B. Courtois, Electrically induced stimuli for MEMS self-test, in *Proc. IEEE VLSI Test Symp.*, April 2001, pp. 210–215.

[Chau 1998] K. H. L. Chau and R. E. Sulouff, Jr., Technology for the High-volume manufacturing of integrated surface-micromachined accelerometer products, *Microelectronics J.*, 29, 579–586, 1998.

[Deb 2002] N. Deb and R. D. (S.) Blanton, Built-in self test of CMOS-MEMS accelerometers, in *Proc. IEEE Int. Test Conf.*, October 2002, pp. 1075–1084.

[Kuehnel 1994] W. Kuehnel and S. Sherman, A surface micromachined silicon accelerometer with on-chip detection circuitry, *Sensors Actuators*, A(45), 7–16, 1994.

[Puers 2002] R. Puers and S. Reyntjens, RASTA: real-acceleration-for-self-test accelerometer—a new concept for self-testing accelerometers, *Sensors Actuators*, A(97–98), 359–368, 2002.

[Rosing 1999] R. Rosing, A. Richardson, A. Dorey, and A. Peyton, Test support strategies for MEMS, in *Proc. 5th Int. Mixed Signal Test Workshop*, June 1999, pp. 345–350.

[Rosing 2000] R. Rosing, A. Lechner, A. Richardson, and A. Dorey, Fault simulation and modeling of microelectromechanical systems, *Comput. Control Eng. J.*, 11(5), 242–250, 2000.

[Tewksbury 2001] S. K. Tewksbury, Challenges facing practical DFT for MEMS, in *Proc. Int. Symp. in Defect and Fault Tolerance in VLSI Systems*, October 2001, pp. 11–17.

[Xiong 2004] X. Xiong, Y. Wu, and W. Jone, A dual-mode built-in self-test technique for capacitive MEMS devices, in *Proc. IEEE VLSI Test Symp.*, April 2004, pp. 148–153.

R12.6—High-Speed I/O Testing

[Cai 2002] Y. Cai, B. Laquai, and K. Luehman, Jitter testing for gigabit serial communication transceivers, *IEEE Des. Test Comput.*, 19(1), 66–74, 2002.

[Cai 2005] Y. Cai, A. Bhattacharyya, J. Martone, A. Verma, and W. Burchanowski, Comprehensive production test solution for 1.5 Gb/s and 3.0 Gb/s serial ATA based on AWG and undersampling techniques, in *Proc. IEEE Int. Test Conf.*, October 2005, Paper 27.1, 8 pp.

[Gillis 1998] P. Gillis, F. Woytowich, K. McCauley, and U. Baur, Delay test of chip I/Os using LSSD boundary scan, in *Proc. IEEE Int. Test Conf.*, October 1998, pp. 83–90.

[Ilkbahar 2001] A. Ilkbahar, S. Venkataraman, and H. Muljono, Itanium™ processor system bus design, *IEEE J. Solid-State Circuits*, 36(10), 1565–1573, 2001.

[Kundu 2004] S. Kundu, T.M. Mak, and R. Galivanche, Trends in manufacturing test methods and their implications, in *Proc. IEEE Int. Test Conf.*, October 2004, pp. 679–687.

[Laquai 2001] B. Laquai, and Y. Cai, Testing gigabit multilane SerDes interfaces with passive jitter injection filters, in *Proc. IEEE Int. Test Conf.*, October 2001, pp. 297–304.

[Li 2004] M. Li, A. Martwick, G. Talbot, and J. Wilstrup, Transfer functions for the reference clock jitter in a serial link: theory and applications, in *Proc. IEEE Int. Test Conf.*, October 2004, pp. 1158–1167.

[Li 2005] M. Li, Statistical and system approaches for jitter, noise and bit error rate (BER) tests for high speed serial links and devices, in *Proc. IEEE Int. Test Conf.*, October 2005, Paper 3.3, 11 pp.

[Lin 2005] M. Lin, K.-T. Cheng, J. Su, M. C. Sun, J. Chen, and S. Lu, Production-oriented interface testing for PCI express by enhanced loop-back technique, in *Proc. IEEE Int. Test Conf.*, October 2005, Paper 27.3, 10 pp.

[Lin, 2003] H.-C. Lin, K. Taylor, A. Chong, E. Chan, M. Soma, H. Haggag, J. Huard, and J. Braat, CMOS built-in test architecture for high-speed jitter measurement, in *Proc. IEEE Int. Test Conf.*, October 2003, pp. 67–76.

[Mak 2004a] T.M. Mak, M. Tripp, and A. Meixner, Testing Gbps interfaces without a gigahertz tester, *IEEE Des. Test Comput.*, 21(4), 278–286, 2004.

[Mak 2004b] T.M. Mak, How do we test for adaptive computing?, *Workshop on Test Resource Partitioning*, April 2004, pp. 17–21.

[Moore 1965] G. Moore, Cramming more components onto integrated circuits, *Electronics*, 38, 114–117, 1965.

[Tripp 2004] M. Tripp, T.M. Mak, and A. Meixner, Elimination of traditional functional testing of interface timings at Intel, in *Proc. IEEE Int. Test Conf.*, October 2004, pp. 1448–1454.

R12.7—RF Testing

[Agilent EVMa-2005] Agilent, *8 Hints for Making and Interpreting EVM Measurements*, 5989-3144EN Appl. Note, Agilent Technologies, Palo Alto, CA, 2005.

[Agilent EVMb-2000] Agilent, *Using Error Vector Magnitude Measurements to Analyze and Troubleshoot Vector-Modulated Signals*, 5965-2898E PN 89400–14, Agilent Technologies, Palo Alto, CA, 2000.

[Agilent Rx-2002] Agilent, *Testing and Troubleshooting Digital RF Communications Receiver Designs (AN 1314)*, 5968-3579E Appl. Note, Agilent Technologies, Palo Alto, CA, 2002.

[Agilent SA-2005] Agilent, *Spectrum Analyzer Basics (AN 150)*, 5952–0292 Appl. Note, Agilent Technologies, Palo Alto, CA, 2005.

[Agilent Tx-2002] Agilent, *Testing and Troubleshooting Digital RF Communications Transmitter Designs (AN 1313)*, 5968-3578E Appl. Note, Agilent Technologies, Palo Alto, CA, 2002.

[Agilent WiMax-2005] Agilent, *WiMAX Concepts and RF Measurements – IEEE 802.16–2004 WiMAX PHY Layer Operation and Measurements*, 5989-2027EN Appl. Note, Agilent Technologies, Palo Alto, CA, 2005.

[Agilent Y-2004] Agilent, *Noise Figure Measurement Accuracy: The Y-Factor Method*, 5952-3706E Appl. Note, Agilent Technologies, Palo Alto, CA, 2004.

[Akbay 2004] S. S. Akbay, A. Halder, A. Chatterjee, and D. Keezer, Low-cost test of embedded RF/analog/mixed-signal circuits in SOPs, *IEEE Trans. on Advanced Packaging*; see also: Advanced packaging, *IEEE Trans. on Comp. Packag. Manuf. Technol., Part B*, 27(2), 352–363, 2004.

[Bhattacharya 2004] S. Bhattacharya and A. Chatterjee, Use of embedded sensors for built-in-test RF circuits, in *Proc. IEEE Int. Test Conf.*, October 2004, pp. 801–809.

[Bhattacharya 2005] S. Bhattacharya, R. Senguttuvan, and A. Chatterjee, Production test technique for measuring BER of ultra-wideband (UWB) devices, *IEEE Trans. Microwave Theory Techn.*, 53(11), 3474–3481, 2005.

[Cherubal 2004] S. Cherubal, R. Voorakaranam, A. Chatterjee, J. McLaughlin, J. L. Smith, and D. M. Majernik, Concurrent RF test using optimized modulated RF stimuli, in *Proc. IEEE Int. Conf. on VLSI Design*, January 2004, pp. 1017–1022.

[Cho 2005] C. Cho, W. R. Eisenstadt, B. Stengel, and E. Ferrer, IIP3 estimation from the gain compression curve, *IEEE Trans. Microwave Theory Techn.*, 53(4, Part 1), 1197–1202, 2005.

[Dabrowski 2003] J. Dabrowski, Loopback BIST for RF front-ends in digital transceivers, in *Proc. Int. Symp. on System-on-Chip*, November 2003, pp. 143–146.

[Ferrario 2001] J. Ferrario, R. Wolf, and H. Ding, Moving from mixed signal to RF test hardware development, in *Proc. IEEE Int. Test Conf.*, October 2001, pp. 948–956.

[Ferrario 2002] J. Ferrario, R. Wolf, and S. Moss, Architecting millisecond test solutions for wireless phone RFICs, in *Proc. IEEE Int. Test Conf.*, October 2002, pp. 1151–1158.

[Ferrario 2003] J. Ferrario, R. Wolf, S. Moss, and M. Slamani, A low-cost test solution for wireless phone RFICs, *IEEE Commun. Mag.*, 41(9), 82–88, 2003.

[Halder 2003] A. Halder, S. Bhattacharya, and A. Chatterjee, Automatic multitone alternate test-generation for RF circuits using behavioral models, in *Proc. IEEE Int. Test Conf.*, September 2003, pp. 665–673.

[Halder 2005a] A. Halder and A. Chatterjee, Low-cost alternate EVM test for wireless receiver systems, in *Proc. IEEE VLSI Test Symp.*, April 2005, pp. 255–260.

[Halder 2005b] A. Halder, S. Bhattacharya, G. Srinivasan, and A. Chatterjee, A system-level alternate test approach for specification test of RF transceivers in loopback mode, in *Proc. Int. Conf. on VLSI Design*, January 2005, pp. 289–294.

[Kasten 1998] J. S. Kasten and B. Kaminska, An introduction to RF testing: device, method and system, in *Proc. IEEE VLSI Test Symp.*, April 1998, pp. 462–468.

[Lau 2002] W. Y. Lau, Measurement challenges for on-wafer RF-SOC test, in *Proc. of Int. Electronics Manufacturing Technology Symp.*, July 2002, pp. 353–359.

[Maxim 2003] Maxim, *Three Methods of Noise Figure Measurement*, Appl. Note AN2875, Maxim Integrated Products, Sunnyvale, CA, 2003.

[Ozev 2001] S. Ozev, C. Olgaard, and A. Orailoglu, Testability implications in low-cost integrated radio transceivers: a Bluetooth case study, in *Proc. IEEE Int. Test Conf.*, October 2001, pp. 965–974.

[Ozev 2004] S. Ozev and C. Olgaard, Wafer-level RF test and DFT for VCO modulating transceiver architectures, in *Proc. IEEE VLSI Test Symp.*, April 2004, pp. 217–222.

[RS 1EF40-1998] Rhode & Schwarz, *Measurement of Adjacent Channel Power on Wideband CDMA Signals*, Appl. Note 1EF40-0E, Josef Wolf, Rhode & Schwarz GmbH & Co. KG, Munich, Germany, 1998.

[RS 1MA40-2002] Rhode & Schwarz, *Testing Power Amplifiers for 3G Base Stations*, Appl. Note 1MA40-0E, Rhode & Schwarz GmbH & Co. KG, Munich, Germany, 2002.

[RS 7BM03-2002] Rhode & Schwarz, *Bit Error Ratio BER in dVb as a Function of S/N*, Appl. Note 7BM03-3E, Rhode & Schwarz GmbH & Co. KG, Munich, Germany, 2002.

[Sylla 2003] I. T. Sylla, Building an RF source for low cost testers using an ADPLL controlled by Texas Instruments digital signal processor (DSP) TMS32OC5402, in *Proc. IEEE Int. Test Conf.*, October 2003, pp. 659–664.

[Valdes-Garcia 2005] A. Valdes-Garcia, R. Venkatasubramanian, R. Srinivasan, J. Silva-Martinez, and E. Sanchez-Sinencio, A CMOS RF RMS detector for built-in testing of wireless transceivers, in *Proc. IEEE VLSI Test Symp.*, April 2005, pp. 249–254.

[Variyam 2002] P. N. Variyam, S. Cherubal, and A. Chatterjee, Prediction of analog performance parameters using fast transient testing, *IEEE Trans. Comput.-Aided Des.*, 21(3), 349–361, 2002.

[Veillette 1995] B. R. Veillette and G. W. Roberts, A built-in self-test strategy for wireless communication systems, in *Proc. IEEE Int. Test Conf.*, October 1995, pp. 930–939.

[Voorakaranam 2002] R. Voorakaranam, S. Cherubal, and A. Chatterjee, A signature test framework for rapid production testing of RF circuits, in *Proc. IEEE Design, Automation and Test in Europe Conf.*, March 2002, pp. 186–191.

[Yin 2005] Q. Yin, W. R. Eisenstadt, R. M. Fox, and T. Zhang, A translinear RMS detector for embedded test of RF ICs, *IEEE Trans. Instrument. Measure.*, 54(5), 1708–1714, 2005.

R12.8—Concluding Remarks

[Gupta 2003] P. Gupta and A. B. Kahng, Manufacturing-aware physical design, in *Proc. IEEE Int. Conf. on Comput.-Aided Des.*, November 2003, pp. 681–687.

[SIA 2003] SIA, *The International Technology Roadmap for Semiconductors: 2003 Edition—Design*, Semiconductor Industry Association, San Jose, CA (http://public.itrs.net), 2003, pp. 30–36.

[SIA 2004] SIA, *The International Technology Roadmap for Semiconductors: 2004 Update*, Semiconductor Industry Association, San Jose, CA (http://public.itrs.net), 2004.

[Zorian 2004] Y. Zorian, Optimizing manufacturing by design for yield, in *Proc. Int. Symp. on Electronics Manufacturing Technology*, July 2004, pp. 255–258.

Index

0-controllability, 153
0-control point, 267
1-controllability, 153
1-control point, 267
5-valued algebra, 196–197
1057 standard. *See* IEEE 1057 standard
1149.1 standard. *See* IEEE 1149.1 standard
1149.5 standard. *See* IEEE 1149.5 standard
1149.6 standard. *See* IEEE 1149.6 standard
1500 standard. *See* IEEE 1500 standard

A
ABM (analog boundary module), 662–664
AC (alternate current) faults, 21
accuracy, 728
AC I/O loopback testing, 725
AC measurement circuitry, 26
AC parametric testing, 635–641
  frequency response measurement, 637–639
  intermodulation distortion measurement, 641
  maximal output amplitude measurement, 636–637
  overview, 635–636
  SNR and distortion measurement, 639–641
acquisition time, 645
adaptive diagnostic approach, 448
adaptive scan, in test compression, 385–386
ADC and DAC circuit structure, 644–647
  ADC, 645–646
  ADC circuit structure, 646–647
  DAC circuit structure, 646
  filter, 644–645
  gain stage, 644
  MUX stage, 645

overview, 644
  S/H stage, 645
ADC/DAC specification and fault models, 647–652
  differential nonlinearity error (DNL), 648–650
  gain error, 648
  hysteresis error, 651
  intermodulation distortion, 652
  load-dependent error, 650–651
  missing code, 651–652
  nonlinearity error, 648
  offset error, 648
  overview, 647–648
  resolution error, 651
  signal-to-noise and distortion, 652
  signal-to-noise ratio, 652
  total harmonic distortion, 652
ADCs (analog-to-digital converters), 29, 269, 619, 641
address (ADDR), 540
address decoder faults (AFs), 464, 518
address faults (AFs), 489
address orders, 467
address remapping unit (ARU), 543
ad hoc approach, in design for testability, 30, 51–53, 96
  overview, 51
  test point insertion, 51–53
adjacent channel power ratio (ACPR), 733
AFs (address faults), 464, 489, 518
AGC (automatic gain control), 622, 629
aggressor (AGR), 418, 477, 519
AGR (aggressor), 418, 477, 519
alias, 290
aliasing probability, 291, 292, 294, 295

aliasing problem, 294, 322
aligned double-capture, 316–317, 323
aligned skewed-load, 312–314
almost full-scan design, 39, 59
alpha-particle radiation, 692
alternate current (AC) faults, 21
alternate test, 735
amplitude modulation (AM), 631
AMS (analog and mixed-signal), 92
AMS (analog and mixed-signal) circuit, 737
analog and mixed-signal (AMS), 92
analog and mixed-signal (AMS) circuit, 737
analog and mixed-signal circuit testing, 29–31
analog blocks, 267–268
analog boundary module (ABM), 662–664
analog boundary scan. *See* IEEE 1149.4 standard
analog circuit properties, 620–623
  absence of suitable fault model, 622–623
  complicated cause–effect relationship, 622
  continuous signals, 621
  feedback ambiguity, 622
  large range of circuits, 621
  nonlinear characteristics, 621–622
  overview, 619–620
analog circuit testing, 627–641
  AC parametric testing, 635–641
  frequency response measurement, 637–639
  intermodulation distortion measurement, 641
  maximal output amplitude measurement, 636–637

overview, 635–636
SNR and distortion
measurement, 639–641
analog test approaches,
627–629
analog test waveforms,
629–631
DC parametric testing,
631–635
common mode rejection
ratio measurement,
634–635
open-loop gain
measurement, 632–633
overview, 631–632
power supply
rejection ratio
measurement, 635
unit gain bandwidth
measurement, 632–633
overview, 627
analog defect mechanisms, and
fault models, 623–627
hard faults, 625
overview, 623–624
soft faults, 625–627
analog test buses, 662
analog test receiver, 580–581
analog test waveforms, 629–631
analog-to-digital
conversion, 645
analog-to-digital converters
(ADCs), 29, 267,
619, 641
aperture time, 645
application-dependent test, 708
application-independent
testing, 708
application-specific integrated
circuit (ASIC), 21, 236,
462, 559
apply command, 429–430
apply event, 595
arbitrary waveform generators
(AWGs), 631, 636
arbitrary waveforms, 631
ARU (address remapping
unit), 543
ascending (⇑) address
sequences, 504
ASICs (application-specific
integrated circuits), 21,
236, 462, 559, 560
asynchronous clock domains,
310, 314
asynchronous set/reset signals,
75–76, 268–269
asynchronous SRAM, 519
ATAP (analog test access
port), 662
ATE. See automatic test
equipment (ATE)

ATPG. See automatic
test pattern
generation (ATPG)
ATPG algorithm, 27, 236
at-speed delay test
technique, 311
at-speed failure, 697
at-speed memory BIST, 737
at-speed scan testing, 737
at-speed testing, 310, 312–313,
315, 697, 737
automatic gain control (AGC),
622, 629
automatic test equipment
(ATE), 25–27, 39,
341, 342, 518, 586,
662, 724
accurate DC and AC
measurement
circuitry, 26
computer, 25
configurability, 26
digital signal processor
(DSP), 26
modularization, 26
overview, 25
parallel test
capabilities, 26
pin electronics and
fixtures, 25
test program, 25–26
third-party components,
26–27
automatic test pattern
generation (ATPG),
27–28, 38, 161, 441
for acyclic sequential
circuits, 247
for non-stuck-at faults,
231–245
ATPG for transition faults,
238–239
bridging fault ATPG,
244–245
designing ATPG that
captures delay defects,
231–237
overview, 231
transition ATPG using
stuck-at ATPG, 240
transition ATPG using
stuck-at vectors,
240–244
automatic test pattern
generation (ATPG)
software, 264
AUXCKn (auxiliary clock),
589–590
average-sum filtering,
438–439
AWGs (arbitrary waveform
generators), 631, 636

B
BAC (BIST activation control)
input, 493
background data sequence
(BDS), 21
background pattern, 473
background word, 473
backtrace algorithm, 407,
408–409
backtrace( ) function, 184
backtracking, 172–173, 175
bad events, 144
bad gates, 143
balanced partial-scan design,
39, 66
bare board, 31
base cell (BC), 468
BC (base cell), 468
BCK (BIST clock) input, 494
BCS (BIST control selection)
input, 493
BDD (binary decision
diagram)-based
approach, 202
BDN (BIST Done), 541
BDS (background data
sequence), 21
BEF (BIST Error Flag)
signal, 525
behavioral level, 22–23, 106
bench test, 627
BER (bit error rate), 722, 727
BERT (bit error rate
tester), 724
BEST (built-in evaluation and
self-test), 297, 299
BGO (BIST Go/No-Go)
signal, 522
BGO (Go/No-Go indicator
signal), 494
BI (burn-in) mode, 495
BID (BIST intermediate
description) data
format, 501, 503
bidirectional I/O ports, 71, 271
BILBO (built-in logic block
observer), 299–300, 301
binary counter, 275
binary decision diagram
(BDD)-based
approach, 202
binary phase shift keying
(BPSK), 631
binary search, 443
BIRA (built-in redundancy
analysis), 517, 518,
542–545
BISD (built-in
self-diagnosis), 462
BISR (built-in self-repair), 517,
518, 537–555, 737

BIRA module, 542–545
BISR architecture and
        procedure, 538–541
BIST module, 541–542
an industrial case, 545–548
overview, 537
redundancy organization,
        537–538
repair rate and yield, 548–555
BIST (built-in self-test), 30, 39,
        96, 341, 397, 442, 679,
        695, 708
BIST (built-in-self-test), 736
BIST (built-in self-test), 737
architectures and functions,
        493–495
BIST module, 541–542
design strategy and RAM
        specification, 489–493
with diagnostic support,
        521–525
    controller, 521–523
    fault site indicator (FSI),
        524–525
    overview, 521
    test pattern generator,
        523–524
implementation in memory
        built-in self-test,
        495–500
logic BIST diagnosis, 442–459
    interval-based methods,
        443–446
    masking-based methods,
        446–459
    overview, 442–443
BIST (built-in self-test)
        features, 266
BIST (built-in self-test)
        generators, 562
BIST (built-in self-test)
        schemes, 488
BIST (logic built-in
        self-test), 265
BIST activation control (BAC)
        input, 493
BIST clock (BCK) input, 494
BIST controller, 495
BIST control selection (BCS)
        input, 493
BIST design rules, 267
Bist for RAm IN Seconds
        (BRAINS), 462, 501
BIST Go/No-Go (BGO)
        signal, 522
BIST intermediate description
        (BID) data format, 501
BIST intermediate description
        (BID) format, 503
BIST length, 445
BIST logic, 463

BIST Mode Select (BMS)
        signal, 521
BIST Normal Selection (BNS)
        signal, 541
BIST pattern generation, 266
BIST ready flag (BRD*), 494
BIST Reset (BRS) signal, 522
BIST scan-in (BSI) input, 493
BIST scan-out (BSO)
        output, 493
BIST Serial-In (BSI)
        terminal, 521
BIST Serial-Out (BSO)
        terminal, 521
BIST-specific design rules, 266,
        267, 320
BIST State Control (BSC)
        input, 522
BIST templates, 501
BIST timing control
        diagrams, 263
bit error rate (BER), 722
bit error rate tester
        (BERT), 724
bit-line short fault (BSF), 473
bitmap, 532
bitwise parallel simulation, 116
blocks under test (BUTs),
        708–709
BMS (BIST Mode Select)
        signal, 521
BNS (BIST Normal Selection)
        signal, 541
board testing, 31
Boolean difference,
        166–168
Boolean operators, 236
borophosphosilicate glass
        (BPSG), 701
bottom-top LFSR, 274
boundary cells, 580
boundary scan, 31, 32–36, 561,
        579–585, 708
    See also IEEE 1149.1
        standard; IEEE 1149.4
        standard; IEEE 1149.6
        standard
    overview, 557–559, 579
boundary-scan cells (BSCs),
        561, 565
boundary-scan description
        language (BSDL), 32,
        558, 574
boundary-scan register (BSR),
        561, 562, 565
BPSG (borophosphosilicate
        glass), 701
BPSK (binary phase shift
        keying), 631
BRAINS (Bist for RAm IN
        Seconds), 462, 501

BRAVES (Built-in Redundancy
        Analysis Verification
        and Evaluation
        System), 535
BRD* (BIST ready flag), 494
bridging fault, 17
bridging fault diagnosis,
        416–418
bridging tests, 697
broadcaster, 363
broadcast mode, 361
broadcast scan, 359, 360
broadcast-scan-based schemes,
        344, 359–364, 377
    broadcast scan, 359–360
    Illinois scan, 360–362
    multiple-input broadcast
        scan, 362
    overview, 359
    reconfigurable broadcast
        scan, 362–363
    virtual scan, 363–364
broad-side capture, 88, 315
broadside-load, 86
broad-side test, 687
BRS (BIST Reset) signal, 522
BSC (BIST State Control)
        input, 522
BSCs (boundary-scan cells),
        561, 565
BSDL (boundary-scan
        description language),
        32, 558, 574
BSF (bit-line short fault), 473
BSI (BIST scan-in) input, 493
BSI (BIST Serial-In)
        terminal, 521
BSO (BIST scan-out)
        output, 493
BSO (BIST Serial-Out)
        terminal, 521
BSR (boundary-scan register),
        561, 562, 565
built-in evaluation and self-test
        (BEST), 297
built-in logic block observer
        (BILBO), 299–300
built-in redundancy analysis
        (BIRA), 517, 518,
        542–545
Built-in Redundancy Analysis
        Verification and
        Evaluation System
        (BRAVES), 535
built-in self-diagnosis
        (BISD), 462
built-in self-repair (BISR), 517,
        518, 537–555, 737
    BIRA module, 542–545
    BISR architecture and
        procedure, 538–541
    BIST module, 541–542
    an industrial case, 545–548

overview, 537
redundancy organization, 537–538
repair rate and yield, 548–555
built-in self-test. *See* BIST (built-in self-test)
built-in-self-test (BIST), 736
built-in self-test (BIST) features, 264
built-in self-test (BIST) generators, 562
built-in self-test (BIST) schemes, 488
burn-in (BI) mode, 495
bus conflict, 114
bus contention, 71
BUTs (blocks under test), 708–709
butterfly algorithm, 469
BYPASS instruction, 569, 665
bypass logic, 267
bypass register, 562, 565, 566
Byzantine general's phenomenon, 427

C
CA (cellular automaton), 280, 281
CAD (computer-aided design), 5, 574
calibration error, 648
candidate test algorithm, 485
capture aligned skewed-load, 313
capture command, 428
Capture-DR state, 568
capture event, 593–594
Capture operation, 566
Capture/Update operations, 561
CaptureWR, 590
CAS (Core Access Switch), 606
CAS-before-RAS (CBR), 499
catastrophic errors, 625
catastrophic faults, 21
cause–effect analysis, 397, 401–405
CBILBO (concurrent BILBO), 302
CBR (CAS-before-RAS), 499
CC (common clock) type, 720
CD (core disconnect), 664
CDR (core data registers), 591
cell fault, 535
cell functional input (CFI), 592
cell functional output (CFO), 592
cell test input (CTI), 592
cell test output (CTO), 592
cellular automata, 278–280
cellular automaton (CA), 278, 279

center test controller (CTC), 608
centralized and separate board-level BIST architecture (CSBL), 296
central processing unit (CPU), 704
CF (clock forwarding) scheme, 720
CFI (cell functional input), 592
CFI (circuit function independent), 366
CFid (idempotent coupling fault), 465, 519
CFin (inversion coupling fault), 465, 518
CFO (cell functional output), 592
CFS (circuit-function-specific), 366
CFs (coupling faults), 21, 465, 489
CFSRs (complete LFSRs), 275–277
CFst (state coupling fault), 465, 519
CG (conversion gain), 731
characteristic polynomial, 273, 293, 374
CHECK_FULL state, 544
CHECK_RMR state, 543, 544
checkerboard algorithm, 467
checkerboard pattern, 467
checking experiment, 50, 53
checkpointing, 703
chip-level strategy, 418–425
chip-level diagnostic flow, 424–425
direct partitioning, 418–420
overview, 418
two-phase strategy, 420–424
aliasing SIC points example, 421
definitions, 420–422
examples, 422–424
observations, 420–421
output group example, 422
chirp, 631
chromosomes, 208
circuit function independent (CFI), 366
circuit-function-specific (CFS), 366
circuit modeling, 198
circuit modification, 198
circuit reduction, 424
circuit structures, IEEE 1149.4
analog boundary module (ABM), 662–664
analog test access port (ATAP), 662

analog test buses (AB1/AB2), 662
test bus interface circuit (TBIC), 662
circuits without scan chains, BIST architectures for, 296–297
built-in evaluation and self-test (BEST), 297
centralized and separate board-level BIST architecture, 296
overview, 296
circuits with scan chains, BIST architectures for, 297–298
LSSD on-chip self-test, 297–298
overview, 297
self-testing using MISR and parallel SRSG, 298
circuit under diagnosis (CUD), 399
circuit under test (CUT), 2, 10, 163, 265, 341, 463, 626
circular BIST architectures, 303
circular self-test path (CSTP), 302–303
CLAMP instruction, 573, 665, 666
CLK (Clock), 541
Clock (CLK), 541
clocked full-scan designs, 62
clocked-scan, 55
clocked-scan cell designs, 56–57
clock forwarding (CF) scheme, 720
clock gating block, 323–325
clock grouping, 78, 93
clock recovery (CR) circuit, 722
clock skew, 82
clock suppression, 323
clock suppression circuit, 313
clock tree synthesis (CTS), 84
CMOS (complementary metal oxide semiconductor) devices, 8
CMRR (common mode rejection ratio), 634
Cocktail–March algorithms, 473, 485–487, 517
code-based schemes, 344, 345–351
dictionary code (fixed-to-fixed), 345–346
Golomb code (variable-to-variable), 350–351
Huffman code (fixed-to-variable), 346–349

overview, 345
run-length code
        (variable-to-fixed),
        349–350
code bins, 654
code generation, 124–125
    approach 1—high-level
        programming language
        source code, 124
    approach 2—native machine
        code, 124
    approach 3—interpreted
        code, 124–125
code level, 654
code transition level, 654
code transition level test, 655
codewords, 345
column-first algorithm, 529
combinational circuits,
        designing stuck-at
        ATPG for, 169–194
    basic ATPG algorithm,
        173–177
    D algorithm, 177–182
    dynamic logic implications,
        191–194
    FAN, 186–187
    naive ATPG algorithm,
        169–173
    overview, 169
    PODEM, 182–186
    static logic implications,
        187–191
combinational controllability
        and observability
        calculation, 41–43
combinational feedback loops,
        74–75, 268
combinational linear
        decompression,
        355, 363
combinational logic diagnosis,
        401–427, 430
    cause–effect analysis,
        401–405
        example 7.1, 402–403
        overview, 401–402
    chip-level strategy, 418–420
        direct partitioning, 418–420
        overall chip-level diagnostic
            flow, 424–425
        overview, 418
        two-phase strategy,
            420–424
    compaction and compression
        of fault dictionary,
        403–405
        detection dictionary, 405
        overview, 403–404
        P&R compression
            dictionary, 404–405
        pass–fail dictionary, 404

diagnostic test pattern
        generation, 425–426
effect–cause analysis,
        405–418
    backtrace algorithm,
        408–409
    inject-and-evaluate
        paradigm, 409–418
    overview, 405–406
    structural pruning, 407–408
overview, 401
combined LFSR/PS approach,
        283–284
combined LFSR/SR
        approach, 283
combined linear and nonlinear
        decompressors,
        357–359
common clock (CC) type, 720
common mode noises, 722
common mode rejection ratio
        (CMRR), 634
common mode rejection ratio
        measurement, 634–635
compaction, 290, 343
compaction filters, 226
COMPARE state, 542
compatibility classes, 287
compatible LFSR, 287
compatible vectors, 246
compiled-code simulation,
        121–125
    code generation, 124–125
    vs. event-driven
        simulation, 129
    logic levelization, 123–124
    logic optimization, 121–123
    overview, 121
compiler engine, 501
complementary metal oxide
        semiconductor
        (CMOS) devices, 8
complete LFSRs (CFSRs),
        275–277
complex network measurement,
        671–672
complex programmable logic
        devices (CPLDs), 710
compression, 290
compression ratio, 361
computer-aided design (CAD),
        5, 574
concurrent built-in logic block
        observer (CBILBO),
        300–302
concurrent checking, 296,
        303–304
concurrent fault simulation, 28,
        143–146, 147, 698
concurrent online BIST, 264
concurrent self-verification
        (CSV), 303–304

condensed LFSR, 284–285
cone intersection, 407
cone union, 407
configurability, automatic test
        equipment, 26
conflict analysis, 173–175
connected component, 420
constant-weight code, 282
constant-weight counter (CWC),
        282–283
continuous signals, 621
control-dominant circuits, 226
controllability, 38, 40
controllability observability
        program (COP)
        testability
        measures, 307
controller (CTR), 495, 501,
        521, 541
controlling case, 408
controlling rule, 408
controlling value, 112
control-only scan point, 267
control point (CP), 51, 305
control point activation, 307
conventional redundancy
        analysis algorithms,
        529–531
converge, 145
conversion gain (CG), 731
conversion range, 642
COP (controllability
        observability program)
        testability
        measures, 309
Core Access Switch (CAS), 606
core-based testing, 559–561,
        585–611
core data registers (CDR), 591
core disconnect (CD), 664
Core Instance block, 603
Core Test Language (CTL), 557,
        560, 587, 601–603
core test supporting, 603–606
correlated signals, 175
correlation, 629
coupling faults (CFs), 21,
        465, 489
CP (control point), 51, 307
CPLDs (complex programmable
        logic devices), 710
CPU (central processing
        unit), 704
CR (clock recovery) circuit, 722
CRC (cyclic redundancy check),
        294, 295, 704, 722
criminal detection, 401
CRIS, 216
critical net, 152
critical path tracing, 152–153,
        408
critical value, 152

crossover, 208
crosstalk, 8, 719
crosstalk delays, 20
crosstalk glitches, 20
crosstalk noises, 680
CSBL (centralized and separate board-level BIST architecture), 296, 298
CSTP (circular self-test path), 302–303
CSV (concurrent self-verification), 303–304, 305
CTC (center test controller), 608
CTI (cell test input), 592
CTL (Core Test Language), 557, 560, 587, 601–603
CTO (cell test output), 592
CTR (controller), 495, 501, 521, 541
CTS (clock tree synthesis) process, 84
CUD (circuit under diagnosis), 399
curable output, 411
curable output count, 415
curable output number, 411
curable outputs, 411, 413, 414
curable-vector-based metric, 413
curable vectors, 413–414, 416
cure injection, 411
current buffers, 672
CUT (circuit under test), 2, 10, 163, 267, 341, 463, 626
CWCs (constant-weight counters), 282–283, 284
cycle-based simulation, 129
cyclic code checker, 293
cyclic codes, 285
cyclic LFSRs, 285–287
cyclic redundancy check (CRC), 292, 293, 704, 722

D
D (digital components), 659
DAC circuit structure, 646
DACs (digital-to-analog converters), 29, 619, 641
DAG (directed acyclic graph), 48, 65
daisy-chain clock triggering, 323
daisy-chained configuration, 604
D algorithm, 27, 177–182
D-Alg-Recursion( ) function, 179
data background, 473
data-dominant designs, 226
data input (D) channels, 489

data input (DI), 55
data output (Q), 503
data pattern (D), 503
data retention fault (DRF), 465
dB (decibel), 731
dBm, 731
DC (direct current) faults, 21
DCG (directed cyclic graph), 65
DC measurement circuitry, 26
DC parametric testing, 631–635
    common mode rejection ratio measurement, 634–635
    open-loop gain measurement, 632–633
    overview, 631–632
    power supply rejection ratio measurement, 635
    unit gain bandwidth measurement, 632–633
DDR (double-data-rate) DRAM, 503
decibel (dB), 731
decision feedback equalizer (DFE), 727
decision tree, 171
decompressor, 342
deductive fault simulation, 28, 139–143, 149–150
defect analysis, 683
defect and error tolerance, 705
defect-based tests, 697–698, 726
defect level, 5, 629, 692
defects, 2, 3, 696
defects per million (DPM) goal, 692, 698
defect tolerance, 705
delay defects, 231–237
    ATPG for path-delay faults, 236–237
    classification of path-delay faults, 233–236
    overview, 231–233
delay fault models, 687–691
delay faults, 19–20, 310
delay fault testing, 288–290
delay testing, 685–691
    delay fault models, 687–691
    overview, 685–686
    test application schemes for testing delay defects, 686–687
dependency graph, 419
dependent faults (DFs), 418, 420
derived clocks, 74
descending ($\Downarrow$) address sequences, 504
designated length, 443
designated seeds, 443
design-for-diagnosis (DFD), 431, 446, 738
design for manufacturability (DFM) method, 398, 681, 737, 738

design for testability (DFT), 5, 29, 50, 95, 692
design for testability (DFT) circuitry, 401
design for testability (DFT) methods, 162, 681, 737
design for testability (DFT) purposes, 224
design for testability (DFT) structures, 686
design for testability (DFT) techniques, 37, 161, 263
design for testability (DFT) tools, 359
design for yield enhancement (DFY), 738
design verification, 106
detectable transition fault, 238
detection dictionary, 405
deterministic activation, 307
deterministic algorithms, 207
deterministic jitter (DJ), 719
deterministic masking, 446, 448–459
deterministic phases, 230
deterministic testability, 45
deterministic test generator, 226
deviation faults, 625–626
device identification register, 565, 665
device-ID register, 562, 566
Device Under Test (DUT), 667, 681, 683, 724, 735
DFD (design-for-diagnosis), 431, 446, 738
DFE (decision feedback equalizer), 727
D flip-flop, 43
DFM (design for manufacturability) method, 398, 681, 737, 738
D-frontier, 177
DFs (dependent faults), 418, 420
DFT (design for testability), 5, 29, 50, 95, 692
DFT (design for testability) circuitry, 401
DFT (design for testability) methods, 162, 681, 737
DFT (design for testability) purposes, 224
DFT (design for testability) structures, 686
DFT (design for testability) techniques, 37, 161, 265
DFT (design for testability) tools, 359

DFY (design for yield enhancement), 738
diagnosability ratio (DR), 526
diagnosis, 8
diagnosis, reliability screens, and yield improvement, 685
diagnostic logic, 265, 325–326
diagnostic resolution, 400
diagnostic test algorithms, 518–521
diagnostic test pattern generation (DTPG), 425–426
diagnostic test procedures, 430, 431
diagnostic test sequences, 434
diagnostic tree, 402
dictionary code (fixed-to-fixed), 345–346
dictionary size problem, 403
difference frequency, 437
difference profile, 437, 438, 439, 440
difference vector set, 350
differential fault simulation, 146–148
differential nonlinearity error (DNL), 648–650
DIGATE, 218–220
digital boundary scan. *See* IEEE 1149.1 standard
digital boundary-scan standard, 558
digital circuit testing, 28–29
digital components (D), 659
digital driver logic, 580, 581–582
digital receiver logic, 582–584
digital signal processors (DSPs), 26, 635, 685
digital-to-analog converters (DACs), 29, 619, 641
direct-access configuration, 604
direct current (DC) faults, 21
directed acyclic graph (DAG), 48, 65
directed cyclic graph (DCG), 65
direct implications, 189
direct partitioning, 418–420
disruptive device technologies, 683
distinguishable under other test algorithms, 527
distortion, 652
distortion measurement, 639–641
distributed delay defects, 691
distributed fault simulation, 150
diverge, 144–145
divide-and-conquer strategy, 419

DJ (deterministic jitter), 719
DNL (differential nonlinearity error), 648–650
dominant-AND/dominant-OR bridging fault, 18
dominant bridging fault model, 18
dominators, 196
double-capture, 88, 315–317
    aligned, 316–317
    one-hot, 315–316
    overview, 315
    staggered, 317
double-capture clocks, 323
double-data-rate (DDR) DRAM, 503
double-fix candidate, 418
double-latch design, 62
double-length LFSR, 288
down-conversion mixer, 729
down transition (DTR), 479
DPM (defects per million) goal, 692, 698
DR (diagnosability ratio), 526
DRAMs (dynamic random-access memories), 270, 499–500, 705
DRF (data retention fault), 465
droop rate, 645
drop-on-k heuristic, 405
drowsy cache, 680
drowsy fault, 680
DSPs (digital signal processors), 26, 635, 685
DTPG (diagnostic test pattern generation), 425–426
DTR (down transition), 479
dual code, 285
dual-mode BIST technique, 714–716
DUT (device under test), 667, 681, 683, 724, 735
duty cycle, 82
dynamic compaction, 246, 363, 377
dynamic fault, 465
dynamic hazard, 130
dynamic hazard detection, 132
dynamic implications, 192
dynamic partial reconfiguration, 706
dynamic random-access memories (DRAMs), 268, 499–500, 705
dynamic reconfiguration, 362

E
EB (extended backward) implications, 190
ECC (error correcting code), 90, 703, 722

EDA (electronic design automation) vendors,50
EDC (error detecting code), 703
edge detection, 434, 438, 439–441
edge placement, 27
edge rate, 630
edge set, 26
edge-triggered muxed-D scan cell design, 55
EDO (extended-data-out), 489
EDRAM (embedded DRAM), 462
EDT (embedded deterministic test), 379–382
EDTC (embedded time to digital converter), 696
effect–cause analysis, 397, 401, 405–418
    backtrace algorithm, 408–409
    inject-and-evaluate paradigm, 409–418
        bridging fault diagnosis using multiplets, 416–418
        computational details, 412–413
        curable-vector-based metric, 413
        ranking heuristic for multiple-fault diagnosis, 414–415
        reward-and-penalty principle, 415
    mismatched output, 406
    overview, 405–406
    structural pruning, 407–408
effective error polynomial, 295
effective fault coverage, 10
effective input sequence, 295
effective number of bits (ENOB), 652
E-frontier (evaluation frontier), 193
eight-valued logic, 132
electromagnetic interference, 694
electronic design automation (EDA) vendors, 50
electronic system level (ESL), 106
electronic system manufacturing process, 6
ELT (embedded logic test) technology, 386
EMA (Export Mask Address), 539, 543, 547
embedded core test standard. *See* IEEE 1500 standard

embedded deterministic test
    (EDT), 379–382
embedded DRAM
    (EDRAM), 462
embedded golden
    signature, 296
embedded logic test (ELT)
    technology, 386
embedded software-based
    self-testing, 737
embedded SRAM
    (ESRAM), 462
embedded time to digital
    converter (EDTC), 696
EN (clock enable signal), 71
encoding efficiency, 354
ending states, 221
end-of-algorithm (EOA)
    command, 498
enhanced-scan, 87–88, 232, 239
ENOB (effective number of
    bits), 652
EOA (end-of-algorithm)
    command, 498
EOP (Error Operation Protocol)
    data, 525
EOP (Error Operation Protocol)
    registers, 523–524
EP (essential pivot), 532
EPG (exhaustive pattern
    generator), 277
equivalent faults, 12–14
ERR (error indicator), 525,
    539, 541
error, 9–10
error bitmaps, 526, 528
error cancellation, 370
error correcting code (ECC), 90,
    703, 722
error detecting code (EDC), 703
error masking, 290, 370
Error Operation Protocol (EOP)
    data, 525
Error Operation Protocol (EOP)
    registers, 523–524
error polynomial, 294
error-resilient scan, 90–92
error sequence, 294
error tolerance, 705
error vector magnitude
    (EVM), 734
escape rate, 697
escaping faults, 423
ESL (electronic system
    level), 106
ESP (essential spare pivoting)
    algorithm, 517,
    531–534
ESP_SA( ) function, 532
ESRAM (embedded
    SRAM), 462
essential line, 532

essential pivot (EP), 532
essential spare pivoting (ESP)
    algorithm, 517,
    531–534
essential vector, 246
etching defects, 624
ETCompression, 386–388
Eulerian trail, 243
evaluation frontier
    (E-frontier), 193
event, 125
even taps, 289
event-driven simulation,
    125–129, 410
  vs. compiled-code
    simulation, 129
  nominal-delay, 126–129
  overview, 125–126
event history, 413
exhaustive pattern generator
    (EPG), 275
exhaustive testing, 10
  binary counter, 275
  complete LFSR, 275–277
  overview, 275
Exit2-DR state, 568–569
Exit-DR state, 568
exponential failure law, 7
exponents, 276
Export Mask Address (EMA),
    539, 543, 547
extended backward (EB)
    implications, 190
extended-data-out (EDO), 489
extended interconnect
    measurement, 667–671
extended interconnects, 660
external BIST, 265
external blocks, 603
external scan cells, 302
external set/reset disable (RE)
    pin, 268
external testing, 708
external-XOR LFSR, 272
EXTEST_PULSE
    instruction, 581
EXTEST_TRAIN
    instruction, 582
EXTEST instruction,
    571–572, 665
extra defects, 624
extra device hard fault
    model, 625

F
fading scheme, 427
failing output, 406
failing output vectors, 405
failing profile, 437, 438
failing test vectors,
    400–401, 414

failure, 9
failure analysis (FA), 461, 683
failure mode analysis (FMA), 4
failure rate, 7, 692
false paths, 88, 233
FAN (Fanout-Oriented TG),
    186–187
fanin cone, 407
Fanout-Oriented TG (FAN),
    186–187
fanout stem, 175
Fast Fourier transformation
    (FFT) transforms, 636
fast synthesis, 93
fault, 9
fault bitmaps, 528
fault candidate area, 407
fault candidates, 402
fault collapsing, 12, 13–14, 134
fault coverage (FC), 10, 108,
    476, 692
fault descriptor, 477
fault detection, 148–149, 294,
    317–319
fault detection efficiency, 10
fault diagnosis, 108, 187
fault dictionary, 401–402, 526
fault dictionary, compaction
    and compression of,
    403–405
  detection dictionary, 405
  overview, 403–404
  P&R compression dictionary,
    404–405
  pass–fail dictionary, 404
fault-dictionary based
    paradigm, 402
fault-driven algorithm, 529
fault dropping, 134
fault effect, 53
fault flag (FF) status flag, 545
fault-free circuit, 136, 166
fault-free profile, 437
fault-free signature, 448
fault-free simulation, 133, 147
fault grading, 38, 108
fault-independent methods,
    200–201
fault injection, 133
fault list propagation, 140, 141
fault models, 10, 11–21,
    623–627
  absence of suitable, 622–623
  analog, 21
  delay faults and crosstalk,
    19–20
  hard faults, 625
  open and short faults, 16–19
  overview, 11–12, 623–624
  pattern sensitivity and
    coupling faults, 20–21
  soft faults, 625–627

stuck-at faults, 12–15
transistor faults, 15–16
fault sampling, 151
fault simulation, 11, 38, 106, 132–159, 187
alternatives to, 151–154
critical path tracing, 152–153
fault sampling, 151
overview, 151
statistical fault analysis, 153–159
toggle coverage, 151
concurrent, 143–146
deductive, 139–143
differential, 146–148
fault detection, 148–149
overview, 132–133
parallel, 135–139
serial, 133–135
techniques, 149–151
for test and diagnosis, 107–108
and VLSI test technology, 28
fault simulator, 108, 208
fault site indicator (FSI), 521, 524–525
fault syndrome (FS), 541
fault syndrome (FS) signal, 539
fault tolerance, 685, 692, 701–704
fault-to-SO, 432, 436
faulty column, 530
faulty line, 529, 530
faulty row, 530
FC (fault coverage), 10, 108, 476, 692
FC (fiber channel), 722
FDR (frequency-directed run-length) code, 350
feature size, 1
feedback ambiguity, 622
feedback bridging fault, 244
feed-forward, 39
feed-forward partial-scan design, 65
FF (fault flag) status flag, 545
FFT (Fast Fourier transformation), 636
FIB (focused ion beam) systems, 399
fiber channel (FC), 722
field programmable gate arrays (FPGAs) testing, 706–711
built-in self-test of logic resources, 708–709
built-in self-test of routing resources, 709–710
impact of programmability, 706–708

overview, 706
recent trends, 710–711
testing approaches, 708
FIFO (first-in first-out) memories, 706
filtering, 434, 438
filters, 226, 421, 644
final-repair phase, 529
final signature, 265
finite state machine (FSM), 248, 322, 475, 493, 495, 543
FIRE, 201–202
first-hit index, 400, 415
first-in first-out (FIFOs) memories, 706
fitness, 208
fitness function, 213–215
fixed-length sequential linear decompressors, 355–356
fixed-to-fixed (dictionary code), 345–346
fixed-to-fixed coding, 345
fixed-to-variable (Huffman code), 346–349
fixed-to-variable coding, 346
flash ADC, 646
flip-flop partition, 448
floating component, 669
floating-point unit (FPU), 705
floating ports, 270–271
flush tests, 85, 427, 432, 433
FM (frequency modulation), 631
FMA (failure mode analysis), 4
focused ion beam (FIB) systems, 399
four nines, 8
FPU (floating-point unit), 705
free variables, 352
frequency analysis, 637
frequency decomposition, 226
frequency-directed runlength (FDR) code, 350
frequency-domain ADC testing, 658–659
frequency modulation (FM), 631
frequency response measurement, 637–639
frequency shift keying (FSK), 631
front-side bus (FSB), 720
FS (fault syndrome), 541
FS (fault syndrome) signal, 539
FSB (front-side bus), 720
FSI (fault site indicator), 521, 524–525
FSK (frequency shift keying), 631
FSM (finite state machine), 248, 324, 325, 475, 493, 495, 543

full power unit gain bandwidth, 634
full-scan design, 39, 59–64
clocked full-scan design, 62
LSSD full-scan design, 62–64
muxed-D full-scan design, 59–61
overview, 59
fully decomposable polynomial, 274
functional circuitry (CUT), 303
functional defects, 696
functional element delay model, 120–121
functional mode, 434
functional offline BIST, 265
functional partitioning, 65
functional pruning technique, 408
functional random (spot) defects, 696
functional sequences, 435–437
functional storage elements, 265
functional testing, 10, 698
functional test patterns, 466–469
definition, 466–467
overview, 466
theorem, 467–469
functional verification, 106

G
GA–HITEC hybrid test generator, 227
gain and offset test, 656–657
gain-bandwidth product, 633
gain error, 648
gain stage, 644
galloping (ping-pong) pattern (GALPAT), 468
galloping column algorithm, 468
galloping diagonal algorithm, 468
galloping row algorithm, 468
Galois field (GF), 274
gated clocks, 71–74, 197–198
gate-delay fault, 19, 688
gate -evel, 23–24
gate-level networks, 109–111
gate-level scan design, 95
gate-level testability analysis, 50
gate outputs, 142
GATEST, 216–217
GATTO, 216
Gauss–Jordan elimination, 353
GBE (gigabit ethernet), 722
generator polynomial, 285
genetic-algorithm-based ATPG, 208–217

CASE studies, 215–217
  issues concerning fitness
      function, 213–215
  issues concerning GA
      parameters, 213
  issues concerning GA
      population, 212–213
  overview, 208–212
genetic algorithm-based test
      generator, 216, 229
genetic algorithms, 216
  crossover, 208
  fitness, 208
  mutation, 208
  parameters, 213
  population, 208, 212–213
  seeding, 208
  selection, 208
getObjective() function,
      184, 196
GF (Galois field), 276
gigabit ethernet (GBE), 722
giga samples per second (Gsps),
      642, 646
global effects, 430
global scan enable (GSE)
      signal, 310
golden signature, 265, 290,
      323, 327
Golomb code
      (variable-to-variable),
      350–351
Go/No-Go indicator signal
      (BGO), 494
good events, 144
Gray code counter, 289
greedy algorithm, 532
gross-delay fault model, 688
group size, 537
GSE (global scan enable)
      signal, 312
Gsps (giga samples per second),
      642, 646

H
Hadamard transform, 226
Hamming distance, 208
hard detected faults, 149
hard faults, 149, 624, 625
hardware-assisted method, scan
      chain diagnosis,
      430–432
hardware description language
      (HDL), 93, 106, 398
hardware partitioning, 288
hazards, 130–132
  dynamic hazard
      detection, 132
  overview, 130–131
  static hazard detection,
      131–132

HCSs (hierarchical control
      signals), 609
HDL (hardware description
      language), 93,
      106, 398
headline, 186
heuristic ranking, for
      multiple-fault
      diagnosis, 414–415
hierarchical architecture, 578
hierarchical control signals
      (HCSs), 609
hierarchical test controller
      (HTC), 606–610
hierarchical test manager
      (HTM), 609
high energy neutrons, 702
high-level programming
      language source
      code, 124
high-performance I/O test
      solutions, 725–726
high-speed compressed
      stimuli, 384
high-speed digital network
      testing. See IEEE
      1149.6 standard
high-speed I/O testing,
      719–728
  challenges of, 724–725
  future challenges of,
      726–728
  high-performance I/O test
      solutions, 725–726
  I/O interface technology and
      trend, 720–724
  overview, 719–720
high-speed networks, 579–580
HIGHZ instruction, 574,
      665, 666
hold cycle, 61
hold-time violation, 429
hot-carriers, 692
hot electron, 700
HTC (hierarchical test
      controller), 606–610
HTM (hierarchical test
      manager), 609
Huffman code
      (fixed-to-variable),
      346–349
Huffman tree, 346
hybrid (H) parameters, 671
hybrid BIST, 305, 343
hybrid deterministic ATPG,
      226–231
hybrid LFSR, 274–275
hybrid methods, 200
hyperactive faults, 149
hypergeometric
      distribution, 151
hysteresis error, 651

I
ICs (integrated circuits), 1, 37,
      344, 397, 623
IDCODE, 573–574
IDDQ testing, 15, 247, 697
idempotent coupling fault
      (CFid), 465, 519
identification (ID) field, 545
IEEE 1057 standard, 652–654
IEEE 1149.1 standard, 561–579
  basic concept, 561–562
  board and system-level
      boundary-scan control
      architectures, 576–579
  hierarchical architecture,
      578–579
  multidrop architecture, 578
  overview, 576
  single-ring architecture
      with separate TMS, 578
  single-ring architecture
      with shared TMS,
      576–577
  star (multi-ring)
      architecture, 578
  data registers and
      boundary-scan cells,
      565–566
  description language, 574
  instruction register, 569–574
  instruction set
      BYPASS, 569
      CLAMP, 573
      EXTEST, 571–572
      IDCODE, 573–574
      INTEST, 572–573
      PRELOAD, 570–571
      RUNBIST, 573
      SAMPLE, 569–570
      USERCODE, 574
  on-chip test support with
      boundary scan,
      574–576
  overall 1149.1 test
      architecture and
      operations, 562–564
  overview, 561
  TAP controller, 567–569
  test access port and bus
      protocols, 564–565
IEEE 1149.2 standard, 559
IEEE 1149.3 standard, 559
IEEE 1149.4 standard, 659–667
  circuit structures, 661–664
      analog boundary module
      (ABM), 662–664
      analog test access port
      (ATAP), 662
      analog test buses
      (AB1/AB2), 662

overview, 661–662
test bus interface circuit
(TBIC), 662
instructions, 665–666
overview, 659–661
test modes, 666–667
complex network
measurement, 671–672
extended interconnect
measurement, 667–671
high-performance
configuration, 672–677
open/short interconnect
testing, 666–667
overview, 666
IEEE 1149.5 standard, 559
IEEE 1149.6 standard, 579–585
analog test receiver, 581
digital driver logic, 581–582
digital receiver logic, 582–584
EXTEST_PULSE instruction,
581–584
EXTEST_TRAIN instruction,
581–584
test access port (TAP), 584
IEEE 1500 standard, 585–618
architecture of, 587–589
core test language (CTL),
601–603
external, 603
internal, 603
overview, 601–603
patterninformation, 603
core test supporting and
system test
configurations,
603–606
hierarchical test control and
plug-and-play, 606–610
instruction set, 597–601
overview, 597
W<S/P/H>_<Command>
{_<Configuration>},
597
WP_EXTEST, 599
WP_PRELOAD, 600
WS_BYPASS, 597
WS_CLAMP, 600
WS_EXTEST, 597–599
WS_INTEST_RING, 600
WS_INTEST_SCAN,
600–601
WS_PRELOAD, 600
WS_SAFE, 600
Wx_EXTEST, 599
Wx_INTEST, 601
overview, 585
SOC (system-on-chip) test
problems, 585–587
deeply embedded
cores, 585

different core
providers, 586
expensive and inefficient
external automatic test
equipment (ATE), 586
hierarchical cores, 585
higher performance core
I/Os than SOC
pins, 586
IP protection/test reuse, 586
large test power, 586
long test application
time, 586
mixing technologies, 585
overview, 585
testable design automation,
586–587
wrapper components,
589–597
AUXCKn, 589–590
CaptureWR, 590
overview, 589
SelectWIR, 590
ShiftWR, 590
TransferDR, 591–592
UpdateWR, 590
WRCK, 589
WRSTN, 590
wrapper functions
functional modes of
WBR, 592
operation events of WBR
and WBC, 592–597
IEEE P1687 (IJTAG
standard), 611
IFs (independent faults), 418
IJTAG (internal JTAG)
Istandard, 611
IL (integrity loss), 693
Illinois scans, 360–362
imbalance fault, 466
IMD (intermodulation
distortion), 641, 652
implication graph, 203
impossible value combination,
203–204
independent faults (IFs), 418
indirect implications, 190
inertial delay, 119
information redundancy,
701, 703
inject-and-evaluate paradigm,
407, 409–418
bridging fault diagnosis using
multiplets, 416–418
computational details,
412–413
curable-vector-based
metric, 413
definitions
curable output, 411
curable vector, 413–414

reproduction principle, 411
resolving mismatched
output, 411
heuristic ranking for
multiple-fault
diagnosis, 414–415
reward-and-penalty
principle, 415
INL (integral nonlinearity
error), 649
Input Address (ADDR), 543
input counting, 116
input data (DI), 519
input referred common mode
voltage, 634
input scanning, 115–116
instruction register (IR), 562,
563, 565, 569–574
instructions, in IEEE 1149.4
standard, 665–666
instruction sets. See IEEE
1149.1 standard; IEEE
1149.4 standard; IEEE
1500 standard
integral nonlinearity error
(INL), 649
integrated circuits (ICs), 1, 37,
344, 397, 623
integrity loss (IL), 693
integrity loss fault model,
693–694
intellectual property (IP), 21,
462, 557
inter-clock-domain fault, 310
interconnects, 694
interconnect test, 660
intermediate logic states, 114
intermittent error, 264
internal BIST, 265
Internal blocks, 603
internal JTAG (IJTAG)
standard, 611
internal test, 660–661
internal-XOR LFSR, 272
International Symposium on
Circuits and Systems
(ISCAS), 27
International Technology
Roadmap for
Semiconductors
(ITRS), 680, 737
interpreted code, 124
inter-scan-chain reordering, 83
interval-based methods,
443–446
interval unloading, 445
INTEST instruction,
572–573, 665
intra-clock-domain fault, 310
intra-scan-chain reordering, 83
inverse-hybrid (G), 672

inversion coupling fault (CFin), 465, 518
inversion operation, 431
inversion value, 115
invisible bad gate, 144
IO_FACE signal, 595
I/O pin cost, 87
I/O subsystems, 728
I/O wrap, 725
IP (intellectual property), 21, 462, 557
IP protection/test reuse, 586
IR (instruction register), 562, 563, 565, 569–574
irreducible primitive polynomial, 274
ISCAS (International Symposium on Circuits and Systems), 27
iterative logic array, 194
ITRS (International Technology Roadmap for Semiconductors), 680, 737

J
JETAG (Joint European Test Action Group), 558
J-frontier, 177
jitter, 645
Johnson counter, 289
Joint European Test Action Group (JETAG), 558
Joint Test Action Group (JTAG), 32, 558
JTAG (Joint Test Action Group), 32, 558
justification sequence, 197
JustifyFanoutFree( ) function, 173–175

K
known good signature, 296

L
large-scale integration (LSI) devices, 1
large signal models, 621
last-read value (LRV), 479
last-write value (LWV), 479
late signal, 429
launch aligned double-capture schemes, 323
launch aligned double-capture waveform, 324
launch aligned skewed-load, 313
launch-on-capture, 88, 232, 239, 315, 687
launch-on-shift, 88, 232, 239, 312, 387, 687

least significant bit (LSB), 642
level of confidence, 163
level order, 40
level-sensitive/edge-triggered muxed-D scan cell design, 55
level-sensitive scan design (LSSD), 30, 55, 267
level-sensitive scan design (LSSD) full-scan designs, 62–64
level-sensitive scan design (LSSD) on-chip self-test (LOCST), 297–298
level-sensitive scan design (LSSD) scan cell designs, 57–59
level-sensitive scan design (LSSD) shift register latches (SRLs), 297
LFSR (linear feedback shift register), 273, 446
LFSR (linear feedback shift register) pattern generation technique, 280
LFSR (linear feedback shift register) properties, 273–274
LFSR (linear feedback shift register) reseeding, 308–309
linear codes, 284
linear decompression, 358
linear-decompression-based schemes, 344, 351–354, 376
  combinational linear decompressors, 355
  combined linear and nonlinear decompressors, 357–359
  fixed-length sequential linear decompressors, 355–356
  overview, 351–354
  variable-length sequential linear decompressors, 356–357
linear decompressors, 351
linear expansion, 367
linear feedback/nlshift registers (LFSRs), 271
linear feedback shift register (LFSR), 446
linear feedback shift register (LFSR) pattern generation technique, 278

linear feedback shift register (LFSR) properties, 273–274
linear feedback shift register (LFSR) reseeding, 308–309
linearity, 619
linearity versus nonlinearity, 366
linear phase shifter, 280
linear phase shifter (PS), 284
line fault, 535
link layer, 722
LNA (low noise amplifier), 729
load-dependent error, 650–651
local bitmaps, 532
local faults, 143
locally controlled configuration, 604
local oscillator (LO) signal, 731
LOCSTEP, 222
logical bridging faults, 697
logic BIST, 343, 708
logic BIST architectures, 96, 296, 304, 320–321
logic BIST controller, 265
logic built-in self-test, 45, 263–331
  architecture, 296–304
    for circuits without scan chains, 296–297
    for circuits with scan chains, 297–298
    overview, 296
    using concurrent checking circuits, 303–304
    using register reconfiguration, 298–303
  design example, 319–329
    design verification and fault coverage enhancement, 326–340
    overview, 319
    RTL BIST synthesis, 326
    rule checking and violation repair, 320
    system design, 320–326
  design rules, 266–267
    overview, 266–267
    re-timing, 271
    unknown source blocking, 267–271
  fault coverage enhancement, 304–309
    hybrid BIST, 309
    mixed-mode BIST, 308–309
    overview, 304–305
    test point insertion, 305–307

output response analysis, 290–295
  ones count testing, 291
  overview, 290
  signature analysis, 292–294
  transition count testing, 291–292
overview, 263–266
test pattern generation, 271–290
  delay fault testing, 288–289
  exhaustive testing, 275–277
  hybrid LFSR, 274–275
  LFSR properties, 273–274
  modular LFSR, 272
  overview, 271–275
  pseudo-exhaustive testing, 280–288
  pseudo-random testing, 277–280
  standard LFSR, 272
timing control, 310–319
  double-capture, 315–317
  fault detection, 317–319
  overview, 310
  single-capture, 310–311
  skewed-load, 311–315
logic built-in self-test (BIST), 263
logic diagnosis, 397–459
  built-in self-test (BIST), 442–459
    interval-based methods, 443–446
    masking-based methods, 446–459
    overview, 442–443
  combinational, 401–427
    cause–effect analysis, 401–405
    chip-level strategy, 418–420
    compaction and compression of fault dictionary, 403–405
    diagnostic test pattern generation, 425–426
    effect–cause analysis, 405–418
    overview, 401
  overview, 397–401
  scan chain diagnosis, 427–442
    hardware-assisted method, 430–432
    modified inject-and-evaluate paradigm, 432–434
    overview, 427
    preliminaries for scan chain diagnosis, 427–430
    signal-profiling-based method, 434–441

logic element evaluation, 114–117
  input counting, 116
  input scanning, 115–116
  overview, 114
  parallel gate evaluation, 116–117
  truth tables, 115
logic levelization, 123–124
logic optimization, 121–123, 187
logic/scan synthesis, 106
logic simulation, 106, 121–132, 187, 215
  compiled-code simulation, 121–125
    code generation, 124–125
    vs. event-driven simulation, 129
    logic levelization, 123–124
    logic optimization, 121–123
    overview, 121
  for design verification, 106–107
  event-driven simulation, 125–129
  hazards, 130–132
    dynamic hazard detection, 132
    overview, 130–131
    static hazard detection, 131–132
  overview, 105–106, 121
logic-simulation-based ATPG, 222–225
logic soft errors, 737
logic symbols, 110–114
  high-impedance state Z, 113–114
  intermediate logic states, 114
  overview, 110–111
  unknown state u, 111–113
logic verification, 187
long test application time, 586
look-up tables (LUTs), 499, 706
lossy compression, 343
lower limit mask, 637
low noise amplifier (LNA), 729
LRV (last-read value), 479
LSB (least significant bit), 642
LSI (large-scale integration) devices, 1
LSSD (level-sensitive scan design), 30, 55, 269
LSSD (level-sensitive scan design) full-scan designs, 62–64
LSSD (level-sensitive scan design) on-chip self-test (LOCST), 297–298, 299

LSSD (level-sensitive scan design) scan cell designs, 57–59
LSSD (level-sensitive scan design)SRLs (shift register latches), 299
LUTs (look-up tables), 499, 706
LWV (last-write value), 479

M
MA (maximum aggressor) fault model, 695
mandatory wrapper clock terminal (WRCK), 589
mandatory wrapper reset terminal resets (WRSTN), 590
manufacturing yield, 692
MAO (Mask Address Output), 539, 543, 547
mapping logic, 287, 309
March commands, 502
March–CW algorithm, 473
March elements, 466, 469
March signature, 526
March syndrome, 526
March template, 480
March test algorithms, 461
March tests, 466, 469–471
Mask Address Output (MAO), 539, 543, 547
masking-based methods, 446–459
masking probability, 291, 292
matching score, 433
material defects, 623
MATS (modified algorithmic test sequence), 470
maximal output amplitude measurement, 636–637
maximal output swing, 636
maximal static error (MSE), 657
maximum aggressor (MA) fault model, 695
maximum-length LFSR, 273, 278
maximum-length sequence, 273, 280
MBILBO (modified BILBO), 302
MCMs (multiple chip modules), 558
mean time between failures (MTBF), 7
mean time to repair (MTTR), 7
MECA (Memory Error Catch and Analysis) system, 517, 527
medical diagnosis, 401
medium-scale integration (MSI) devices, 1

mega samples per second (Msps), 642

memories and non-scan storage elements, 268

memory BIST compiler, 462

memory BIST mode, 495

memory block, 530

memory built-in self-test, 461, 488–515
  BIST architectures and functions, 493–495
  BIST implementation, 495–500
  overview, 488–489
  RAM BIST compiler, 500–515
  RAM specification and BIST design strategy, 489–493

memory diagnosis, 461
  BIST with diagnostic support, 521–525
    controller, 521–523
    fault site indicator (FSI), 524–525
    overview, 521
    test pattern generator, 523–524
  overview, 517–518
  RAM defect diagnosis and failure analysis, 526–529
  RAM redundancy analysis algorithms, 529–537
    conventional redundancy analysis algorithms, 529–531
    essential spare pivoting algorithm, 531–534
    overview, 529
    repair rate and overhead, 535–537
  refined fault models and diagnostic test algorithms, 518–521

Memory Error Catch and Analysis (MECA) system, 517, 527

memory fault simulator, 461

memory library, 501

memory scan (MSCAN) algorithm, 467

memory specifications, 503, 504

MEMS. *See* microelectromechanical system (MEMS) testing

metal oxide semiconductor (MOS) transistor, 692

MFSD (multiple faulty subwords detector), 543

microelectromechanical system (MEMS) testing, 711–719

basic concepts for capacitive MEMS devices, 711–712

BIST example for MEMS comb accelerometers, 716–719

dual-mode BIST technique, 714–716

example for MEMS comb accelerometers, 716–719

overview, 711, 713

sensitivity BIST scheme, 713

symmetry BIST scheme, 713–714

minimum distance, 285

minimum heap structure, 412

min–max delay model, 118

mismatched outputs, 407, 411, 415

mismatched outputs reachable, 420

MISR (multiple-input signature register), 68, 296, 360, 374, 386, 442

missing code, 651

missing device hard fault model, 625

mixed-mode BIST, 305, 308–309
  embedding deterministic patterns, 309
  LFSR reseeding, 308–309
  overview, 308
  ROM compression, 308

mixed-signal components (M), 659

mixed-signal testing, 641–658
  ADC and DAC circuit structure, 644–647
    ADC, 645–646
    ADC circuit structure, 646–647
    DAC circuit structure, 646
    filter, 644–645
    gain stage, 644
    MUX stage, 645
    overview, 644
    S/H stage, 645
  ADC/DAC specification and fault models, 647–652
    differential nonlinearity error (DNL), 648–650
    gain error, 648
    hysteresis error, 651
    load-dependent error, 650–651
    missing code, 651–652
    nonlinearity error, 648
    offset error, 648
    overview, 647–648

resolution error, 651

signal-to-noise ratio, signal-to-noise and distortion, total harmonic distortion, intermodulation distortion, 652

analog–digital conversion overview, 642–644

frequency-domain ADC testing, 658–659

IEEE 1057 standard, 652–654
  overview, 652
  purpose, 652–654
  scope, 652

IEEE 1149.4 circuit structures, 661–664
  analog boundary module (ABM), 662–664
  analog test access port (ATAP), 662
  analog test buses (AB1/AB2), 662
  overview, 661–662
  test bus interface circuit (TBIC), 662

IEEE 1149.4 instructions, 665–666

IEEE 1149.4 overview, 659–661

IEEE 1149.4 test modes, 666–667
  complex network measurement, 671–672
  extended interconnect measurement, 667–671
  high-performance configuration, 672–677
  open/short interconnect testing, 666–667
  overview, 666

overview, 641–642

time-domain ADC testing, 654–658
  code bins, 654
  code transition level test, 655
  gain and offset test, 656–657
  linearity error and maximal static error, 657
  overview, 654
  sine wave curve-fit test, 658

MO (MISR Observe) pins, 379

Mode signal, 595

modified algorithmic test sequence (MATS), 470

modified built-in logic block observer (MBILBO), 300

modified inject-and-evaluate paradigm, 432–434
modularization, 26
modular LFSR, 272, 293
MONITOR state, 544
Moore's law, 1, 701
MOS (metal oxide semiconductor) transistor, 692
most significant bits (MSBs), 538
mouse-bites, 399
M-out-of-N code, 283
MOVI (moving inversion) algorithm, 469
moving inversion (MOVI) algorithm, 469
MP-LFSR (multiple-polynomial LFSR), 310
MSBs (most significant bits), 538
MSCAN (memory scan) algorithm, 467
MSE (maximal static error), 657
MSI (mediumscale integration) devices, 1
Msps (mega samples per second), 642
MT (multiple transition) model, 695
MTBF (mean time between failures), 7
MTTR (mean time to repair), 7
multidrop architecture, 578
multiple chip modules (MCMs), 558
multiple-cycle paths, 270
multiple-fault diagnosis, 419
multiple-fault model, 11
multiple faulty subwords detector (MFSD), 543
multiple-input broadcast scans, 362
multiple-input signature register (MISR), 68, 294, 360, 374, 386, 442
multiple-line conflict analysis, 203–207
    ATPG for transition faults, 238–239
    definition 1, 206
    definition 2, 206
    definition 3, 206
    lemma 2, 206–207
    overview, 203–206
    proof, 207
multiple-pass fault simulation approach, 150
multiple-polynomial LFSR (MP-LFSR), 308
multiplet, 417

multiple transition (MT) model, 695
multiplexed configuration, 604
multiplexer (MUX)-based modifications, 199
multi-port memory, 473–475
multi-ring (star) architecture, 578
must-repair faulty line, 530
must-repair phase, 529
mutation, 208
mutation operator, 212
muxed-D full-scan designs, 59–61
muxed-D scan cell designs, 55–56

N
naive ATPG algorithm, 169–173
nanoelectromechanical system (NEMS), 711
nanometer age test technology trends, 679–749
    defect and error tolerance, 705
    defect-based test, 697–698
    delay testing, 685–691
        delay fault models, 687–691
        overview, 685–686
        test application schemes for testing delay defects, 686–687
    fault tolerance, 701–704
    FPGA testing, 706–711
        built-in self-test of logic resources, 708–709
        built-in self-test of routing resources, 709–710
        impact of programmability, 706–708
        overview, 706
        recent trends, 710–711
        testing approaches, 708
    high-speed I/O testing, 719–728
        future challenges, 726–728
        high-performance I/O test solutions, 725–726
        I/O interface technology and trend, 720–724
        I/O testing and challenges, 724–725
        overview, 719–720
    MEMS testing, 711–719
        basic concepts for capacitive MEMS devices, 711–712
        BIST example for MEMS comb accelerometers, 716–719

MEMS built-in self-test, 713–716
        overview, 711
    overview, 679
    parametric defects, 697
    physical failures, coping with, 692
    process variations, 697
    reliability issues, coping with, 692
    RF testing, 728–749
        core RF building blocks, 729–730
        overview, 728–729
        RF test specifications and measurement procedures, 730–733
        tests for system-level specifications, 733–734
        trends, 735–749
    signal integrity and power supply noise, 692–696
        integrity loss fault model, 693–694
        location, 694
        overview, 692–693
        pattern generation, 694–695
        sensing and readout, 695–696
    soft errors, 698–701
    test technology overview, 680–685
national semiconductor (NS), 578
native machine code, 124
ND (noise detector), 696
N-detect, 266, 318, 697
N-detect ATPG, 247
N-detect single stuck-at fault test vectors, 14
N-detect stuck-at tests, 697
neighborhood pattern-sensitive faults (NPSFs), 471
neighboring parasitic, 697
NEMS (nanoelectromechanical system), 711
netlist, 76
newly visible bad gates, 145
NO_SP_ROW state, 544
NO_SP_SCS state, 544
noise detector (ND), 696
noise factor, 733
noise figure, 733
noise figure analyzer (NFA), 733
nominal-delay event-driven simulation, 126–129
nominal delay model, 118
nonadaptive diagnostic approach, 448
nonbinary coding, 212
nonchronological backtracking, 173

nonconcurrent online BIST, 266
noncontrolling case, 409
noncontrolling gate inputs, 142
noncontrolling rule, 408
non-feedback bridging
    fault, 244
nonlinear decompression, 358
nonlinearity error, 648
nonrecurring transient
    errors, 692
nonrobust sensitization, 235
nonterminating necessary
    condition
    set (NTC), 206
normal/capture mode, 55
normal mode operation, 53, 566
NP-complete problem, 288
NPSFs (neighborhood
    pattern-sensitive
    faults), 471
NS (national
    semiconductor), 578
NTC (nonterminating necessary
    condition set), 206
Nyquist rate, 645

O
observability, 38, 40, 153
observation boundary scan cell
    (OBSC), 696
observation point, 305
observed image, 427
odd taps, 289
OE (Output-Enable) signals,
    519, 523
off-input, 232
offline BIST, 264, 296
offline testing, 8
offset error, 648
OIF (optical internetworking
    forum), 722
on-chip test support, 574–576
one-hot clocking scheme, 78
one-hot decoder, 269
one-hot double-capture,
    315–316
one-hot single-capture, 310–311
one-hot skewed-load, 312
one-pass synthesis, 79
one-point crossover, 211
ones count testing, 290, 291
ones count test technique, 291
on-input, 232
online BIST, 264
online testing, 8
open and short faults, 16–19
open-circuit impedance (Z), 672
open hard fault model, 625
open-loop gain measurement,
    632–633
open/short interconnect testing,
    666–667

operation, 466
OPMISR, 377–379
optical internetworking forum
    (OIF), 722
ORAs (output response
    analyzers), 31, 265,
    290, 708–710
orthogonal faulty cell, 530
oscillation faults, 149
OSR (oversampling ratio), 644
output (Q) channels, 489
output column, 405
output data (DO), 519
Output-Enable (OE) signals,
    519, 523
output response analysis,
    290–295
    ones count testing, 291
    overview, 290
    signature analysis, 292–294
        overview, 292
        parallel, 294–295
        serial, 292–294
    transition count testing,
        291–292
output response analysis
    technique, 290
output response analyzers
    (ORAs), 31, 265, 290,
    708–710
output response
    compaction, 290
output space, 351
over-killing effect, 423
overlapping populations, 213
overlapping relationship, 420
oversampling ADCs, 647
oversampling ratio (OSR), 644
overshoot, 694
over-test problem, 88

P
P&R compression dictionary,
    404–405
$P_{1parameters}$, 231
$P_{2parameters}$, 231
$P_{3parameters}$, 231
$P_{4parameters}$, 231
PA (power amplifier), 729
packaging failures, 624
parallel fault simulation, 28,
    135–139, 149–150, 230
parallel gate evaluation,
    116–117
parallel pattern fault
    simulation, 135
parallel-pattern single-fault
    propagation
    (PPSFP), 137
parallel signature analysis
    (PSA), 292, 295

parallel simulation, 116
parallel test capabilities, 26
parametric defects, 696, 697
parametric faults, 21, 625–626
parametric test, 660
parity-check polynomial, 285
partially curable vector, 420
partial reproduction, 411
partial-scan design, 39, 59,
    64–67
parts per million (PPM), 38
parts per million (PPM) chips, 6
pass-band ripple models, 637
pass bands, 637
pass–fail dictionary, 404
path-delay fault, 19, 232, 688
path-delay tests, 697
path filtering phase, 691
path sensitization, 691
pattern counter, 443
pattern decoding logic, 307
pattern generation, 694–695
PatternInformation blocks, 603
pattern sensitivity and coupling
    faults, 20–21
Pause-DR state, 568
PCA (printed circuit
    assembly), 659
PCBs (printed circuit boards),
    2, 106, 558
PDFs (probability density
    functions), 691
PE (process element), 543
peak harmonic, 641
performance degradation
    cost, 87
period, 273
phase error, 734
phase interpolator (PI), 722
phase-locked loop (PLL), 74,
    382, 622, 629, 690, 722
PHY (physical layer), 722, 727
physical bridging faults, 697
physical defects, 696
physical failures, 692
physical fault modeling, 698
physical layer (PHY), 722, 727
PI (phase interpolator), 722
pilot-run stage, 397
pin electronics and fixtures, 25
ping-pong (galloping) GALPAT
    (pattern), 468
pipelined, 39, 65
pipelined ADCs, 646, 647
pipelining register, 271
PIs (primary inputs), 12, 61,
    109, 269,
    272, 298
PLAs (programmable logic
    arrays), 710
PLBs (programmable logic
    blocks), 706–707

PLL (phase-locked loop), 74, 382, 622, 629, 690, 722
plug-and-play (PnP) protocol, 587
plug-and-play test control, 606–610
PMU (precision measurement unit), 495
PnP (plug-and-play) protocol, 587
PODEM, 27, 182–186
Poisson model, 550
Polya–Eggenberger distribution, 535
polynomial identifier, 309
polynomial remainder, 293
Pomeranz, I., 216
Pomeranz and Reddy (P&R), 404
POR (Power-On Reset) signal, 539, 541
POs (primary outputs), 12, 61, 109, 272, 406
potentially detected, 149
power amplifier (PA), 729
power integrity problem, 680
Power-On Reset (POR) signal, 539, 541
power supply noise (PSN), and signal integrity, 692–696
    integrity loss fault model, 693–694
    location, 694
    overview, 692–693
    pattern generation, 694–695
    sensing and readout, 695–696
power supply rejection ratio measurement, 635
PPIs (pseudo primary inputs), 61, 109, 194
PPOs (pseudo primary outputs), 61, 109, 194
PPSFP (parallel-pattern single-fault propagation), 137
PRAS (progressive random-access scan) design, 68
precision measurement unit (PMU), 495
PRELOAD instruction, 570–571
primary inputs (PIs), 12, 61, 109, 267, 270, 296
primary outputs (POs), 12, 61, 109, 270, 406
prime candidate, 420, 421, 424
primitive polynomial, 274, 354
printed circuit assembly (PCA), 659
printed circuit boards (PCBs), 2, 106, 558

probabilistic fault simulation, 307
probability-based testability analysis, 45–47
probability-based testability measures, 45
probability density functions (PDFs), 691
PROBE instruction, 665
process element (PE), 543
processing faults, 623
process variations, 680, 694, 697
profiling-based analysis, 437–441
    average-sum filtering, 438–439
    edge detection, 439–441
programmable logic arrays (PLAs), 710
programmable logic blocks (PLBs), 706–707
progressive random-access scan (PRAS) design, 68
progress limit, 209
PROOFS sequential circuit fault simulator, 216
PropagateFanoutFree( ) function, 173–175
propagation delay, 120
protocol layer, 727
PRPG (parallel shift register sequence generator [SRSG]), 298
PRPGs (pseudo-random pattern generators), 279, 280, 378, 442, 695
PSA (parallel signature analysis), 294, 297
pseudo-exhaustive pattern generation, 281, 282
pseudo-exhaustive pattern generation (EPG/PEPG), 301
pseudo-exhaustive testing, 166, 272, 280–288
    overview, 280–281
    segmentation testing, 287–288
    verification testing, 281–287
        combined LFSR/PS, 284
        combined LFSR/SR, 283–284
        compatible LFSR, 287
        condensed LFSR, 284–285
        constant-weight counter, 282–283
        cyclic LFSR, 285–287
        syndrome driver counter, 281–282

pseudo primary inputs (PPIs), 61, 109, 194
pseudo primary outputs (PPOs), 61, 109, 194
pseudo-random masking, 446
pseudo-random pattern generators (PRPGs), 277, 278, 378, 442, 695
pseudo-random testing, 272, 277–280, 304
    cellular automata, 278–280
    maximum-length LFSR, 278
    overview, 277–278
    weighted LFSR, 278
pseudo RTL scan synthesis, 94
PSN. *See* power supply noise (PSN), and signal integrity
PSN monitor circuit, 696

Q
q-compactor, 375
q-output zero-aliasing space compactor, 368
quadrature phase shift keying (QPSK), 631
quantization error, 642
quantization noise, 642
quantization steps, 642
quiescent power supply current, 28

R
R-2R ladder DAC, 646
RA (redundancy analysis), 518, 529
RA (redundancy analysis) algorithms, 517
radiofrequency (RF), 9, 737
radiofrequency (RF) amplifier, 620
radiofrequency (RF) circuit, 679
radiofrequency (RF) devices, 728
radiofrequency (RF) testing, 728–749
    Core RF building blocks, 729–730
    future trends, 735–736
    overview, 728–729
    RF test specifications and measurement procedures, 730–733
        conversion gain, 731
        gain, 730–731
        noise figure, 733
        overview, 730
        third-order intercept, 731–733

tests for system-level
   specifications, 733–734
adjacent channel power
   ratio, 733
error vector magnitude,
   magnitude error, and
   phase error, 734
overview, 733
radiofrequency identification
   (RFID), 728, 736
RAID (redundant array of
   inexpensive disk), 704
RAM (random access memory),
   1, 67, 464
RAM BIST compiler, 500–515
Rambus DRAM, 503
RAM defect diagnosis and
   failure analysis,
   526–529
RAM dynamic faults, 465–466
RAM fault simulation, 475–488
   overview, 475–477
   RAMSES, 477–480
RAM functional fault models,
   461–463
   coupling faults, 465
   data retention fault
      (DRF), 465
   idempotent coupling fault
      (CFid), 465
   inversion coupling fault
      (CFin), 465
   overview, 461–462
   read disturb fault (RDF), 465
   stuck-at fault (SAF), 464
   stuck-open fault (SOF), 464
   transition fault (TF), 464
ramp waveforms, 630
RAM redundancy analysis
      algorithms, 529–537
   conventional, 529–531
      definition 9.1, 530
      definition 9.2, 530
      definition 9.3, 530–531
      overview, 529–530
   essential spare pivoting
      algorithm, 531–534
      example, 531–534
      overview, 531–533
   overview, 529
   repair rate and overhead,
      535–537
RAMSES, 528
RAMSES (random access
   memory simulator for
   error screening), 461,
   477–480
RAM test algorithm generation,
   480–488
RAM test algorithms
   functional test patterns and,
      466–469

definition, 466–467
overview, 466
theorem, 467–469
March tests, 469–471
multi-port memory, 473–475
RAM test patterns,
   comparison of,
   471–472
word-oriented memory, 473
random-access memory (RAM),
   1, 67, 464
random access memory
   simulator for error
   screening (RAMSES),
   461, 477–480
random-access scan (RAS), 67
random-access scan design, 59,
   67–70
random activation, 307
random defects, 696
random fill operation, 377
random imperfection, 697
random jitter (RJ), 719
random localized
   manufacturing
   imperfection, 697
random-pattern resistant
   (RP-resistant), 47, 278
random-pattern resistant
   (RP-resistant) faults,
   163, 304, 343
random resistant fault analysis
   (RRFA) method, 48
random-resistant faults, 163
random testability, 45
random test generation
   (RTG), 163
random test generators, 208
random vectors, 163, 208
ranked list of fault
   candidates, 440
ranking metrics, 410, 411
rank ordering, 124
RAS (random-access scan), 67
RCV (recoverer), 478
RDF (read disturb fault), 465
reachable output group, 422
read disturb fault (RDF), 465
read-only memory (ROM), 308
Read/Write-mode
   implementations, 485
real-time errors, 264
reciprocal polynomial, 274
reconfigurable broadcast scan
   method, 362, 363
reconfigurable broadcast scans,
   362–363
reconfigurable counter, 288
reconfiguration, 706
reconfigured storage
   elements, 39
recoverer (RCV), 478

recovery fault, 465
redundancy analysis (RA),
   518, 529
redundancy analysis (RA)
   algorithms, 517
redundancy organization,
   537–538
redundancy repair, 705
redundant array of inexpensive
   disk (RAID), 704
redundant elements, 518
redundant multithreading
   (RMT), 704
REF (Repair End Flag)
   signal, 539
refined fault models, 518–521
refresh (retention) fault, 465
register reconfiguration, BIST
   architecutres using,
   298–303
   built-in logic block observer,
      299–300
   circular self-test path (CSTP),
      302–303
   concurrent built-in logic
      block observer,
      300–302
   modified built-in logic block
      observer, 300
   overview, 298–299
register-transfer level, 22–23
register-transfer level (RTL), 4,
   37, 39, 319, 489
register-transfer level (RTL)
   BIST-ready core, 320
register-transfer level (RTL)
   design for testability,
   92–103
   overview, 92–93
   RTL scan design rule
      checking and repair,
      93–94
   RTL scan extraction and scan
      verification, 95–103
   RTL scan synthesis, 94–95
register-transfer level (RTL)
   scan design, 95
register-transfer level (RTL)
   testability analysis, 40
reject rate, 5–6
Relation block, 603
reliability, 7, 692
reliability screens, 682
repair efficiency, 531
repair end flag (REF), 543
Repair End Flag (REF)
   signal, 539
Repair Flag (RF) status flag, 545
repair-most (RM)
   algorithm, 529
repair-most rules, 541

repair rate, 535–537, 548–555
repair rate (RR), 7, 531,
        548, 551
repair time (R), 7
repeatability, 728
reproduction principle, 410
reset signal masking, 224
residue, 647
residue amplifier, 647
RESIST, 237
resistance–inductance–
        capacitance (RLC)
        model, 24
resistive open, 626
resistive short, 626
resolution, 642
resolution error, 651
resolution function, 114
resolving mismatched
        output, 411
response graph, 368
Result bus, 321
retention (refresh) fault, 465
re-timing, 271, 325
re-timing logic, 271
reward-and-penalty basis, 440
reward-and-penalty
        principle, 415
rewards, 415
RF. See radiofrequency (RF)
RF (Repair Flag) status flag, 545
RFID (radiofrequency
        identification),
        728, 736
ring counter, 289
rise/fall delay model, 118
RJ (random jitter), 719
RLC (resistance–inductance–
        capacitance)
        model, 24
RM (repair-most)
        algorithm, 529
RMT (redundant
        multithreading), 704
roadmapping, 681
robustly testable, 234
robustly untestable paths, 235
rollback, 264
ROM (read-only memory), 310
ROM compression, 308
routing BIST, 708
row-first algorithm, 529
row-repair phase, 541
RP-resistant (random pattern
        resistant), 47, 278–279,
        281, 304–309
RR (repair rate), 7, 531,
        548, 551
RRFA (random resistant fault
        analysis) method, 48
RTL. See register-transfer level
        (RTL)

rule 151, 279–281
rule 90, 279–281
rule of ten, 2
run-and-scan test application,
        434–435
RUNBIST instruction, 573,
        665–666
run-length code
        (variable-to-fixed),
        349–350
run-length coding, 349
run-length encoding, 372
Run-Test/Idle state, 568, 584
runtime reconfiguration, 706

S
S (sequential circuit), 214–215,
        246, 298
SA0 (stuck-at-0), 12, 373
SA1 (stuck-at-1), 12, 368
SAFs (stuck-at faults), 12–15,
        23, 464, 489, 518
sample/hold (S/H), 645, 647
SAMPLE instruction, 569–570
SAMPLE/PRELOAD
        instruction, 665
Sandia Controllabil-
        ity/Observability
        Analysis Program
        (SCOAP), 40, 41–45
        combinational controllability
        and observability
        calculation, 41–43
        overview, 41
        probability-based testability
        analysis, 45–47
        RTL testability analysis,
        48–50
        sequential control lability and
        observability
        calculation, 43–45
        simulation-based testability
        analysis, 47–48
scan architectures, 59–70
        full-scan design, 59–64
        clocked full-scan design, 62
        LSSD full-scan design,
        62–64
        muxed-D full-scan design,
        59–61
        overview, 59
        overview, 59
        partial-scan design, 64–67
        random-access scan design,
        67–70
scan–capture–scan scenario, 432
scan–capture–scan test
        procedure, 434
scan cell designs, 55–59
        clocked, 56–57
        LSSD, 57–59

muxed-D, 55–56
        overview, 55
scan cells, 39, 53, 91–92
scan chain diagnosis, 427–442
        hardware-assisted method,
        430–432
        modified inject-and-evaluate
        paradigm, 432–434
        overview, 427
        preliminaries for, 427–430
        signal-profiling-based
        method, 434–441
        diagnostic test sequence
        selection, 434
        functional sequences,
        435–437
        overview, 434
        profiling-based analysis,
        437–441
        run-and-scan test
        application, 434–435
        summary, 441–442
scan chains, 39, 53, 55
scan configuration, and scan
        synthesis, 79–82
scan design costs, 86–87
        area overhead cost, 86
        design effort cost, 87
        I/O pin cost, 87
        overview, 86
        performance degradation
        cost, 87
scan design flow, 76–87
        overview, 76–77
        scan design costs, 86–87
        area overhead cost, 86
        design effort cost, 87
        I/O pin cost, 87
        overview, 86
        performance degradation
        cost, 87
        scan design rule checking and
        repair, 77–78
        scan extraction, 83–84
        scan synthesis, 78–84
        overview, 78–79
        scan configuration, 79–82
        scan reordering, 82–83
        scan replacement, 82
        scan stitching, 83
        scan verification, 84–86
        overview, 84
        verifying scan capture
        operation, 86
        verifying scan shift
        operation, 85–86
scan design rules, 39, 70–76,
        266, 320
        asynchronous set/reset
        signals, 75–76
        bidirectional I/O ports, 71

checking and repair, 77–78, 93–94
combinational feedback loops, 74–75
derived clocks, 74
gated clocks, 71–74
overview, 70
tristate buses, 71
scan enable (SE), 55, 268
scan extraction, 83–84, 95–103
scan hold time, 85
scan-input (SI) pin, 379, 427
scan input (SI) port, 39
scan inputs (SIs), 55, 322
ScanInternal block, 603
scanning electronic microscopy (SEM), 399
scan output (SO) pin, 427
scan output (SO) port, 39
scan outputs (SOs), 322
scan point, 267
scan-ready design, 82
scan slice, 345, 446
scan synthesis
    overview, 78–79
    scan configuration, 79–82
    scan reordering, 82–83
    scan replacement, 82
    scan stitching, 83
scan test mode, 495
scan vector, 433
scan verification, 84–86
    overview, 84
    verifying scan capture operation, 86
    verifying scan shift operation, 85–86
SCGs (spare column groups), 537
SCOAP. *See* Sandia Controllability/Observability Analysis Program (SCOAP) testability analysis
SD (skew detector) cells, 696
SDC (silent data corruption), 700
SDCs (syndrome driver counters), 281–282, 284
SDF (standard delay format), 84
SE (scan enable), 55, 270
seeding, 218
segmentation testing, 280, 287–288
segment delay fault, 232
segments, 361
Select-DR-Scan state, 568
selection, 208
selection pressure, 210
selective Huffman code, 347
SelectWIR, 590
self-checking checker, 701, 702

self-test cells, 302
self-testing using MISR and parallel SRSG (STUMPS), 298
SEM (scanning electronic microscopy), 399
Semiconductor Industry Association (SIA), 679, 680, 737
semiconductor memory testing, 461
sense amplifier recovery fault, 465
sensitivity BIST mode, 713
sensitized partitioning, 288
sequence controller, 498
sequencer, 495, 501
sequential circuit (S), 109–110, 214–215, 246, 296
sequential control lability and observability calculation, 43–45
sequential depth, 48, 65
sequential fault simulation, 147
sequential feedback loops, 39
sequential linear decompressors, 355
SER (soft error rate), 90, 701
serial fault simulation, 133–135
serializer–deserializer (SERDES) devices, 736
serial scan design, 67
serial scan mode, 361–362
serial signaling, 722
serial signature analysis (SSA), 292, 294
serial simulation, 86
set/reset clock point (SRCK), 268
set/reset signals, 75–76
setup time violation, 429
SEUs (single-event upsets), 90
SF (Solid Flag) status flag, 545
S/H (sample/hold), 645, 647
shift clock, 82
Shift-DR state, 568
shift event, 593
shift mode, 53, 55
shift operation, 566
shift register (SR), 283
shift register latch (SRL), 57, 267
ShiftWR, 590
short-circuit admittance (Y), 672
shortened cyclic LFSR (linear feedback shift register), 287
short hard fault model, 625
S/H stage, 645
SI. *See* signal integrity (SI), and power supply noise

SI (scan input) pin, 379, 427
SI (scan input) port, 39
SIA (Semiconductor Industry Association), 679, 680, 737
SIC (single-input change), 291
SIC (structurally independent fault candidate) point, 420
sigma-delta ADC, 646
signal-1 frequency, 434
signal distortion element, 428
signal flipping, 434
signal integrity (SI), and power supply noise, 692–696
    integrity loss fault model, 693–694
    location, 694
    overview, 692–693
    pattern generation, 694–695
    sensing and readout, 695–696
signal-profiling-based method, in scan chain diagnosis, 434–441
    diagnostic test sequence selection, 434
    functional sequences, 435–437
    overview, 434
    profiling-based analysis, 437–441
        average-sum filtering, 438–439
        edge detection, 439–441
        example, 438
    run-and-scan test application, 434–435
signal-value pair, 412
signature, 290–295
signature analysis, 292–294
    overview, 292
    parallel, 294–295
    serial, 292–294
silent data corruption (SDC), 700
simple interconnects, 660
simulation-based ATPG, 207–217, 218–226
    genetic-algorithm-based ATPG, 208–217
        CASE studies, 215–217
        issues concerning fitness function, 213–215
        issues concerning GA parameters, 213
        issues concerning GA population, 212–213
        overview, 208–212
    logic-simulation-based ATPG, 222–225
    overview, 207–208, 218

seeding GA with helpful
sequences, 218–222
spectrum-based ATPG,
225–226
in test generation, 226–231
ALT-TEST hybrid, 228–231
overview, 226–228
simulation-based design
verification, 24
simulation-based testability
analysis, 40, 47–48
simulation-based test
generators, 208, 226
simulation models, 108–121
gate-level network, 109–110
logic element evaluation,
114–117
input counting, 116
input scanning, 115–116
overview, 114
parallel gate evaluation,
116–117
truth tables, 115
logic symbols, 110–114
high-impedance state Z,
113–114
intermediate logic
states, 114
overview, 110–111
unknown state u, 111–113
overview, 108–109
timing models, 118–121
functional element delay
model, 120–121
inertial delay, 119
overview, 118
transport delay, 118–119
wire delay, 119–120
simulation passes, 134, 137, 138
simultaneous self-test
architecture, 303
sine wave curve-fit test, 658
single-capture, 310–311, 323
single-event upsets (SEUs), 90
single-excitation patterns, 442
single-fault assumption, 11
single-fault candidate, 413
single-fault model, 11, 679
single-fix candidate, 417
single-input change (SIC), 289
single-input signature register
(SISR), 292
single-latch design, 62
single-line-conflict
algorithm, 203
single location at a time (SLAT)
pattern, 413
single location at a time (SLAT)
table, 416
single location at a time (SLAT)
vectors, 416
single-pass synthesis, 79

single-path sensitizable, 233
single-ring architecture
with separate TMS, 578
with shared TMS, 576–577
single-time-frame vector, 221
sinusoidal waveforms, 629
SIP (system-in-package), 737
SIs (scan inputs), 55, 324
SISR (single-input signature
register), 294
six sigma manufacturing, 6
six-valued logic, 131
skew detector (SD) cells, 696
skewed-load, 88, 311–315,
387, 687
aligned, 312–314
one-hot, 312
overview, 311–312
staggered, 314–315
SLAT (single location at a time)
pattern, 413
SLAT (single location at a time)
table, 416
SLAT (single location at a time)
vectors, 416
slew rate (SR), 633
slew rate limitation, 633
sliding diagonal/row/column
algorithm, 468–469
slow-speed testing, 310
slow-to-fall faults, 429, 687
slow-to-fall node, 238
slow-to-rise faults, 429, 687
slow-to-rise node, 238
slow-to-rise transition fault, 238
small-scale integration (SSI),
1, 38
small signal model, 621
small signal unit gain
bandwidth, 633
smoothed difference profile,
438, 440
snapshot image, 427–428, 432
snapshot scan, 88–90
SNDR (signal-to-noise and
distortion ratio), 640
SNR (signal-to-noise ratio),
639–641, 652, 722
SO (scan output) pin, 427
SO (scan output) port, 39
SOC. See system-on-chip (SOC)
test problems
SOCRATES, 27, 188
SOF (stuck-open fault), 464, 518
soft error rate (SER), 90, 701
soft errors, 39, 692
soft faults, 624
soft repair scheme, 539
software-based self-testing, 738
solid background, 467, 485
Solid Flag (SF) status flag, 545
SONET (synchronized optical
network), 722

SOs (scan outputs), 324
source synchronous (SS)
scheme, 720
space compaction, 367–374
overview, 367
X-blocking, 371–372
X-compact, 369–371
X-impact, 373–374
X-masking, 372
zero-aliasing linear
compaction, 367–369
space dimension, 365
space information, 443
spare column groups
(SCGs), 537
spare elements, 518
spatial redundancy, 701, 702
special-purpose scan designs,
39, 87–92
enhanced scan, 87–88
error-resilient scan, 90–92
overview, 87
snapshot scan, 88–90
specification-oriented
testing, 627
spectrum-based ATPG, 225–226
spectrum-based test generation
procedure, 226
spread networks, 376
spread polynomials, 376
SPT (suspect), 478
SR (shift register), 285
SR (slew rate), 633
SRAM (static random access
memory), 68, 270
SRCK (set/reset clock
point), 270
Srinivas, M., 215
SRL (shift register latch),
57, 269
SS (source synchronous)
scheme, 720
SSA (serial signature analysis),
294, 296
SSI (small-scale integration),
1, 38
STA (static timing analysis), 85
STAFAN (statistical fault
analysis), 48, 153–159
staggered clocking scheme, 78
staggered double-capture, 317,
323, 324
staggered single-capture,
311, 323
staggered skewed-load, 314–315
standard delay format (SDF), 84
standard LFSR, 272
Standard Test Interface
Language (STIL)
standard, 601

star (multi-ring)
     architecture, 578
starting pattern, 293
Start signal, 321
state coupling fault (CFst),
     465, 519
static, 696, 697
static 0-hazard, 130
static 1-hazard, 130, 132
statically sensitizable, 233
statically unsensitizable
     path, 233
static compaction, 246
static fault, 465
static hazard detection, 131–132
static logic implications,
     187–191
static random access memory
     (SRAM), 68, 268
static reconfiguration, 362
static timing analysis (STA), 85
statistical coding, 346
statistical delay testing and
     diagnosis, 691
statistical fault analysis
     (STAFAN), 48, 153–159
statistically constrained ATPG
     method, 691
statistically long paths, 691
statistically sensitizable, 233
statistically unsensitizable, 233
statistical path selection, 691
statistical process
     variations, 689
statistical sampling, 47
statistical timing analysis, 691
step waveforms, 629
STIL (Standard Test Interface
     Language)
     standard, 601
stop-band rejection ratio, 637
stop bands, 637
stop condition, 224
storage elements, 80
STRATEGATE test generators,
     218–219, 221
structural faults, 310
structurally dependent
     fault, 418
structurally independent fault,
     418, 420
structurally independent fault
     candidate (SIC)
     point, 420
structural offline BIST, 265, 266
structural offline test
     techniques, 317
structural pruning,
     407–408, 424
structural testing, 10, 698
structured approach, 50
structure graph, 48, 65

stuck-at-0 (SA0), 12, 373
stuck-at-1 (SA1), 12, 368
stuck-at faults (SAFs), 12–15,
     23, 464, 489, 518
stuck-at tests, 697
stuck-off, 15
stuck-on, 15
stuck-open fault (SOF), 464, 518
stuck-short, 15–16, 24
STUMPS (self-testing using
     MISR and parallel
     SRSG), 300
STUMPS architecture, 263, 442
subword, 540
success rate, 401
superposition principle, 448
surface-micromachined comb
     accelerometer, 716
surround disturb
     algorithms, 469
suspect (SPT), 478
suspicion profile, 438, 440
sweep sine, 631
switch level, 24
switch-level model, 106
symbolic simulation, 427
symbols, 345
symmetry BIST scheme,
     713–714
synchronized optical network
     (SONET), 722
synchronizing sequence, 496
synchronous clock domains,
     310, 317
synchronous SRAM, 519
syndrome analysis, 403
syndrome driver counters
     (SDCs), 281–282
synthesized waveforms, 631
systematic defects, 696
system availability, 7
SystemC, 106
system clock (CK), 269
system failure, 625
system-in-package (SIP), 737
system-level operation, 6–8
system-level specifications, tests
     for, 733–734
   adjacent channel power
     ratio, 733
   error vector magnitude, 734
   magnitude error, 734
   overview, 733
   phase error, 734
system-on-chip (SOC) test
     problems, 585–587
   deeply embedded cores, 585
   different core providers, 586
   expensive and inefficient
     external automatic test
     equipment (ATE), 586
   hierarchical cores, 585

higher performance core I/Os
     than SOC pins, 586
   IP protection/test reuse, 586
   large test power, 586
   long test application
     time, 586
   mixing technologies, 585
   overview, 585
   testable design automation,
     586–587
system test configurations,
     603–606
SystemVerilog, 106

T

TAGS (test algorithm generator
     by simulation), 461,
     480, 528
TAM (test access mechanism),
     585, 588
TAP (test access port), 32, 325,
     387, 562, 584–585, 610,
     660–661
TAPC (TAP controller), 562,
     567–569
TAP controller (TAPC), 562,
     567–569
TARO (transition fault
     propagated to all
     reachable outputs),
     320, 697
TBIC (test bus interface
     circuit), 662
TCK (test clock), 325, 562, 611
TDDB (time-dependent
     dielectric
     breakdown), 692
TDDM (time-division
     demultiplexer),
     382–383
TDI (test data input), 325, 562,
     563, 565
TDM (time-division
     multiplexer), 383
TDO (test data output), 325,
     562, 565
TDR (test data register), 563
TE (Test Enable) signal, 714
temperature-dependent
     error, 649
temperature sensitivity, 649
temporal (time)
     redundancy, 702
terminating necessary condition
     set (TNC), 206
ternary logic, 111
testability, 37–103
   analysis of, 40–50
   overview, 40
   SCOAP testability analysis,
     41–45

design for, 50–54
    ad hoc approach, 51–53
    overview, 50–54
    structured approach, 53–54
overview, 37–40
RTL design for, 92–103
    overview, 92–93
    RTL scan design rule
        checking and repair,
        93–94
    RTL scan extraction and
        scan verification,
        95–103
    RTL scan synthesis, 94–95
scan architectures, 59–70
    full-scan design, 59–64
    overview, 59
    partial-scan design, 64–67
    random-access scan design,
        67–70
scan cell designs, 55–59
    clocked-scan cell, 56–57
    LSSD scan cell, 57–59
    muxed-D scan cell, 55–56
    overview, 55
scan design flow, 76–87
    overview, 76–77
    scan design costs, 86–87
    scan design rule checking
        and repair, 77–78
    scan extraction, 83–84
    scan synthesis, 78–84
    scan verification, 84–86
scan design rules, 70–76
    asynchronous set/reset
        signals, 75–76
    bidirectional I/O ports, 71
    combinational feedback
        loops, 74–75
    derived clocks, 74
    gated clocks, 71–74
    overview, 70
    tristate buses, 71
special-purpose scan designs,
    87–92
    enhanced scan, 87–88
    error-resilient scan, 90–92
    overview, 87
    snapshot scan, 88–90
testability-driven factoring, 307
testability measures, 38
testable design, 76
test access mechanism (TAM),
    585, 586, 588, 604–606
test access port (TAP), 32, 323,
    387, 562, 584–585, 610,
    660–661
test algorithm generator by
    simulation (TAGS),
    461, 480, 528
test bus interface circuit
    (TBIC), 662

test clock (TCK), 323, 562, 611
test command (CMD), 543
test compaction, 108
test compression, 341–389,
    698, 737
    industry practices, 376–396
        adaptive scan, 385–386
        embedded deterministic
            test, 379–382
        ETCompression, 386–396
        OPMISR, 377–379
        overview, 376–377
        virtualscan and ultrascan,
            382–384
    overview, 341–344
    test response compaction,
        364–376
        mixed time and space
            compaction, 375–376
        overview, 364–367
        space compaction, 367–374
        time compaction, 374–375
    test stimulus compression,
        344–364
        broadcast-scan-based
            schemes, 359–364
        code-based schemes,
            345–351
        linear-decompression-
            based schemes,
            351–354
        overview, 344
test controller, 322–323
test cubes, 344, 377
test data input (TDI), 323, 562,
    563, 565
test data output (TDO), 323,
    562, 565
test data register (TDR), 563
Test Enable (TE) signal, 714
tester memory requirement, 240
test escapes, 317
test generation, 11, 161–262
    ATPG for non-stuck-at faults,
        231–245
        ATPG for transition faults,
            238–239
        bridging fault ATPG,
            244–245
        overview, 231
        transition ATPG using
            stuck-at ATPG, 240
        transition ATPG using
            stuck-at vectors,
            240–244
    combinational circuits,
        designing stuck-at
        ATPG for, 169–194
        basic ATPG algorithm,
            173–177
        D algorithm, 177–182

dynamic logic implications,
    191–194
FAN, 186–187
naive ATPG algorithm,
    169–173
overview, 169
PODEM, 182–186
static logic implications,
    187–191
hybrid deterministic and
    simulation-based
    ATPG, 226–231
    ALT-TEST hybrid,
        228–231
    overview, 226–228
multiple-line conflict
    analysis
    definition 1, 206
    definition 2, 206
    definition 3, 206
    lemma 2, 206–207
    overview, 203–206
    proof, 207
other topics in, 246–262
    ATPG for acyclic sequential
        circuits, 247
    designing high-level ATPG,
        248–262
    IDDQ testing, 247
    N-Detect ATPG, 247
    overview, 246
    test set compaction, 246
overview, 161–163
random, 163–166
    exhaustive testing, 166
    lemma 1, 165
    overview, 163–165
    proof, 165
sequential ATPG, 194–200
    5-valued algebra is
        insufficient, 196–197
    gated clocks and multiple
        clocks, 197–200
    overview, 194
    time frame expansion,
        194–196
simulation-based ATPG,
    207–217, 218–226
    genetic-algorithm-based
        ATPG, 208–217
    logic-simulation-based
        ATPG, 222–225
    overview, 207–208, 218
    seeding GA with helpful
        sequences, 218–222
    spectrum-based ATPG,
        225–226
theoretical background,
    166–168
untestable fault
    identification, 200–207

multiple-line conflict analysis, 203–207
overview, 200–202
in VLSI testing, 9–11
testing delay defects, 690
test input (TDI), 576
test interface unit (TIU), 725
test invalidation, 289
TestKompress compaction scheme, 381
Test–Logic–Reset state, 568
test methodologies, 683
test mode (TM), 51, 566
TestMode parameter, 603
test modes, IEEE 1149.4, 666–667
  complex network measurement, 671–672
  extended interconnect measurement, 667–671
  high-performance configuration, 672–677
  open/short interconnect testing, 666–667
  overview, 666
test mode select (TMS), 323, 562, 565
test-oriented design methodology, 53
test pattern generation, 271–290
  delay fault testing, 288–289
  exhaustive testing, 275–277
    binary counter, 275
    complete LFSR, 275–277
    overview, 275
  hybrid LFSR, 274–275
  LFSR properties, 273–274
  modular LFSR, 272
  overview, 271–275
  pseudo-exhaustive testing, 280–288
    overview, 280–281
    segmentation testing, 287–288
    verification testing, 281–287
  pseudo-random testing, 277–280
    cellular automata, 278–280
    maximum-length LFSR, 278
    overview, 277–278
    weighted LFSR, 278
  standard LFSR, 272
test pattern generators (TPGs), 31, 187, 265, 288, 321–322, 501, 521, 541, 708
test patterns, 133
test-per-clock BIST system, 298
test-per-scan BIST system, 297
test point insertion, 30, 305–307

control point activation, 307
overview, 305–306
test point placement, 306–307
test point placement, 306–307
test programs, 25–26, 39
test quality, 163
test requirements, 504
test reset (TRST*), 562, 565
TestResourceConstraint block, 603
test response compaction, 364–376
  mixed time and space compaction, 375–376
    example 6.6, 376
    overview, 375–376
  overview, 364–367
  space compaction, 367–374
    overview, 367
    X-blocking, 371–372
    X-compact, 369–371
    X-impact, 373–374
    X-masking, 372
    zero-aliasing linear compaction, 367–369
  time compaction, 374–375
    example 6.5, 375
    overview, 374–375
test reuse, 586
Test Selection (TS) signal, 715
test sequence, 195
test set compaction, 246
test stimulus compression, 344–364
  broadcast-scan-based schemes, 359–364
    broadcast scan, 359–360
    Illinois scan, 360–362
    multiple-input broadcast scan, 362
    overview, 359
    reconfigurable broadcast scan, 362–363
    virtual scan, 363–364
  code-based schemes, 345–351
    dictionary code (fixed-to-fixed), 345–346
    Golomb code (variable-to-variable), 350–351
    Huffman code (fixed-to-variable), 346–349
    overview, 345
    run-length code (variable-to-fixed), 349–350

linear-decompression-based schemes, 351–354
  combinational linear decompressors, 355
  combined linear and nonlinear decompressors, 357–359
  fixed-length sequential linear decompressors, 355–356
  overview, 351–354
  variable-length sequential linear decompressors, 356–357
overview, 344
test vectors, 10, 11, 14, 133, 238, 405
TFs (transition faults), 464, 489, 518
THD (total harmonic distortion), 640, 652
third-order intercept (TOI), 732
third-party components, 26
TIAs (timing interval analyzers), 724
tie-breaker outputs, 415
time (temporal) redundancy, 702
time and space compaction, 375–376
time compaction, 374–375
time compactor compacts, 365
time-dependent dielectric breakdown (TDDB), 692
time-dependent failures, 624
time dimension, 365
time-division demultiplexer (TDDM), 382–383
time-division multiplexer (TDM), 383
time-domain ADC testing, 654–658
  code bins, 654
  code transition level test (dynamic), 655
  code transition level test (static), 655
  gain and offset test, 656–657
  linearity error and maximal static error, 657
  overview, 654
  sine wave curve-fit test, 658
time information, 443
timing control, 310–319
  double-capture, 315–317
    aligned double-capture, 316–317
    one-hot double-capture, 315–316

overview, 315
staggered
    double-capture, 317
fault detection, 317–319
overview, 310
single-capture, 310–311
    one-hot single-capture,
        310–311
    overview, 310
    staggered
        single-capture, 311
skewed-load, 311–315
    aligned skewed-load,
        312–314
    one-hot skewed-load, 312
    overview, 311–312
    staggered skewed-load,
        314–315
timing-driven test point
    insertion
    techniques, 307
timing fault test mode, 495
timing interval analyzers
    (TIAs), 724
timing models, 118–121
timing physical failures, 696
timing wheel, 127
TIU (test interface unit), 725
TM (test mode), 51, 566
TMR (triple modular
    redundancy), 703
TMS (test mode select), 325,
    562, 565
TNC (terminating necessary
    condition set), 206
toggle coverage, 151
TOI (third-order intercept), 732
token-ring clock enabling, 323
too-early signal change, 429
top-10 hit, 400
top-bottom LFSR, 274
topology-based testability
    analysis, 40
totally self-checking two-rail
    checker, 304
TPGs (test pattern generators),
    31, 187, 267, 290,
    321–322, 501, 521,
    541, 708
TransferDR, 591–592
transfer event, 595
transient error, 264
transient power supply
    current, 29
transient responses, 630
transients, 698
transistor fault models, 24
transistor-level, 106
transition ATPG, using stuck-at
    ATPG, 240–244
transition bands, 637

transition count testing,
    291–292
transition fault models, 19, 687
transition fault propagation to
    all reachable outputs
    (TARO), 320, 697
transition faults (TFs), 464,
    489, 518
transition faults, ATPG for,
    238–239
transition faults propagated to
    all reachable outputs
    (TARO), 318
transition-independent delay
    model, 118
transition test chains, 241–244
transition tests, 697
transparent space compactor,
    367–368
transport delay, 118–119
triangular waveforms, 631
triple modular redundancy
    (TMR), 703
tristate buses, 71, 269
TRST* (test reset), 562, 565
true at-speed testing,
    315–316, 319
true critical path selection, 691
TS (Test Selection) signal, 715
two-phase chip-level strategy,
    420–424
    aliasing SIC points
        example, 421
    definitions, 420–422
    examples, 422–424
    observations, 420–421
    output group example, 422
two-point crossover, 211

U
UDL (user-defined logic),
    587, 597
ultrascan, 382–384
ultrawideband (UWB), 736
unimportant states, 223
unit gain bandwidth
    measurement, 632–633
universal CA, 280
universal injection model, 434
unknown (X) values, 266
unknown source blocking,
    267–271
    analog blocks, 267–268
    asynchronous set/reset
        signals, 268–269
    bidirectional I/O ports, 271
    combinational feedback
        loops, 268
    critical paths, 270
    false paths, 270
    floating ports, 270–271

memories and non-scan
    storage elements, 268
multiple-cycle paths, 270
overview, 267
tristate buses, 269
unknown test response bits, 366
unmodeled-fault problem, 403
untestable fault identification,
    187, 200–207
    multiple-line conflict
        analysis, 203–207
    overview, 200–202
up-conversion mixer, 729
Update-DR state, 569
update event, 595
Update operation, 566
UPDATE signal, 88
UpdateWR, 591
upper limit mask, 637
up transition (UTR) flag, 479
USERCODE instruction, 574
user-defined logic (UDL),
    587, 597
UTR (up transition) flag, 479
UWB (ultrawideband), 736

V
valid fault multiplets, 416, 418
value–change–dump (VCD)
    file, 435
value-change event, 410
variable-length-input Huffman
    code (VIHC), 351
variable-length sequential linear
    decompressors,
    356–357
variable-to-fixed coding, 349
variable-to-variable coding, 350
VCD (value–change–dump)
    file, 435
VCO (voltage-controlled
    oscillator), 631
vectors, 240
verification test approach, 280
verification testing, 281–287
    combined LFSR/PS, 284
    combined LFSR/SR, 283–284
    compatible LFSR, 287
    condensed LFSR, 284–285
    constant-weight counter,
        282–283
    cyclic LFSR, 285–287
    syndrome driver counter,
        281–282
very-large-scale integration. See
    VLSI
VHDL (VHSIC hardware
    description language),
    4, 106, 154, 574
victim (VTM), 477
victim signal, 418
VIHC (variable-length-input
    Huffman code), 351

virtual instrumentation, 636
virtual scan, 363–364, 382
visible bad gates, 144
VLSI testing, 8–21
  fault models, 11–21
    analog fault models, 21
    delay faults and crosstalk, 19–20
    open and short faults, 16–19
    overview, 11–12
    pattern sensitivity and coupling faults, 20–21
    stuck-at faults, 12–15
    transistor faults, 15–16
  historical review of technology, 25–36
    analog and mixed-signal circuit testing, 29–31
    automatic test equipment, 25–27
    automatic test pattern generation, 27–28
    board testing, 31
    boundary scan testing, 32–36
    digital circuit testing, 28–29
    fault simulation, 28
    overview, 25
  levels of abstraction in, 22–24
    gate level, 23–24
    overview, 22
    physical level, 24
    register-transfer level and behavioral level, 22–23
    switch level, 24
  overview, 8–9
  test generation, 9–11
voltage buffers, 672
voltage-controlled oscillator (VCO), 631
VTM (victim), 477

W
walking pattern (WALPAT), 468
waveform-oriented testing, 627
WBCs (wrapper boundary cells), 592
WBR (wrapper boundary register), 592
WBR_Cntrl signal, 596
WBY (wrapper bypass register), 591–592
WCSs (wrapper control signals), 608
WDR (wrapper data registers), 591
WE (Write-Enable) signals, 523
weighted, 287
weighted pattern generation technique, 278

weighted pattern graph, 244
weighted pseudo-random LFSR reseeding, 372
weighted transition graph algorithm, 241
weighted transition pattern, 241, 243
white-noise fault, 437
WiMAX, 736
WIR (wrapper instruction register), 588, 591
wired-AND/wired-OR bridging fault model, 17
wire delay, 119–120
word-level prime candidate, 418, 422, 425
wordline short fault (WSF), 473
word-oriented memory, 473
WP_EXTEST instruction, 599
WP_PRELOAD instruction, 600
WPC (wrapper parallel control), 588, 591
WPI (wrapper parallel input), 588, 591
WPO (wrapper parallel output), 588, 591
WPP (wrapper parallel port), 591
wrapper boundary cells (WBCs), 592
wrapper boundary register (WBR), 592
wrapper bypass register (WBY), 591–592
wrapper components, 589–597
  AUXCKn, 589–590
  CaptureWR, 590
  overview, 589
  SelectWIR, 590
  ShiftWR, 590
  TransferDR, 591–592
  UpdateWR, 591
  WRCK, 589
  WRSTN, 590
wrapper control signals (WCSs), 608
wrapper data registers (WDR), 591
wrapper functional mode, 592
wrapper instruction register (WIR), 588, 591
wrapper parallel control (WPC), 588, 591
wrapper parallel input (WPI), 588, 591
wrapper parallel output (WPO), 588, 591
wrapper parallel port (WPP), 591
wrapper serial control (WSC) terminal, 588, 589

wrapper serial input (WSI) terminal, 588, 589
wrapper serial output (WSO) terminal, 588, 589
wrapper serial port (WSP), 588–592
  AUXCKn, 589–590
  Capture WR, 590
  overview, 589
  SelectWIR, 590
  ShiftWR, 590
  TransferDR, 591–592
  Update WR, 591
  WRCK, 589
  WRSTN, 590
wrapper TAP (WTAP), 387
WRCK (wrapper clock) terminal, 589
Write-Enable (WE) signals, 523
write recovery fault, 465
write-through operation, 519
WRSTN (wrapper reset) terminal, 590
WS_BYPASS instruction, 597
WS_BYPASS naming convention, 597
WS_CLAMP instruction, 600
WS_EXTEST instruction, 597–599
WS_EXTEST naming convention, 597
WS_INTEST_RING instruction, 600
WS_INTEST_SCAN instruction, 600–601
WS_PRELOAD instruction, 600
WS_SAFE instruction, 600
WSC (wrapper serial control) terminal, 588, 589
WSF (wordline short fault), 473
WSI (wrapper serial input) terminal, 588, 589
WSO (wrapper serial output) terminal, 588, 589
WSP (wrapper serial port), 588, 589
WTAP (wrapper TAP), 387
Wx_EXTEST instruction, 599
Wx_INTEST instruction, 601
Wx_PRELOAD instruction, 597

X
X (unknown) values, 268
X-blocking, 267, 343, 371–372
X-bounding, 267, 343, 371
X-compact, 369–371
X-impact, 343, 373–374
X-masking, 343, 372
XOR network, 357, 365, 387, 390

X-tolerant, 343, 369, 387, 390
X-tolerant XOR network,
        387, 390

Y
yield, 4, 5–6, 11, 33, 200, 255,
        397–399, 450, 461, 517,

518, 548–552, 684, 685,
        692, 696–698

Z
zero-aliasing linear compaction,
        367–369
zero defects, 6

zero-delay event-driven
        simulation, 125
zero-delay simulation, 125
zero-detector, 275
zero-one algorithm, 467

Printed and bound by CPI Group (UK) Ltd, Croydon, CR0 4YY

03/10/2024

01040314-0006